Proceedings of the 2016 International Conference on

Biotechnology & Medical Science

Proceedings of the 2016 International Conference on

Biotechnology & Medical Science

Edited by

Yong Zhang

NUS, Singapore

World Scientific

NEW JERSEY • LONDON • SINGAPORE • BEIJING • SHANGHAI • HONG KONG • TAIPEI • CHENNAI • TOKYO

Published by

World Scientific Publishing Co. Pte. Ltd.
5 Toh Tuck Link, Singapore 596224
USA office: 27 Warren Street, Suite 401-402, Hackensack, NJ 07601
UK office: 57 Shelton Street, Covent Garden, London WC2H 9HE

Library of Congress Cataloging-in-Publication Data
Names: International Conference on Biotechnology and Medical Science (2016 : Nanjing, China) | Zhang, Yong, (Professor) editor.
Title: Proceedings of the 2016 International Conference on Biotechnology and Medical Science / Yong Zhang, National University of Singapore, Singapore.
Description: New Jersey : World Scientific, 2016. | Includes bibliographical references.
Identifiers: LCCN 2016023599 | ISBN 9789813145863 (hardcover : alk. paper)
Subjects: LCSH: Medicine--Congresses. | Biotechnology--Congresses.
Classification: LCC R106 .I516 2016 | DDC 610--dc23
LC record available at https://lccn.loc.gov/2016023599

British Library Cataloguing-in-Publication Data
A catalogue record for this book is available from the British Library.

Typeset by Stallion Press
Email: enquiries@stallionpress.com

Printed in Singapore

Preface

BMS2016 is held in Nanjing, China from April 16-17, 2016. Aiming at providing an excellent international academic forum for all the researchers and practitioners, the conference enjoys a wide spread participation among all over the universities and research institutes. Any paper and topic concentrates on related fields of biotechnology and medical science is broadly welcomed. Besides that we also look forward to making the conference perfect circumstances to share resources, exchange opinions and inspire studying.

BMS2016 proceeding tends to collect the most up-to-date, comprehensive, and worldwide state-of-art knowledge on biotechnology and medical science. All the accepted papers have been submitted to strict peer-review by 2-4 expert referees, and selected based on originality, significance and clarity for the purpose of the conference. The conference program is extremely rich, profound and featuring high-impact presentations of selected papers and additional late-breaking contributions. We sincerely hope that the conference would not only show the participants a broad overview of the latest research results on related fields, but also provide them with a significant platform for academic connection and exchange.

The Technical Program Committee members have been working very hard to meet the deadline of review. The final conference program consists of 96 papers divided into 3 sessions. The proceedings would be published in a volume by World Scientific Publishing Company.

We would like to express our sincere gratitude to all the TPC members and organizers for their hard work, precious time and endeavor preparing for the conference. Our deepest thanks also go to the volunteers and staffs for their long-hours work and generosity they've given to the conference. The last but not least, we would like to thank each and every of the authors, speakers and participants for their great contributions to the success of BMS2016.

Yong Zhang

Committees

Editor
Yong Zhang, Professor & Deputy Chair, National University of Singapore, Singapore
Technical Program Committee
Savita Yadav, Professor, All India Institute of Medical Sciences, New Delhi, India
Collins Collins, Professor, University of British Columbia, Canada
Chris Chatwin, Professor, University of Sussex, UK
Jie Wu, Professor, Barrow Neurological Institute, Phoenix, Arizona, USA
Tariq Mahmood, Assistant Professor, University Of Central Punjab, Lahore, Pakistan
Yangchao Chen, Assistant Professor, The Chinese University of Hong Kong, China
Prashant N. Kumta, Edward R. Weidlein Chair Professor, University of Pittsburgh, USA
Ke Zheng, Professor, Anhui University of Science and Technology, China
Qi Zheng, Associate Professor, Guilin University Of Electronic Technology, China
Zhaoyun Zhong, Associate Professor, Beijing Normal University, China
Miguel Angel Fernandez Granero, Professor, University of Cadiz, Spain
Yiming Feng, Professor, Guangdong Ocean University, China
Bei Li, Professor, Shantou University, China
Shang-You Yang, Associate Professor, Wichita State University, USA
Mei Cheng, Professor, China Science & Technology University, China
Yuzhuo Wang, Professor, University of British Columbia, Canada
Ghulam Murtaza, Assistant Professor, COMSATS Institute of Information Technology, Pakistan
Erzheng Su, Associate Professor, Nanjing Forestry University, China
Hebron C. Chang, Associate Professor, Asia University, Taiwan, China
Chang-Hwan Im, Associate Professor, Hanyang University, Korea
Farid Dorkoosh, Associate Professor, Tehran University of Medical Sciences, Iran
Gylles Ricardo Ströher, Professor, Federal University of Technology Paraná, Brazil
Guifen Qian, Professor, Fujian Normal University, China
Guai Gou, Professor, Guangxi University, China
Geetha Manivasagam, Professor (SMBS) and Assistant Director, CBMST, VIT University, India
Chin-Yih Hong, Professor, National Chung Hsing UniversityTai Wan, China

Contents

Session 2: Medical Science 145

Session 3: Food Hygiene, Environment and Human Health 491

Session 1

Biotechnology

STUDY ON THE OCULAR BLOOD PRESSURE BASED ON HARMONIC ANALYSIS

Jintao Yu[a], Fang Feng, and Zhiping Yang

School of Computer and Information Engineering, Harbin University of Commerce, Harbin, 150028, China

E-mail: [a]hityjt@sina.com

Lots of researches have shown that ocular blood pressure (OBP) of healthy volunteers is much different from glaucoma patients. Increased OBP is the most well-known risk factor for the induction of glaucoma. What's more, our results indicated that the calculations of the first through fifth harmonics of the OBP pulse are superior in distinguishing between healthy people and glaucoma patients. This paper mainly studies on the relationship between the OBP and Age based on the harmonic content of envelope waveform aimed at the healthy volunteers (age: 41.5±12.5years), here we randomly choose 27 (SBP: 115.9±8.6, DBP: 66.0±7.9) without the history of eye disease. A laser speckle flowgraphy (LSFG, NAVI LSFG system, Soft-care, Japan) device is utilized to estimate the BF in the ONH. OBP is recorded around one of the fingers by a noninvasive blood recorded at 200Hz sampling rate (Finometer Pro®, Holland). Then, calculated the Frequency Coefficient (FC: 0.45±0.447) and Frequency Index (FI: 0.05±0.032).The results indicated that age in the healthy body has great influence on FC and FI ($P<0.001$). We considered that the FC of the first five harmonics is less suitable for calculations independent from age, FI is not related to the blood pressure, which changes with aging. So FI is used to be an indictor to assess ocular cycle in the healthy volunteers. This study has important significance for the diagnosis in clinical ophthalmology.

1. INTRODUCTION

Glaucoma is the second most cause of blindness in the world. Although the pathogenesis of glaucoma remains unclear, some references indicated that increased OBP is the most risk factor [1]. Nowdays, majority of studies have shown that abnormal ocular blood flow (OBF) will cause some physiological disease [2-4].OBP has obvious difference along with the slight changes caused by each heartbeat in Pulsation cycle .The pulse waveform of OBP has been used to be a method for assessing the vascular status of glaucoma [5-7].The present study indicated that it's so important for glaucoma patients to decrease the amplitude of OBP, compared to the healthy volunteers.

In the circulation of system, the pulse waveform of ocular blood pressure has entered into a more detailed research stage. Many studies have focused on the ocular blood pressure pulse based on fast Fourier transform (FFT). In short, the analysis principle of FFT is also used in the periodic function [8-10], e.g. arterial pressure pulse, which composed of sine wave that the amplitude and phase change repeatedly [11]. Fourier analysis, which decomposes the signal into the sum of different frequency sine wave functions [12], Compared to the time-domain waveform, the former technologies are considered to be on behalf of the frequency. The frequency domain waveform contains the highest power spectral components and the frequency of the arterial pressure has direct relationship with the heartbeat [13-15].The result of the waveform analysis is that the spectrum component part lower than the fundamental frequency, which relates to respiratory system and cardiovascular system [16].

Some studies have shown that the higher harmonic of the arterial pressure represents the stronger vascular resistance [17] .The vascular resistance is effected by size, elastic, physical properties of the blood flow and wave reflection from distal arterial tree part .What's more, the risk of vascular, which is defined by the resistance of the oscillation system [18-20], have great relationship with vascular resistance. The resistance is defined by the non-oscillatory or stable movement, so it is regarded as a risk factor in zero frequency [21].

Although the pulses derived from blood vessel originally, it relies mainly on vascular quantitative change, the size of the eye, eye external elastic, rather than the changes of their eyes' OBF [22]. This research ignores the effect caused by other factors, e.g. height, weight, blood type and so on.

From the perspective of OBP, if the harmonic components of OBP have relationship with some other ocular measures, waveform analysis of FFT will provide a new method for the eye state of its own blood vessels. Therefore, the purpose of this research is to analyze the harmonic components of OBP waveform based on FFT for the healthy volunteers, to achieve the relationship between OBP and age.

2. RESEARCH METHOD OF THE RELATIONSHIP BETWEEN OBP AND AGE

2.1. Data Collection

Participants (age: 41.5±12.5 years, n=27) that just take account of the healthy which are randomly selected from society. We collect the data in the Devers Eye Institute United States. The participants have a normal ocular blood pressure (systolic ocular blood pressure [SBP]:≤130mmHg and diastolic ocular blood pressure [DBP]:≤80mmHg). Before the collection of data, all of the participants received a questionnaire, including age, gender, height, known cardiovascular risk factors (e.g. arterial hypertension, diabetes, and smoking history), atherosclerosis and so on. Then, the study is conducted in accordance with the Declaration of Helsinki on Biomedical Research Involving Human Subjects [23-25]. What's more, the participants sign informed consent. They must remain at least 20min rest in an ambient temperature at 23°C, pupil expansion without drug. OBP and BF data are recorded for 2 minutes during spontaneous respiration. LSFG-NAVI is a updated model, the special area sensor is replaced by the ordinary charge-coupled device camera, to describe a more clear spatial resolution, and the measurement has good reproducibility. LSFG is used to collect BF data of the ONH, utilizing the laser speckle phenomenon. Finometer is used to achieve the OBP data at the same time as shown in Fig. 1.

Fig.1. LSFG and Finometer pictorial diagram of the system: Universal serial bus, PC, LCD.

The heartbeat is estimated by home visit. The method takes the instrument to collect the data of OBP and BF for 2 minitues. Since the rapple waveform frequency is 1Hz, the frequency of the rapple has direct relationship with the pulse of OBP. In order to meet the sampling theorem, the data sampling frequency is larger than 2Hz, and then the data collection is considered to be feasible.

2.2. Cubic Spline Interpolation

Since collecting the data of the OBP, some data lost during blinking (eg.Fig.2) In order to reduce error in the process of calculation, this method adopts cubic spline interpolation. Cubic spline interpolation is very simple in the calculation, stable enough and good convergence. What's more, it has good smoothness.

Although the time domain figure is based on the cubic spline interpolation, the time domain methods sometimes cannot satisfy the basic requirement of the analyst. Many research methods in favor of frequency domain, which is deemed to process data easily.

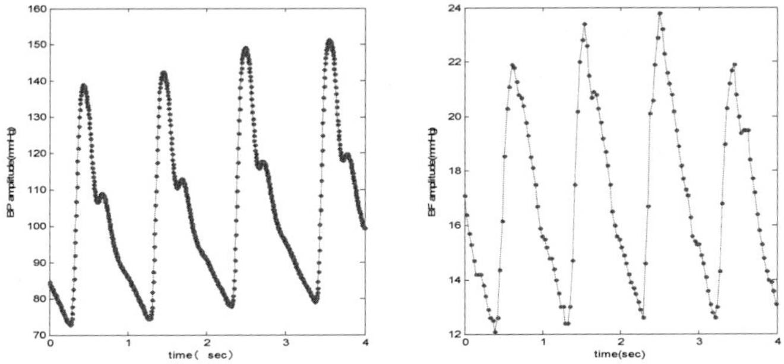

Fig.2. Voluteers averaged ocular blood pressure levels(left)and averaged blood flow levels(right) in time domain.

2.3. Fast Fourier Transform

On account of the analysis is very complicated in time domain, we conduct the research in frequency domain. FFT is a method of signal processing as a mathematical branch. In 1807, Fourier proposed a theory that periodic function change along with time repeatly, expressed by the sum of different frequency of sine and cosine function. In the case of simple and uncertain sine waveform, only fundamental frequency can be used to represent the data. For more complex waveform, the fundamental frequently contains a large number of spectrum energy which represents the heart rate of the patient, e.g. OBP pulse. FFT can decompose the signal into different frequency of waveforms to analyze the characteristics of a single wave conveniently.

The frequency of collecting the data of OBF (16.7±4.8) and OBP by LSFG and Finometer(LSFG and Finometer are instruments which are used to collect the data of OBF and OBP)is 200Hz, continuous acquisition for 10s, providing 2000 data points available for analysis. Fourier analysis by FFT algorithm required 2^n data points as data sample. For example, the sample need contain the following data points:2,4,8,16….512,1024,2048 and so on[22]·Select the sample size of 1024 data the sample data size of 1024 to analyze ,because it can provide the biggest data point, capture it during the sample period in 10s. Then, the data of OBP is downloaded to the computer to be analyzed by a form. Except the three differences in overlap part, 1024 data points of OBP pulse waveform are assessed by FFT. After the FFT, the power percentage of each harmonic in Eq:

$$p(i)=\frac{a_i^2}{a_1^2+a_2^2+a_3^2+a_4^2+a_5^2} \tag{1}$$

$$FC=\frac{\sum_i^n ip(i)}{\sum_i^n i}=\frac{1*p1+2*p2+3*p3+4*p4+5*p5}{1+2+3+4+5} \tag{2}$$

$$FI=P(3)/P(1) \tag{3}$$

i indicates the numbering of the harmonic, p(i) indicates the power percentage value.

The frequency composition of FFT is effected by the truncation of the signal and part of datasets waveform perfusion. If the sample datasets do not contain an accurate cycle period, it can reduce some harmonics. Because the data collection in our study occurred in a relatively short time, each patient has different heart rate and different harmonic power percentage, the result of truncation may make data deviation.

2.4. Statistical Analysis

Analyses are performed using MATLAB tool (release 2010.b, MATLAB). In the subjects, one-way ANOVA is applied to validate the model and hypotheses, and age is a variable and it displayed a relative statistical weight in FC (71%, $P<0.01$) and FI (73%,$P<0.01$). FI as an indictor, the calculation is based on the first and the third harmonic power percentage; it is also influenced by age. Based on the theory of the preceding discussion, establishment simple linear regression model to get the conclusion. All of the values expressed as mean±SD.$P<0.05$ were considered to be significant. Cronbach-a at the 95% confidence interval was 31%-52%, 20%,15%,13%,11% for FC and 39%-51% for FI. Therefore, the intraclass correlation coefficients of reliability derived from intraocular pressure's envelope waveform are within a very suitable rang of reliability.

3. EXPERIMENT RESULT

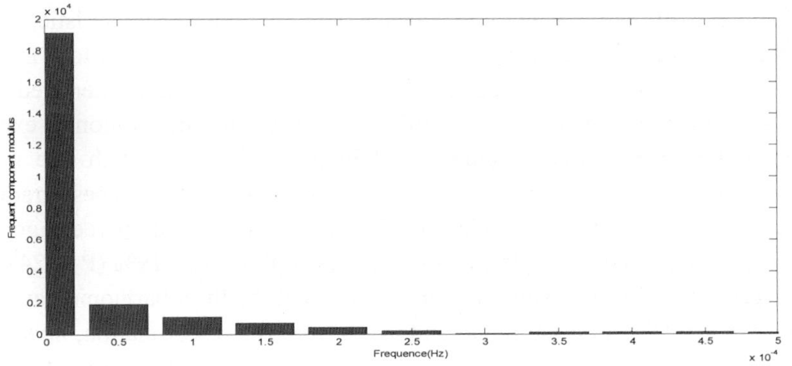

Fig.3. The oscillation amplitude average of the separate signal components in all voluteers, as shown in the picture, ie, the first through fifth oscillation is the main frequent component modulus by FFT. Althought the fundamental frequency effected by heartbeat has the lowest frequency, the amplitude is largest, ie, sound wave.

From Fig.3, we can suggest that the first through fifth harmonics (the amplitude is $1.92* 10^4$, $1.89* 10^3$, $1.1* 10^3$, 673,440 respectively) have significant affects on studying healthy voluteers' eye and has an important theoretical basis for the analysis of OBP. After it, the other harmonic amplitude is not obvious, we ignored it. In addition, the relationship of the power percentage of the harmonic on OBP and age is shown in Fig. 4.

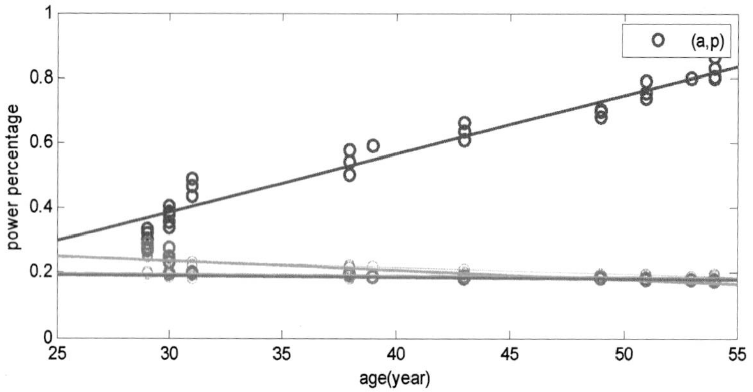

Fig.4. The first through fifth harmonics oscillation amplitude of the separate signal components is indicated by the power absolute values. Oscillation frequencies of the five harmonics are by dividing the power in each frequency component by the total power.

The harmonic of participants' OBP has a very obvious characteristic; the first through fifth power percentage of harmonic has a great influence. Owing to it influenced by the arterial pressure pulse associated with the heartbeat and interfered by the respiratory and vasoactive cycles of the cardiovascular system, the harmonics except the first harmonic power percentage values is failing slowly along with age growing. However, the first and the third harmonic oscillation (H) have the steepest gradient (b) and highest square regression coefficient (R^2). The average power percentage of five harmonics is 58.1% (P=0.027), 23.9% (P=0.04), 22.1% (P=0.015), 19% (P=0.24), 19.4% (P=0.19) respectively. As the results shown, the first and the third harmonic have a great correlation with age. The first harmonic and age have positive correlation, however, the third harmonic and age have negative correlation, and the other harmonic do not have obvious change along with the age growing. In addition, it indicated that data populations are similar among groups. The similarities of variable coefficient among groups provide some similarity for the measurement of participants. However, from the Fig. 4, we found that the power percentage values of the first and the third harmonics achieved the steepest gradients over ages and the greatest squared coefficients of determination: R^2, which is why their ratio is considered for definition of the FI. The relationship between FI and age is shown in Fig. 5

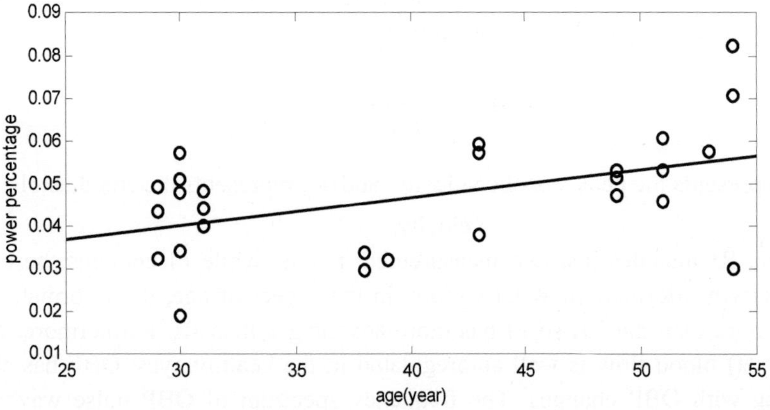

Fig.5. The ratio of percentage power of the third and the first harmonic components, it indicated that most voluteers (P=71%)OBP is more and more higher along with the age growing, especially the order voluteers(age>40(year)) with thicker blood vessel walls lead to a high-input resistance, the high-input resistance cause a high OBP. So we can consider that the OBP is increasing along with the growth of age.

From the Fig. 5, we concluded that the FI of the OBP increase along with the growth of age and they are positive correlation. In other words, as the growth of the age, the risk of eye diseases increased.

FC is influenced significantly by age(71%,$p<0.01$). However, we have considered the FC of the first five harmonics less suitable for calculations independent from age, FC was related to the blood pressure, while FI(73%,$p<0.01$) were not related to the blood pressure, which changes with aging. Therefore, FI was preferred for further calculations.

In conclusion, our results have shown that the amplitudes from the first harmonic to the five harmonic provided obvious difference. FI will be a key indicator on the changes of OBP for patients with eye disease compared to the healthy bodies in the future. The paper provides the statistics of healthy voluteers' bodies in Table 1.

Table 1. The statistics of healthy voluteers' bodies (n=27, mean ±SD,*P<0.001)

Age(years)	41.5±12.5
Sex(F)	27
Systolic ocular blood pressure(mmHg)	115.9±8.6
Diastolic ocular blood pressure(mmHg)	66.0±7.9
Optic nerve head blood flow	20.9±5.3
RI	0.43±0.447
heartbeat	71.5±18.5
height	163±12
FC	0.45±0.443
FI	0.05±0.032

We render that FI is a better indictor than RI which used broadly as a marker of peripheral vascular resistance, because the FI of the ophthalmic artery allows an assessment of the ocular circulation in consideration of age and arterial ocular blood pressure. The computing formula of RI as follows:

$$RI = \frac{(V_s - V_d)}{V_d} \tag{4}$$

V_s represents the peak systolic velocity, and V_d represents the end diastolic velocity.

In addition, RI includes just two measurement points, while FI contains the entire envelope waveform information. What's more, in the aspect of age, the information FI contained far more than the RI, so FI has more advantages than RI. Furthermore, optic nerve head(ONH) blood flow is well autoregulated in the healthy eyes, OBF has slight influence along with OBP changes. The frequency spectrum of OBP pulse waveform provided a good mathematic method for the development of clinical ocular diseases.

4. CONCLUSIONS

First, the outcome of research show that the pulse of OBP is very easy to be analyzed in the frequency domain, FFT can provide the entire information for the dynamic characteristic of cardiovascular system. It just used one indicator to describe the relationship between OBP and age in detail in the frequency domain.

Second, our result also show that FC and FI were not affected by mean arterial pressure ($p<0.001$). However, FI has obvious difference between people; FI is much lower for the young compared to the value of the older. RI is effected by arterial pressure and widely used as an important marker of peripheral resistance. A small artery has higher vascular resistance than the large arteries, so the movement of the waveform is also reflecting the size of resistance. Related literature has also shown that age can change the endothelium dependent relaxation of the artery; the waveform amplitude is gradually smaller along with the growth of the age. But many interference factors were caused, during we study the relationship between RI and age, the factors will effect the result. On the contrary, the first harmonic and the third harmonic has more important impact on making the decision. so we consider FI as the indictor to judge the relationship between OBP and age. Beyond that, the result of experiment explains that FI of the OBP and age is positive correlation.

Third, the heart rate decides the number of OBP pulse occurred in 2 minutes. From the point of average level, the same level between the sample period and truncation show the same cycle number. Previous studies have shown that the OBP pulse of glaucoma have different with the healthy eyes. The heart rate has important relevance with the OBP measurement, e.g. the higher the heart rate, the lower the pulse amplitude, pulse volume and pulse BF, so we select the data that the heart rate is not much of a difference.

In a word, the results show that the harmonic feature of OBP pulse waveform has important value for studying the characteristic of healthy voluteers' eye vascular. What's more, the outcome suggests that the relevance between OBP and age will bring significant development on clinical ophthalmology.

References

1. Harris, A, Chung, H. S Ciulla, T. A. and Kagemann, L. Progress in measurement of ocular blood flow and relevance to our understanding of glaucoma and age-related macular degeneration. Prog. Retin. *Eye Res 1999;* **18**, 669–687.
2. Nichols WW. Clinical measurement of arterial stiffness obtained from noninvasive pressure waveforms. *Am J Hypertense*. 2005; **18**: 3S–10S.
3. Lehman ED, Gosling RG, Parker JR, da Silva T, Taylor MG. An ocular ocular blood pressure in dependent index of aortic distensibility. Br J Radiol. 1993; **66**:126-131.
4. Kelly, R., Hayward, C., Avolio, A. and O'Rourke, M. Noninvasive dete rminationof age-related changes in the human arterial pulse. *Circulation.1989;* **80**, 1652–1659.
5. Stanley S.Franklin,MD. Hemodynamic Pattern of Age-Related Changes in Ocular ocular blood pressure. UCT Heart Disease Prevention Program, C240 Medical Sciences, University of California Irvine, 1948;**2**(9):1165-1169.
6. Kannel WB, Stokes J III. Hypertension as a cardiovascular risk factor. In: Robertson JIS, ed. Handbook of Hypertension. Clinical Aspects of Hypertension. NewYork, NY: Elsevier Science Publishing Co, Inc; 1985; **6**:15-34.
7. Savage DL, Levy D, Dannenberg AL, Garrison RJ, Castelli WP. Association of schocardiographic left ventricular mass with body size, ocular ocular blood pressure and physica l activity (TheFramingham study). *Am J Cardiol*. 1990; **65**:371-376.
8. Quick, C. M., Berger, D. S. and Noordergraaf, A. Constructive and destructive addition of forward and reflected arterial pulse waves. Am Physiol. *Heart Circ. Physiol*. 2001;**280**, H1519–H1527.
9. Cooley JW, Tukey OW. An algorithm for the machine calculation of complex Fourier series. *Math Comput.* 1965; **19**: 297–301.
10. London GM, Guerin AP, Pannier BM, Marchais SJ, Metivier F. Body height as a determinant of carotid pulse contour in humans. *Hypertens*. 1992;**10**(suppl6):S93-S95.
11. London GM, Guerin AP, Pannier B, Marchais SJ, Benetos A, Safar ME. Increased systolic pressure in chronic uremia: role of arterial wave reflections. *Hypertension*. 1992; **20**:10-19.
12. Kelly R, Hayward C, Ganis J, Daley J, Avolio A, O'Rourke M. Noninvasive registrationof the arterial pressure pulse wave form using high-fidelity applanation tonometry. *J Vasc Med Biol*. 1989; **1**:142-149.
13. London GM, Pannier B, Guérin AP, Marchais SJ, Safar ME, Cuche [J-L]. Ca rdia c hypertrophy, aortic compliance, peripheral resistance, and wave reflection in end - stage renaldisease: comparative effects of ACE inhibition and calcium channel blockade. *Circulation*. 1994; **90**:2786-2796.
14. Nakano Eye Clinic of Kyoto Medical Co-operative,Basic Technology and ClinicalApplications of the Updated Model of Laser Speckle Flowgraphy to Ocular Diseases. *Circulation*.2014; **99**:622-631.
15. Webster IW, Logie AR. A relationship between functional age and health status in female subjects. *J Gerontol*. 1976; **31**: 546–550.

16. Njemanze PC, Beck OJ, Gomez CR, Horenstein S. Fourier analysis of the cerebrovascular system. *Stroke*. 1991; **22**: 721–726.
17. Attinger EO, Anne A, McDonald DA. Use of Fourier series for the analysis of biological systems. *Biophys J*. 1966; **6**: 291–304.
18. Burattini R, Campbell KB. Assessment of aortic pressure power components andtheir linkto overall elastic and resistive arterial properties. *Med Biol Eng Comput.* 1999; **37**: 366–376.
19. Michelson G, Warntges S, Harazny J, Oehmer S, Delles C, Schmieder RE. Effect of nos inhibition on retinal arterial and capillary circulation in early arterial hypertension. *Retina*. 2006; **26**: 437–444.
20. Baskett JJ, Lewis RR, Beasley MG, Gosling RG. Red blood cell velocity changes with age and sex in the common carotid and supra-orbital arteries. Age Ageing.1985; **14**: 102–108.
21. Korner PI, Bobik A, Angus JA, Adams MA, Friberg P. Resistance control in hypertension. *Hypertens*. 1989; **7** (suppl): S125–S134.
22. Folkow B, Grimby G, Thulesius O. Adaptive structural changes of the vascular walls in hypertension and their relation to the control of the peripheral resistance. *Acta Physiol Scand*. 1958; 44: 255–272.
23. Adamson, S. L., Morrow, R. J., Langille, B. L., Bull, S. B. and Ritchie, J. W.(1990) Site-dependent effects of increases in placental vascular resistance on the umbilical arterial velocity waveform in fetal sheep. *Ultrasound Med*. Bio l. **16**, 19–27.
24. Bude, R. O. and Rubin, J. M. Relationship between the resistive index and vascular compliance and resistance. *Radiology* 1999; **211**:411–417.
25. Pieper, G. M. Review of alterations in endothelial nitric oxide productionin diabetes: protective role of arginine on endothelial dysfunction. *Hypertension* 1998; 31:1047–1060.
26. McVeigh, G. E., Allen, P. B., Morgan, D. R., Hanratty, C. G. and Silke, B. Nitricoxide modulation of blood vessel tone identified by arterial waveform analysis. *Clin. Sci* 2001*;* **100**:387–393.
27. Nakatsuka, M., Takata, M., Tada, K. and Kudo, T. Effect of a nitric oxide donoron the ophthalmic artery flow velocity waveform in preeclamptic voluteers. *Ultrasound Med* 2002*;* **21**: 309–313.
28. 24 Kouvidis, G. K., Benos, A., Kyriakopoulou, G., Anastopoulos, G. and Triantafyllou, D.(2000) Colour Doppler ultrasonography of the ophthalmic artery: flow parameters in normal subjects[J]. The significance of the resistance index. Int. Angiol 2000; **19**:319–325.
29. Hochschild R. Validating biomarkers of aging – mathematical approaches and results of a2462–person study. In: Balin AK, ed. Practical Handbook of Human Biologic Age Determination. Boca Raton, FL: CRC Press.1994; 93–144.
30. PC Njemanze,OJ Beck,CR Gomez and S Horenstein. Fourier analysis of the cerebrovascular system. *Stroke*.2010; **22**:721-726.

THE PROSPECT ON APPLICATION OF NANOTECHNOLOGY IN THE TRAINING OF COLLEGE STUDENTS MAJOR IN PHYSICAL EDUCATION

Gang Li
Institute of Physical Education, Shandong University of Finance and Economics, Shandong, China
E-mail: Lg0224shi@163.com

Nanotechnology as a kind of new and high technology, it has many excellent properties. Physical education in college has become a new area for the application of Nanotechnology. This article adopts Literature Study, System Analysis and other research methods to describe the main aspects of Nanotechnology. Through studying from scientific training, equipment and facilities, sports injury and nutrition supplements, sports fatigue and recovery, instruments and equipment for sports scientific research, etc. , this paper discusses the prospect on application of Nanotechnology in the training of college students major in physical education, to promote the rapid development of physical education in colleges.

1. INTRODUCTION

With the development and extension of Nanotechnology in the field of international science and technology, Nanotechnology is increasingly becoming the focus of application in physical training. It is quite mature in operation of nanotechnology industrialization at abroad, and widely applied to sports, while it is still infancy in our country. Especially the weak initiative consciousness of the combination between college sports training and nanotechnology, which affect the further development of sports level in colleges.

This article mainly studying from the perspective of demand for training of physical education major, introduces Nanotechnology into the sports field. Once Nanotechnology fusing with training of physical education major, it will provide a good opportunity for the development of physical education major in colleges in the future. In order to further promote Nanotechnology and products so that they can be fully applied by students of physical education major in college and to do an excellent result in the training. It is believed that Nanotechnology will be more and more applied to physical education major in college with its development in the future.

2. NANOTECHNOLOGY

Nanometer is a length unit, expressed by nm, 1nm = 10 - 9m, or 1 nm is billionths of a meter. Concept of Nanometer comes from the fundamental unit ----- control material in atom and molecule level, combination of extremely small new materials and new machines. 1NM is the sum of a number of nanometer metal atoms diameter in general. The size of Red blood cell is 200 ~ 300nm. The length of the Escherichia coli is 200 ~ 600nm. The viruses that cause human disease are usually nanometers [1]. So nanometer is smaller than red blood cells and bacteria, and same or slightly smaller than viruses.

Nanotechnology mainly includes the following four aspects:

(1) Nanometer Materials: when material becoming to nanometer scale which is in the size range of 0.1-100 nm, the performance of material will mutate and some special performance will appear. It is not only different from the original composed of atoms or molecules, but also different from the macro material consist of materials with special performance. This is the nanometer materials.

(2) Nanometer Dynamics: It is mainly the micro mechanical and micro-motor, or Micro Electromechanical System (MEMS), which used for micro sensors and actuators of mechanical transmission, fiber communication system, special electronic equipment, medical and diagnostic instruments, etc. It is a kind of new technology that similar to the design and manufacture of integrated appliances. Its characteristic is very small, etching depth tends to be required tens to hundreds of micrometers, and error of width is very small. In terms of research and testing, micro deformation and micro friction of atomic should be tested accordingly, etc.

(3) Nanometer Biology and Pharmacology: such as fasten DNA particles in the mica surface with colloidal gold that as same as nanoparticles, doing experiments on interdigitated electrode on the surface of silica which is about biomolecular interaction, phospholipids and aliphatic acid double deck plane biological film, the fine structure of DNA, etc. With nanotechnology, we can also put parts or components into cell to form new materials by using self-assembly method. New drugs, even the micron particle powder, which is about half of insoluble in water; But as for nanoscale particles (that is the superfine particles), it can dissolve in water.

(4) Nanoelectronics: including nanoscale electronic devices based on quantum effects, optical/electrical properties of nanometer structure, the characterization of nano electronic materials, as well as atoms manipulation and assembly, etc. Now, the trend of current electronic technology requires smaller, faster and colder devices and system. Smaller means faster response speed while cooler means smaller power consumption of single device. But smaller is not without limits. Nanotechnology is the last field for constructer and it will have enormous effect.

3. THE PROSPECT ON APPLICATION OF NANOTECHNOLOGY IN THE TRAINING OF COLLEGE STUDENTS MAJOR IN PHYSICAL EDUCATION

There are more and more current applications of nanotechnology in sports, especially those nano products related to sports which take almost 50% of nanometer technology application. With the continuous application of nanotechnology in every field, it is believed that application of nanotechnology in the training of college students major in physical education will be more and more widely, and the application prospect is also very optimistic.

3.1 Application of Nanotechnology in the training of college students major in physical education

Nanotechnology promotes the training of college students major in physical education more scientific. There are some incorrect postures in training progress, some details even can't find by coach while it is perceptive of the slightest by using high technology supplementary means. Such as diving athletes using computer in training in Tsinghua University's diving training base [2]. There are sensor, speed table and miniature electrocardiograph that record students heart load on college students, which can provide information without missing. After entering the water, they can analyze data that provided by graphic method and find any defects of technology or physical condition.

Research has shown that the efficiency of nano structure microprocessor which made by nanotechnology can be improved 100 times, and density of nano-memory can be improved 100 times. The combination of nanotechnology and integration technology could make system of nanometer sensor, to achieve on-line monitor for the small system and complex system. Magnetic recording materials made of nano powder can increase the recording density of tapes and floppy by 10 times, and can significantly improve the fidelity of their performance. All of this provides powerful technical support for the training of college students major in physical education, and it has also provided guarantee for accuracy of the records in competition.

3.2 Application of nanotechnology in required for college students majored in physical training equipment, facilities and equipment etc.

3.1.1. *Nanotechnology and Sports Equipment, Facilities*

College students in the required training equipment and facilities, nanotechnology can be used as a key research direction, through the research method of the invention effectively applied to nano materials, sports equipment and facilities, to help the university sports specialized university obtain better effect in the training, such as the application of nanotechnology in university athletics track, nano plastic runway than general stadiums in use polyurethane plastic runway

More wear resistant, flame retardant, anti - mildew is better, and the environmental performance has reached the international standard. Nano track not only adhering to the general polyurethane plastic of high strength, good elasticity, abrasion resistance, aging resistance, moderate hardness and durable characteristic more showed good mechanical properties, the tensile strength and elongation at break into more than ordinary polyurethane material, and has better wear resistance, flame retardant and anti - mildew properties, service life will be extended. And its excellent rebound value and the compression resilience is highly respected Sports Professional College students.

With a strong, super elastic nano table tennis racket made of carbon fiber composite material with glass steel pole, nylon fabrication of nanometer materials, the application of nano materials production less resistance swimsuit, nano soft volleyball, "nano sports shoes", "nano ball" and other sports equipment. These nanotechnology applications sports equipment and facilities will be more in line with sports and athletes play requirements, more conducive to college students' level of professional sports.

3.1.2. *Nanotechnology and Sports Equipment*

Sports clothes and shoes is the most basic training, college sports specialized university sports training puts forward higher requirements for the equipment better. At present, great progress has been made in the use of various functional fabric production of nanometer technology. As in the aspects of antibacterial, deodorant. The nanometer Tio2, ZnO, Sio2 etc. photocatalytic antibacterial agent and chemical fiber blended spinning to obtain lasting antibacterial fiber, far beyond the traditional antibacterial properties of antibacterial agent. In recent years, has been able to mass production of submicron and nanometer antibacterial agent, through the introduction of fiber blend, designed as antibacterial sportswear [3]. In terms of waterproofing. Waterproof fabric nanotechnology treated not only has the singularity of super hydrophobic, oleophobic, and can keep the original fabric any properties and characteristics, and the chemical composition does not contain harmful to human body. Running shoes materials such as elastic, weight, strength, and number of spike size, installation angle has strict requirements. But the project of different pairs of shoes with different requirements, such as high jump half sole with a nail, heel also has two nails, this design is mainly in order to facilitate the high jump athlete doing take-off action, can brake, in order to obtain the driving force, and not easy to slip. By comparison, gymnastics shoes requires soft and comfortable, on the way of agreement, to match the gymnastics athletes foot leaves the ground straight action.

3.3 Application of nanotechnology in sports college students majored in physical training injury and sports nutrition supplements on

The motion process of agency training college sports majors in university can hardly be avoided damage, repair the injured body is more important than prevention. Chinese medicine plaster is a commonly used method of treatment of sports injury, one of the difficulties it is how to make more of the drug through the skin, into the blood circulation, thus improving the efficacy of the drug. Research shows that: the nanotechnology may function in promoting the drug through the skin barrier plays. Nano Chinese medicine due to the small particle size and high selective adsorption ability, may have a stronger ability to penetrate, so that the nano traditional Chinese medicine more penetrate the skin barrier, enter the blood circulation.

Very important to improve nutrition for sports performance training. Sports nutrition food raw material mainly is the natural flora and fauna, especially traditional Chinese medicine treasure -- traditional Chinese medicine provides a larger research space for sports nutrition, the nutritional ingredients of nano powder or suspension of the use of nanotechnology, can increase the body's absorption rate, better play the role of nutritional supplement biological activity [4].

3.4 Nanotechnology in sports college students majored in physical training fatigue mechanism and the recovery applications

Prevent exercise-induced fatigue of college students majoring in sports training in Colleges and universities and speeding up the recovery process using nanotechnology. The breakthrough of nano technology in medicine, the mechanism of sports fatigue especially in central nervous system and the target organ and target cells will be more in-depth, people can directly study the metabolic process of skeletal muscle, cardiac muscle during exercise in liver and nerve tissue using nano bio chip, to investigate the biological mechanism of the central and peripheral sports fatigue and its recovery, and by some means (such as nano drug) suppression leads to exercise fatigue induced gene expression or accelerating the recovery of gene expression.

3.5 Application of nanotechnology in required for college students majored in physical training in the sports scientific research instruments and equipment

College students majored in physical training process will often use some of the sports scientific research instruments and equipment, the use of nanotechnology in the electronics and computer technology to enable the computer to improve the efficiency of millions of times. Nano machining technology progress makes nano disk appears possible, nanoscale optical disk storage capacity will be 106 times the existing disc, can store 300 trillion Chinese characters [5]. Performance of the electronic instrument products of nanotechnology manufacturing will be greatly improved, and the volume can be greatly reduced. For example, nanotechnology will be very large scale integrated circuit capacity, speed increased by 100 times, and the volume is reduced to 1 / 1000. These products if used for training will be the design innovation, new movements of the traditional action of consolidation, the error action is found to provide strong technical support to correct and formulate specific tactics, to help improve athletic performance accuracy, coaches to have training details, provide technical guarantee to improve the work efficiency and the quality of judges[6].

4. CONCLUSION

With the development of science and technology, the application of nanotechnology in school physical education will become increasingly important, the school sports as an important part of modern social culture, the college sports level, and improve the health level of college students plays an important role[7], but at present we also need to find a combination point nanotechnology and college physical education professional training of the college students, to really make the nanotechnology and school sports combine with each other, and make contributions to promoting the students specialized in sports training level of colleges and universities.

References

1. Yue HL. Effect of new high tech materials on the development of sports. *Journal of Xi'an Physical Education University*, 1999; **2**:5-8.
2. Xin YL. The progress of application and research of nanometer materials. *Anhui chemical industry* 2002; **5**:11-13.
3. Zhan L. The nanotechnology and nano traditional Chinese medicine research progress of. *Shanghai Journal of traditional Chinese medicine*2003; **1**:17-19.
4. Zhang Lide, Mou LD. Nano materials and nano structure. *Science and Technology Press*2001; **2**:21-25.
5. Hui BX, ang Xiang LY, Chang SX et al. Nanotechnology in medicine research and applications. *Journal of China Medicine University* 2001; **32**:16-19.
6. Xia PW. The research of nanotechnology in Sports Biological Sciences in the application prospect of. *Xi'an bodyEducation Journal*2002; **4**:8-12.
7. Hua D. New high-tech materials in sports by means of investigation and analysis of temperature and excellent. *Sports scientific research*2001; **4**:15-18.

THE EFFECTS OF LOW TEMPERATURE ON FLUCTUANTING ASYMMETRY OF CHINA-LINDIAN NATIVE CHICKENS

Changping Wang[a], Bo Wang[b], Qingbo Zhou, Hongsheng Yang, Guangdong Cheng, Tianlin Miao, Chengwu Han, and Bin Wang

College of Life Science, Jiamusi University, Jiamusi, 154007, China
Jiamusi University Animal Husbandry and Veterinary Medicine, Jiamusi,154007,China
E-mail:[a]wcp100200@163.com,[b]408081288@qq.com

In the study, 448 China-Lindian Native Chickens were conducted on three factors (interval days, decrease temperature and stimulation time) .we identified the influence of cold stimuli on welfare of CLNC through detecting fluctuanting asymmetry welfare indexs. The fluctuanting asymmetries were carried out after slaughter. The main fluctuanting asymmetry welfare indexes of detection were asymmetry of wings and asymmetry of legs. It was concluded from the study that the interval days of cold stimuli had great effect on asymmetry of legs ($P<0.05$), while had no impact on asymmetry of wings ($P>0.05$); the decrease temperature of cold stimuli had no impact on asymmetry of wings and asymmetry of legs ($P>0.05$); and the time of cold stimuli had great effect on asymmetry of legs ($P<0.05$), while had no impact on asymmetry of wings ($P>0.05$).

1. INTRODUCTION

Cold stress is a common stress factors to livestock in the cold area of the world, especially to newborn livestock [1]. Cold stress is also the main factors to restrict the development of animal husbandry in the cold area of the world [2].

Breeding the cold resistant varieties, Adding cold stress resistance material, nutrition regulation, improving livestock and poultry breeding management and environmental control facilities could significantly enhance the broiler ability to extreme cold environment, but the cost is higher, the effect is not very good. If we can find a simple and economical method, making the broiler increase its ability to adapt to the cold environment.

2. MATERIALS AND METHODS

2.1. Animal Selection and Daily Feeding Management

In this study, China-Lindian Native Chickens were randomly selected as experimental animals, they were reared from 0 week age, and the chickens were reared in artificial climate chamber. Formal trial started from fourth weeks.

They were reared in 28 cages; whose rearing density was 10 birds / m^2.Feed and water were provided ad libitum. The diet composition of test group and the control group of are same.1-3 weeks of age Commercial complete feed standards respectively are crude protein21.00% and energy 12.10MJ/kg.4-6 weeks of age Commercial complete feed standards respectively are crude protein19.00% and energy 12.60MJ/kg. In strict accordance with the China-Lindian Native Chickens breeding management, we control the temperature, humidity and light in the shed. At the aspect of temperature, the control group at first week with 32-35°C, second weeks with 29-32 °C, third weeks with 26-29°C, fourth weeks with 23-26 °C, fifth weeks with 20-23 °C, the sixth weeks temperature has been controlled at 20°C.At the aspect of humidity, relative humidity inside the house is 60-70%.At the aspect of light, we adopt artificial illumination in n artificial climate chamber. Experiment illumination per day of 1-3 days is 23h; experiment illumination per day of 4-42days is 16h. Water tank cleaned daily to ensure water supply and clean. Daily observating chicken drinking water and faeces, and make a good record everyday.

2.2. Artificial Climate Chamber

We used the artificial climate chamber for temperature control. The artificial climate chamber can be controlled automaticlly by computer program. Thus the chamber's temperature, humidity and illumination can be controlled automaticlly by computer program according to low temperature stimulation test design.

2.3. Low Temperature Stimulation Test Design

There was not low temperature stimulation test to China-Lindian Native Chickens from 1 age in days to 21 age in days. We started to give the low temperature stimulation test to randomly selected 448 birds whose age are three weeks from 22 age in days to 42 age in days. These stimulated CLNC were still reared in normal brooding conditions after stimulating test. In this study, repeat experiments (4 repeations in groups)were conducted on three factors (interval days, decrease of temperature and stimulation time) and three levels (interval of 0 day, interval of 1 day, interval of 2 days; with temperature decreasing 3°C, temperature decreasing 5°C, with temperature decreasing 7°C respectively in different tested groups; stimulating time of 1h, stimulating time of 3h, stimulating time of 6h).3- week-old China-Lindian Native Chickens, physically healthy and weighing evenly, were selected. 448 chickens were divided into 28 groups according to the design, 16 chickens in one group, four repeat experiments in one group, and four chickens for each repetition.

2.4. Fluctuanting Asymmetry Index Determination and Method

Four birds were randomly selected for slaughter from each repeat at one day before the end of the cold stimulus test. The fluctuanting asymmetries were carried out after slaughter. We use vernier caliper to accurately measure the length of left and right sides of the legs (tarsometatarsus), flange width (tarsometatarsus and tibial articulation) and wings length (ulna) when we measured the fluctuanting asymmetry and calculated the FA value. The formula is as follows:FA(%)=2×|R-L|/R+L×100.

3. RESULTS AND ANALYSIS

3.1. Effect of Cold Stimulation on the Wings Fluctuanting Asymmetry of China-Lindian Native Chickens

3.1.1. *The Effects of Interval Days of Cold Stimulus on the Wings Fluctuanting Asymmetry in China-Lindian Native Chickens*

Table 1. The effects of interval days on the wings fluctuanting asymmetry in China-Lindian Native Chickens

interval days	average	difference significant	
		α=0.05	α=0.01
0 d	2.67±0.25	a	A
1 d	2.32±0.32	a	A
2 d	2.55±0.23	a	A
control group	2.03±0.42	a	A

As shown in Table1, the wings fluctuanting asymmetry on all experimental group were higher than the control group, the difference was not significant (P>0.05). The wings fluctuanting asymmetryon interval of 0 day was higher than interval of 1 day and 2 day in cold stimulus trial group, the difference was significant(P>0.05).Through the variance analysis, interval days had no significant effect on the wings fluctuanting asymmetry (P>0.05).

3.1.2. *The Effects of Decreased Temperature of Cold Stimulus on Wings Fluctuanting Asymmetry in China-Lindian Native Chickens*

Tab 2. The effects of decreased temperature on wings fluctuanting asymmetry in China-Lindian Native Chickens

decreased temperature	average	difference significant	
		α=0.05	α=0.01
3°C	2.35±0.26	a	A
5°C	2.77±0.19	a	A
7°C	2.64±0.34	a	A
control group	2.03±0.42	a	A

As shown in Table 2, wings fluctuanting asymmetry on all experimental group were higher than the control group, the difference was not significant (*P*>0.05).Wings fluctuanting asymmetry on decreased temperature 5°C was higher than decreased temperature 3°C and 7°C in cold stimulus trial group, the difference was not significant (*P*>0.05).Through the variance analysis, decreased temperature had no significant effect on wings fluctuanting asymmetry (P>0.05).

3.1.3. *The Effects of Stimulation Time of Cold Stimulus on Wings Fluctuanting Asymmetry in China-Lindian Native Chickens*

Table 3. The effects of the time of cold stimulus on wings fluctuanting asymmetry in China-Lindian Native Chickens

stimulation time	average	difference significant	
		α=0.05	α=0.01
1 h	2.20±0.25	a	A
3 h	2.72±0.37	a	A
6 h	2.84±0.28	a	A
control group	2.03±0.42	a	A

As shown in Table 3, wings fluctuanting asymmetry on all experimental group were higher than the control group, the difference was not significant (P>0.05).Wings fluctuanting asymmetry on stimulation time 6h was higher than stimulation time 1h and 3h in cold stimulus trial group, the difference was not significant (P>0.05).Through the variance analysis, cold stimulation time had no significant effect on wings fluctuanting asymmetry (P>0.05).

3.2. Effect of Cold Stimulation on the Legs Fluctuanting Asymmetry of China-Lindian Native Chickens

3.2.1. *The Effects of Interval Days of Cold Stimulus on the Legs Fluctuanting Asymmetry in China-Lindian Native Chickens*

Table 4. The effects of interval days on thelegs fluctuanting asymmetry in China-Lindian Native Chickens

interval days	average	difference significant	
		α=0.05	α=0.01
0 d	2.38±0.33	a	A
1 d	2.52±0.29	a	A
2 d	3.11±0.41	a	A
control group	2.12±0.36	a	A

As shown in Table 4, the legs fluctuanting asymmetry on all experimental group were higher than the control group, the difference was not significant(P>0.05). The legs fluctuanting asymmetryon interval of 2 day was higher than interval of 0 day and 1 day in cold stimulus trial group, the difference was significant(P>0.05).Through the variance analysis, interval days had significant effect on the legs fluctuanting asymmetry (P<0.05),there was no significant difference among the groups(P>0.05).

3.2.2. *The Effects of Decreased Temperature of Cold Stimulus on Legs Fluctuanting Asymmetry in China-Lindian Native Chickens*

As shown in Table 5, legs fluctuanting asymmetry on all experimental group were higher than the control group, the difference was not significant(P>0.05). Legs fluctuanting asymmetry on decreased temperature 5°C was higher than decreased temperature 3°C and 7°C in cold stimulus trial group, the difference was not significant (P>0.05).Through the variance analysis, decreased temperature had no significant effect on legs fluctuanting asymmetry (P>0.05).

Table 5. The effects of decreased temperature on legs fluctuanting asymmetry in China-Lindian Native Chickens

decreased temperature	average	difference significant	
		α=0.05	α=0.01
3°C	2.31±0.19	a	A
5°C	2.87±0.31	a	A
7°C	2.53±0.29	a	A
control group	2.12±0.36	a	A

As shown in Table 5, legs fluctuanting asymmetry on all experimental group were higher than the control group, the difference was not significant (P>0.05).Legs fluctuanting asymmetry on decreased temperature 5°C was higher than decreased temperature 3°C and 7°C in cold stimulus trial group, the difference was not significant (P>0.05).Through the variance analysis, decreased temperature had no significant effect on legs fluctuanting asymmetry (P>0.05).

3.2.3. *The Effects of Stimulation Time of Cold Stimulus on Legs Fluctuanting Asymmetry in China-Lindian Native Chickens*

Table 6. The effects of the time of cold stimulus on legs fluctuanting asymmetry in China-Lindian Native Chickens

stimulation time	average	difference significant	
		α=0.05	α=0.01
1 h	2.48±0.20	a	A
3 h	2.75±0.37	a	A
6 h	2.21±0.33	a	A
control group	2.12±0.36	a	A

As shown in table 6, legs fluctuanting asymmetry on all experimental group were higher than the control group, the difference was not significant (P>0.05). Legs fluctuanting asymmetry on stimulation time 3h was higher than stimulation time 1h and 6h in cold stimulus trial group, the difference was not significant(P>0.05).Through the variance analysis, cold stimulation time had significant effect on legs fluctuanting asymmetry (P<0.05), there was no significant difference among the groups(P>0.05).

4. DISCUSSION

Fluctuanting asymmetry reflects that the growth condition is good or not under the condition of certain raising environment. It is a behavior parameters to general evaluate the environmental condition and genetic stress. It is also an index to evaluate animal welfare and comfort [4][5].The greater the value of FA, the more asymmetry of the traits of both sides of the poultry body. The two different parts of wings and legs were selected in this experiment according to the characteristics of broiler bone development. The results of this experiment show that interval days decreased temperature and cold stimulation time had no significant effect on the wings fluctuanting asymmetry($P>0.05$).The wings fluctuanting asymmetry of cold stimulation training chickens were higher than the control group without low temperature stimulation. The results of this experiment show that interval days decreased temperature and cold stimulation time had no significant effect on the legs fluctuanting asymmetry($P>0.05$).The legs fluctuanting asymmetry of cold stimulation training chickens were higher than the control group without low temperature stimulation.

Acknowledgments

This work was funded by Natural science foundation of China (31302008) and Natural science foundation of Heilongjiang province (QC2012C104).

References

1. Chun-yu Li. Study on the effect of low temperature on broiler chickens.master thesis: Northeast Agricultural University 2001. (In Chinese).
2. Wen-dian Cui, Guo-xian Zhao, Xiao-lei Zuo,et al.The relationship between environmental temperature and feed intake, dietary energy and protein level. Heilongjiang animal husbandry and veterinary medicine 2005; **9**:11 – 12. (In Chinese).
3. Dan Xia. Low temperature and energy on production performance and biochemical index of laying ducks. Master thesis. Northeast Agricultural University 2007. (In Chinese).
4. BRUNO L D G, FURLAN R L, MALHEIROS E B, MACARI M. Influence of early quantitative feed restriction on long bone growth at different environmental temperatures in broilers chickens. Br. Poultry Sci 2000; **41**:389 – 394.
5. MOLLER A P, MANNING J T. Growth and developmental instability.Vet.J 2003; **166**:19 – 27.

FACILE GREEN FABRICATION OF ANTIBACTERIAL ZNO NANOPARTICLES

Wei Shao[a,*], Shuxia Wang, Jimin Wu, Hui Liu, Min Huang, and Zhiyin Guo

College of Chemical Engineering, Nanjing Forestry University, Nanjing 210037, China
E-mail: [a]w.shao@njfu.edu.cn
[*]Corresponding author

ZnO nanoparticles were fabricated by a facile green method. The effects of $Zn(NO_3)_2{\cdot}6H_2O$ concentration, reaction temperature, calcining temperature and calcining time on the diameters of ZnO nanoparticles were investigated. The effect of ZnO nanoparticles on antibacterial ability was also evaluated. The results demonstrated that the optimum $Zn(NO_3)_2{\cdot}6H_2O$ concentration and reaction time in this system are 0.1 mol/L and 95°C, and the best calcining temperature and time are 450°C and 3h, respectively. ZnO nanoparticles exhibited good antibacterial activity against *S. aureus* and *C. albicans*.

1. INTRODUCTION

Nanocrystalline transition metal oxides have attracted great attention due to their unique properties, which are very useful to fabricate nano devices [1]. Zinc oxide (ZnO) is a unique inorganic engineered nanomaterial since it has its distinctive and fascinating properties which have various applications in the science and technology areas[2]. ZnO has been widely used in many fields such as textiles[3], solar cells[4], catalyst[5], light emitting devices[6], industrial coatings[7], paintings[8], inorganic antibacterial agents[9] and so on. ZnO has demonstrated to possess various kinds of morphologies, such as nanospheres, nanorods, nanosheets, nanowires, nanowhiskers and nanotubes[10]. Researches on ZnO nanospheres have also been widely investigated due to their diverse excellent properties of large surface interactions and low densities[10]. Hitherto, conventional templating methods were used the most frequently to prepare ZnO nanospheres[11-13]. However, these methods have some disadvantages that they not only consume lots of energy and time to remove the templates, but also could bring in heterogeneous impurities[10]. To overcome these problems, template-free synthesis methods and green synthesis strategies have been intensely developed [14-16].

In this study, we report a facile, convenient and template-free method for synthesis of ZnO nanoparticles. The effects of preparation concentrations including Zn $(NO_3)_2{\cdot}6H_2O$ concentration, the reaction time, the calcining temperature and time on the diameters of ZnO nanoparticles were investigated. Furthermore, the antibacterial activities of ZnO nanoparticles against *S. aureus* and *C. albicans* were studied.

2. EXPERIMENTAL

2.1. Production of Zno Nanoparticles

ZnO nanoparticles were prepared by eco-friendly green synthesis method. Suitable $Zn(NO_3)_2 \cdot 6H_2O$ and urea were dissolved in 10 mL of de-ionized water to achieve concentrations of 0.2 mol/L and 0.6 mol/L, respectively. Then the mixture was stirred constantly at 97°C for 6 h and naturally cooled to room temperature before filtering, washing with 1% ammonia until the filtrate was neutral and then rinsing with ethanol, drying at 80°C for 4 h. Finally, the obtained specimens were kept in a pre-heated muffle furnace at 450°C for 3 h.

2.2. Characterization

The diameters of fabricated ZnO nanoparticles were analyzed by a Melvin Nano ZS. All tests were proceeded in triplicate and the average values were calculated.

2.3. Antibacterial Activities

The antibacterial activities of the prepared ZnO nanoparticles were investigated against *S. aureus* ATCC 6538 and *C. albicans* CMCC (F) 98001 by disk diffusion method.

Lawns of test bacteria (about 1×10^5 CFU/mL) were prepared on TSA. The sterilized ZnO nanoparticles were then carefully placed upon the lawns. The plates were placed in a 37°C incubator for 24 h. Then inhibitory action of tested samples on the growth of the bacteria was determined by observing diameter of inhibition zone.

3. RESULTS AND DISCUSSION

The effect of Zn $(NO_3)_2 \cdot 6H_2O$ concentrations in reaction system on the diameter of ZnO nanoparticles is illustrated in Figure 1a. Figure 1a showed that for a given temperature, the diameter of fabricated ZnO nanoparticles increased with increasing $Zn(NO_3)_2 \cdot 6H_2O$ concentration. The effect of reaction temperature in reaction system on the diameter of ZnO nanoparticles was shown in Figure 1b. For a given Zn $(NO_3)_2 \cdot 6H_2O$ concentration, firstly, the diameter of ZnO nanoparticles increased with increasing reaction temperature. Then it achieved the maximum value at 95°C. After that, the diameter of ZnO nanoparticles decreased dramatically and it reached a minimal value at 97°C.

The effects of calcining temperature and calcining time on the diameters of ZnO nanoparticles were displayed in Figure 1c and 1d, respectively. From Figure 1c, there existed a minimum diameter (184 nm) of ZnO nanoparticles with calcining temperature at 450°C. For a given Zn $(NO_3)_2 \cdot 6H_2O$ concentration, reaction temperature and calcining temperature, the diameter of ZnO nanoparticles decreased with increasing calcining time at first. Then the diameter reached the minimum with calcining time for 3h. The temperature increased again with increasing calcining time.

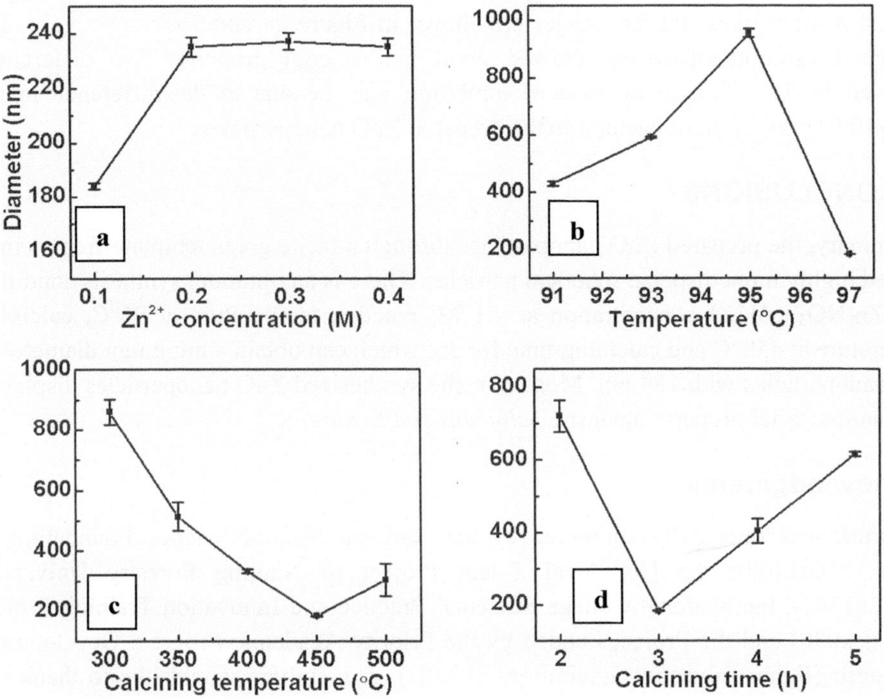

Fig. 1. The effects of Zn^{2+} concentration (a), reaction temperature (b), calcining temperature (c) and calcining time (d) on the diameter of ZnO nanoparticles.

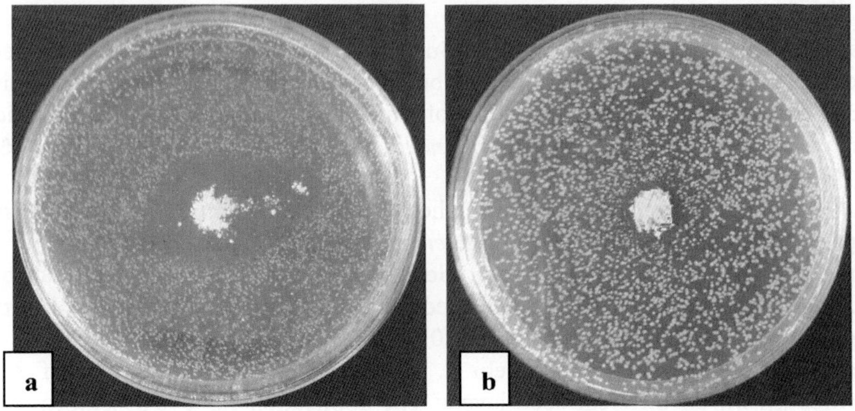

Fig. 2. Inhibition zones of ZnO nanoparticles: *C. albicans* (a), *S. aureus* (b).

Two strains including fungal *C. albicans* CMCC (F) 98001 and Gram-positive *S. aureus* ATCC 6538 were selected for antibacterial tests because they are usually associated with the infections. The antibacterial activities of ZnO nanoparticles against *C. albicans* and *S. aureus* were investigated by disk diffusion method. The prepared ZnO nanoparticles were placed on a lawn of tested bacteria in TSA, respectively. The antibacterial activity is determined by the clear zone of inhibition around the samples after 24 h incubation and the images are shown in Figure 2a and 2b, respectively. The fabricated ZnO nanoparticles showed good antibacterial property. The differences observed in the diameter of zone of inhibition may be due to the difference in the susceptibility of different bacteria to the prepared ZnO nanoparticles.

4. CONCLUSIONS

In summary, the prepared ZnO nanoparticles through a facile green template-free method showed highly monodisperse spherical particles. There is an optimum synthesis condition with $Zn(NO_3)_2{\cdot}6H_2O$ concentration at 0.1 M, reaction temperature at 95°C, calcining temperature at 450°C and calcining time for 3h, which can obtain a minimum diameter of ZnO nanoparticles with 184 nm. Moreover, the synthesized ZnO nanoparticles displayed good antibacterial property against *C. albicans* and *S. aureus*.

Acknowledgments

The work was financially supported by the National Natural Science Foundation of China (51401109), the High-level Talent Project of Nanjing Forestry University (GXL201301), the National College Students' Practice and Innovation Training Project (2015sjcx030) and the Project Funded by the Priority Academic Program Development of Jiangsu Higher Education Institutions (PAPD). The authors would like to thank the Advanced Analysis & Testing Center of Nanjing Forestry University.

References

1. Madan HR, Sharma SC, Udayabhanu, Suresh D, Vidya YS, *et al.* Facile green fabrication of nanostructure ZnO plates, bullets, flower, prismatic tip, closed pine cone: Their antibacterial, antioxidant, photoluminescent and photocatalytic properties. *Spectrochimica acta Part A, Molecular and biomolecular spectroscopy* 2016; **152**: 404-416.
2. Saikia I, Hazarika M and Tamuly C. Synthesis, characterization of bio-derived ZnO nanoparticles and its catalytic activity. *Materials Letters* 2015; **161**: 29-32.
3. Petkova P, Francesko A, Perelshtein I, Gedanken A and Tzanov T. Simultaneous sonochemical-enzymatic coating of medical textiles with antibacterial ZnO nanoparticles. *Ultrasonics sonochemistry* 2016; **29**: 244-250.
4. Nayeri FD, Soleimani EA and Salehi F. Synthesis and characterization of ZnO nanowires grown on different seed layers: The application for dye-sensitized solar cells. *Renewable Energy* 2013; **60**: 246-255.
5. Suwanboon S, Amornpitoksuk P, Sukolrat A and Muensit N. Optical and photocatalytic properties of La-doped ZnO nanoparticles prepared via precipitation and mechanical milling method. *Ceramics International* 2013; **39**: 2811-2819.
6. Shabnam, Kant CR and Arun P. White-light emission from annealed ZnO:Si nanocomposite thin films. *Journal of Luminescence* 2012; **132**: 1744-1749.

7. John S, Joseph A, Jose AJ and Narayana B. Enhancement of corrosion protection of mild steel by chitosan/ZnO nanoparticle composite membranes. *Progress in Organic Coatings* 2015; **84**: 28-34.
8. El-Feky OM, Hassan EA, Fadel SM and Hassan ML. Use of ZnO nanoparticles for protecting oil paintings on paper support against dirt, fungal attack, and UV aging. *Journal of Cultural Heritage* 2014; **15**: 165-172.
9. Liu Y, He L, Mustapha A, Li H, Hu ZQ*, et al.* - Antibacterial activities of zinc oxide nanoparticles against Escherichia coli O157:H7. 2009; - 107: - 1201.
10. Wang Y, Yang J, Kong J, Jia H and Yu M. ZnO microspheres: Controllable preparation and optical properties. *Superlattices and Microstructures* 2015; **86**: 228-235.
11. Han J, Liu Z, Guo K, Zhang X, Hong T*, et al.* AgSbS2 modified ZnO nanotube arrays for photoelectrochemical water splitting. *Applied Catalysis B: Environmental* 2015; **179**: 61-68.
12. Li X, Chen Y, Kumar A, Mahmoud A, Nychka JA*, et al.* Sponge-Templated Macroporous Graphene Network for Piezoelectric ZnO Nanogenerator. *ACS applied materials & interfaces* 2015; **7**: 20753-20760.
13. Zang Z, Wen M, Chen W, Zeng Y, Zu Z*, et al.* Strong yellow emission of ZnO hollow nanospheres fabricated using polystyrene spheres as templates. *Materials & Design* 2015; **84**: 418-421.
14. Kadam AN, Dhabbe RS, Kokate MR, Gavade NL, Waghmare PR*, et al.* Template free large scale synthesis of multi-shaped ZnO nanostructures for optical, photocatalytical and antibacterial properties. *Journal of Materials Science: Materials in Electronics* 2015; **26**: 8367-8379.
15. Thema FT, Manikandan E, Dhlamini MS and Maaza M. Green synthesis of ZnO nanoparticles via Agathosma betulina natural extract. *Materials Letters* 2015; **161**: 124-127.
16. Suresh D, Nethravathi PC, Udayabhanu, Pavan Kumar MA, Raja Naika H*, et al.* Chironji mediated facile green synthesis of ZnO nanoparticles and their photoluminescence, photodegradative, antimicrobial and antioxidant activities. *Materials Science in Semiconductor Processing* 2015; **40**: 759-765.

EFFECTS ON T7 DNA LIGASE BY 2'-FLUORINATED NUCLEOTIDE SUBSTITUTION

Hua Ding[a], Bin Zhao[b], Qingyun Niu, Songlin Wu, and Jirui Song

Key Laboratory of National Sport Bureau, Department of Human Movement Science, Shenyang Sport University, Shenyang 110102, China

E-mail:[a]707813077@qq.com,[b]firelion2006@126.com

To further understand the ligation mechanism, we investigated the effects of 2′-fluorinated Nucleotide on the T7 DNA ligation efficiency. Molecular beacon (MB) based fluorescence resonance energy transfer assay was used to monitor the nick-joining process. Results showed that the substitutions on 5′-terminus of the nick inhibited ligation efficiency by 21.8±1.6 %, and substitution on 3′-terminus of the nick inhibited ligation by 53.9±10.8 %, Dual-substitution at both 5′- and 3′-termini decreased the ligation efficiency by 45.1±12.8 %. Corresponding kinetic parameters, V_{max}, K_m and k_{cat} were determined. Mismatches on 3′-terminus and 5′-terminus of the nick influenced the ligation differently. Results and conclusions in the study demonstrated for the first time the capability of regulating the ligation efficiency of T7 DNA ligase by 2′-FN substitutions. These results could be used to uncover the mechanism of ligation and to facilitate other biological applications.

1. INTRODUCTION

Researches on nucleic acids have gained much remarkable progress since the nucleotide derivatives were discovered [1]. Chemical modification on the 2'-position of the ribose linked with base can indirectly improve the stability of phosphodiester bond and increase the Tm value [2]. 2'-fluorinated nucleotide (2'-F) is the most typically modified nucleotide derivatives [3], which has been used to improve the stability of RNA such as siRNA [4]. 2'-F substitution is described as -H of the -OH groups on the ribose's C2' positions are substituted by fluorinated group [5]. It is one of naturally mutational RNA derivatives which were found in mammalian ribosomal and transfer RNAs [6]. The lighting point of 2'-F is that it has also effectively constrained the ribose in the C3'-endo pucker conformation, which leads an A-type DNA or RNA duplexes [7].

T7 DNA ligase is widely used in genetic engineering research, which catalyzing mechanism and structure have been known clearly. It is a representative ligase which depends on ATP [8] performing very well in joining the blunt-ended of dsDNA [9]. T7 DNA ligase is an Mg2+ and ATP dependent enzyme that seals DNA nicks in three steps: it covalently binds AMP, transadenylates the nick phosphate, and catalyses formation of the phosphodiester bond releasing AMP. The ligation efficiency influenced by Mg2+, ATP and Mg2+-ATP compound have been reported [10]. In a specific condition, the ligation reached its maximum[11]. Recent years, influences on T7 DNA ligases by mismatching and mutation oligos were well studied [12].

Although the catalytic mechanism of T7 DNA ligase has been delineated, little is known about the interaction change of DNA-enzyme after chemical modification of oligonucleotide. This lights our interest in studying the influences on the ligation reaction of T7 DNA ligase after target sequence were substituted by 2'-F. We previously proved that chemical modification especially LNA (Locked nucleic acid) indeed improved the discrimination capability of oligonucleotides by increasing the melting temperature [13]. And we have studied the relationship between short primers and DNA polymerase [14].

2. MATERIALS AND METHODS

2.1. Chemicals

T7 DNA ligase with the concentration of 350 U/μl and then prepared with ligation buffer. All these oligonucleotides (Table 1) were purchased from TaKaRa Biotechnology Co., Ltd. (Dalian, China). The stock solutions of oligonucleotides at the concentration of 100 μM.

Table 1. Oligonucleotides used in this study

Name		Sequence (3′ end to 5′ end)
MB		5′-F-AATACACACTCTGCTGTGATGTCTCATCTGTGTATT-Q-3′
Complementary	L-ON	GAGACGACAC
	R-ON	TACAGAGTAG
	C-ON	GAGACGACACTACAGAGTAG
length of R-ON	R-7	TACAGAG
	R-8	TACAGAGT
	R-9	TACAGAGTA
	R-11	TACAGAGTAGT
	R-12	TACAGAGTAGTG
	R-13	TACAGAGTAGTGT
FN	5'-F	GAGACGACAC
substitutions	3'-F	TACAGAGTAG
Mismatches	L1-G:g	GAGACGACAg
	L1-G:t	GAGACGACAt
	L1-G:a	GAGACGACAa
	R1-A:g	gACAGAGTAG
	R1-A:c	cACAGAGTAG
	R1-A:a	aACAGAGTAG
	R2-T:c	TcCAGAGTAG
	R3-G:a	TAaAGAGTAG
	R4-T:c	TACcGAGTAG
	R5-C:c	TACAcAGTAG
	R6-T:c	TACAGcGTAG
	R7-C:c	TACAGAcTAG
	R8-A:c	TACAGAGcAG

Note: The nucleotide sequences are written from 3'-end to 5'-end except MB. The underlined and the lower cases represent the 2'-OMeN and mismatched nucleotides, respectively

2.2. Preparation of Oligonucleotides

Molecular beacon (MB) of 36 nucleotides was labeled with fluorescein isothiocyanate (FITC) at the 5'-end and DABCYL at the 3'-end (Table 1). Considering the stability and structure of MB.

Left oligonucleotide (L-ON) and right oligonucleotide (R-ON), each being 10 nucleotides long, respectively. L-ON and R-ON are complementary to the left portion and the right portion of the loop, respectively. Furthermore, the nucleotide at either 5'-end of the L-ON or the nucleotide at 3'-end of the R-ON have also been replaced with mismatched nucleotides (Table 1).

2.3. FRET analysis of ligation efficiency

Equal molar ratio of MB, L-ON and R-ON at the same concentration of 50 nM was pre-incubated at 37 ℃ in 100 μl ligation buffer. 2.0 U T7 DNA ligase was added into the reaction solution, the florescence intensity was recorded every 20 seconds for 120 circles.

Ligation initial velocity (V0) was determined from the changing of fluorescence intensity (ΔFI) with respect to the reaction time. ΔFI was calculated using formulation △FI = (Ft-F1) [15], where F1 was the initial fluorescence intensity at the time T7 DNA ligase was added. All the data presented following are the averaged results of three independent measurements.

2.4. Enzymatic kinetics of T7 DNA ligation

To compare the ligation efficiency, kinetic parameters of T7 DNA ligation were determined. The substrate (MB, L-ON, R-ON) was prepared at the concentrations range from 50 nM to 400 nM. Ligation initial velocities with different substrate concentrations were calculated. The substrate concentrations and the initial velocities were plotted in the Lineweaver-Burk graph. Km, Vmax and Kcat were got in the units of nM, nM/s and s-1 respectively.

3. RESULTS

3.1. FRET assay of ligation

As shown in Figure 1, there is significant difference in fluorescence intensity when oligonucleotides were mixed with MB. The fluorescence spectrum of MB alone represents a background having a small peak at 518 nm of FITC (Figure 1a). The hybridization of L-ON and R-ON with the whole loop of MB through complementary base-pairing resulted in a slight increase in fluorescence intensity (Figure 1b). However, the fluorescence was enhanced substantially when T7 DNA ligase is added into the reaction solution (Figure 1c). When C-ON is added into the reaction solution, its hybridization with MB generates a fluorescence spectrum similar to Figure 1C (Figure 1d).

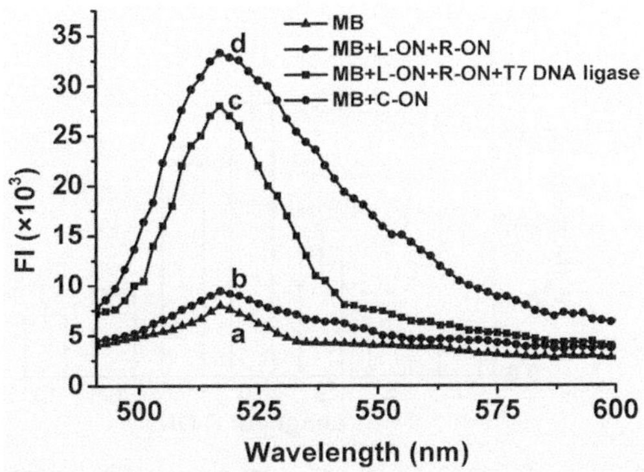

Fig. 1. Fluorescence spectra of MB hybridized with different oligonucleotides.

3.2. Optimization of ligation reaction

T7 DNA ligase in the concentration range of 0.455 U/ml to 8.75 U/ml was added into the reaction solution and the fluorescence intensity was monitored (Figure 2). The initial ligation velocity V0 of the ligation reaction was directly proportional to the concentration of T7 DNA ligase.

We repeated the FRET using the same method described above. Figure 3 shows clearly that when R-7 was hybridized with MB, the initial velocity was relatively low.

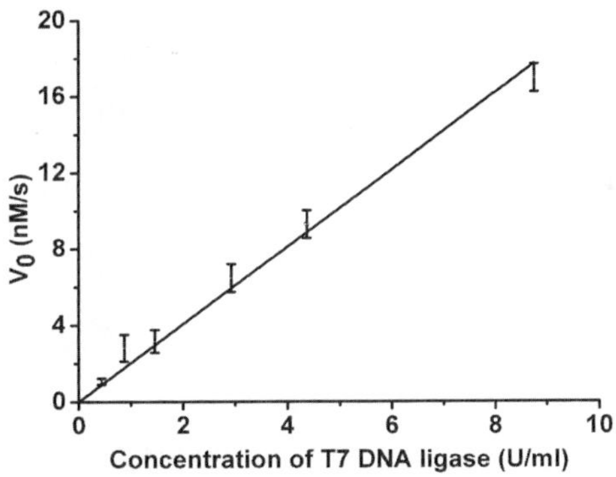

Fig. 2. Fluorescence intensity changes of FRET-based DNA ligase assay for different T7 DNA ligase's concentrations.

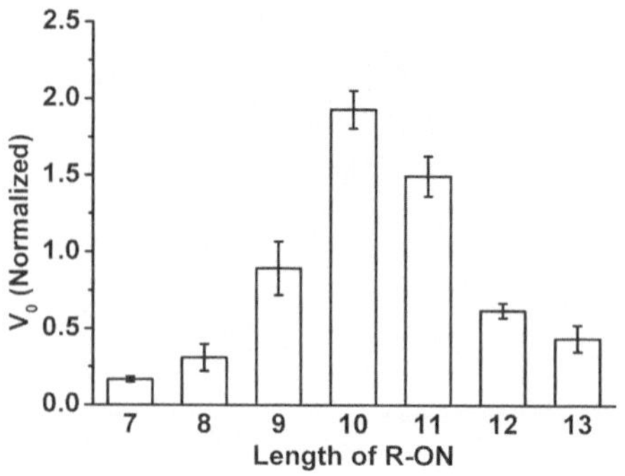

Fig. 3. Effects of the length of R-ON on ligation efficiency in terms of ligation velocity (V_0).

3.3. Effect of the substitution and mismatched on the ligation efficiency

As shown in Figure 4A, when the 2'-F substitution was made on the 5'-end of L-ON, the initial ligation velocity was reduced by 21.8±1.6 % compared with no substitution, while the 2'-F substitution on the 3'-end of R-ON decreased the initial ligation velocity by 53.9±10.8 %. When both sides of the nick were substituted, the corresponding initial velocity was less inhibited (by 45.1±12.8 %) compared with the 3'-FN substitution.. Further study was conducted to understand the effects of mismatches on the nick (Figure 4B, 4C, 4D).

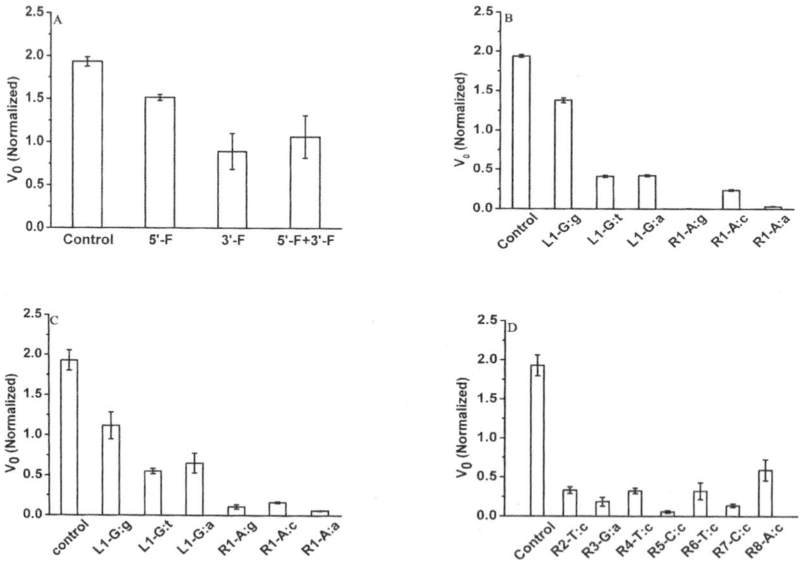

Fig. 4. Influences of 2′-FN substitutions and mismatches on ligation efficiency.

3.4. Influence on Kinetics after 2'-F substitution

The kinetic analysis of the F-substitution was also conducted. By varying the concentration of the substrates in the range from 50 nM to 400 nM were prepared and adding 2.0 U T7 DNA ligase in every well, the initial velocity of the ligation process was graphed as the Lineweaver-Burk plots (Figure 5), and consequently, kinetic parameters of V_{max}, k_m and K_{cat} were calculated (Table 2).

Table 2. Kinetic parameters for different 2′-FN substituted substrates

Substitution site	V_{max} (nM/s)	K_m (nM)	k_{cat} (s^{-1})	k_{cat}/K_m $(nMs)^{-1}$
Control	2.3±0.1	5.7±0.4	4.0±0.8	0.7×103
5′-F	5.8±0.2	17.5±0.2	11.2±0.3	0.6×103
3′-F	264.5±35.9	911.9±0.2	361.5±13.5	0.4 × 103
5′-F+3′-F	19.6±1.9	60.0±0.2	16.3±2.5	0.3 × 103

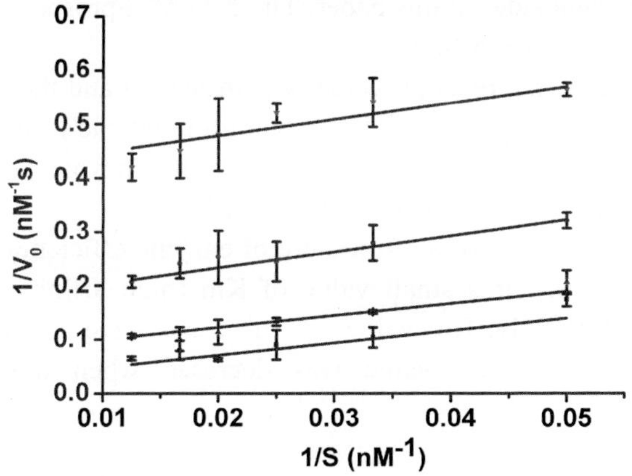

Fig. 5. Libeweaver-Burk plots of control and 2′-FN substituted substrates.

4. DISCUSSION

A certain concentration of T7 DNA ligase catalyzing the ligation process was important for the latter research. Proved by Michaelis-Menten equation, the initial reaction velocity catalyzed by enzymes is directly proportional to the concentration of the enzyme within a certain concentration range. It demonstrated that the initial ligation velocity was directly proportional to the T7 DNA ligase concentrations range examined.

The reasons caused the lower rate of MB open could be supposed as the following: the first was that the ligation rate of T7 DNA ligase was low, and the second was that the ligation products couldn't make the MB open when the length of R-ON were shorter than 10. In contrast, when the optimum length of complementary strands for the ligation process is 10 nucleotides as expected, the ligation velocities of other lengths are all lower than 10 nucleotides.

Both 2'-F substituted and base mismatched substrates inhibited the ligation process; After 2'-F substitution, C2'-endo conformation of DNA was constrained to C3'-endo, which indicates that the change of conformation probable to be responsible for the inhibition; Inhibiting more when changes were happened on 3'-OH side, which means that the effect seems to be position dependent.

The catalysis mechanism of T7 DNA ligase is to link the nick between phosphate and hydroxyl termini [11]. Reports has shown that F modification on the 2'-position of the ribose linked with base can indirectly improve the stability of the inter-nucleotide phosphate bond and increase the Tm value [2]. This may cause the enzyme and substrate binding efficiency becoming lowe. It was reported that mismatched bases around the ligation positions indeed inhibited the ligation velocity of DNA ligase, both on the terminal positions and other positions [16].

The above researches suggested that the binding between T7 DNA ligase and nick will be influenced more if bases were modified or mismatched on the 3'-OH side, which was defined as the 'right side' in this paper. The delighted prospect of the results was the application of 2'-F on the RCA biology.

In enzyme kinetics, the reaction velocity is measured and the effects of varying the conditions of the reaction investigated. Studying an enzyme's kinetics in this way can reveal the catalytic mechanism and efficiency of the enzyme [17]. Enzyme becomes saturated with substrate and the rate reaches Vmax [18].

kcat/Km is often thought of as a measure of enzyme efficiency. Either a large value of kcat (rapid turn over) or a small value of Km (high affinity for substrate) makes kcat/Km large. In this study, Km values of substrates were small, which caused large kcat/Km values, but kcat/Km value was decresed when the substrate was 2'-F substituted.

References

1. Kurreck J. Antisense technologies. Improvement through novel chemical modifications. Eur J Biochem 2003; **270**: 1628–1644.
2. Braasch DA, Jensen S, Liu Y. RNA interference in mammalian cells by chemically–modified RNA. *Biochemistry* 2003; **42**: 7967–7975.
3. Grunweller A, Wyszko E, Bieber B. Comparison of different antisense strategies in mammalian cells using locked nucleic acids, 2'–O–methyl RNA, phosphorothioates and small interfering RNA. *Nucleic Acids Res* 2003; **31**: 3185–3193.
4. Freier SM, Altmann KH. The ups and downs of nucleic acid duplex stability: structure–stability studies on chemically–modified DNA:RNA duplexes. *Nucleic Acids Res* 1997; **25**: 4429–4443.
5. Majlessi M, Nelson NC, Becker MM. Advantages of 2'–O–methyl oligoribonucleotide probes for detecting RNA targets. *Nucleic Acids Res* 1998; **26**: 2224–2229.
6. Czauderna F, Fechtner M, Dames S. Structural variations and stabilising modifications of synthetic siRNAs in mammalian cells. *Nucleic Acids Res* 2003; **31**: 2705–2716.

7. Harborth J, Elbashir SM, Vandenburgh K. Sequence, chemical, and structural variation of small interfering RNAs and short hairpin RNAs and the effect on mammalian gene silencing. *Antisense Nucleic Acid Drug Dev* 2003; **13**: 83–105.
8. Rossi R, Montecucco A, Ciarrocchi G. Functional characterization of the T4 DNA ligase: a new insight into the mechanism of action. *Nucleic Acids Res* 1997; **25**: 2106–2113.
9. Subramanya HS, Doherty AJ, Ashford SR. Crystal structure of an ATP–dependent DNA ligase from bacteriophage T7. *Cell* 1996; **85**: 607–615.
10. Cherepanov AV, de Vries S. Kinetic mechanism of the Mg2+–dependent nucleotidyl transfer catalyzed by T4 DNA and RNA ligases. *J Biol Chem* 2002; **277**: 1695–1704.
11. Tang Z, Wang K, Tan W. Real–time monitoring of nucleic acid ligation in homogenous solutions using molecular beacons. *Nucleic Acids Res* 2003; **31**: e148.
12. Cherepanov A, Yildirim E, de Vries S. Joining of short DNA oligonucleotides with base pair mismatches by T4 DNA ligase. *J Biochem* 2001; **129**: 61–68.
13. Piao X, Yan Y, Yan J. Enhanced recognition of non–complementary hybridization by single–LNA–modified oligonucleotide probes. *Anal Bioanal Chem* 2009; **394**: 1637–1643.
14. Zhao G, Guan Y. Polymerization behavior of Klenow fragment and Taq DNA polymerase in short primer extension reactions. *Acta Biochim Biophys Sin (Shanghai)* 2010; **42**: 722–728.
15. Liu L, Tang Z, Wang K. Using molecular beacon to monitor activity of E. coli DNA ligase. *Analyst* 2005; **130**: 350–357.
16. Pritchard CE, Southern EM. Effects of base mismatches on joining of short oligodeoxynucleotides by DNA ligases. *Nucleic Acids Res* 1997; **25**: 3403–3407.
17. Tseng SJ, Hsu JP. A comparison of the parameter estimating procedures for the Michaelis–Menten model. *J Theor Biol* 1990; **145**: 457–464.
18. S. Schnell PKM. A century of enzyme kinetices: reliability of the Km an Vmax estimates. *Comments on theoretical biology* 2003; **8**: 169–187.

THE STUDY OF COLD STIMULATION EFFECTING ON IMMUNE WELFARE INDEX OF CHINA-LINDIAN NATIVE CHICKENS

Changping Wang[a], Bo Wang[b], Guangdong Cheng, Hongsheng Yang, Qingbo Zhou, Chengwu Han, Tianlin Miao, and Bin Wang

College of Life Science, Jiamusi University, Jiamusi,154007, China
Jiamusi University Animal Husbandry and Veterinary Medicine, Jiamusi, 154007, China
E-mail:[a]wcp100200@163.com,[b]408081288@qq.com

3-week-old 448 China-Lindian Native Chickens were selected as experimental animals in this experiment. They were conducted on cold stimuli at interval days, decrease temperature and stimulation time. We identified the influence of cold stimuli on welfare of CLNC through detecting immune welfare indexs. The main immunel welfare index of detection were the levels of interleukin-2(IL-2) and complement (C_3).The content change of interleukin-2(IL-2) and complement (C_3) were detected by enzyme-linked immunosorbent assay (ELISA). It was concluded from the study that the interval days of cold stimuli had great effect on the levels of IL-2($P<0.05$), while had no impact on complement (C_3)($P>0.05$); the decreasing temperature of cold stimuli had no impact on the level of IL-2, ($P<0.05$) and complement (C_3) ($P>0.05$); and the time of cold stimuli had no impact on the level of complement (C_3) and IL-2 in immune welfare aspect.

1. INTRODUCTION

Cold stress influences the broiler chicks in the breeding process. Cold stress can make the broiler chick grow slowly, resist disease poorly, even die. It will have a significant impact on the northern winter cold area of our country poultry production. The workgroup therefore make broilers suitable cold stimulation training.

We try to find the effect of low temperature stimulation in the process of chicken brooding by immune aspect. The main immune welfare index of detection were the levels of interleukin-2(IL-2) and complement (C_3).The aim of the study was to determine the effects of cold stress on China-Lindian Native Chickens immune index.

2. MATERIALS AND METHODS

2.1. Test Design

In the study, China-Lindian Native Chickens were randomly selected as experimental animals, they were reared from 0 week age, and the chickens were reared in artificial climate chamber.

The specific schemes of cold stimulus are shown in Table 1, Table2 and Table 3.

Table 1. The design of cold stimulus(1)

cold stimulus-interval of 0 day								
temperature decreasing 3 °C			temperature decreasing 5 °C			temperature decreasing 7 °C		
1h	3 h	6h	1h	3 h	6h	1h	3 h	6h

Table 2. The design of cold stimulus(2)

cold stimulus-interval of 1 day								
temperature decreasing 3°C			temperature decreasing 5°C			temperature decreasing 7°C		
1h	3 h	6h	1h	3 h	6h	1h	3 h	6h

Table 3. The design of cold stimulus(3)

cold stimulus-interval of 2 day								
temperature decreasing 3 °C			temperature decreasing 5 °C			temperature decreasing 7 °C		
1h	3 h	6h	1h	3 h	6h	1h	3 h	6h

2.2. Slaughter and Sampling Method

The chicken were fasting for 12h prior to slaughter, one healthy chick was selected to blood sampling Per the Jugular Vein in each repeat which close to the average weight. Each chick was collected 10 ml blood. The blood was put in the refrigerator for 4 hours at 4°C, again with 3500r / min centrifugation for 10 minutes. The extracted serum was put in 1.5ml centrifuge tube, then put in the refrigerator, freezed at -20°C,used for the immunel index determination.

2.3. Immune Indexes Determination Method

The content change of interleukin-2(IL-2) and complement (C_3) were detected by enzyme-linked immunosorbent assay (ELISA).

3. RESULTS AND ANALYSIS

3.1. Effect of Cold Stimulation on C_3 of China-Lindian Native Chickens

3.1.1. *The Effects of Interval Days of Cold Stimulus on C_3 in China-Lindian Native Chickens*

As shown in Table 4, Serum C_3 content on all experimental group was higher than the control group, the difference was not significant($P>0.05$).Serum C_3 content on interval of 1 day was higher than interval of 0 day and 2 day in cold stimulus trial group, the difference was not significant($P>0.05$).

Through the variance analysis, interval days had no significant effect on the content of C_3 (P>0.05).

Table 4. The effects of interval days of cold stimulus on C_3 in China-Lindian Native Chickens (Unit:ug/ml)

interval days	average	difference significant	
		α=0.05	α=0.01
0 d	7.41±0.94	a	A
1 d	7.82±1.04	a	A
2 d	7.70±0.74	a	A
control group	7.13±0.38	a	A

3.1.2. *The Effects of Decreased Temperature of Cold Stimulus on C_3 in China-Lindian Native Chickens*

Table 5. The effects of decreased temperature of cold stimulus on C_3 in China-Lindian Native Chickens (Unit:ug/ml)

decreased temperature	average	difference significant	
		α=0.05	α=0.01
3°C	7.66±0.93	a	A
5°C	7.84±1.02	a	A
7°C	7.44±0.78	a	A
control group	7.13±0.38	a	A

As shown in Table 5, Serum C_3 content on all experimental group was higher than the control group, the difference was not significant(*P*>0.05). Serum C_3 content on decreased temperature 5°C was higher than decreased temperature 3°C and 7°C in cold stimulus trial group, the difference was not significant(*P*>0.05).Through the variance analysis, decreased temperature had no significant effect on the content of C_3 (P>0.05).

3.1.3. *The Effects of Cold Stimulation Time on C_3 in China-Lindian Native Chickens*

Table 6. The effects of time of cold stimulus on C_3 in China-Lindian Native Chickens (Unit:ug/ml)

stimulation time	average	difference significant	
		α=0.05	α=0.01
1 h	7.71±0.89	a	A
3 h	7.49±0.95	a	A
6 h	7.72±0.93	a	A
control group	7.13±0.38	a	A

As shown in Table 6, Serum C_3 content on all experimental group was higher than the control group, the difference was not significant(*P*>0.05).Serum C_3 content on stimulation time 6h was higher than stimulation time 1h and 3h in cold stimulus trial group, the difference was not significant(*P*>0.05).Through the variance analysis, cold stimulation time had no significant effect on the content of C_3 (*P*>0.05).

3.2. Effect of Cold Stimulation on IL-2 of China-Lindian Native Chickens

3.2.1. *The Effects of Interval Days of Cold Stimulus on IL-2 in China-Lindian Native Chickens*

Table 7. The effects of interval days of cold stimulus on IL-2 in China-Lindian Native Chickens(Unit:pg/ml)

interval days	average	difference significant	
		α=0.05	α=0.01
0 d	113.75±7.97	a	A
1 d	112.40±13.13	a	A
2 d	113.17±10.12	a	A
control group	110.35±8.93	a	A

As shown in Table 7, Serum IL-2 content on all experimental group were higher than the control group, the difference was not significant(P>0.05).Serum IL-2 content on interval of 0 day was higher than interval of 1 day and 2 day in cold stimulus trial group,the difference was not significant(P>0.05).Through the variance analysis, interval days had significant effect on the content of IL-2 (P<0.05),but the difference between groups was not significant.

3.2.2. *The Effects of Decreased Temperature of Cold Stimulus on IL-2 in China-Lindian Native Chickens*

Table 8. The effects of decreased temperature of cold stimulus on IL-2 in China-Lindian Native Chickens (Unit:pg/ml)

decreased temperature	average	difference significant	
		α=0.05	α=0.01
3°C	112.86±9.58	a	A
5°C	109.08±11.34	b	B
7°C	117.39±9.12	a	A
control group	110.35±8.93	a	A

As shown in Table 8, Serum IL-2 content on decreased temperature 3°C and 7°C were higher than the control group, Serum IL-2 content on decreased temperature 5°C was lower than the control group, the difference was significant(P<0.01).Serum IL-2 content on decreased temperature 7°C was higher than decreased temperature 3°C and 5°C in cold stimulus trial group, the difference was significant(P<0.01). Through the variance analysis, decreased temperature had significant effect on the content of IL-2 (P<0.01).

3.2.3. *The Effects of Cold Stimulation Time on IL-2 in China-Lindian Native Chickens*

Table 9. The effects of the time of cold stimulus on IL-2 in China-Lindian Native Chickens (Unit:pg/ml)

stimulation time	average	difference significant	
		α=0.05	α=0.01
1 h	112.08±8.24	a	A
3 h	113.38±11.43	a	A
6 h	108.86±11.79	a	A
control group	110.35±8.93	a	A

As shown in Table 9, Serum IL-2 content on stimulation time 1h and 3h were higher than the control group, Serum IL-2 content on stimulation time 6h were higher than the control group, the difference was not significant($P>0.05$).Serum IL-2 content on stimulation time 3h was higher than stimulation time 1h and 6h in cold stimulus trial group, the difference was not significant($P>0.05$).Through the variance analysis, cold stimulation time had no significant effect on the content of IL-2 ($P>0.05$).

4. DISCUSSION

The complement system is widely involved in the immune response and immune regulation, also can mediate immune pathological damage response.It is an effective system of biological significance in the body[1][2].C_3 is a a key component in the process of complement activation. Whether it is the classical activation pathway or alternative activation pathways, the chain reaction of the subsequent complement component can be promoted after C_3 was activated. The level of C_3 was parallel to the content of total complement, so it can Determinate the levels of total complemen by measuring C_3,and its level was an important index to measure the humoral immunity[3]. The results of this experiment show that interval days decreased temperature and stimulation time had no significant effect on the content of C_3($P>0.05$).IL-2 is produced mainly by the activate T cells. It has very important physiological and immune regulation effect in the immune system, can activate and enhance the immune cells functions, such as improving the ability of their killing cancer cells and viral or bacterial infected cells, promoting the antibody formation and secretion. The level of IL-2 is an important symbol of cell immunity in the body. Lv Caixia (2008) studies confirmed that long-term cold stimulation can stimulate the body to generate multiple serum immunosuppressive factor, inhibition interleukin -2, glucocorticoid and prostaglandin[4]. The results of this experiment show that interval days had significant effect on the content of IL-2 ($P<0.05$).Serum IL-2 content on all experimental group were higher than the control group, the difference was not significant($P>0.05$).This conclusion is consistent with the Paul' idea that the mild stress can enhance the immune response and strong stress can suppress the immune response. There is also evidence that the appropriate cold stimulus can improve the body's immune system.

Acknowledgments

This work was funded by Natural science foundation of China (31302008) and Natural science foundation of Heilongjiang province (QC2012C104).

References

1. Xue-jun Yuan, Jing-hua Niu, Hai-yan Wu. Effect of cold stress on peripheral blood lymphocyte amount in Red brown chicken. *Journal of Heilongjiang Bayi Agricultural University* 2002;**1**:61–63.(In Chinese).
2. Hong Lei, Cheng-nian Qi, Shi-ning Guo. The effect of Several kinds of traditional Chinese medicine prescriptions on immunoglobulin and complement in chicken serum. *The Chinese Journal of veterinary medicine* 2007;**43** (1): 33–34.(In Chinese).
3. HANGALAPURA B N, NIEUWLM G, BUYSE J,et al. Effect of duration of cold stress on plasma adrenal and thyroid hormone levels and immune responses in chicken lines divergently selected for antibody responses. *Poult Sci* 2004;**83**(10):1644–1649.
4. Cai-xia Lv, Yong-jie Wang, Rui-gang Li,et al. The effect of cold stress on neural mechanism and the immune system. *veterinary herald* 2008;**3**:27.(In Chinese).

EFFECT OF COPPER, ZINC, LEAD AND SILVER STRESS ON ANTIOXIDATIVE ACTIVITY IN EICHHORNIA CRASSIPES

Minghui Fu[a,*], Xiaodong Yu, Hailin Chen, and Jinping Peng

Department of Bioengineering, Guangdong University of Technology, Guangzhou, China

E-mail:[a]mhfugd@126.com

[*]Corresponding author

This study investigated the effect of copper, zinc, lead and silver over a concentration gradient of 0 – 0.20mmol/L on catalase activity, superoxide dismutase (SOD) activity and malondialdehyde (MDA) content in root and leaf of *Eichnornia crassipes*. The results showed that the catalase activity in root of E. crassipes cultured in copper and zinc was firstly decreased and then increased with the culture time. For the plants cultured in lead and silver, the catalase activity was increased all the experiment time. The change of SOD activity of plants cultured with copper, zinc and lead was firstly increased, and then decreased with the culture time. The SOD activity of plants cultured with silver was decreased all the experiment time. MDA content in root and leaf of plants cultured in copper, zinc, lead and silver was firstly decreased, and then increased with the culture time. This implied that E. crassipes could induce its antioxidative enzyme such as catalase, SOD etc. to decrease the accumulation of the oxidative product such as MDA to protect itself from oxygen damage when cultured in copper, zinc, lead and silver. But too much exposed to these heavy metals could make this self-protection not enough to avoid the oxygen damage so that MDA content was increased greatly. Moreover, the results also indicated that the antioxidative enzyme of E. crassipes responded more slowly to lead and silver than copper and zinc. They could not protect the plants from oxidation damage more quickly so that these two heavy metals were more toxic.

1. INTRODUCTION

Copper, zinc, lead and silver are belong to heavy metals, which are defined as metals with a density higher than $5gcm^{-3}$ [1]. Zinc and copper are important to plants with very low concentrations as trace elements, but they are toxic with high concentrations. On the other hand, silver and lead have no known functions as nutrients [2]. Toxic levels of these metals usually appear as a result of environment pollution due to the removal technology of mining, heavy automobile traffic, smelting, manufacturing and agricultural wastes in natural and agricultural areas [3]. Heavy metals can cause plants oxidative damage by the inducement of reaction oxygen species such as hydroxyl radical, superoxide anion and others which may attack many biological macromolecules such as proteins, membrane lipids and nucleic acids [4,5]. Malondialdehyde (MDA) is a cytotoxic product of lipid peroxidation and an indicator of free radical production and consequent tissue damage [6]. So detecting the level of MDA in plants can show the degree of the oxidative damage.

Eichnornia crassipes is an aquatic plant originated from South America. It is one of the most productive plants in the world. Due to its rapid reproduction and accumulation of heavy metals [7,8], it can be used in phytoremediation of waste-water [9,10]. In this study, we detected the catalase activity, SOD activity and malondialdehyde content in root and leaf of E. crassipes exposed to the environment with copper, zinc, lead and silver. The result can illustrate the reason why E. crassipes can tolerate with heavy metal stress to some extent from the angle of oxygen damage and can provide the theory basis for further development and utilization of E. crassipes as a kind of water remediation plant.

2. MATERIALS AND METHODS

2.1. Preculture

E. crassipes was collected from the lake in the west area of Guangdong University of Technology. The plant roots were washed with water and sterilized with sodium hypochlorite, and then the sodium hypochlorite on the surface was washed with distilled water. Finally, the plant was cultured in 1/4 Hoagland solution for 20 days. The solution was replaced every two days.

2.2. Culture with Heavy Metals

After preculture, copper (as $CuSO_4$), zinc (as $ZnSO_4$), lead (as $Pb(NO_3)_2$) and silver (as $AgNO_3$) were added respectively to the growth medium in calculated amount to obtain the final concentrations of 0mmol/L, 0.04mmol/L, 0.08mmol/L, 0.12mmol/L, 0.16mmol/L, 0.20mmol/L. Every two days, the root or the leaf were fetched and prepared for evaluation.

2.3. Preparation of Samples

2.5g root or leave was homogenized with 5mL 0.05mol/L phosphate buffer solution (pH7.8), and then the homogenate was centrifuged at 8000rpm for 10 minutes. The supernatant was prepared for subsequent evaluation.

2.4. Evaluation of the Antioxidant Activity

The evaluation of every antioxidative activity was elaborated as below. Every evaluation was doubled. The data were recorded as means±standard deviations and analyzed by SPSS 13.0. One-way analysis of variance (ANOVA) and Student-Newman-Keuls (S-N-K) multiple comparisons were carried out to test for any significant differences between the means. Differences between means at 5% ($P<0.05$) level were considered significant.

2.5. Evaluation of the Catalase Activity

Evaluation method was according to the method of Peng et al. [11]. Firstly, 0.1mL of the supernatant and 0.3mL of 50mmol/L H2O2 were mixed to react for 120s, and then 1mL saturated NaCl solution and 2.6mL 50mmol/L ammonium molybdate were added. The mixture was shaken and the absorbance at 405nm was determined. The controlled added the supernatant at last, just before the determination. The protein concentration of the supernatant was determined by the Bradford method [12]. 1 unit catalase activity was defined as 1 μmol/L of H2O2 decomposed every second and the activity referred to micrograms of protein.

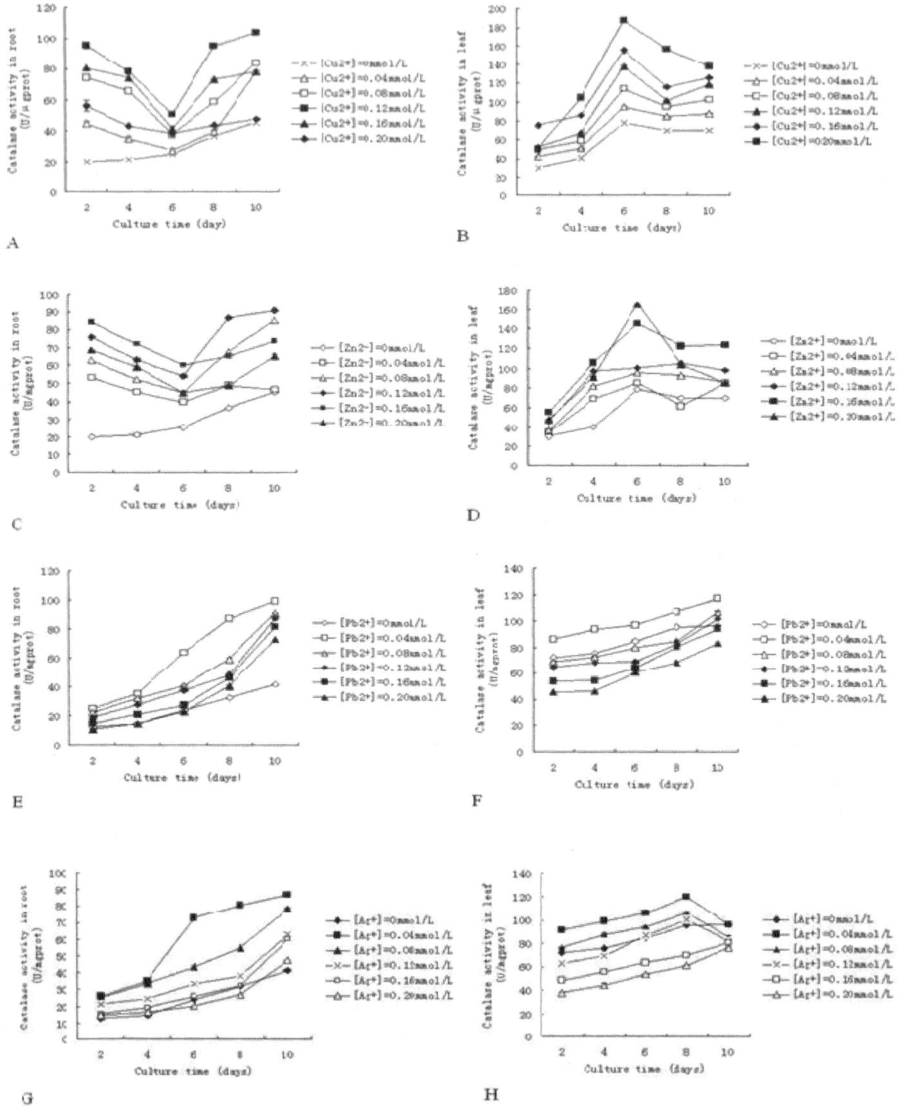

Fig. 1. The effect of culture time on catalase activity in roots (A, C, E and G) and leaves (B, D, F and H) of *Eichnornia crassipes* cultivated in different concentrations of copper, zinc, lead and silver respectively.

2.6. Evaluation of the SOD Activity

Evaluation method was according to Calatayud et al. [13]. The reaction mixture included 1.5mL of 0.05mol/L potassium phosphate buffer (pH 7.8), 0.3mL of 130mmol/L methionine, 0.3mL of 750μmol/L nitroblue tetrazolium (NBT), 0.3mL of 100μmol/L EDTA-Na2, 0.3mL of 20μmol/L riboflavin, 0.05mL of supernatant and 0.25mL of distilled water. The controlled used 0.05mL of potassium phosphate buffer instead of the supernatant in the reaction mixture. The mixture was thoroughly shaken and illuminated for 20min at 4000Lx light. The absorbance at 560nm was determined. 1uint of suoeroxide dismutase activity was defined as a 50% inhibition of the NBT phytoreduction.

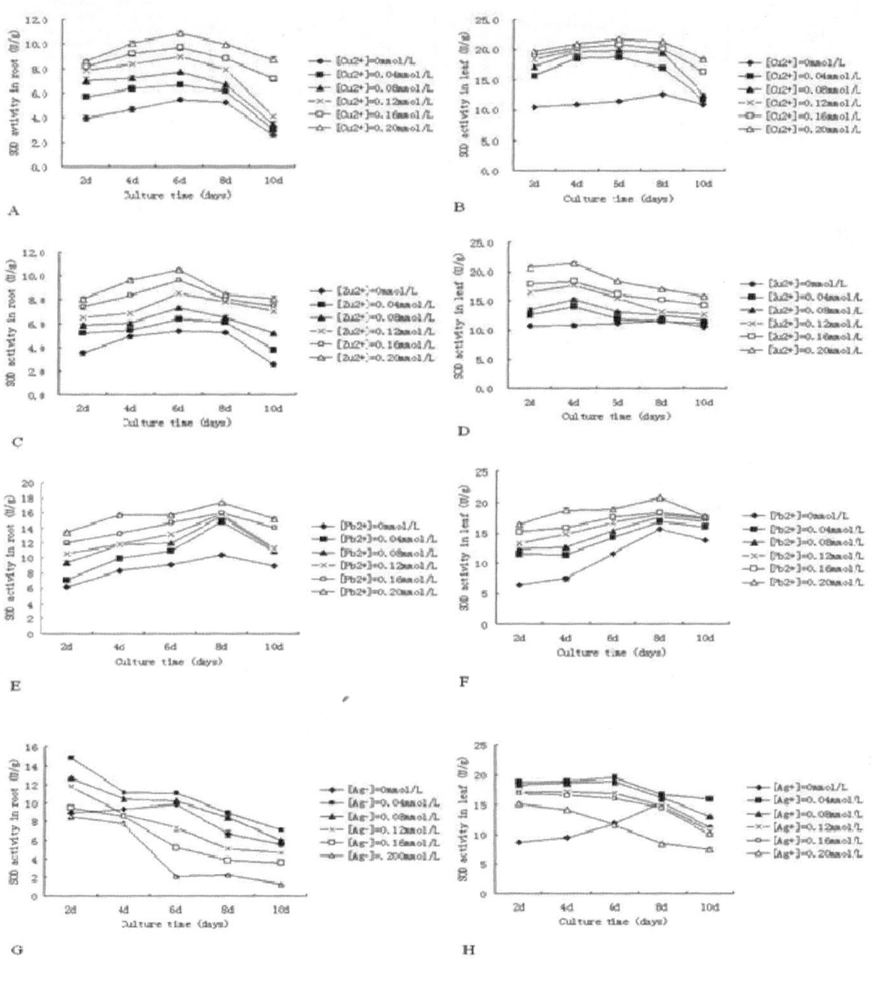

Fig. 2. The effect of culture time on superoxide dismutase (SOD) activities in roots (A, C, E and G) and leaves (B, D, F and H) of *Eichnornia crassipes* cultivated in different concentrations of copper, zinc, lead and silver respectively.

2.7. Evaluation of MDA

Evaluation method was according to Li et al. [14]. 4mL (Vs) of the supernatant and 6mL of 0.5% 2- thiobarbituric acid (TBA) were mixed and then heated in boiled water for 10 minutes. Cooled immediately in an ice bath, centrifuged at 4500rpm for 10 minutes. The absorbance at 532nm, 450nm and 600nm of the supernatant were determined. The concentration of MDA (CMDA) was defined as 6.452*(OD532-OD600)-0.559*OD450. The total content of MDA was calculated as CMDA*Vt/(Vs*fresh weight of sample). Vt was the total volume of the supernatant.

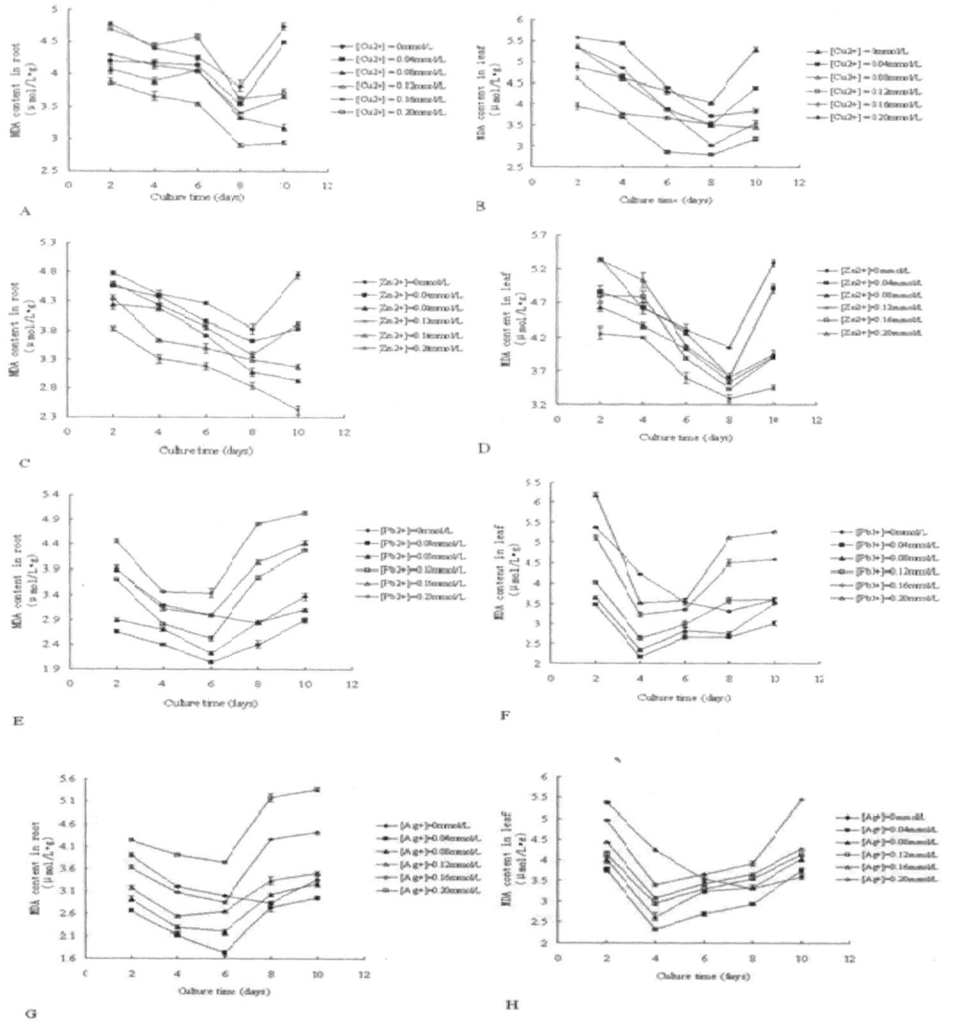

Fig. 3. The effect of culture time on malondialdehyde (MDA) content in roots (A, C, E and G) and leaves (B, D, F and H) of *Eichnornia crassipes* cultivated in different concentrations of copper, zinc, lead and silver respectively.

3. RESULTS

3.1. Evaluation of the Catalase Activity

The change of catalase activity with the culture time was showed in Fig. 1. The catalase activity in root of E. crassipes cultured in copper and zinc was firstly decreased and then increased with the culture time. The minimum was at the sixth day. While in leaf it was firstly increased, and then decreased. The maximum was at the sixth day. This indicated the root has quicker response to the copper and zinc. At the second day, the catalase activity of plants cultured in copper and zinc had been increased a lot compared with those cultured with no copper and zinc. For the plants cultured in lead and silver, the catalase activity was increased all the experiment time except for the leaf of plants cultured in silver, at the tenth day there was a little decrease in catalasse activity.

3.2. Evaluation of the SOD Activity

The change of SOD activity of plants cultured with copper, zinc and lead was firstly increased, and then decreased with the culture time (Fig. 2). The maximum was at the sixth day except for those cultured with lead and those in leaf of plants cultured with zinc. The maximum of those cultured in lead was at eighth day, and the maximum of the leaf of plants cultured with zinc was at the fourth day. The SOD activity of plants cultured with silver was decreased all the experiment time although the decreased tendency of leaf was slowly.

3.3. Evaluation of MDA Content in Plant Cultured in Copper

The evaluation results of MDA in leaf and root of E. crassipes cultivated in culture solution with copper, zinc, lead and silver were depicted in Fig. 3. As shown in Fig. 3, both in root and leaf, the content of MDA was firstly decreased, and then increased with the increasing of culture time. When E, crassipes was cultured in copper and zinc, the content of MDA of both root and leaf arrived at the minimum after eight days. But when E, crassipes was cultured in lead and silver, MDA of root arrived at the minimum after six days, while MDA of leaf arrived at the minimum after four days.

4. DISCUSSION

MDA is a main product of lipid peroxidation. Its accumulation has been used as an important indicative of membrane lipid peroxidation [15]. Many reports [16-18] had showed that when plants lived in poor condition or at aging process, the balance of intracellular reactive oxygen could be destroyed, which would cause the accumulation of reactive oxygen species, and then trigger or exacerbate membrane lipid peroxidation, at last result in serious membrane damage, even lead to death of plant cells.

The results of our study showed that MDA content in the root and leaf of E. crassipes would be firstly decreased, and then increased with the increasing of culture time or the concentration of copper, zinc, lead and silver. When these heavy metals acted on E. crassipes at first, they may induce the plant's defense system, perhaps by enhancing the activity of some antioxidation enzymes such as catalase activity, SOD activity and peroxidase, which could decrease the MDA content in plants. Some previous reports [19] had confirmed that when E. crassipes encountered with heavy metals, its activities of SOD, peroxidase and catalase in leaf would be improved greatly. Our study also confirmed that most catalase and SOD activity were increased when E. crassipes was exposed to heavy metals. But for different metals and for different tissues, the respond speeds of these antioxidative enzymes were different. Copper and zinc could induce more quickly than lead and silver, and root responded more quickly than leaf which might result in more bearing time of MDA content responding to copper and zinc than lead and silver, and more bearing time of MDA content in root than in leaf. Our study showed that the highest concentration and the longest culture time of the least MDA content in copper and zinc were 0.12mmol/L and 8 days, higher or longer than those in lead and silver. This was also consistent with that copper and zinc were important to plants as trace elements at low concentration, but were toxic to plants at high concentration, while lead and silver were toxic to plants at low concentration [2]. When the culture time or the concentration of heavy metals went on increasing, antioxidation mechanism of plant could not inhibit the function of the reactive oxygen completely, MDA content would be increased again. This had been proved in our experiments.

Acknowledgments

This work was funded by the national science foundation under grant 21177029, innovation and entrepreneurship training program for college students under grant 201411845184 and technology innovation program of educational commission of Guangdong Province of grant 2012KJCX0043.

References

1. Weast RC. *CRC Handbook of chemistry and physics.* Florida: Boca Raton, 1984.
2. Neis DH. Microbial heavy-metal resistance. *Applied Microbiol Biotech* 1999; **51**: 730-750.
3. Oncel I, Kele Y, Ustun AS. Interactive effects of temperature and heavy metal stress on the growth and some biochemical compounds in wheat seedlings. *Environ Poll* 2000; **107**: 315-320.
4. Matsumura T, Tabayashi N, Kamagata Y, Souma C, Saruyama H. Wheat catalase expressed in transgenic rice can improve tolerance against low temperature stress. *Physiol Plant* 2002; **116**: 317-327.
5. Choudhary M, Jetley UK, Khan MA, Zutshi S, Fatma T. Effect of heavy metal stress on proline, malondialdehyde, and superoxide dismutase activity in the cyanobacterium Spirulina platensis-S5. *Ecotoxicology and Environmental Safety* 2007; **66**: 204-209.
6. Ohkawa H, Ohishi N, Yagi K. Assay for lipid peroxidation in animal tissues by thiobarbituric acid reaction. *Anal Biochem* 1979; 95: 351.

7. Deng H, Ye ZH, Wong MH. Accumulation of lead, zinc, copper and cadmium by 12 wetland plant species thriving in metal-contaminated sites in China. *Environmental Poll* 2004; **132**: 29-40.
8. Caldelas C, Iglesia-Turino S, Araus JL, Bort J, Febrero A. Physiological responses of *Eichnornia crassipes* (Mart.) Solms to the combined exposure to excess nutrients and Hg. *Braz J Plant Physiol* 2009; **21**: 01-12.
9. Casabianca MLD, Laugier T, Posada F. Pertoliferous wastewater treatment with water hyacinth: experimental statement. *Waste Management* 1995; **15**: 651-655.
10. Chua H. Bio-accumulation of environmental residues of nlre earth elements in aquatic flora *Eichnornia crassipes* (Mart.) Solms in Guangdong Provence of China. *The science of the total Environment* 1998; **214**: 79-85.
11. Peng J, Wang DY, Xu CM, Chen LP, Deng F, Zhang XF. Ammonium molybdate method for detecting the activities of rice catalase. *Chinese agricultural science bulletin* 2009; **16**: 61-64.
12. Bradford MM. A rapid and sensitive method for the quantification of microgram quantities of protein, utilizing the principle of protein-dye binding. *Anal Biochem* 1976; **72**: 248-254.
13. Calatayud A, Ramirez JW, Iglesias DJ, Barreno E. Effects of ozone on photosynthetic CO2 exchange, chlorophyll a fluorescence and antioxidant system in lettuce leaves. *Physiol Plant* 2002; **116**: 308-316.
14. Li HS, Sun Q, Zhao SJ. *Principles and techniques of plant physiology and biochemistry.* Beijing: Higher education press, 2000.
15. Flohr L, Fuzinatto CF, Melegari SP, Matias WG. Effects of exposure to soluble fraction of industrial solid waste on lipid peroxidation and DNA methylation in erythrocytes of Oreochromis niloticus, as assessed by quantification of MDA and m5dC rates. *Ecotoxicology and Environmental Safety* 2012; **76**: 63-70.
16. Zhirov VK, Meryzlyak MN. Cultivation of pea plants (Pisum sativum L.) at low temperature decreases lipid peroxidation induced by freezing-thawing. *Plant Science Letters* 1983; **30**: 185-191.
17. Hung KT, Kao CH. Lipid peroxidation in relation to senescence of maize leaves. *Journal of Plant Physiology* 1997; **150**: 283-286.
18. Ghnaya AB, Charles G, Hourmant A, Hamida JB, Branchard. M. Physiological behaviour of four rapeseed cultivar (Brassica napus L.) submitted to metal stress. *Comptes Rendus Biologies* 2009; **332**: 363-370.
19. Odjegba VJ, Fasidi, IO. Changes in antioxidant enzyme activities in Eichhornia crassipes (Pontederiaceae) and Pistia stratiotes (Araceae) under heavy metal stress. *Rev Biol Trop* 2007; **55**: 815-823.

PROBLEMS AND SUGGESTIONS FOR QUALITY AND SAFETY OF AQUATIC PRODUCTS IN CHINA

Yuan Wang[a], Xiaosheng Shen[b], Youqiong Cai[c], Cong Kong[d], and Xuan Zhang[e]

Laboratory of Quality and Safety Risk Assessment for Aquatic Products (Shanghai), Ministry of Agriculture, East China Sea Fisheries Research Institute, Chinese Academy of Fishery Sciences, Shanghai, 200090, China

E-mail: [a]wangyuan81@163.com, [b]foodsmc98@126.com, [c]caiyouqiong@163.com, [d]kongcong@gmail.com, [e]zhangxuan0430@163.com

With the increased aquaculture production and the continuous expansion of the scale of aquaculture, deterioration of the environment, expanded culture density and other factors lead to serious diseases of aquaculture species, and the problems such as drug abuse in aquaculture and poor implementation of withdrawal in accordance with the norms are very serious. The above issues pose a threat to quality and safety of aquaculture products and seriously affect sustainable development of fisheries. This article explored the impact factors of quality and safety of aquatic products from the aspects of aquaculture facilities and modes, disease prevention and control, detection and supervision system construction, etc, and further made some suggestions.

1. INTRODUCTION

In recent years, aquaculture industry of China has developed rapidly with the continuous expansion of the aqua-scale and fast increase of output. China has been the first largest aquaculture country in the world, and the export amounts of aquatic products are in the first place for more than ten years [1, 2]. According to China Fishery Statistical Yearbook from 2010 to 2014 [3-7], total output, culture production output and catch production output of aquatic products between 2009 and 2013 are increased year by year (Table 1). The percentage of culture production occupied a main proportion of the total output in present situation.

Table 1. Comparisons of total output, culture production output and catch production output of aquatic products from Year 2009 to 2013

Unit: million ton

Year	2009	2010	2011	2012	2013
total output	51.17	53.73	56.03	59.07	61.72
culture production output	36.22	38.29	40.23	42.88	45.42
catch production output	14.95	15.44	15.80	16.19	16.30

However, with the increased aquaculture production, the problems of quality and safety of aquatic products have limited the further development of aquaculture. Because of the deterioration of aquaculture environments and immoderately expanded culture density, the illegal use of drugs in aquaculture has still been a serious problem. With the changes of situation on supply and demand of aquatic products and the continuous improvements of people's living conditions, the residents show strong requirements of cleaning-up "dining-table contamination" and the guarantee of food safety. To improve the quality and competitiveness of aquatic products in the international markets and to ensure the safety and quality of aquatic products have become the urgent task of economic restructuring of fisheries and increasing the income of fishermen.

2. PROBLEMS AFFECTING THE QUALITY AND SAFETY OF AQUATIC PRODUCTS

2.1. Deterioration of Culture Environment

Fishery ecological environment is an important condition for the development of aquaculture. Generally, the major farming areas locate in lakes, coastal lough and waters near the estuaries, which are the main receiving sites of land-based sources of pollutant and sewage at sea. With the rapid development of economy and urbanization, a large number of industrial waste and urban sewage untreated according to standards have been discharged into rivers and sea, which is a direct threat to the survival of aquaculture and the safety and quality of aquatic products. On the other hand, self-pollution of aquaculture industry cannot be ignored. At present, some practitioners of aquaculture are pursuing short-term economic benefits through random expansion of aquaculture area and farming layout and density. Besides, poor quality of feed materials, low levels of culture managements and illegal use of antibiotics, disinfectants and water conditioner also exist in the process of aquaculture. These issues exacerbate the pollution of culture waters and make contamination cumulated in aquatic products [8, 9].

2.2. Frequent Occurrence of Fishery Diseases

The excessive expansion of culture density and the pollution of aquaculture environment resulted in frequent occurrence of fishery diseases. It's reported that the diseases amounted to more than 200 species caused by viruses, bacteria, parasites, fungi and other pathogenic and non-biological factors [10]. The country and multilevel governments are very concerned about the diseases of aquaculture and have made great achievements in prevention and treatment on diseases of aquaculture by organizing cooperation of many institutes. However, the basic scientific research is little, and most reasons for fishery diseases, such as the pathogen, pathology and epidemiology of culture species have to be further studied. Now, we lack of not only the effective early rapid monitoring techniques, but also those drugs that can prevent and treat diseases effectively with no toxic side effects. Once the diseases occurred, the farmers still used illegal drugs because of no better substituting drugs to minimize losses, which resulted in drug residues in aquatic products [11].

2.3. Illegal Use of Drugs

In the culturing, processing and transporting fields of aquatic products, drug abuse and residues are still serious problems. The quality and safety incidents of aquatic products often occurred, such as the malachite green residues in freshwater fish in 2005, the Nitro furan metabolites residues in turbot in 2006, and the crawfish event in 2009 etc., which seriously affected the development of aquaculture industry. From the current monitoring results of Ministry of Agriculture (MOA) in China, the malachite green and Nitro furan metabolites and other types of prohibited drugs are still the main reasons for unqualified aquatic products. It is also showed that the phenomenon of drug abuse still exists and it's a tremendous job to enhance the quality of aquatic products by developing safety and effective culture ways.

2.4. Lack of Traceability System of Aquatic Products

In recent years, the supervision regulations and system of fishery have made great achievements [12]. But decentralized organizations of aquaculture are commonplace and a comprehensive traceability system is not be well established, the supervising departments cannot require enough information about the producers and transporters and how the drug used in production and distribution of aquatic products. These above reasons make unqualified aquatic products into the markets and consumers' tables.

3. SUGGESTIONS ON SOLVING THE PRESENT PROBLEMS OF AQUATIC PRODUCTS

In order to ensure the quality and safety of aquatic products and sustainable development of China's aquaculture, we should take measures not only in the culture stage, such as from fry to adult, feeding phase, breeding density, diet quality and drug use, but also in transport and sale stages, such as aquaculture facilities, water treatment, and aquaculture managements. When culture techniques are reasonable and scientific, the quality and safety of aquatic products can be insured.

3.1. Speeding Up the Research on Aquaculture Facilities and Ecological Environment

Aquaculture facilities are important substance basis on the well development of aquaculture, the structure and design of which have direct impact on aquaculture and ecological benefits. Aquaculture facilities of China, especially the main way of farming-aquaculture ponds [13], followed the traditional structure and layout only with the spaces for culture species and basic entrance and drainage function. The facilities are lack of effective water control system and the nutrient-rich salts and other wastes can pollute the water and have a negative impact on environment, which in turn affect the aquaculture industry itself.

In order to ensure the well development of aquaculture, it's necessary to study and reconstruct the current aquaculture facilities, and then guide the industry to the direction of fishery facilities [14]. Besides the spaces for culture species and basic entrance and drainage function, the facilities should also have a strong regulation and control of water quality and purification, so that the culture water can meet the demands of in-house recycling. By the recycle of waste in farming inside system, many kinds of resources can be well used and waste in the process of aquaculture can be minimized, which can realize a good combination of economic and ecological benefits and promotion of the quality and safety of aquatic products.

3.2. Developing High-Quality Aquatic Feeds

Feed is the main inputs of aquaculture, and high-quality feeds play a crucial role for improving quality of aquaculture products, such as lowing costs, reducing diseases, preventing environmental pollution and increasing economic benefits. Using and popularization of quality feed with general nutrition are the signs of technological progress in aquaculture. The scientific and technological workers recently carry out many systematic researches on the nutrition demands of aquaculture species and compositions of feed, but compared with the needs of industry development, the technical levels of fishery feeds are still low, which mainly reveal in the aspects of poor quality of compositions and laggard processing technology leading to the high cost of feeds, low feed efficiency, and causing pollution of culture water [15]. Research and development of environment-friendly feeds, which are of high quality, good stability, bait-eating and easy-digestion, can provide an important guarantee for quality and safety of aquatic products.

3.3. Strengthening Prevention and Control of Diseases

At present, the problems of diseases have become an important factor restricting development of aquaculture. On the one hand, aquaculture production units pursue the output and economic benefits blindly, resulting in aquaculture technology and mode unreasonable. On the other hand, China's aquaculture health management and disease control techniques are far behind the production development. Besides, the deterioration of ecological environment and increased probability of the occurrence of diseases result in drug abuse, which reduced the drug efficiency and whereas increased the dosage, then form a vicious circle. In this way, the cost of aquaculture increases and the benefits decline. Also, drug abuse has a very negative impact on ecological environment and quality and safety of aquatic products.

Disease control techniques in aquaculture are the key technology, including the following aspects.

First is the health management in the process of culture production. Aiming at the different growth phases under the specific types of culture and the specialties of production management, based on the reasonable aquaculture techniques and modes, it should adopt reasonable water quality management and control techniques (including biological and chemical techniques) to maintain a good culture environment and control the occurrence of diseases.

Second is the ecological prevention and treatment technology of diseases. Many diseases in aquaculture relate to not only the existence of biological pathogens, but also ecological balance of microorganism in culture water. In another word, the compositions of microorganism community in waters determine if the occurrence of diseases would happen by pathogens. As a result, through the research on relationship between physical and chemical factors in water and the compositions of microorganism community, to maintain the micro-ecological balance of water for eliminating occurrence of diseases is an effective way to prevent and treat diseases in healthy aquaculture.

The third is the rational use of drugs. Presently there are 179 kinds of aquaculture drugs recommended by Chinese government [16], which are able to meet the current needs for prevention and treatment of common disease. However, from the monitoring results of MOA, the main reasons for unqualified aquatic products have been due to not using the drugs in strict way accordance with the norms of use, especially for the illegal use of prohibited drugs. Therefore, we should strengthen the basic research on drug efficiency, metabolites and residues in aquaculture, regulate the rational use of drugs and prescribe the list of prescription and non-prescription drugs, strictly enforce the regulations of off-drug period, and severely strike on the illegal use of prohibited drugs, so that we can completely improve the quality and safety of aquatic products.

3.4. Improving Standard System for Quality and Safety of Aquatic Products

It should be important problems that establishing and improving standard system of aquaculture and food safety to meet the demands of international and domestic market requirements [17]. Focus on famous, unique and excellent seafood, it's important to establish and improve kinds of standards of aquaculture and quality and safety of aquatic products as soon as possible. From the existing problems of quality and safety of aquatic products, we should study and develop the crucial detecting, controlling and monitoring technology for aquatic products. To establish and improve the standard system of quality and safety of aquatic products according with national conditions is to make our standard system adapt to international trade market and protect domestic market, which can improve the competitiveness of the aquatic products and increase incomes of farmers and serve for the sustainable development of fisheries.

3.5. Enhancing the Supervision and Detection System of Quality and Safety of Aquatic Products

To ensure the overall safety of aquatic products, it's not enough to only rely on self-discipline of farmers. Since Nov. 2006, the implementation of "Law of Agro-food Quality and Safety of People's Republic of China" clarified the agricultural (fishing) department is responsible for quality and safety of primary agro-products. For the last ten years, the construction of monitoring system of agro-products' quality and safety processed by the MOA has made great achievements, which formed a system of ministry, provincial, municipal and county level to administrate supervision responsibilities. Because of some restrictions, the development of supervision and detection system remains uneven. In the future, the fishing department should seize the opportunities to train the professionals and strengthen technical training in order to establish a comprehensive supervision and detection system as soon as possible. It's necessary to encourage the enterprises, professional associations and markets to establish their own professional detecting stations to control quality of products in the whole processing under the guidance of authorities, which can provide a guarantee for sustainable development of fisheries.

4. CONCLUSION

In short, there is a close relationship between quality and safety of aquatic products and sustainable development of fisheries. To enforce the quality and safety of aquatic products, the supervising departments and the research institutes and the production units should play good roles in their own fields, including the supervising system, science and technology input, developing scientific research, the mode of aquaculture and process control, the basic facilities' promotion and use of high-quality drug and feeds an so on. These above measures can ensure the quality and safety of aquatic products and consumption safety of consumers, and increase the output of aquaculture and incomes of farmers, and further achieve the real sustainable development of fisheries.

References

1. Glitnir Bank. China is the world's largest producer and exporter of seafood, 2007, http://www.acnnewswire.com/press_release/english/52/china_is_world%27s_largest_producer_and_exporter_of_seafood_according_to_the_latest_glitnir_report
2. Edward I, Broughton, Damian G, etc. Policies and practices for aquaculture food safety in China, *Food Policy* 2010; **35**: 471–478
3. Ministry of Agriculture, Fisheries Bureau. *China Fishery Statistical Yearbook.* Beijing: China Agriculture Press, 2010: 1–5
4. Ministry of Agriculture, Fisheries Bureau. *China Fishery Statistical Yearbook.* Beijing: China Agriculture Press, 2011: 1–5
5. Ministry of Agriculture, Fisheries Bureau. *China Fishery Statistical Yearbook.* Beijing: China Agriculture Press, 2012: 1–5
6. Ministry of Agriculture, Fisheries Bureau. *China Fishery Statistical Yearbook.* Beijing: China Agriculture Press, 2013: 3–7
7. Ministry of Agriculture, Fisheries Bureau. *China Fishery Statistical Yearbook.* Beijing: China Agriculture Press, 2014: 6–10
8. Fan YH. The analysis and suggestions for the role of technology in the quality and safety of aquatic products. *China Fisheries* 2012; **9**: 36–37

9. Wang YT. Drug use in aquaculture and the quality and safety of aquatic products. *China Fisheries* 2012; **5**: 54–58
10. Yan DC, Wei FQ. Research on virus transmission routes in aquaculture animals. Reservoir Fisheries 2006; 26(1): 110–112
11. Lu JS, Ou ZD, Li YP. Problems and solutions of the quality and safety of aquatic products. *Fishery Guide to Be Rich* 2013; **13**: 16–18
12. Zhu YL, Luo TL. The status and countermeasures of the quality and safety of aquatic products in China. *Journal of Quality of Goods* 2012; **1**: 189
13. Ren XL, Ling WH. Problems and countermeasures of the quality and safety of aquatic products. *Journal of Aquaculture* 2011; **12**: 48–50
14. Han XF, Han LM. Status and suggestions of fishery facilities in Shandong Province of China. *Modern Fisheries Information* 2011; **26**(11): 6–8, 29
15. Zhang C, Jiang QJ. Problems and countermeasures and suggestions of the quality and safety of aquatic products in China. *Journal of Shanxi Agricultural Sciences* 2010; **38**(3): 61–64
16. National Fishery Technical Extension Center. Current Situation and Development of drug use in aquaculture in China. China Fisheries 2005; 11: 14–17
17. Li Q. Problems and measures of the quality and safety of aquatic products. *Science and Technology* 2012; **41**(6): 51–52, 115

ESTABLISHMENT OF EFFICIENT CALLUS INDUCTION AND THE REGENERATION SYSTEM OF KIYOMI TANGOR

Jun Wang[a], Ke Huang[b], Xia Qiu, Zhihui Wang, and Xun Wang[c,*]

College of Horticulture, Sichuan Agricultural University

No. 211, Hui Min Road, Wenjiang, Chengdu, Sichuan, P.R.China

E-mail: [a]4692670@qq.com, [b]872945120@qq.com,[c]wangxun0104@hotmail.com

**Corresponding author*

An efficient in vitro culture system, callus induction and regeneration, has been established in Kiyomi tangor (*Citrus unshiu* × *Citrus sinensis*). Seedling epicotyls, geminated from sterile seeds in germination medium, were cut off to short fragments and used to induce callus. The effects of seedling epicotyls fragments length and wounded fragment on callus induction were analyzed. It indicated that the fragment 0.3 cm without wounding was the best material for callus induction. The optimal medium for callus induction, callus proliferation and shooting regenerating form callus were studied.

1. INTRODUCTION

Developing tissue culture methods for citrus crop species has practical benefits. Most notable among these are techniques for obtaining virus-free and mycoplasma-free stocks using *in vitro* grafting of apical meristems from infected plants onto decapitated seedlings[1]. To improve techniques for Citrus tissue culture, there have been many efforts to investigate conditions that enhance callus culture growth. It was certified that MT (Milk-Tween) culture medium was better than MS (Murashige and Skoog) medium[2]. *In vitro* citrus shoots formation and proliferation were controlled primarily by the type of explant material and supplemented components in culture medium. In previous studies, hypocotyl and young embryo were proved to the preferred explant materials[3], and frequently callus turning brown appeared during callus culturing[4].

Hybrid citrus fruit is becoming more popular because it is easy to peel, seedless (low seediness), fragrant, and with special flavor[5,6]. Hybrid citrus breeding is one of important research fields. Using cell engineering and gene engineering is necessary in citrus breeding, and efficient system of citrus culture is the foundation to realize cell genetic engineering. However, *in vitro* culturing for hybrid citruses is still little reported. It is important to understand the optimal culture system for an interested citrus variety, but not just accepting that has been established on other varieties.

In the present research, we adopted the breed variety of Kiyomi tangor (*Citrus unshiu* × *Citrus sinensis*) to establish its *in vitro* culture system. Kiyomi tangor is a

Japanese citrus fruit that is a hybrid of a *Miyagawa Wase* mikan and a *Citrus sinensis*[7] orange. This breed variety was spread to plant rapidly in south and west of China[8]. We used seedling epicotyls as the explant material and compared different operated treatments on explant material. In addition, we investigated the suitable medium for callus induction, callus proliferation, and shoot regeneration on Kiyomi tangor. This research is expected to provide an efficient *in vitro* culture system for genetic engineering on Kiyomi tangor.

2. MATERIALS AND METHODS

2.1. Materials

Kiyomi fruits were provided by tangor orchard in Suining city, Sichuan province, China.

2.2. Seed Sterilization

Seeds were taken out of the fruits, and soaked in 1 mol/L NaOH solution for 10 min for cleaning. Cleaned seeds were sterilized in 75% alcohol disinfection 2-3 min, and stirred in sodium hypochlorite (1% active chlorine) for 20 min, followed by rinsing three times with sterile water. Sterilized seeds were soaked in sterile water over night till to swell. Seeds were removed the exopleura and endopleura by pincets, and inoculated in germination medium. Seeds germination was incubated in darkness at 25±1°C. Germination medium was MT (Milk-Tween) medium supplemented with 50 g/L sucrose, and solidified by 8 g/L agar, pH 5.8. At the same time, the swelled seeds without husking were inoculated directly. The test were performed in germination medium and as well incubated in darkness at 25± 1°C.

2.3. The Explant Materials for Callus Induction

Seeds geminated several days after inoculating. When Seedling epicotyl grew to 4-6cm, it was cut off to short fragments with different length, 0.3cm, 0.5cm, and 1.0cm. In addition, seedling epicotyl fragment (0.5cm long) was cutting into two parts longitudinally, or wounded by knife longitudinally. All explant materials were inoculated in callus induction medium. Six explant materials were in one medium dish, with three replications. Callus medium, confirmed in previous test, was MT basal medium supplemented with 50 g/L of sucrose, and with a combination of 1.0 mg/L 2,4-D and 0.5 mg/L 6-BA, pH 5.8. Callus induction was incubated in darkness at 25± 1°C.

2.4. Callus Proliferation Medium Optimization

Callus induced in callus medium for one month was divided into small pieces 0.5cm diameter, and inoculate in proliferation medium. Proliferation media with different combination of plant growth regulators were tested the efficiency. Supplemented plant growth regulators were 1.0 mg/L or 2.0 mg/L 2,4-D (2,4-dichlorophenoxyacetic acid) combined with 0.5 mg/L or 1.0 mg/L 6-BA (6-benzyladenine). The combination were added in MT basal medium with sugar 50 g/L, and solidified with rubber plant 2 g/L, pH5.8. All media were maintained in in darkness at 25± 1°C, with three replications.

2.5. Shoot Regeneration Medium Optimization

Friable embryogenic callus was divided into 0.5-cm pieces. Approximately 6 callus pieces were inoculated in each medium with three replications. Media were MT basal medium (pH5.8) in glass bottles, added with sugar 50 g/L and solidified with agar 8 g/L, supplemented different combination of plant growth regulators, 6-BA (1.0 mg/L and 2.0 mg/L) and 1-Naphthaleneacetic acid NAA (0.01 mg/L and 0.1 mg/L). Culture for plant regeneration was kept in a 16:8 h light:dark (LD) photoperiod at 25± 1 °C.

3. RESULTS

3.1. The Effect of Seed Husking On Seed Germination

In order to understand the effect of seed husking on seed germination, we inoculated directly swelled seeds without husking in germination medium. Seven days after inoculation, seeds husked began to germinate, and seedling epicotyls got to 7-10 cm few days later (Fig. 1). On the other hand, the seeds without husking began to germinate in 15th day after inoculation. The results showed that existence of seed coats (exopleura and endopleura) inhibited seed germination. Husking before inoculating would short the time to germinate and improve the germination efficiency.

Fig. 1. Seed germination *in vitro.*

a: Seeds inoculated in generation medium; b: Seeds rooting; c: Seeds shooting

3.2. The Effect of Treatment on Explant Materials on Callus Induction

The seedling epicotyls were cut off to short fragments with different length, 0.3cm, 0.5cm, and 1.0cm, and inoculated in callus induction medium. After few days' in darkness culture, yellow callus emerged at both ends of epicotyl fragments. Among different lengths of epicotyl fragment, 0.3 cm was the best treatment. Epicotyl fragments 0.3 cm long produced greater number of callus, and more friable embryogenic calli were observed (Fig. 2). The epicotyl fragments 1.0 cm long produced callus slowest. Two weeks later, only little callus emerged, and after four weeks, in some epicotyl fragment only one end produced callus (Fig. 2).

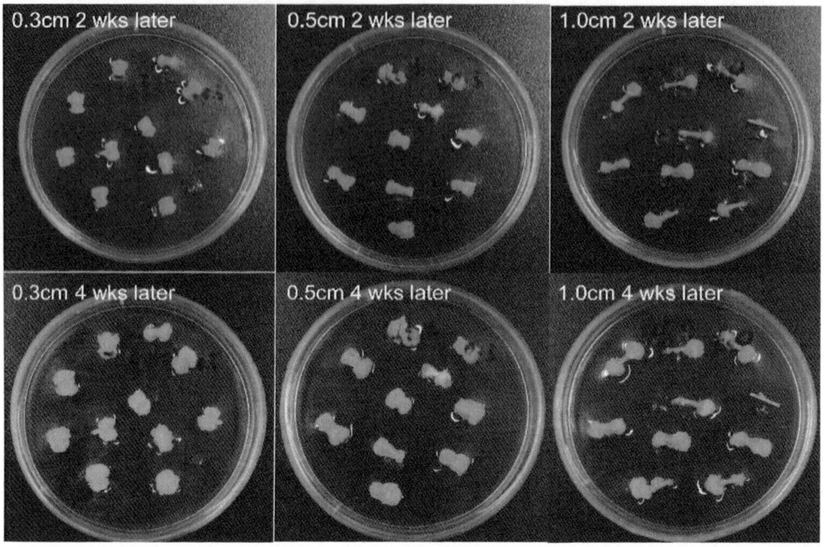

Fig. 2. Callus formation from different lengths of epicotyl fragment in 2 weeks and 4 weeks.

In order to understand whether wounding would promote callus production, seedling epicotyl fragment (0.5cm long) was cutting into two parts longitudinally, or wounded by knife longitudinally. In the wounded longitudinally epicotyl fragments, callus produced at both ends of wound of epicotyl fragments, but the callus production was slower than that without wounds (Fig. 3c). In another treatment, epicotyls were cutting into two parts longitudinally. Because the diameter of epicotyl is too small to separate symmetrically, the two separated parts were generally in different size. The smaller part was difficult to form callus (Fig. 3b), and another part didn't showed advantage in callus formation efficiency (Fig. 3b). As a result, seedling epicotyl wounding didn't showed advantage in callus formation efficiency comparing to that without wounding (Fig. 3a), as the epicotyl materials were too small to bear helpful wounding.

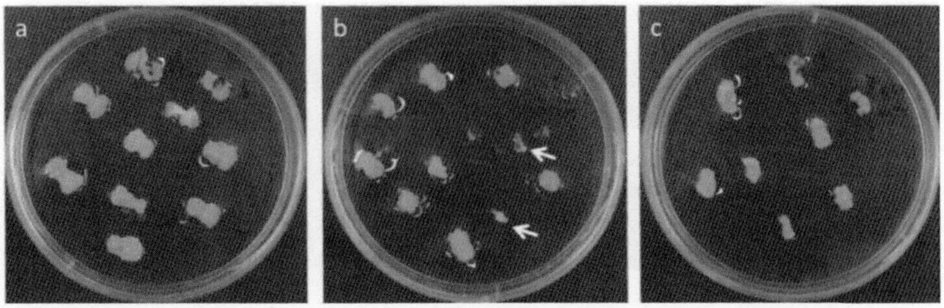

Fig. 3. Callus induction in 4 weeks from injured/intact epicotyls.

a: Intact epicotyls (no injured); b: Injured epicotyls with longitudinal cutting (white arrow showing the smaller parts cut off which was too small to form callus); c: Injured epicotyls with longitudinal scratching

3.3. Callus Proliferation Medium Optimization

Calli formed in callus medium for one month were divided into small pieces 0.5cm diameter, and used to optimize callus proliferation medium. One month after inoculating, in two tested media callus proliferated better. One contained the combination of 2.0 mg/L 2,4-D and 0.5 mg/L 6-BA, and the other one contained 2.0 mg/L 2,4-D and 1.0 mg/L 6-BA (Fig. 4). Furthermore, the callus produced in the medium containing 2.0 mg/L 2,4-D and 0.5 mg/L 6-BA were mostly friable embryonic callus (Fig. 5b). However, in the medium containing 2.0 mg/L 2,4-D and 1.0 mg/L 6-BA, callus were mostly dry, yellow-brown, and nonembryonic (Fig. 5a). In other two media, containing 1.0 mg/L 2,4-D and 0.5 or 1.0 mg/L 6-BA, callus were generally white, loose and watery nonembryonic callus (Fig. 5c).

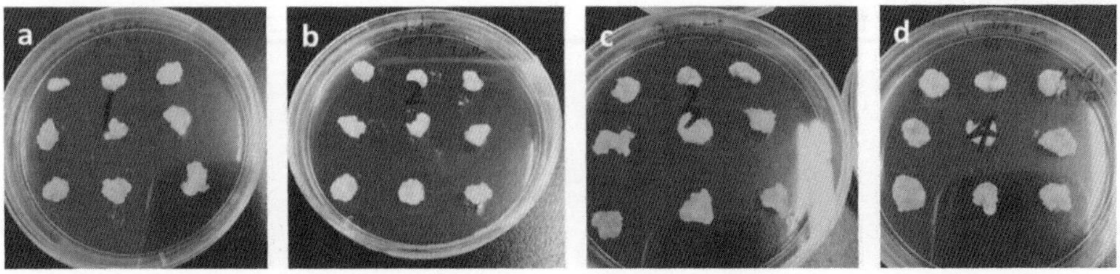

Fig. 4. Callus proliferation in 30 days in test proliferation media.

a: Treat 1 (1.0 mg/L 2,4-D + 0.5 mg/L 6-BA); b: Treat 2 (1.0 mg/L 2,4-D + 1.0 mg/L 6-BA); c: Treat 3 (2.0 mg/L 2,4-D + 0.5 mg/L 6-BA); d: Treat 4 (2.0 mg/L 2,4-D + 1.0 mg/L 6-BA).

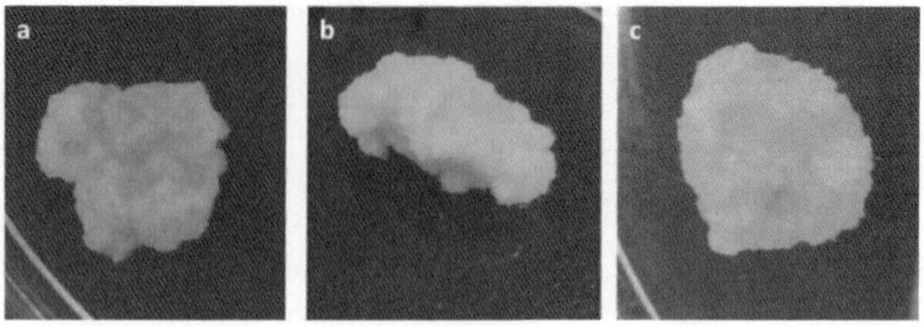

Fig. 5. Three callus types.

a: Dry callus (Yellowish-brown and dry); b: Embryonic callus (light white and friable); c: Soft callus (white, loose and watery)

3.4. Shoot Regeneration Medium Optimization

By calculating the number of callus producing green tissues in 50 days (Fig. 6), shoot regeneration medium were optimized. In the media containing 1.0 mg/L 6-BA and 0.01 or 0.1 mg/L NAA, there were same numbers of callus forming green tissues (Table 1). However, in the media containing less NAA (0.01 mg/L), the area of green tissue in one callus was less than that containing higher NAA (0.1 mg/L). As the result, the medium was the best one for shoot regeneration for now.

Table 1. The effect of different concentrations of plant growth regulator on shoot regeneration

6-BA (mg/L)	*NAA (mg/L)*	*Number of Callus*	*Number of callus turning green*	*Ratio of green callus (%)*
1.0	0.01	15	5	33.33
1.0	0.1	15	5	33.33
2.0	0.01	15	4	26.67
2.0	0.1	15	2	13.33

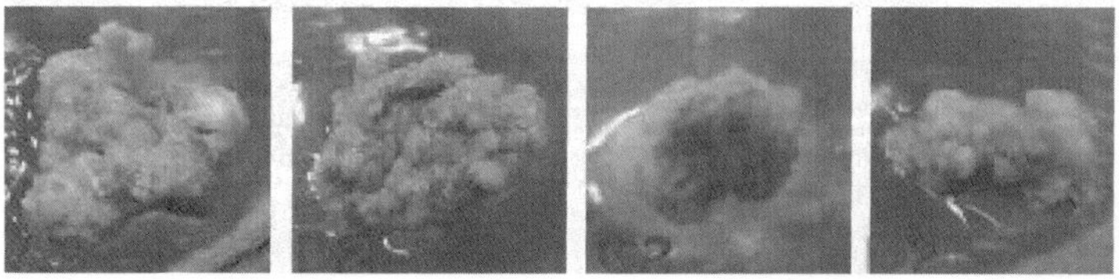

Fig. 6. Green callus in the bud regeneration medium in 50 days.

A: Treat 1 (1.0 mg/L 6-BA + 0.01 mg/L NAA); B: Treat 2 (1.0 mg/L 6-BA + 0.1 mg/L NAA); C: Treat 3 (2.0 mg/L 6-BA + 0.01 mg/L NAA); D: Treat 4 (2.0 mg/L 6-BA + 0.1 mg/L NAA)

4. DISCUSSION

Husking was an important factor to restrain seeds germination *in vitro*, as the result of this research. The seeds which were removed exopleura and endopleura germinated 8 days earlier than that without husking. Moreover, husked seeds were mostly germinated but the seeds without germinated with lower efficiency.

In this research, seedling epicotyls were cut off to short fragments with different length, 0.3 cm, 0.5 cm, and 1.0 cm, to test the best cutting length of explant for callus induction. It showed that 0.3 cm of epicotyl fragments produced better amount and higher quality callus. Besides, adopting 0.3 cm of epicotyl fragments is an economical way to use explant materials, as 1.0 cm of epicotyl fragments can separated to three 0.3 cm of fragments.

In plant biology, callus cells are those cells that cover a plant wound[9]. This irritated the wounding treatment on seedling epicotyls in this research. We cutting off the seedling epicotyl fragment (0.5cm long) to two parts longitudinally, or wounded seedling epicotyl fragment (0.5cm long) by knife longitudinally. The results indicated that cutting and wounding didn't help to produce more and better callus. The reason was that the diameter of epicotyl is too small to separate symmetrically, which resulted the two separated parts were generally in different size, and the smaller part was difficult to form callus.

Generally in the process of plant tissue culture, the frequency of medium change was once a month. Callus continuously absorbed nutrition from medium to sustain their growth, as well as discharge metabolic substances to the medium. During culturing, the nutrient in the medium would be consumed gradually, and metabolic waste would be accumulated. In this experiment we found that, during culture in proliferation medium, callus first turned to pale and yellow, but in the later stage of culturing, callus became to brown and watery. It was due to nutrient consuming and metabolic substances accumulating in medium.

References

1. Roistacher CN, Navarro L, Murashige T. Recovery of Citrus Selection Free of Several Viruses, Exocortis Viroid, and Spiroplasma citri by Shoot-Tip Grafting in vitro. *Seventh IOCV conference* January 1976: 186-193.
2. Lu LX, Lin SQ. *In vitro* culture of nucelli in 3 species of citrus subfamily plants. *Fujian Agriculture and Forestry University* 2003; **32**(1):193-195, (In Chinese).
3. Chen DM, Xu CJ, Zhang SL. Studies on a direct shoot regeneration system of *Citrus Sinensis* L. Osbeck with a high frequency *in vitro*. *Journal of Zhejiang Agricultural University (Agric.& Life Sci.)* 2000; **26**(05): 493-499, (In Chinese).
4. Zhang JE, Guo WW, Deng XX. Relationship between ploidy variation of citrus calli and competence for somatic embryogenesis. *Acta Genetica Sinica* 2006; **33**(7): 647-654.
5. Wei ZX. The Hybrid Citrus Cultivation Techniques. *Guangxi Gardening* 2005; **16**(3): 20-22, in Chinese.
6. Wang ZH. Research of Frost Injury Mechanism and Control Techniques in "Skiranui Tangerine" Hybrid Citrus Fruit. *Sichuan: Sichuan Agricultural University*, 2008, (In Chinese).
7. Yao LH, Jiang YM, Shi J, et al. Flavonoids in food and their health benefits. *Plant Foods for Human Nutrition* 2004; **59**: 113-122.

8. Weng FL, Xu JG, Ke PZ. The biological characteristics of Kiyomi tangor citrus cultivar and its utilization in citrus breeding. *South China Fruits* 2004; **33** (5): 8-9, (In Chinese).
9. Liang YC, Yang H. Advances of plant tissue culture. *Journal of Fujian College of Forestry* 2002; **22**(1): 1-3, (In Chinese).

CLONING AND EXPRESSION ANALYSIS OF *ZOMNHX2* GENE FROM EELGRASS (*ZOSTERA MARINA L.*)

Changlu Guo[1,a,*], Yanhong Zhao[2,b], Yanfang Wang[2,c], Wenjie Li[1,d], Xipeng Tian[1,e], Meifeng Lu[1,f], Xiangyu Zhang[1,g], and Xin Liu[1,h]

[1]*School of Marine Science and Technology, Harbin Institute of Technology (Weihai), Weihai, Shandong Province, China*

[2]*College of Life Sciences, Ludong University, Yantai, Shandong Province, China*

E-mail: [a]guochanglu@163.com, [b]Zyhbob@163.com, [c]wyf863@126.com, [d]1933818491@qq.com, [e]913826637@qq.com, [f]lanshlu@163.com, [g]zxy@163.com, [h]hitliuxin@126.com

[*]Corresponding author

The vacuolar Na^+/H^+ antiporter (*ZomNHX2*) gene from eelgrass was cloned by RT-PCR successfully, and we got related information by bioinformatics analysis. Then Semi-quantitative RT-PCR was preformed to analyse the tissue specificity of *ZomNHX2* gene expression. The results showed that *ZomNHX2* gene of eelgrass was cloned by RT-PCR, length of 1626 bp, encoding 541 amino acids. In addition, the *ZomNHX2* from eelgrass displayed high homology with Na^+/H^+ antiporters from other species. Protein secondary structure prediction showed that alpha helix was the main secondary structure of *ZomNHX2*. Hydrophobic analysis indicated that the N-terminal of *ZomNHX2* was hydrophobic, while the C-terminal was hydrophilic and that the whole protein was hydrophobic. By TMpred prediction, it was presented that the *ZomNHX2* had 10 transmembrane domains. Phylogenetic tree indicated eelgrass had close genetic relationship with Iris lactea. Analyzing the specific expression of *ZomNHX2* gene in stems and leaves by semi-quantitative RT-PCR technology, we found that the gene expression in leaves was higher than in stems.

1. INTRODUCTION

At present, improving the plant salt tolerance and breeding salt-tolerance varieties by plant genetic engineering can improve its economic value and ecological value. With the research on the salt-tolerance genes, the genes which have the function of maintaining ion balance play an important role in plant salt-tolerance mechanisms. Among these genes, the Na^+/H^+ antiporter gene plays a key role in reducing salt stress. Some researchers had found that the Na^+/H^+ antiporter gene in the vacuole membrane could reduce the damage of Na^+ to the cytoplasm by promoting Na^+ to be transferred into vacuoles, thereby enhancing the plant salt tolerance[1].

The Na^+/H^+ antiporter gene encodes a polypeptide containing about 550 amino acid residues which forms about nine to twelve transmembrane domain, and the C-terminal of the peptide chain was stronger hydrophilic. Na^+/H^+ antiporter gene was a member of multi-copy gene family. Along with Na^+/H^+ antiporter genes from *Spartina*, *Populus*, *Limonium gmelinii*, and other plants being cloned by RT-PCR (Reverse transcription-polymerase chain reaction), RACE (Rapid-amplification of cDNA ends) , it was showed that this protein commonly existed in plants[2-6]. In addition, some studies indicated that Na^+/H^+ antiporter genes in different plants had high homology. Guo et al. analyzed the Na^+/H^+ antiporter gene sequence from *Bruguiera gymnorrhiza*, and found that this gene had high homology with Na^+/H^+ antiporter genes from *Populus*, *Vitis vinifera*, *Lotus corniculatus* and other plants[7].

Numerous studies indicated that overexpression of the Na^+/H^+ antiporter gene in vacuole membrane could improve plant salt tolerance. Eelgrass is a marine higher angiosperm, which lives in the sea water with the salinity of 600mmol/L. Its salt-tolerance is obvious and therefore eelgrass is one of the absolutely good objects for the research on the salt-tolerance mechanisms of the higher plants. In this study, eelgrass was used as the material for cloning, expression and bioinformatics analysis of the Na^+/H^+ antiporter gene. It lays the foundations for further researching the salt-tolerance mechanisms of this gene.

2. MATERIALS AND METHODS

2.1. Plant Material

Eelgrass was from Xiaoshidao of Weihai.

2.2. CDNA Cloning of *ZomNHX2* Gene

Total RNA was extracted from eelgrass by Trizol method. The cDNA synthesis kit (Takara, Dalian, China) was employed to synthesize cDNA for the following RT-PCR. The specific set, *ZomNHX*-F (5'-ATG GAT TTG GGT GTC ATT ACG G-3') and *ZomNHX*-R (5'-CTA AGA TGG AGG CTC ACT TGG TTC-3'), were designed, according to the published mRNA sequence of the Na^+/H^+ antiporter gene in Genbank. Reactions were subjected to the following conditions: 5 min denaturation at 94 °C, 33 cycles of 94 °C for 45 s, 65.8 °C for 50 s and 72 °C for 120 s, and a final extension of 10 min at 72 °C. Amplified fragment was recovered from agarose electrophoresis with an MiniBEST Agarose Gel DNA EXtraction Kit Ver.3.0 (TaKaRa, Dalian, China) and cloned into the pMD®19-T Simple Vector (TaKaRa, Dalian, China), followed by DNA sequencing for confirmation.

2.3. Structure Analysis of *ZomNHX2*

The nucleotide sequences and deduced amino acid sequences of *ZomNHX2* were analyzed by the BLAST algorithm (NCBI, http://www.ncbi.nlm.nih.gov/BLAST/). The ORF was identified by the ORF Finder tool (http://www.ncbi.nlm.nih.gov/gorf/gorf.html). The protein secondary structure was predicted by SOPMA. Protein transmembrane structure and protein hydrophobicity were analyzed by TMHMM and expasy (http://web.expasy.org/cgi-bin/protscal/protscale.pl).

2.4. Phylogenetic Analysis of *ZomNHX2*

Phylogenic tree was built by the Clustal X tool with neighbor-joining (N-J) method of the MEGA5.0.

2.5. Expression of *ZomNHX2* Gene in Different Tissues

Specific primers were designed as follows: *ZomNHX*-F2 (5'-ATC AAT GGC ACT TGC TTA-3') and *ZomNHX*-R2 (5'-AAC AAA TGG AAC AAA ACC TC-3'), according to the sequencing result of *ZomNHX2* gene. The size of the amplified fragments is 393 bp. Specific primers were designed as follows:18SF (5'-GTT AGA TAC AAG GTT GAC ACG AC-3') and 18SR (5'-CTG ACT ATG AAA TAC GAA TGC C-3'), according to the sequence of 18SrRNA of eelgrass in Genebank. The size of amplified fragment was 695 bp. Total RNA from stems and leaves of eelgrass was extracted and reversely transcribed into cDNA. Regarding 18sRNA as the reference gene, the expression of *ZomNHX2* gene in stems and leaves of eelgrass can be detected by semi-quantitative RT-PCR.

3. RESULTS AND ANALYSIS

3.1. CDNA Cloning of *ZomNHX2* Gene

Total RNA was extracted from eelgrass, then reversely transcribed into cDNA. The specific product was amplified from cDNA of eelgrass by RT-PCR. As shown in Fig. 1, agarose gel electrophoresis demonstrated PCR product was located between 1000 bp and 2000 bp, about 1600 bp. PCR product was cloned by T-A, and recombinant transformation was verified by colony PCR.

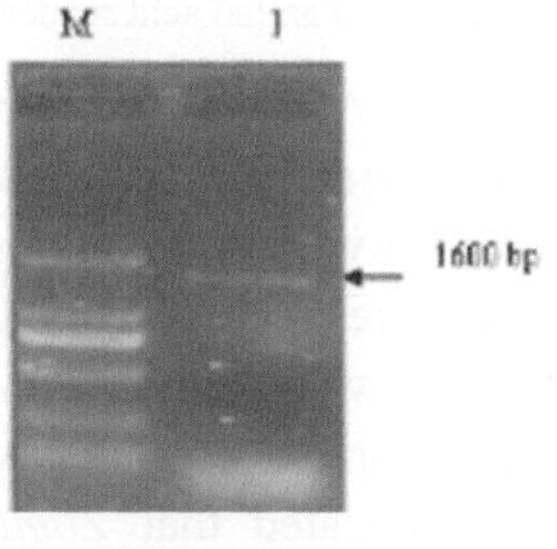

M: DNA Molecular Weight Markers (DL2000);1: PCR products of *ZomNHX2* gene

Fig. 1. Electrophoresis result of PCR product.

3.2. Structure Analysis of *ZomNHX2*

Blastn comparison of the sequencing result showed that the gene was the Na^+/H^+ antiporter gene from eelgrass. The full length of the gene was analyzed by NCBI ORF finder, which was 1626 bp and encoded 541 amino acids, as shown in Fig. 2. The molecular weight of the protein was 59470.7 and the isoelectric point was 6.67, which was analyzed by ProtParam of ExPASy database.

```
   1 atggatttgggagtgattacggagatggcaaatcaagtgacggtctccacctccggccatgcttcggtatcgtcgataagcttgttctcg
     M  D  L  G  V  I  T  E  M  A  N  Q  V  T  V  S  T  S  G  H  A  S  V  S  S  I  S  L  F  S
  91 gctcttctgtgcggctgcatcgtgattggtcatctgttggaagagaaccgatgggtcaacgaatccatcaccgcacttctcattgggttg
     A  L  L  C  G  C  I  V  I  G  H  L  L  E  E  N  R  W  V  N  E  S  I  T  A  L  L  I  G  L
 181 gttaccggagttgtcatccttctaacgacaaagggaaaaagttcacatattctggtgttcagtgaggatctgttcttcatatacttacta
     V  T  G  V  V  I  L  L  T  T  K  G  K  S  S  H  I  L  V  F  S  E  D  L  F  F  I  Y  L  L
 271 cctcctattatcttcaacgctgggttccaagtcaagaagaagcagtttttcagaaatttcatgaccatcacgttatttggcgtagtcggc
     P  P  I  I  F  N  A  G  F  Q  V  K  K  K  Q  F  F  R  N  F  M  T  I  T  L  F  G  V  V  G
 361 acattgatatcattttcaattatttcatttggcgctattattctgcttaaaaaaatggacgtaggcctcactgatatcggagattaccta
     T  L  I  S  F  S  I  I  S  F  G  A  I  I  L  L  K  K  M  D  V  G  L  T  D  I  G  D  Y  L
 451 gccattggagccatctttgcagctacagattctgtatgcaccttacaggttcttaatcaagatgagactcctctgctttacagtttggtg
     A  I  G  A  I  F  A  A  T  D  S  V  C  T  L  Q  V  L  N  Q  D  E  T  P  L  L  Y  S  L  V
 541 tttggagaaggtgtggtcaatgatactacttcagttgtgctgttcaatgcaatccagaacttcgatcttgttcacatagacggcactatt
     F  G  E  G  V  V  N  D  T  T  S  V  V  L  F  N  A  I  Q  N  F  D  L  V  H  I  D  G  T  I
 631 gccctgaaattttcggaaactttctctatctgtttcttgctagcaccgttcttggtgttgcaactggattattaagtgctttcgtcacc
     A  L  K  F  F  G  N  F  L  Y  L  F  L  A  S  T  V  L  G  V  A  T  G  L  L  S  A  F  V  T
 721 aagaaattgtacttcggcaggcattctacagaccgtgaagtttctttaatgatgctcatggcctacctctcctacttgctggcagagtta
     K  K  L  Y  F  G  R  H  S  T  D  R  E  V  S  L  M  M  L  M  A  Y  L  S  Y  L  L  A  E  L
 811 ttctatctgagtggtatactgactgtgttcttctgtaggattgtaatgtctcactacacatggcacaatgtaacagagagttcaagaata
     F  Y  L  S  G  I  L  T  V  F  F  C  R  I  V  M  S  H  Y  T  W  H  N  V  T  E  S  S  R  I
 901 acaaccaagcatactttcgcaacactgtcatttatcgctgaaatcttcatctttctgtatgttgggatggatgcattggatatcgacaag
     T  T  K  H  T  F  A  T  L  S  F  I  A  E  I  F  I  F  L  Y  V  G  M  D  A  L  D  I  D  K
 991 tggagattcatcagtgatagtccaggaaaatcgattggaataagctccattctgctgggattggttatggtaggaagggcagcattcgtg
     W  R  F  I  S  D  S  P  G  K  S  I  G  I  S  S  I  L  L  G  L  V  M  V  G  R  A  A  F  V
1081 tttccactgtcatttctgtccaacttgtccaagaaattctcacatgaaaaaattgaattcaagcagcaggttacaatatggtgggcggga
     F  P  L  S  F  L  S  N  L  S  K  K  F  S  H  E  K  I  E  F  K  Q  Q  V  T  I  W  W  A  G
1171 ctaatgagaggtgcagtatcaatggcacttgcttacaatcagtttaccagcgcaggtcacacggaattacccgcaaatgcaattatgata
     L  M  R  G  A  V  S  M  A  L  A  Y  N  Q  F  T  S  A  G  H  T  E  L  P  A  N  A  I  M  I
1261 accagtaccataacagttgttcttttcagcacagtggtatttggattgctgactaaacccttggtatttcttttacttcctcaacaagga
     T  S  T  I  T  V  V  L  F  S  T  V  V  F  G  L  L  T  K  P  L  V  F  L  L  L  P  Q  Q  G
1351 ttaaagcagcagatgagcagtattgatgacccaacaagcccaaaatcatttctagctgcactcctggggagtgtgcatgggtctgaaatt
     L  K  Q  Q  M  S  S  I  D  D  P  T  S  P  K  S  F  L  A  A  L  L  G  S  V  H  G  S  E  I
1441 gaagaacccactagtattcctcgaccgtccagtctacgtctactcgtcacttcacctactcgcaccgtgcattactattggcgcaaattt
     E  E  P  T  S  I  P  R  P  S  S  L  R  L  L  V  T  S  P  T  R  T  V  H  Y  Y  W  R  K  F
1531 gatgacgcagtcatgagacctatgtttggtggtagaggttttgttccatttattccaggatcaccaacagaggaaccaagtgtccatcct
     D  D  A  V  M  R  P  M  F  G  G  R  G  F  V  P  F  I  P  G  S  P  T  E  E  P  S  V  H  PP
1621 ttttag 1626
     F  *
```

Fig. 2. The nucleotide sequence and amino acid sequence encoded by *ZomNHX2* gene.

The secondary structure of *ZomNHX2* was predicted by SOPMA software and the results showed that α helix was the main secondary structure of *ZomNHX2*, which accounted for about 36.48%, followed by random coil and main chain extension, accounting for 27.41% and 27.22% respectively; the last one was β corner, about 8.89%. *ZomNHX2* transmembrane structure was analyzed By TMHMM, and it has been shown that *ZomNHX2* had 10 transmembrane domains, which differed from the *AtNHX1* transmembrane protein, and it indicated that *ZomNHX2* and *AtNHX1* had different functions. Hydrophobicity analysis indicated that the N-terminal of *ZomNHX2* was highly hydrophobic, while the C-terminal had a strong hydrophilicity. Overall, the entire peptide chain was uniformly distributed hydrophobic amino acids, and hydrophobic amino acids were more than hydrophilic amino acids, so the whole protein expressed as hydrophobicity. The C-terminal had hydrophilic tail with particularity, presuming that the C-terminal was related with transporter activity. The amino acid sequence of the present study was analyzed by NCBI database and we found that *ZomNHX2* had conserved functional domains: NhaP, Asp-Al-Ex Superfamliy, Na-H-Exchanger, Endosomal/prevacuolar_sodium/hydrogen_exchange, COG3263 and PRK05326 etc. Na-H-Exchanger domains belonged to Na^+/H^+ exchange protein family, which showed that *ZomNHX2* was Na^+/H^+ antiporter family member.

3.3. Phylogenetic Analysis of *ZomNHX2*

Eelgrass *ZomNHX2* was aligned with Na^+/H^+ antiporter proteins of some species. Eelgrass *ZomNHX2* showed a similarity of 77% with *Suaeda pruinosa*, 77% with *Theobroma cacao*, 79% with *Iris lactea*, 75% with *Atriplex gmelinii* and 78% with *Oryza sativa Japonica Group*, respectively. To investigate the evolutionary relationships of the *ZomNHX2* and other plant Na^+/H^+ antiporter proteins, a phylogenetic tree was constructed from the deduced amino acid sequences of Na^+/H^+ antiporter proteins from 15 species, as shown in Fig. 3. Eelgrass *ZomNHX2* and *Iris lactea* were located on a branch, indicating that eelgrass *ZomNHX2* had closer phylogenetic relationships with *Iris lactea*.

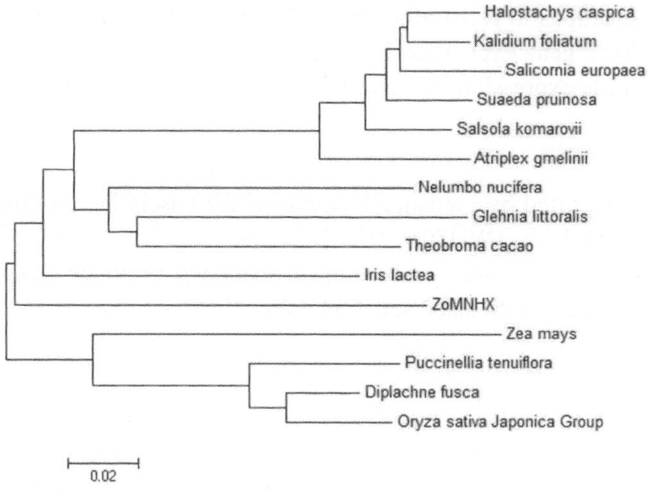

Fig. 3. Phylogenetic tree of Na^+/H^+ antiporter proteins from 15 species.

3.4. Expression of *ZomNHX2* gene in different tissues

The expression of *ZomNHX2* gene in stems and leaves of eelgrass was detected by semi-quantitative RT-PCR. The expression of *ZomNHX2* gene both in stems and leaves was not high, as can be seen from Fig. 4, however the expression of reference gene was high. It also showed that the expression of *ZomNHX2* gene in leaves was higher than that in stems relatively.

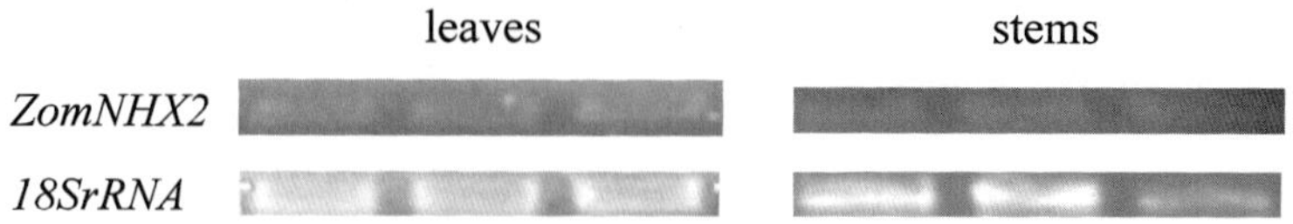

Fig. 4. *ZomNHX2* gene specific expression.

4. CONCLUSIONS

In this study, the eelgrass *ZomNHX2* gene was cloned by gene engineering technology successfully. The amino acid sequence deduced from the DNA sequence by software showed that the eelgrass *ZomNHX2* gene had typical Na^+/H^+ antiporter protein family's conserved region, and it had a high similarity with *oryza sativa, atriplex gmelinii, Theobroma cacao, suaeda* and *iris lacteal*, and it had a close genetic relationship with *iris lacteal* in evolution. Additionally, a large number of studies showed that Na^+/H^+ antiporter had functions to resist other stress. Semi-quantitative RT-PCR was preformed to detect *ZomNHX2* gene expression in stems and leaves of eelgrass. The results showed that *ZomNHX2* gene expression in leaves was higher than that in stems.

Acknowledgments

The study was supported by guidance Foundation for Discipline construction of Harbin Institute of Technology (WH20140205), Yantai Science and Technology Project (2015ZH065), Natural Science Foundation of Shandong Province (ZR2013CL013) and Key Project of Coal Based Science and Technology in Shanxi Province (FT2014-01-14).

References

1. Li M, Zhang J, Li YJ, et al. The research progress of Plant Physiology and tolerance gene [J]. *Jiangsu Agricultural Sciences*, 2012, **40**(10): 45-49(In Chinese).
2. Gaxiola RA, Rao R, Sherman A, et al. The Arabidopsis thaliana Proton Transporters, *AtNhxl* and *AvP1*, can Function in Cation Detoxification in Yeast[J]. *Proceedings of the National Academy of Sciences of the United States of America*, 1999, **96**(4): 1480-1485.
3. Fukuda A, Nakamura A, Tanaka Y. Molecular Cloning and Expression of the Na^+/H^+ Exehanger Gene in *Oryza Sativa*. *Biochimica et BioPhysica Acta*. 1999; **1446**(1): 149-155.
4. Uji T, Monma R, Mizuta H, et al. Molecular Characterization and Expression Analysis of two Na^+/H^+ Antiporter Genes in the Marine Red Alga *Porphyra Yezoensis*. *Fisheries Science* 2012; **78**: 985-991.
5. Ren L. Analysis of *Populus* Na^+/H^+ antiporter *PeNHX3* biological function [D]. Gansu: Lanzhou University, 2014(In Chinese).

6. Zhou LL. Preliminary study of *Limonium gmelinii* Na^+/H^+ antiporter gene cloning and function [D]. Xinjiang: Shihezi University, 2009(In Chinese).
7. Guo QS, Xu LX, Yu W. Cloning of Na^+/H^+ Antiporter Gene from *Bruguiera gymnorrhiza* (L.) LAM. *Chinese Journal of Tropical Crops*. 2010; **1**(2): 105-109(In Chinese).

THE CHARACTER OF NEXT-GENERATION TECHNOLOGIES IN PERENNIAL GRASS

Maobo Zheng[1,2,a], Jidi Gao[4,b], Fengqi Liu[3,c], Hailing Zhang[3,d], Ping Chen[4,e], Jilin Li[4,f], and Yanming Zhang[4,g,*]

[1]*Institute of Crop Breeding, Heilongjiang Academy of Agricultural Sciences, Harbin, 150086, China*

[2]*School of Science, Harbin University, Harbin, 150086, China*

[3]*Institute of Prataculturai Sciences, Heilongjiang Academy of Agricultural Sciences, Harbin, 150086, China*

[4]*Key Laboratory of Molecular Cytogenetics and Genetic Breeding of Heilongjiang Province, College of Life Science and Technology, Harbin Normal University, Harbin, 150025, China*

E-mail: [a]biotech0451@163.com, [b]1204620356@qq.com, [c]l_fq2003@163.com, [d]hailingzhang2014@163.com, [e]1113565172@qq.com , [f]jilinli2013@163.com, [g]blueright@163.com

**Corresponding author*

DNA sequencing technology is undergoing a revolution and the same time fuelling a revolution in genetics and genomics. To understand a plant's heritable traits, it is important to understand its genome. If a plant has a very complex genome it may be a challenge to correlate genome variation with important agronomic traits. The genomics extended by high-throughput sequencing now impacts the disciplines of agronomy, biochemistry, physiology and microbiology. This review summarizes advances of next generation technologies and introduces the development of next generation sequence (NGS) for perennial grass as well as perspectives the important role of NGS in research of genomics for perennial grass in the future.

1. INTRODUCTION

With rapidly increasing efficiencies from DNA and RNA high-throughput sequencing, next-generation technologies are quickly becoming a mainstay in plant genetics, genomics and biochemistry[1]. In present study, multiple genomes from several species have been sequenced including: Arabidopsis[2], rice[3], maize[4], soybean[5], and Medicago truncatula[6]. Next-generation sequencing (NGS) of plant genomes is providing new insight into the genomic variation within and between species, such as single nucleotide polymorphisms (SNPs), presence–absence variations (PAVs), copy number variations (CNVs) and other types of structural variations (SVs), the introgression of genetic elements such as quantitative trait loci (QTLs), the evolution of organs and pathways, and the status of epigenetic modifications[4,6,7]. Likewise, sequencing of plant transcriptomes (RNA-seq) is providing atlases of gene expression, demonstrating alternative splicing events, defining QTL introgressions, characterizing small RNAs and identifying genes and pathways involved in acclimation to biotic and abiotic stresses[8,9]. The family of grasses (Poaceae, also known as Gramineae) contains 10000 species including rice, wheat, barley, maize and forage grasses; it is the only family of plants more important to humans than legumes in in economic and nutritional value. For grasses, the primary source of genetic information is rice. The availability of genome sequences for these species has enhanced our ability to understand their physiology, with examples such as quantitative trait loci (QTL) analysis for genes associated with domestication and drought tolerance in rice[10], shoot fly resistance in sorghum[11], as well as disease resistance in maize[12]. In this article, we will discuss advances of next generation technologies in present and outline the development of next generation sequence (NGS) for perennial grass as well as prospect the important role of NGS in research of genomics for perennial grass in the future.

2. ADVANCES OF NEXT-GENERATION TECHNOLOGIES

The genome sequence provided the foundation to better understand important agronomic traits and assist researchers seeking to improve varieties. The first commercially available NGS system was developed by 454 and commercialized by Roche as the GS20, capable of sequencing over 20 million base pairs in just over 4h[13]. While traditional Sanger-based sequencing remains popular for some applications, the majority of DNA and expressed gene sequence data produced today comes from the range of next- or second-generation sequencing (NGS/2GS) technologies. Current systems can produce around 400 Mbp of sequence data with reads of up to 400 bp (http://www.454.com), such as Roche/454, Illumina/Solexa and ABI/SOLiD, has made it possible to generate genome resources at a large scale and relatively low cost[14].

NGS technologies are now used routinely for transcript profiling. Some whole-genome sequencing projects were accompanied by transcriptome analyses via RNA-seq[15,16]. RNA-seq data allowed for in-depth mining of splice variants, SNPs, simple sequence repeats (SSRs) and small interfering RNA (siRNA) species, the identification of sub-/neo-functionalized genes and transcript. It can be applied for analysis of gene expression, the structure of genomic loci, and sequence variation present at expressed gene loci[17], without prior knowledge of gene sequence. Roche 454 technology was the first of the second-generation sequencing technologies, and Illumina platform is becoming increasingly popular for transcriptome studies due to the power of vast read depth and read pair technology, which provides more readily than 454 sequencing[18,19].

3. DEVELOPMENT OF NGS FOR PERENNIAL GRASS

The perennial grass is of fundamental importance for the development of environmentally friendly biofuels and platform chemicals. The value of grass will continue to increase in line within an increase in the requirement for meat and milk at a rate of rapid per annum.

Perennial ryegrass (*Lolium perenne*, 2n = 2x =14; DNA content =2034 Mbp)[20] is one of the most conomically and environmentally important grass species. For example, in the UK it accounts for 50% of all land use and 70% of all agricultural land use. It is a member of a group of interfertile, largely outbreeding, related species which encompass a number of ryegrasses and fescues (*Lolium* and *Festuca* spp.) including, Italian ryegrass (*L. multiflorum*) and meadow and tall fescue (*F. pratensis* and *F. arundinacea*). Because of their economic significance and evolutionary relationships, there has been interest in characterising these grasses in molecular terms both as individual species and as ancestral and recent hybrids[21]. The number of microsatellites for use in the *Lolium* that are publicly available has started to increase in the last few years, which was the aim of developing a genomic microsatellite library for *L. perenne*.[22,23], and was combined with a set of EST-based microsatellites produced from a *L. perenne* GeneThresher database[24]. Moreover, The ryegrass transcriptome map displays the genetic location of 732 expressed genes putatively underlying specific biochemical or physiological functions that control variation for agronomically important traits. For the same traits, the present transcriptome map contains additional candidate genes such as the TERMINAL FLOWER1-like gene (LpTFL1) that is a well characterised repressor of flowering and a con-troller of axillary meristem identity in ryegrass[25], and a homologue of the Triticum monococcum L. gene TmVIL3, that is up-regulated by vernalization[26].

Leymus is a genus of the grass family (Triticeae; Poaceae). Approximately 34 species of Leymus have been identified[27]. As using for forage and conservation, Leymus is one of several perennial Triticeae genera used in wide-hybridization wheat breeding[28], and some of the resulting introgression lines display potentially useful traits, including salt tolerance[29].

Sheepgrass [*Leymus chinensis* (Trin.) Tzvel.] is an important perennial grass species in the Leymus genus. It is widely distributed on the eastern Eurasian steppe, including the outer Baikal area of Russia, western North Korea, Mongolia, the Northeast Plain, and the Inner Mongolian Plateau of China[30]. These grasslands play important roles in soil and water conservation, ecological construction and support of livestock farming, especially in northern China[31]. Recently, a transcriptome data was generated in sheepgrass under saline-alkaline treatment using Roche-454 massive parallel pyrosequencing technology and a large number of saline-alkaline responsive differentially expressed genes (DEGs) were obtained[32].

4. CONCLUSION

At present, in plant genetics and biochemistry, using next-generation sequencing have resulted in profound paradigm shifts, which lead to our fundamental understanding for genome architecture was expanded, such as transcript variants[33], species evolution[34], gene networks[35] and biochemical pathways[36]. Likewise, perennial grass as important role in the earth for environmentally friendly, agriculture, and human life, will need pay more attention to research its hereditary basis and take full advantage of NGS for genome analysis, and in the finally, to improve grass breeding and genetics.

Acknowledgements

This work was financially supported by Scientific Research Fund of Heilongjiang Provincial Education Department (No:12531198).

References

1. Chia JM, Ware D. Sequencing for the cream of the crop. *Nature Biotechnology* 2011; **29**: 138–139.
2. Cao J, Schneeberger K, Ossowski S, et al. Whole-genome sequencing of mulitple Arabidopsis thaliana populations. *Nature Genetics* 2011; **43**: 956–963.
3. Huang X, Lu T, Han B. Resequencing rice genomes: an emerging new era of rice genomics. *Trends in Genetics* 2013; **29**: 225–232.
4. Jiao Y, Zhao H, Ren R, et al. Genomewide genetic changes during modern breeding of maize. *Nature Genetics* 2012; **44**: 812–815.
5. Lam HM, Xu X, Lui X, et al. Resequencing of 31 wild and cultivated soybean genomes identifies patterns of genetic diversity and selection. *Nature Genetics* 2010; **42**: 1053–1059.
6. Branca A, Paape TD, Zhou P, et al. Whole-genome nucleotide diversity, recombination, and linkage disequilibrium in the model legume Medicago truncatula. *Proceedings of the National Academy of Sciences USA* 2011; **108**: E864–E870.
7. Varshney RK, Song C, Saxena RK, et al. Draft genome sequence of chickpea (Cicer arietinum) provides a resource for trait improvement. *Nature Biotechnology* 2013; **31**: 240–246.
8. Meyer K, Stecca KL, Ewell-Hicks K, Allen SM, Everard JD. Oil and protein accumulation in developing seeds is influenced by the expression of a cytosolic pyrophosphatase in Arabidopsis. *Plant Physiology* 2012; **159**: 1221–1234.
9. Atwood SE, O'Rourke JA, Peiffer GA, et al. Replication protein A subunit 3 and the iron efficiency response in soybean. *Plant, Cell and Environment* 2014; **37**: 213–234.
10. Li C , Zhou A, Sang T. Genetic analysis of rice domestication syndrome with the wild annual species, Oryza nivara. *New Phytologist* 2006; **170**: 185-194.
11. Satish K, Srinivas G, Madhusudhana R, Padmaja P, Nagaraja Reddy R, Murali Mohan S, Seetharama N. Identifi cation of quantitative trait loci for resistance to shoot fly in sorghum. [Sorghum bicolor; (L.)Moench] *Theoretical and Applied Genetics* 2009; **119**: 1425–1439.
12. Chung CL, Jamann T, Longfellow J, Nelson R. Characterization and fine-mapping of a resistance locus for northern leaf blight in maize bin 8.06. *Theoretical and Applied Genetics* 2010; **121**: 205 – 227.

13. Margulies M, Egholm M , Altman WE, Attiya S, Bader JS, Bemben LA, Berka J, et al. Genome sequencing in microfabricated high-density picolitre reactors. *Nature* 2005; **437**: 376–380.
14. Duran C, Appleby N, Vardy M, Imelfort M, Edwards D, Batley J. Single nucleotide polymorphism discovery in barley using autoSNPdb. *Plant Biotechnology Journal* 2009b; 7: 326–333.
15. Young ND, Debelle F, Oldroyd GE, et al. The Medicagogenome provides insight into the evolution of rhizobial symbioses. *Nature* 2011; **480**: 520–524.
16. Varshney RK, Song C, Saxena RK, et al. Draft genome sequence of chickpea (Cicer arietinum) provides a resource for trait improvement. *Nature Biotechnology* 2013; **31**: 240–246.
17. Morozova O, Marra MA. Applications of next-generation sequencing technologies in functional genomics. *Genomics* 2008; **92**: 255–264.
18. Barski A, Cuddapah S, Cui K, Roh TY, Schones DE, Wang Z, Wei G, et al. High-resolution profi ling of histone methylations in the human genome. *Cell* 2007; **129**: 823–837.
19. Varshney RK, Nayak SN, May GD, Jackson SA. Next-generation sequencing technologies and their implications for crop genetics and breeding. *Trends Biotechnol* 2009, **27**: 522–530.
20. Bennett MD, Smith JB. Nuclear DNA amounts in angiosperms. Philosophical Transactions of the Royal Society of London. Series B. *Biological Sciences* 1976; **274**: 227–274.
21. King J, Thorogood D, Edwards KJ, Armstead IP, Roberts L, Skøt K, Hanley Z, King IP. Development of a Genomic Microsatellite Library in Perennial Ryegrass (Lolium perenne) and its Use in Trait Mapping. *Annals of Botany* 2008; **101**: 845–853.
22. Jensen LB, Muylle H, Arens P, Andersen CH, Holm PB, Ghesquiere M, et al. Development and mapping of a public reference set of SSR markers in Lolium perenne L. *Molecular Ecology Notes* 2005; **5**: 951–957.
23. Studer B, Widmer F, Enkerli J, Kölliker R. Development of novel microsatellite markers for the grassland species Lolium multiflorum, Lolium perenne and Festuca pratensis. *Molecular Ecology Notes* 2006; **6**: 1108–1110.
24. Gill GP, Wilcox PL, Whittaker DJ, Winz RA, Bickerstaff P, Echt CE, et al. A framework linkage map of perennial ryegrass based on SSR markers. *Genome* 2006; **49**: 354–364.
25. Jensen CS, Salchert K, Nielsen KK. A terminal flower1-like gene from perennial ryegrass involved in floral transition and axillary meristem identity. *Plant Physiol* 2001; **125**: 1517–1528.
26. Fu D, Dunbar M, Dubcovsky J. Wheat VIN3-like PHD finger genes are up-regulated by vernalization. *Mol Gen Genomics* 2007; **277**: 301–313.
27. Liu ZP, Chen ZY, Pan J, Li XF, Su M, et al. Phylogenetic relationships in Leymus (Poaceae: Triticeae) revealed by the nuclear ribosomal internal transcribed spacer and chloroplast trnL-F sequences. *Mol Phylogenet Evol* 2008; **46**: 278–289.
28. Jiang J, Friebe B, Gill BS. Recent advances in alien gene transfer in wheat. *Euphytica* 1994; **73**: 199–212.

29. Liu X, Shi J, Zhang XY, Ma YS, Jia JZ. Screening salt tolerance germplasms and tagging the tolerance gene(s) using microsatellite (SSR) markers in wheat. *Acta Bot Sinica* 2001; **43**: 948–954.
30. Bai YF, Han XG, Wu JG, Chen ZZ, Li LH. Ecosystem stability and compensatory effects in the Inner Mongolia grassland. *Nature* 2004; **431**: 181–184.
31. Bai WM, Xun F, Li Y, Zhang WH, Li LH. Rhizome severing increases root lifespan of Leymus chinensis in a typical steppe of Inner Mongolia. *PLoS One* 2010; **5**: e12125.
32. Sun Y, Wang F, Wang N, Dong Y, Liu Q, et al. Transcriptome exploration in Leymus chinensis under saline-alkaline treatment using 454 Pyrosequencing. *PLoS One* 2013; **8**(1): e53632.
33. Libault M, Stacey G. Evolution of FW2.2-like(FWL)andPLAC8genes in Eukaryotes. *Plant Signaling and Behavior* 2010; **5**: 1226–1228.
34. Lam HM, Xu X, Lui X, et al. Resequencing of 31 wild and cultivated soybean genomes identifies patterns of genetic diversity and selection. *Nature Genetics* 2010; **42**: 1053–1059.
35. Tomato Genome Consortium. The tomato genome sequence provides insights into fleshy fruit evolution. *Nature* 2012; **485**: 635–641.
36. Li H, Peng Z, Yang X, et al. Genome-wide association study dissects the genetic architecture of oil biosynthesis in maize kernels. *Nature Genetics* 2012; **45**: 43–50.

COMPARATIVE PROTEOMIC ANALYSIS OF THE HIGH-YIELDING STRAIN *AUREOBASIDIUM PULLULANS* CGMCC NO.3337 AND THE ORIGINAL STRAIN *AUREOBASIDIUM PULLULANS* TKPM10017 ON SYNTHESIS OF POLY (Β-L-MALIC-ACID) METABOLISM

Xujia Fan[a], Yanhui Bian, and Changsheng Qiao[b,*]

School of Biological Engineering, Tianjin University of Science and Technology, Tianjin economic and technological development zone 13th avenue 29, Tianjin, 300457, China

E-mail addresses: [a]gudusafa@163.com, [b]qiaochangsheng@tust.edu.cn

[*]Corresponding author

Poly(β-L-malic-acid) (PMLA), a sort of polyester and absolute biodegradable material ,has attracted an increasing amount of investment in the areas of pharmacy and medical industry. PMLA is showing a promising application foreground while little information concerning the molecular basis and the mechanism of biosynthesis was accumulated. In this paper, two strains which have different production-- thehigh-yieldingstrain Aureobasidium pullulans CGMCC No.3337 and the original strain Aureobasidium pullulans TKPM10017 were studied. In order to study the significant differences, we carried out the proteomics containing protease digestion and LC-MS/MS detection. The protein spectrum results show that many proteins had change, involveing energy-generating enzymes, antioxidant family and members function in different biochemistry procedures such as transcription, selective splicing, translation and so on. These studies have verified our previous hypothesis, which would provide a molecular understanding of the physiological response of A. pullulans for optimizing the performance of industrial pullulan fermentation.

1. INTRODUCTIONS

Poly (β-L-malic-acid)(PMLA)is a functional aliphatic polyester which is composed of malic acid repeating units[1]. Its pendant carboxylic groups allows the introduction of functional groups and molecules, even make it possible to bring in drugs via chemical modification. These advantages make us to realize it's of vital importance to find out how to escalate its yield and purity. As a new type of nontoxic macromolecular material with favorable biodegradation and biocompatibility[2,3], the PMLA has gained great industrial interest for its potential applications in medicine[4]and other industries. For example, it was reported that a PMLA-based drug delivery system and biodegradable nano particles could be used for Tumor-targeting therapy [5]. Therefore, PMLA can be used as prodrug or for drug delivery system, which has attracted a great deal of interest in industries[6-7].According to the reports, *Aureobasidium sp.* is the best strain to product PMLA. However, the productivity of PMLA is still at a low level under natural conditions, as is shown in some publishment about the researches on the fermentation by *Aureobasidium pullulans* [8-9].

Complex protein mixture separation and protein identification can be solved by proteomics techniques. A large number of proteins, are qualitatively and quantitatively measured by proteomic involving morphological and biochemical process, so as to provide exact analysis of alternative cell proteins in the course of growth, differentiation, and response to environmental factors[10-12].

Although PMLA has been used in different fields in industry, the molecular basis of PMLA biosynthesis of *A.pullulans* is still poorly understood. To control the fermentation process, it is important to realize the occurrence and development in the cell at a molecular level. The objective of the present paper is to describe the comparative global protein profiles in *A. pullulans* cells in the high-yielding strain *Aureobasidium pullulans* CGMCC No.3337 and the original strain *Aureobasidium pullulans* TKPM10017, thus extending our understanding of intracellular regulation mechanism in relation to cell biomass growth and PMLA production.

2. MATERIALS AND METHODS

2.1. Strain and Medium

Aureobasidium pullulans CGMCC No.3337 was isolated by the Key Laboratory of Industrial Microbiology, Ministry of Education, Tianjin University of Science and Technology. *A.pullulans* CGMCC No.3337 were selected from *A.pullulans* TKPM10017 by UV and He-Ne laser. The cells were grown on LB mediums for 3 days at 25 °C and stored at 4 °C.

The medium for inoculum preparation (seed medium) contained the following components (g/L): glucose 140 , yeast extract 3, $(NH_4)_2SO_4$ 1 , succinic acid 2 , K_2CO_3 0.4, KH_2PO_4 0.1, $CaCO_3$ 20, $ZnSO_4 \cdot 7H_2O$ 5×10^{-3} , $MgSO_4 \cdot 7H_2O$ 0.1, corn steep liquor.

The fermentation medium for flask culture was consisted of the following components (g/L): glucose 180, peptone 35, KCl 0.5, KH_2PO_4 0.1, $NaNO_3$ 2, $MnSO_4$ 0.05, $MgSO_4 \cdot 7H_2O$ 0.3, $CaCO_3$ 20 . The initial pH of the medium was adjusted by using 5 M NaOH and 6 M HCl to 7.0.

2.2. Cultivation Conditions

2.2.1. *Shake-flask Culture*

The cells growing on newly prepared were transferred to 500 ml Erlenmeyer flask containing 50 ml seed medium and incubated at 25°C and 200 rpm for inoculums preparation. The cells used for experiment was carried out by 500 ml sharking flasks containing 50 ml fermentation medium and incubated at 25°C and 200 rpm.

2.2.2. *Fermentation Cylinder Culture*

The seed prepared for the bioreactor was cultured in the Erlenmeyer flask each time. When its absorbance measured at 600 nm using a ultraviolet spectrophotometer

(UV-1800 PC, China), was at 0.6-0.7, transferred it to a 10.0 L bioreactor (Shanghai, China) containing 7 L fermentation culture medium.

2.3. Biomass Concentration

Two milliliter of broth mixed 3 mol/L hydrochloric acid(aim to wipe off calcium carbonate) was also centrifuged at 2000xg at 4°C for 20 min(Microfuge 18, Beckman Coulter Co., Fullerton, CA). Then, the precipitate were washed twice with distilled water and centrifuged again to remove impurities. After the precipitate were dried at 80 °C overnight(>8 h), dry weight amortized computation with fermentation broth was determined as biomass.

2.4. Quantitative Analysis of PMLA

The concentration of PMLA was determined by a reversed phase high performance liquid chromatography method on a column of prevail C18(4.6 mm×250 mm, 4 μm) with a mobile phase containing acetonitrile and 0.025 mol/L KH2PO4 buffer (pH 2.0) (5:95, v:v), the flow rate of mobile phase was 1.0 mL/min, the detection wavelength was 210 nm and the column temperature was 25°C. Before sampling, the fermentation broth was centrifuged and hydrolysed. L-malic acid was completely separated and determined in 4 minutes.

2.5. Protein Spectrum Experiment

LYSIS BUFFER was added to cells firstly.After repeated freezing and thawing and ultrasonic methods. As used lysate containing 7 mol / L urea, 2 mol / L thiourea, 4% m / V of CHAPS, 1% m / V of DTT, 2% V / V of the IPG buffer (pH 4-7), 1 % m / V of PMSF and Tris-HCl 15 mmol / L, pH, 8.0, ultrapure water configuration. The microbial cells collected after the addition of the above lysate full vortexed, frozen in liquid nitrogen for 10 min, then at 37 ° C to thaw. Freezing and thawing process was repeated three times. After repeated freezing and thawing, using ultrasound (30% power) for 5 s pulse ultrasonic method ultrasonic 15 min full cleavage. After sonication was allowed to stand on ice for 15 min was cooled, the mixture was added a nuclease 1% V / V, and allowed to stand on ice for 30 min. At 4 ° C to 13200 r / min centrifugation 30 min, the supernatant. After protein quantification assay to Brandford which contains protein concentration, placed in -80 ° C to save for later use.

In the resulting protein sample, added to a final concentration of 10 mmol / L of dithiothreitol (DTT), a water bath at 37 ° C reduction reaction 3h, and then added to a final concentration of 20 mmol / L of iodoacetamide (IAA) alkylation process for 1 h at room temperature, dark place. Adding 4 volumes of acetone at -20 ° C for protein precipitation for 12 h and at 4 ° C, to 12000 r / min centrifugation 40 min collect protein, 1 mL ethanol rinse twice. Was added to the protein precipitation 50 mmol / L of bicarbonate, the volume is used so that the concentration of the protein dissolved in 5 g / L of the amount. After the ice bath sonicator protein precipitation will break, by an enzyme: protein mass ratio of the amount of added trypsin 1:50, were digested at 37 ° C the reaction effect 20 h. End of the reaction after the vacuum drying apparatus manipulation resulting dried peptide hydrolysis.

HPLC mobile phase: liquor A H_2O (mass spectrometry), liquor B acetonitrile (mass spectrometry).flow rate: 200 ul/min (after shunt 2 ul/min).The elution gradient:120 min (5%B to 35%B in 40 min, to 95% in 20 min, balance for 20 min).Sample: sample automatically. Sample quantity: 20 ul.

Mass spectrometry: followed by lysate (containing 0.5% of CAN, 0.1% formic acid) to dissolve peptides after digestion by the automatic sampler like to Trap desalting column above, after C18 reverse phase column (Michrom, Bioresources , Auburn, CA) directly after the separation LTQ-Orbitrap XL mass spectrometer detection. In the reverse peptide elution gradient column with 0-35% CAN (containing 0.1% formic acid) to give 150 min at a flow rate of 250 nL / min of. Setting the main parameters and mass data collection method in, ion transfer tube temperature is 200°C; Electrospray voltage is 1.85 kV, stage Orbitrap mass spectrometry in the scan, complete with two CID LTQ in collision mode. A mass scan range is 400-2000 D, two mass collision energy of standardization, activation q value and the activation time is 35 percent, respectively, 0.25 and 30 ms. With BioWorks 3.3.1 (Thermo Finnigan, San Jose, CA) for all of the raw data, to give Peak list file, and then retrieve SEQUEST software in NCBI yeast database, complete identification of proteins.

3. RESULTS

3.1. Differences between *A.pullulans* CGMCC No.3337 and *A.pullulans* TKPM10017 on PMLA Fermentation

As is shown in the Fig.1, the High-yield strain *A.pullulans* CGMCC No.3337 and the original strain *A.pullulans* TKPM 10017 showed the same trend. 0-36 h is lag phase, 36-120h is logarithmic growth phase, 120 h-144 h is stationary phase. Especially in the logarithmic growth phase of the two strains exhibited significant differences, *A.pullulans* TKPM 10017 biomass increased from 3.7g / L to 23.2 g / L, residual sugar from 65 g / L down to 21 g / L, while the high-yield strain of biological volume increased from 3.8 g / L to 35.8 g / L, residual sugar by 62 g / L down to 8 g / L, respectively, the original strain of 1.64 times and 1.23 times, indicating that this stage can make better use of high-yielding strains of culture based synthetic cells.Producing strain of higher substrate utilization, better PMLA synthesis. The stage for the main product PMLA synthesis stage, the product PMLA with cell growth and large-scale synthesis, showing the same conclusion growth phase coupling, Weifeng Cao[13],in the study found *A.pullulans* ipe, which stated PMLA in log phase of growth is coupled with the cells, only in the late logarithmic growth phase occurs partially dissociated, and when the cells stop growing PMLA synthesis also terminated. At the same time can be found from the figure, the consumption of high-yield strains before and 72 h starting strain sugar are basically the same, but PMLA yield and cell amount less than the original strain will have high-yield strains, suggesting that consumption of sugar this time starting strain used part pluronic synthesis of polysaccharides.

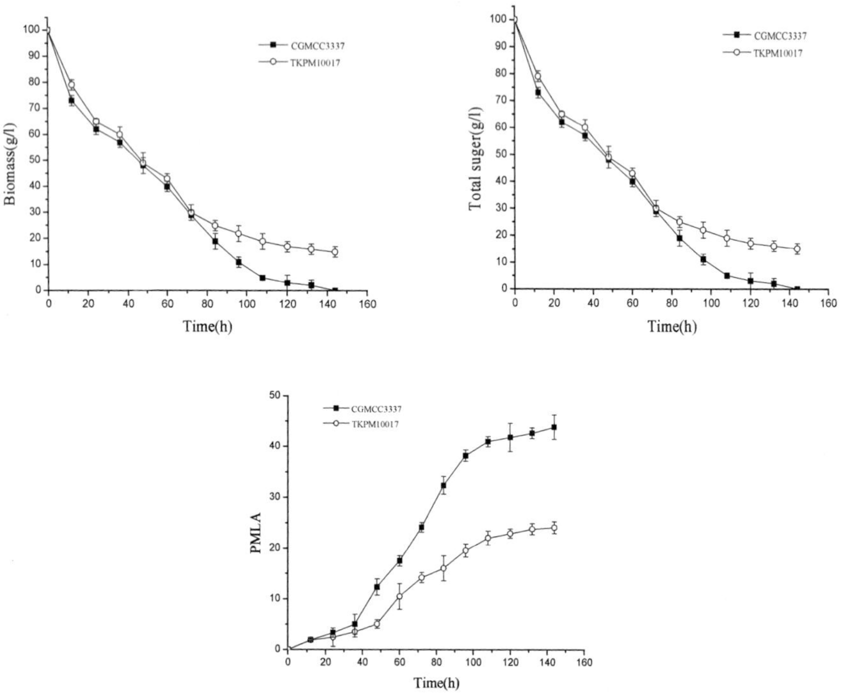

Fig.1. Comparison of high-yield strains and the original strain during fermentation of biomass, total sugar and synthesis PMLA.

3.2. Protein Identification

In previous studies, we found that exopoly saccharide accumulation was tightly related to cell growth. When *A. pullulans* cells reached a logarithmic growth phase and the cell growth wasvigorous, PMLA could be synthesized more as a secondary metabolite. To search for differences in physiological metabolism and its administration between the cell biomass growth and secondary metabolite PMLA biosynthesis in *A.pullulans* CGMCC No.3337 and *A.pullulans*TKPM10017, proteins extracted from the cells at 60h were subjected to proteome analysis.After digestion of the sample by LC - MS/MS analysis, the results are shown in Fig. 2 to Fig.4. From the chromatogram (Fig. 2), 20 min-120 min between sample was cleared, can be used for further analysis. From primary mass spectra (Fig. 3), peak is normal, can be used for further analysis. Peptides from secondary mass spectrum matching figure (Fig. 4), matching rate is higher, shows that the results to be trusted.

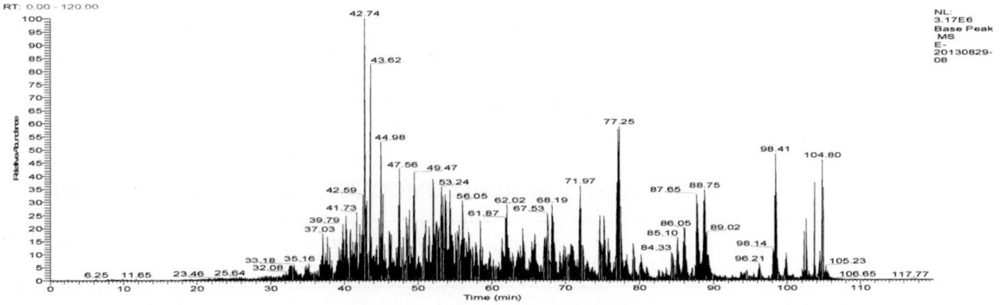

Fig.2. Results of peptides chromatograms.

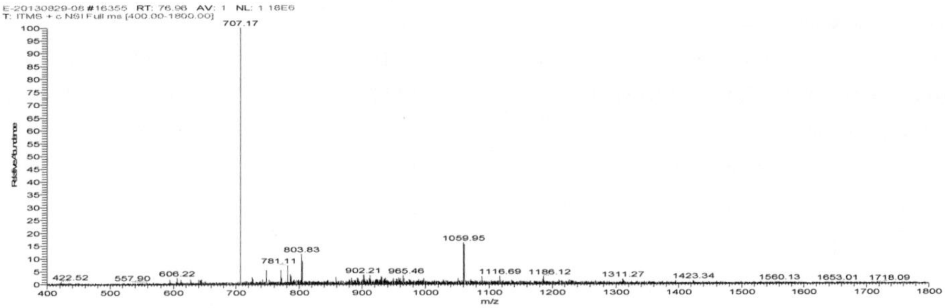

Fig.3. Results of peptides mass spectrometry analysis.

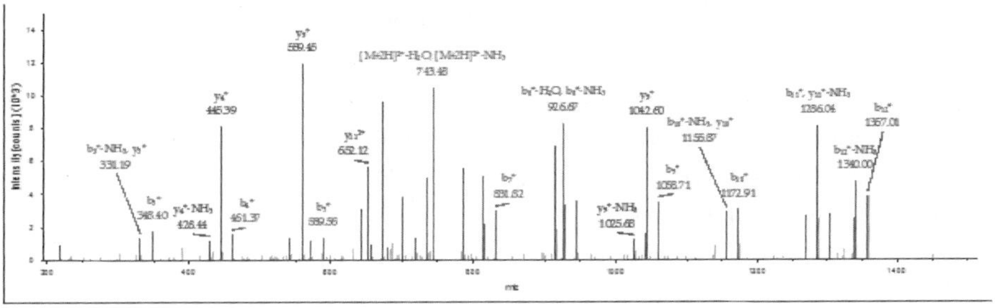

(Blue: b ion;Red: y ion;Green: neutral loss)

Fig.4. Results of peptides Mass Spectrometer – Mass Spectrometeranalysis.

Protein spectrum according to the results: 51 protein were detected in *A.pullulans* TKPM 10017; while, 55 protein were detected in *A.pullulans* CGMCC No.3337. Meanwhile, the identified proteins were matched to known proteins from the fungus species and other microorganisms. Of the proteins identified, 9 were up-regulated under *A.pullulans* TKPM10017and 7 were up-regulated under that is shown in Table 1.

The up-regulated proteins in *A.pullulans* CGMCC No.3337 sample contained ATP synthase β subunit, glyceraldehyde-3-phosphate dehydrogenase, enolase, glucose-6-phosphate dehydrogenase, heat shock protein, 26s proteasome like protein, actin similar protein, GTP-binding protein, F1F0-ATP enzyme. Contrast significantly down-regulated proteins contained ATP synthase α subunit, turn dihydroxyacetone group enzyme, UDP-glucose pyrophosphorylase, glucose-6-phosphate isomerase, tubulin α-1 chain, predict protein MYCTH_2316642, predict protein PTT_12738. These proteins are involved in many areas of *Aureobasidium sp* intracellular energy metabolism, antioxidant, amino acid synthesis and metabolism, glycolysis, the pentose phosphate pathway, protein trafficking and transcriptional control.

Table 1. Protein coverage contrasts between *A.pullulans* CGMCC3337 and *A.pullulans* TKPM10017

Protein	*A.pullulans* TKPM10017	*A.pullulans* CGMCC No.3337
ATP synthase subunit beta	*	41.13
ATP synthase beta chain, mitochondrialprecursor	*	21.02
ATP synthase subunit alpha	41.09	*
UDP-glucose pyrophosphorylase	25.19	11.82
hypothetical protein MYCGRDRAFT_105895	17.57	16.1
heat shock protein 70	15.47	30.36
glucose-6-phosphate isomerase	11.57	10.03
glyceraldehyde-3-phosphate dehydrogenase	13.02	51.48
hypothetical protein MYCTH_75824	12.36	13.51
Glucose-6-phosphate dehydrogenase	10.43	22.19
similar to transaldolase	11.15	3.19
enolase	12.98	17.58
tubμLin alpha-1 chain	13.84	6.79
hypothetical protein FOXB_14076	10.32	10.84
hypothetical protein MYCGRDRAFT_70588	10.37	10.37
similar to S-adenosyl-L-homocysteine hydrolase	10.24	5.32
hypothetical protein PTT_12738	13.33	10.1
hypothetical protein MYCTH_2316642	45.35	9.58
U3 small nucleolar ribonucleoprotein IMP3	6.33	14.87
mitochondrial peroxiredoxin PRX1	4.18	16.52
similar to 26S protease regμLatory subunit S10b	3.46	10.18
ATP2-F1F0-ATPase complex, F1 beta subunit	9.75	20.37
actin-like protein	8.48	29.33
similar to GTP-binding protein YchF	2.49	10.94

3.3. Proteome Changes

Glyceraldehyde-3-phosphate dehydrogenase and enolase are the two proteins involved in the glycolytic pathway (EMP). Glyceraldehyde-3-phosphate dehydrogenase (GAPDH) is a sulfhydryl enzyme in glycolysis, Catalyzes glyceraldehyde 3-phosphate into 1,3-diphosphoglycerate, this process requires NAD^+ and inorganic phosphate (Pi) participation. Glyceraldehyde 3-phosphate dehydrogenase (GAPDH) catalyses the conversion of glyceraldehyde 3-phosphate to D-glycerate 1,3-bisphosphate. This is the 6th step in the glycolytic breakdown of glucose, an important pathway of energy and carbon molecule supply which takes place in the cytosol of eukaryotic cells. In this catalysis, glyceraldehyde-3-phosphate oxidative dehydrogenation, while phosphorylation to produce 1,3-diphosphoglycerate containing a high-energy phosphate bonds, the reaction off of two H^+ accepted by GAPDH coenzyme NAD^+, generating NADH, and H^+. This is a very important process of glycolysis reaction, because after the formation of glyceraldehyde-3-phosphate carboxy, energy will be stored to the oxidation process produces ATP molecule[13]. Enolase, also known as phosphopyruvate hydratase, is a metalloenzyme responsible for the catalysis of the conversion of 2-phosphoglycerate (2-PG) to phosphoenolpyruvate (PEP), the ninth and penultimate step of glycolysis. The chemical reaction catalyzed by enolase is:2-phospho-D-glycerate⇌phosphoenolpyruvate $^+$ H_2O. Enolase belongs to the family of lyases, specifically the hydro-lyases, which cleave carbon-oxygen bonds[14]. Extensive literature confirmed that enolase is a multifunctional protein, especially plays a very important role in the glycolytic flux level of transcription regulation. For example, in K. lactis, the enzyme glucose permease Rag1 by inducing expression to regulate the flux of glucose metabolism[15]. Obviously, glyceraldehyde-3-phosphate dehydrogenase and enolase raised, leaded to increasing carbon flux flow in the glycolytic pathway, which on the one hand would result in biomass growth, and more importantly, there was more carbon used for pyruvate synthesis. Pyruvate reduction pathway, a major pathway polymalic acid precursor L-malic acid synthesis, thus presumably Aureobasidium Poly malate main routes include pyruvate carboxylase pathway, which pre-metabolic flux analysis same result. The strain mutagenesis of *A.pullulans* CGMCC3337, may be due to its high yield expression of enzymes in the glycolytic pathway mutagenesis makes glyceraldehyde-3-phosphate dehydrogenase and enolase, increase, so that more carbon used in the synthesis of pyruvate, the final product was synthesized more precursors L- malic acid via pyruvate carboxylase pathway.

Glucose-6-phosphate dehydrogenase, dihydroxyacetone group transfer enzyme are the proteins in the pentose phosphate pathway (ppp), also known as hexose monophosphate pathway (EMP). EMP pathway [16] is the main source way of microbial cytoplasmic NADH. NADH play an important role in the cells growth and metabolism, which provide reducing power to synthetic amino acids, lipids and nucleotides. Meanwhile study found that erythrose-4-phosphate, sedoheptulose 7-phosphate compounded in the HMP pathway is the metabolic precursor for the synthesis of fat acids and nucleic acids. Glucose 6-phosphatedehydrogenase(G6PD) is the major regulatory enzyme of the EMP pathway, $NADP^+$as a coenzyme catalytic glucose-6-phosphate dehydrogenation of 6-phosphorgluconolaetone, which make an important role to maintain the balance of intracellular NADPH and redox reaction. Glucose-6-phosphate dehydrogenaseupregulation makes EMP pathway enhancing to provide more adequate NADH and ATP to intracellular anabolism, which is the reason that cell growth rate and product synthesis rate of *A.pullulans* CGMCC3337 is significantly higher than*A.pullulans* TKPM10017.Transaldolasecatalyzes thedihydroxyacetone of 7-sedoheptulose phosphate transfer to the first carbon atom of glyceraldehyde 3-phosphate, resulting D-erythrose-4-phosphate and D-fructose-6-phosphate. However , D-erythrose-4-phosphate is the intermediate products of phosphopen- tose pathway and the precursor of biosynthesis of aromatic amino acid of tyrosine and phenylalanine. D-erythrose-4-phosphate and phosphoenolpyruvate reactionto produce tyrosine. Through a series of reactions,tyrosine is changed into melanin, then the carbon flow to the melanin synthesis pathway, affecting the synthetic product, while reducing consumption of phosphoenolpyruvate, glyceraldehyde-3-phosphate.On the one hand,Cutting level of transaldolasewill reduce the synthesis of D-erythritol 4-phosphate, thereby reducing the synthesis of melanin, on the other hand it reduces consumption of phosphoenolpyruvate, glyceraldehyde-3-phosphate. It will make more carbon flows pyruvate synthesis of EMP pathway, and thenthrough pyruvate carboxylase L- malic acid synthesis pathway.

UDPG pyrophosphorylase is related enzyme of glycogen synthesis in vivo, before glucose synthesis of glycogen, glucose needs activation, This process the glucose-1-phosphate and UTP molecules synthesized as UDP- glucose (UDPG) in UDP- glucose pyrophosphorylase catalysis. Duan et al[17] proposed a possible synthesis route of pullulan, from glucose units into pullulan requires three key enzymes, which are α glucose phosphate mutase, UDPG pyrophosphorylase and glucose glycosyltransferase. Meanwhile Ono et al found that adding ATP and UDPG to Aureobasidium cell lysate, there under the presence of sucrose can also be synthesized pullulan, and when replacing UDPG use ADPG, found it impossible to synthesize pullulan, indicating precursors pullulan is UDPG[18]. The decline of UDPG pyrophosphorylase will inevitably lead to the synthesis of byproducts pullulan decreased in high-yielding strains, so that more carbon flow to PMLA synthesis and the product PMLA yield higher.

Glucose-6-phosphateisomerase[19] is a kind of multifunctional protein, Catalytic a reversible reaction between glucose-6-phosphate, fructose 6-phosphate and glucose-1-phosphate in carbohydrate metabolism. Glucose-1-phosphate is the precursor for synthesis UDPG and UDPG are important precursors in pullulan synthesis. GPI decline in high-yield strains, indicating that byproduct pullulan synthesis pathway weakened within high-yielding strains, reducing its substrate competition with PMLA synthesis, so that more carbon flows to PMLA product synthesis pathway.

The 70 kDa heat shock proteins (Hsp 70) belonged to a family of widespread heat shockproteins. Proteins with similar structures were found in almostall organisms. Hsp 70 played a crucial role in protein folding inthe cells and meanwhile helped to protect the cells fromenvironmental pressure. Hsp 70 seemed to be involved intreating defective and damaged proteins. Hsp70 could be ableto protect protein from aggregating through close connectionto fractional synthesized peptide sequences[20]. As the nitrogensource was consumed, the environment outside the cellsbecame rigorous; the expression of more Hsp 70 could protectA. pullulans cells from oxidative stress, which generally hurtsproteins, leading to aggregation and partial unfolding.

26 s proteasome is responsible for degradation proteins in eukaryotic cells, which has ATP-dependent and substrate specificity and can be specifically degraded the target protein. Production and degradation of proteins is the most fundamental and important metabolic activity in vivo and participated in virtually all life activities within biological, therefore the 26 s proteasome has a crucial role in organisms. 26 s proteasome protein

were detected only in the high-yield strains, suggesting that may be mutagenized the strain to enhance the environmental adaptability[21].

GTP-binding protein, also known as G protein, was widespread in the biosphere and played an extremely important role in signal transduction in cells. It's can regulate highly conserved transmembrane signaling pathways in eukaryotes[22]. G protein generally refers to the membrane receptor coupling of a heterologous trimer protein, but the latest results shown that G protein is in addition to coupled with membrane receptors, is also played an important role in cell communication. GTP-binding protein growth is beneficial to intracellular signaling.

Actin is a Micro filament structural protein and exists in two forms of monomer and polymer in cells. Microfilament involved in the maintenance of cell shape and polarity, endocytosis, intracellular transport, cell contraction and movement, cell division and many other features. Many scholars have reported that actin existed on chromosomes and nuclear skeleton, reported that it is a non-histone on nuclear matrix and chromatin. Some study results also showed that actin is involved in a number of important physiological activities in the nucleus, such as chromatin condensation, DNA transcription, RNA transport[23]. Producing strain A. pullulans CGMCC 3337 biomass was significantly higher than the starting strain A. Pullulans TKPM 10017,A large number of freshmen protein and intermediate metabolites are synthesized, regulation of actin protein in the nucleus is beneficial to intermediate metabolites transmembrane transport, protein transport and transcriptional control.

F1F0-ATPase is a transmembrane protease complex .It not only can regulate intracellular pH of lactic acid bacteria, it can also enhance the acid resistance of cells, promoting the conversion of ATP .When the pH of the extracellular environment to reduce, F1F0-ATPase on membrane made intracellular proton pumped out to the extracellular which took advantage of the energy provided by ATP hydrolysis the, thereby forming a pH gradient difference, maintain the stability of the intracellular pH and avoid affecting intracellular metabolic, so that the cell can low pH environment that can continue to grow and produce acid[24]. Thus, the up-regulation of F1F0-ATPase in high-yield strains of *A.pullulans* CGMCC 3337 is beneficial to increase acidic environment tolerance.

4. DISCUSSION

This issue provide a distinct perspective of the biosynthesis mechanism of PMLA,which may be helpful for improving PMLA production by *A.pullulans* in practice.According to the total protein spectrum analysis of Aureobasidium pullulans, we could find significant differences between *A.pullulans* CGMCC No.3337 and *A.pullulans* TKPM 10017 in various aspects of typical biochemistry properties ,such as the glycolysis pathway,the pentose phosphate pathway, the biosynthesis of pullulan as well as the energy metabolism. The higher enzyme activity of the High-yield strain functions in the glycolysis pathway and pentose phosphate pathway cause a higher-level synthesis of PMLA.Likewise, the down-regulation of UDP-glucose pyrophosphorylase and glucose-6-phosphate isomerase make it possible to decrease the synthesis of byproduct. Compared with the original strain, High-yield strain also has a better performance in energy metabolism. To sum up, the differences mentioned are closely related to the increasing yield of *A.pullulans* CGMCC No.3337.

Acknowledgments

This work was financially supported by the Project of Tianjin University of Science and Technology Support Program and the Company of Tianjin Peiyang Biotrans. The authors would like to thank these organizations for their generous financial support.

References

1. Holler M, Cardinali F, Mamlouk H, et al. Synthesis of Fullerohelicates and Fine Tuning of the Photoinduced Processes by Changing the Number of Addends on the Fullerene Subunits. *J Cheminform*; 2006; **37**(18): 2060-2073.
2. Fujita M, Khazenzon NM, Ljubimov AV, et al. Inhibition of laminin-8 in vivo using a novel poly(malic acid)-based carrier reduces glioma angiogenesis. *J Angiogenesis* 2006; **9**(4): 183-91.
3. Gödde C, Liebergesell M, Steinbüchel A. Isolation of poly(β-l-malic acid)□degrading bacteria and purification and characterization of the PMA hydrolase from Comamonas acidovorans strain 7789. *J Fems Microbiology Letters* 1999; **173**(2): 365-372.
4. Osanai S, Nakamura K. Effects of complexation between liposome and poly(malic acid) on aggregation and leakage behaviour. *J Biomaterials* 2000; **21**(9): 867-76.
5. Fujita M, Lee B S, Khazenzon N M, et al. Brain tumor tandem targeting using a combination of monoclonal antibodies attached to biopoly(β-l-malic acid). *J Journal of Controlled Release* 2007; **122**(3): 356-363.
6. Ljubimova JY, Manabu F, Ljubimov AV, et al. Poly(malic acid) nanoconjugates containing various antibodies and oligonucleotides for multitargeting drug delivery. *J Nanomedicine* 2008; **3**(2): 247-65.
7. Ding H, Portilla-Arias J, Patil R, et al. The optimization of polymalic acid peptide copolymers for endosomolytic drug delivery.J Biomaterials 2011; **32**: 5269–5278.
8. Rathberger K, Reisner H, Willibald B, et al. Comparative synthesis and hydrolytic degradation of poly (L-malate) by myxomycetes and fungi. *J Mycological Research* 1999; **103**: 513-520.
9. Weifeng Cao, Jianquan Luo, Juan Zhao,et al. Intensification of β-poly(l-malic acid) production by Aureobasidium pullulans ipe-1 in the late exponential growth phase. *J Journal of Industrial Microbiology & Biotechnology* 2012, **39**(7): 1073-1080.
10. Bhadauria V, Banniza S, Wang L X, et al. Proteomic studies of phytopathogenic fungi, oomycetes and their interactions with hosts. *J European Journal of Plant Pathology* 2009; **126**: 81-95.
11. Maik B, Thomas C, Christian B, et al. Proteomic analysis of dimorphic transition in the phytopathogenic fungusUstilago maydis. *J Proteomics* 2007; **7**(5): 675-685.

12. Brigitte W L, Hanns-Rüdiger G, Thomas P. Two-dimensional gel electrophoresis as tool for proteomics studies in combination with protein identification by mass spectrometry. *J Proteomics* 2006; **6**(17): 4688–4703.
13. Díaz-Ramos A, Roig-Borrellas A, García-Melero A, et al. α-Enolase, multifu - nctional protein: its role onpathophysiological situations. *J Journal of Biomedicine and Biotechnology* 2012; **2012**: 156795
14. Lemaire M, Wesolowski-Louvel M. Enolase and glycolytic flux play a role in the regulation of the glucose permease gene RAG1 ofKluyVeromyces lactis. *J Genetics* 2004; **168**: 723-731
15. Cohan N, Karimi M, Khalili AH, et al. The efficacy of a neonatal screening programme in decreasing the hospitalization rate of patients with G6PD deficiency in southern Iran. *J Med Screen* 2010; **17**: 66-67.
16. Lemaire M, Wesolowski-Louvel M. Enolase and glycolytic flux play a role in the regulation of the glucose permease gene RAG1 ofKluyVeromyces lactis. *J Genetics* 2004; **168**: 723-731
17. Duan X H, Chi Z M, Wang L, et al. Influence of different sugars on pullulan production and activities of α-phosphoglucose mutase, UDPG-pyrophosphorylas and glucosyltransferase involved in pullulansynthesis in Aureobasidium pullulansY68. *J Carbohydrate Polymers* 2008; **73**: 587-593
18. Ono K, Kawahara Y, Ueda S. Effect of pH on pullulan elaboration by Aureobasidium pullulans S-1. *J Agric Biol Chem* 1977; **44**: 2113-2118
19. Gaalen, Floris A. Van, et al. Association of Autoantibodies to Glucose-6-Phosphate Isomerase with Extraarticular Complications in Rheumatoid Arthritis. *J Arthritis & Rheumatology* 2004; **50**: 395-399.
20. Morano K A. New Tricks for an Old Dog: The Evolving World of Hsp70. *J Annals of the New York Academy of Sciences* 2007; **1113**: 1-14.
21. Ding MZ, Wang X, Yang Y, et al. Comparative metabolic profiling of parental and inhibitors-tolerant yeasts during lignocellulosic ethanol fermentation. *J Metabolomics* 2011; **8**(2): 232-243.
22. Kaziro Y, Itoh H, Kozasa T, et al. Structure and function of signal-transducing GTP-binding proteins. *J Annual Review of Biochemistry* 1991; **60**(1):349-400.
23. Kumar B A, Paily K P. Actin protein up-regulated upon infection and development of the filarial parasite, Wuchereria bancrofti (Spirurida: Onchocercidae), in the vector mosquito, Culex quinquefasciatus (Diptera: Culicidae). *J Experimental Parasitology* 2008; **118**(3): 297-302.
24. Walker J E, Lutter R , Dupuis A ,et al. Identification of the subunits of F1F0-ATPase from bovine heart mitochondria. *J Biochemistry* 1991; **30**(22): 5369-5378.

THE EFFECTS OF COLD STIMULATION ON PHYSIOLOGICAL INDEX OF CHINA-LINDIAN NATIVE CHICKENS

Changping Wang[a], Bo Wang[b], Qingbo Zhou, Guangdong Cheng, Chengwu Han, Hongsheng Yang, Tianlin Miao, and Bin Wang

College of life science, Jiamusi university,Jiamusi,154007,China
Jiamusi university animal husbandry and veterinary medicine, Jiamusi,154007,China
E-mail: [a]wcp100200@163.com, [b]408081288@qq.com

448 China-Lindian Native Chickens were given the low temperature stimulation from 22 age in days to 42 age in days. They were reared in artificial climate chamber. The influence of cold stimuli were identified through detecting physiology welfare indexs. The main physiological welfare index were the levels of three triiodothyronine (T_3) and adrenocorticotropic hormone(ACTH).The content change of three triiodothyronine (T_3) and adrenocorticotropic hormone (ACTH) were detected by enzyme-linked immunosorbent assay (ELISA).It was concluded from the study that the cold stimuli time had impact on the level of ACTH($P<0.05$), the interval days of cold stimuli had impact on the level of T_3 ($P<0.05$),the interval days of cold stimuli had no impact on the level of ACTH ($P> 0.05$); the decreasing temperature of cold stimuli had no impact on the level of ACTH and T_3($P> 0.05$); and the time of cold stimuli had no impact on the level of T_3($P> 0.05$).

1. INTRODUCTION

Thermoregulation is crucial for survival of homeotherms in warm and cold environments. During cold exposure adult birds actively regulate their body temperature and generate heat by up-regulating their energy intake and metabolism. Although adult birds have a high tolerance for cold, neonatal chicks are particularly vulnerable to low-temperature and cannot survive in cold environments without adequate shelter and heating.Stress under low-temperature can disrupt the balance in oxidant/antioxidant system and cause oxidative damage to several tissues by altering the enzymatic and non-enzymatic antioxidant status, protein oxidation and lipid peroxidation (Sahin and Gumuslu, 2004)[1].Cold ambient temperatures increase the oxygen requirement, cardiac output, and blood flow and may result in increased pulmonary arterial pressure overload on the right ventricle (Julian et al., 1989)[2]. Exposure of growing chickens (21-d old) to low-temperature (12-14°C) has been reported to increase the plasma malondialdehyde (MDA, Pan et al., 2005)[3]. Young chicks are more sensitive to cold exposure than adult chickens. Chicks older than 6 days can tolerate a cold temperature (4°Cfor 24 h) as compared to neonatal chicks (2-d old) who are unable to maintain homeothermy even on lowtemperature (20°C) exposure (Hirabayashi et al., 2005; Mujahid and Furuse, 2008a, in press)[4][5].

As soon as we find a useful means of stimulation, we can make the chicken receive a good low temperature stimulation in the process of brood. Thereby improving the immunity of the late broilers, to provide a theoretical basis for the healthy breeding of new technology.

In this study, China-Lindian Native Chickens(CLNC)were given cold stimuli from four to six weeks in brood period. These stimulated CLNC were still reared in normal brooding conditions after stimulating test. Method of cold stimuli is controlling the chambers temperature by artificial climate chamber from the following three aspects: stimuli extent (decrease of temperature), stimuli intensity (stimulation time) and stimuli frequency(interval days).

We will identify the influence of cold stimuli on welfare of CLNC through detecting physiology welfare indexs. The main physiological welfare index of detection were the levels of three triiodothyronine (T_3) and adrenocorticotropic hormone(ACTH).The aim of the study was to determine the effects of cold stress on China-Lindian Native Chickens physiological index.

2. MATERIALS AND METHODS

The artificial climate chamber can be controlled automaticlly by computer program. These stimulated CLNC were still reared in normal brooding conditions after stimulating test. In this study, repeat experiments (4 repeations in groups) were conducted on three factors (interval days, decrease of temperature and stimulation time) and three levels (interval of 0 day, interval of 1 day, interval of 2 days; with temperature decreasing 3°C, temperature decreasing 5°C, with temperature decreasing 7°C respectively in different tested groups; stimulating time of 1h, stimulating time of 3h, stimulating time of 6h).

The content change of three triiodothyronine (T_3) and adrenocorticotropic hormone (ACTH) were detected by enzyme-linked immunosorbent assay (ELISA).

3. RESULTS AND ANALYSIS

3.1. Effect of Cold Stimulation on ACTH of China-Lindian Native Chickens

3.1.1. *The Effects of Interval Days of Cold Stimulus on ACTH in China-Lindian Native Chickens*

As shown in Table 1, Serum ACTH content on interval of 2 day was higher than the control group, the serum ACTH content on interval of 0 day and 1 day were lower than the control group, the difference was not significant($P>0.05$).

Serum ACTH content on interval of 2 day was higher than interval of 0 day and 1 day in cold stimulus trial group, the difference was not significant($P>0.05$).

Through the variance analysis, interval days had no significant effect on the content of ACTH ($P>0.05$).

Table 1. The effects of interval days of cold stimulus on ACTH in China-Lindian Native Chickens (Unit:pg/ml)

interval days	average	difference significant	
		α=0.05	α=0.01
0 d	83.99±12.95	a	A
1 d	84.82±15.12	a	A
2 d	87.30±9.78	a	A
control group	85.55±3.85	a	A

3.1.2. *The Effects of Decreased Temperature of Cold Stimulus on ACTH in China-Lindian Native Chickens*

As shown in Table 2, Serum ACTH content on decreased temperature 7°C was higher than the control group, the serum ACTH content on decreased temperature 3°C and 5°C were lower than the control group, the difference was not significant($P>0.05$).Serum ACTH content on decreased temperature 7°C was higher than decreased temperature 3°C and 5°C in cold stimulus trial group, the difference was not significant(P>0.05).

Through the variance analysis, decreased temperature had no significant effect on the content of ACTH (P>0.05).

Table 2. The effects of decreased temperature of cold stimulus on ACTH in China-Lindian Native Chickens (Unit:pg/ml)

decreased temperature	average	difference significant	
		α=0.05	α=0.01
3°C	84.51±15.54	a	A
5°C	84.61±9.99	a	A
7°C	86.99±12.36	a	A
control group	85.55±3.85	a	A

3.1.3. *The Effects of Cold Stimulation Time on ACTH in China-Lindian Native Chickens*

Table 3. The effects of time of cold stimulus on ACTH in China-Lindian Native Chickens(Unit:pg/ml)

stimulation time	average	difference significant	
		α=0.05	α=0.01
1 h	82.55±13.71	a	A
3 h	88.51±12.50	b	A
6 h	85.04±11.63	ab	A
control group	85.55±3.85	ab	A

As shown in Table 3, Serum ACTH content on stimulation time 3h was higher than the control group, the serum ACTH content on stimulation time 1h and 6h were lower than the control group, the difference was significant($P<0.05$).Serum ACTH content on stimulation time 3h was higher than stimulation time 1h and 6h in cold stimulus trial group, the difference was significant($P<0.05$).

Through the variance analysis, cold stimulation time had significant effect on the content of ACTH ($P<0.05$).

3.2. Effect of Cold Stimulation on T_3 of China-Lindian Native Chickens

3.2.1. *The Effects of Interval Days of Cold Stimulus on T_3 in China-Lindian Native Chickens*

Table 4. The effects of interval days of cold stimulus on T_3 in China-Lindian Native Chickens (Unit:ng/ml)

interval days	average	difference significant	
		α=0.05	α=0.01
0 d	1.19±0.17	a	A
1 d	1.11±0.21	ab	A
2 d	1.08±0.21	b	A
control group	1.04±0.20	ab	A

As shown in Table 4, Serum T_3 content on all experimental group were higher than the control group, the difference was significant($P<0.05$).Serum T_3 content on interval of 0 day was higher than interval of 1 day and 2 day in cold stimulus trial group, the difference was significant($P<0.05$).

Through the variance analysis, interval days had significant effect on the content of T_3 ($P<0.05$).

3.2.2. *The Effects of Decreased Temperature of Cold Stimulus on T_3 in China-Lindian Native Chickens*

Table 5. The effects of decreased temperature of cold stimulus on T_3 in China-Lindian Native Chickens (Unit:ng/ml)

decreased temperature	average	difference significant	
		α=0.05	α=0.01
3°C	1.12±0.19	a	A
5°C	1.12±0.20	a	A
7°C	1.14±0.21	a	A
control group	1.04±0.20	a	A

As shown in Table 5, Serum T_3 content on all experimental group were higher than the control group, the difference was not significant($P>0.05$).Serum T_3 content on decreased temperature 5°C was higher than decreased temperature 3°C and 7°C in cold stimulus trial group, the difference was not significant($P>0.05$).Through the variance analysis, decreased temperature had no significant effect on the content of T_3 ($P>0.05$).

3.2.3. *The Effects of Stimulation Time of Cold Stimulus on T_3 in China-Lindian Native Chickens*

Table 6. The effects of the time of cold stimulus on T_3 in China-Lindian Native Chickens (Unit:ng/ml)

stimulation time	average	difference significant	
		α=0.05	α=0.01
1 h	1.17±0.20	a	A
3 h	1.11±0.20	a	A
6 h	1.10±0.20	a	A
control group	1.04±0.20	a	A

As shown in Table 6, Serum T_3 content on all experimental group were lower than the control group, the difference was not significant(P>0.05).Serum T_3 content on stimulation time 1h was higher than stimulation time 3h and 6h in cold stimulus trial group, the difference was not significant(P>0.05).Through the variance analysis, cold stimulation time had no significant effect on the content of T_3 (P>0.05).

4. DISCUSSION

Adrenocorticotropic hormone(ACTH) Promotes the growth of adrenal cortex,stimulates steroid hormone production and maintains the secretion function of adrenal cortex, secreted by adenohypophysis posterior pituitar ACTH cells. Adrenocorticotropic hormone is the most important secretion regulator of the cortisol synthesis. zhao jianrong et al(1998)research shows that cold stress significantly increased the Concentration of cortisol in the plasma of layers[6]. Thyroid hormones almost acts on all the organs and tissues of the chick body, and have a certain effect on the growth, development, metabolism and tissue differentiation. Fuller (1980) observed the effect of ambient temperature on the size of thyroid gland in broiler chickens. Under normal physiological conditions, thyroid hormone levels remain within a certain range. This study results showed that interval days decreased temperature and stimulation time had no significant effect on the content of T_3(P>0.05).

Acknowledgments

This work was funded by Natural science foundation of China (31302008) and Natural science foundation of Heilongjiang province (QC2012C104).

References

1. Sahin, E., Gumuslu, S. Cold-stress-induced modulation of antioxidant defence:role of stressed conditions in tissue injury followed by protein oxidation and lipid peroxidation. Int. *J. Biometeorol* 2004;**48**:165–171.
2. Julian, R. J., I. Mcmillan and M. Aquinton. The effect of cold and dietary energy on right ventricular hypertrophy, right ventricular failure and ascites in meat type chickens. *Avian Pathol* 1989;**18**:675–684.

3. Pan, J.Q., Tan, X., Li, J.C., Sun, W.D.,Wang, X.L. Effects of early feed restriction and cold temperature on lipid peroxidation, pulmonary vascular remodelling and ascites morbidity in broilers under normal and cold temperature.Br.*Poult.Sci* 2005; **46**:374–381.
4. Hirabayashi, M., Ijiri, D., Kamei, Y., Tajima, A., Kanai, Y. Transformation of skeletal muscle from fast- to slow-twitch during acquisition of cold tolerance in the chick.*Endocrinology* 2005;**146**:399–405.
5. Mujahid, A., Furuse, M. Homeothermy in neonatal chicks exposed to low environmental temperature with or without intracerebroventricular administration of corticotropin-releasing factor. *FEBS Lett* 2008a;**20**:3052–3060.
6. Jian-rong Zhao, et al. Effect of cold and heat stress on physiological and biochemical indexes of laying hens. *Chinese Journal of Animal Science* 1998;**3**(6):6–8.(In Chinese).

THE STABILITY OF GINSENG WHEY PROTEIN POLYPEPTIDE BEVERAGE

Pingping Hou[1,a], Gang Wang[1,b], Dongxue Li[1,c], Ting Zhang[2,d], Xudong Li[1,e], Xinhui Zhou[1,f], and Tiehua Zhang[1,g,*]

[1]*College of Food Science and Engineering, Jilin University, Jilin Changchun, China*
[2]*Tourism Institution, Jilin Business and Technology College, Jilin Changchun, China*
E-mail: [a]m15754304718@163.com, [b]18844547553@163.com, [c]18843866598@163.com, [d]147834156@qq.com, [e]15688965021@163.com, [f]651065148@qq.com, [g]zhangth@jlu.edu.cn
[*]Corresponding author

The ginseng whey protein polypeptide beverage was mainly made from ginseng extract and whey protein concentrate hydrolyzate. Single factor experiments were carried out to analyze the stability of ginseng whey protein polypeptide beverage utilizing precipitation rate, stability coefficient, and viscosity as study indicators. The results showed the best stabilizer formula was consisted of 0.12% pectin, 0.12% STPP, and 0.12% CMC. Under such circumstance, the precipitation rate of the beverage was 0.167% and the stability coefficient was 80.38%, which resulted to have the best beverage stability.

1. INTRODUCTION

Ginseng is a rare traditional Chinese medicine in China, which belongs to Araliaceae perennial herb[1]. Many published literature, confirmed that ginseng had biological activities in immune regulation, anti-tumor, antioxidant and hypoglycemic. Whey protein polypeptide could significantly improve the stability and thermal gel properties of milk beverage including hypoglycemic, antioxidant and other biological activities[2]. The ginseng whey protein beverage was rich in nutrition with unique flavor. However, protein denaturation and precipitation were the main problems existed during manufacturing. Current studies showed pectin, tripolyphosphates (STPP), carboxymethyl cellulose (CMC) and other good stabilizers could enhance beverage stability.

Pectin is considered to be one of the most effective and stable milk beverage hydrocolloid. Through adsorbing on the surface of the protein to use the electrostatic repulsion and steric hindrance to prevent aggregation and precipitation of protein, which resulted to improve the stability of milk beverage products[3]. Aqueous solution of sodium tripolyphosphate is weakly alkaline, which could form an emulsion in specific PH range (4.3-14) through dispersion. CMC is a kind of colloid with water holding capacity that could suspend in acid system. It can form network structure with other food gum. It can suspend stability in low apparent viscosity and can effectively prevent precipitation layer[4]. In present research paper, we aimed to study beverage stability to solve quality issues of products.

2. MATERIALS AND METHODS

2.1. Materials and Reagents

Ginseng extract (the total saponin content for 20%) was purchased from Jilin Hongjiu Biotechnology Co., Ltd., China; White sugar in line with GB 317.1-2010 health standards; Citric acid, Danisco pectin, STPP and CMC were purchased from Henan Fenglu Food Additive Co., Ltd., China; WPC (protein content of 82%) was purchased from Galaxy Albert Tianjin Import and Export Co. The company, China; Flavor protease and neutral protease were purchased from Novozymes Biotech Co., Ltd., China.

2.2. Instruments and Equipment

DHR-3 rheometer was purchased from Instrument American Ta; 752 UV spectrophotometer was purchased from Shanghai Precision Scientific Instruments; DSY-1-6 shaped hole electric heated water bath was purchased from Beijing Guohua medical instrument factory China; SW- CJ- 2D type clean bench was purchased from Suzhou Purification Equipment Co., China; TGL- 16G high-speed desktop centrifuge was purchased from Shanghai Anting Scientific Instrument Factory; GYB60- 6S-pressure homogenizer was purchased from Shanghai Donghua high pressure homogenizer Factory, China; TC10KB electronic balance was purchased from Test Instrument Factory Changshu Shuangjie, China.

2.3. Process

Whey powder → rehabilitation → homogenization → pasteurization → enzymolysis →enzyme inactivation→ enzymolysis liquid whey→ mix → homogenization → pasteurization → bottling → sealing → secondary sterilization → cooling → products.

2.4. Operating Points

Firstly, the fully hydrated material was homogenized by high-pressure homogenizer, the homogenization pressure was 18-20MPa. And then pasteurization at 85±2°C, sterilization of 30min. 4% neutral protease and 4% flavourzyme were added into the material at 55°C. After 4 hour's enzymolysis[5], it would be heated to 85°C to kill the enzyme for 30min. Then all materials were added in turn and mixed well. The material after mix was homogenized and pasteurized again. Then the liquid material was filled to 2 cm away from the bottle, screw on the lid. Finally, bottling finished products should be sterilized at 90-95°C some time. The products were required to be preserved at 2-28°C.

2.5. Screening Test of Stabilizer

2.5.1. *Screening of Stabilizer Type*

By single factor experiments, the total quality of the product was calculated by 100g, in the case of 0.7% whey, 0.2% ginseng extract, 8.0% sugar, 0.1% citric acid, to analyze effects of stabilizers were 0.1% pectin; 0.1% STPP; 0.1%CMC; 0.1% pectin and STPP; the group combining 0.1% pectin with CMC; the group combining 0.1% STPP with CMC; the group combing 0.1% pectin, STPP with CMC on the stability of products utilizing precipitation rate, stability coefficient and viscosity as study indicators.

2.5.2. *Determination of Stabilizer Concentration*

By single factor test, based on the screening of the stabilizer type, the concentration of the stabilizer was further determined. The total quality of the product was calculated by 100g, in the case of 0.7% whey, 0.2% ginseng extract, 8.0% sugar, 0.1% citric acid, to analyze effects of stabilizer concentration were 0.06%, 0.08%, 0.10%, 0.12%, 0.14%, 0.16% on products utilizing precipitation rate, stability coefficient and viscosity as study indicators.

2.6. Measurements of Stability Indexes

2.6.1. *Measurement of Sedimentation Rate*

Take a sample of products to centrifuge (3000 rpm, 15 min, 4 °C), and discard the supernatant, dry the centrifuge tube and weigh it, according to the following formula to calculate the precipitation rate.

Precipitation rate $X = (W-X) / Z \times 100\%$

Wherein:

X represents the weight of the centrifuge tube, accurate to 0.01g;

Z represents the weight of the material liquid in centrifuge tube, accurate to 0.01g;

W represents the weight of the centrifuge tube after centrifuging, accurate to 0.01g.

2.6.2. *Measurement of Stability Coefficient*

Measure the stability factor of beverage by UV spectrophotometer[6]. Take the same amount of stock solution and the supernatant to dilute 50-fold, respectively, measure the absorbance at the wavelength of 500nm. According to the following formula to calculate the stability coefficient. The greater the stability coefficient, the better the stability of the drink.

Stability coefficient $K=T_2/T_1$

Wherein:

T_1 represents the absorbance of the diluted stock solution;

T_2 represents the absorbance of the diluted supernatant.

2.6.3. *Measurement of Viscosity*

Measure the beverage viscosity[7] by DHR rheometer to obtain the viscosity change curve with shear rate. The greater the viscosity, the better the stability of the beverage.

2.7. Data Processing

Using SPSS and Origin Software for data processing and mapping.

3. RESULTS AND DISCUSSION

3.1. Analysis of Screening Results of Stabilizer Type

3.1.1. *Effect of Stabilizer Type on Precipitation Rate*

Fig. 1 showed that the group combining pectin, STPP with CMC had the lowest precipitation rate and the best stability. When there was only one kind of stabilizer, the precipitation rate of beverage added pectin was the lowest, which illustrated pectin has a great effect on beverage stability. Among the three groups with two kinds of stabilizers, pectin and STPP had the greatest effect.

3.1.2. *Effect of Stabilizer Type on Stability Coefficient*

Fig. 2 showed that the group combining pectin, STPP with CMC had the highest stability coefficient and the best stability. When there was only one kind of stabilizer, the stability coefficient of beverage had little difference. The stability coefficient substantially increased with two kinds of stabilizers.

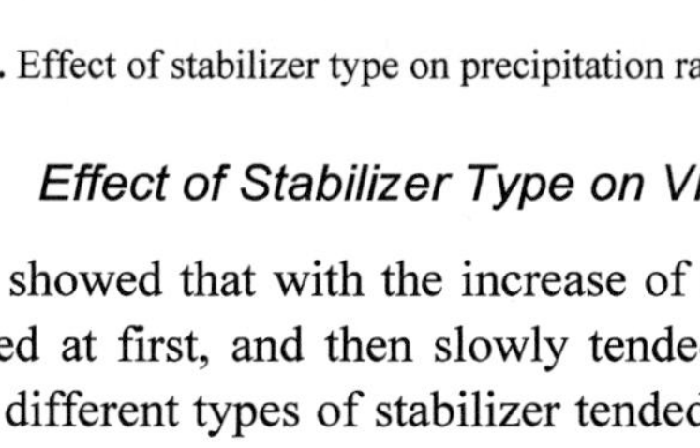

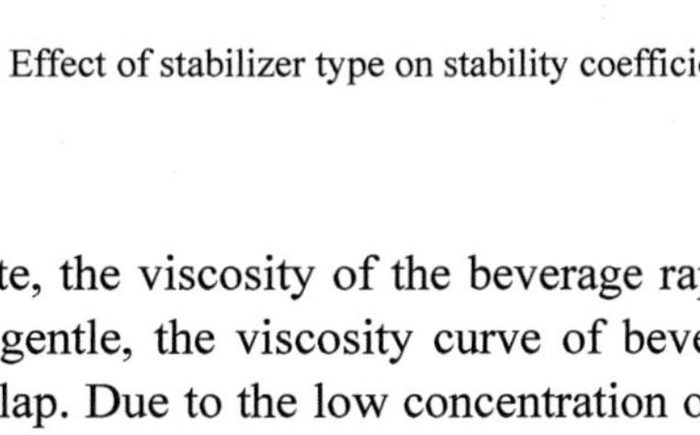

Fig.1. Effect of stabilizer type on precipitation rate. **Fig.2.** Effect of stabilizer type on stability coefficient.

3.1.3. *Effect of Stabilizer Type on Viscosity*

Fig. 3 showed that with the increase of shear rate, the viscosity of the beverage rapidly declined at first, and then slowly tended to be gentle, the viscosity curve of beverage added different types of stabilizer tended to overlap. Due to the low concentration of the beverage, the change of the stabilizer type had little influence on the viscosity of the beverage.

3.2. Analysis of Screening Results of Stabilizer Concentration

After selecting the stabilizer type, we carried out next test using the mixtures of pectin, STPP and CMC. Based on above findings, the effect of stabilizer concentration on the stability of the beverage was further analyzed.

3.2.1. *Effect of Stabilizer Concentration on Precipitation Rate*

Fig. 4 showed that the precipitation rate of beverage was the lowest when the stabilizer concentration was 0.12% and its value was 0.167%. With the increase of the concentration of stabilizer, the precipitation rate generally decreased first and increased later. When the stabilizer concentration was too low, the stability was not good, tended to precipitation, stratification and other phenomena. When the stabilizer concentration was too high, it would influence the taste and flavor of beverage.

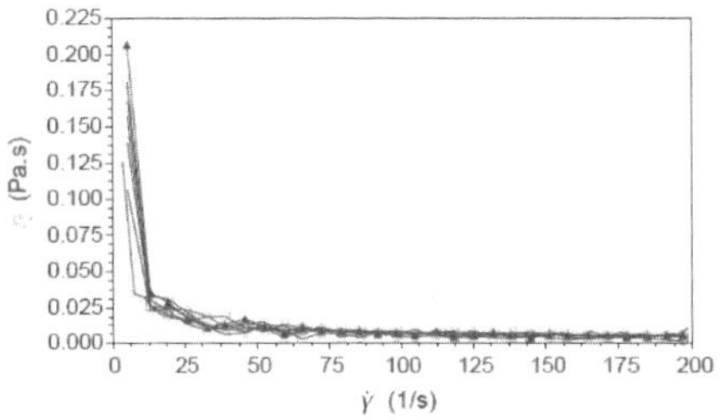

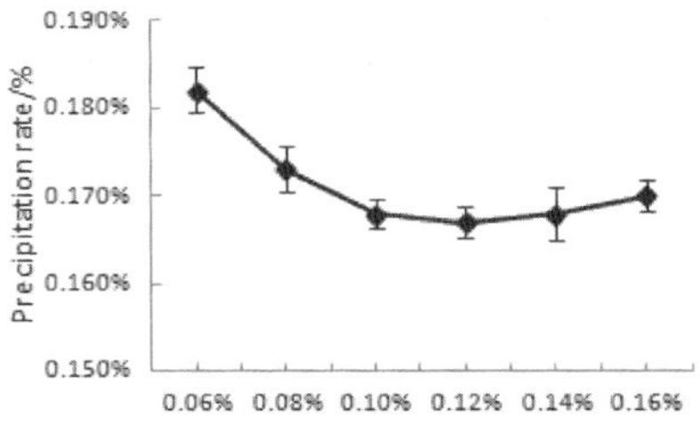

Fig.3. Effect of stabilizer type on viscosity. **Fig.4.** Effect of stabilizer concentration on precipitation rate.

3.2.2. *Effect of Stabilizer Concentration on Stability Coefficient*

Fig. 5 showed that the stability coefficient was the highest and its value was 80.38% when the concentration of stabilizer was 0.12%. With the increase of the stabilizer concentration, the stability coefficient generally increased first and decreased later.

3.2.3. *Effect of Stabilizer Concentration on Viscosity*

Fig. 6 showed that the viscosity curve of beverage added different concentrations of stabilizers tended to overlap in the end. Due to the low concentration of beverage, the change of the stabilizer concentration had little influence on the viscosity of beverage.

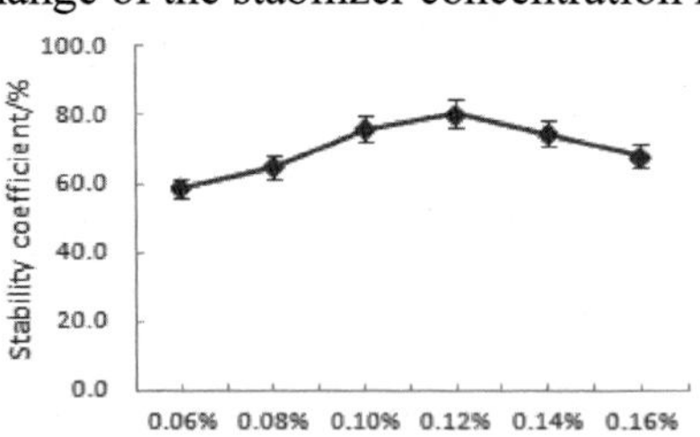

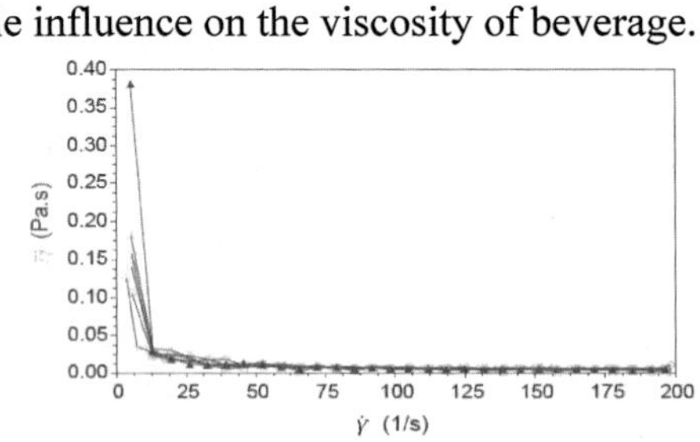

Fig.5. Effect of stabilizer concentration on coefficient. **Fig.6.** Effect of stabilizer concentration on viscosity.

4. CONCLUSIONS

The optimum stabilizer formulation of ginseng whey protein polypeptide beverage was consisted of 0.12% pectin, 0.12% STPP and 0.12% CMC. Under such circumstance, the beverage had no sediment stratification with a good stability.

Acknowledgments

This project was funded by Jilin University "Students' Innovation and Entrepreneurship Training Program" (Grant no. 2015831299). The authors thank The National Twelfth Five-Year Grant (2013BAD18B07) for financial support of this study.

References

1. Hongping Wang, Xinbao Yang, et al. Study on the chemical constituents of root and rhizome of Panax ginseng in Jilin in Chinese [J]. *China journal of Chinese meteria medica*, 2013; **9**:2809-2817.
2. Jing Liu, Qingbo Han. Characteristics and Application of Whey Protein in Chinese [J]. *Food Science*, 2007; 7:535-537.
3. A. Maroziene, C.G. de Kruif. Interaction of pectin and casein micelles [J]. *Food Hydrocolloids*. 2000; **14**(4):391-394.
4. Zhiyun Yan, Yuying Chen, Xuesong Zhou. Effect of Combination of CMC-Na and other Colloids on the Stability of an Active Lactobacillus Drink Containing Double Proteins in Chinese [J]. *Modern Food Science and Technology*, 2008; **7**:664-666.
5. Luyun Wen, Lijuan Zhang, Bin Zhang, et al. Development of Chromium-rich Germinated Brown Rice Biscuit in Chinese [J]. *Food research and development*, 2015; **4**:91-93.
6. Junjun Zou, Yanli Xie. Rheological properties of proteins, polysaccharides and hydrocolloids in Chinese [J]. *Journal of Henan University of Technology Natural Science Edition*, 2015; **6**:12-16.
7. Ping Hou, Wenshui Xia. Research on Optimization of Emulsifier in Chinese [J]. *Food Research and Development*, 2005, **26(3)**:53-56.

BIOACTIVITIES OF BLUEBERRY ANTHOCYANINS

Qing Xu[1,a], Yuanjing Zhou[2,b], Yuxiang Wu[1], Qiang Jia[1], Fei Nie[1], and Guilong Gao[1]

[1]Institute of Biology, Guizhou Academy of Sciences, Guiyang 550009, China

[2]Guizhou Academy of Test and Analysis, Guiyang 550002, China

E-mail: [a]xqhaoy@163.com, [b]gd8298123@126.com

Blueberry anthocyanins are the major active ingredients of blueberry with a variety of biological activities and have widespread applications in the fields of drugs, cosmetics, food additives and so on. The deep investigation of their biological activities has been made in recent year and this article reviewed the advance. This work would provide useful references for further application of blueberry anthocyanins.

1. INTRODUCTION

Anthocyanins are a kind of natural water-soluble colorant and widely present in plants, which belong to flavonoids. Research on anthocyanins has raised a growing interest due to their beneficial health effects with great potential applications in food, cosmetic, medicine, etc [1]. Bluebery (or bilberry), also known as Vaccinium vitisidaea L., is rich in anthocyanidins [2]. The deep investigation of their bioactivities has been made in recent year, however, a systematic report on the subject appears to be lacking. This paper reviewed the advance.

2. BIOACTIVITIES OF BLUEBERRY ANTHOCYANINS

Blueberry anthocyanins are formed mainly by glycosylated five anthocyanidins (cyanidin, delphinidi, peonidin, petunidin, and malvidin) with three sugars (glucose, galactose, and arabinose). And the glycosides may be acylated. As natural small molecules and water soluble substances, blueberry anthocyanins can be rapidly absorbed by human body and enter tissues and organs in 45min after oral administration [3], and have a wide range of biological activities as follows.

2.1. Antioxidant Activity

Numerous studies have shown that blueberry anthocyanins act as strong antioxidants, which is also the reason for their beneficial health benefits [4, 5]. Unlike other antioxidants, anthocyanins can pass through the blood-brain barrier, directly protect the brain and nervous system. They show a higher antioxidant activity than vitamins C and E and can capture free radicals by donation of phenolic hydrogen atoms [6, 7]. The antioxidant activity of blueberries varies greatly by origin, variety, maternity, and so on, because they contain different amounts and types of anthocyanins [8-13]. Sun et al assessed the antioxidant activity of nine rabbiteye blueberry cultivars in China. Their results indicated that Gardenblue and Powderblue showed stronger antioxidant activities, and delphinins and anthocyanidin-3-glucosides were the main contributive components to the antioxidant activities of blueberries [8]. Kim et al. analyzed the total anthocyanin content and antioxidant activity of 45 commercial cultivars blueberry fruits (39 northern highbush and 6 half highbush blueberry) grown in Suwon, Korea. In different blueberry cultivars, the antioxidant activity ranged from 0.7 to 2.1 mg of quercetin equivalents per gram of fresh berries. Among the 45 blueberry cultivars, high amounts of anthocyanins and polyphenols, and high antioxidant activity were observed in Elliott, Rubel, Rancocas, and Friendship [9].

2.2. Anticancer Activity

Hou et al summarized the mechanisms of cancer chemoprevention by anthocyanins and subdivided them into three aspects: 1) the antioxidation; 2) the molecular mechanisms involved in anticarcinogenesis; 3) the molecular mechanisms involved in the apoptosis induction of tumor cells [14]. They also reviewed the molecular mechanisms behind the chemopreventive effects of anthocyanidins [15]. As natural pigments, the potential anticancer activities (eg. stomach cancer, prostate cancer, intestine cancer, breast cancer, etc.) of blueberry anthocyanins have been evaluated in vitro. [16, 17]. Chen et al. investigated the ability of anthocyanins from the wild blueberries of Inner Mongolia to suppress the growth of the oral cancer cell line KB, and found the blueberry anthocyanins suppressed the proliferation of KB cells in a dose-dependent manner, as well as induced G2/M cell cycle arrest and apoptosis of oral cancer KB cells [17].

2.3. Antimicrobial and Antiinflammatory Properties

Recently, there has been renewed interest in the antimicrobial and antiinflammatory properties of blueberry extracts [18-20]. Burdulis et al revealed that Toro blueberry cultivars and bilberry fruits had antimicrobial properties to eight tested Gram-negative and Gram-positive bacteria [18]. Taverniti et al. showed that the blueberry anthocyanins could reduce the activation of NF-κB, induced by IL-1β in intestinal epithelial Caco-2 cells. Specifically, concentrations of 50 and 100 $\mu g\ mL^{-1}$ decreased NF-κB activation by 68.9 and 85.2%, respectively ($p \leq 0.05$).

2.4. Anti-aging Function

Some evidences indicate that blueberry supplementation are able to ameliorate age-related declines in neuronal and cognitive function, common in disorders such as Alzheimer disease, neurodegenerative diseases, and so on [21-23]. Microglial proinflammatory activation is widely known to cause neuronal and synaptic damage that correlates with cognitive impairment in Alzheimer disease. The aggregation of amyloid-beta (A beta) into fibrillar amyloid plaques is a key pathological event in the development of the disease. Zhu et al found that blueberry extract significantly enhanced microglial clearance of amyloid-beta, inhibited aggregation of amyloid-beta (1-42), and suppressed microglial activation, and so should help alzheimer's patients [21]. Thangthaeng et al believed that daily consumption of blueberries might reverse some age-related deficits in cognition, as well as preserve function among those with intact cognitive ability [23].

2.5. Protection Cardiovascular

Lots of studies have shown that blueberry has the function of lowering blood pressure, lowering cholesterol, and protecting the cardiovascular [24-28]. Riso et al. obtained that the consumption of the wild blueberries drink for 6 weeks significantly reduced the levels of oxidized DNA bases and increased the resistance to oxidatively induced DNA damage in subjects with risk factors for cardiovascular disease [24]. Kalea et al showed that incorporating wild blueberry in the diet could affect the endothelium-dependent vasorelaxation by modulating alternative metabolic pathway (such as affecting the production/activity of COX-derived products) in the young spontaneously hypertensive rats aorta [26]. Matsunaga et al. found that Vaccinium myrtillus (Bilberry) extracts could reduce angiogenesis in vitro and in vivo [27].

2.6. Improving Vision

It has long been recognized intaking buleberry can keep eye healthy and prevent the diseases of eyes [29, 30]. Song et al obtained that bilberry anthocyanin extract could protect cells from light damage and inhibit vascular endothelial growth factor level secreted from damaged cells [29]. Meng et al studied that blueberry anthocyanins had a significant protective effect against rat retinal light damage and thought it might be due to the anti-lipid peroxidation activity of anthocyanins [30].

2.7. Other Bioactivities

In addition to the biological activity mentioned above, other physiological effects of blueberry anthocyanins have been reported and continually investigated, such as diabetes prevention, protection against obesity and hypoglycemia, protection cardiac, antiproliferation and antiadhesion, improving learning and memory, and so on [31-35].

3. CONCLUSION

Overall, blueberry anthocyanins have caused wide attentions due to their nutritional and health benefits. And more physiological effects of blueberry anthocyanins are also expected to be found soon.

Acknowledgments

The authors are grateful for the financial support of Youth Fund Project of Guizhou Academy of Sciences, Guizhou province, China (J [2014]01) and Science and Technology Project of Guizhou Province, China (J Z [2015]2006, Z [2015]6013).

References

1. Tsuda T. Dietary anthocyanin-rich plants: biochemical basis and recent progress in health benefits studies. *Molecular Nutrition & Food Research* 2012; **5:** 159–170.
2. Hosseinian FS, Beta T. Saskatoon and wild blueberries have higher anthocyanin contents than other manitoba berries. *Journal of Agricultural and Food Chemistry* 2007; **26:** 10832–10838.
3. Del Bò C, Ciappellano S, Klimis-Zacas D, et al. Anthocyanin absorption, metabolism and distribution from a wild blubeery-enriched diet (*Vaccinium angustifolium*) is affected by diet duration in the Sprague-Dawley rat. *Journal of Agricultural and Food Chemistry* 2010; **4:** 2491–2497.
4. Bornsek SM, Ziberna L, Polak T, et al. Bilberry and blueberry anthocyanins act as powerful intracellular antioxidants in mammalian cells. *Food chemistry* 2012; **4:** 1878–1884.
5. Chen J, Sun H, Sun A, et al. Studies of the protective effect and antioxidant mechanism of blueberry anthocyanins in a CCl_4-induced liver injury model in mice. *Food and Agricultural Immunology* 2012; **4:** 352–362.
6. Mazza G, Kay CD, Cottrell T, Holub BJ. Absorption of anthocyanins from blueberries and serum antioxidant status in human subjects. *Journal of Agricultural and Food Chemistry* 2002; **26:** 7731–7737.
7. Li J, Solval KM, Alfaro L, et al. Effect of blueberry extract from blueberry pomace on the microencapsulated fish oil. *Journal of Food Processing and Preservation* 2015; **2:** 199–206.
8. Sun L, Ding X, Qi J, et al. Antioxidant anthocyanins screening through spectrum-effect relationships and DPPH-HPLC-DAD analysis on nine cultivars of introduced rabbiteye blueberry in China. *Food Chemistry* 2012; **2:** 759–765.
9. Kim JG, Kim HL, Kim SJ, Park K. Fruit quality, anthocyanin and total phenolic contents, and antioxidant activities of 45 blueberry cultivars grown in Suwon, Korea. *Journal of Zhejiang University Science B* 2013; **9:** 793–799.
10. Li YC, He YT. Anthocyanin content and antioxidant activity of different varieties blueberries. *Advanced Materials Research* 2013; **610:** 3421–3427.
11. Wang SY, Chen H, Camp MJ, Ehlenfeldt MK. Genotype and growing season influence blueberry antioxidant capacity and other quality attributes. *International Journal of Food Science & Technology* 2012; **7:** 1540–1549.
12. Requea PM, Steffensa RS, Jablonskib A, et al. Cold storage of blueberry (Vaccinium spp.) fruits and juice: Anthocyanin stability and antioxidant activity. *Journal of Food Composition and Analysis* 2014; **1:** 111–116.

13. Pertuzattia PB, Barciaa MT, Rodriguesa D, et al. Antioxidant activity of hydrophilic and lipophilic extracts of Brazilian blueberries. *Food Chemistry* 2014; **164:** 81–88.
14. Hou DX. Potential mechanisms of cancer chemoprevention by anthocyanins. *Current Molecular Medicine* 2003; **2:** 149–159.
15. Hou DX, Fujii M, Terahar N, Yoshimoto M. Molecular mechanisms behind the chemopreventive effects of anthocyanidins," *Journal of Biomedicine and Biotechnology* 2004; **5:** 321–325.
16. Grace MH, Ribnicky DM, Kuhn P, et al. Anthocyanins extracted from Chinese blueberry (*Vaccinium uliginosum* L.) and its anticancer effects on DLD-1 and COLO205 cell. *Chinese Medical Journal* 2010; **19:** 2714–2719.
17. Chen Q, Li S, Jia Y, Wang L. Blueberry anthocyanins induce G2/M cell cycle arrest and apoptosis of oral cancer KB cells through down-regulation methylation of p53. *Yi Chuan* 2014; **6:** 566–573.
18. Burdulis D, Šarkìnas A, Jasutìenè I, et al. Comparative study of anthocyanin composition, antimicrobial and antioxidant activity in bilberry (*Vaccinium myrtillus* L.) and blueberry (*Vaccinium corymbosum* L.) fruits. *Acta Poloniae Pharmaceutica-Drug Research* 2009; **4:** 399–408.
19. Taverniti V, Fracassetti D, Del Bo' C, et al. Immunomodulatory effect of a wild blueberry anthocyanin-rich extract in human Caco-2 intestinal cells. *Journal of Agricultural and Food Chemistry* 2014; **33:** 8346–8351.
20. Esposito D, Chen A, Grace MH, et al. Inhibitory effects of wild blueberry anthocyanins and other flavonoids on biomarkers of acute and chronic inflammation in vitro. *Journal of Agricultural and Food Chemistry*2014; **29:** 7022–7028.
21. Zhu Y, Bickford PC, Sanberg P, et al. Blueberry opposes beta-amyloid peptide-induced microglial activation via inhibition of p44 /42 mitogenactivation protein kinase. *Rejuvenation Research* 2008; **5:** 891–901.
22. Devareddy L, Hooshmand S, Collins JK, et al. Blueberry prevents bone loss in ovariectomized rat model of postmenopausal osteoporosis. *The Journal of Nutritional Biochemistry* 2008; **10:** 694–699.
23. Thangthaeng N, Miller M, Poulose S, et al. Differential effects of blueberry polyphenols on age-associated neuroinflammation and cognition. *The FASEB Journal* 2015; **1:** 756–758.
24. Riso P, Klimis-Zacas D, Del Bo' C, et al. Effect of a wild blueberry (*Vaccinium angustifolium*) drink intervention on markers of oxidative stress, inflammation and endothelial function in humans with cardiovascular risk factors. *European journal of nutrition* 2012; **3:** 949–961.
25. Basu A, Du M, Leyva MJ, et al. Blueberries decrease cardiovascular risk factors in obese men and women with metabolic syndrome. *The Journal of nutrition* 2010; **9:** 1582–1587.
26. Kalea AZ, Clark K, Schuschke DA, et al. Dietary enrichment with wild blueberries (*Vaccinium angustifolium*) affects the vascular reactivity in the aorta of young

spontaneously hypertensive rats. *The Journal of Nutritional Biochemistry* 2010; **1:** 14–22.

27. Matsunaga N, Chikaraishi Y, Shimazawa M, et al. Vaccinium myrtillus (bilberry) extracts reduce angiogenesis in vitro and in vivo. *Evidence-Based Complementary and Alternative Medicine* 2010; **1:** 47–56.
28. Roopchand DE, Kuhn P, Rojo LE, et a1. Blueberry polyphenol-enriched soybean flour reduces hyperglyeemia, body weight gain and serum cholesterol in mice. *Pharmacological Research* 2013; **1:** 59–67.
29. Song X, Han Y, Ji B, et al. Protective effect of bilberry anthocyanin extract (BAE) against photo-oxidative damage of human retinal pigment epithelial cells. *Food Science* 2010; **21:** 242–245.
30. Meng X, Bi W, Zhang Q, Li B. Protective effect and mechanism of blueberry anthocyanins against retinal photodamage in rats. *Food Science* 2013; **11:** 242–245.
31. Wu T, Tang Q, Gao Z, et al. Blueberry and mulberry juice prevent obesity development in C57BL/6 mice. *Plos One* 2013; **10:** e77585.
32. Meng X, Song D, Shi L. et al. Protective effect of blueberry anthocyanins on cyclophosphamide-induced cardiac injury in rats. *Chinese Journal of Public Health* 2015; **2:** 187–190.
33. Ma P, Ji Z, Ji C, et al. Protective effect of blueberry anthocyanins pretreatment on myocardial infarction in rats. *Chinese Journal of comparative Medicine* 2013; **23:** 49–52.
34. Diaconeasa Z, Leopold L, Rugină D, et al. Antiproliferative and antioxidant properties of anthocyanin rich extracts from blueberry and blackcurrant juice. *International Journal of Molecular Sciences* 2015; **2:** 2352–2365.
35. Pranprawit A, Heyes JA, Molan AL, Kruger MC. Antioxidant activity and inhibitory potential of blueberry extracts against key enzymes relevant for hyperglycemia. *Journal of Food Biochemistry* 2015; **1:** 109–118.

PROPERTIES OF LACCASE FROM *BACILLUS* SP. C2 AND ITS USE IN DYE DECOLORIZATION

Chunxia Lou[a], Te Wang, and Linghua Zhang[b,*]

Environmental Science and Engineering College, Dalian Maritime University, 1 Linghai Road, Dalian, China

E-mail: [a]lcx1511867657@163.com, [b]dlzlh2008@163.com

[*]Corresponding author

A bacteria which had capability of laccase production was screened from the sludge of sludge wastewater treatment plant. The strain was affirmed as Bacillus sp. C2 by microscopy observation and 16S rDNA sequences. The propeties of laccase and the dye decolorization had been carried on preliminary research. Results were concluded as follows: The optimum temperature and pH of laccase activity were respectively 30 °C and 3, laccase was induced well by 3 mM Cu^{2+}. The optimum concentration of ABTS (2,2'-azino-bis (3-ethylbenz-thiazoline-6-sulfonate) was 1 mM, Km and Vmax value of michaelis-menten equation was 9.992 mM and 76.923 mM/min. The laccase could degrade the acid green dye, the acid green decolorization rate was 83.1%.

1. INTRODUCTION

Dyes are widely used in textile, leather industries and so on, which has been reported are more than 10000 kinds, and the dyes that produced every year are more than 70 million[1-2]. In the production and usage of dye, a lot of direct dye wastewater is discharging into the environment, not only affect the transparency of water and DO, but also can produce carcinogens, such as benzidine, so people pay more attention to the treatment of dye wastewater[3]. In practical appilication, the traditional physical and chemical methods have some limitations because of its high cost, large energy consumption, and easy to form toxic by-products[4]. The biodegradable technology of dye, which is nonpoisonous and harmless, and has features such as specificity, high efficiency and mildness, it can effectively remove the color of dyes and can degrade the dye. Biological enzymatic treatment of dye wastewater has good application prospects because of its characters: low cost, no secondary pollution, well ecological recovery[5].

Laccase (Laccase, benzene diol: oxygen oxidoreductase, EC1.10.3.2) is a kind of polyphenol oxidase containing copper[6], can degrade dyes. Laccase is the most potential biological enzyme in the treatment of dye wastewater in recent years, it can apply to a varity of substrates, and can catalytic degradate different structure of organic compounds including dyes, and some small molecule redox mediators can help to improve the laccase activity[7], which will be conducive to the development and application of laccase.

A bacterium with laccase activity was isolated and screened in this paper. It was analyzed and identified as *bacillus*. Laccase activity properties were studied. And the decolorization of dye was conducted a preliminary study.

2. MATERIALS AND METHODS

2.1. Materials

Isolation and screening medium (g/L): peptone 10, yeast powder 5, NaCl 0.5, ABTS 1 mM, pH 7.2.

Laccase production medium (g/L): sucrose 20, sodium glutamate 20, Sodium citrate 10, $(NH_4)_2SO_4$ 10, yeast powder 1, KH_2PO_4 3, K_2HPO_4 9, $MgSO_4 \cdot 7H_2O$ 0.4, $MnSO_4 \cdot H_2O$ 0.01, NaCl 5, pH 7.2.

2.2. Determination Method of Laccase Enzyme Activity

The system of laccase enzyme activity contained 2.73 mL 0.1 M phosphate buffer (pH 3), 60 μL 1 mM ABTS, 150 μL 3 mM $CuCl_2 \cdot 2H_2O$ and 60 μL crude laccase. The absorbance changes were determined within 4 min at 420 nm value. Laccase enzyme-activity unit was defined that the quantity of enzyme that required for oxidizing 1 μmol ABTS in one minute[8].

2.3. 16S rDNA Analysis and Identification.

The strain was identified by 16S rDNA PCR using TaKaRa 16S rDNA Bacterial Identification PCR Kit (TaKaRa Biotechnology (Dalian) Co., Ltd.). The PCR product was sequenced by the TaKaRa Biotechnology (Dalian) Co., Ltd.

2.4. Determination METHOD of the K_m and V_{max} Value

Different final concentration (0, 1, 2, 3, 4, 5, 6) of ABTS solutions were configurate with phosphate buffer which pH was 3, then the laccase enzyme activity was determinated under 30 °C . Enzymatic kinetics equation (michaelis-menten equation) (1):

$$V = V_{max} \times [S]/(K_m + [S]) \quad (1)$$

According to the double inverse method of Lineweaver-Burk[9], The transformation for the equation (2):

$$\frac{1}{V} = \frac{K_m}{V_{max}} \times \frac{1}{S} + \frac{1}{V_{max}} \quad (2)$$

A linear equation was obtained when plotting with 1/V and 1/S, then the K_m and V_{max} value could acqurie under the conditions.

2.5. Dyes Decolorizing and Chromaticity Measurement Method

The 4 mL system contained pH 3 phosphate buffer, ABTS, acid green dye and culture broth, at the same time the same system without laccase was took as blank control. The sample was taked out after a certain time and was centrifuged at 1200 rpm for 5 min, then the absorbance value in the maximum absorption wavelength of dye was determinated. Calculation formula of decoloring rate was: decolorizing ratio = $(A_0-A)/A_0\times$ 100%, A_0 was the initial dye absorbance value, A was the absorbance values when sampiling[10].

3. RESULTS AND DISCUSSION

3.1. The Separation and Selection of Laccase Production Strain Selection

Strains were screened from the sludge of sludge wastewater treatment plant and strains were cultivated in the ABTS filter medium with the marking method. The green colonies were picked out and separated by the marking method until there were all single colonies from the microscopic identification of gram staining and the form[11], four strains were screened and cultivated then laccase activities were determinated according to the method of 2.2. The strain C2 had the highest laccase activity, and was used as the research object.

While the distribution of laccase activity was studied, the distribution was divided into three parts: fermented liquid, supernatant and bacteria suspension laccase activity was determinated according to the method 2.2, the result was shown in the Fig.1, 90% laccase activity existed in bacteria cells, so the strain C2 laccase was surface enzymeand. All cells can be used laccase enzymatic reaction.

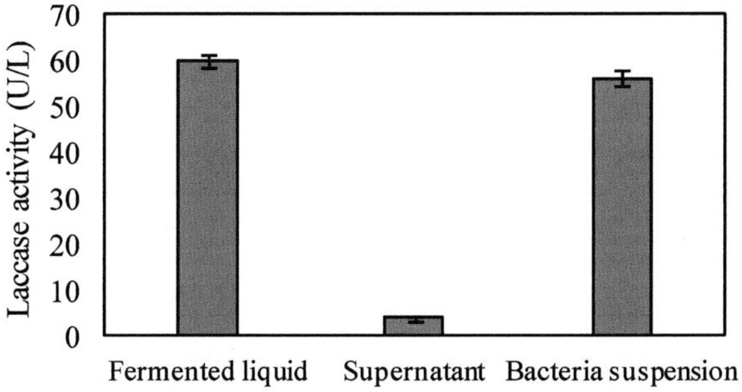

Fig.1. The strain C2 laccase activity distribution.

3.2. 16S rDNA Identification of Laccase Strains C2

Characteristics of C2 bacteria colony: milky white, smooth surface, smooth edges. C2 strain was a short rod under microscope, gram positive bacteria, spore staining showed generating spore. 16S rDNA identification was performed by the method of 2.3.

Comparison of 16S rDNA sequence of the strain C2 (GenBank, KM016238) with the sequences in GenBank database showed 99.9% similarity to *Bacillus pumilus* strain TCCC11014 (GenBank, KJ722434.1). The strain was named as *Bacillus* sp. C2.

3.3. Laccase Properties of *Bacillus* sp. C2

3.3.1. *Influence of PH on Laccase Activity*

At room temperature (25 °C), laccase activity was determinate under different phosphate buffer with pH (1.4, 2.2, 3, 2.2, 4.6, 5.4, 6.2), the result was showed in Fig.2.(a). The optimal pH of *Bacillus* sp.C2 was 3.

3.3.2. *Influence of Temperature on Laccase Activity*

Laccase activity was determinated under pH 3 phosphate buffer and different temperature of 10, 20, 30, 40, 50, 60 °C , the result was showed in Fig.2.(b). The optimal temperature of *Bacillus* sp. C2 was 30 °C .

3.3.3. *Influence of Concentration of Cu^{2+} on Laccase Activity*

Laccase is a polyphenol oxidase containing copper, this paper discussed the influence of the concentration of Cu^{2+}. Different final concentration (0, 1, 2, 3, 4, 5 mM) of Cu^{2+} soltion were configurated, laccase activity were determinated under 30 °C and pH 3. The result was showed in Fig.2.(c). The optimal concentration of Cu^{2+} of *Bacillus* sp. C2 was 3 mM.

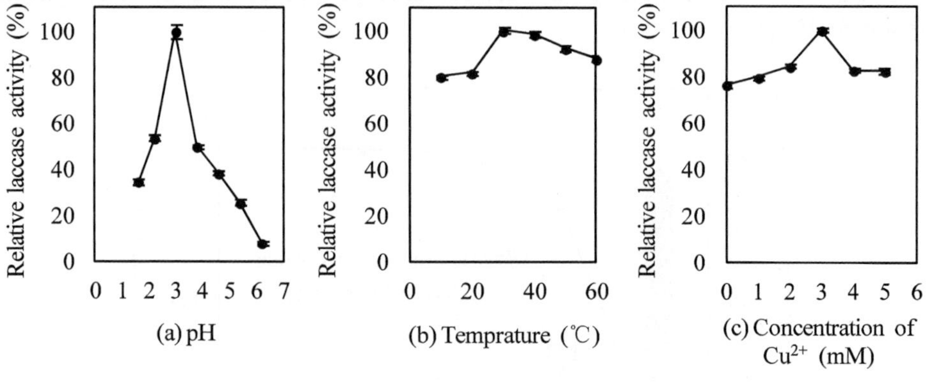

Fig.2. Different factors ((a): pH; (b): temperature; (c): Concentration of Cu^{2+})) influence on *Bacillus* sp. C2 laccase activity.

3.3.4. *Kinetics of Bacillus sp. C2 Laccase Enzyme Reaction*

The influence result on laccase activity of the substrate concentration (ABTS) was shown in Fig.3.(a). As the method 2.4, the regression line was a straight line when plotting with

1/V and 1/S, the result was shown in Fig.3.(b), the equation was y=0.1299x+0.013, R^2=0.9958, from the above equation, V_{max} was 76.923 mM/min, K_m was 9.992 mM. *Bacillus* sp. C2 michaelis-menten equation was V=76.923× [S]/(9.992+[S]).

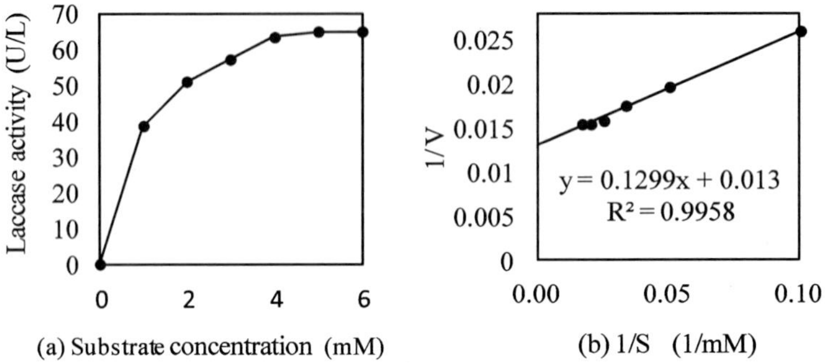

Fig.3. (a)The influence of the substrate concentration on *Bacillus* sp. C2 laccase ctivity; (b) Enzymatic reaction kinetics fitting.

3.4. Laccase Decolorization of Dye Wastewater

4 mL fermented liquid was centrifuged, and then the cell was suspended with 3 mL pH 3 phosphate buffer, 250 μL ABTS, 750 μL 50 mg/L acid green dye, which was the decoloring system, at the same time the same system without bacteria was took as blank control. Sample was taked out each 6 h and centrifuged at 1200 rpm for 5 min, then the absorbance value under 636 nm was determinated. The result was shown in Fig.4, the decolorization of dye acid green was balanced after 12 h processing, the decolorization rate was 83.1%.

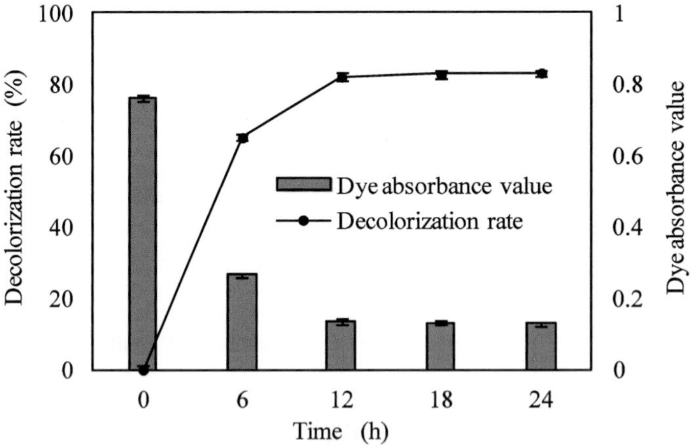

Fig.4. *Bacillus* sp. C2 laccase decolorization of acid green under different time.

4. CONCLUSIONS

(i) Bacteria C2 which had capability of laccase production was screened from the sludge of sludge wastwater treatment plant. The strain was affirmed as *Bacillus* sp.C2 by microscopy observation and 16S rDNA sequences.
(ii) The most suitable conditions with ABTS as the substrate of *Bacillus* sp. C2: pH of 3, the temperature of 30 °C , the substrate concentration of 1 mM, and Cu^{2+} had a promoting effect to laccase. The enzymatic reaction of *Bacillus* sp.C2 laccase was in line with the michaelis-menten equation, Vmax was 76.923 mM/min, Km was 9.992 mM.
(iii) *Bacillus* sp. C2 laccase decolorization of dye acid green was basic balanced after 12 h processing, the decolorization rate was 83.1%.

References

1. Ayed L, Mahdhi A, Cheref A, et al. Decolorization and degradation of azo dye methyl red by an isolated Sphingomonas paucimobilis: Biotoxicity and metabolites characterization. *J Desalination* 2011; **274**(1–3): 272–277.
2. Kadam AA, Telke AA, Jagtap SS, et al. Decolorization of adsorbed textile dyes by developed consortium of *Pseudomonas* sp. SUK1 and Aspergillus ochraceus NCIM-1146 under solid state fermentation. *J Hazardous Materials* 2011; **189**(1–2): 486–494.
3. Li J, Li GF, Lu L, et al. Application of spore laccase from *Bacillus amyloliquefaciens* in dye decolorization. *Beijing Forestry University* 2013; **2**: 125–129. (In Chinese).
4. Yu Y, Zheng P, Jin RC, et al. Biotechnology for dyeing and printing wastewater treatment. *Chemical Industry and Engineering Progress* 2008; **27**(11): 1724–1727. (In Chinese).
5. Sun QF, Zhao LL, Li YR, et al. Decolorization Effect of Anthraquinone Dye with Laccase/ABTS Mediator System. *Guizhou Agricultural Sciences* 2011; **39**(1): 190–193. (In Chinese).
6. Liu YX, Yan KL, Xiong Z, et al. Experimental Study on Decolorizing Malachite-green in Laccase-mediator System. *Environmental Science and Technology* 2008; **31**(7): 37–39. (In Chinese).
7. Baldrian P. Fungal laccases-occurrence and properties. *J FEMS Microbiol Rev* 2006; 30: 215–242.
8. Li W. The optimization of fermentation conditions and decolorization properties of dye of the *Bacillus thuringiensis* ME506 cells with surface-displayed laccase. *Hubei University*, Hubei 2014. (In Chinese).
9. Hou HM. White-rot fungus *Pleurotus ostreatus* laccase and the research of decoloring of anthraquinone dyes and alkali lignin. *Dalian University of Technology*, Dalian 2003. (In Chinese).
10. Li TL. Properties of spore laccase from *Bacillus subtilis* WN02 and its use in dye decolorization. *Northaest Forestry University*, Harbin 2010. (In Chinese).
11. Li J. Produce laccase bacteria strain screening and its application. *Northaest Forestry University*, Harbin 2013. (In Chinese).

OPTIMIZATION OF FERMENTATION CONDITIONS OF BIOCONTROL STRAIN *BACILLUS SUBTILIS* HS5B5

Peng Song[1,a], Xiaogai Hou[1], Xiupu Guo[1], Guozhan Fu[1], Xiaomei Zhang[1], Zhipeng Huo[2], and Shaodan Liu[2]

[1]*College of Agriculture, Henan University of Science and Technology, Luoyang, Henan Province, China*

[2]*Flower Company of Luoyang, Luoyang, Henan Province, China*

E-mail: [a]songpeng0826@126.com

Bacteria of *Bacillus* can produce antibacterial peptide and protein, etc. They have an inhibitory action on harmful microorganisms. *Bacillus subtilis* HS5B5 was screened and identified previously. It had obvious antibacterial activity against *Cladosporium paeoniae*. In order to make better use of this biocontrol strain, its fermentation medium and condition were studied through orthogonal test involving 4 factors at 3 levels and single-factor calibration test. Results showed the optimal fermentation conditions were corn flour 2.0%, bean flour 1.0%, fermentation time 24h and original pH 7.0. Under the optimal conditions, bacterial quantity and the transparent zone diameter were 9.2×10^8 CFU/ml and 6.21mm respectively after fermentation.

1. INTRODUCTION

Bacillus is the dominant population in the micro-ecosystem of soil and plants, which has strong resistance ability and antimicrobial effect[1-2]. *Bacillus* not only can inhibit a variety of fungal and bacterial diseases of plants, but also is of certain growth promoting effect on growth[3-4]. In addition, it can produce many kinds of stable antibacterial substances widely used in industrial and agricultural production. Many researches showed the antimicrobial mechanism of *Bacillus* was varied, including antibiotic, dissolve hypha, antimicrobial protein and competition[5-7]. *Bacillus* strain named HS5B5 was screened and identified from the topsoil of peony garden previously. It had obvious antibacterial activity against *Cladosporium paeoniae*, which was pathogenic microorganism of peony leaf spot disease. Based on *Bacillus subtilis* HS5B5, the fermentation medium and condition of the strain were optimized through orthogonal and single-factor calibration test, in order to provide theoretical support for industrial and agricultural application.

2. MATERIALS AND METHODS

2.1. Strains

Bacillus subtilis HS5B5, a newly isolated strain with obvious antagonistic activity against *Cladosporium paeoniae*, pathogenic microorganism of peony leaf spot disease, was used in this study. Its fermentation liquid including the highest concentration of 7.1×10^8 CFU/ml can cause transparent zone with a diameter of 5.75mm through oxford cup method. It was identified and deposited in College of Agriculture, Henan University of Science and Technology, Luoyang, China.

2.2. Medium

Bacillus subtilis HS5B5 was preserved on NA medium (peptone 10 g l^{-1}; beef extract 3 g l^{-1}; NaCl 5 g l^{-1}; agar 18 g l^{-1}; pH 7.2) at 4°C. Strain 36063 of *Cladosporium paeoniae* from ACCC was preserved on PDA medium (potato 200 g l^{-1}; dextrose 20 g l^{-1}; agar 20 g l^{-1}). Corn flour and bean flour were studied and taken as the main carbon and nitrogen sources of fermentation medium, and the rest was beef extract 3 g l^{-1}, peptone 2 g l^{-1} and $MgSO_4\cdot7H_2O$ 0.5 g l^{-1}.

2.3. Orthogonal Test

Bacillus subtilis HS5B5 was cultured in the NA medium at 30 °C for 12 h. Then it was inoculated into 200 ml fermentation medium under the conditions of 30°C and 1% incubation in the shaker (180 r/min). A $L9(3^4)$ orthogonal test was designed by SPSS 19.0 and four factors were selected: corn flour (A), bean flour (B), fermentation time (C) and original pH (D). Three levels were set for each selected factor, and bacterial quantity in fermentation medium was counted[8-10]. The test was repeated and design was shown in Table 1.

Table 1. Factors and levels for orthogonal test

Factor / Level	A Corn flour (%)	B Bean flour (%)	C Fermentation time(h)	D Original pH
1	1.0%	0.5%	18	6.5
2	1.5%	1.0%	24	7.0
3	2.0%	1.5%	30	7.5

2.4. Single-factor Calibration Test

Based on the result of orthogonal test, single-factor calibration test was carried out on fermentation time (18, 24, 30 and 36 h) and original pH (6.0, 6.5, 7.0 and 7.5) to get better fermentation conditions. Both bacterial quantity and antagonistic activity were determined to optimize the result of orthogonal test.

2.5. Bacterial Quantity and Antagonistic Activity Assay in Vitro

The amount of bacteria was counted by blood cell counting chamber after diluted with sterile water. Antagonistic activity was determined by the transparent zone diameter of growth inhibition against *Cladosporium paeoniae*. Strain of *Cladosporium paeoniae* was inoculated at the centre of the plate. The plates were cultured at 30°C for 5-7 d until the fungus of the control plate had grown to 30mm in diameter. Then oxford cups filled with fermentation liquid of *Bacillus subtilis* HS5B5 were placed at four points (10mm from the edge of the fungus). The transparent zone diameter was measured within 24h.

3. RESULTS

3.1. Orthogonal Test

The impacts of corn flour, bean flour, fermentation time and original pH on fermentation were shown in Table 2, Table 3 and Table 4. As can be seen from Table 3, the impacts of corn flour and bean flour on the bacterial quantity in fermentation medium was remarkably significant ($p < 0.01$), while that of fermentation time and original pH reached a significant level ($0.01 < p < 0.05$). The mean values in Table 4 indicated following conclusions of each level: A3> A2> A1, B3> B2> B1, C1> C2> C3 and D1> D2> D3. Thus, the optimal combination came to $A_3B_3C_1D_1$, viz. corn flour 2.0%, bean flour 1.5%, fermentation time 18 h and original pH 6.5.

Table 2. $L_9(3^4)$ orthogonal test

No.	Factor				Bacterial quantity($\times 10^8$CFU/ml)	
	A	B	C	D	Repeat I	Repeat II
1	1	1	1	1	3.5	3.3
2	1	2	2	2	4.5	4.3
3	1	3	3	3	4.2	4.3
4	2	1	2	3	5.5	5.6
5	2	2	3	1	7.0	7.2
6	2	3	1	2	7.4	7.1
7	3	1	3	2	6.5	6.3
8	3	2	1	3	8.1	8.2
9	3	3	2	1	8.3	8.6

Table 3. Tests of between-subjects effects

Source	Type III Sum of Squares	*df*	Mean Square	*F*	*Sig.*
Corrected Model	52.085[a]	9	5.787	251.012	0.000
Intercept	671.001	1	671.001	29103.639	0.000
Corn flour	42.474	2	21.237	921.133	0.000
Bean flour	8.831	2	4.416	191.518	0.000
Fermentation time	0.374	2	0.187	8.120	0.012
Original pH	0.404	2	0.202	8.771	0.011
Repeating group	0.001	1	0.001	0.024	0.880
Error	0.184	8	0.023		
Total	723.270	18			
Corrected Total	52.269	17			

a. R Squared =0.996 (Adjusted R Squared =0.993); Dependent Variable: bacterial quantity.

Table 4. Results of each factor

	Mean	*Std. Error*	95% Confidence Interval	
			Lower Bound	Upper Bound
Corn flour				
1	4.017	0.062	3.874	4.160
2	6.633	0.062	6.490	6.776
3	7.667	0.062	7.524	7.810
Bean flour				
1	5.117	0.062	4.974	5.260
2	6.550	0.062	6.407	6.693
3	6.650	0.062	6.507	6.793
Fermentation time				
1	6.267	0.062	6.124	6.410
2	6.133	0.062	5.990	6.276
3	5.917	0.062	5.774	6.060
Original pH				
1	6.317	0.062	6.174	6.460
2	6.017	0.062	5.874	6.160
3	5.983	0.062	5.840	6.126

Dependent Variable: bacterial quantity.

3.2. Single-factor Calibration Test

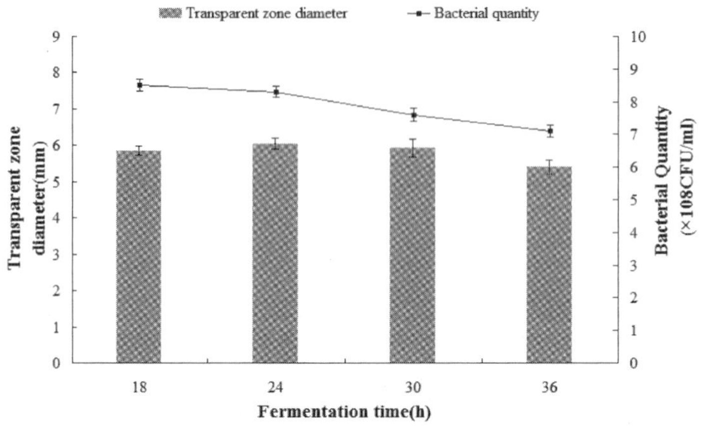

Fig. 1. Result of fermentation time in single-factor calibration test.

Based on the result of orthogonal test, single-factor calibration test of fermentation time and original pH was carried out. As shown in Fig. 1, although the bacterial quantity dropped down after 18 h, the transparent zone diameter reached its max level at 24h. As fermentation time went on, antibacterial effect decreased continually. As shown in Fig. 2 similarly, the increase of original pH of fermentation medium was not conducive to bacteria after pH 6.5. The amount of bacteria was higher under the fermentation condition of pH 6.5 than that of pH 7.0, but the transparent zone diameter was relatively smaller under the former condition. Therefore, based the above, the fermentation time and original pH were corrected to 24 h and 7.0 respectively. Under the final result, viz. corn flour 2.0%, bean flour 1.5%, fermentation time 24 h and original pH 7.0, the bacterial quantity and the transparent zone diameter were 9.2×10^8 CFU/ml and 6.21mm respectively.

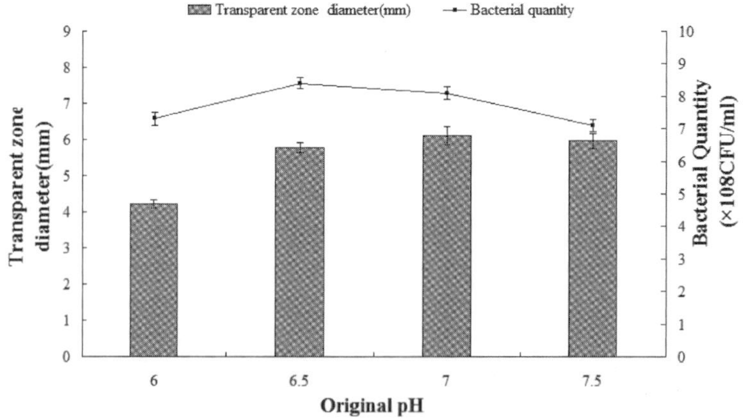

Fig. 2. Result of original pH in single-factor calibration test.

4. CONCLUSION

Through orthogonal test involving 4 factors at 3 levels and single-factor calibration test, the results showed that the impacts of corn flour and bean flour on the bacterial quantity in fermentation medium was remarkably significant ($p < 0.01$), while that of fermentation time and original pH reached a significant level ($0.01 < p < 0.05$). The final optimal combination was corn flour 2.0%, bean flour 1.5%, fermentation time 24 h and original pH 7.0. The bacterial quantity and the transparent zone diameter were 9.2×10^8 CFU/ml and 6.21mm respectively under optimized fermentation medium and condition.

5. DISCUSSION

The higher the amount of bacteria is, the more antibacterial substance is produced. Is that true? As shown in this study, the bacterial quantity reached a maximum at 18h and then dropped down, while antagonistic ability was not higher than that at 24h. There was similar situation of original pH in single-factor calibration test. The reason may be that substances produced from microbial metabolism or medium degradation had an inhibitory action on increase of antibacterial substances[11-12]. Further research is needed to reveal the structure, properties and antibacterial mechanism of antibacterial substance produced by *Bacillus subtilis* HS5B5.

Acknowledgments

This study was supported by the Key Scientific Research Foundation for University of Henan Province (Study on the Antagonistic Mechanism of *Bacillus subtilis* against *Cladosporium paeoniae*, **15A180003**) and the Scientific and Technological Project of Henan Province (Research and Development of Composite Microbial Fertilizer for Peony, **162102110131**).

References

1. Alvarez F, Castro M, Principe A, *et al.* The plant-associated *Bacillus amyloliquefaciens* strains MEP2 18 and ARP2 3 capable of producing the cyclic lipopeptides iturin or surfactin and fengycin are effective in biocontrol of sclerotinia stem rot disease. *J Appl Microbiol* 2012; **112**(1): 159-174.
2. Chen Y, Yan F, Chai Y, *et al.* Biocontrol of tomato wilt disease by *Bacillus subtilis* isolates from natural environments depends on conserved genes mediating biofilm formation. *Environ Microbiol* 2013; **15**(3): 848-864.
3. Grover M, Nain L, Singh SB, *et al.* Molecular and biochemical approaches for characterization of antifungal trait of a potent biocontrol agent *Bacillus subtilis* RP24. *Curr Microbiol* 2010; **60**(2): 99-106.
4. Huang CJ, Wang TK, Chung SC, *et al.* Identification of an antifungal chitinase from a potential biocontrol agent, *Bacillus cereus* 28-9. *J Biochem Mol Biol* 2005; **38**(1): 82-88.
5. Ji SH, Paul NC, Deng JX, *et al.* Biocontrol Activity of *Bacillus amyloliquefaciens* CNU114001 against Fungal Plant Diseases. *Mycobiology* 2013; **41**(4): 234-242.

6. Li XY, Mao ZC, Wang YH, *et al.* Diversity and active mechanism of fengycin-type cyclopeptides from *Bacillus subtilis* XF-1 against Plasmodiophora brassicae. *J Microbiol Biotechnol* 2013; **23**(3): 313-321.
7. Zhan Y, Xu Q, Yang MM, *et al.* Screening of freeze-dried protective agents for the formulation of biocontrol strains, *Bacillus cereus* AR156, *Burkholderia vietnamiensis* B418 and *Pantoea agglomerans* 2Re40. *Lett Appl Microbiol* 2012; **54**(1): 10-17.
8. Zhang JD, Yang Q. Optimization of solid-state fermentation conditions for *Trichoderma harzianum* using an orthogonal test. *Genet Mol Res* 2015; **14**(1): 1771-1781.
9. Wang LY, Cheong KL, Wu DT, *et al.* Fermentation optimization for the production of bioactive polysaccharides from *Cordyceps sinensis fungus* UM01. *Int J Biol Macromol* 2015; **79**: 180-185.
10. Zhu T, Wang L, Wang W, *et al.* Enhanced production of lipstatin from *Streptomyces toxytricini* by optimizing fermentation conditions and medium. *J Gen Appl Microbiol* 2014; **60**(3): 106-111.
11. Tian Y, Fan Y, Zhao X, *et al.* Optimization of fermentation medium for acetoin production by *Bacillus subtilis* SF4-3 using statistical methods. *Prep Biochem Biotechnol* 2014; **44**(5): 529-543.
12. Zhang P, Luo XY, Wu Y, *et al.* Optimization of submerged fermentation of *Thelephora ganbajun zang*. *J Basic Microbiol* 2014; **54**(8): 866-872.

EFFECT OF *JUGLANS MANDSHURICA* MAXIM EXTRACT ON IMMUNITY RESPONSE AND OXIDATIVE STRESS IN MICE

Jingjing Li, Yan Li*,a, and Kai Zhu

School of Life Science, Beijing Institute of Technology, Beijing, China
E-mail: a2120141331@bit.edu.cn
*Corresponding author

The present study examined the potential application of Juglans mandshurica Maxim extracts (JMME) for immune regulation and antioxidation in mice. From the perspective of using JMME as a herbal medicine, sheep red blood cell model and low-immunity mice model were utilized to characterize the effect of JMME on immune regulation and antioxidation. The 50% hemolysin value (HC50), the concentration of IgM antibody in splenic cells, index of thymus and spleen, serum lysozyme (LZM), malondialdehyde (MDA), total antioxidationability (T-AOC), contents of reduced glutathione (GSH) in thymus, content of nitrogen oxide (NO), total supperoxide dismutase activity (T-SOD) and anti-superoxide anion ability(ASAA) in spleen were measured. The results showed that that JMME had a significant effect on immune regulation and antioxidation in mice. HC50, IgM, T-AOC, LZM, T-SOD and ASAA were increased in the model-dosed group, MDA and NO levels were decreased in the model-dosed group. In conclusion, the present study demonstrated that JMME exhibited immune regulation and antioxidation effects comprising the down-regulation of MDA as well as NO levels and the up-regulation of HC50 and IgM T-AOC, LZM, T-SOD and ASAA in mice.

1. BACKGROUD

The Juglans mandshurica maxim JMM is a fast-growing deciduous tree that is widely distributed in China, Siberia and Korea[1]. It has been used as a folk medicine plant for the treatment of esophageal, gastric, cardiac, and lung cancer[1-3]. Its primary chemical constituents include juglone, b-sitosterol, naphthoquinones, naphthalenyl glycosides, a-tertalonyl glucopyranosides, flavonoids, and diarylheptanoyl glucopyranosides[4]. Additionally, JMM, has been shown to have anti-tumor, anti-oxidative, antibacterial, and anti-inflammatory effects[3, 5, 6]. The extracts of juglans mandshurica maxim(JMME) has significant lipoxygenase inhibitory activity and Oxy-radical inhibition[7]. JMME has a protective effect to the immune cells in irradiated mice[8].This paper studies the effects of JMME on mice with different immune status and the antioxidant capacity to explore whether there are correlations between the immune-modulatory effects of JMME and the body's free radical reactions. To this purpose we adopted dexamethasone sodium phosphate injection to establish low immunity model and built a mice immune stimulation model induced by sheep red blood cells.

2. MATERIALS AND METHODS

2.1. Materials

All animals were approved by the Institutional Animal Experiments Committee. Adult KM mice (20 ± 2 g body weight, all male, SPF level) were purchased from Department of Laboratory 90Animal Science of Peking University Health Science center (Beijing, China). Animals were maintained under 12 h light/dark cycles at a temperature of approximately 24 ± 1°C with food and water ad libitum. JMME was extracted from Juglans mandshurica Maxim in the Pharmacology Lab. Coomassie brilliant blue protein assay kit, LZM detection kit, malondialdehyde assay kit (thiobarbituric acid, TBA method), T-AOC kit, reduced GSH detection kit, NO test kit, T-SOD kit and anti-superoxide anion radicals test kit were purchased from Nanjing Jiancheng Bioengineering Institute.

2.2. Effect of JMME on SRBC Induced Generation of Hemolysin Antibody

The 32 mice were divided into 4 groups according to their weight randomly and there were 8 mice in each group. They were the Normal control group, Normal JMME group, SRBC model control group and SRBC model JMME group. Normal JMME group and SRBC model JMME group were given JMME (1g·kg-1·d-1, 0.5% sodium carboxymethyl cellulose preparation) for 7 days. Normal control group and SRBC model control group were administered intragastrically with 0.5% sodium carboxymethyl cellulose (1g·kg-1·d-1) for 7 days. SRBC model control group and SRBC model JMME group were administered intraperitoneal injection of 2% (V / V, prepared with saline) SRBC 0.2 ml on the third day.

2.3. Determination of Half Hemolytic Value (HC_{50}) of Serum

0.2 mL blood per mice was obtained by the orbital blooding on the seventh day and then they were placed about 1h. Serum was collected after 2000rpm centrifugation. Using PBS buffer serum was diluted for 400 times by PBS buffer. Adding serum in an 1mL EP tube, followed by adding 0.5 mL 10% (prepared with PBS buffer, v/v) SRBC volume, 1 mL (complement with PBS at 1:10 dilution). Control tubes were added PBS instead of serum. The constant temperature of 37°C water bath heat for 30 min and then using ice bath to terminate the reaction. Supernatant were obtained by centrifugal 2000r/m for 10min. Adding 3 mL doushi reagent in 1mL supernatant. The absorbance value were measured by spectrophotometer (λ=540 nm). HC50= sample absorbance value / (SRBC half hemolysis×400).

2.4. Induction and Detection of IgM Antibody in Spleen Cells

Spleen cell suspension was prepared (1×1010mL-1), 0.5mL spleen cell suspension, 0.2%SRBC and0.1mL guinea pig serum were added in each tube. 37°C water bath for 1 h, 3000r/m centrifuge for 5min. The absorbance value were measured by spectrophotometer(λ=413 nm).

2.5. Effects of JMME on Immune Organ Index and Antioxidant Capacity in Mice

The 32 mice were divided into 4 groups according to their weight randomly and there were 8 mice in each group. They were the Normal control group, Normal JMME group, Low-immunity control group and Low-immunity JMME group. The dexamethasone sodium phosphate by intraperitoneal injection (100mg·kg-1·d-1) for 7 days and immunosuppressed mice model was established. Normal JMME group and Low-immunity JMME group were administered intragastrically with JMME (1g·kg-1·d-1) from the eighth day. Normal control group and Low-immunity control group were given by 0.5% CMC-Na. Continuous administration for 10 days, the mice were measured and recorded before administration. On the eleventh day, all the mice were weighed. Blood were obtained by the orbital blooding and the spleen and thymus were taken and weighed.

2.6. Immune Organ Index Detection in Mice

The thymus and spleens were rapidly removed and washed by cold saline. The calculation formula is as follows: Thymus (spleen) index = thymus (spleen) weight (g) / body weight (100g).

2.7. Detection and Method of Oxidation Resistance Index

All the indexes were measured by using their own assay kits respectively.

3. RESULTS

3.1. Effect of JMME on The HC_{50} and IgM Antibodies in The Serum of Mice

As shown in Table 1, Compared with the normal control group, HC_{50} significantly increased ($P < 0.01$), spleen cell IgM antibody levels were slightly increased in the normal JMME group. Compared with SRBC model control group, HC_{50} and spleen cell IgM antibody level significantly improve ($P < 0.01$) in the SRBC model JMME group.

Table 1. The effect of JMME on HC_{50} and IgM of laboratory mice ($\overline{x}$ ±s, n=8)

group	HC50	IgM
Normal control group	95.8±7.9	0.489±0.036
Normal JMME group	128.7±12.3**	0.502±0.048
SRBC model control group	83.2±8.7*	0.428±0.048*
SRBC model JMME group	143.6±14.7**##	0.531±0.057#

Note: Compared with normal control group, * P<0.05, **P<0.01; compared with Low-immunity Model group, #P<0.05, ##P<0.01

3.2. Effect of JMME on The Immune Organ Index of Mice

The result is shown in Table 2, compared with normal control group, thymus index and spleen index of low-immunity control group were significantly decreased (P < 0.01), the thymus index and spleen index of normal JMME group although improved, but the difference was not statistically significant. Compared with model group, the thymus index and spleen index of model group were slightly improved, but there was no significant difference between the model group and the model-dosed group.

Table 2. The effect of JMME on immune organ of laboratory mice ($\overline{x}$ ±s, n=8)

Groups	immune organ index (g/100g weight)	
	Thymus index	Spleen index
Normal control group	0.373±0.044	0.397±0.038
Normal JMME group	0.377±0.032	0.525±0.017
Low-immunity control group	0.096±0.021**	0.244±0.042**
Low-immunity JMME group	0.097±0.028**	0.256±0.021**

Note: Compared with normal control group, * P<0.05, **P<0.01

3.3. Effect of JMME on The Contents of LZM, MDA and T-AOC in Serum of Mice

As shown in Table 3, compared with the normal control group, the T-AOC significantly reduced (P < 0.01), MDA increased significantly (P < 0.01) in the serum of model group. MDA significantly reduced (P < 0.01), while LZM (P < 0.05) and T-AOC (P < 0.01) increased significantly in Normal JMME group. Compared with model group, the content of MDA in serum of model JMME group was significantly lower (P<0.01), while LZM and T-AOC were significantly higher (P<0.01).

Table 3. The effect of JMME on the content of LZM, MDA, T-AOC in serum ($\overline{x}$ ±s,n=8)

Groups	LZM(μg/mL)	MDA(nmol/mL)	T-AOC(U/mgprot)
Normal control group	0.690±0.263	18.342±1.240	7.914±1.058
Normal JMME group	2.894±0.790*	6.891±0.834**	11.959±1.504**
Low-immunity control group	2.130±0.857*	22.257±1.260**	6.681±0.877*
Low-immunity JMME group	12.546±2.163**##	7.456±0.383**##	20.843±2.416**##

Note: Compared with normal control group, * P<0.05, **P<0.01; compared with Low-immunity Model group, #P<0.05, ##P<0.01.

3.4. Effect of JMME on The Content of GSH in Thymus of Mice

Table 4 shows the result, compared with the normal control group, the model group was significantly lower GSH (P<0.05), the Low-immunity model was successfully established, and the GSH content in the thymus of the normal group was significantly increased (P<0.05). Compared with the model group, the content of GSH in thymus of model mice increased slightly, but there was no significant difference between the two groups.

Table 4. The effect of JMME on the content of GSH in thymus ($\bar{x}$ ±s, n=8)

Groups	GSH (mg/gprot)
Normal control group	44.837±2.754
Normal JMME group	49.958±1.366*
Low-immunity control group	18.493±1.946**
Low-immunity JMME group	20.845±1.572**

Note: Compared with normal control group, * P<0.05, ** P<0.01

3.5. Effect of JMME on The Contents of NO, T-SOD and Anti-O^{2-} in Spleen of Mice

As shown in Table 5, compared with the normal control group, the content of no NO in spleen in Low-immunity control group increased significantly (P < 0.01), T-SOD content was significantly decreased (P < 0.05). NO in normal JMME group decreased significantly (P < 0.05), while anti-O^{2-} significantly increased (P < 0.01). Compared with Low-immunity control group, the content of NO in spleen of low-immunity JMME group was significantly decreased (P<0.01), T-SOD was significantly higher (P<0.01), and anti-superoxide anion ability (ASAA) was significantly higher (P<0.05).

Table 5. The effect of JMME on the content of NO, T-SOD, anti-O2- in Spleen($\bar{x}$ ±s,n=8)

Groups	NO(mol/gprot)	T-SOD (U/mgprot)	ASAA (U/mgprot)
Normal control group	0.647±0.090	28.381±2.345	119.870±7.141
Normal JMME group	0.422±0.162*	26.248±1.086	153.477±8.245**
Low-immunity control group	6.901±0.645**	20.196±2.348*	123.508±5.277
Low-immunity JMME group	0.378±0.124*##	38.697±3.110**##	148.196±15.412#

Note: Compared with normal control group, * P<0.05, **P<0.01; compared with Low-immunity Model group, # P<0.05, ## P<0.01.

4. CONCLUSIONS

The results showed that that JMME had a significant effect on immune regulation and antioxidation in mice. HC50, IgM, T-AOC, LZM, T-SOD and ASAA were increased in the model-dosed group, MDA and NO levels were decreased in the model-dosed group. In conclusion, the present study demonstrated that JMME exhibited immune regulation and antioxidation effects comprising the down-regulation of MDA as well as NO levels and the up-regulation of HC50 and IgM T-AOC, LZM, T-SOD and ASAA in mice.

In conclusion, JMME had significant immunomodulatory and antioxidant effect on mice by enhancing the hemolysin, antibody level and lysozyme activity and increasing the immunity. When body was immunosuppressed, JMME may through eliminate free radicals and reactive oxygen species, reduce peroxide in living cells, reduce lipid peroxidation, and through improve the antioxidant capacity of the body, thereby improving the body immunity.

Acknowledgments

The authors would like to express the thanks to all the fund support of pharmacology laboratory of Beijing Institute of Technology.

References

1. Son J K. Isolation and structure determination of a new tetralone glucoside from the roots of Juglans mandshurica[J]. Archives of Pharmacal Research, 1995, 18(3):203-205.
2. J. Guo, L. Cui, S. Li, L. Liu, R. He, B. Liu. Inhibitory effects and immunoregulation of Juglans mandshurica maxim extract on S-(180) mouse sarcoma, *Journal of Jilin University (Medicine Edition)* 2007; **2**: 286–289.
3. C. Wang, Y. Bao, Y. Duan, Y. Luan, L. An. Study on anti-tumor of Juglans mandshurica. *Chinese Traditional Patent Medicine* 2003; **8**: 285–286.
4. C.L.S. Chuan-Ling Si, Y.S.B. Young-Soo Bae. Antioxidant effect of Juglans mandshurica bark gallotannins, *Journal of Forest Science* 2007; **23**: 1–4.
5. L. Yan, L. Jing. Studies on anti-oxidation effects of alcohol extract of Juglans Mandshurica leaves, *IEEE Reviews in Biomedical Engineering* 2011; 463–465.
6. S. Xu, Q. Zhang, W. Song, Y. Meng, Q. Xu, Effects of ethanol extract of the walnut leaf on anti-inflammatory and index of immune organ in mice, *Journal of Medical Science Yanbian University* 2008;**3**: 285.
7. Tran Minh N, Tran Manh H, Phuong Thien T, et al. Antioxidative activities of galloyl glucopyranosides from the stem-bark of Juglans mandshurica. *Bioscience Biotechnology & Biochemistry*, 2008; **8** :2158-63.
8. Yang S, Zhao X Y, Jing L I, et al. Effects of juglans mandshurica maxim on immune function of ionizing irradiated mice. Journal of Jilin University, 2008;10:542-846.

THE BMSPATZLE5 GENE ENCODES A COMPONENT OF TOLL-SIGNALING PATHWAY AGAINST FUNGUS AND GRAM-POSITIVE BACTERIA INVASION IN THE INTEGUMENT OF SILKWORM, *BOMBYX MORI*

Ji Liu[a], Jie Yang, Xiaonan Wang, Yang Zhang, Zhongyuan Shen, and Jinmei Wu[b,*]

[1]*College of Biotechnology, Jiangsu University of Science and Technology, Zhenjiang, 212018, China*

[2]*The Sericultural Research Institute, Chinese Academy of Agricultural Sciences, Zhenjiang, 212018, China*

E-mail: [a]1141756029@qq.com, [b]jwuus@hotmail.com

[*]*Corresponding author*

Spatzle (spz) plays an important role in the innate immunity of insects. In this study, the Open Reading Frame(ORF)of Bmspz5 gene and its 3 'end non-coding region were cloned from Bombyx mori (B. mori). The gene encodes a protein of 405 amino acids, of which 56 amino acids with the C-terminal fragment were shown to be different from the corresponding predicted ORF in silkworm genome database. Phylogenetic tree and sequence analyses showed that Bmspz5 has a common conserved Cysteine Knot region and is closely associated with spz5 of several other insects. The expression of Bmspz5 was upregulated in the integument 24 h after Yeast and Bacillus subtilis invasion, In addition, RNAi knockdown of Bmspz5 caused an obvious reduction in the transcription expression of the transcription factors Cactus, Rel and in the antimicrobial peptides Attacin, Moricin, and Gloverin in the integument of B. mori. These results indicate that Bmspz5 participated in the Toll signal transduction pathway in the integument of B. Mori.

1. INTRODUCTION

Innate and acquired immune system are the two major components of multicellular organism defense system. However, contrary to the vertebrates, insects have no acquired immunity, and to resist the invasion of external microbes, they can only rely on their own natural immunity[1]. In the innate immune response, a group of germline-encoded pattern recognition receptors (PRR) can recognize and bind to the conserved pathogen-associated molecular pattern (PAMP) in bacteria and fungi when microbes invade[2]. In response to microbial infection, the host activates multiple signaling pathways to induce the expression of immune effector molecules (various transcription factors and antimicrobial peptides), in which the main signaling pathway has Toll, IMD, and JAK / STAT pathways.

Toll signaling pathway is the most studied signaling pathway. The Toll pathway in Drosophila and the interleukin-1 receptor (IL-1R)-NF-kB pathway in mammals are homologous signal transduction pathways that mediate several different biological responses[3]. The Drosophila spz protein is involved in the embryonic development[4] and the immune response[5]. In the early embryonic development of Drosophila, the serine protease Easter cleaves proSpatzle into active spz[6, 7]. Fungal and gram-positive bacteria infections can also activate spz. Active spz binds to and activates the transmembrane receptor Toll, causing cytoplasmic Toll / interleukin-1 receptor dimerization and autophosphorylation. The phosphorylated dimer is then combined with the protein Tube and cytoplasmic serine / threonine kinase Pelle. The activation of Pelle controls the degradation of Cactus protein. Once Cactus response signal is degraded, Dorsal is free to move into the nucleus to regulate the transcription of specific target genes, thus regulating the expression of AMP genes.

B. mori spz family have six members amongst which only Bmspz1 has been studied. Upregulation of Bmspz1 can induce the expression of AMPs, indicating that Bmspz1 also known as Toll receptor ligands is involved in regulating the expression of AMPs in the immune response[8]. Notably, Bmspz1 full-length cDNA cannot induce the expression of AMPs, but truncated Bmspz1 can significantly increase the level of expression of AMPs, suggesting that only the activation of the spz1 cleavage can induce an immune response and not prospatzle1[9].

Integument as the first line of defense in an organism plays an important role in the immune response. The integument is not only the organism's physical barriers but also an important immune organ. Brey, et al. showed that lepidopteran larval integument mounts an immune response when challenged with live bacteria or their cell wall components[10]. Hence, the insect integument, which is often considered as an inert protective armor, is indeed actively participating in defense.

Whether other members of B. mori spz family have a similar immune function with Bmspz1, or the signaling pathway that involves Bmspz, or the molecular mechanism of spzs in the induction of AMP expression is unclear. In this study, Bmspz5 gene was cloned from B. mori. Its structure was analyzed using bioinformatics. A phylogenetic tree was constructed, and the results showed that Bmspz5 is similar to Drosophila spz5, has the same conserved regions, and thus presumably have similar biological functions to Drosophila spz5. Microbial induced expression and RNA interference show that Bmspz5 can be up-regulated in integument at 24 h by Yeast and Bacillus subtilis challenge, and the gene silencing of Bmspz5 can inhibit the expression of Toll pathway transcription factor Cactus, Rel and antimicrobial peptides Attain, Gloverin and Moricn in the integument. The results indicate that Bmspz5 plays a significant role in the innate immunity of silkworm.

2. MATERIALS AND METHODS

2.1. Cloning the Bmspz5 gene from B. mori

RNAiso^{TM} Plus kit was used to extract RNA from the integument of B. mori larvae. Using total RNA as a template, oligo dT as primers for reverse transcription reaction, first strand cDNA was synthesized using M-MLV reverse transcriptase. Using Bmspz5-F (5'-AGTTCCTTATGGGACGATGC-3') and Bmspz5-R (5'-GTTCCTCTTCCTCCGCTTCT-3') primers for the PCR reaction, the 25μl volume reaction system containing 2.0μl cDNA, 1.0μl Bmspz5-F, 1.0μl Bmspz5-R, 2.5μl 10 × PCR buffer, 4.0μl dNTP, 0.25μl Taq polymerase, and RNase Free ddH2O was employed. PCR reaction conditions were 95 °C 3.0min, 95 °C 30 s, 55 °C 30 s, 72 °C 1.0min for 30 cycles, and 72 °C 10 min. Using total RNA as a template, first strand cDNA was synthesized using 3'RACE adapter primer. Using GSP1 (5'-CAAGAAGCGGAGGAAGAGGA-3 ') and 3'RACE Outer Primer (5'-CGCGGATCCTCCACTAGTGATTTCACTATAGG-3') for the first round (1stPCR) reaction, a 25μl volume reaction system containing 2μl reverse transcription cDNA, 2μl 1×cDNA Dilution Buffer, 2.5μl 10×LA PCR Buffer II (Mg2 + Plus), 1μl 3'RACE Outer Primer (10μM), 1μlGSP1, 0.25μlTaKaRa LA Taq® (5 U / μl), and RNase Free ddH2O was employed. The reaction conditions were 95 °C 3.0min, 95 °C 30 s, 57 °C 30 s, 72 °C 1.0min for 20 cycles, 72 °C 10 min. The first PCR reaction failed to give the desired product, and then GSP2 (5'- GTTTCCGACCAGGACCACGAC-3 ') and 3'RACE Inner Primer (5'-TACCGTCGTTCCACTAGTGATTT-3') were used for a second round (2nd PCR) reaction. A 25μl volume reaction system containing 2μl of 1stPCR product, 4μl dNTP Mixture (2.5 mM each), 2.5μl 10 × LA PCR Buffer II (Mg2 + Plus), 1μl 3'RACE Inner Primer (10μM), 1μl GSP2, 0.25μl TaKaRa LA Taq® (5 U / μl), ands RNase Free ddH2O was employed at the following conditions: 95 °C 3.0min, 95 °C 30 s, 56 °C 30 s, 72 °C 1.0min for 30 cycles, 72 °C 10 min. PCR product was separated using 1% agarose gel electrophoresis, and after gel DNA recovery, it was then connected to PMD18-T vector and transformed into competent E. coli to be amplified and sent to Shanghai Sangon Biotech for sequencing.

2.2. Bioinformatics Analysis of Bmspz5

SignalP (http://www.cbs.dtu.dk/services/SignalP/) was used to find signal peptide online, and TMHMM (http://www.cbs.dtu.dk/services/TMHMM-2.0/) was used to predict the cross membrane structure of the protein. ProtScale (http://us.expasy.org/cgi-bin/protscale.pl) was used for hydrophobicity analysis while SMART (http://smart.embl-heidelberg.de/) was used for protein domain prediction. ClustalX1.83 was used for Protein sequence comparison and conserved sequence alignment, and then MEGA6.06 was used for phylogenetic tree construction.

2.3. Expression of Bmspz5 in different Tissues

Semi-quantitative RT-PCR of the mRNA extracted from the head, integument, midgut, fat body, Malpighian tubules, testis, ovary, and silk gland of fifth-instar day 3 Bombyx mori larvae was conducted. eTIF4A-F (5'-TGGTGGGCACTCCAGGTCGT-3 ') and eTIF4A-R (5'-AGCCAATCCACCTTGCGACGG-3') as internal reference primers, and Bmspz5-F (5'-ACACCAAGTTTCACACCCGA-3 ') and Bmspz5-R (5'-CTTCCTCCGCTTCTTGAGGG- 3 ') as primers were used for semi-quantitative PCR analysis. The PCR method was the same as 2.1.

2.4. Induction of the Expression of Bmspz5 by Microorganisms

Semi-quantitative RT-PCR of the mRNA extracted from the head, integument, midgut, and fat body of the fifth-instar day 3 Bombyx mori larvae was conducted 24 h after they were injected with inactivated yeast (5.6 × 10^9cells / ml, 10μl), Escherichia coli (2.9 × 10^9cells / ml, 10μl), Bacillus subtilis (7.2 × 10^8cells / ml, 10μl) and sterile water (10μl). First strand cDNA was synthesized by reverse transcription. The PCR method was the same as 2.1.

2.5. RNA Interference

RNA was extracted from the integument of the fifth-instar day 3 B. mori larvae. Two pairs of primers, Bmspz5-T7-F/Bmspz5-R and Bmspz5-F/Bmspz5-T7-R were used for PCR respectively. The reaction conditions were 94 °C 3.0 min, 94 °C 30 s, 55 °C 30 s, 72 °C 1.0 min for10 cycles, 94 °C 30 s, 60 °C 30 s, 72 °C 1.0 min for 35 cycles, 72 °C 10 min. PCR products were purified and used as templates for the synthesis of double-stranded RNA. T7 RiboMAX Express RNAi System (P1700, Promega Corp., Madison, WI) was used to synthesize dsRNA.

Table 1. The primer pairs of transcription factor and AMPs

Primer name	Oligonucleotide sequence(5'-3 ')
BmAttacin-F	GGCTCGCTCTAGACAACGTA
BmAttacin-R	ACATTGGGCTCCCACGAAG
BmMoricin-F	GCAAAAACAGTAAACCGCGCA
BmGloverin-F	TACTCGATCAGCGGGCAATC
BmGloverin-R	GACCATACCACCGGCAGAAA
BmCactus-F	TAGGATGTTGGTCATCGCCG
BmCactus-R	CAGACTCTCGTTGTCGCTGT
BmRel-F	CAGATGCGAACGATACCGGA
BmRel-R	TCGCTTGTGGAGAGCTTGTT
BmRelish-F	TACAAAGTCGGGTTCGGTGG
BmRelish-R	TGAAACTCCCGTTAGCCGTC

Newly exuviated (day 0) fifth instar Bombyx mori larvae were divided into eight groups including four experimental groups and four control groups. Experimental groups were injected with 5.0 μg ds-RNA specific for Bmspz5 and 6h later with yeast (5.6 × 10^9cells / ml, 5μl), Escherichia coli (2.9 × 10^9cells / ml, 5μl), Bacillus subtilis

(7.2 × 10^8cells / ml, 5μl) and phosphate-buffered saline PBS (5μl). In contrast, control groups were injected with 5.0 μl PBS and 6h later with the same yeast, Escherichia coli, Bacillus subtilis and PBS. mRNA was extracted from the integument 6h after the injection of microorganisms. Using eTIF4A-F/eTIF4A-R as internal reference primers. The primer pairs of transcription factor and AMPs (Table 1) were used respectively for Semiquantitative RT-PCR[11]. The semiquantitative RT-PCR method was the same as that used in 2.4.

3. RESULTS

3.1. Nucleic Acid and Amino Acid Sequence Analysis of Bmspz5

ORF of Bmspz5 was deposited in GenBank under accession number KT943515. After cloning Bmspz5, the nucleic acid sequence was located on chromosome 7 of scaffold3055 that have only one copy in the genome, and a total of six exons and five introns, exon /intron boundary, were in line with GT-AG rules. It encodes 405 amino acids (Fig.1) with a protein molecular weight of 45.93 kD and an isoelectric point of 8.96. There was no N terminal signal peptide predicted, nor transmembrane structure as extracellular proteins; hydrophobic and hydrophilic regions were staggered.

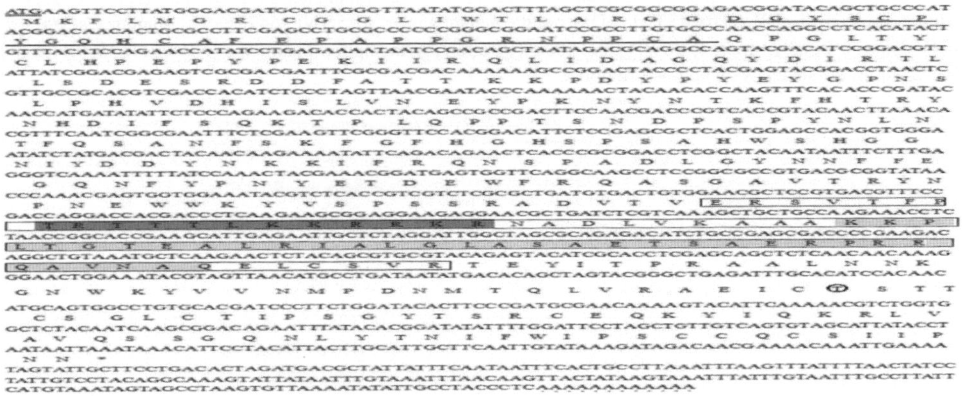

Fig. 1. Nucleic acid and amino acid sequence of Bmspz5.

The initiation codon is double underlined, the star is the stop codon, and the broken line is ployA tail. The N-terminal EGF-like domain is underlined; the Ribosomal_L31e domain is boxed; the Low complexity region is marked dark gray; the helix_turn_helix, lux Regulon (HTH_LUXR) domain is light gray; the wavy line is AWS domain. 56 amino acids with C-terminal fragment more than the predicted ORF in silkworm genome database. The last amino acid of the predicted ORF is indicated by a circle.

NCBI BLAST online comparison found that Bmspz5 protein sequences have high homology with spz5 from Drosophila, Tribolium castaneum, Papilio xuthus, Anopheles gambiae, Aedes aegypti, Acyrthosiphon pisum, Apis mellifera, Manduca sexta, and Apis cerana. Sequence alignment showed that it has a highly conserved C-terminal seven cysteine domains (Fig.2). A phylogenetic tree (Fig. 3) was constructed using the neighbor joining method with a 1,000 bootstrap test based on the multiple alignments of Bmspz5 and spz5s from Drosophila, Tribolium castaneum, Papilio xuthus, Anopheles gambiae, Aedes aegypti, Acyrthosiphon pisum, Apis mellifera, Manduca sexta, and Apis cerana. The result showed that Bmspz5 was most similar to Papilio xuthus spz5 and relatively close to Drosophila spz5 and Manduca sexta spz5.

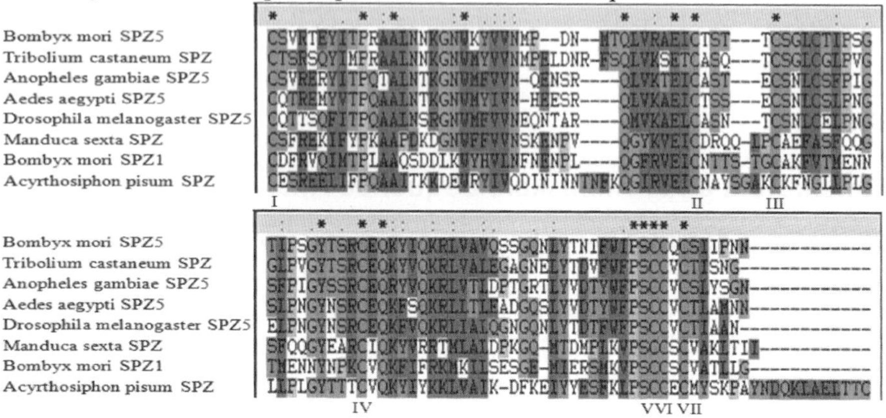

Fig. 2. The conserved amino acid sequence of the Bombyx mori and other insects' spz.

Bombyx mori spz5 and Tribolium castaneum (XP_008193940.1), Anopheles gambiae spz5 (XP_308593.4), Aedes aegypti spz5 (XP_001654338.1), Drosophila melanogaster spz5 (NP_647753.1), Manduca sexta spz (ACU68553.1), Bombyx mori spz1(NP_001108066.1), and Acyrthosiphon pisum spz (NP_001153589.1) were aligned within the seven cysteine I-VII region.

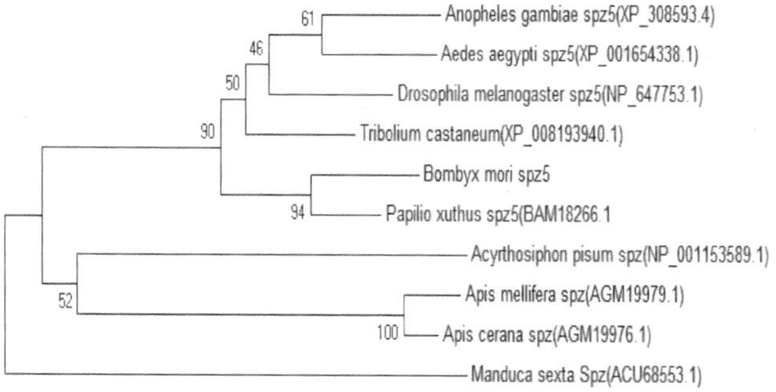

Fig.3. Phylogenetic tree of Bmspz5 with spzs from other insects.

3.2. Tissue Expression of Bmspz5

Semi-quantitative RT-PCR results showed that Bmspz5 gene expression was different in various tissues of the fifth-instar day 3 larvae. The highest expression was found in the head followed by the integument; there was virtually no expression found in other tissues (Fig. 4). This result is consistent with the Bmspz5 gene expression profile in the silkworm microarray database.

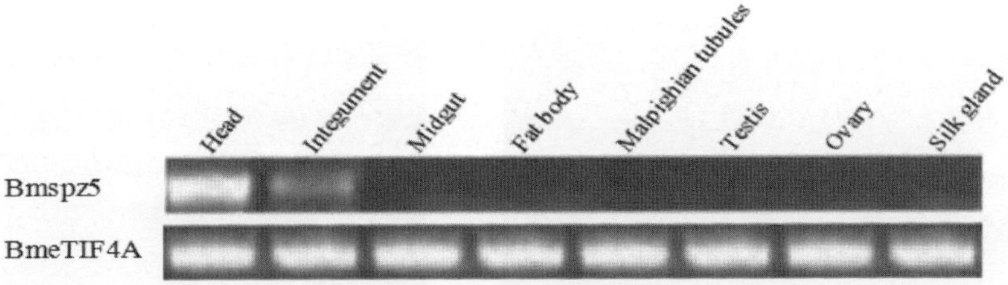

Fig. 4. Bmspz5 gene expression profile in the fifth-instar day 3 larvae.

3.3. Bmspz5 Was Upregulated by Microbial Induction

Bmspz5 was upregulated in the integument 24h after yeast and Bacillus subtilis injection but presented no changes after the injection of E. coli (Fig. 5). This was attributed to spz5, which participated in the immune response against fungi and gram-positive bacteria invasion.

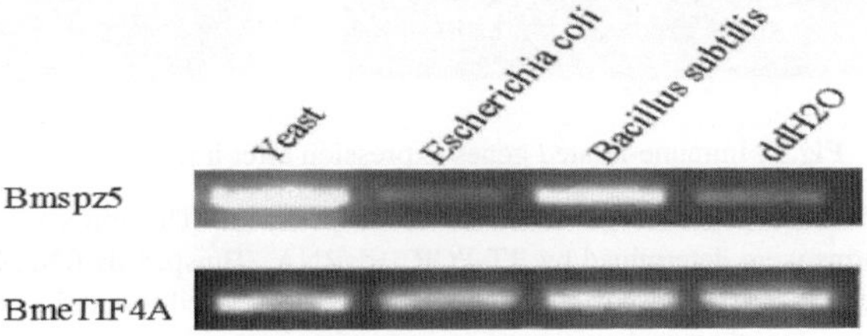

Fig.5. Bmspz5 gene expression in the integument of the fifth-instar day 3 larvae after microbial injection.

3.4. Interference of Bmspz5 Gene Resulted in the Downregulation of Transcription Factor Cactus, Rel and AMPs Attain, Gloverin and Moricin in the Integument

Experimental results showed that AMPs Attain, Moricin and Gloverin could be detected in the integument after injection of PBS. The injection of dsRNA specific for Bmspz5 did not down-regulate the Bmspz5 expression (Fig. 6D). After injection of yeast, silencing Bmspz5 expression could inhibit the expression of the transcription factor Cactus, Rel, and AMPs Attain, Gloverin and Moricin (Fig. 6A). Also, a similar result was obtained after the injection of bacillus (Fig.6C). However, after the injection of E. coli, silencing Bmspz5 expression could not inhibit the expression of these genes (Fig. 6B). From the results, the yeast had the highest knockdown efficiency, followed by the bacillus; E. Coli had no knockdown efficiency.

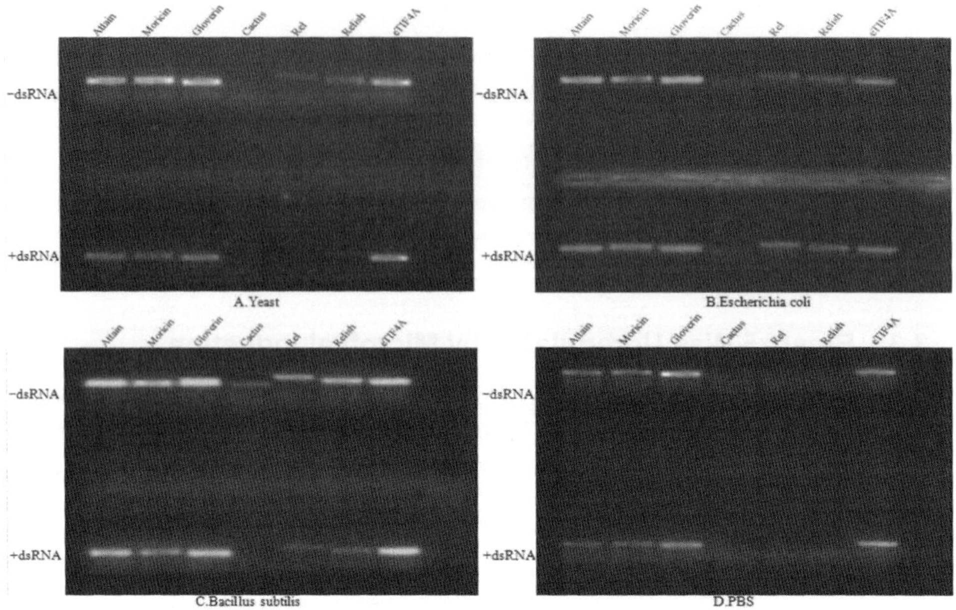

Fig. 6. Immune-related genes expression after injection of dsRNA.

Newly exuviated fifth-instar larvae were used for the experiment. The mRNA levels of the genes in the integument of silkworms were determined by RT-PCR. +dsRNA, (Bmspz5 ds-RNA injection) or –dsRNA (no Bmspz5 ds-RNA injection) before yeast (A), Escherichia coli (B), Bacillus subtilis (C) and PBS (D) injection. eTIF4A was used as an internal control gene.

4. DISCUSSION

In this study, we found that the expression of Bmspz5 was higher in the head and integument than in the other tissues, and plays an important role in innate immunity. Yeast and Bacillus subtilis infection can induce the expression of Bmspz5. RNA interference experiments show that Bmspz5 can regulate the expression of antimicrobial peptides in the integument, and may be involved in Toll signaling pathway.

In Drosophila, seven AMPs with different spectra of activities were synthesized by the fat body during the immune response and secreted into the hemolymph[12]. Various microorganisms activate different signaling pathways to produce different AMPs. Several AMP genes such as Attacin[13], Moricin[14] and Gloverin[15] have been identified from B. mori. In this study, it was discovered that yeast and bacillus injection could upregulate the expressions of Bmspz5 and AMPs Attacin, Moricin, and Gloverin. However, injection of E. coli could not upregulate Bmspz5 expression. However, silencing of Bmspz5 expression inhibited the expression of the AMPs when yeast and Bacillus subtilis invaded instead of E. Coil. These results indicate that Bmspz5 participated in the immune response induced by fungi and gram-positive bacteria, which is similar to Drosophila spz5.

In Drosophila, Cactus and Rel is the transcription factor in Toll pathway, and Relish is the transcription factor in IMD pathway[1]. BmCactus was involved in the Toll signal transduction pathway in B. Mori[16]. In this study, injection of yeast and Bacillus subtilis upregulated the expression of Cactus and Rel but silencing of Bmspz5 expression inhibited the expression of Cactus and Rel. However, E. coil injection failed to induce upregulation of Cactus and Rel expression, and silencing Bmspz5 expression could not inhibit the expression of Cactus and Rel. These results suggest that Bmspz5 played a role in the expression of Cactus and Rel induced by fungi and gram-positive bacteria.

In Drosophila, AMPs expression was induced in surface epithelia during the local response through imd pathway[12,17], and was regulated through the Toll pathway during the systemic response. In our experiment, the Toll pathway was found in the integument after yeast and bacillus were injected into the body of the silkworm, but the imd pathway existed in the surface epithelia in response to pathogens in Drosophila.

In conclusion, the Bmspz5 expression could be induced and upregulated in the integument after fungi and gram-positive bacterial invasion. The induced Bmspz5 could in turn induce the expression of AMPs, and participate in Toll signaling pathway in the integument. This study demonstrates that Bmspz5 was involved in the Toll signaling pathway present in the Silkworm Integument.

Acknowledgments

This work was supported by the National Natural Science Foundation of China (31272508) to J. Wu.

References

1. Hoffmann JA. The immune response of Drosophila. Nature. Nov 6 2003;426(6962):33-38.
2. Janeway CA, Jr. Pillars article: approaching the asymptote? Evolution and revolution in immunology. Cold spring harb symp quant biol. 1989. 54: 1-13. *Journal of immunology (Baltimore, Md : 1950).* Nov 1 2013;**191**(9):4475-4487.
3. Belvin MP, Anderson KV. A conserved signaling pathway: the Drosophila toll-dorsal pathway. Annual review of cell and developmental biology. 1996;12:393-416.
4. Halfon MS, Hashimoto C, Keshishian H. The Drosophila toll gene functions zygotically and is necessary for proper motoneuron and muscle development. *Developmental biology.* May 1995;**169**(1):151-167.
5. Lemaitre B, Nicolas E, Michaut L, Reichhart JM, Hoffmann JA. The dorsoventral regulatory gene cassette spatzle/Toll/cactus controls the potent antifungal response in Drosophila adults. *Cell.* Sep 20 1996;**86**(6):973-983.
6. Chasan R, Anderson KV. The role of easter, an apparent serine protease, in organizing the dorsal-ventral pattern of the Drosophila embryo. *Cell.* Feb 10 1989;**56**(3):391-400.
7. Weber AN, Tauszig-Delamasure S, Hoffmann JA, et al. Binding of the Drosophila cytokine Spatzle to Toll is direct and establishes signaling. *Nature immunology.* Aug 2003;**4**(8):794-800.
8. Taniai K, Ishii T, Sugiyama M, Miyanoshita A, Yamakawa M. Nucleotide sequence of 5'-upstream region and expression of a silkworm gene encoding a new member of

the attacin family. *Biochemical and biophysical research communications.* Mar 27 1996;**220**(3):594-599.
9. Wang Y, Cheng T, Rayaprolu S, et al. Proteolytic activation of pro-spatzle is required for the induced transcription of antimicrobial peptide genes in lepidopteran insects. *Developmental and comparative immunology.* 2007;**31**(10):1002-1012.
10. Brey PT, Lee WJ, Yamakawa M, et al. Role of the integument in insect immunity: epicuticular abrasion and induction of cecropin synthesis in cuticular epithelial cells. *Proc Natl Acad Sci U S A.* Jul 1 1993;**90**(13):6275-6279.
11. Yang J, Wang X, Tang S, Shen Z, Wu J. Peptidoglycan recognition protein S2 from silkworm integument: characterization, microbe-induced expression, and involvement in the immune-deficiency pathway. *Journal of insect science (Online).* 2015;15.
12. Tzou P, Ohresser S, Ferrandon D, et al. Tissue-specific inducible expression of antimicrobial peptide genes in Drosophila surface epithelia. *Immunity.* Nov 2000;**13**(5):737-748.
13. Sugiyama M, Kuniyoshi H, Kotani E, et al. Characterization of a Bombyx mori cDNA encoding a novel member of the attacin family of insect antibacterial proteins. *Insect biochemistry and molecular biology.* Mar 1995;**25**(3):385-392.
14. Furukawa S, Tanaka H, Nakazawa H, Ishibashi J, Shono T, Yamakawa M. Inducible gene expression of moricin, a unique antibacterial peptide from the silkworm (Bombyx mori). *The Biochemical journal.* May 15 1999;**340** (Pt 1):265-271.
15. Mrinal N, Nagaraju J. Intron loss is associated with gain of function in the evolution of the gloverin family of antibacterial genes in Bombyx mori. *The Journal of biological chemistry.* Aug 22 2008;**283**(34):23376-23387.
16. Furukawa S, Tanaka H, Ishibashi J, Imanishi S, Yamakawa M. Functional characterization of a cactus homolog from the silkworm Bombyx mori. *Bioscience, biotechnology, and biochemistry.* Dec 2009;**73**(12):2665-2670.
17. Davis MM, Engstrom Y. Immune response in the barrier epithelia: lessons from the fruit fly Drosophila melanogaster. *Journal of innate immunity.* 2012;**4**(3):273-283.

THE STUDY OF COLD STIMULATION EFFECTING ON PRODUCTION PERFORMANCE OF CHINA-LINDIAN NATIVE CHICKENS

Changping Wang[a], Bo Wang[b], Tianlin Miao, Chengwu Han, Qingbo Zhou, Guangdong Cheng, Hongsheng Yang, and Bin Wang

College of life science, Jiamusi university,Jiamusi,154007,China

Jiamusi university animal husbandry and veterinary medicine, Jiamusi, 154007,China

E-mail: [a]wcp100200@163.com,[b]408081288@qq.com

448 China-Lindian Native Chickens were given the cold stimulation from from fourth weeks to sixth weeks. The three cold stimulation interval days levels were 0 day 1 day and 2 days. The three decrease temperature levels were temperature decreasing 3 °C, temperature decreasing 5 °C and temperature decreasing 7 °C. The three stimulation time levels were 1h 3h and 6h.we identified the influence of cold stimuli on welfare of CLNC through detecting production performance welfare indexs. The main production performance welfare index of detection were the average daily weight gained, the ratio of meat to feed. It was concluded from the study that interval days had no significant effect on the average daily weight and ratio of meat to feed($P>0.05$);decrease temperature had no significant effect on the average daily weight and ratio of meat to feed($P>0.05$);stimulation time had no significant effect on the average daily weight and ratio of meat to feed($P>0.05$).

1. INTRODUCTION

Exposure to low temperature provokes physiological responses related to oxygen consumption and induces changes in energy production and subsequent reactive oxygen species (ROS) production (Blagojevic, 2007)[1]. The acute cold stress reaction in rodents involves initial muscle shivering thermogenesis that is later replaced by a nonshivering state of the whole organism (Himms-Hagen, 1990)[2]. Cold temperature is closely associated with the ability of the broiler to produce heat.Extreme selection pressure towards either growth rate and feed conversion ratio (FCR) puts high demands on the metabolic processes in broilers (Decuypere et al., 2000)[3]. A reduction in environmental temperature from 20 to 2°C for example, almost doubles the oxygen requirement in White Leghorn hens (Gleeson, 1986)[4].Chicks on exposure to low environmental temperature also show increase in oxygen consumption (Mujahid and Furuse, unpublished observations). Cool temperatures are the primary triggers for ascites during commercial broiler production (Wideman, 2001)[5].If chicks are chilled before day 6, it may affect their metabolic rate for several weeks and increase the ascites. However, Malan et al. (2003) found that ascites can relate to a low metabolic rate[6].The main production performance welfare index of detection were thethe average daily weight gained, the ratio of meat to feed.The aim of the study was to determine the effects of cold stress on China-Lindian Native Chickens performance.

2. MATERIALS AND METHODS

In our study, there were four repeations in test groups, we conducted on three factors in our experiment. The three factors were interval days, decrease of temperature and stimulation time. The factor of interval days had three levels, which were interval of 0 day, interval of 1 day, interval of 2 days. The factor of decrease temperature had three levels, which were temperature decreasing 3 °C, temperature decreasing 5 °C, temperature decreasing 7 °C respectively. The factor of stimulation time had three levels, which were stimulating time of 1h, stimulating time of 3h, stimulating time of 6h.We used the artificial climate chamber for temperature control.

We weighed experimental chicken with empty stomach in the next morning 8:00, at the same time, we weighed the amount of the remaining diet. Average daily weight, daily feed intake and ratio of feed to meat were calculated.

The formula is as follows: Average daily weight(g/bird)=final weight-Initial weight/experimental days; ratio of feed to meat=Daily gain/daily feed intake.

3. RESULTS AND ANALYSIS

3.1. Effect of Cold Stimulation on Average Daily Weight of China-Lindian Native Chickens

3.1.1. *The Effects of Interval Days of Cold Stimulus on Average Daily Weight in China-Lindian Native Chickens*

Table 1. The effects of interval days of cold stimulus on average daily weight in China-Lindian Native Chickens (Unit:g/d)

interval days	average	difference significant	
		α=0.05	α=0.01
0 d	92.70±52.89	a	A
1 d	98.11±51.46	a	A
2 d	90.66±42.38	a	A
control group	107.87±75.21	a	A

As shown in Table 1, average daily weight on all experimental group was lower than the control group, the difference was not significant($P>0.05$).

Average daily weight on interval of 1 day was higher than interval of 0 day and 2 day in cold stimulus trial group, the difference was not significant($P>0.05$).

Through the variance analysis, interval days had no significant effect on the average daily weight ($P>0.05$).

3.1.2. *The Effects of Decreased Temperature of Cold Stimulus on Average Daily Weight in China-Lindian Native Chickens*

Table 2. The effects of decreased temperature of cold stimulus on average daily weight in China-Lindian Native Chickens (Unit:g/d)

decreased temperature	average	difference significant	
		α=0.05	α=0.01
3°C	91.20±50.22	a	A
5°C	93.28±48.29	a	A
7°C	96.99±48.99	a	A
control group	107.87±75.21	a	A

As shown in Table 2, average daily weight on all experimental group was lower than the control group, the difference was not significant (P>0.05). Average daily weight on decreased temperature 7°C was higher than decreased temperature 3°C and 5°C in cold stimulus trial group, the difference was not significant (P>0.05).

Through the variance analysis, decreased temperature had no significant effect on the average daily weight (P>0.05).

3.1.3. *The Effects of Cold Stimulation Time on Average Daily Weight in China-Lindian Native Chickens*

Table 3. The effects of time of cold stimulus on average daily weight in China-Lindian Native Chickens (Unit:g/d)

stimulation time	average	difference significant	
		α=0.05	α=0.01
1 h	97.60±55.42	a	A
3 h	92.78±46.46	a	A
6 h	91.10±44.94	a	A
control group	107.87±75.21	a	A

As shown in Table 3, average daily weight on all experimental group was lower than the control group, the difference was not significant (P>0.05).

Average daily weight on stimulation time 1h was higher than stimulation time 3h and 6h in cold stimulus trial group,the difference was not significant (P>0.05).

Through the variance analysis, cold stimulation time had no significant effect on the average daily weight (P>0.05).

3.2. Effect of Cold Stimulation on the Ratio of Meat to Feed of China-Lindian Native Chickens

3.2.1. *The Effects of Interval Days of Cold Stimulus on the Ratio of meat to Feed in China-Lindian Native Chickens*

As shown in Table 4, the ratio of meat to feed on all experimental group were lower than the control group, the difference was not significant (P>0.05).the ratio of meat to feed on interval of 1 day was higher than interval of 0 day and 2 day in cold stimulus trial group, the difference was not significant (P>0.05).Through the variance analysis, interval days had no significant effect on the ratio of meat to feed (P>0.05).

Table 4. The effects of interval days of cold stimulus on the ratio of meat to feed in China-Lindian Native Chickens

interval days	average	difference significant	
		α=0.05	α=0.01
0 d	0.45±0.14	a	A
1 d	0.48±0.17	a	A
2 d	0.47±0.15	a	A
control group	0.49±0.16	a	A

3.2.2. *The Effects of Decreased Temperature of Cold Stimulus on the Ratio of Meat to Feed in China-Lindian Native Chickens*

Table 5. The effects of decreased temperature on the ratio of meat to feed in China-lindian native chickens

decreased temperature	average	difference significant	
		α=0.05	α=0.01
3°C	0.45±0.16	a	A
5°C	0.46±0.09	a	A
7°C	0.47±0.11	a	A
control group	0.49±0.13	a	A

As shown in Table 5, the ratio of meat to feed on all experimental group were lower than the control group, the difference was not significant (P>0.05).The ratio of meat to feed on decreased temperature 7°C was higher than decreased temperature 3°C and 5°C in cold stimulus trial group, the difference was not significant (P>0.05).Through the variance analysis, decreased temperature had significant effect on the ratio of meat to feed (P>0.05).

3.2.3. *The Effects of Cold Stimulation Time on the Ratio of Meat to Feed in China-Lindian Native Chickens*

Table 6. The effects of the time of cold stimulus on the ratio of meat to feed in China-Lindian Native Chickens

stimulation time	average	difference significant	
		α=0.05	α=0.01
1 h	0.43±0.15	a	A
3 h	0.45±0.15	a	A
6 h	0.52±0.16	a	A
control group	0.49±0.16	a	A

As shown in Table 6, the ratio of meat to feed on all experimental group were lower than the control group, the difference was not significant(P>0.05).The ratio of meat to feed on stimulation time 6h was higher than stimulation time 1h and 3h in cold stimulus trial group, the difference was not significant(P>0.05).

Through the variance analysis, cold stimulation time had no significant effect on the the ratio of meat to feed (P>0.05).

4. DISCUSSION

The results show that average daily weight of cold stimulation training chickens were lower than the control group without low temperature stimulation. Interval days decreased temperature and cold stimulation time had no significant effect on average daily weight. The results of this study are consistent with the results of Yang Chengli et al. (2000),The study concluded that there was no significant difference in the weight gain between the experimental groups at initial temperature 30°C,33°Cand 35°C .

Deaton(1977)showed that the maximum weight gain of the test group was 21.1°Ccompared with 15.6 °C and 23.9 °C when we adopt constant temperature from 4 to 7 weeks. The results of this experiment show that interval days decreased temperature and cold stimulation time had no significant effect on the ratio of meat to feed(P>0.05).

The ratio of meat to feed of cold stimulation training chickens were higher than the control group without low temperature stimulation. The results of this study are consistent with the results of Deaton et al(1996) and Li yuchun(2001)[7][8].

Acknowledgments

This work was funded by Natural science foundation of China (31302008) and Natural science foundation of Heilongjiang province (QC2012C104).

References

1. Blagojevic, D.P.Antioxidant systems in supporting environmental and pro-grammed adaptations to low temperatures. *Cryo Letters* 2007;28: 137–150.
2. Himms-Hagen, J. Brown adipose tissue thermogenesis: interdisciplinary studies.*FASEB J* 1990;**4**:2890–2898.
3. Decuypere,E., J.Buyse and N.Buys. Ascites in broiler chickens:exogenous and endogenous structural and functional causal factors. *World's Poult.Sci* 2000; **56**(4):367–377.
4. Gleeson, M.. Respiratory adjustments of the unanaesthetized chicken, gallus domesticus, to elevated metabolism elicited by 2,4 dinitrophenol or cold exposure. Comp. *Biochem. Phys* 1986; 83a: 283–289.
5. Wideman, R.F.Pathophysiology of heart/lung disorders: pulmonary hypertension syndrome in broiler chickens. *World's Poult. Sci* 2001;**57**(3):289–307.

6. Malan, D. D., C. W. Scheele, J. Buyse, C. Kwakernaak, F. K. Siebrits, J. D. Van Der Klis and E. Decuypere. Metabolic rate and its relationship with ascites in chicken genotypes. *Br. Poult. Sci* 2003;**44**:309–315.
7. DEATON J W, BRANTON S L,et al. The effect of brooding temperature on broiler performance. *Poult Sci* 1996;**75**:1217–1220.
8. Chun-yu Li. Study on the effect of low temperature on broiler chickens.*master thesis: Northeast Agricultural University* 2001.(In Chinese).

Session 2

Medical Science

OBJECTIFIED STUDY ON TONGUE MANIFESTATION OF PATIENTS WITH ECZEMA

Zhifeng Yu[a], Haifang Zhang[b], Linjie Fu[c], and Xiaozuo Lu[d]

School of Chinese Medicine and Engineering, Tianjin University of Traditional Chinese Medicine, Tianjin, 300193, China

E-mail: [a]fengzhiyu86@hotmail.com, [b]zhanghf@163.com, [c]flinjie@163.com, [d]13702093979@163.com

Objective: To describe the characteristic of tongue images of patients with eczema of different types and to reveal the elemental rule on the changes of the tongue images. Methods: two hundred patients with eczema were recruited and divided into three groups according to the diagnostic criteria. Acute group had 47 patients, subacute group had 82 patients, and chronic group had 71 patients. The tongue parameters were detected by computerized tongue image digital analysis device, and the tongue images were described with qualitative, tongue color and quantitative analysis, respectively. The International Commission on Illumination (CIE) L*a*b* (CIELAB) color model was used for the quantitative classification. Results: There was a significant statistical difference between different types of eczema on tongue color, tongue shape, coating color, and thickness of tongue coating ($P<0.05$), and there was significant statistical difference between different types of eczema on Lab values of the tongue and coating ($P<0.05$). The acute eczema patients usually had pale or red tongue with yellow coating, while the subacute eczema patients usually had pale puffy tongue with teeth marks and the tongue coating was thin or thick with white color. Conclusion: The changes of tongue manifestation might be associated with the type of the patients with eczema. The computerized tongue image digital analysis instrument can objectively describe the tongue features of patients with eczema.

1. BACKGROUND

Inspection of the tongue is one of the four diagnostic methods in Chinese medicine (CM), which include looking, listening and smelling, asking, and touching. According to Traditional Chinese Medicine (TCM) theory, the tongue mirrors the viscera. It can not only diagnose the health status of the body but also objectively manifest the states of the disease. Recently there were some clinical reports demonstrate that tongue manifestation was closely associated with diseases[1-4]. We can observe disease progression, evaluate prognosis, definite clinical stages, and make syndrome differentiation by inspecting tongue manifestation[5-6].

However, observation diagnosis is often biased by subjective judgment, especially for tongue observation, which is more dependent on light sources, personal knowledge, experience, thinking patterns, and diagnostic skills. Moreover, there are no precise or existing quantifiable standards. Different practitioners may pass varying judgments on the same tongue, while a practitioner may even reach different diagnoses on the identical tongue if examined at different time. Therefore, it has become an important issue to sum up tongue diagnosis theory and clinical experience systematically and to establish objective standards for inspection of the tongue by using modern scientific methods, which can assist TCM practitioners to establish reliable diagnoses by providing them with standardized procedures as well as objective, reliable, and quantified data. In order to solve these problems, researchers have been developing methods and systems for computerized tongue diagnosis[7-8]. Among these means, computerized imaging technology is one of the latest methods for tongue diagnosis.

Eczema is a common hypersensitive and inflammatory dermatosis which may attack people of all ages, either occurring all over the body or being confined in certain parts. Clinically, it is classified into three types: acute eczema, subacute eczema and chronic eczema. In order to avoid the side effect of west medicine, Chinese medicine was widely used in the clinic. Because all the Chinese medicine treatment were based on syndrome differentiation, so how to make a correct diagnosis is very important. Inspection of tongue manifestation is a mainly used diagnostic method in clinical and plays an important role in syndrome differentiation. In this study, computerized tongue image digital analysis device was used to acquire tongue images and calculate the images. Tongue color and coating of different types of eczema patients were compared, in an attempt to establish a tongue diagnosis standard of eczema based on tongue images and analysis.

2. METHODS

2.1. Diagnostic Criteria

Inclusion criteria was referred to the "Standardized Diagnosis and Treatment of Chinese Medicine Disease and Syndrome"[9] set by the State Administration of Traditional Chinese Medicine, P. R. C.

2.2. Clinical Data and Grouping

From September 2011 to March 2012, a total of 200 eczema patients were recruited from the first Hospital Affiliated to Tianjin University of Traditional Chinese Medicine. There were 117 male and 83 female cases. The youngest was 8 years old, while the oldest was 87, with the mean age 54.7±16.9 years old.

The patients were divided into three groups according to diagnostic criteria, including acute eczema group (acute, 47 cases), subacute eczema group (subacute, 82 cases), and chronic eczema group (chronic, 71 cases).

2.3. Qualitative Indices

The tongue images were categorized by tongue color (pale tongue, pink tongue, red tongue, crimson tongue and blue purple tongue), tongue shape (size and teeth marks), coating color (white coating, light yellow coating and yellow coating), and texture of tongue coating (no coating, thin coating and thick coating).

2.4. Quantitative Indices

The computerized tongue image digital analysis instrument was supplied by the Department of Chinese Medicine and Engineering, Tianjin University of Traditional Chinese Medicine, which consisted of a Nikon D80 digital single-lens reflex camera (referred to as a digital camera Nikon D80), Nikon F-135 lens, laptops and tongue image analysis software.

The International Commission on Illumination (CIE) L*a*b* (CIELAB) color model was set up based on the standard color measurement formulated in 1931.In this three-dimensional model, "L" stands for the lightness of the color with a diffusion range of 0 (black) to 100 (white), "a" stands for the changes from redness to greenness, while "b" stands for the variation from yellowness to blueness.

2.5. Statistical Analysis

All data were analyzed by SPSS 11.5 software package for windows and adopted two-sided test at a common level of significance $\alpha=0.05$. Measurement data were expressed in the form of meanlstandard deviation. The statistical differences were accepted as $P<0.05$.

3. RESULTS

3.1. Overall Distribution of All Kinds of Tongue Manifestations

Groups of different tongue colors: pink was 11% (22/200), pale 43.5% (87/200), red 36% (72/200), crimson 5.5% (11/200), blue purple 4% (8/200); Groups of different tongue shape: enlarged tongue 49% (98/200), normal size tongue 33% (66/200), small size tongue 18% (36/200), teeth marks 51% (102/200); Groups of no coating 2.5% (5/200), thin coating 46.5% (93/200), group of thick coating 51% (102/200); Groups of different tongue coating colors: white 61% (122/200), light yellow 21.5% (43/200), yellow 17.5% (35/200).

3.2. Comparison of Tongue Body Colors in Different Eczema Types

Result is shown in Table 1. Among 87 patients of pale tongue, 15 had acute eczema, 31.91% of acute eczema group, while 39 patients with subacute eczema for 47.56% of subacute eczema group and 33 patients with chronic eczema for 46.48% of chronic group, showing significant difference between acute eczema group and the other two groups ($P<0.05$). Among 72 patients of red tongue, 21 had acute eczema, 44.68% of acute eczema group, while 26 patients with subacute eczema for 31.71% of subacute eczema group and 25 patients with chronic eczema for 35.21% of chronic group, showing significant difference between acute eczema group and the other two groups ($P<0.05$).

Table 1. Comparison of Tongue Body Colors in Different Eczema Types (Case,%)

Group	Case	Pale	Pink	Red	Crimson	Blue Purple
Acute	47	15(31.91)	2(4.88)	21(44.68)	6(12.77)	3(6.38)
Subacute	82	39(47.56)[a]	11(13.41)	26(31.71)[a]	3(3.66)	3(3.66)
Chronic	71	33(46.48)[a]	9(12.68)	25(35.21)[a]	2(2.82)	2(2.82)

Note: Table notes.

[a]Sample footnote: P<0.05, compared with acute

3.3. Comparison of Tongue Shape in Different Eczema Types

Result is shown in Table 2. Among 98 patients of enlarged tongue, 56 had subacute eczema, 68.29% of subacute eczema group, and 21 patients with acute eczema for 44.68% of acute eczema group, while 21 patients with chronic eczema for 29.58% of chronic group, showing significant difference between the former two groups and chronic eczema group (P<0.05). Among 102 patients of teeth marks tongue, 68 had subacute eczema, 82.93% of subacute eczema group, while 13 patients with acute eczema for 27.66% of acute eczema group and 21 patients with chronic eczema for 29.58% of chronic group, showing significant difference between subacute eczema group and the other two groups (P<0.05).

Table 2. Comparison of Tongue Shape in Different Eczema Types (Case,%)

Group	Case	Normal	Enlarged	Small	Teeth Mark
Acute	47	17(36.17)	21(44.68)[a]	9(19.15)	13(27.66)
Subacute	82	18(21.95)	56(68.29)[a]	8(9.76)	68(82.93)[b]
Chronic	71	31(43.66)	21(29.58)	19(26.76)	21(29.58)

Note: Table notes.

[a]Sample footnote: P<0.05, compared with chronic

[b]Sample footnote: P<0.05, compared with acute and chronic

3.4. Comparison of Tongue Coating Colors in Different Eczema Types

Result is shown in Table 3. Among 122 patients of white tongue coating, 3 had acute eczema, 6.38% of acute eczema group, while 63 patients with subacute eczema for 76.83% of subacute eczema group and 56 patients with chronic eczema for 78.87% of chronic group, showing significant difference between acute eczema group and the other two groups (P<0.05). Among 35 patients of yellow tongue coating, 32 had acute eczema, 68.09% of acute eczema group, while 2 patients with subacute eczema for 2.44% of subacute eczema group and 1 patient with chronic eczema for 1.41% of chronic group, showing significant difference between acute eczema group and the other two groups (P<0.05).

Table 3. Comparison of Tongue Coating Colors in Different Eczema Types (Case,%)

Group	Case	White	Light Yellow	Yellow
Acute	47	3(6.78)	12(25.53)	32(68.09)
Subacute	82	63(76.83)[a]	17(20.73)	2(2.44)[a]
Chronic	71	56(78.87)[a]	14(19.72)	1(1.41)[a]

Note: Table notes.

[a]Sample footnote: P<0.05, compared with acute

3.5. Comparison of Texture of Tongue Coating in Different Eczema Types

Result is shown in Table 4. Among 93 patients of thin coating, 42 had chronic eczema, 59.15% of chronic eczema group, while 16 patients with acute eczema for 34.04% of acute eczema group and 35 patients with subacute eczema for 42.68% of subacute group, showing significant difference between chronic eczema group and the other two groups ($P<0.05$). Among 102 patients of thick coating, 27 had chronic eczema, 38.03% of chronic eczema group, while 30 patients with acute eczema for 63.83% of acute eczema group and 45 patients with subacute eczema for 54.88% of subacute group, showing significant difference between chronic eczema group and the other two groups ($P<0.05$).

Table 4. Comparison of Texture of Tongue Coating in Different Eczema Types (Case,%)

Group	Case	No Coating	Thin	Thick
Acute	47	1(2.13)	16(34.04)[a]	30(63.83)[a]
Subacute	82	2(2.44)	35(42.68)[a]	45(54.88)[a]
Chronic	71	2(2.82)	14(59.15)	27(38.03)

Note: Table notes.

[a]Sample footnote: P<0.05, compared with chronic

3.6. Comparison of Lab Values of Tongue Body Colors in Different Eczema Types

The results in Table 5 indicated that tongue colors presented in different eczema types had different "Lab" values. "L" values of tongue presented in acute group showed significant difference compared with those in subacute group and chronic group ($P<0.05$). "a" values of tongue presented in acute group had significant difference with the other two groups ($P<0.05$).

Table 5. Comparison of Lab Values of Tongue Body Colors in Different Eczema Types ($\bar{x}\pm s$)

Group	Case	L	a	b
Acute	47	59.47±2.15	15.81±1.13	7.58±1.34
Subacute	82	67.38±2.31[a]	11.99±1.22[a]	6.99±1.45
Chronic	71	66.55±2.09[a]	12.68±1.17[a]	7.13±1.27

Note: Table notes.

[a]Sample footnote: P<0.05, compared with acute

3.7. Comparison of Lab Values of Tongue Coating Colors in Different Eczema Types

The results in Table 6 indicated that tongue coating colors presented in different eczema types had different "Lab" values. "b" values of tongue coating presented in acute group showed significant difference compared with those in subacute group and chronic group ($P<0.05$).

Table 6. Comparison of Lab Values of Tongue Coating Colors in Different Eczema Types ($\bar{x}\pm s$)

Group	Case	L	a	b
Acute	47	61.45±2.77	7.26±1.23	7.97±1.15
Subacute	82	69.26±2.91	7.52±1.09	5.44±1.09[a]
Chronic	71	67.81±2.51	8.03±1.36	5.60±1.29[a]

Note: Table notes.

[a]Sample footnote: $P<0.05$, compared with acute

4. DISCUSSION

This study enrolled 200 eczema patients, collected information on the tongue with computerized tongue image digital analysis system to ensure its objectivity. The results of this study showed that tongue color, coating color, texture of tongue, thickness of coating, and objective parameters in different types of eczema had differences, although the patients had the same disease. The results also showed that tongues of patients with different types had various features which were inherent to the pathogenesis of syndromes according to TCM theory.

This study demonstrated that most patients with acute eczema had pale or red tongue but with yellow coating which indicates damp heat as the main pathogenic factor, while the patients with subacute eczema usually had pale puffy tongue with teeth marks and the tongue coating was thin or thick with white color, which indicates spleen deficiency with dampness retention. For the chronic eczema, the tongue manifestation had some features like normal or puffy or small size of the tongue body, pale or red color of tongue body, white color of tongue coating, which shows blood deficiency may be involved during the development of the disease.

In this study, Lab values of the tongue were analyzed and showed significant difference in three eczema types. "L" and "a"values of tongue body colors presented in acute group showed significant difference compared with those in subacute group and chronic group. These results were in accordance with those of tongue body colors in different eczema types. "b" values of tongue coating presented in acute group showed significant difference compared with those in subacute group and chronic group. In L*a*b* model, "b" stands for the variation from yellowness to blueness, "b" value in acute eczema group is higher than the other two groups showed the color of tongue coating in acute eczema group is more yellowish than the other two groups. This result was quite similar to that of tongue coating colors in different eczema types. By analyzing these results, we can draw a conclusion that L*a*b* model can reflect the tongue manifestation objectively and can be used in computerized tongue image digital analysis system.

In this study, the computerized tongue image digital analysis system was used to observe the tongue manifestation of eczema patients and provided a reference for computerized technique to apply in Chinese medicine diagnostic field. In order to develop the study of Chinese medicine objectification and standardization, more diseases and cases are needed in further research to enhance the validity and credibility of computerized diagnostic technique.

Acknowledgments

This study was supported by National "Twelfth Five-Year" Plan for Science & Technology Support Project (No.2012BAI25B05).

References

1. Xue-Qin Wei, Quan Zhang, Hong Wei, Ling P. Observation on the helicobacter pylon and the tongue fur of chronic gastritis. *Sichuan J Tradit Chin Med* (In Chinese) 1999;**11**:7-8.
2. Shi-Kun Wu, Bo-Qiang Tan. Clinical observation on endoscope manifestation and tongue manifestation in 311 patients of stomach-duodenum pathological changes. *Guangxi J Tradit Chin Med* (In Chinese) 1990;**1**:1-3.
3. Hai-Yin Huang, Jin-Mei Su, Ming-Quan Han, Wei-Hong Long. Comparative study on picture of the tongue in patients of lung cancer and other respiratory diseases. *J Tradit Chin Med* (In Chinese) 2002;**7**:535-536.
4. Xiao-Qiang Yue, Jing-Dong Gao, Xin-Feng Zhai, Qing Liu, Dong Jiang, Chang-Quan Ling. Correlation between the width of lingual vein and the changes of hemodynamics of portal system in patients with primary liver cancer. *J Chin Integr Med* (In Chinese) 2006;**4**:482-484.
5. Fen-Qiao Chen, Yan-Ru Du, Qi-Quan Liu. Correlation between tongue proper and gastroscope, pathology in 900 patients of chronic gastritis. *Hebei J Tradit Chin Med* (In Chinese) 1999;**6**:349-350.
6. Yong-Wei Zhao, Shu-Ming Yang. Investigation of tongue manifestation in 360 patients of digestive system cancer. *J Pract Tradit Chin Med* (In Chinese) 2008:**5**:318.
7. Liang-Min Tao, Ge-Hui Xu. Color problem and application in computer vision. *Chin Sci Bull* (In Chinese) 2001;**46**:178-190.
8. Chui CC. The development of a computerized tongue diagnosis system. *Biomed Eng Appl Basis Commun* 1996;**8**:342-350.
9. State Administration of Traditional Chinese Medicine, P. R. C. *Standardized Diagnosis and Treatment of Chinese Medicine Disease and Syndrome*. Nanjing: Nanjing University Publishing House, 1994:146.

CONTENT DETERMINATION OF CICHORIC ACID IN ONYCHIUM JAPONICUM (THUNB.) KUNZE. BY HPLC

Hongli Teng[a], Licheng Guo, and Xiangpei Zhao

Guangxi Institute of Minority Medicine/Traditional Zhuang Hospital of Guangxi Zhuang Autonomous Region, Nanning, 530001, China

E-mail: [a]THL555@163.com

Objective: To develop a method to determine the content of Cichoric Acid in Onychium japonicum by HPLC. Methods: Agilent XDB C18 (5μm, 4.6mm×150mm) was adopted with mobile phase of acetonitrile-phosphate buffer (38:62), the flow rate was 1.0mL/min, the detection wavelength was at 326 nm, the column temperature was at 25 °C and the flow rate was 1.0 mL/min^{-1}. Results: A good line arcorrelation of Cichoric Acid was observed within the range of0.16~0.48μg/mL (r=0.999 8) and the average recovery was 99.79% with RSD 1.27% (n=6). Conclusion: The method is convenient, sensitive and accurate. It can be used for the quality control of Onychium japonicum.

1. INTRODUCION

Onychium japonicum (Thunb.) Kunze (Chinese name Jieducao) is widely used as medicinal plant for a long time in Zhuang Autonomous Region. It is mainly distributed in Guangxi, Guangdong, Fujian, Taiwan, etc [1].With effects of toxic materials, anti-inflam-matory, removing the Necrotic Tissue and Promoting Granulation, clearing heat, removing dampness and other effects, it is used for treatment of wind-heat type common cold, acute enteritis, dysentery, jaundice, etc [2]. Researches show that Onychium japonicum have anti-inflammatory and analgesic effect [3]. And there are many reported constituents in this plant, among them Cichoric Acid are regarded as the most active anti-inflammatory and enhance immune function. In recent years, more and more attention has been paid to the quality assessment of the row materials of traditional Chinese medicines in order to guarantee efficacy and safety when they are utilized in clinical practices. However, there is no report on the quality control methods for the cichoric acid in Onychium japonicum. Therefore, it is essential to establish an analytical method to control the quality by monitoring the active compound. In this study, the cichoric acid in Onychium japonicum were determined by HPLC.

2. METHODS AND RESULTS

2.1. Chemicals and Materials

Onychium japonicum were collected in Mashan County (Nanning China) and other county. The botanical origins of the materials were identified by Prof. Zhu Yi-lin

(Guangxi University of Chinese Medicine). Cichoric acid was purchased from National Institute for food and drug control. Acetonitrile was purchased from Fisher, Fair Lawn, NJ, USA. Redistilled water was made in our own lab. Phosphoric acid (≥85%), was purchased from Shanghai Chemical Co. The methanol for samples preparation was of analytical-reagent grade and purchased from Nanjing Chemical Factory (Nanjing, China). Other chemicals were of analytical grade. All solvents and samples were filtered through 0.45 μm filters before they were injected into HPLC.

2.2. Apparatus and Conditions

An Agilent series1100 HPLC instrument (Agilent, Waldbronn, Germany) equipped with a quaternary pump, a diode-array detector, an auto sampler and a column compartment was used. The samples were separated on an Agilent XDB column 18 (150 mm×4.6 mm I.D., 5μm, Agilent). The mobile phase was acetonitrile-phosphate buffer (38:62), with the flow-rate of 1.0 ml/min. The detection wavelength was set at 326nm. The column temperature was set at 30°C and the injection volume was 10μl. Chromato graphic peaks were identified by comparing the retention time (tR). In addition spiking samples with the standards further confirmed the identities of peaks.

2.3. Preparation of Standard Solution

Some reference cichoric acid was weighed accurately and then added into a 10 mL volumetric flask and dissolved with methanol to make a solution containing 0.1980 mg/ml barbaloin. The solution was filtered through 0.45μm micropore filter before HPLC analysis.

2.4. Sample Preparation

A 0.4 g crushed sample of Onychium japonicum was accurately weighed and Ultrasonic extracted with 50ml of Methanol -0.5% phosphoric acid (4:1) for 40 min. After cooling, weighed again, add the extraction solvent to make up the weight loss. Shook well, filtered and took 5ml of the filtrate into 10ml volumetric flask, then brought to volume with the extraction solvent. The sample solutions were filtered through a 0.45 μm micropore filter before HPLC analysis.

2.5. Stability Test

Reference cichoric acid solution was analyzed by HPLC for 0, 1, 2, 4, 6, 8, 10 h respectively under room temperature. The average of peak area integral values was 1694382. RSD% was 0.83%. The experimental results indicated that samples were stable in 10 h.

2.6. Determination of the linear relationship

Injected 4.0, 6.0, 8.0, 10.0, 12.0 μl standard solution into HPLC. Determined the cichoric acid peak area integral values according to the above chromatograph condition. Cichoric acid standard curve: Y=2.0×10-7X-0.0162 (x was thecichoric acid and y was the peak area integral value of cichoric acid). The linear range of cichoric acid was between 0.16~0.48μg, with a mean correlation coefficient r=0.999 8.

2.7. Precision Test

10μl of cichoric acid solution was injected into the HPLC system five times successively. The peak areas were calculated and The average of peak area integral values was 1692576. The RSD was found to be 0.2%.

2.8. Repeatability Test

Five portions of the same batch of Onychium japonicum were sampled and extracted for analysis. The content of cichoric acid was determined and its RSD was3.3%.

2.9. Recovery Test

We have examined the recovery property of cichoric acid by adding certain amount of its standard solution to the plant samples prior to the extraction, and all the samples were processed with the same method as described in Section 2.4. The results were illustrated in Table 1.

Table 1. Result of recovery test of cichoric acid

No.	Added amount (μg)	Sample amount(μg)	Found amount (μg)	Recovery(%)	Mean recovery(%)	RSD(%)
1	1.400	1.421	2.850	102.01	99.79	1.27
2		1.423	2.798	98.21		
3		1.422	2.817	99.64		
4		1.420	2.808	99.14		
5		1.420	2.822	100.14		
6		1.422	2.816	99.57		

2.10. Sample Analysis

The content of cichoric acid was determined by the above method in Onychium japonicum from the extraction solution. The results are shown in Table 2, Fig.1 and Fig.2. From Tables 2, we can conclude that the contents of cichoric acid in Onychium japonicum are 0.0538~0.0676%.

Table 2. Contents of cichoric acid in Onychiumjaponicum

No.	Sample	Habitat	Time of collection	Cichoric acid(%)
1		Mashan	2013.4.12	0.0647
2		Jinxiu	2013.4.9	0.0606
3		Jingxi	2013.4.17	0.0676
4	Onychiumjaponicum	Liuzhou	2013.7.5	0.0583
5		Guilin	2013.4.5	0.0627
6		Hezhou	2013.3.13	0.0638
7		Yongning	2013.6.10	0.0619
8		Nanning	2013.7.13	0.0607

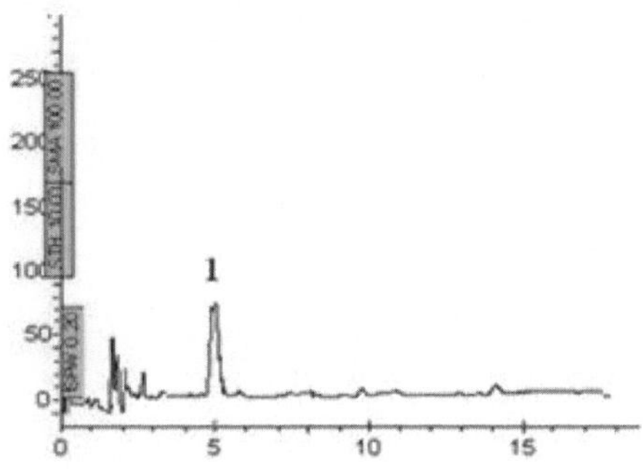

Fig. 1. HPLC chromatogram of referencecichoric acid.

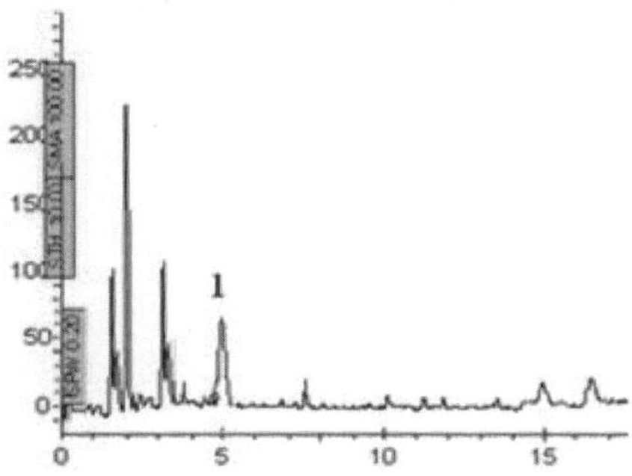

Fig. 2. HPLC chromatogram of Onychium japonicum.

3. CONCLUSION

It is the first time to establish a quantitative method to investigate the content of cichoric acid in Onychium japonicum. It documents that the verified HPLC method can be success fully applied to quantification of cichoric acid in Onychium japonicum with good sensitivity, repeatability and reproducibility.Our results suggested that this method can be helpful for the further study on Onychium japonicum.

Acknowledgments

This work was supported by the National Science & Technology Pillar Program during the 12th Five-year Plan Period (Grant No. 2012BAI27B06-1, 2013BAI11B06-7), Guangxi science and technology development plan(14124002-14), Nanning science and technology development plan(20133158).

References

1. Health department of Guangxi Zhuangautonomous region. Standard of Chinese medicinal materials in Guangxi(Vol. II). Nanning, Guangxi science and technology press. 1996; 32.
2. Feng YP, Guo LC, Li B, et al. Studies on Pharmacological Action of Onychium japonicum(Thunb.) Kunze. *Chin J of ExpTrad Med Formulae* 2011; **17**: 197-199.
3. Chen B, Pan ZQ, Fan YC, et al. Study of preparation on cichoricacid standard by preparative high performance liquid chromatography. *Practical Preventive Medicine* 2001; **8**:15-17.

COMPARATIVE METABOLISM CHARACTERISTICS RESEARCH OF TETRAHYDROPALMATINE IN FIVE SPECIES (DOG, HUMAN, MICE, MONKEY, AND RAT) LIVER MICROSOMES BY UHPLC/ESI-QTOF-MS/MS

Hui Ouyang[1,#], Huiping Liao[1,2,#], Junmao Li[1], Mingzhen He[1], Yan Li[1], Xiaoyong Rao[2], Qi Wang[1], Shilin Yang[1,2], Zhifeng Li[1,a,*], and Yulin Feng[2,b,*]

[1]Jiangxi University of Traditional Chinese Medicine, N0.818Yunwan Road, Nanchang 330002, China

[2]National Pharmaceutical Eengineering Center for Solid Preparation in Chinese HerbalMedicine, No.56 Yangmin Road, Nanchang 330006, China

E-mail: [a]fengyulin2003@126.com, [b]wangqilizhifeng@126.com

[*]Corresponding author. Tel.: +8679187119632

[#]These authors contributed equally to this work.

Tetrahydropalmatine (THP) is a natural product with a wide spectrum of pharmacological activities such as analgesic, resisting arrhythmia and dropping blood pressure etc. THP is now under investigation for the treatment of drug addiction, a chronic disease characterized by compulsive drug use and loss of control over intake. As we know, the metabolic fate of drug candidates is an essential requirement in assessing their safety and efficacy. Based on this requirement, a simple, rapid and reliable ultra-high performance liquid chromatography combined with electrospray ionization quadrupole time of flight mass spectrometry (UHPLC/ESI-QTOF-MS/MS) method was developed and successfully applied to the study on the metabolites of THP in liver microsomes of dog (DLM), human (HLM), mice (MLM), monkey (MKLM) and rat (RLM) incubation systems. As a result, a total of 32 metabolites (17 metabolites in DLM, 15 metabolites in HLM, 12 metabolites in MLM, 21 metabolites in MKLM and 18 metabolites in RLM) were proposed and identified in liver microsomes of five species *in vitro* respectively. The results also showed that the metabolic patterns and products are different among liver microsomes within five species. Furthermore, the present study also provided vital data for further pharmacokinetic and in *vivo* metabolism study.

1. INTRODUCTION

Tetrahydropalmatine (THP) is a tetrahydroprotoberberine isoquinoline alkaloid that widely distributed in the herbal plants *Corydalis decumbens(Thunb.)Pers* and THP has been used for clinic in China for more than 40 years [1].Recently, pharmacological investigation revealed that THP had multiple pharmacological effects such as analgesic, anti-cerebral ischemia, resisting arrhythmia, resisting myocardial ischemia and anticoagulant as well as dropping blood pressure etc [2].The unique pharmacological property of THP distinguishes it from other dopamine antagonists. THP is now under new drug development for drug addiction, a chronic disease characterized by compulsive drug use and loss of control over intake. More and more researchers have recognized bioactive components from herbal medicine are often subjected to extensive phase I and II metabolism [3-5]. Consequently, the metabolism of an investigational new drug should be defined during drug development and that its interactions with other drugs should be explored as part of an adequate assessment of its safety. In recent year, liver microsomes have been widely used to determine the drug metabolism of metabolic pathways, preparation of metabolites and comparing the metabolic differences between species for their convenience and effectiveness [6-9]. In addition, no detailed information in regarding liver microsomes of THP to the best of our knowledge, especially to the metabolic of the characteristics between different species of liver microsomes *in vitro* [10-12].

Ultra high performance liquid chromatography combined with electrospray ionization quadrupole time of flight mass spectrometry (UHPLC/ESI-QTOF-MS/MS) has recently become a powerful and reliable analytical technique for metabolites identification for higher mass resolution, sensitivity and accuracy can provide accurate masses of ions and molecular formulas [13-16]. In this research, 32 metabolites were detected and identified in DLM, HLM, MLM, MKLM and RLM. The fragmentation patterns and chromatographic behaviors were also examined to confirm the metabolic identification. The principles of the *in vitro* metabolic pathways of THP were also proposed in this research.

2. MATERIALS AND METHODS

2.1. Chemicals and Reagents

THP with the purity of >98% was obtained in our laboratory. The structure was confirmed through MS, ^{13}CNMR, and ^{1}H NMR methods. HPLC grade methanol, ammonium formate and acetonitrile were purchased from Fisher Scientific (NJ, USA). Deionized water was produced with A10 equipment (Millipore, Bedford, MA, USA). DLM, HLM, MLM, MKLM and RLM were purchased from Becton, Dickinson and Company(BD, USA),NADPH(Roche), PBS buffer(WuHan DingGuo, China). All the other chemicals and reagents were of analytical grade.

2.2. Sample Preparation

Firstly, PBS buffer (pH7.4) and THP were performed at 1mmol/L; DLM, HLM, MLM, MKLM and RLM were performed at 1 mg/mL respectively. All of five species liver microsomes and blank samples were added PBS buffer (150 μL), THP (20 μL) and liver microsomes (10 μL). Then incubation bags at 37°C in a shaking water bath for 5 min. After the incubation, 1 mmol/L (20 uL) NADPH was added to the incubation of liver microsomes samples which in order to initiate the metabolic reaction, as well as 1 mmol/L (20 μL) PBS buffer was added to blank samples instead of NADPH. Incubation experiment was performed at 37°C for 1 h and terminated reaction by adding acetonitrile (400 μL). The incubation samples were centrifuged at 14,000 g for 10 min. Then supernatant was pooled and dried at 40°C using a nitrogen evaporator. The residue was redissolved in 200 μL of 20% acetonitrile-water, centrifuged at 14,000 g for 10 min, then 20 μL supernatant were injected with for UHPLC/ESI-QTOF–MS /MS analysis.

2.3. Instrumental and Analytical Conditions

The blanks and samples were analyzed on a triple TOF™5600+ MS/MS system (AB SCIEX, Canada) coupled to a Prominence™ UHPLC system (Shimadzu, Japan).The UHPLC system consisted of a solvent delivery system LC-30AD, autosampler SIL-30AC, column oven CTO-30AC, degasser DGU-20A3, and controller CBM-20A from Shimadzu (Kyoto, Japan). Separations were accomplished on a Welch C_{18} column (1.7 mm × 100 mm, 1.7 μm, Milford, USA). The column oven was maintained at 25 °C. The mobile phase was composed of 5 mmol/L ammonium formate in water (*v/v*) as mobile phase A (MA) and acetonitrile as mobile phase B (MB) using a gradient elution of 5% of MB (0–2.5 min), from 5% to 50% of MB (2.6–10.0 min), keeping 50% of MB (10–12.5 min) ,from 50% to 90% of MB (12.6–17.0 min), and 5% of MB (17.1–20.0 min).The flow rate was 0.25 mL/min, the temperature of the column and autosampler were maintained at 45°C and 4°C, respectively.

In the MS analysis, positive ionization mode was used for metabolite detection with a DuoSpray™ source (AB SCIEX, Canada) and mass range was set at *m/z* 100–1000. The parameters throughout this work were set as the following: ion spray voltage: 5500 V; ion source heater: 600 °C; curtain gas: 30 psi; nebulizing gas (GS1), 50 psi; and TIS gas (GS2), 50 psi. For the TOF-MS-IDA-MS/MS analysis, the survey scan TOF/MS and MS/MS experiments were run with 200 ms of accumulation time for TOF/MS and 100 ms of accumulation time in MS/MS experiments. Collision energy (CE) is 50 eV; collision energy spread (CES) is 20eV in MS/MS experiment. In addition, information dependent acquisition (IDA) was used to trigger acquisition of MS/MS spectra for ions matching the IDA criteria. A realtime multiple mass defect filter (MMDF) was used in IDA criteria. IDA priority was given to ions that match the mass defect window to acquire MS/MS spectra.

3. RESULTS AND DISCUSSION

3.1. UHPLC/ESI-QTOF-MS/MS conditions

The acetonitrile-5 mmol/L ammonium formate solution in the water system performed more powerful separation ability for the detected compounds. For the ESI-QTOF-MS conditions, a full scan of analysis in positive ionization modes form ESI source proved a higher intensity and supplied sufficient abundance. The positive ionization modes were used to MS/MS analysis toobtain more structure information [17-18].

3.2. Fragmentation Studies of Parent

The fragmentation of the parent (THP) is useful in metabolite identified by UHPLC/Q-TOF-MS/MS. In this case, the fragmentation pattern of THP was subsequently used for the confident structural identification of metabolites. Under the chromatographic and mass spectrometry conditions, THP was eluted at 12.74 min (Fig.1.A). Full scan mass spectrum analysis of THP at $[M+H]^+$ (*m/z*356.1865) and two isotope peaks (*m/z* 357.1900, 358.1923) were proposed in Fig.1.B. The MS/MS product spectrum of THP was presented in Fig. 1.C. Fragmentation of THP in the ion trap lead to six product ions at *m/z* 192.1010, 176.0698, 165.0897 and 150.0667. The abundant product ion at *m/z* 192.1010 and 165.0897 was produced by the loss of 164 Da ($C_{10}H_{12}O_2$) and 191 Da ($C_{11}H_{13}O_2$), respectivly (table1). The product ion at *m/z* 176.0698 was 16 Da (CH_4) less than *m/z* 192.1010, indicating the loss of methyl and two of hydrogen. The product ion at *m/z* 150.0667 were formed by loss of 15 Da (CH_3) comparing with *m/z* 165.0897, involving the loss of methyl and hydrogen. Accordingly, the ions at *m/z* 192.1010, 176.0698, 165.0897 and 150.0667 were the characteristic product ions of THP. However, 164 Da, 191 Da, 16 Da and 15 Da were the characteristic fragment ions losses of THP. All characteristic product ions and fragment ions losses were essential for the identification of THP metabolites. Accurate mass measurements of molecular and fragment ions of THP and 32 metabolites were presented in Table 1 [19-20].

Table 1. Accurate mass measurements of protonated molecular and fragment ions of THP

No.	Formula	Ion mode	Rt(min)	Experimental *m/z*	Calculated *m/z*	error (ppm)
Parent	$C_{21}H_{25}NO_4$	$[M+H]^+$	12.75	356.1865	356.1868	0.3
Fragment ions 1	$C_{20}H_{21}NO_4$	$[M+H]^+$	ND	340.1512	340.1549	3.5
Fragment ions 2	$C_{11}H_{13}NO_2$	[M+H]+	ND	192.1010	192.1025	1.5
Fragment ions 3	$C_{11}H_{11}NO_2$	$[M+H]^+$	ND	190.0856	190.0868	3.2
Fragment ions 4	$C_{10}H_9NO_2$	$[M+H]^+$	ND	176.0698	176.0712	2.4
Fragment ions 5	$C_{10}H_{12}O_2$	$[M+H]^+$	ND	165.0897	165.0916	1.9
Fragment ions 6	$C_9H_9O_2$	$[M+H]^+$	ND	150.0667	150.0675	0.9

Note: ND: Not detected

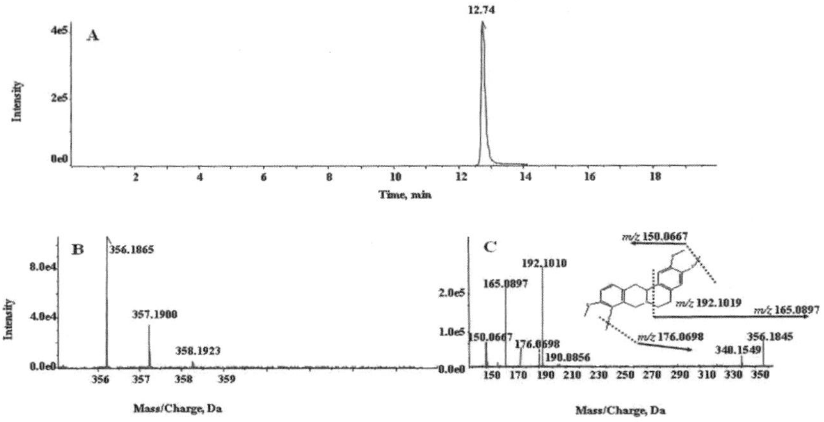

Fig.1. XICs (A), the MS (B) and MS/MS (C) spectrum of THP.

3.3. Identify of Metabolites

In this study, in order to find out the possible metabolites, the full scan mass spectrum of fragment ions with all samples which adding THP to liver microsomes in DLM, HLM, MLM, MKLM and RLM. Based on the method which mentioned above, by compared with blank samples, 32 metabolites were found as potential metabolites, among them 17 metabolites in DLM (M1,M2b-2e,M2g,M3b-3c,M4a-4d,M5a-5b, and M6d),15 metabolites in HLM (M2b-2d,M3a-3b,M4d-4e,M6b-6e, and M7a-7c), 12 metabolites in MLM (M2a-2b,M2d,M2f,M3a-3b and M6b-6f), 21 metabolites in MKLM (M2a-2b,M2d, M2f,M3a-3c,M4e-4f,M5a-5b,M6a-6f, and M7a-7d), 18 metabolites in RLM (M2b,M2d,M2f,M3a-3c,M6a-M6e, M6g, M7a-7d and M8a-8b). Their MS/MS ions along with their relative abundance and retention time which were performed by comparing with THP. Therefore, these metabolites were supposed that undergo further phase I and/or II metabolism in *vivo* by analysis their fragmentation ions. At the same time, coupled with the maximum mass errors between measured and calculated values were all less than 10 ppm which guaranteed high resolution and good accuracy. From all the above analysis, their structure could be identification effective and rapidly. Meanwhile, there main biotransformation were hydroxylation, demethylation, dehydrogenation reaction and so on of THP. Finally, the proposed structures and their mass fragmentation pathways were presented (Table 2).

Table 2. Accurate mass measurement for the protonatedmolecules of metabolites in 5 liver microsomes samples [19-20]

No	Metabolite description	Formula	Ion mode	RT (min)	Fragment ions (m/z)	Caculated (m/z)	DLM	HLM	MLM	M1LM	RLM
M	Parent	$C_{21}H_{25}NO_4$	[M+H]+	12.75	340.1549,192,1010,190.0556,165.0897,150.0667	356.186	√	√	√	√	√
M1	Dehydrogenation a× hydroxylation	$C_{21}H_{23}NO_5$	[M+H]+	10.88	192.1009,165.0909,150.0667	370.165	√	×	×	×	×
M2a	Hydroxylation	$C_{21}H_{25}NO_5$	[M+H]+	8.73	190.0861,165.0905	372.181	√	×	√	√	×
M2b	Hydroxylation	$C_{21}H_{25}NO_5$	[M+H]+	9.09	190.0861,165.0905	372.181	√	√	√	√	√
M2c	Hydroxylation	$C_{21}H_{25}NO_5$	[M+H]+	10.71	190.0861,165.0905	372.181	√	√	×	×	×
M2d	Hydroxylation	$C_{21}H_{25}NO_5$	[M+H]+	10.89	190.0861,165.0905	372.181	√	√	√	√	√
M2e	Hydroxylation	$C_{21}H_{25}NO_5$	[M+H]+	11.19	190.0861,165.0905	372.181	√	×	×	×	×
M2f	Hydroxylation	$C_{21}H_{25}NO_5$	[M+H]+	11.55	190.0861,165.0905	372.181	×	×	√	√	√
M2g	Hydroxylation	$C_{21}H_{25}NO_5$	[M+H]+	12.76	190.0861,165.0905	372.181	√	×	×	×	×
M3a	demethylation	$C_{20}H_{23}NO_4$	[M+H]+	10.45	192.1024,165.0909	342.17	√	√	√	√	√
M3b	demethylation	$C_{20}H_{23}NO_4$	[M+H]+	10.9	192.1024,165.0909	342.17	√	√	√	√	√
M3c	demethylation	$C_{20}H_{23}NO_4$	[M+H]+	11.19	192.1024,165.0909	342.17	√	√	√	√	√
M4a	Dehydrogenationa× demethylation	$C_{20}H_{21}NO_4$	[M+H]+	8.77	190.0834	340.154	×	√	×	√	×
M4b	Dehydrogenation a× demethylation	$C_{20}H_{21}NO_4$	[M+H]+	9.2	190.0834	340.154	×	×	×	√	×
M4c	Dehydrogenation a× demethylation	$C_{20}H_{21}NO_4$	[M+H]+	9.97	190.0834	340.154	√	×	×	×	×

M4d	Dehydrogenation a× demethylation	$C_{20}H_{21}NO_4$	[M+H]+	10.44	190.0834	340.154	√	×	×	×	×
M4e	Dehydrogenation a× demethylation	$C_{20}H_{21}NO_4$	[M+H]+	10.91	190.0834	340.154	√	×	×	×	×
M4f	Dehydrogenationn a× demethylation	$C_{20}H_{21}NO_4$	[M+H]+	11.2	190.0834	340.154	√	√	×	×	×
M5a	Di-demethylation	$C_{19}H_{21}NO_4$	[M+H]+	9.16	178.0858	328.155	√	×	×	√	×
M5b	Di-demethylation	$C_{19}H_{21}NO_4$	[M+H]+	9.61	178.0858	328.155	√	×	×	√	×
M6a	Demethylation a× hydroxylation	$C_{20}H_{23}NO_5$	[M+H]+	8.05	340.1546,190.0862	358.165	×	×	×	√	√
M6b	Demethylation a× hydroxylation	$C_{20}H_{23}NO_5$	[M+H]+	9.13	340.1546,190.0862	358.165	×	√	√	√	√
M6c	Demethylation a× hydroxylation	$C_{20}H_{23}NO_5$	[M+H]+	9.86	340.1546,190.0862	358.165	×	√	√	√	√
M6d	Demethylation a× hydroxylation	$C_{20}H_{23}NO_5$	[M+H]+	10.04	340.1546,190.0862	358.165	√	√	√	√	√
M6e	Demethylation a× hydroxylation	$C_{20}H_{23}NO_5$	[M+H]+	10.49	340.1546,190.0862	358.165	×	√	√	√	√
M6f	Demethylation a× hydroxylation	$C_{20}H_{23}NO_5$	[M+H]+	11.07	340.1546,190.0862	358.165	×	×	√	×	×
M6g	Demethylation a× hydroxylation	$C_{20}H_{23}NO_5$	[M+H]+	11.25	340.1546,190.0862	358.165	×	×	×	√	√
M7a	Dehydrogenation	$C_{21}H_{23}NO_4$	[M+H]+	9.39	190.0853,174.0553,165.0910	354.17	×	√	×	√	√
M7b	Dehydrogenation	$C_{21}H_{23}NO_4$	[M+H]+	9.67	190.0853,174.0553,165.0910	354.17	×	√	×	√	√
M7c	Dehydrogenation	$C_{21}H_{23}NO_4$	[M+H]+	10.88	190.0853,174.0553,165.0910	354.17	×	√	×	√	√
M7d	Dehydrogenation	$C_{21}H_{23}NO_4$	[M+H]+	11.57	190.0853,174.0553,165.0910	354.17	×	√	×	√	√
M8a	Dehydrogenation a× di-demethylation	$C_{19}H_{19}NO_4$	[M+H]+	7.93	190.0790,176.0633	326.139	×	×	×	×	√

M8b	Dehydrogenation a× di-demethylation	$C_{19}H_{19}NO_4$	[M+H]+	10.51	190.0790,176.0633	326.139	×	×	×	×	√

"√" : Detected "×" : Not detected

In this study, all metabolites peaks could be observed in these liver microsomes samples of different time, among them, 17 peaks (Fig.2.A), 15 peaks (Fig.2.B), 12 peaks (Fig.2.C), 21 peaks (Fig.2.D), 18 peaks (Fig.2.E) were in DLM, HLM, MLM, MKLM and RLM, respectively. All these peaks were showed as follow in Fig.2.

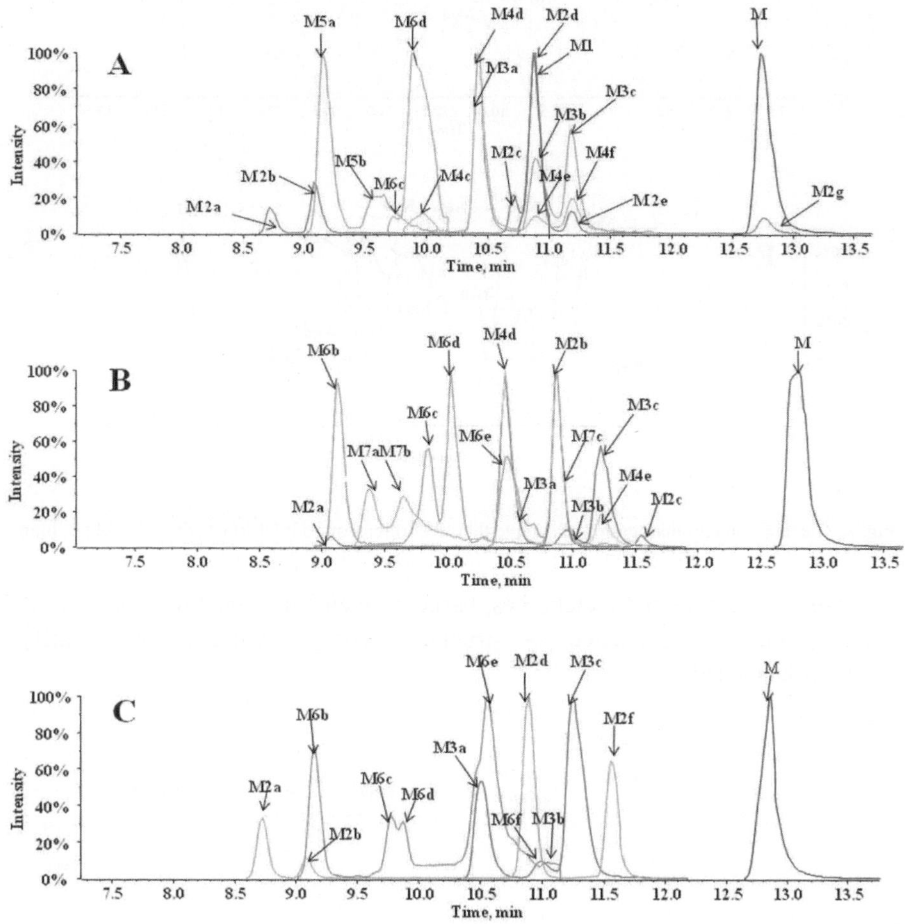

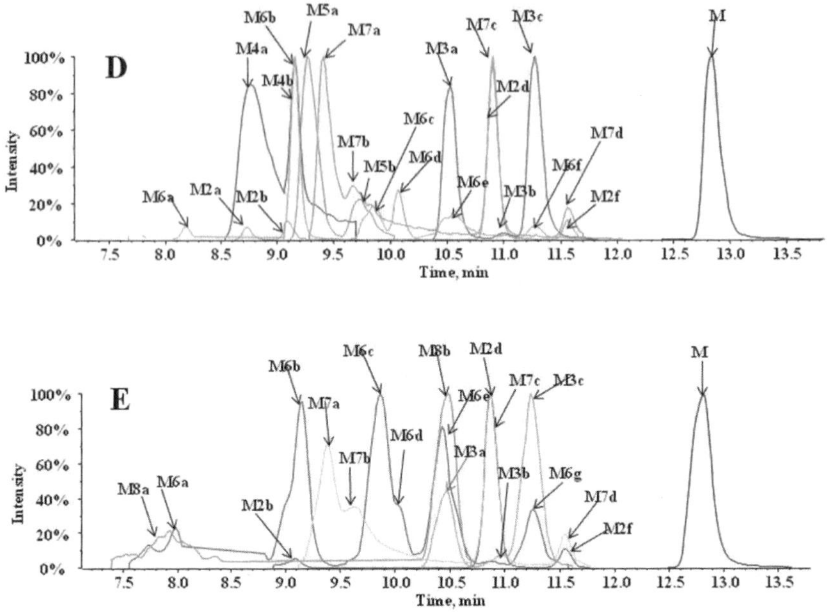

Fig. 2. The total ion chromatograms of sample that liver microsomes inDLM (A, dog), HLM (B, human), MLM (C, mice), MKLM (D, monkey) and RLM (E, rat).

To assess these potential metabolites, based on their fragmentation ions scan enrich concentration that we speculated the structures. The following we will clarify 32 metabolitesas below (Fig.3)

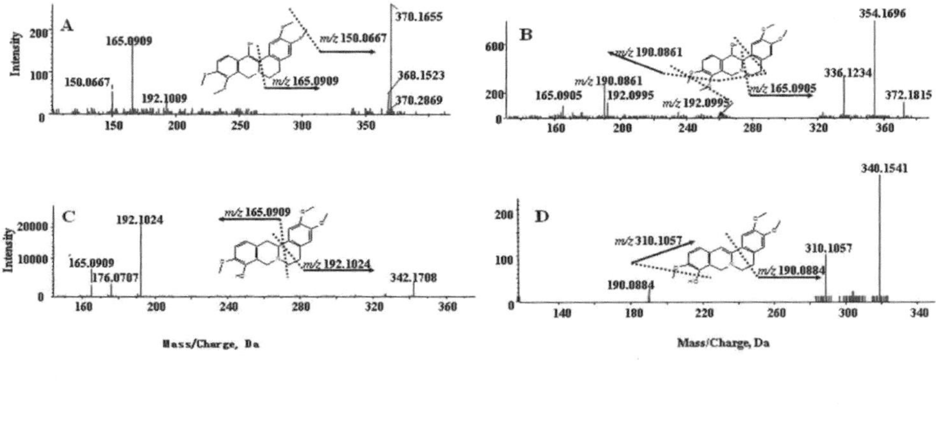

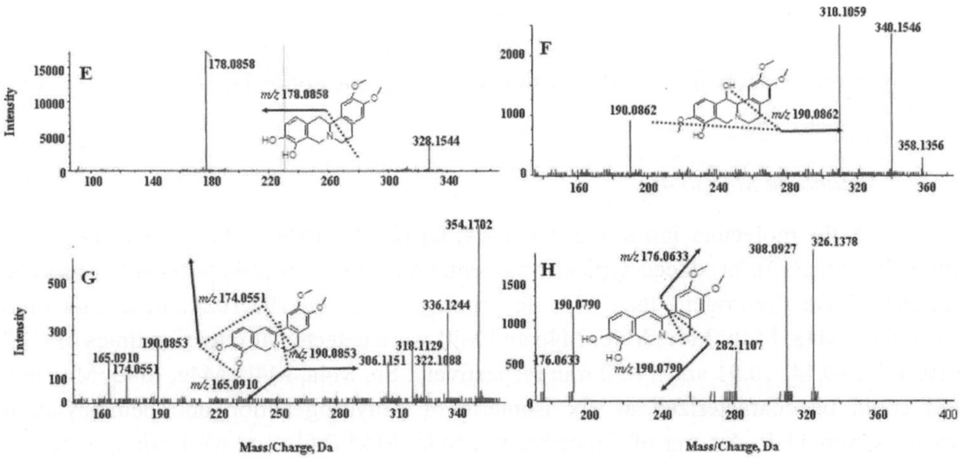

Fig. 3. Mass spectrum (MS/MS) of metabolites of THP.

3.3.1. *Metabolite M1.*

The molecular ion at *m/z* 370.1653, $C_{21}H_{23}NO_5$ (Fig.3.A) was eluted at 10.88 min. M1 generated from increasing 14 Da (loss of 2 hydrogen and increased hydroxyl) comparing with the THP, produced two typical fragmentation ions at *m/z* 192.1009 and *m/z* 165.0909 were the same as the two product ions of parent. Therefore, M1 was characterized as dehydrogenation and hydroxylation product from THP. In five species, M1 was only survived in DLM.

3.3.2. *Metabolite M2a to 2g.*

The molecular ion at *m/z* 372.1817, $C_{21}H_{23}NO_5$ (M2a, M2b, M2c, M2d, M2e, M2f and M2g) were eluted at 8.73, 9.09, 10.71, 10.89, 11.19, 11.55 and 12.76 min (Fig.3.B), showed the same MS and MS/MS but their retention times were differences. The seven metabolites were produced by increasing 16 Da (hydroxyl) from THP. The fragment ion were displayed at *m/z* 192.0995 and 165.0905 were the same as the product ion from THP. Thus, M2a to 2g were characterized as seven isomerides of hydroxylation products from THP. The potential mechanism for the formation of isomeric and hydroxylation position requires further study. However, M2a were detected in DLM, MLM and MKLM, M2b and M2d in five species, M2c in DLM, HLM and MLM, both M2e and M2g only in DLM, M2f in MLM, MKLM and RLM.

3.3.3. *Metabolite M3a to 3c.*

The molecular ion at *m/z* 342.1716, $C_{20}H_{23}NO_4$ (M3a,M3b and M3c) were eluted at 10.45, 10.90 and 11.19 min respectively (Fig.3.C), showed the same MS and MS/MS but their retention times were differences, produced by loss of 14 Da (methyl) from THP. The fragment ion at *m/z* 192.1024, 176.0707 and 165.0909 were the same as the three product ions of THP. Thus, M3a, M3b and M3c was identified as three isomerides of demethylation products from THP. As showed, metabolites of M3a, M3b and M3c were found in five species.

3.3.4. *Metabolite M4a to 4f.*

In the case, the molecular ion at *m/z* 340.1544, $C_{20}H_{21}NO_4$ (M4a, M4b, M4c, M4d, M4e and M4f) (Fig.3.D), produced typical fragmentation ions at *m/z*190.0884 was observed eliminate 2 Da (hydrogen) than M3a-3c product ion at *m/z*192.1024. In a total of 6 metabolites M4a, M4b, M4c, M4d, M4e and M4f were detected at retention times of 8.77, 9.20, 9.97, 10.44, 10.91 and 11.20 min respectively. So, M4a, M4b, M4c, M4d, M4e and M4f could be characterized as six isomeric of dehydrogenation and demethylation products from THP. Among of 6 metabolites, M4c, M4d, M4e and M4f primly detected in DLM, M4a and M4f detected in HLM, M4a and M4b were found in MKLM, but both in MLM and RLM were not detected in these metabolites.

3.3.5. *Metabolite M5a and M5b.*

The molecular ion at *m/z* 328.1549, $C_{19}H_{21}NO_4$ (M5a and M5b) (Fig.3.E) were observed at a retention time of 9.16 and 9.61 min, exhibited the same MS and MS/MS. Their fragmentation ions at *m/z* 178.0858 were all less than 14 Da of M3a-3c fragmentation ions at *m/z* 192.1024, suggested it may be produced via demethylation of M3a-3c. Therefore, M5a and M5b were identified as two isomeric of di-demethylation products from THP. In addition, M5a and M5b were only detected in DLM and MLM.

3.3.6. *Metabolite M6a to 6g.*

The molecular ion at *m/z* 358.1665, $C_{20}H_{23}NO_5$ (M6a,M6b, M6c, M6d, M6e, M6f and M6g) (Fig.3.F) were eluted at a retention time of 7.98, 9.13, 9.86, 10.04, 10.49 and 11.07 min respectively. Their main fragment ions at *m/z* 194.0706 were all 2 Da (loss of methyl and add hydroxyl) more than THP fragment ions at *m/z* 192.1010. So, M6a, M6b, M6c, M6d, M6e, M6f and M6g could be characterized as seven isomeric of demethylation and hydroxylation products from THP. However, in combination with the detailed table 2, it was showed that except M6a detected in RLM, the rest of four species were not detected.M6b, M6c and M6e were detected in despite of DLM, but M6d was detected in five species.M6f was observed in MKLM,MLM and RLM, the rest of M6g only detected in RLM.

3.3.7. *Metabolite M7a to 7d.*

The molecular ion at *m/z* 354.1705, $C_{21}H_{23}NO_4$ (M7a, M7b, M7c and M7d) (Fig.3.G) along with the main characterized fragment ions at m/z 190.0853 and *m/z* 174.0551 were all 2 Da (two hydrogen) less than THP. The other characteristicsof *m/z* 165.0901 showed the same MS and MS/MS spectra with THP, indicating the loss of two hydrogen at m/z 190.0853. A set of metabolites M7a, M7b, M7c and M7d were detected at retention times of 9.39, 9.67, 10.88 and 11.57 min, respectively. Therefore, M7a, M7b, M7c and M7d were identified as four isomeric of dehydrogenation products from THP. According to table2, the metabolites of M7a, M7b and M7c were detected in HLM, MKLM and RLM. In addition, M7d were only detected in MLM and RLM.

3.3.8. *Metabolite M8a and M8b.*

The molecular ion at *m/z* 326.1387, $C_{19}H_{19}NO_4$ (M8a, M8b) (Fig.3.H) were at the retention time of 7.91 and 10.91 min. M8a and M8b.were all 2 Da less than M5a and M5b. The product ions at *m/z* 176.0633 and 190.0709 were all 2 Da less than the product ions of M5a and M5b. Accordingly, M8a and M8b were identified as dehydrogenation and di-demethylation products from THP. Based on the table2, we can observe the metabolites of M8a and M8b only existed in RLM of five species liver microsomes.

In general, 32 metabolites were detected in THP among DLM, HLM, MLM, MKLM and RLM, such as, dehydrogenation and hydroxylation, dehydrogenation and demethylation, demethylationand hydroxylation. The possible metabolic pathways of THP were subsequently proposed (Fig. 4).

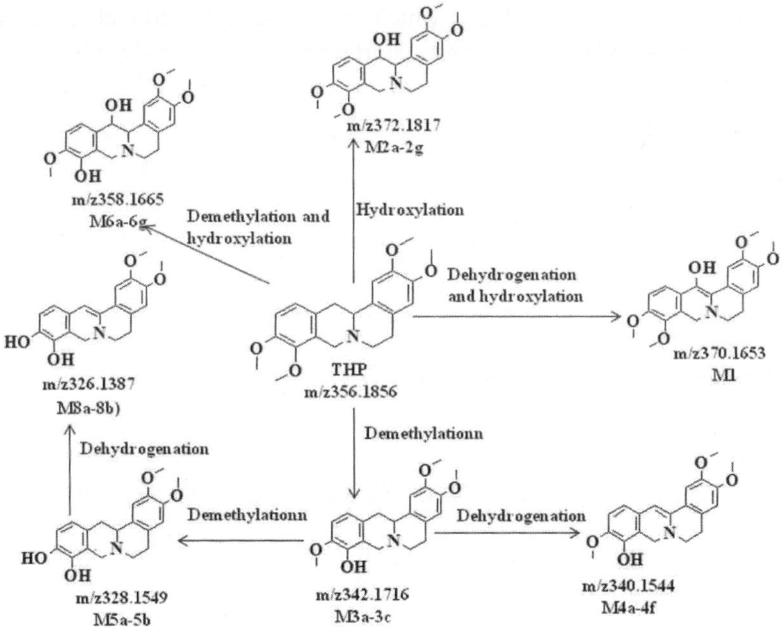

Fig.4. Proposed metabolic pathway of THP in DLM, HLM, MLM, MKLM and RLM.

4. CONCLUSIONS

In this study, we discussed the differences of THP metabolites among DLM, HLM, MLM, MKLM and RLM *in vitro* metabolism by using a sensitive and high efficiency UHPLC/ESI-QTOF-MS/MS method. A total of 32 metabolites in liver microsomes sample were detected and tentative confirmed. On the basis of fragmentation information, not only their structure were identified, but also their mainly biotransformation were demonstrated as hydroxylation, dehydrogenation and hydroxylation, dehydrogenation and demethylation, demethylationand hydroxylation. However, several groups of isomers cannot be discriminated from each other and the accurate position remains to be determined, in order to solve this problem, more advanced technology such as NMR-MS should be used in further study.

Comparing with fragmentation information and retention times which in different species of liver microsomes, we acquire a significant and interesting conclusion that monkey and rat metabolism were similar to human metabolism. The available data indicated monkey and rat were suitable for further *in vivo* research. For the first time, the metabolism studies employing five species liver microsomes *in vitro* were demonstrated to be a suitable tool for the THP metabolism [21-22]. In a nutshell, the present investigation adds information to the existing knowledge on *in vitro* and *in vivo* biotransformation of THP metabolites, which is particularly critical used safely pharmacokinetic and *in vivo* metabolism. Furthermore, this study also provides significant information about other active alkaloids for its potential clinical applications with the traditional medicines.

Acknowledgments

The authors gratefully acknowledge the award from the Young Scientists Culture Project of Jiang Xi Province (20142BCB23022), Postdoctoral Research Project of Jiang Xi Province (2012-195) and School Research Project of Jiangxi University of Traditional Chinese Medicine (ZX1002, ZX1004).

The authors declare no conflict of interest.

References

1. Xiao, W. B., Zhuang, X. M., Shen, G. L., Zhong, Y. H., J Sep Sci. 2014, **37**: 696–703.
2. Li, P. F., Mu, C. H., Wang, T., Lei, F. H., S.Afr.J.Chem. 2014, **07**: 67–70.
3. Zhang, Y. Y., Dong, X., Le, J., Wen, J., Lin, Z. B., J Pharmaceut Biomed. 2014, **94**: 152–162.
4. Marquesa, L. M., Juniorb, E. S., Gouveaa, D. R., Vessecchic, R., J Pharmaceut Biomed. 2014, **95**: 113–120.
5. Qi, M., Zhong, A. Z., Li, P. F., Yang, Q. M., Agric.Food Chem. 2013, **141**: 357–365.
6. Akama, T., Bu, W., Chanda, S., Sullivan, D., J Pharmaceut Biomed. 2012, **70**: 344–353.

7. Hu, Z., Yang, J., Cheng, C., Huang, Y., Du, F., Drug Metab.Dispos. 2013, **41**: 1457–1469.
8. Varkhede, N. R., Jhajra, S. L., Ahire, D. S., Singh, S., Rapid Commun Mass Sp. 2014, **28**: 311–331.
9. Caslavska, J., Thormann, W., J.Chromatogr.A. 2011, **1218**: 588–601.
10. Shen, J., Yang, C., Wu, C., Feng, P., Wang, Z., Li, Y., Li, Y., Zhang, S., Rapid Commun.Mass Sp. 2010, **24**: 375–383.
11. Han, Y. L., Yu, H. L., Li, D., Meng, X. L., Phytother Res. 2011, **25**: 1660–1665.
12. Takizawa, T. T., Shimizu, M., Kume, T., Yamazaki, H., DMPK. 2014, **6**: 1–6.
13. Ma, C. H., Dastmalchi, K., Flores, G., Wu, S. B., Agric.Food Chem. 2013, **61**: 3548–3559.
14. Rao, R. N., Prasad, K. G., J Pharmaceut Biomed. 2014, **94**: 92–98.
15. Shen, J., Yang, C., Wu, C., Feng, P., Wang, Z., Li, Y., Li, Y., Zhang, S., Rapid Commun. Mass Sp. 2010, **24**: 375–383.
16. Zhang, Y., Wu, W. H., Han, F. M.. Chen, Y., Biomed Chromatogr. 2008, **22**: 1360–1367.
17. Deng, P., Yong, T., Chen, X., Yuan, T., Huang, H., Zhong, D., Drug Metab Dispos. 2011, **39:** 1058–1066.
18. Shayeganpour, A., El-Kadi, A. O. S., Brocks, D. R., Drug Metab. Dispos. 2006, **34**: 43–50.
19. Li, S. P., Liu, W., Tenga, L., Cheng, X. M., J Pharmaceut Biomed. 2014, 92, 53–62.
20. Huang, M., Ho, P. C., Biochem Pharmacol. 2009, **77**: 1418–1428.
21. Ouyang, H., Zhou, M. F., Guo, Y. C., He, M. Z., Huang, H. S., Fitoterapia. 2014, **96**: 152–158.
22. Liu, H. X., Hu,Y., He, Y. Q., Liu, Y., Talanta. 2009, **79**: 1433–1440.

FEATURE EXTRACTION ON XINJIANG HIGH MORBIDITY ESOPHAGUS CANCER BASED ON WAVELET TRANSFORM

Ximei Kong[a], Murat Hamit[b,*], Chuanbo Yan, Jing Sun, and Juan Yao
College of Medical Engineering Technology, Xinjiang Medical University, Urumqi 830011,China
E-mail: [a]326481822@qq.com, [b]murat.h@163.com
[*]Corresponding author

The feature extraction is the basis of image interpretation and analyzing. It is also the main problem of pattern recognition and classification. In this paper, we proposed extracting the characteristic of Xinjiang high morbidity esophagus cancer of X-ray barium angiogram by using wavelet transform. First, normalized the image, removing the noise by using the median filter and enhancing. Then, using the texture feature extraction method to extract mushroom umbrella, coarctate esophageal of mean and variance, respectively by using wavelet-based transform. Those characters are texture features. Finally, the feature vector was using descriptive statistical analysis to obtain the best characteristics of classification ability, and we used the Bayes discriminant analysis to evaluate feature classification ability. The results showed that mushroom umbrella esophageal classification accuracy rate reached to 79%, coarctate esophageal cancer classification accuracy rate reached to 74%. And feature classification ability was different when classifying different images through wavelet transform, which could discriminate the different esophageal disease, and which would help the clinician to diagnosis the esophageal disease. As for a certain extent could lay a foundation for computer diagnosis system of Kazakh in Xinjiang, also lay a foundation for the subsequent content-based image retrieval system.

1. INTRODUCTION

Esophageal cancer is one of the most common clinical tumor, ranking first in the gastrointestinal tract[1]. It was due to the esophageal epithelium or glands malignant. Esophageal cancer is also one of the most common cancer in China. The morbidity and mortality of Esophageal Cancer was ranked first in the world. Esophageal cancer still has a high morbidity in the Xinjiang Uygur Autonomous Region[2]. We mainly put emphasis on the prevention and treatment of locality esophageal cancer. In Xinjiang, ethnic variance is one of the trait of ethnic distribution among esophageal cancer. The morbidity of Kazakh population ranked the first in nationalities. Since the early clinical features of esophageal cancer is performed not obviously. Therefore, it's hard to rely on conventional diagnostic methods to detect the early lesions. Medical technology have sustainable evolved in Xinjiang in recent years. Large amount of medical imaging techniques have become an effective way of detecting esophageal disease. Especially the Kazakh residents already have the awareness of prevention and treatment of esophageal cancer and this awareness is further improved. The phenomenon promoted the demand for screening and diagnosis of various medical images associated with esophageal cancer substantial growth within the next few years.

Image feature extraction is separated from the key techniques and methods of the computer vision technology and image processing technology. We mainly use computers to process and analyze the image information in order to determine the effective feature of image. Thus the application for the extraction of those useful features could solve real technical problem. Image feature extraction is the foundation of image recognition, image data mining, and content-based image retrieval. And it is also the key step of pattern recognition and classification[3]. Different attributes of images feature and different feature extraction methods will get different characteristic parameters. Chen[4]and D.B had put forth a gray level co-occurrence matrix(GLCM) method to analysis texture features of liver CT images. It mainly described the difference in image gray space. Then they study computer-aided diagnosis system for liver disease. Wavelet transform needs texture features. Those features show various features of wavelet domain provided a more accurate method for feature extraction[5].

Texture is a common existence in a property of the nature object. Compared with other image characteristics, it can be a batter balance between the macroscopic properties and fine structure of the image, thus texture feature analysis has become a significant means of image analysis. However, the texture is indescribable the characteristics. Early studies scholars texture, such as Muerle and Hawkins both agreed that the texture is not present accurate or precise definition. After 30 years, all generally accepted view is that the texture is composed of many mutual close and mutual woven of unit. It is the pixel brightness or gray in local space through a certain variation generated mode, which can get very rich visual scene information from texture. Texture feature analysis provided an important way for image perception, scene identification and target location and so on research. And it is widely used in the field of content-based image retrieval, image fusion, image matching, and medical image analysis etc. At present, most scholars consider that commonly texture feature analysis consists of four methods, such as statistical analysis, structural analysis, model analysis and spectrum analysis[6]. In this paper, we proposed spectrum analysis process feature extraction in frequency domain for esophageal cancer image. First, we normalized scale, removing the noise by median filter, enhancing by histogram equalization, and acquiring clearly gradation picture. Secondly, we used wavelet transform method to calculate the mean and variance of mushroom umbrella esophageal cancer and coarctate esophageal cancer respectively, and constitute the texture feature vector. Finally, we processed descriptive statistical analysis for feature vector, and obtain the classification ability of the best features, in order to evaluate the feature's classification ability by Bayes discriminant analysis.

2. MATERIALS AND METHODS

2.1. Image Acquisition

The performance of the method has been tested on a real image set from Xinjiang medical university first affiliated hospital. Clinicians were to guide the position, size and extent of the esophagus were defined in X-ray barium angiogram images by experienced radiologist. A total of 200 free-hand regions of interest(ROI) were sampled,100 of which corresponded to mushroom umbrella esophageal cancer and 100 to coarctate esophageal [7] cancer. Using free-hand way to extract ROI is a result of our emphasis focused on the esophageal disease classification. At present, feature extraction of literature is very little for esophageal X-ray barium angiogram. Therefore, this paper aimed at the characteristics of esophageal disorders, for mushroom umbrella esophageal cancer and coarctate esophageal cancer images will be analyzed. We proposed a wavelet-based texture analysis method and carried out effective feature extraction for esophageal lesions. First, we normalized scale, removing the noise by median filter, enhancing by histogram equalization, and acquiring clearly gradation picture[8].Second, we used wavelet transform method to respectively calculate the mean and variance of mushroom umbrella esophageal cancer and coarctate esophageal cancer, to constitute the texture feature vector; Finally, we used descriptive statistical analysis for feature vector, and obtain the classification ability of the best features to evaluate the feature's classification ability by Bayes discriminant analysis.

2.2. Wavelet Transform

Image processing is an important area of wavelet analysis and application. In recent years, with the improvement of people's demands for image compression, edge, feature detection and texture analysis. Multi-scale and time-frequency analysis of wavelet-based transform has become one of the powerful tools for image processing [9-10]. According to different levels of wavelets, the wavelet decomposition can make the image expanded, and can also be separated from approximation and details. So the wavelet-transformation can be used to de-noise, reconstructed and feature-extracted the image.

Since the image is two-dimensional signal, so the first need is to be extended to two-dimensional wavelet transform. f(x1,x2)represents a two-dimensional signal, x1,x2 denote horizontal and vertical axis, respectively. ϕ(x1,x2)represents a basic two-dimensional wavelet, scaling function corresponding to ϕ(x1,x2).If the scaling function could be separated, That's ϕ(x1,x2)=ϕ(x1)*ϕ(x2).Make φ(xi) is a one-dimensional wavelet function[11],and with corresponded to ϕ(xi). In that way, the two-dimensional dyadic wavelet can be expressed as the following three separable of orthogonal wavelet function:

$$\phi^1(x_1,x_2)=\varphi(x_1)\phi(x_2) \tag{1}$$

$$\phi^2(x_1,x_2)=\phi(x_1)\varphi(x_2) \tag{2}$$

$$\phi^3(x_1,x_2)=\phi(x_1)\phi(x_2) \tag{3}$$

It shows that the separable, two-dimensional wavelet can divide into two steps. First along with the direction x1, used $\phi(x_1)$and $\varphi(x_2)$as analysis respectively. The $f(x_1,x_2)$ is represent by approximation and details. The image process wavelet-transform, that is to say, the rows and columns of the images are convolved with both a low-pass filter h and high-pass filter g. Then the sample is processed from one out of two. Thus a two-dimensional wavelet decomposition of image is performed that results in four sub-bands, including one approximation and three detail images. It represents the low-frequency and high-frequency contents image, respectively [12].The wavelet two-dimensional process is shown in Fig.1. I represents original image;I_{LL},I_{LL1} are lower frequency parts of image, and represent approximation at first and second level,respectively;I_{LH},I_{HL},I_{HH},I_{LH1},I_{HL1},I_{HH1} are high frequency part of image, and represent the details of horizontal, diagonal and vertical directions at first and second level, respectively[13].

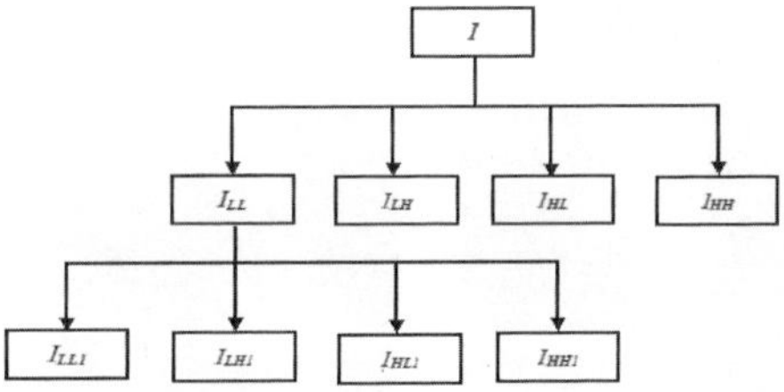

Fig.1. The discrete wavelet transformation at two levels diagram.

2.3. Feature Extraction

The purpose of feature extraction is to get the difference texture features. two-level wavelet decomposition reflects frequency conversion in different directions, thus it also reflect texture feature of the image [14]. High frequency prat of the original image is detailed sub-images, it contains a main texture information. Therefore, this research select 12 sub-channels and I_{LL4} sub-channels of four level wavelet transform, a total of 13 sub-channels. The mean and variance of wavelet coefficients, they are employed as the characteristic value set.

During feature values are choosing, the experiment reference to literature[15]about the calculation of the wavelet coefficients, it can be concluded 26 characteristic as feature set.

1)Average

$$M_{LC}=\frac{1}{i\times j}\sum_{i=0}\sum_{j=0}M_{ij} \tag{4}$$

2)Variance

$$V_{LC}=\frac{1}{i\times j}(M_{ij}-M_{LC})^2 \tag{5}$$

3. EXPERIMENTAL RESULTS AND ANALYSIS

3.1. Image Preprocessing Results

The purpose of image preprocessing is processing for image obtained from medical equipment, removing outside interference appears of the noise, enhancing the contrast between normal tissue and the pathological tissue, and getting visibility and contrast strong of ROI. Thus the doctor's clinical diagnosis and subsequent image processing provides better support. In this paper, our preprocessing methods consists of normalizing scale, removing the noise by median filter, enhancing by histogram equalization etc.

The preprocessing results is preformed on two different types of mushroom umbrella and coarctate esophageal cancer images. From Fig.2, Fig.3, Fig.2(a) and Fig.3(a) represent ROI images; Fig.2(b) and Fig.3(b) represent normalizing scale images; Fig.2(c) and Fig.3(c) represent removing the noise images, and Fig.2(d) and Fig.3(d) represent senhancing by histogram equalization.

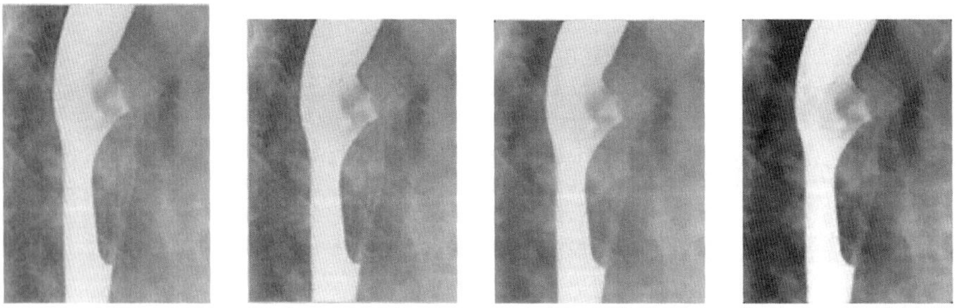

(a) ROI image (b) normalized image (c)Image Denoising (d)Image enhancement

Fig.2. Preprocessing results of mushroom umbrella esophageal X-ray barium angiogram image.

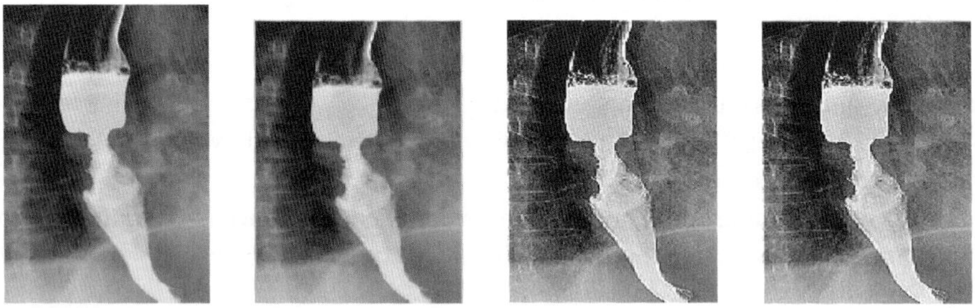

(a) ROI image (b) normalized image (c)Image Denoising (d)Image enhancement

Fig.3. Preprocessing results of coarctate esophageal cancer X-ray barium angiogram image.

3.2. Wavelet-based Transform Feature Extraction Results

We used symlet wavelet to process four level decomposition for mushroom umbrella esophageal images, calculated 13 sub-channels of the mean and variance of wavelet coefficients. HL sub-channel set first level is decomposed to get the mean of wavelet coefficients, it represents HL1-M;HL1-V represents the variance of wavelet coefficient, during HL sub-channels at first level is decomposed, and so on. Part of feature results for mushroom umbrella and coarctate esophageal cancer are tabulated in Table 1.

Table 1. Feature results of mushroom umbrella and coarctate esophageal cancer

Feature	Image Type							
	mushroom umbrella esophageal cancer				coarctate esophageal cancer			
HL1-M	2.452	3.225	...	2.007	2.148	1.853	...	1.828
HL1-V	4.090	7.593	...	2.5150	4.010	2.671	...	2.547
LH1-M	10.018	12.886	...	7.513	9.472	7.663	...	8.136
LH1-V	13.388	28.170	...	7.899	16.098	9.550	...	9.923
HH1-M	29.104	33.1761	...	29.096	28.027	20.805	...	24.038
HH1-V	31.394	69.155	...	31.497	34.779	25.324	...	29.014
....	....	....	...	....	....	....	...	....
LH4-V	16.820	23.977	...	18.910	16.492	15.775	...	14.325
HH4-M	35.2033	23.216	...	34.2673	29.303	21.128	...	21.028
HH4-V	42.707	32.380	...	43.500	45.460	29.077	...	32.969
LL4-M	1998.148	1907.0421	...	1750.991	2008.733	1993.883	...	1721.957
LL4-V	967.569	1109.158	...	1147.108	1193.272	1023.038	...	1237.699

3.3. Feature Selection

Original feature data obtained directly from the image feature extraction, generated large amounts of data. To effectively carry the pattern classification and recognition needed is that original data analyzed and selected, obtained classification ability of strong features, thereby constitute a feature vectors. Through feature extraction and selection, the experiment is not only can reduce classification errors, but also can reduce the processing time. In this paper, based on independent samples t-test of feature selection has been applied to the feature vectors estimated from wavelet transform, in order to produce redundant feature vectors. According to Table 1, the feature selection results are tabulated in Table 2.

Table 2. The feature selection results for two types of esophageal cancer

	95% Confidence Interval of the Difference		F	P
	Lower	Upper		
HL1-M	0.053	0.385	7.003	0.009
HL1-V	0.110	0.889	6.859	0.010
LH1-M	0.382	1.589	4.964	0.002
LH1-V	0.751	2.786	5.623	0.001
HH1-M	1.725	5.703	9.202	0.000
HH1-V	3.123	9.673	10.436	0.000
HL2-M	3.604	14.413	7.992	0.001
HL2-V	5.397	19.524	9.271	0.001
LH2-V	-1.242	-0.229	0.348	0.005
HH2-V	-2.909	-0.199	0.636	0.025
HL3-V	-7.029	-0.323	2.462	0.032
LH4-M	0.060	2.020	5.119	0.025
HH4-M	0.151	4.618	1.602	0.037

From table 2,HL1-M, HL1-V, LH1-M, LH1-V, HH1-M, HH1-V, HL2-M, HL2-V, LH2-V, HH2-V, HL3-V, LH4-M, HH4-M proceed independent samples t-test.P values were less than 0.05. The analyzing results showed that mushroom umbrella and coarctate esophageal cancer images appeared on gray distribution uniformity and the degree of complexity texture in different. In this case, it may provide a certain gist for pathogenesis, tracking disease, and efficacy monitoring. So this 13 feature values can be used to distinguish mushroom umbrella and coarctate esophageal cancer reference feature.

3.4. Bayes Discriminant Analysis

The analyzed results can be seen in table 2, obtained 13 with better characteristics of the classification ability. Therefore, with this 13 features to classify, the classified results had shown in Table 3.

Table 3. Classification results of mushroom umbrella and coarctate esophageal cancer feature vector

Image of the original classification	Discrimination and Classification	
	Mushroom umbrella esophageal cancer	Coarctate esophageal cancer
Mushroom umbrella esophageal cancer	79	21
Coarctate esophageal cancer	26	74

We processed the classification for 200 pictures of mushroom umbrella esophageal cancer and coarctate esophageal cancer respectively. The experimental results showed that using Bayes discriminant analysis and independent samples t-test is classifying, mushroom umbrella esophageal classification accuracy rate reached to 79%,esophageal ulcer type classification accuracy rate reached to 74%.

Moreover, we don't use independent samples t-test for the classification of X-ray barium angiogram esophageal into mushroom umbrella esophageal cancer and coarctate esophageal cancer images. Table 4 provided a summary of the experimental results.

Table 4. Classification results of original feature vector

Image of the original classification	Feature set	Classification accuracy (%)
Mushroom umbrella esophageal cancer	26	71
Coarctate esophageal cancer	26	73

From Table 4, mushroom umbrella esophageal classification accuracy rate reached to 71%, esophageal ulcer type classification accuracy rate reached to 73%.Compared Table 3 with Table 4, the experimental results can be also concluded that the proposed method is more suitable for mushroom umbrella esophageal classification.

4. CONCLUSION

As a conclusion, we have proposed a method for wavelet-based feature extraction and selected the optimal characteristics using independent samples t-test, and evaluated the Bayes discriminant analysis to classify two kinds of Xinjiang Kazakh esophageal cancer X-ray barium angiogram images into mushroom umbrella esophageal and coarctate esophageal cancer. According to the Xinjiang Kazakh high morbidity of esophageal types, there were differences in the texture distribution, combined with the characteristic of esophageal disorders, and presented wavelet-based feature extraction method. This method is effectively operation well for the classification of the Kazakh esophageal cancers into mushroom umbrella esophageal and coarctate esophageal cancer images with high accuracy. From the results of experimental analysis, the wavelet-based feature selection was yielding better results as compared to the feature extraction directly from the image without applying independent samples t-test. The testifies the choice of using wavelet transform and independent samples t-test. Therefore, based on wavelet - transformation of feature extraction algorithm can get a certain extent improved accurate classification. Meanwhile, the method can be extended to other organs and the lesion area as a future work.

Acknowledgments

The authors are very grateful to support of department of radiology of the first affiliated hospital of Xinjiang Medical University. This research was partially supported by National Science Foundation under grant number(81460281,81160182,81560294) .

References

1. Parkin DM, Bray FI,Devesa SS.Cancer burden in the year 2000. *Eur J Cancer* 2001;**37**:S4-66.
2. Qiu YD, Li JC.Molecular targeted therapy for esophageal cancer.*Practical Oncology Journal* 2015;**29**:57-60.(In Chinese).
3. Zhai JH, Zhao WX, Wang XZ. image feature extraction for reserch.*journal of hebei university* 2009;1885-89. (In Chinese).
4. E.L. Chen, P.C. Chung, C. L. Chen, etal. An automatic diagnosis system for CT liver image classification. *IEEE trans. Biomed. Eng* 1988;**45**:783-794.
5. Yan FH Xu PJ, Ling ZQ, etc.sclerosis type hepatocellular carcinoma of spiral CT multi-phase scan manifestation. *Chinese journal of medical imaging* 2003;**l9**:417-420. (In Chinese).
6. Zhu BY, Chen H. Research on methods and application of texture analysis of medical image. *the capital university of medical of biomedical engineering* 2003. (In Chinese).
7. S.Gr.Mougiakakou, I.Valavanis, K.S.Nikita, A.Nikita, etc. Characterization of CT Liver Lesions Based on Texture Features and a Multiple Neural Network Classification Scheme.*EMBC* 2003;1287-1290.
8. Murat Hamit, Yang F, Kong DW,etc. Classification on Xinjiang high morbidity of Kazakh Esophageal Disease based on texture feature. *Journal of Xinjiang Medical University* 2014;**37**: 273-276. (In Chinese).
9. F.Abramovich,T.C.Bailey,T.Sapatinas.Wavelet analysis and its statistics applications. *The Statistician* 2000;49:1-29.
10. Liu Y.H. Wavelet feature extraction for high-dimensional microarray data. *Neurocomputing* 2009;**72**:985-990.
11. Duo HQ, Wang YC, Xue HR. Frequency features extraction and analysis of wood texture based on the 5/3 wavelet transform. *College of Material science and art design, Inner Mongolia agriculture university* 2011. (In Chinese).
12. A.Padma Nanthagophal R, Sukanesh Rajamony. Automatic classification of brain computed tomography images using wavelet-based statistical texture features. *The Visualization Society of Japan* 2012;1-10.
13. Yu HP, Liu YX, Sun JP. Separated frequency features extraction and analysis of wood texture based on wavelet.*Key Laboratory of Bio-based Material science and technology of ministry of education, Northeast Forestry University* 2005;41. (In Chinese).
14. Yang ZS, Li CF, Zhou KY, etc. Diagnosis of Prostate Cancer and texture feature extraction of ultrasound images based on wavelet transform. *Space Medicine & Medical Engineering* 2009;**22**:281-285. (In Chinese).
15. Wang SZ, Ji XP, Yan WJ. The extraction of texture features based on wavelet transform, sci-tech information development & economy 2009;**18**:149-150. (In Chinese).

ELECTROCHEMICAL IMPEDANCE SPECTROSCOPIC TO REFLECT THE CYTOTOXICITY IN THE DOXORUBICIN-INDUCED HELA CELLS APOPTOTIC MODEL ON DIFFERENT GRAPHENE/GELATION ASSEMBLED LAYER INTERFACING LIVE CELLS

Lingzhi Zhao[1,3,a], Liu Zhao[2], Yanqing Miao[1], and Chunye Liu[1]

[1]*Department of pharmacy, Xi'an Medical College, Xi'an 710021, China*

[2]*Beijing Research Center of Agricultural Standards and Testing, Beijing 100097, China*

[3]*Key Laboratory of Analytical Chemistry for Life Science of Shaanxi Province, School of Chemistry and Chemical Engineering, Shaanxi Normal University, Xi'an 710062, China*

E-mail: [a]oldskyhappy_zlz@163.com

This study was designed to investigate the cytotoxicity in the doxorubicin-induced Hela cell apoptotic model on different reduced grapheme(rGO)/gelation assembled layer interfacing live Hela cells. Respectively with gelatin/GC, rGO/GC, and gelatin/rGO/GC as the growth substrates interfacing Hela cells, the change of the charge-transfer resistance for cell apoptosis was studyed by using electrochemical impedance spectroscopic when Hela cells were treated with a dose of the anticancer drug doxorubicin. The results showed that the same concentration of drugs can cause a large number of cells death growing on the gelatin /GC substrate, while the cytotoxicity of drugs weakened for cells growing on the rGO /GC and gelatin /rGO/GC substrate. It is assumed that these results could be attributed to the distinct nanotopographic features of these two kinds of substrates. but it need further exploration for the possible mechanisms that graphene nanomaterial as a cell growth substrate can weaken the cytotoxic effects of drugs.

1. INTRODUCTION

Grapheme, as one new kind of nanostructure in the carbon family, with extraordinary structure and unique electronic properties, which are different from those of other kinds of carbon structures such as diamond, fullerene, and carbon nanotubes, has attracted considerable attention[1-3]. Good biocompatibility of graphene in the biomedical field also caused a great deal of interest, related research and applications, such as the in vitro studies about grapheme biotoxicity, the in vivo biotoxicity, and the research of graphene as drug carrier, which have been repeatedly reported[4-8]. Chen *et al*[6] studied the ability of two kinds of nanocarbon substrates (SWCNT and rGO) to interface different types of cells. They demonstrated that rGO film, the flat 2D nanocarbon structure analogous to a 2D cell membrane, is more biocompatible with the tested cells, indicating its potential applications in biology. On the basis of the study mentioned above, our study was designed to investigate the cytotoxicity in the doxorubicin-induced Hela cell apoptotic model on different graphene/gelation assembled layer interfacing live Hela cells by using electrochemical impedance spectroscopic. This comparative study suggests that Hela cells with doxorubicin treatment respond differently to these three types of substrates. Furthermore, to verify that the graphene as a cell growth substrate can weaken the cytotoxicity of doxorubicin which can induced Hela cells apoptosis, bright field picture was applied for directly reflecting the cytotoxicity of doxorubicin with grapheme and gelatin nanocomposite interfacing live cells, which had good agreement with the results of electrochemical impedance spectroscopic. These observations could be attributed to the distinct nanotopographic features of these two kinds of substrates. but it need further exploration for the possible mechanisms that graphene nanomaterial as a cell growth substrate can weaken the cytotoxic effects of drugs.

2. EXPERIMENTAL

2.1. Chemicals and Reagents

Dulbecco's modified Eagle Medium (DMEM, high glucose) was purchased from Gibco Invitrogen, Carlsbad, CA, USA. Fetal bovine serum (FBS, South American Origin) was obtained from Hyclone. Gelatin (from porcine skin, type A), and doxorubicin (DOX) were purchased from Sigma. $K_3Fe(CN)_6$,$K_4Fe(CN)_6$, NaH_2PO_4,Na_2HPO_4, and NaCl were purchased from Beijing Chemical Co. (Beijing, China). Phosphate-buffered saline (PBS, pH 7.4) was prepared containing 136.7mM NaCl, 2.7mM KCl, 0.087M Na_2HPO_4, and 0.014M KH_2PO_4 into water. All reagents were at least analytical grade and used as received. Aqueous solutions were prepared with Milli-Q water.

2.2. Cell Culture

The human cervical carcinoma cells (Hela cells, Peking Union Medical College, Tsinghua University) were cultured in DMEM medium supplemented with 10% fetal bovine serum, penicillin (100μg/mL), and streptomycin (100μg/mL) in an humid incubator (5% CO_2, 37°C). After incubation for some time (typically, 2 days), the cells were trypsinized with 0.25% (v/v) trypsin solution dissolved in phosphate buffer (pH 7.4) containing 0.02% EDTA when the adherent cells approached confluence. To grow Hela cells onto electrode surface for the EIS measurements, a thin layer of gelatin was first confined onto glassy carbon (GC) electrode surface by immersing the electrodes into the aqueous solution of gelatin (0.2%) for 90 min and then rinsing the electrodes with PBS to remove unstable adsorbed gelatin (gelatin-modified electrodes, gelatin/GC). 10mL of 0.1 $mg \cdot mL^{-1}$ rGO aqueous solution was drop-coated onto the surface of the pretreated GC electrode, dried under the infrared lamp (rGO-modified electrodes, rGO/GC). The rGO-modified electrode was inserted into 0.5 mL of 1.0 $mg \cdot mL^{-1}$ gelatin aqueous solution for 30 min, and then rinsed with water thoroughly to remove the weakly adsorbed gelatin molecules (gelatin/rGO/GC). The gelatin/rGO/GC was obtained and stored in a refrigerator (4°C) before use. Hela cells were grown onto the modified electrodes by immersing the electrodes into the cell suspension of 50 μ L 1×10^6 cells mL^{-1}cell suspension for 6 h in a humid incubator (5% CO_2, 37°C). Cell number was determined using a traditional countering chamber (Thoma).

2.3. Apparatus and EIS Studies

Electrochemical impedance spectroscopy (EIS) were carried out in 5 mL of 10 mM PBS containing 2 mM $K_4Fe(CN)_6$ and 2 mM $K_3Fe(CN)_6$ (1:1) and 0.10 M NaCl as electrolyte solution at the equilibrium potential of $[Fe(CN)_6]^{3-/4-}$ (0.25 V). The open circuit potential was used as a DC bias potential, on which amplitude of the AC potential was 5 mV (Popkirov and Schindler, 1995). And the EIS was recorded over the frequency from 100 kHz to 1 Hz with a sampling rate of 12 points per decade. 2 s as quiet time was performed before each measurement. All EIS measurements were repeated for a minimum of three times with an electrode. The modified GC electrodes with diameter of 2 mm were used as working electrode, a Ag/AgCl (KCl-saturated) electrode as reference electrode, and stainless steel plate (2 cm×3 cm) as counter electrode. GC electrodes used as the substrate for cell growth were first polished with emery paper and then with aqueous slurries of fine alumina powder (0.3 and 0.05 mm) on a polishing cloth. The electrodes were finally rinsed with acetone and doubly distilled water under an ultrasonic bath, each for 5 min, and allowed to dry at room temperature. Prior to the measurements, the Hela cells /modified electrodes were taken out from the DMEM medium and rinsed twice with the PBS. doxorubicin-induced apoptotic model was constructed by immersing the Hela cells/modified GC electrodes in DMEM medium containing 15 μg/mL doxorubicin for different time. All electrochemical experiments were performed at ambient temperature.

3. RESULTS AND DISCUSSION

3.1. Characterization of Different Graphene/Gelation Assembled Layer Interfacing Live Cells

Electrochemical impedance spectroscopy(EIS) is employed to monitor the interfacial properties of electrode during the modification process. Fig. 1A compares the Nyquist plots recorded with bare (blue line), gelatin /GC (black line), rGO/GC (red line), and gelatin/rGO/GC electrodes (green line) in 0.10 M NaCl solution buffered with 10 mM phosphate (pH 7.4) in the presence of 2 mM $K_4Fe(CN)_6$ and 2 mM $K_3Fe(CN)_6$.. At these modified GC electrodes, Nyquist plots include a semicircle portion and a linear line portion, corresponding to the charge-transfer process and diffusion process, respectively. The diameter of the semicircle represents the charge-transfer resistance (R_{et}) of the $[Fe(CN)_6]^{3-/4-}$redox probe at the electrode surface. As shown, the confinement of gelatin onto the electrode surface (Fig. 1A, black line) clearly increases the charge-transfer resistance, as compared with that at the bare electrode, presumably due to the fact that the gelatin layer was not conductive and could thus block the interfacial charge-transfer of the $[Fe(CN)_6]^{3-/4-}$ redox probe. While the R_{et} at rGO/GC electrode greatly decreased (Fig. 1A, red line), This is mainly attributed to the fact that rGO can accelerate electron transfer of $[Fe(CN)_6]^{3-/4-}$ at the electrode surface owing to its excellent two-dimensional electron-conductive network. Since gelatin/rGO/GC electrodes was rGO and gelatin nanocomposite, the R_{et} at gelatin /rGO/GC electrodes should be in between rGO/GC and gelatin /GC. Different from those obtained with bare and above-mentioned three kinds of modified GC electrodes, the Nyquist plot recorded with the Hela cells/gelatin/GC(Fig. 1B, black line), the Hela cells/rGO/GC(Fig.1C, black line), and the Hela cells/gelatin/rGO/GC(Fig. 1D, black line) electrodes displays two semicircles with no linear portion. These results imply that at least two time constants were involved at the Hela cells-modified GC electrode and that the diffusion process of redox probe was not the limiting process within the frequency range employed here. In the EIS, the semicircle at high and low frequency presumably reflects the charge-transfer resistance from the above-mentioned three kinds of modified layer and cell layer, respectively. These result demonstrates that, on one hand, Hela cells could be cultured onto three kinds of modified GC electrode surface and, on the other hand, the attachment of Hela cells onto the electrode surface hinder the access of the redox probe to the electrode surface from the solution.

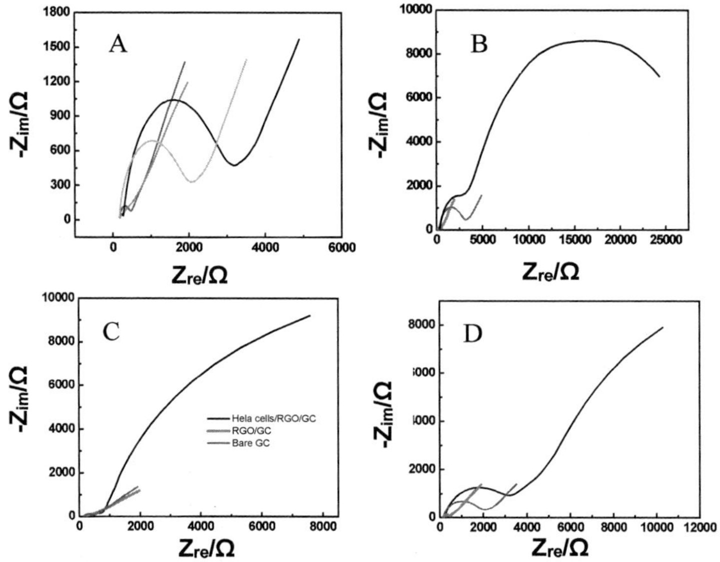

Fig. 1. (A)Nyquist plots obtained at bare (blue line), gelatin/GC (black), rGO/GC (red), and gelatin/rGO/GC electrodes (green). (B) Nyquist plots obtained at the Hela cells/gelatin/GC (black), gelatin /GC(blue), and bare (red).(C) Nyquist plots obtained at the Hela cells/ rGO /GC (black), rGO/GC(blue), and bare (red) (D) Nyquist plots obtained at the Hela cells/gelatin/ rGO/GC (black), gelatin/rGO/GC(blue), and bare (red). EIS was recorded in 0.10 M NaCl solution buffered with 10 mM phosphate buffer (pH 7.4) containing 2 mM $K_4Fe(CN)_6$ and 2 mM $K_3Fe(CN)_6$ within a frequency range of 1 to 10^6 Hz at the open circuit potential of $[Fe(CN)_6]^{3-/4-}$ couple. The amplitude of the alternate voltage was 5 mV.

3.2. Electrochemical Impedance Spectroscopic to Reflect the Cytotoxicity of Doxorubicin on Different rGO/gelation Assembled Layer Interfacing Live Cells

The electrochemical impedance method was also applied to describe the viability of cells and to evaluate the effectiveness of antitumor drug doxorubicin on Hela cells. Theoretically, the cells can have good attachment, grow and proliferate on our as-established electrode surface. Since the conductivity of cell membrane is quite poor, cells can be considered to create a barrier for the charge transfer process of the redox probe. These demonstrated that the charge-transfer resistance (R_{et}) of the redox probe can reflect the state of cells immobilized on the surface of electrodes and the cell membrane permeability, which had been verified in our previous report[9]. In this work, we interface living cells with chemically reduced graphene oxide (rGO) film, biomacromolecule gelatin film, and a composite of rGO and gelatin assembled layer, respectively. The rGO was employed as a substrate to be interfaced with cells, on the surface are in favor of adhesion of cells, implying its potential application in biology. And the use of biomacromolecule gelatin, which is useful in promoting cell adhesion, was made in building relative friendly matrix for immobilized cells. From Fig. 2, it can be seen that EIS measurements were conducted at three kinds of modified GC electrodes with $[Fe(CN)_6]^{3-/4-}$ as the redox probe, in which Hela cells/modified GC electrodes were immersed into 15μg/mL doxorubicin for different times. The EIS results show that the presence of doxorubicin in the cell culture clearly decreases the R_{et} of the $[Fe(CN)_6]^{3-/4-}$ redox probe in different degrees at the Hela cells/modified GC electrodes, and the R_{et} decreases with increasing time employed for immersing the Hela cells/modified GC electrodes into the cell culture containing doxorubicin. The R_{et} at Hela cells/gelatin/GC electrodes decreased sharply with increasing the immersing time, while the R_{et} at Hela cells/rGO/GC electrodes slightly decreased. In summary, Hela cells with doxorubicin treatment respond differently to these three types of substrates by electrochemical impedance method, likely due to their distinct topographic features. The mechanism that rGO as a cell growth substrate can weaken the cytotoxicity of doxorubicin was discussed later.

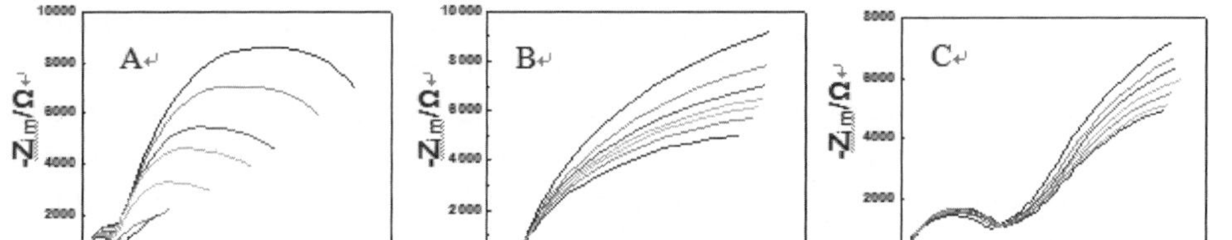

Fig. 2. Nyquist plots recorded at the Hela cells/gelatin/GC(A), Hela cells/rGO/GC(B), and Hela cells/gelatin/rGO/GC(C) electrode in 0.10 M NaCl solution buffered with 10 mM phosphate buffer (pH 7.4) containing 2 mM $[Fe(CN)_6]^{3-/4-}$. Prior to the measurements, the electrodes were immersed into the DMEM medium containing 15μg/mL for different time of 0 min (black line), 10 min (red), 20 min(blue), 30 min (green), 40 min (cyan), 50 min (magenta), and 60 min (navy).

3.3. Bright Field Picture to Reflect Cytotoxicity on rGO/glass and Glass Substrate Interfacing Live Cells

To further verify that the graphene as a cell growth substrate can weaken the cytotoxicity of doxorubicin which can induced Hela cells apoptosis., As shown in Fig. 3, bright field picture can directly reflect the cytotoxicity of doxorubicin on rGO/glass (Fig.3B, Fig.3D) and gelation/glass substrate (Fig.3A, Fig.3C) interfacing Hela cells. Picture A displays that normal cell lines without doxorubicin treatment have clear cytoplasm on gelation/glass substrate. while Picture C showed the pathological state of Hela cells on the gelation/glass substrate incubating 15μg/mL doxorubicin after 1 h, it can been seen cell body shrinkage, round and detachment from the substrate, indicating that most of that programmed cell death has been launched with loss of cellular three-dimensional structures, which illustrates that 15μg/mL doxorubicin is enough to induce Hela cells apoptosis. Compared to gelation/glass substrate, on rGO/glass substrate, Hela cells without doxorubicin treatment grew well and have good cell attachment and rapid proliferation, indicating rGO nanomaterials has good biocompatibility as a cell growth substrate, which had good agreement with other reports[4-7]. After 15μg/mL doxorubicin treatment, on rGO/glass, the growing state of Hela cells did not change significantly, cells was still in good condition and seemed to have no cell death programme runing. This phenomenon had entirely different from the state of Hela cells on the glass substrate after the same dose of doxorubicin treatment, which had good agreement with the results of electrochemical impedance spectroscopic, it need further exploration for the possible mechanisms that graphene nanomaterial as a cell growth substrate can weaken the cytotoxic effects of drugs. A preliminary deduction is: on the one hand, since rGO with great conjugated structure produced π-π stacking and hydrophobic interactions with drug molecules(e.g. doxorubicin) and the surface of rGO had some physical adsorption of drug molecules, resulting in the actual concentration of drug into cells is lower, so that drug toxicity decreased at some degree. On the other hand, any drugs are only required to undergo the biorecognition of the cell membrane and adopt certain way to enter into the cells, they can exert the efficacy of drugs. When graphene nanomaterials as cell growth substrates, changed the state of the cell adhesion, cell membrane permeability and the transport mechanisms of cell membrane for transporting a variety of bioactive molecules and so forth, causing that some drugs can not be effectively passed the cell membrane, so that the efficacy of drugs were sharply weakened. Specific inhibition mechanism of these needs to be further verified by other experiments.

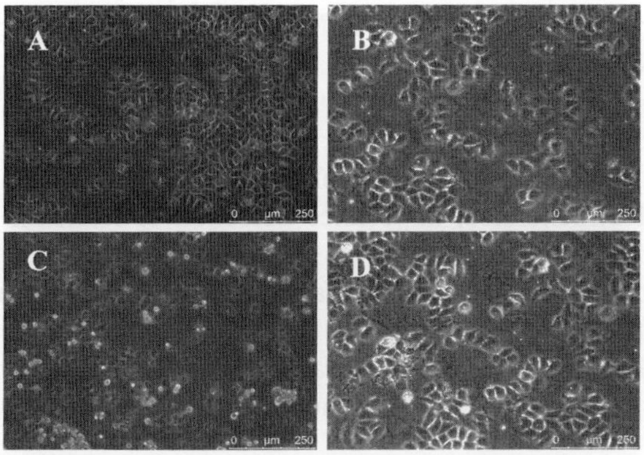

Fig. 3. Microscopic images of Hela cells on (A) gelation/glass; (B) rGO/glass without 15μg/mL doxorubicin treatment. Microscopic images of Hela cells on (C) gelation/glass; (D) rGO/glass with 15μg/mL doxorubicin treatment for 1 h.

4. CONCLUSIONS

The present study here described a novel electrochemical impedance method for reflecting the cytotoxicity in the doxorubicin-induced Hela cells apoptotic model on different graphene/gelation assembled layer interfacing live Hela cells. Cell apoptosis was studied when Hela cells were treated with 15μg/mL anticancer drug doxorubicin. The results showed that the same concentration of drugs can cause a large number of cells death growing on the gelatin/GC substrate, while the cytotoxicity of drugs weakened for cells growing on the rGO /GC and gelatin /rGO/GC substrate. Bright field picture was also applied for directly reflecting the cytotoxicity of doxorubicin with grapheme and gelatin nanocomposite interfacing live cells, which had good agreement with the results of electrochemical impedance spectroscopic. It is assumed that two factors can explain this phenomena, one is that graphene adsorbs drugs physically; another is that graphene changes cell membrane state, it need further exploration for specific protection mechanism that graphene nanomaterial as a cell growth substrate can weaken the cytotoxic effects of drugs.

Acknowledgements

This work is financially supported by NSF of China (Grant Nos. 21305109 for L. Zhao, 81202492 for C. Liu), by the PhD Startup Foundation of Xi'an Medical University of China (No. 2012DOC09) and by the Scientific Research Plan Projects Foundation of Shaanxi Science and Technology Department of China (No. 2014JQ2073).

References

1. Li D, Kaner RB. Graphene-based materials. *Nat Nanotechnol* 2008; 3: 101.
2. Allen MJ, Tung VC, Kaner RB. Honeycomb carbon: a review of graphene. *Chemical reviews* 2009; **110**: 132-145.

3. Chen H, Müller MB, Gilmore KJ, et al. Mechanically strong, electrically conductive, and biocompatible graphene paper. *Adv. Mater* 2008 ; **20**: 3557-3561.
4. Chang Y, Yang ST, Liu JH, et al. In vitro toxicity evaluation of graphene oxide on A549 cells. *Toxicology letters* 2011;**200**: 201-210.
5. Zhang D, Zhang Y, Zheng L, et al. Graphene oxide/poly-l-lysine assembled layer for adhesion and electrochemical impedance detection of leukemia K562 cancer cells. *Biosensors and Bioelectronics* 2013;**42**: 112-118.
6. Agarwal S, Zhou X, Ye F, et al. Interfacing live cells with nanocarbon substrates. *Langmuir*, 2010; **26**: 2244-2247.
7. Sasidharan A, Panchakarla LS, Chandran P. Differential nano-bio interactions and toxicity effects of pristine versus functionalized graphene. *Nanoscale* 2011; **3**: 2461–2464.
8. Yang K, Wan JM, Zhang S. In vivo pharmacokinetics long-term biodistribution, and toxicology of PEGylated grapheme in mice. *ACS Nano* 2011; **5**:516–522.
9. Zhao L, Li X, Lin Y, et al. Electrochemical impedance spectroscopic measurements of FCCP-induced change in membrane permeability of MDCK cells. *Analyst* 2012; **137**: 2199-2204.

EFFECT OF MECHANICAL STRETCH ON APOPTOSIS AND BAX/BCL-2 PROTEIN EXPRESSION OF HUMAN ORAL SQUAMOUS CELL CARCINOMA KB CELLS *IN VITRO*

Yang Wang[a] and Hua Qu[b,*]

College of Science, Tianjin Polytechnic University, Tianjin, China

E-mail: [a]wangy0419@163.com, [b]*quhua-930@163.com*

[*]*Corresponding author*

Objective: To investigate the effect of mechanical stretch force on the apoptosis of Human oral Squamous Cell Carcinoma KB Cells cultured in vitro. Method: The KB cells were subjected to cyclic equibiaxial stretch forces for different regimens (squarewave, 10% and 15% elongation amplitude, 0.5Hz, 24h) using Flexcell4000TM strain unit. Apoptotic levels of the KB cells were quantified by flowcytometer right after the stretch application. The ratios of bax/bcl-2 was analysed by RFQ-PCR. Result: different stretch force (10% or 15% elongation) led to a increase of apoptosis rate of KB cells compared with the control, the apoptosis rate of 15% elongation was significantly higher than 10% elongation ($P<0.05$), different stretch force led to a decrease of bcl-2 mRNA expressing of KB cells compared with the control, the bcl-2 mRNA expressing of 15% elongation was significantly lower than 10% elongation ($P<0.05$), different stretch force led to an increase of bax mRNA expressing of KB cells compared with the control, the bax mRNA expressing of 15% elongation was significantly higher than 10% elongation ($P<0.05$). Conclusion: Stretching force can affect apoptosis in the KB cells culture. Different force loading magnitudes could induce the KB cells apoptosis by altering bax/bcl-2 protein expression.

1. INTRODUCTION

The mechanical stretch affect many respects of function and metabolism of tissues and organs, cell proliferation, apoptosis, differentiation, etc. Many researchers have studied the biologic effects of mechanical forces on tissues and cells, but most research focus on osteoblasts[1] , human lung adenocarcinoma A549 cells and Human Squamous Carcinoma of Tongue Tca8113 Cell [2,3] , etc. Very few reports about the effects of mechanical force on tissue of buccal cavity and else tumor cells can be found from correlative reference. Except some reports about periodontal ligament cells [4], The study of mechanical stimulus on Human oral Squamous Cell Carcinoma KB Cells have never been reported.

Operation and radiation-therapy are important therapeutic tool of malignant tumors of oral and maxillofacial region for years. However, the serious surgical trauma and toxic and side-effect of radiation and chemotherapy go against anticancer treatment ability into full play of patient. It's necessary to develop adjuvant therapy or new methods. Exploring the practicability of adjuvant therapy by mechanics means to oral cancer and providing the theoretical and experimental base for the therapy of oral cancer by physiotherapy, have significance in the treatment of Human oral Squamous Cell Carcinoma therapy.

In the study, the effects of cyclic stretch of different Loading project (squarewave, 0.5Hz, 24h) on apoptosis and apoptotic gene expression of the KB Cells were studied by FX-4000 strain unit. the apoptosis rates were detected by flow cytometry, the Bcl-2 and Bax mRNA expression were detected by Real-time fluorogenic quantitative polymerase chain reaction (RFQ-PCR).

2. MATERIALS AND METHODS

2.1. Cell Culture

Human oral Squamous Carcinoma (KB) Cell comes from the cell repository of Chinese Academy of Science ,The KB cells were grown on culture plates of six-well, The culture plates consisted of a flexible, collagen I-impregnated elastic membrane of 5 cm^2 (Flexcell international). the KB cells were cultured in RPMI1640 medium supplemented with 10% FBS and 100 U /ml streptomycin at 37 °C in 5% CO_2 humidified incubator.

2.2. The FX-4000 Strain System

The FX-4000 strain unit (Flexcell International) is used for mechanical loading experiment. It provides regulated strain to cells in culture, i.e., different waveforms, amplitude frequency, and duration of strain are controlled by computer. The instrument consisted of a Flexcell4000 FlexCentral system controller, a CO_2 incubator in which there's a Flexcell vacuum baseplate and gaskets on which the culture plates reside.

The baseplate contains loading stations, the plate consisted of six wells. The cells are grown on the six-well culture plates. a negative pressure was applied to the flexible wells through air pump.

2.3. Experiment Regimens

The KB cells which density is 1×10^4 cells/ml were grown on six-well culture plates 24 hours before the experiment. The base of culture plates consisted of a collagen I-impregnated elastic membrane which surface area is 5 cm^2.

The experimental regimens:

- square wave(Fig.1)

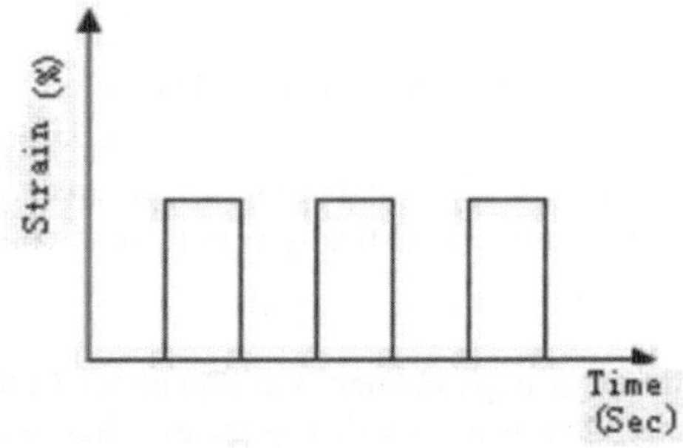

Fig. 1. Square wave.

- Elongation amplitude: 10 %, 15 %.
- Frequency: 0.5 Hz.
- Duration of strain: 24 h.

2.4. Examination

The cells were collected 24h after experiment. The percentage of apoptotic cells was determine d by flow cytometry using FITCAnnexin V/PI stain method. FITC-Annexin V could specifically bind to phosphatidylserine which exposed when apoptosis occurred. The KB cells of staining with FITCAnnexin V were shown as green-fluorescence by inverted fluorescence microscope, and could be detected by flow cytometry. The FITC-Annexin V could be used to test cell apoptosis. The apoptosis rate was tested and analyzed by flow cytometry; Control group cells were detected at the same condition.

The KB cells which density is 1×10^4 cells/ml were grown on six-well culture plates 24 hours before the experiment. The cells were harvested 24h after experiment, the mRNA expression of Bcl-2 and Bax were detected by Real-time fluorogenic quantitative polymerase chain reaction (RFQ-PCR). Control group cells were detected at the same condition.

The RNA was attained with reagent box(Gibco company), the DNA/ RNA tester (Britain Pharmarcia) was used to mensurate the purity of RNA quantificationally, Bcl-2, Bax and internal reference β-actin were specifically amplified by PCR after reverse transcriptase reaction, the upstream primer and downstream primer of Bcl-2, Bax and β-actin were designed by Primer3 Output(Table 1).

Table 1. The primer and sequence of BCL-2, BAX and B-actin

	upstream	downstream	The length of amplified product
β-actin	5'-gatgagattggcatggcttt-3'	5'-caccttcaccgttccagttt -3'	100bp
Bcl-2	5'-ggatgcctttgtggaactgt-3'	5'-agcctgcagctttgtttcat- 3'	236bp
Bax	5'-tttgcttcagggtttcatcc-3'	5'-cagttgaagttgccgtcaga-3'	246bp

3. STATISTICAL ANALYSIS

Results are analysed as mean ± SD of three independent determination. a single factor analysis of variance was used to decide statistically significant differences from the controls. Comparisons between the given experimental groups were made using t-test. Usually, $P<0.05$ was considered as the statistically significant level.

4. RESULTS

The apoptosis rates of different groups Human oral Squamous Cell Carcinoma KB Cells were showed in the Table 2. Data was recorded as mean ± SD. Independent-Samples T-test were used to determine-significant-differences between different groups. The level of significance was 5%. Apoptosis rate of different amplitude were shown in Fig.2.

Table 2. Apoptosis rates of the different groups KB cells

Square wave 0.5Hz, 24h	control	10% Amplitude	15% Amplitude
Apoptosis rate(%)	1.17±0.34	6.06±1.19	13.32±2.58

4.1. The Effect on Apoptosis of the KB Cell by Square Wave

Human oral Squamous Cell Carcinoma KB Cell was stretched by Squarewave (10% or 15%, 0.5Hz, 24h), The apoptosis rate of the KB cells stretched by Squarewave (elongation amplitude 0~10%) increased significantly compared with control group, The apoptosis rate of the KB cells stretched by Squarewave (elongation amplitude 0~15%) was significantly higher than Squarewave (elongation amplitude 0~10%) group($P<0.05$) (Fig.2). It indicated that elongation amplitude was very important for promoting the KB cells apoptosis.

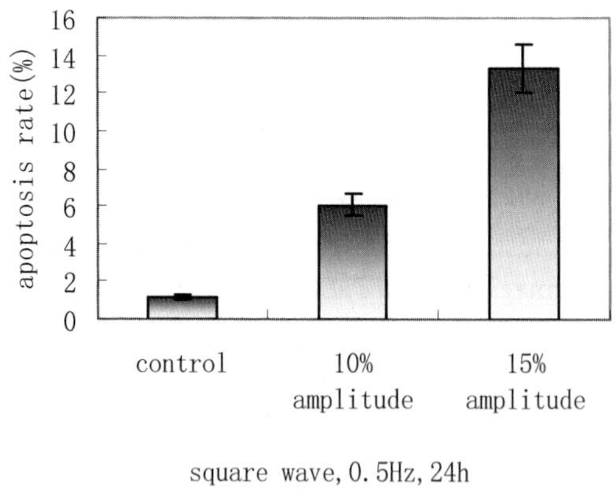

Fig. 2. The apoptosis rates of square wave group.

4.2. The Effect on BCL-2 MRNA of the KB Cell by Square Wave

Human oral Squamous Cell Carcinoma KB Cell were stretched by Square wave(10% or 15%, 0.5Hz, 24h), The bcl-2 mRNA of the 10% KB cells group was reduced significantly compared with the control group ($P<0.05$), the Bcl-2 mRNA expression of the 15% KB cells group was reduced significantly compared with the 10% KB cells group ($P<0.05$) (Fig.3). It indicated that elongation amplitude was very special to reducing the bcl-2 mRNA of KB cell.

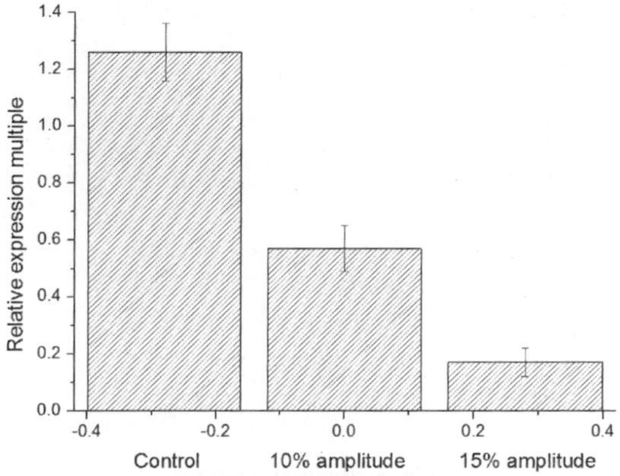

Fig.3. Bcl-2 mRNA expression.

4.3. The Effect on BAX MRNA of the KB Cell by Square Wave

Human oral Squamous Carcinoma(KB)Cell was stretched by Square wave(10% or 15%, 0.5Hz, 24h), The Bax mRNA expression of the 10% KB cells group was increased significantly compared to control group ($P<0.05$), the Bax mRNA expression of the 15% KB cells group was increased significantly compared with the 10% KB cells group ($P<0.05$) (Fig.4). It indicated that elongation amplitude was very special to reducing the Bax mRNA of KB cell.

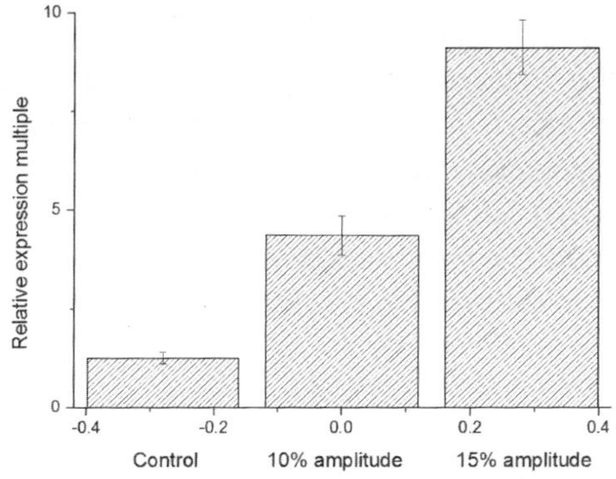

Fig. 4. Bax mRNA expression.

4.4. The Effect on BAX/BCL-2 of the KB Cell by Square Wave

The experimental results indicated that different elongation amplitude had different effects on bax/bcl-2 of the KB cells, all regimens could induce the apoptosis of the KB cells, and the apoptosis rate increased significantly with the stretch amplitude. The bax/bcl-2 increased significantly with the stretch amplitude(Fig.5). The experimental results indicated that the amplitude and waveform played important role with regard to apoptosis of the KB cells.

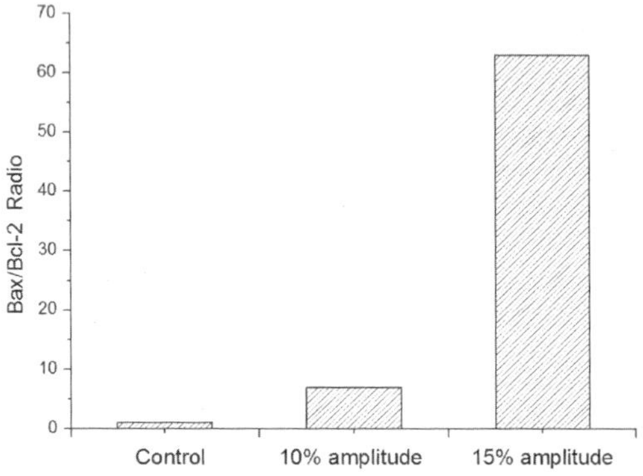

Fig. 5. Bax/Bcl-2 ratio.

5. DISSCUSSION

Recently, to research apoptosis of tumor cell is one of the important aspects. Previous studies on apoptosis of human squamous cell carcinoma of tongue focus on biochemistry, gene therapy, immunotherapy, molecular biology [5-8],etc. Mechanical stretching could induce many responses in different biological systems. It induces growth of vascular endothelium, bone, cardiac muscle and skeletal muscle. Strain regulates extra-cellular matrix molecule formation and morphological [9]. Very few reports to discuss the apoptosis of tumor cells by using mechanical stress. The research of the apoptosis of A549 cells by different stretch shows that apoptosis rate increases with strain (4 h ,0.5 Hz), time (15 % ,0.5 Hz) and frequency (15 %, 4 h). Research shows that the apoptosis of the cancer (KB) cell is induced after cells are subjected to stretch. Different stretch amplitude have different effects on the apoptosis of the cancer (KB) cell, square wave stretch have the higher apoptosis rate of the cancer (KB) cell, All groups can induce the apoptosis of the cancer (KB) cell. The higher the amplitude is given, the higher apoptosis rate of the cancer (KB) cell. The results are consistent with Ning [10].

The effects of mechanical stretch on proliferation of human squamous (Tca8113)cell carcinoma of tongue were studied under the same conditions [11]. The results shown that the proliferation rate of Tca8113 cells reduced significantly with square wave group compared to the control group, The higher the amplitude is given, the lower proliferation rate of the cancer (Tca8113) cell. The cellular proliferation rate reduced significantly with 10% elongation amplitude compared to the control group.

The paper showed, the square wave could increase apoptosis rate of the KB cells, the apoptosis rate increased with the stretch amplitude. The waveform indicates mode of mechanical stress, The elongation amplitude is proportional to the stress intensity. The results indicate that the waveform and amplitude play important role in apoptosis of the KB cells. The intrinsic pathway of apoptosis is activated by the cell itself, such as in response to the cyclic strain of different stretch amplitude. Apoptosis mediated by intrinsic pathway involves the release of cytochrome c which comes from mitochondria, an essential factor required the activation of caspases which regulated by the Bcl-2 proteins upstream family of caspase activation [12]. Bcl-2 are apoptosis suppressing factors. When Bcl-2 are overexpression, cells are protected against-apoptosis. when Bax is overexpression, cells are susceptible to programmed death. the ratios of Bax/Bcl-2 determine the fate of a cell consequently [13]. The stretch cultures of the KB cell induce a significant ratios ascent of Bax/Bcl-2. the imbalance between Bcl-2 and Bax by mechanical stretch play a significant role as observed here [3]. Square wave can raise apoptosis rate, it is used to inhibit or kill Human oral Squamous Carcinoma cancer (KB) cell. It is of practical and referential importance to the treatment of oral Squamous Cell Carcinoma and new treatment methods of cancer.

It was reported that the cells had sensitive receptor respond to mechanical signal, these receptors could transmission to intracellular components by special molecule channels, and the cellular physiological significance were adjusted [14]. The KB cells may have sensitive receptors of waveform and amplitude. Mechanical strain could change cytoskeleton, alter synthesis of DNA and many biological function by mechanics chemistry coupling of the receptors. The cellular mechanical signal transduction and the other experiment regimens need study continually.

References

1. J.S.Liu ,J.F.Ma , Q.Wang et al. Effects of fluid shear stress on apoptosis of osteoblasts and expression of bcl-2,Bax and caspase-3. *J Shanghai Journal of Stomatology*, 2008,**17**:501-504.
2. H.Qu, P.Gao. The Effect of Squarewave Stretching on Apoptosis of Human Oral Squamous Cell Carcinoma KB Cells. *Congres Proceedings-2012 5th International Conference on Biomedical Engineering and Informatics*, 2012, p1548-1550.
3. G.H.Wang, H.Qu, X.L.Gao, Y.F.Xie, Effects of Cyclic Squarewave Stretch on Proliferation and Apoptosis of Human Squamous Carcinoma of Tongue Cell Line Tca8113, *J Taiyuan University of Technology*, 2009, **40**:192-194.
4. Y. F. Zhang, Y. Zh. Duan, N. Wang, Effect of elastic stress on expression of corebinding factor a 1 mRNA in human periodontal ligament cells in vitro, *Chin J Stomatol*, 2003, **38**: 374-376.
5. J.Shen, C. Huang. Enhancement of cisplatin induced apoptosis by suberoylanilide hydroxamic acid in human oral squamous cell carcinoma cell lines, *J Biochemical Pharmacology*, 2007 ,**73**:1901–1909.

6. Y. Jun, D. Li, NF-kappaB signaling pathway is involved in growth inhibition, G2/M arrest and apoptosis induced by Trichostatin A in human tongue carcinoma cells, *J Pharmacological Research*, 2006 ,**54**: 406–413.
7. S.F. Zhao, X.Y.Tong. F.D.Zhu, Nitric oxide induces oral squamous cell carcinoma cells apoptosis with p53 accumulation, *J Oral Oncology*, 2005,**41**:785–790.
8. Z.N.Gao , Y.Q. Wei, P.S.Yang, Combined effects of soluble vascular endothelial growth factor receptor FLT-1 gene therapy and cisplatin chemotherapy in human tongue carcinoma xenografts, *J Oral Oncology*, 2007,**43**:477–483.
9. T H Korver, R J van de Stadt, E Kiljan, Effects of loading on the synthesis of proteoglycans in different layers of anatomically intact articular cartilage in vitro, *J Rheumatology*, 1992,**19**:905-912.
10. Q. M. Ning, R. X. Wang, Early apoptosis induced by mechanical stretch in human alveolar type Ⅱ epithelia l cells, *J Shanghai Jiaotong University*, 2007,**27**:961-964.
11. H.Qu, P. Gao. The Effect of Mechanical Strain on Proliferation of Lung Cancer A549 Cells And Squamous Carcinoma of Tongue Tca8113 Cells. *Congres Proceedings-2011 4th International Conference on Biomedical Engineering and Informatics*, 2011,**v3**, p1545-1548.
12. B C Baliga and S Kumar, Role of Bcl-2 family of proteins in malignancy.*J Hematological Oncology*, 2002, **20**: 63-74.
13. A Gross, Bcl-2 proteins: regulators of the mitochondrial apoptotic program. *J IUBMB Life*. 2001, **52**: 231-236.
14. H. J. Zhang, S. X. Cai, X. Lu, Effects of Mechanical strain on the proliferation of pulmonary Eithelial cells and Reorganization of Integrins, *Acta Biochemical et Biophysical sinic*, 2002, **34**:748- 752.

THE SURGICAL TREATMENT OF RENAL CELL CARCINOMA WITH RENAL VEIN AND/OR INFERIOR VENA CAVA TUMOR EMBOLI

Tianli Shi[1,a], Xinquan Gu[1], Zhuang Li[1], and Cao Xia[2,b]

[1]*Department of Urology, China-Japan Union Hospital Affiliated to Jilin University, Changchun Jilin, 130033, China*

[2]*School of Pharmaceutical Sciences, Jilin University, Changchun Jilin, 130021, China*

E-mail: [a]345488525@qq.com, [b]caoxia@jlu.edu.cn

**Corresponding author*

Objective: Discuss the diagnosis and treatment of renal cell carcinoma with renal vein and/or inferior vena cava tumor emboli; Method: 10 patients with renal cell carcinoma with renal vein tumor emboli, tumor emboli levels according to the (Mayo Clinic) five-level taxonomy, divided into 6 level 0, 1 level I, 2 level II and 1 level IV. The surgery adopts Open Radical Nephrectomy and Vein Tumor Thrombus Extraction, using Vein Suture Repair directly in 8 patients and Venous Blood Vessel Replacement in 2 patients. Results: All 10 patients were successed in operation, 9 cases get follow-up, 9-56 months follow-up time, 1 case was lost to follow-up. Three-year survival rate is 60% (Lost to follow-up cases calculated at death), including the survival rate with the renal vein tumor emboli is 67%, the survival rate with the inferior vena cava tumor emboli is 50%. Conclusion The Open Radical Nephrectomy and Vein Tumor Thrombus Extraction is an effective way to diagnosis and treatment of renal cell carcinoma with renal vein and/or inferior vena cava tumor emboli.

1. INTRODUCTION

Renal carcinoma is the second most common cancer in urinary system. Renal carcinoma have the characteristics that can invade the renal vein and the inferior vena cava, 4% - 10% of cases could be simultaneously have the tumor emboli in renal vein and inferior vena cava[1]. My hospital recives and cures 10 patients of renal carcinoma with the renal vein and/or inferior vena cava tumor emboli in January 2009 to September 2015. All cases adopt Open Radical Nephrectomy and Vein Tumor Thrombus Extraction, presently reports as follows.

2. MATERIALS AND METHODS

2.1. General Information

10 patients with renal carcinoma with renal vein and/or inferior vena cava tumor emboli. The course of 2 months to 4 years, the average is 13 months; Male 6 cases, female 4

cases; The median age is 62 (51-73). 6 cases are right renal carcinoma, and the other 4 cases are left. 7 cases are limited in the kidney, 3 cases infringe the ipsilateral adrenal gland, including 1 case that the chest CT revealed nodules, consider transfer. The tumor diameter is 40-180 mm. Judging the level of tumor emboli by imaging, 6 cases are tumor emboli in renal vein and the other 4 are in the inferior vena cava. Five-level taxonomy (Mayo Clinic) [2]: level 0 (tumor emboli is limited in the renal vein) 6 cases; level I (in inferior vena cava, the top of the tumor emboli apart to the opening of the renal vein≤2cm) 1 case; level II (under the hepatic vein, the top of the tumor emboli apart to the opening of the renal vein≥2cm) 2 cases; level III (intrahepatic inferior vena cava, under the diaphragm) 0 case; level IV (in the inferior vena cava ,above the diaphragm) 1 case.

2.2. Surgical Treatment

Level 0, ligaturing the renal pedicle close to the wall of inferior vena cava in the radical nephrectomy. Level I, radical nephrectomy should separation and exposing of renal pedicle, ligaturing the ureter before the fifth lumbar vertebra. Level II, the inferior vena cava near and far end are free to normal position. Cut out tumor emboli after blocking-up. Then suture the inferior vena cava. Level II, 1 case incision via a ventral midline, 1 case "T" incision. First of all , free the kidney and the inferior vena cava. Then ligature the renal pedicle to expose the liver. Blocking the vena cava, vena cava and the offside renal vein of the tumor emboli up and down. Level IV, cut off the kidney and use artificial vascular replacement.

3. RESULT

10 cases are successfully completed, pulling up temporary vena cava filter immediately after the surgery. The average operation time is 265(75-710) min; blood loss 2140(1200-4700) ml; the length of the tumor emboli is 3.6(0.7-5.1) cm; the average hospitalization time 17 day. Postoperative pathological reports: Renal clear cell carcinoma 8 cases, papillary carcinoma 1 case, chromophobe cell tumor 1 case; TNM stage: T3aN0M0 6 cases, T3cN0M0 1 case, T3bN0M0 1 case, T4N0M0 1 case, T4N0M1 1 case. 9 patients get follow-up, 1 case was lost to follow-up, follow-up time 9-56 months, average follow-up time (35±4) months. 7 cases alive without tumor, 1 case diagnosis lung transfer before the surgery, 1 case appears the pulmonary nodules reviewing the CT after half a year. Three-year survival rate is 60% (Lost to follow-up cases calculated at death), including the survival rate with the renal vein tumor emboli is 67%, the survival rate with the inferior vena cava tumor emboli is 50%.

4. DISCUSSION

Most of the inferior vena cava tumor emboli patients without clinical symptoms, so it should be paid attention when the diagnosis of renal cancer. CT can clearly show imaging features of renal carcinoma, to provide reliable basis for the diagnosis and staging of the renal carcinoma [3]. MRI performance high signal in the vein. It can image in different levels to know the range of the tumor emboli. In particular, it has high degree of sensitivity and specificity to show whether the tumor emboli infringe the vein [4]. To the patients with venous tumor emboli and no distant metastases, open radical nephrectomy and vein tumor thrombus extraction were the major treatment. The complications of these surgeries have a large influence on patients, if in the surgery found blood oxygen saturation, blood pressure decline faster should be alert to pulmonary embolism. It should be carefully enough to the surgery operation, avoiding milk out and pursuit of quantity when get out the emboli. After getting out the emboli, it should try to do the repair of inferior vena cava. If the emboli can't cut off clearly, or vessel wall rough of the wound that easy to form the blood clots, to do the artificial vascular replacement could be considered. Without extracorporeal circulation, to reduce the blocking time of the superior vena cava, it can be used to suture the vein what close to the heart before blocking. Complex surgery can use extracorporeal circulation technology, performer can have enough time and clean surgical field to complete removal of tumor emboli and further the operation of inferior vena cava [5]. But reducing the time of extracorporeal circulation is necessary.

What should be paid attention to in the surgery: ligaturing the renal pedicle close to the wall of inferior vena cava; level II tumor emboli above carcinoma to block belt, traction of the inferior vena cava to prevent the backflow of blood, cutting out vein wall to take out tumor emboli; level III tumor emboli which should blocking the portal vein, the time of blocking should less than 20 min[6], avoiding affecting liver function; level IV tumor emboli could take the extracorporeal circulation to protect the important viscera from ischemia and to reduce the damage of the low blood pressure. We hold the opinion that, to the no lymph nodes invasion and distant metastasis of renal carcinoma in patients with venous tumor emboli, should be positive surgical treatment. Current research suggests that tumor emboli grading may influence the recurrence rate, but does not affect overall survival rate[7]. Preoperative detailed planning scheme is the key, and the skilful operation is also necessary. Resection of tumor emboli must be carefully. The complete resection can avoid the death from the pulmonary embolism or tumor proliferation caused by emboli loss.

References

1. Sweeney P, Wood CG, Pisters LL, et al. Surgical management of renal cell carcinoma associated with complex inferior venacaval thrombi [J]. *Urolo Oncol*, 2003, **21**(5):327-333.
2. Blute ML, Leibovich BC, Lohse CM, et al. The mayo clinic experience with surgical management, complications and outcome for patients with renal cell carcinoma and venous tumor thrombus [J].BJU Int, 2004, **94**(1):33-41.
3. Xiaosong Wang, Wenhua Chen. 16 layers spiral CT in the diagnosis of renal cancer preoperative value. *Jiangsu Yiyao*. 2011, **37**(18), 2166-2168.
4. Aslam Sohaib SA, The J, Nargund VH, et al. Assessment of tumor invasion of the vena caval wall in renal cell carcinoma cases by magnetic resonance imaging [J]. *Journal of Urology*, 2002, **167**(3):1271-5.

5. Parissis H, Akbar MT, Tolan M, et al. Surgical resection of a renal cell carcinoma involving the inferior vena cava: the role of the cardiothoracic surgeon [J]. *Journal of Cardiothoracic Surgery*, 2010, **5**(1):1-6.
6. Wotkowicz C, Wszolek M F, Libertino J A. Resection of renal tumors invading the vena cava.[J]. *Urologic Clinics of North America*, 2008, **35**(4):657-671.
7. Otaibi M A, Youssif T A, Alkhaldi A, et al. Renal cell carcinoma with inferior vena caval extention: impact of tumour extent on surgical outcome.[J]. Bju International, 2009, 104(10):1467–1470.

BASED ON FUZZY ANALYTIC HIERARCHY PROCESS TO STUDY STRESS EVALUATION METHOD

Zhangyong Li[a], Dewei Lv[b], Shengrong Liu, Peng Ran, Wei Wang, Yu Pang, and Jinzhao Lin

Research Center of Biomedical Engineering, Chongqing University of Posts and Telecommunications, Chongqing, 400065, China

E-mail: [a]lizy@cqupt.edu.cn, [b]lv_dewei@163.com

The latest study found that balance, stability and regulation of the autonomic nervous system can reflect the changes of individual mental closely and objectively, while heart rate variability (HRV))able reflect function of the human autonomic nervous system. Therefore, this paper applies analytic hierarchy process(AHP)to determine the weight of each evaluation factors qualitatively and quantitatively; after that, building fuzzy comprehensive evaluation model based on AHP, then using a fuzzy evaluation method make a comprehensive evaluation to the mental pressure of the subjects; experiments show that assessment of individual mental stress based on the model is scientific and feasible.

1. INTRODUCTION

In recent years, stress problems cause widespread concern in the world. The World Health Organization calls it the "Western world catastrophe," which not only endanger the individual's physical and mental health, but also harm society[1]. People begin to live a happy life with the development of society, which is rich material, a wide choice, information flow. However, due to intense competition and unbalanced society, mental state suffers a severe blow and erosion, the task concerned about people's mental health becomes particularly important[2].

The autonomic nervous system is divided into sympathetic (SNS) and parasympathetic (PNS) nerve(also known as the vagus nerve), shown in Fig.1. They are distributed in nerves of the internal organs, blood vessels, glands and other organs, which control breathing, heart rate, perspiration, body temperature and other functions together. For example heartbeat, its speed does not depend on individual consciousness, but SNS will make it accelerate in motion naturally, or slow at rest or sleep state, which is a natural result of the regulation of the body's autonomic nervous system[3].SNS and PNS play a balancing role in the body, like a seesaw. When one is more active, another will be suppressed. In this case, heart beat, blood pressure, rapid breathing and other stress condition will be reflected. However, when PNS is more active, SNS will be suppressed. The body will gradually enter a resting state. Thence, the autonomic nervous system in normal and stable operation ensure mental health.

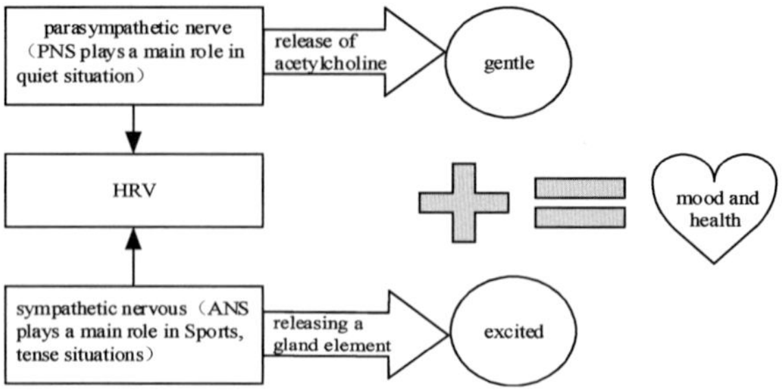

Fig.1. Relationship between the autonomic nervous system, HRV.

HRV is a dynamic pointer generated by the autonomic nervous activity, and reflects blance between SNS and PNS[4]. Many factors affecting the autonomic nervous activity will affect or reduce HRV, such as: high blood pressure, shortness of breath, physical and mental stress tension, hyperthyroidis -m and medication[5].HRV of patients with diabetes or hypertension is low. Although HRV is a complex mechanism, but both the pointer which is the health reference of self mental health care in daily life and regular health checks will maintain individual mental health at any time[6]. Therefore, the paper makes use of some parameters of HRV to analyze the activity of the autonomic nervous system, and uses the balance of SNS, PNS and some HRV relevant factors to determine the subject's stress load[7].Using fuzzy AHP identifies several indexes and index weight affecting stress initially with experimental data, then creating a mathematical model based on the ratio between indicators, which makes the stress assessment visual[8].

2. STRESS INDICATORS DETERMINATION

HRV analysis is the essence of quantitative analysis of sinus rhythm. Its analysis methods have been constant progress and development, but now the main research methods are sometimes domain analysis, frequency domain analysis and nonlinear analysis, which can evaluate balance, tension between the cardiac sympathetic and parasympathetic quantitatively, and its effects on the cardiovascular system activity, so consistent changes in certain parameters of HRV obtained by experiment are seen as stress evaluation index. Experimenters who are 20 graduate students meet the test conditions, their specific details as follows:(1) Age 22 to 27 year-old male or female, healthy;(2) no diagnosis of cardiovascular disease, mood surgical history;(3) Day of the experiment do not take affect cardiovascular drugs, do not drink alcoholic beverages. Detecting the body signs under different emotions by changing the measurement position or movement, in which one group is in a state of quiet flat; another group on a treadmill to 2m/s uniform running 20min,third group in a standing posture. Before measurement, the subjects in the measurement posture keep quiet for 10min to make the body be calm and stable, which can avoid fluctuations caused by postural mutations. During measurement, the subjects try to make mental relaxation, breathing steadily, and do not speak, can not move the body. Three measurements are averaged each person each position to reduce the measurement error. According to the above described experiments and experimental considerations, the experimental data is recorded with the experimental procedure strictly,meanwhile the data is analyzed, shown in Table 1 and Table 2.

Table 1. Statistics temporal parameters before and after exercise

Sample	HR(time/min)		SDNN(ms)		RMSSD(ms)		NN50(ms)		PNN50(%)	
	Before exercise	After exercise	Before exercise	After exercise	Before exercise	After exercise	Before exercise	After exercise	Before exercise	After exercise
1	80	97	49.58	35.26	49.57	51.45	5	5	13.11	0.2
2	88	102	42.63	21.43	28.10	41.09	39	3	14.09	0
3	77	99	43.8	25.48	54.36	37.54	20	25	13.5	0
4	71	75	34.01	24.36	24.85	47.49	31	4	5.28	0.2
5	68	76	40.52	23.28	22.40	47.27	21	7	8.61	9.39

Table 2. Frequency domain parameters before and after exercise Statistics

Sample	TP(ms^2)		VLF(ms^2)		LF(ms^2)		HF(ms^2)		LF/HF	
	Before exercise	After exercise	Before exercise	After exercise	Before exercise	After exercise	Before exercise	After exercise	Before exercise	After exercise
1	2922	812.84	1307.64	173.07	374.56	299.25	1081.16	340.52	0.35	0.88
2	2210	532.28	515.27	350.71	390.42	88.18	1096.70	93.39	0.36	0.94
3	5899	4714.14	2622.16	2434.94	1700.62	1681.78	970.11	521.67	1.75	3.22
4	4744	2045.74	1884.17	437.41	1329.75	1008.90	615.68	191.72	2.16	5.26
5	2303	772.03	365.41	312.31	246.22	328.78	1221.30	130.94	0.20	2.51

As the table, with the stress increasing, the heart rate HR,LF / HF value trend to increase; and phase plan, SDNN,PNN50,TP value tends to decrease. The index range is shown in Table 3:

Table 3. Statistics temporal parameters before and after exercise

Pressure Indicator	before experiment	after experiment
HR	73±4	100±6
SDNN	46±12	30±12
PNN50	1526±710	1650±623
TP	2.74±0.63	3.06±0.52
LF/HF	2.10±1.53	4.26±2.68
Phase plan	0.946±0.132	0.2245±0.0573
The relative dispersion	5.48±0.72	6.58±0.32
Lee index	5.11±0.46	6.35±0.96

3. STEP APPLICATION OF FUZZY ANALYTIC HIERARCHY PROCESS

To gain pairwise comparison matrix, the relative importance of each stage factor is described quantitatively by parameter changes and 1-9 ratio scale method based on experimental data[9].

• Setting the objectives and the constituent elements,then to establish evaluation index

$$u = \{u_1, u_2, \cdots, u_p\}.$$

• Constructing judgment matrix A=(a_{ij}),Where a_{ij} is the importance of the element i compared with j element.

• Evaluating weight of each element on the same level by the summation method or rooting method, This paper selects summation method to calculate the relative weights.

$$d_{ij} = \frac{a_{ij}}{\sum_{k=1}^{n} a_{kj}}, i, j = 1, 2, \ldots, n \tag{1}$$

$$\overline{w} = \sum_{j=1}^{n} d_{ij}, i = 1, 2, \ldots, n \tag{2}$$

• Obtaining the weight of each main element of the matrix by matrix normalization

$$w_i = \frac{\overline{w}_i}{\sum_{i=1}^{n} \overline{w}_i} \tag{3}$$

• Consistency test

$$\lambda_{\max} = \frac{1}{n}\sum\left(\frac{(A \bullet w)}{w_i}\right) \tag{4}$$

$$C.I. = \frac{\lambda_{\max} - n}{n - 1} \tag{5}$$

$$C.R. = \frac{C.I.}{R.I.} \tag{6}$$

where $\lambda_{\max}$ is the maximum eigenvalue of judgment matrix. $C.I.$ is consistency index of the judgment matrix. $C.R.$ is the random consistency ratio. $R.I.$ is the mean random consistency index, which found in Table 4. $R.I.$ overcomes $C.I.$ increasing significantly to some extent, as n increase. It is generally believed that as long as $C.R. < 0.1$,the judge matrix can accept or can be re-amend[10].

Table 4. Mean Random Consistency Index

Order	1	2	3	4	5	6	7	8	9	10
$R.I.$	0	0	0.58	0.90	1.12	1.24	1.32	1.41	1.45	1.49

• Determining Reviews Ratings domain

$v = \{v_1, v_2, \cdots\cdots, v_p\}$, which is class set. Each level can correspond to a fuzzy subset.

• Fuzzy relation matrix R establishment

After structuring the rating fuzzy subset, each factor be quantified individually, then get the fuzzy relation matrix.

$$v_i = \sum_{j=1}^{n} d_{ij}, i = 1,2,3,\ldots,n \tag{7}$$

Element r_{ij} located in i-th row j column of Matrix R is expressed the degree of membership $(R \mid u_i)$ u_i to v_j in level fuzzy subset.u_i influencing assessed things is shown by fuzzy vector $(R \mid u_i) = (r_{i1}, r_{i2}, \cdots\cdots, r_{im})$,while is mostly represented by one simple value in other evaluation methods.Thus,fuzzy comprehensive evaluation request more information from this perspective[11].

- Synthesis of fuzzy comprehensive evaluation result vector

To get fuzzy comprehensive evaluation Vector B, judgment matrix A and each rated things R are synthesized by the appropriate operator. Which is:

$$A \times R = (a_1, a_2, \cdots\cdots, a_p)\begin{bmatrix} r_{11} & r_{12} & \cdots & r_{1m} \\ r_{21} & r_{22} & \cdots & r_{2m} \\ \cdots & \cdots & \cdots & \cdots \\ r_{p1} & r_{p2} & \cdots & r_{pm} \end{bmatrix} = (b_1, b_2, \cdots\cdots, b_m) = B \tag{8}$$

Where b_j getting from arithmetic operation result that A and a j-th column of R, which represents the degree of membership something assessed to v_j in level fuzzy subset from the whole.

- Vector analysis of fuzzy comprehensive evaluation results

The most commonly used method is maximum membership degree principle, but it is some reluctance in some cases, which leads to many information loss, or even gets unreasonable evaluation results. So weighted average seek membership grade method is proposed. It makes many evaluated things be sorted according to their level location[12].

4. FUZZY COMPREHENSIVE EVALUATION MODEL BASED ON MENTAL STRESS

4.1. Stress evaluation system establishment

To achievement stress evaluation system, a hierarchy model needs to be established based on AHP. The hierarchical analysis model has three centimeters in the paper, shown in Table 5:

- The first layer represents stress problems to be solved, called the target layer.
- The second layer shows the middle parts involving in the intended target, called the guidelines layer.
- The third layer represents a number of evaluation criteria for each layer, called the index layer.

Table 5. Stress index system

Target layer(A)	Guidelines layer(B)	Index layer(C)
Mental stress	Nonlinear parameters(B_1)	Phase plan area(C_{11})
		The relative dispersion(C_{12})
		Lee index(C_{13})
	Time Domain Parameters(B_2)	SDNN(C_{21})
		PNN50(C_{22})
		HR(C_{23})
	Frequency domain parameters(B_3)	TP(C_{31})
		LF/HF(C_{32})

4.2. Judgment matrix, the index weight calculation and consistency check

Each element of each hierarchy is pairwise comparison judgment to determine their weight, then construct judgment matrix (shown in Table 6-Table 9).The formula (3) can calculate w and w_i, where w and w_i are the weight vector in the target layer (A) and the weight vector in the guidelines layer(B) respectively. While λ_{max}, C.I. and C.R. can be obtained from the formula (4) to (6). C.R. <0.1 indicates that judgment matrix is more than acceptable.

Table 6. A-B_i judgment matrix

A	B_1	B_2	B_3	w
B_1	1	4	5	0.68
B_2	$\frac{1}{4}$	1	3	0.22
B_3	$\frac{1}{5}$	$\frac{1}{3}$	1	0.1
λ max=3.086, C.R.=0.07<0.10				

Table 7. B_3-C_{3i} judgment matrix

B_3	C_{11}	C_{12}	C_{13}	W_3
C_{11}	1	3	4	0.62
C_{12}	$\frac{1}{3}$	1	2	0.24
C_{13}	$\frac{1}{4}$	$\frac{1}{2}$	1	0.14
λ max=2, C.R.=0<0.10				

Table 8. B_2-C_{2i} judgment matrix

B_2	C_{21}	C_{22}	C_{23}	W_2
C_{21}	1	1	3	0.43
C_{22}	1	1	3	0.43
C_{23}	$\frac{1}{3}$	$\frac{1}{3}$	1	0.14
λ max=3, C.R.=0<0.10				

Table 9. B_1-C_{1i} judgment matrix

B_1	C_{31}	C_{32}	W_1
C_{31}	1	3	0.75
C_{32}	$\frac{1}{3}$	1	0.25
λ max=3.083, C.R.=0.07<0.10			

4.3. Evaluation Set Establishment

Mental stress is more subjective evaluation. In order to make it more accurate apply, which is divided into four levels according to semantics scales: relaxation, general pressure, greater pressure, very big pressure. Then let semantics scale of the subjective evaluation quantify by Likert amount method, and the quantifed result is that 1,2,3,4 represent relaxation, general pressure, greater pressure, very big pressure respectively. The quantitative evaluation grading standards are shown in Table 10.

Table 10. Evaluation of quantitative grading

Evaluation value	Comment	Grading
$x_i \le 1.5$	Relax	E_1
$1.5 \le x_i \le 2.5$	General pressure	E_2
$2.5 \le x_i \le 3.5$	Greater pressure	E_3
$x_i > 3.5$	Very much pressure	E_4

4.4. Fuzzy Relation Matrix Structure

10 evaluation experts need to grade the indicators, and give evaluation grades referring to Table 11, When assessing the examinees' mental stress situation. Statistically, getting each index membership to evaluation set, then setting fuzzy relationship matrix R_1.

In order to describe the process that the fuzzy relation matrix is structured, taking Non-linear (B_1) as an example. R_1 is given according to (7):

$$R_1 = \begin{bmatrix} 0.2 & 0.4 & 0.2 & 0 \\ 0.6 & 0.1 & 0.3 & 0 \\ 0.3 & 0.1 & 0.5 & 0.1 \end{bmatrix}$$

Table 11. Nonlinear parameter fuzzy relation matrix

Evaluation	Relax	General pressure	Greater pressure	Very much pressure
C11	0.2	0.4	0.2	0
C12	0.6	0.1	0.3	0
C13	0.3	0.1	0.5	0.1

4.5. Vector Calculation Based on Fuzzy Comprehensive Evaluation Results

Specific steps are as follows :

- Seeking vector evaluation results criteria of layer (Bi):

$$B_i = w_i \times R \tag{9}$$

Where W_i is the weight vector of important elements in criterion level (Bi), shown in Table 6 to Table 9. Then criterions layer (B) evaluation matrix is formed with B_i:

$$B = [B_1 \quad B_2 \quad \dots \quad B_n]^T \tag{10}$$

$$B = \begin{bmatrix} 0.31 & 0.28 & 0.27 & 0.14 \\ 0.25 & 0.35 & 0.30 & 0.1 \\ 0.34 & 0.15 & 0.23 & 0.28 \end{bmatrix}$$

- Seeking evaluation vector in target layer(A): $A = w \times B$ (11), Where w is the weight vector of the elements of the target layer (A).

$$A = \begin{bmatrix} 0.29 & 0.28 & 0.27 & 0.15 \end{bmatrix}$$

- seeking the final evaluation score by the fuzzy comprehensive evaluation results, then rate evaluation: $K = 1\times 0.15 + 2\times 0.27 + 3\times 0.28 + 4\times 0.29 = 2.7$, which explains the mental state is in greater pressure, belong to E_3 level.

5. CONCLUSIONS

To analyze mental pressure accurately and quantitatively, and to visualize it. This paper introduces the fuzzy analytic hierarchy process in mental stress assessment system, as an aid decision-making tool of stress evaluation phase. This evaluation method makes qualitative analysis and quantitative calculation combined. Examples show that the fuzzy analytic hierarchy process achieves a comprehensive trade-off analysis for each factor, effectively makes a comprehensive assessment of mental stress. However, the method has limited places, such as:it is suitable for people between 20-30 years of age, but with age increasing, the stress index change selected is not obvious, even without consistent trend. Therefore, the signal can be added to blood pressure, skin electrical and EEG number, etc. [13-15]to achieve a more accurate analysis of mental stress. Mental stress assessment is only a means, not an end. Mental stress assessment is the need for physical and mental health development, and only taking stress assessment seriously, people could make their spiritual stay healthy to make better contribution to the society.

References

1. Deng Lifang. Positive Students mental stress and Mental Health Analysis [J].*Journal of national institute of education administration*(NAEA OF),2009,**3**(3): 39-40.
2. M.Vargas-Luna, J.B.Montes. Evaluation of the Cardiac Response to Psychological Stress by Short-Term ECG Recordings: Heart Rate Variability and Detrended Fluctuation Analysis[J]. Leno Guanajuato: Departamento de Ingenieria Fisica DCI CL Universidad de Guanajuato,2012,**20**(3): 80-83.
3. Johan Ormel, Harritte Riese, Judith G.M. Rosmalen. Interpreting neuroticism scores across the adult life course: immutable or experience-dependent set points of negative affect. *Journal of Clinical Psychology*. 2012.
4. Naoko Kudo, Hitomi Shinohra, Hideya Kodama. Heart Rate Variability Biofeedback Intervention for Reduction of Psychological Stress During the Early Postpartum Period[J]. *Appl Psychophysical Biofeedback*,2014,**39**(3): 203-205.
5. Naoko Kudo, Hitomi Shinohra, Hideya Kodama. Heart Rate Variability Biofeedback Intervention for Reduction of Psychological Stress During the Early Postpartum Period[J].*Appl Psychophysiol Biofeedback*, 2014 ,**39**(3):203-205.
6. Lin Xiaofeng, Liu Ze, Liu Ling, Dongfeng Ying, Wang Luni, Wu Ziqiang, Yuan Hui. Effect of Individualized aerobic exercise on elderly patients with coronary heart rate recovery [J]. *Chinese Journal of Cerebral Blood Heart Tube Diseases*. 2013 (06).
7. M.Vargas-Luna, J.B.Montes. Evaluation of the Cardiac Response to Psychological Stress by Short-Term ECG Recordings: Heart Rate Variability and Detrended Fluctuation Analysis[J].Leno Guanajuato: Departamento de Ingenieria Fisica DCI CL Universidad de Guanajuato,2012,**20**(3):80-83.
8. Guiwu Wei. Hesitant fuzzy prioritized operators and their application to multiple attribute decision making[J].*Knowledge-Based Systems* . 2012.
9. Nian Zhang, Guiwu Wei. Extension of VIKOR method for decision making problem based on hesitant fuzzy set[J]. *Applied Mathematical Modelling*.2012.
10. LI Weixia1, ZHANG Hao, ZHANG Chengyi. Drag use risk evaluation problem of comprehensive hospital based on Intuitionistic fuzzy analytic hierarchy process[J]. CEA, 2013, **49**(18): 242-244.
11. Zhang Chengyi, Li Weixia, Wang Leng. AHP under the intuitionistic fuzzy environment[C]// 2011 8th International Conference on Fuzzy Systems and Knowledge Discovery,2011: 583 - 587.
12. Lin Jian Qiao, Huangde Chun. Fuzzy analytic hierarchy process to select investment bank stock valuation model [J]. *Statistics and Decision*, 2011 (14): 55-59.
13. Hanqing Peng. Makeing the use of EEG signal based on wavelet packet transform and nonlinear analysis to determine mental fatigue [J]. *Vibration and shock*, 2013,**32** (2): 182-184.
14. Ramesh Sahoo, Srinvas Sethi. Functional Analysis of Mental Stress Based on Physiological Data of GSR Sensor[C].Suresh Chandra Satapath. Advances in

Intelligent Systems and Computing. Springer Cham Heidelberg New York Dordrecht London. Springer International Publishing Switzerland,2015: 120-130.
15. Zheng Fei, Chen Jie, Mo Xiong, Wei Wenqi, Zhou Chao. Correlation between circadian rhythm of blood pressure in patients and silent myocardial is chemia [J / OL]. *Anhui Medicine*, 2015 (04).

INFRARED IMAGING DEVICE FOR *EX VIVO* VISUALIZATION THROUGH DILUTED BLOOD

Lujie Chen

Singapore University of Technology and Design, 487372, Singapore
E-mail: chenlujie@sutd.edu.sg

An imaging device is developed for real-time IR medical imaging systems that are able to penetrate the diluted blood, visualize the region in front of the device, and potentially reconstruct the 3D surface profile of tissue in the field of view. A series of benchtop, endoscope, and catheter-based prototype systems are constructed to validate these principles, and to demonstrate their benefits in guiding medical procedures.

1. INTRODUCTION

Physicians need to see the inside surface of heart chambers in minimally invasive medical procedures (e.g. transapical aortic valve replacement, catheter-based ablation) but the lack of transparency of blood in the visible region prevents the use of conventional visible-light imaging. Furthermore, while anatomical structures are three dimensional, current infrared (IR) imaging systems are able to acquire only low-resolution 2D images that lack depth information. This fundamental restriction greatly limits clinicians' ability to perceive and understand the complex intracardiac anatomy.

IR imaging is rapidly advancing and has shown tremendous promise as the next generation imaging tool. IR light (wavelength 650 – 2500 nm), with its deep penetration in living tissues shown in Fig. 1 and high signal-to-background ratio, has been exploited for biomedical imaging, photo-ablation and photodynamic therapy [1-3]. The extent of IR light propagation would be governed by absorption and scattering properties of tissues. Major tissue absorbers of light are haemoglobin, melanin and water, while composition, size and morphology of tissue components control the light scattering [4]. The absorption and scattering properties would influence volumetric energy distribution induced by laser irradiation. Near IR (NIR) light was reported to travel through 10 cm of breast tissue and 4 cm of skull tissue using microwatt sources.

Optical imaging between 1,000 and 1,350 nm has also been impeded by the lack of sensitive and low-cost CCD (charge-coupled device) cameras [3]. Silicon, the material most commonly used in CCD cameras, is not sensitive at wavelengths longer than 1,000 nm. Instead, near-infrared cameras use semiconductor alloys with narrower band gaps such as InGaAs and HgCdTe. InGaAs cameras have a high quantum efficiency in the second near-infrared window, whereas HgCdTe cameras are available in higher-resolution arrays, but with a lower overall quantum efficiency. In general, these types of sensors are becoming more affordable with adequate sensitivity and resolution (0.1–0.3 megapixels) for most imaging applications [3].

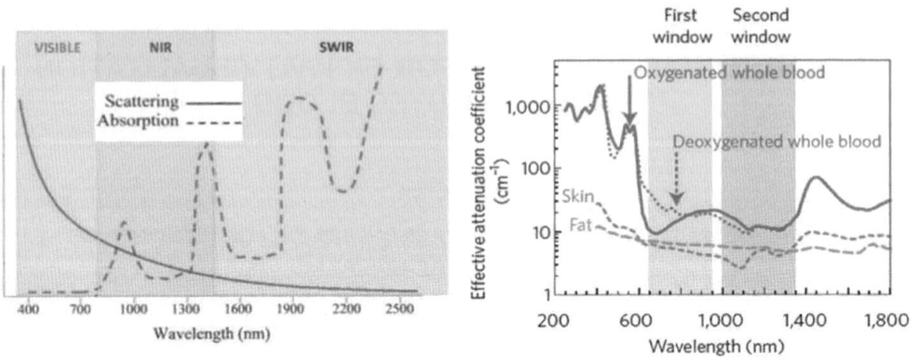

Fig. 1. Left: Schematic illustration of the infrared wavelength variation of scattering and absorption in blood showing three wavelength windows; imaging ability is related to the product of these two parameters. Right: Experimental measurement of infrared attenuation for blood, skin, and fat [3].

3D information at a working distance of a few centimeters can be obtained by a number of non-contact methods, such as structured light projection [5] and stereo vision [6]. The structured light projection approach exploits triangulation between the projected light and the imaging light to infer 3D surface projection. The basic principle can be implemented a variety of ways: random dot projection as in the Kinect video game interface (Microsoft USA), sinusoidal fringe projection [5], etc. It is a relatively mature technique for visible light and many commercial products exist. The stereo vision approach mimics human binocular vision to extract depth information [6]. It makes use of the disparity in two images taken by two cameras separated by some distance to infer the 3D information. In order to retrieve the Euclidian distance with sufficient accuracy, the cameras must be accurately calibrated [7, 8]. In the past ten years, computer vision theory and algorithms have also reached maturity and various commercial products are available on the market (Polaris by NDI Technology).

At IR wavelengths (800 ~ 2000nm), the 3D reconstruction methodology applicable to visible light has not been investigated fully, although a few consumer products, like the Kinect (Microsoft USA), have successfully incorporated near IR (around 800nm) 3D sensing technology. This system is aimed at sensing human limb positions, and is limited to solid surfaces in air within a workspace volume of approximately (2 m^3).

The propagation of light through scattering media has been extensively studied due to its importance in applications such as atmospheric imaging and astrophysical modeling [9]. Recent work has begun to incorporate scattering into analyses of structured light systems [10, 11], with particular application to underwater applications [12]. This analysis will form the basis for the required modeling of IR imaging through blood, but the near-close-packed configuration of red cells in blood will require modification of the model. In addition, the application in year 3 to imaging subcutaneous tumors will require an original formulation of the model. In particular, we must take into account the fact that the tumor target “surface” will be composed of attached nanoparticles at a range of depths and densities, rather than a well-defined reflective surface. This can be implemented as scattering from a statistically-defined ensemble of reflectors. Recovery of the reflecting statistics will be useful to differentiate tumors with well-defined boundaries (likely static) from tumors with diffuse boundaries (likely growing).

2. METHODS

2.1. Bench-Top 2D Imaging System

Several experiments have been conducted to test IR imaging in open-air and under diluted blood. We have also tried to enhance the resulting image quality by implementing image-processing methods, such as frequency domain de-noising, time averaging, and real-time surface segmentation. These experiments form an initial understanding of the imaging capability at IR range based on fiber optics (Fig. 2).

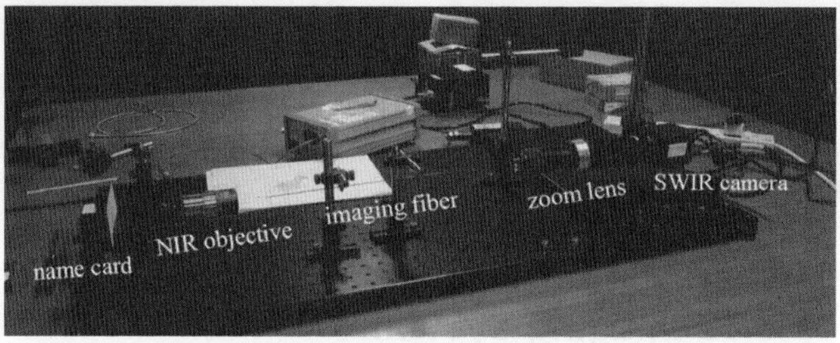

Fig. 2. Lab experiment testing IR imaging based on an optical fiber.

2.2. Bench-Top 3D Imaging System

To measure true three dimensional shapes we developed structured light projection and stereo vision techniques. For optimal display for end-users we have investigated 2D and 3D display techniques including segmented surface rendering, rotation, multiple view, and stereo approaches. Prototype systems have been constructed for 3D shape measurement in confined space (Fig. 3).

2.3. Ex-Vivo 2D Imaging System

2D imaging and coaxial triangulation experiment for 3D reconstruction were implemented. An artificial blood circular system was setup at an animal experiment lab, shown in Fig. 4. This enabled us to perform 2D imaging experiment in a controlled environment.

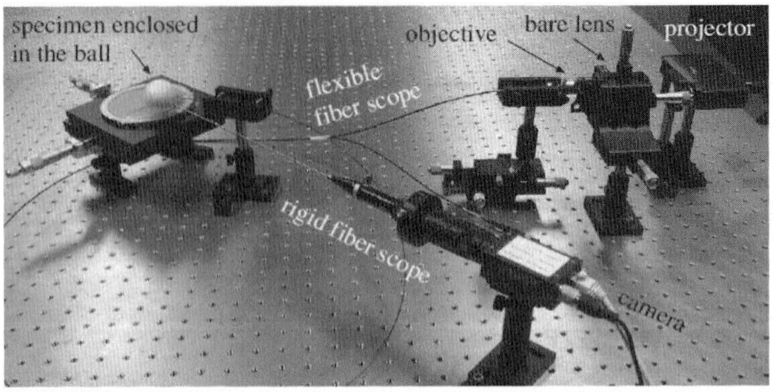

Fig. 3. Experimental setup for 3D shape measurement in confined space. The specimen, a MEMS component, is enclosed in a ball.

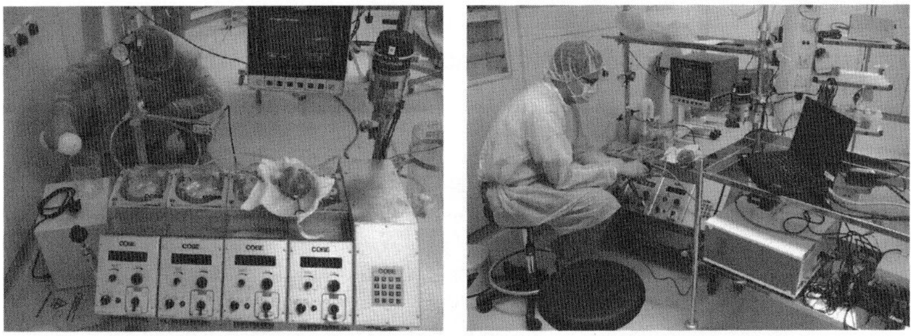

Fig. 4. Artificial blood circular system.

2.4. Endoscopic 3D Imaging System

To measure true three dimensional shapes we developed a novel coaxial structured light projection method. Prototype system is shown in Fig. 5.

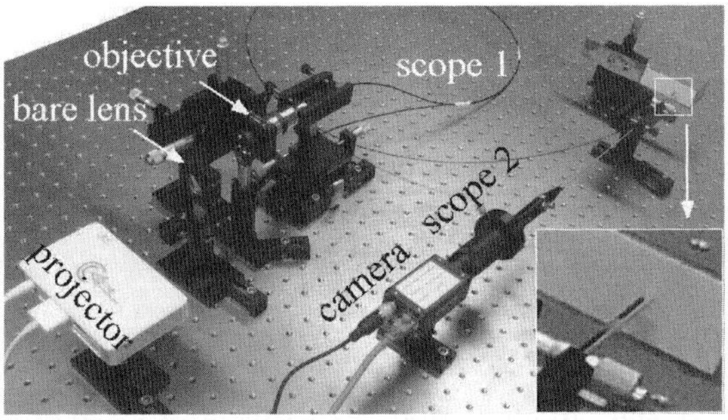

Fig. 5. Experimental setup for 3D shape measurement based on a coaxial structured light projection system.

3. RESULTS

3.1. Ex-Vivo 2D Imaging

Fig. 6 shows the 2D images obtain through water at illumination wavelength from 1000 to 1690nm (left image). The right image indicates quantitatively the intensity variation with wavelength. The results suggest an issue with laser power. At longer wavelength, the absorption is stronger and much stronger power is needed. This issue is currently under investigation.

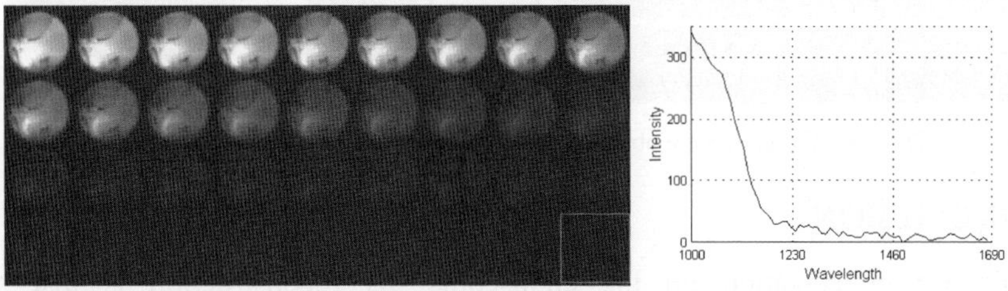

Fig. 6. Left: a particular heart issue structure under illumination wavelength from 1000 to 1690nm. Right: Quantitative intensity variation of a point.

3.2. 3D Imaging

Investigation on the 3D reconstruction in confined space has led to promising findings. We have incorporated structured light projection in a fiber optic system based on an off-the-shelf digital projector. This enables flexible generation of structured light on a specimen and subsequently retrieval of its 3D shape, as shown in Fig. 7.

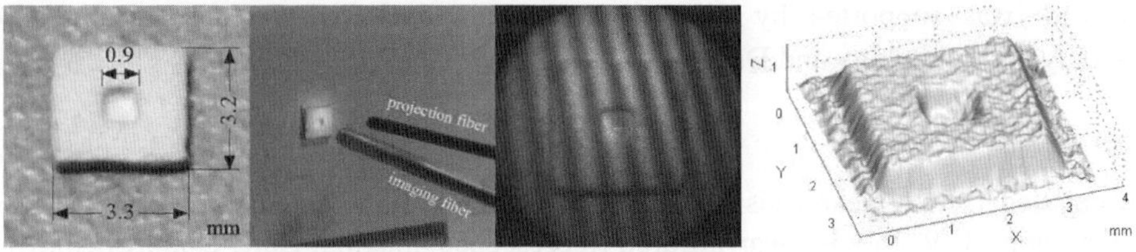

Fig. 7. A MEMS component, visualized under a fiber optic projection system.

We have noticed that 3D information can even be retrieved based on a counter-intuitive principle: on-axis 3D reconstruction. Usually, it is not possible to obtain 3D shape when the projection and imaging views are parallel (on-axis); however, with different lens embedded in the projection and imaging optics, an off-axis situation is created at each local region of the specimen. It is as if each point on a specimen followed a different angular relationship and thus possible to produce 3D perception. Despite the non-uniformity over the surface, 3D information can still be obtained except at the two optical centers. Fig. 8 illustrates this effect. The optical centers coincide at the top-left corner where 3D sensitivity is near zero. The third image shows noisy data in that region. Other regions have obtained reasonably well reconstructed 3D surface data.

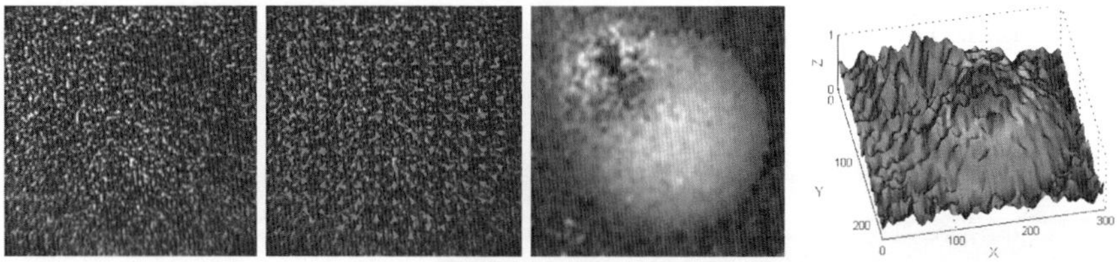

Fig. 8. 3D shape of a semisphere obtained by an on-axis projection imaging geometrical setup.

4. CONCLUSION

Although our investigation on the endoscopic 3D reconstruction is just started, experiments have shown promising results that would help to verify the new approach. It is inevitable that the accuracy and resolution will be lower than the normal off-axis setup but the impact to the research community and to 3D sensing device design could be significant. The principle of on-axis 3D reconstruction would enable a series of new designs leading to compact 3D sensors potentially useful in diverse areas, especially in endoscopes and catheters for medical applications.

Acknowledgments

This work was supported by the International Design Center (IDC) at Singapore University of Technology and Design.

References

1. Weissleder, R. A clearer vision for in vivo imaging, Nat. Biotechnol. 2001, **19**: 316.
2. Frangioni, J. V. In vivo near-infrared fluorescence imaging, Curr. Opin. Chem. Biol. 2003, **7**: 626.
3. Smith, A.M., Mancini M. C., Nie, S. Second window for in vivo imaging, Nat Nanotechnol. 2009, **14**: 710.
4. Niemz, M.H. Laser-tissue interactions: Fundamentals and applications, Springer-Verlag: Berlin, 2002.
5. Geng, J. Structured-light 3D surface imaging: a tutorial, Advances in Optics and Photonics, 2011, **3**: 128–160.
6. Multiple View Geometry In computer vision, Richard Hartley and Andrew Zisserman, Cambridge University Press, 2003.
7. Roger Y. Tsai, A versatile camera calibration technique for high-accuracy 3D machine vision metrology using off-the-shelf TV cameras and lenses, IEEE Journal of Robotics and Automation, 1987, RA-3, 4.
8. Zhengyou Zhang, A flexible new technique for camera calibration, IEEE Transactions on Pattern Analysis and Machine Intelligence, 22, 11, November, 2000.
9. Kokhanovsky, A., Light Scattering Media Optics, 3rd ed., Springer, 2004.
10. S.G. Narasimhan, S.K. Nayar, B. Sun, and S.J. Koppal, Structured Light in Scattering Media, in Proceedings of the Tenth IEEE International Conference on Computer Vision, 2005.

11. T. Treibitz and Y.Y. Schechner, Instant 3Descatter, Proceedings of IEEE Computer Vision and Pattern Recognition, vol. II, 1861-1868, 2006.
12. F.M. Caimi, D.M. Kocak, F. Dalgleish, J. Watson, Underwater imaging and optics: Recent advances, OCEANS 2008.

THE STUDY OF FREQUENCY SPECTRUM CHANGE ON EEG UNDER +GZ ACCELERATION

Yifeng Li[a,*], Lihui Zhang, Baohui Li, Guiding Yan, Xichen Geng, Zhao Jin, and Xiaoyang Wei

Institute of Aviation Medicine, Air Force, Beijing 100142, China

E-mail:[a]liyf8886@sina.com

[*]Corresponding author

The EEG data under +Gz acceleration is obtained, frequency spectrum data graphs and frequency spectrum curve graphs are made by using method of frequency spectrum analysis, and the EEG data under +Gz acceleration and at quiet state are made comparison and study. Result shows that, energy of every frequency band has increasing of different degree. The energy percentage of beta rhythm accounts is the most, secondly theta rhythm. Under +Gz acceleration, energy percentage of delta rhythm to all rhythms increases, energy percentage of beta rhythm to all rhythms decreases, and among this, energy percentage of beta2 rhythm to all rhythms decreases obviously. For all leads, the leads which changes are obvious mostly distribute at all sides of brain and at right rear side of brain. Rhythms changes reflect the ischemia state change of brain, and the distribution of leads which change is obvious is relevant to the special body position which human body lies at. Power spectrum analysis can better help to find EEG waveform characteristic, and it is the effective index to observe EEG change under +Gz acceleration.

1. INTRODUCTION

As a reflection of brain function state and activity law, EEG is widely used in basic research and clinical diagnosis and treatment. In aerospace field, in order to prevent pilots occurring air syncope in flying, many scholars began to study and monitor human physiological signals under +Gz acceleration[1-2]. There are scholars specially monitoring rhythms activity of brain to make a judgment for the spirit state of human body[3].Because that EEG has the defects of long recording time, complex waveform, easy to cause errors and fatigue, in recent dozens of years, all kinds of digital signal processing technologies have been widely used in EEG analysis and extraction of EEG feature, such as wavelet analysis [4-8], artificial neural network [9] and so on, but the application of these techniques is not very common. Among this the frequency spectrum of EEG is a kind method which application is earlier and more mature [10-12]. EEG analysis method includes time domain analysis and frequency domain analysis. Frequency domain analysis is the main analysis method in EEG analysis and clinical application. Frequency domain analysis is the power spectrum estimation on EEG signal. It can calculate firstly autocorrelation function of a lead EEG, then and makes a Fourier transformation for the autocorrelation function, so getting the power spectrum of EEG. Also the square of a lead EEG which is directly after FFT transformation can be the power spectrum estimation of EEG. Having EEG spectrum can determine the main rhythm of this segment of EEG. This method is based on the quasi stationary segment of EEG. Because the spectral estimation is used to analyze the stationary process, and although EEG is non-stationary process, it is considered to be quasi stationary for one segment in a short time. Frequency domain analysis method smoothes the transient information of EEG. The application of computer technology on EEG has overcome the subjective interference produced by traditionally visual analysis, making that the result of EEG can be expressed by using quantitative data. Therefore, based on EEG naked eye analysis and combining with quantitative analysis, it is necessary to further improve the diagnostic level of EEG. In quantitative analysis methods, power value data graph [13-18] and power spectrum curve graph [19-24] are all the main analytical technique and method. In general, the application of EEG quantitative analysis technique is not very common, and has a great development prospect. EEG power spectrum curve graph is the amplitude frequency curve graph shown in 30 frequency bands, longitudinal coordinate is power value, and transverse coordinate is EEG frequency of 0Hz~30Hz. It can clearly see the shape of graph and the change of main frequency.

In this study, EEG data under +Gz acceleration is first obtained, then power spectrum data graph and power spectrum curve graph are made by using the method of frequency spectrum analysis method, and EEG change characteristic under +Gz acceleration is observed. Results show that, under +Gz acceleration, the percentage value that each frequency band power value to this frequency band power value of the whole brain and the percentage value that each frequency band power value to the whole power value of this position all have significant change. The energy percentages of few leads show obvious change. Power spectrum data graph and power spectrum curve graph can intuitively and better help to find characteristic of EEG, and the individual parameter has the potential that becoming the effective indicator to measure the ischemia extent of brain and whether being the ischemic syncope. However, the method also has insufficiency, such as analysis time is longer, and it is based on the quasi stationary segment of EEG, and it is easy to smooth the transient information of EEG and so on, so it still has distance apart from the monitoring and analysis of EEG and needs sequential deeper study.

2. DATA ACQUISITION

The experimental data is acquired from 5 healthy volunteers. The subjects are all healthy, no human centrifuge contraindications. The main equipment is domestic, new type human centrifuge, and subjects are exposed under +Gz acceleration according to the set curve. The set curve starts from 1G, reaches the maximum G value at a certain growth rate, lasting for 10-15 seconds, and then drops to 1G at a certain growth rate. The maximum G value starts from 2.5G, increasing at the rate of 0.5G/r, until the subject gets to the endurance end. The doctor real-time monitors ear pulse, ECG and other physiological signals, and comprehensively judges if the subject gets to the endurance end, according to the level state of ear pulse, combining with the subjective narration and the expression of the subject. EEG data when running, and before and after running is recorded by using a portable EEG recorder. Electrode signal of 16 leads is recorded in accordance with the electrodes positions of 10-20 system of international EEG standard electrodes installation method[25]. All electrodes take the electrode at ipsilateral earlobe (A1 or A2) as the reference electrode[26]. Experiment requires that the subject should try to keep the natural state. Measuring method of single electrode lead is selected.

16 leads signal of EEG goes to amplify box after going to EEG interference box to remove interferes with the sampling rate of 128Hz, then the signal is recorded onto the internal SD card with the specific format. The obtained data make filtering, playing back and so on by the matching software of EEG physiological parameters recorder produced by Shanghai NUOCHENG Company, then realize frequency spectrum analysis and other functions. 16 leads data of EEG first pass pretreatment, namely the low pass digital filtering below 30Hz, then make a fast Fourier transformation, namely FFT transformation with the sampling frequency of 128Hz, from time domain transformating to frequency domain, getting cycle diagram of each segment, and the amplitude of cycle diagram that is the power spectrum function is the square of model of each point after FFT transformation. Data graph and curve graph of frequency spectrum analysis can be got according to power spectrum data. For every subject,10s data of during +Gz acceleration, and 10s data before +Gz acceleration that is at quiet state are selected, then data graph and curve graph of frequency spectrum analysis of 16 leads are made, in order to better observe the energy change of all kinds of rhythms at different parts of brain. In this study, the segment standard of all rhythms takes as: delta 0~3.5Hz, theta 4~7.5Hz, alpha1 8~11Hz, alpha2 11.5~13.5Hz,, beta1 14~18.5Hz, beta2 19~30Hz, and the six kinds of rhythms are made a spectral analysis and are observed the change.

3. RESULTS

Power spectrum curve graphs of 16 leads during +Gz acceleration and at quiet state are made and analyzed the change feature. Figure 1 and Figure 2 are the EEG power spectrum curve graph and EEG power value data graph of a subject at quiet state that is when 1G. Figure 3 and Figure 4 are the EEG power spectrum curve graph and EEG power value data graph of the subject when 7G. EEG power value data graph is also known as EEG frequency spectrum percentage graph , which is that EEG power values converted by six frequency bands of 16 brain regions are listed out by charts. For the numbers of brain regions, the first column values are the absolute power values, the second column values are the percentage values that each frequency band power value to this frequency band power value of the whole brain, and the third column values are the percentage values that each frequency band power value to the whole power value of this position. EEG power spectrum curve graph is the amplitude frequency curve graph shown in 30 frequency bands, longitudinal coordinate is power value, and transverse coordinate is EEG frequency of 0Hz~30Hz. It can clearly see the shape of graph and the change of main frequency.

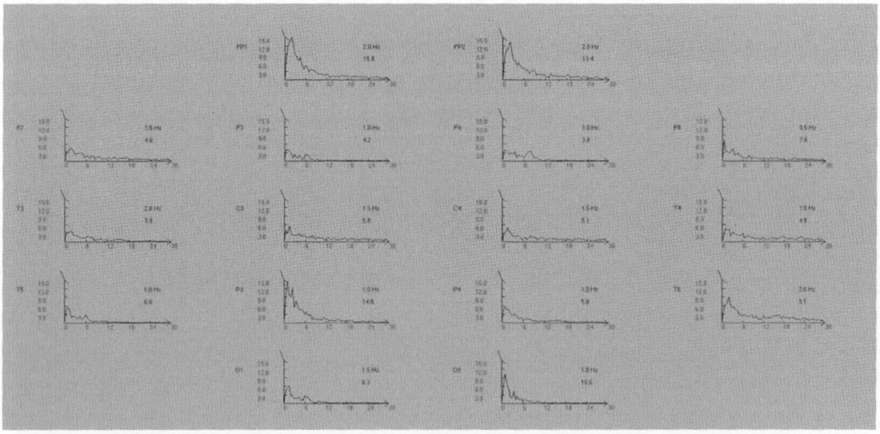

Fig. 1. EEG power spectrum curve graph of subject at quiet state that is when 1G.

	FP1	FP2	F3	F4	C3	C4	P3	P4	O1	O2	F7	F8	T3	T4	T5	T6
	79.6	67.4	20.1	21.8	23.9	23.0	68.4	27.7	26.4	36.9	26.1	28.5	21.8	24.5	24.0	44.5
δ	61.5%	49.0%	48.4%	41.3%	35.4%	36.5%	58.3%	49.3%	50.2%	62.4%	38.5%	43.9%	41.7%	33.0%	50.2%	37.3%
	14.1%	11.9%	3.6%	3.9%	4.2%	4.1%	12.1%	4.9%	4.7%	6.5%	4.6%	5.0%	3.9%	4.3%	4.2%	7.9%
	34.1	29.9	11.7	18.2	13.7	12.9	24.3	10.6	14.5	10.9	14.2	12.0	12.4	17.1	12.7	22.0
θ	22.1%	21.8%	28.2%	34.5%	20.2%	20.4%	20.8%	18.9%	27.6%	18.4%	21.0%	18.5%	23.7%	23.0%	26.6%	18.7%
	12.6%	11.0%	4.3%	6.7%	5.0%	4.7%	9.0%	3.9%	5.4%	4.0%	5.2%	4.4%	4.6%	5.8%	4.7%	8.1%
	14.3	13.6	2.7	5.2	9.2	8.8	8.6	5.4	4.0	3.4	8.3	7.2	6.2	8.7	4.2	13.5
α 1	9.3%	9.9%	6.5%	9.9%	13.5%	14.0%	7.4%	9.6%	7.6%	5.8%	12.2%	11.0%	11.9%	11.7%	8.7%	11.5%
	11.6%	11.0%	2.2%	4.2%	7.5%	7.1%	7.0%	4.4%	3.5%	2.6%	6.7%	5.8%	5.0%	7.0%	3.4%	11.0%
	7.0	6.9	1.5	1.8	5.4	4.5	4.6	2.5	1.7	2.0	4.8	4.6	3.2	4.4	1.8	8.2
α 2	4.6%	5.0%	3.7%	3.4%	8.1%	7.2%	3.9%	4.5%	3.3%	3.4%	7.1%	7.1%	6.1%	5.9%	3.4%	7.0%
	10.8%	10.7%	2.3%	2.8%	5.4%	7.0%	7.1%	3.9%	2.7%	3.1%	7.4%	7.1%	4.9%	6.8%	3.5%	12.6%
	10.1	10.3	2.6	2.7	7.4	6.4	6.5	5.8	2.8	3.0	7.6	6.2	5.1	10.0	2.4	16.3
β 1	6.6%	7.5%	6.4%	5.1%	11.0%	10.1%	5.6%	10.2%	6.9%	5.0%	11.1%	9.6%	9.8%	13.4%	6.1%	13.9%
	9.6%	9.7%	2.5%	2.6%	7.1%	6.1%	6.2%	5.5%	2.6%	2.8%	7.1%	5.9%	4.9%	9.5%	2.3%	15.5%
	9.4	9.4	2.9	3.0	7.8	7.6	4.7	4.1	3.1	2.9	6.9	6.5	3.5	9.7	2.9	12.8
β 2	6.1%	6.9%	6.9%	5.8%	11.6%	11.8%	4.0%	7.4%	5.9%	4.8%	10.1%	10.0%	6.7%	13.0%	6.1%	10.9%
	9.7%	9.7%	3.0%	3.1%	8.1%	7.7%	4.9%	4.3%	3.2%	2.9%	7.1%	6.7%	3.6%	10.0%	3.0%	13.1%

Fig. 2. EEG power value data graph of subject at quiet state that is when 1G.

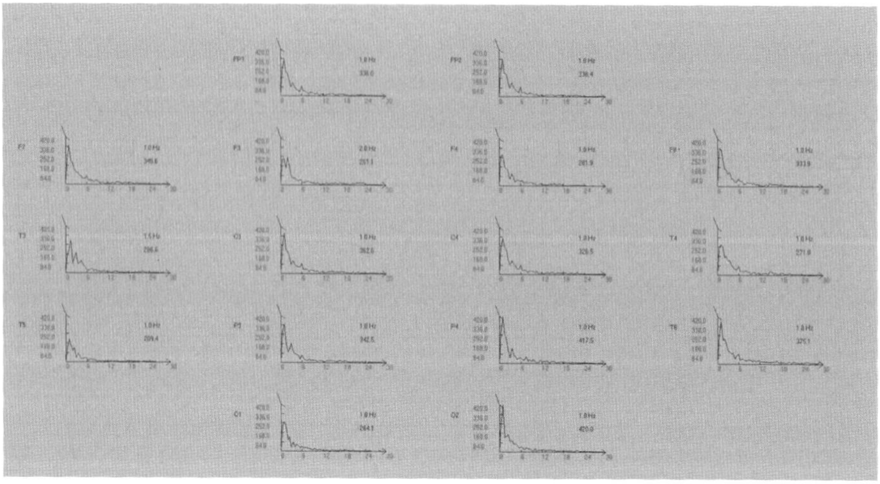

Fig. 3. EEG power spectrum curve graph of subject when 7G.

	FP1	FP2	F3	F4	C3	C4	P3	P4	O1	O2	F7	F8	T3	T4	T5	T6
	1256.8	1214.5	1078.1	1246.2	1331.1	1359.7	1405.5	1647.4	1179.6	1283.7	1279.4	1496.8	1203.9	1372.0	884.9	1506.9
δ	57.9%	56.4%	57.1%	61.0%	59.4%	59.8%	59.9%	60.4%	69.3%	61.3%	60.2%	64.8%	61.4%	60.5%	63.2%	55.9%
	6.1%	5.9%	6.2%	6.0%	6.4%	5.6%	6.8%	7.9%	6.7%	6.2%	6.2%	7.2%	5.8%	6.6%	4.3%	7.3%
	352.1	362.2	351.9	339.4	366.7	363.5	397.4	515.5	243.4	366.0	323.7	328.8	371.9	337.4	267.7	422.6
θ	16.2%	16.8%	18.6%	16.6%	16.8%	16.0%	15.9%	18.9%	14.3%	17.0%	15.2%	14.2%	19.0%	14.9%	18.4%	15.7%
	6.2%	6.4%	6.2%	6.0%	6.4%	6.4%	7.0%	9.1%	4.3%	6.2%	5.7%	5.8%	6.6%	5.9%	4.5%	7.4%
	213.4	221.2	166.3	185.6	203.7	227.4	183.2	204.8	105.4	160.7	184.7	177.0	132.0	195.2	96.0	251.1
α 1	9.8%	10.3%	8.8%	9.1%	9.1%	10.0%	7.8%	7.5%	6.2%	7.7%	8.7%	7.6%	6.7%	8.6%	6.9%	9.3%
	7.3%	7.6%	5.7%	6.4%	7.0%	7.8%	6.8%	7.0%	3.6%	5.5%	6.4%	6.1%	4.6%	6.7%	3.8%	8.6%
	95.1	92.7	75.2	72.8	84.7	79.6	109.3	108.4	46.7	84.5	91.4	76.3	60.9	91.8	42.3	169.7
α 2	4.4%	4.3%	4.0%	3.6%	3.8%	3.5%	4.7%	3.9%	2.7%	4.0%	4.3%	3.2%	3.1%	4.0%	3.0%	6.3%
	6.9%	6.7%	5.5%	5.3%	6.1%	5.8%	7.9%	7.7%	3.4%	6.1%	6.6%	5.5%	4.4%	6.6%	3.1%	12.3%
	149.7	150.6	105.1	125.7	149.0	143.7	154.1	149.9	74.6	114.2	161.1	158.9	96.7	157.6	74.9	181.7
β 1	6.9%	7.0%	5.6%	6.2%	6.6%	6.3%	6.8%	5.5%	4.4%	5.5%	7.1%	6.6%	4.9%	7.0%	5.3%	6.7%
	7.0%	7.1%	4.9%	5.9%	7.0%	6.7%	7.2%	7.0%	3.5%	5.4%	7.1%	7.2%	4.5%	7.4%	3.5%	8.5%
	104.7	111.6	112.1	72.6	107.4	101.2	95.2	105.4	51.9	95.2	95.9	86.8	84.7	113.5	45.1	162.2
β 2	4.8%	5.2%	5.9%	3.6%	4.8%	4.4%	4.1%	3.9%	3.0%	4.5%	4.5%	3.7%	4.8%	5.0%	3.2%	6.0%
	6.7%	7.2%	7.2%	4.7%	6.9%	6.5%	6.1%	5.8%	3.3%	6.1%	8.2%	5.6%	6.1%	7.3%	2.9%	10.4%

Fig. 4. EEG power value data graph of subject when 7G.

Through spectrum analysis, it can be seen that EEG signal under +Gz acceleration has obvious difference from the EEG signal at static state that is when 1G. From the data graph and frequency spectrum curve graph, it can be seen that the low frequency components, such as delta and theta components, are both the main components no matter when 1G or under +Gz acceleration, especially the low frequency component below 9Hz manifests obviously and the amplitude is larger. The low frequency component below 2Hz takes the maximum value of the amplitude of frequency spectrum curve. While the high frequency component above 25Hz is less, and the amplitude is smaller. Especially under high G acceleration, the amplitude of high frequency component of this part is close to 0. Under +Gz acceleration, the amplitude of power spectrum curve of 16 leads of brain all increases for all frequency bands, and the absolute value of energy has increasing of different degrees, and the values are different from several times to dozens of times. For the maximum value of increasing time, the maximum value of increasing time of delta rhythm is about 3 times that of beta rhythm. Under +Gz acceleration, the percentage value that each frequency band power value to this frequency band power value of the whole brain, and the percentage value that each frequency band power value to the whole power value of this position have both significant change. Among this, the percentage values that delta and theta frequency bands power values to this frequency band power value of the whole brain increase, and the percentage value of delta rhythm increasing is the most high, about 30%. The percentage values that alpha and beta frequency bands power values to this frequency band power value of the whole brain decrease, and among this the percentage value of beta2 rhythm decreases significantly, the maximum decreasing value about 10%. That is, the maximum change value of percentage value that delta frequency band power value to this frequency band power value of the whole brain is about 3 times of that of beta2 frequency band. For the energy change of each lead, changes of energy percentages of F4, C3,C4,F7, F8,T3,T4,P4,O1,T5,T6 leads are bigger, and changes of C3,C4,F7,F8,T4 leads are more obviously. Among this, the max value of energy percentage increasing of T4 lead is about28%,and the max value of energy percentage decreasing of F4 lead is about 18%.

4. RESULT ANALYSIS

Under +Gz acceleration, the change of EEG is a complex process, and EEG waveform change can only reflect the external manifestation of electrical activity inside brain, and spectrum analysis can help better to find the change characteristic of each frequency band. The results of spectrum analysis show that, under +Gz acceleration the low frequency rhythms, the energy percentages that delta and theta rhythms to all frequency bands increase, and the energy percentages that alpha and beta rhythms to all frequency bands decrease, and among this, the energy percentage of beta2 rhythm decreases mostly. This is because that EEG alpha rhythm, between states of awake and sleeping, more common in awake state with eyes closed. Alpha rhythm is one of the main components of spontaneous EEG, and corresponding to idle rhythm of visual cortex it is the component with obvious characteristic in rhythms. The inhibited extent of alpha wave indicates the awake degree of a person. While beta rhythm marks that human is at the "alert" state, and this rhythm transformating for slow wave indicates and show that the alert state of human is changing [27]. There is scholar who finds the significant phenomenon that power percentage index of β_2 (20~29. 8 Hz) suddenly drops in 1 minute before syncope in human lower body negative pressure experiment in 2000, so it is suggested that if further studying the variation characteristics of power spectrum of EEG beta frequency band, it is possible to more know the change of brain function state before G-LOC or before syncope which cannot be understood and known only through the recognition and judgment of naked eyes [28].In this study, the conclusion about beta2 rhythm is also consistent with this point. This parameter has the potential to become an effective indicator to measure ischemia degree of brain and whether human body at the state of ischemic syncope. The energy percentages of delta and theta rhythms increasing, and the energy percentages of alpha and beta rhythms decreasing, manifest that under the action of centrifugal force, body blood transfers from head to lower body part, brain starts ischemia and hypoxia, the inhibited degree of brain increases, consciousness degree decreases, and the ischemia state of brain is increasing. The spectrum change curve helps to better reflect physiological change process of human body. In the leads, the changes of the central point, anterior temporal point, and middle temporal point are relatively more obvious, and the changes of right leads are relatively more obvious. Analyzing the cause, this is related to the special body position of human body under +Gz acceleration. During +Gz platform, after rotation the body is in the half-lying position with the left side of body down, the right side of body up, and head toward the center of axis, feeling the action of centrifugal force from head to foot, and affected by blood circulation and other body circulation, the impact of ischemia subjected by the right side of body is greater than that subjected by the left side of body, so the change of leads of right side is more obvious than that of left side. Due to the cockpit, after rotation body is not completely lie low, there is a small slope between front body and rear body, and influenced by gravity, the impact of ischemia subjected by front side of brain is greater than that subjected by rear side of brain, so the change of leads of brain front side is more obvious than that of brain rear side.

EEG power value data graph and EEG power spectrum curve graph both can better reflect EEG change under +Gz acceleration and help to find EEG change feature under +Gz acceleration. They both can help to play the biggest role in quantitative analysis of EEG. Power value data graph is often used to study the effect of brain function and a variety of functional areas, that is, a variety of operations causing change of power value for the activation of some brain area. Power spectrum curves graph can help to clearly see the shape of diagram and the change of main frequency. For the diagnose of lazy wave, the difficulty of naked eye analysis is very big, and naked eye analysis often causes missed diagnosis, while power spectrum curve graph is convenient and sensitive. Lazy wave is mainly manifested as frequency slows down and waveform amplitude decreases, and if the frequency of main frequency slows down more than 0.5 Hz or above or the power value decreases more than 1/3 of itself, the waveform can be considered as an exception. But analysis time of power value data graph method is longer, so it is rarely used in the clinical practice, and it is seldom used in real-time monitoring of EEG, and many scholars believe that it is not as intuitive, catching the eye and rapid as brain atlas. If making power spectrum curve graph by using less frequency bands such as 6 frequency bands, it is not of this advantage that easy to see the shape of graph and the change of main frequency. EEG power value data graph and EEG power spectrum curve graph reflect the energy distribution of EEG in different frequency bands, but this method is based on the quasi stationary segmentation of EEG, thus smoothing the transient information of EEG. Therefore, at present the application of EEG quantitative analysis technology such as power value data graph and power spectrum curve graph and so on still is not very common, and there is a long distance to realize monitoring and analysis of EEG using them. There are researchers who have proposed a number of new methods for EEG analysis from different angles. It is believed that, with the continuous improvement of all kinds of EEG analysis methods and the continuous improvement of modern analysis technology, still there are many work needs to be done in the research of quantitative relationship between brain functional status and EEG characteristic signal.

5. CONCLUSIONS

In this study, human EEG data is first obtained by using a new type of human centrifuge, then power value data graph and power spectrum curve graph are made by using the method of frequency spectrum analysis, and the EEG data under +Gz acceleration and at quiet state are made a comparison and study. It is found under +Gz acceleration, the amplitude of power spectrum curve of each frequency band for 16 leads has a increasing, and the absolute value of energy has increasing of different degrees, value differing from several times to dozens of times. The percentage value that each frequency band power value to this frequency band power value of the whole brain, and the percentage value that each frequency band power value to the whole power value of this position both have significant change. Among this, the percentage values that delta and theta frequency bands power values to this frequency band power value of the whole brain increase, and the percentage value of delta rhythm increasing is the most high. The percentage values that alpha and beta frequency bands power values to this frequency band power value of the whole brain decrease, and among this the percentage value of beta2 rhythm decreases significantly. This is related to that, under +Gz acceleration, under the action of centrifugal force , human blood transfers from head to lower body , the ischemic state of brain aggravates , and the hypoxia extent of ischemic aggravates . For the energy change of leads, the change of energy percentage of lead C3, C4, F7, F8, T4 is more obvious. In the leads, the change of central point, anterior temporal point, middle temporal point and the right side leads is comparatively more obvious. Analyzing the cause, this is related to that, under +Gz acceleration, human body is at a special position and is affected by body cycle. Power spectrum data graph and power spectrum curve graph both can better help to find characteristic of EEG under +Gz acceleration, and the individual parameter has the potential that becoming the effective indicator to measure the ischemia extent of brain and whether being the ischemic syncope. Power value data graph is often used to make special study for a variety of functional areas of brain, while power spectrum curves graph can help to clearly see the shape of diagram and the change of main frequency. However, they also have shortages, such as analysis time of power value data graph method is longer, and if making power spectrum curve graph by using less frequency bands such as 6 frequency bands, it is not of this advantage that easy to see the shape of graph and the change of main frequency, and the two methods are based on the quasi stationary segmentation of EEG, thus easy to smooth the transient information of EEG and so on.

Therefore, at present the application of EEG quantitative analysis technology still is not very common, and there is not a short distance to realize monitoring and analysis of EEG using them. There have been many researchers who have proposed a number of new methods for EEG analysis. It is believed that, with the continuous improvement of all kinds of EEG analysis methods and the continuous raising of modern computer analysis technology, in the research of realizing monitoring and analysis of EEG by using power frequency spectrum analysis, and in the research of brain functional status and quantitative analysis of EEG characteristic signal, still there is work needed to be done and needed to make further and deeper study.

References

1. Morikawa K, Matsumoto A, Patki S, Grundlehner B,Verwegen, Xu JW, Srinjoy Mitra and Julien Penders. Compact Wireless EEG System with Active Electrodes for Daily Healthcare Monitoring. *IEEE International Conference on Consumer Electronics (ICCE)* 2013;**6**:204-205.
2. Xu JW, Yazicioglu RF, Grundlehner B, Harpe P, Makinwa KAA, Hoof CV. A 160μW 8-Channel Active Electrode System for EEG Monitoring. *J IEEE TRANSACTIONS ON BIOMEDICAL CIRCUITS AND SYSTEMS* 2011;**5**(6):555-567.
3. Lin CT, Chang CJ, Lin BS, Hung SH, Chao CF, Wang IJ. A Real-Time Wireless Brain–Computer Interface System for Drowsiness Detection. *J IEEE TRANSACTIONS ON BIOMEDICAL CIRCUITS AND SYSTEMS* 2010;**4**(4):214-222.
4. Pan JS, Yue L.A ridge extraction algorithm based on partial differential equations of the wavelet transform.2014 *IEEE Symposium on Computational Intelligence for Multimedia, Signal and Vision Processing (CIMSIVP), Orlando, FL* 2014;1 – 5.
5. Kumar, K, Mustafa N, Li JP, Shaikh RA, Khan SA. Khan A.Image edge detection scheme using wavelet transform.*2014 11th International Computer Conference on Wavelet Active Media Technology and Information Processing (ICCWAMTIP), Chengdu* 2014;261 – 265.
6. Ramesh K, Sushama M. Power transformer protection using wavelet packet transform.*2014 International Conference on Computer Communication and Informatics (ICCCI),Coimbatore* 2014;1 – 7.
7. Bharkad S, Kokare M. Fingerprint matching using discreet wavelet packet transform, *2013 IEEE 3rd International Advance Computing Conference (IACC), Ghaziaba* 2013;1183 – 1188.
8. Naqvi SAR, Touqir I, Raza SA. Adaptive geometric wavelet transform based two dimensional data compression. *2013 7th IEEE GCC Conference and Exhibition (GCC), Doha* 2013;583 – 588.
9. Borenovic M, Neskovic A. ANN Based Models for Positioning in Indoor WLAN Environments. *19th Telecommunications forum TELFOR 2011,Belgrade* 2011;305-312.
10. Lv WT, Shen C, Gui F, Tian ZS, Jiang DZ. Real-Time Spectrum Analyzer Based on All Phase FFT Spectrum Analysis. *2013 Fourth International Conference on Digital Manufacturing and Automation (ICDMA),Qingdao* 2013; 966 – 969.

11. He CH, Hou B, Wang LW,En YF, Xie SF.A failure physics model for hardware Trojan detection based on frequency spectrum analysis.*2015 IEEE International Reliability Physics Symposium (IRPS), Monterey, CA* 2015; PR.1.1 - PR.1.4.
12. Zabalza, Ren JJC, Zheng JB, Han JW, Zhao HM, Li ST , Marshall S. Novel Two-Dimensional Singular Spectrum Analysis for Effective Feature Extraction and Data Classification in Hyperspectral Imaging. *J IEEE Transactions on Geoscience and Remote Sensing* 2015; **53**(8):4418 – 4433.
13. Jang IS, Kim JW, Kim JS. Makeup color reproduction based on spectrum data. *2013 19th Korea-Japan Joint Workshop on Frontiers of Computer Vision (FCV), IEEE, Incheon* 2013:233 – 236.
14. Liu JX, Sun QS. Mass spectrum data processing based on compressed sensing recognition and sparse difference recovery. *2012 9th International Conference on Fuzzy Systems and Knowledge Discovery (FSKD), Sichuan* 2012;1413-1417.
15. Gardel P, Morinigo-Sotelo D, Duque-Perez O, Perez-Alonso M, Garcia-Escudero LA. Neural network broken bar detection using time domain and current spectrum data.*2012 XXth International Conference on Electrical Machines (ICEM), Marseille* 2012; 2492 -2497.
16. Feng L,Qiu RC, Browning JP. Spectrum Sensing With Small-Sized Data Sets in Cognitive Radio: Vehicular Technology. *IEEE Transactions on Algorithms and Analysis* 2015; **64** (1) :77 – 87.
17. Stamatakis G, Tragos EZ, Traganitis A. Periodic collection of spectrum occupancy data by energy constrained cognitive IoT devices, *2015 International Wireless Communications and Mobile Computing Conference (IWCMC), Dubrovnik, Croatia* 2015; 880-885.
18. Sheng SP,Liu MY, Saigal R. Data-Driven Channel Modeling Using Spectrum Measurement. *J IEEE Transactions on Mobile Computing* 2015; **14**(9): 1794-1805.
19. Aram Z, Setarehdan SK. R-R interval simulation based on power spectrum curve fitting. *2013 20th Iranian Conference on Biomedical Engineering (ICBME), Tehran* 2013; 132-136.
20. Shuman DI, Wiesmeyr C, Holighaus N, Vandergheynst P. Spectrum-Adapted Tight Graph Wavelet and Vertex-Frequency Frames. *J IEEE Transactions on Signal Processing* 2015;**63**(16):4223- 4235.
21. Xu Z, He Y, Xu XD, Peng F. QoS graph coloring spectrum allocation for femtocell in Macro/Femto heterogeneous network. *2013 8th International ICST Conference on Communications and Networking in China (CHINACOM), Guilin* 2013; 374-378.
22. Yang M, Tang CY. Distributed estimation of graph spectrum, *American Control Conference (ACC), 2015 , Chicago, IL* 2015;2703-2708.
23. Wang YJ, Tan ZH. Graph-Based and QoS Guaranteed Spectrum Allocation for Dense Local Area Femtocell Networks. *2014 IEEE Military Communications Conference (MILCOM), Baltimore, MD* 2014; 1556-1561.
24. Bera D, Chakrabarti I, Ray P, Pathak SS. Factor Graph Based Cooperative Spectrum Sensing in Cognitive Radio over Time-Varying Channels. *2013 IEEE 77th Vehicular Technology Conference (VTC Spring), Dresden* 2013; 1-5.
25. Okuma T. *Clinical Electroencephalogram (5th ed), Beijing: Tsinghua University Press* 2005; pp.1-112.

26. Nakamura M, Chen Q, Sugi T, Ikeda A, H. Shibasaki. Technical quality evaluation of EEG recording based on electroencephalographers' knowledge.*J Medical Engineering & Physics* 2005; **27**:93–100.
27. Nakamura M, Sugi T, Ikeda A, Kakigi R, Shibasaki H. Clinical application of automatic integrative interpretation of awake background EEG: quantitative interpretation, report making, and detection of artifacts and reduced vigilance level. *J Electroencephalography and clinical Neurophysiology* I996; **98**: 103-112.
28. Chen YS, Jiao ZG, Yan GD, Jin ZR. The characteristics of cerebral functional indices during occurrence of presyncopal symptom induced by lower body negative pressure. *J Chinese Journal of Aerospace Medicine* 2000;**11**(3):145-148.

ASYMMETRIC SYNTHESIS ATOMOXETINE INTERMEDIATE (R)-ETHYL 3-HYDROXY-3-PHENYLPROPIONATE WITH MICROORGANISMS

Zhimin Ou[a,*] and Wangping Zhao[b]

Pharmaceuticals College, Zhejiang University of Technology, Hangzhou, Zhejiang, 310014, China
E-mail: [a]oozzmm@163.com, [b]506776730@qq.com
[*]*Corresponding author*

(R)-ethyl 3-hydroxy-3-phenylpropionate is an important intermediate in the synthesis of (R)-Atomoxetine. It can be gained by reduction of ethyl benzoylacetate with Candida tropicalis ZJUTSS 4.1.2 screened from soil samples as catalyst. The optimal bioconversion conditions are as follows: pH6.0, 30°C, 120 rpm/min and reaction time 24 h. The yield and enantiomeric excess of (R)-ethyl 3-hydroxy-3-phenylpropionate can get to 100% when 4.8 g/L of substrate and 18.4 g/L (cell dry weight) of cellswere used.

1. INTRODUCTION

Attention deficit hyperactivity disorder(ADHD) is a neurodevelopmental disorder associated with metabolic disorders of dopamine (DA) and norepinephrine (NE). Atomoxetine isa selective inhibitor of NE transporter and a non-irritating drug first used in the treatmentof ADHD. The efficacy of (R)-Atomoxetine is 2 times than racemic Atomoxetine, nine times than the efficacy of (S)-Atomoxetine. (R)-ethyl 3-hydroxy-3-phenylpropionate is a key chiral building block of (R)-Atomoxetine. It can be obtained by the biotransformationmethod[1-6].

Reaction conditions of biotransformation are mild and environmentally friendly compared with the chemical method. The asymmetric reduction of (R)-ethyl 3-hydroxy-3-phenylpropionate with a microorganism as the catalyst was recognized as an effective method because conversion can theoretically reach 100% and the microorganism can be cultured on a large scale[7, 8].

In this paper, (R)-ethyl 3-hydroxy-3-phenylpropionate was synthesized by the asymmetric reduction of ethyl benzoylacetate with C. tropicalis ZJUTSS 4.1.2. The reaction scheme was shown in Fig. 1. And the biotransformation conditions on reduction were investigated in detail[9,10].

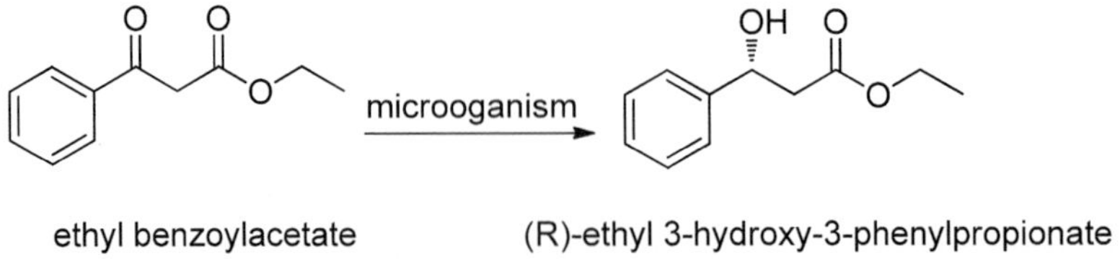

Fig. 1. Transformation of ethyl benzoylacetate to (R)-ethyl 3-hydroxy-3-phenylpropionate.

2. MATERIALS AND METHODS

2.1. Reagents and Instruments

(R)-ethyl 3-hydroxy-3-phenylpropionate and (S)-ethyl 3-hydroxy-3-phenylpropionate were supplied by J&K SCIENTIFIC Co., Ltd., China. Other chemical reagents were purchased commercially. GC(Shimadzu GC-2014) and HPLC(Shimadzu-LC-20AP) were used inthe analysis.

2.2. Microorganism and Cultivation

Microorganisms were screened from soil samples (5-15 cm deep) collected from Zhejiang University of Technology. The solid medium for the preservation of the strain was composed of 20 g/L glucose, 3 g/L NaCl, 1 g/L $(NH_4)_2SO_4$, 0.1 g/L $MgSO_4$, 1.5 g/L $K_2HPO_4 \cdot 3H_2O$, 0.5 g/L KH_2PO_4, and 20 g/L agar. The liquid medium for the cultivation of cells on alarge scale was composed of 20 g/L glucose, 3 g/L NaCl, 1 g/L $(NH_4)_2SO_4$, 0.1 g/L $MgSO_4$, 1.5 g/L $K_2HPO_4 \cdot 3H_2O$, and 0.5 g/L KH_2PO_4. The strain was inoculated into 25 ml liquid culture medium and cultivated in 30°C shaker (200 rpm/min) for 24 h. The 2.5 ml cell suspension gained above was inoculated into 100 ml liquid medium. After cultivationfor 24 h, the free cells were harvested by centrifugation (4000 ×g, 20 min) and used in reduction.

2.3. Biotransformation Process

0.24 g substrates were added into a 100 ml shaking flask contained 25 ml distilled water with cells. The flask was shaken at 200 rpm/min and 30°C. After 72 h, the cells were collected by centrifugation (4000 ×g, 20 min) and discarded, and the (R)-ethyl 3-hydroxy-3-phenylpropionate were extracted with ethyl acetate.

2.4. Analytical Methods

Ethyl 3-hydroxy-3-phenylpropionate and ethyl benzoylacetate were analyzed in a chiral column supelco β-120 (0.25 mm diam, 25 m long, 0.25 μm film). GC detection conditions:injection volume: 1 μL, column temperature: 125°C, injection port temperature: 250°C, detecting port temperature: 255°C, carrier gas was N_2, carrier gas flow rate: 1.62 ml/min, split ratio: 1:15, and detector: chydrogen flame ionization detector (FID). The retention time of ethyl 3-hydroxy-3-phenylpropionate and ethyl benzoylacetate were 17.815 min and27.182 min, respectively(Fig. 2). The enantiomeric excess of ethyl 3-hydroxy-3-phenylpropionate were analyzed using a Chiracel OB-H column (4.6 mm diam, 250 mm long; Diacel). GC detection conditions: mobile phase: n-hexane/isopropanol (80/20, V/V), UV detection wavelength: 254 nm, injection volume: 3 μl, flow rate: 1.0 ml/min, and column temperature: 35 °C. The retention time of (R)-ethyl 3-hydroxy-3-phenylpropionate and (S)-ethyl 3-hydroxy-3-phenylpropionate were 6.182 min and 7.398 min, respectively(Fig. 3).

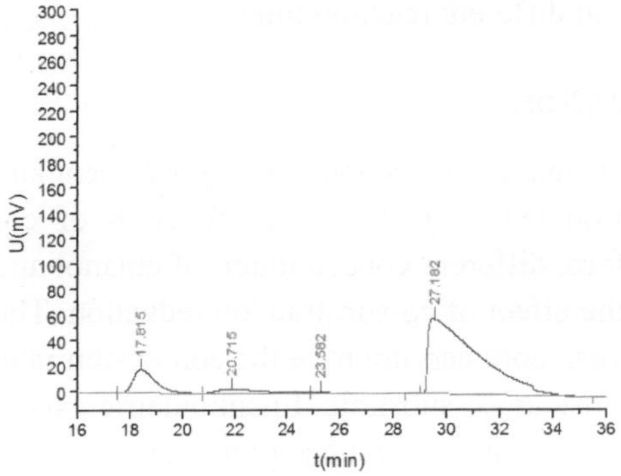

Fig. 2. Gas Chromatography of ethyl 3-hydroxy-3-phenylpropionate(17.815 min) and ethyl benzoylacetate (27.182 min).

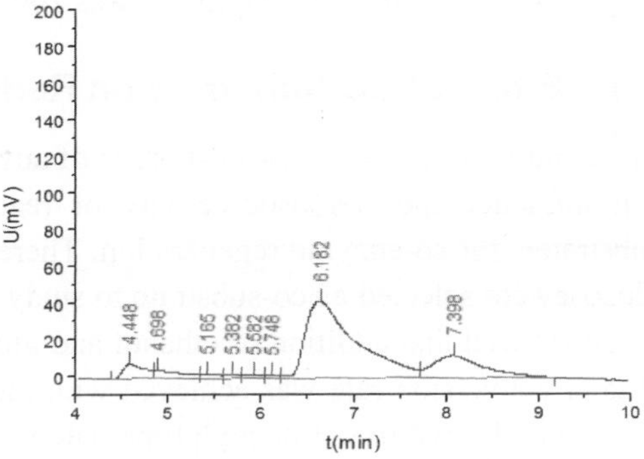

Fig. 3. Liquid Chromatography of (R)-ethyl 3-hydroxy-3-phenylpropionate (6.182min) and (S)-ethyl 3-hydroxy-3-phenylpropionate (7.398 min).

3. RESULTS AND DISCUSSION

3.1. Select the Optimum Strain

Six strains were selected from soil samples to find our objective strain in research. Table 1 show that conversion rate and enantiometric excess of product can get to 51.6% and100% with *C. tropicalis ZJUSS* 4.1.2 used as catalyst. *C. tropicalis ZJUSS* 4.1.2 was selected as the optimal microorganism and was used in further study.

Table 1. Selection of cells which can transform ethyl benzoylacetate to (R)-ethyl 3-hydroxy-3-phenylpropionate

Strain name	Product configuration	Conversion rate (%)	Enantiomeric excess (e.e.%)
S. cerevisiae ZJUTSS 2.1.1	R	96.2	54.6
C. tropicalis ZJUTSS 4.1.2	R	51.6	100
S. cerevisiae ZJUTSS 1.1.1	S	32.4	84.2
C. tropicalis ZJUTSS 1.2.1	R	28.6	50.1
S. cerevisiae ZJUTSS 2.2.2	S	30.0	70.5
S. cerevisiae ZJUTSS 5.4.1	S	10.2	69.1

3.2. Effect of Transformation Time on Reduction

The conversion rate significantly increased from 12 h to 24 h (Fig. 4). The optimum reduction time was 24 h. The conversion rate did not rise after 24 h as expected but drop obviously. Probably, the reduction activity of cells was inhibited by products, while some by-products had reacted with the substrates or the products. Enantiometric excess of (R)-ethyl 3-hydroxy-3-phenylpropionate was 100% at different reaction time.

3.3. Effect of Co-Substrate on Reduction

The addition of suitable co-substrate is of advantage to conversion rate improvement and can influence the stereosodrectivity of reaction (Fig. 5). The main function of co-substrateis for co-enzyme regeneration. Therefore, different concentration of ethanol and glucosewere selected as co-substrate to study the effect of co-substrate on reduction. The result showed that addition of ethanol and glucose both can improve the conversion rate. Higher conversion rate was achieved with glucose as co-substrate. Enantiometric excess of (R)-ethyl3-hydroxy-3-phenylpropionate was 100% at different reaction time.

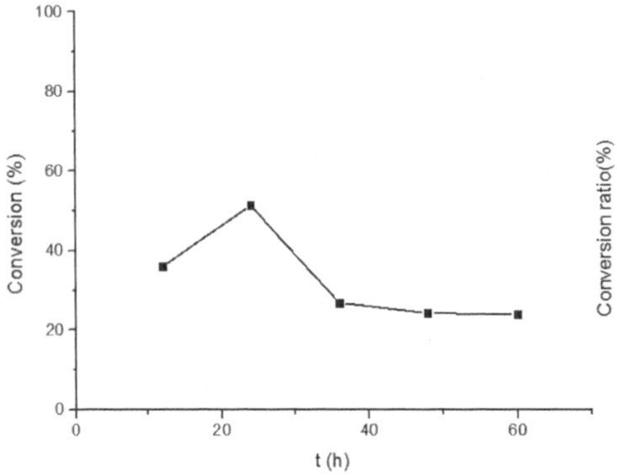

Fig. 4. Effect of time on reduction.

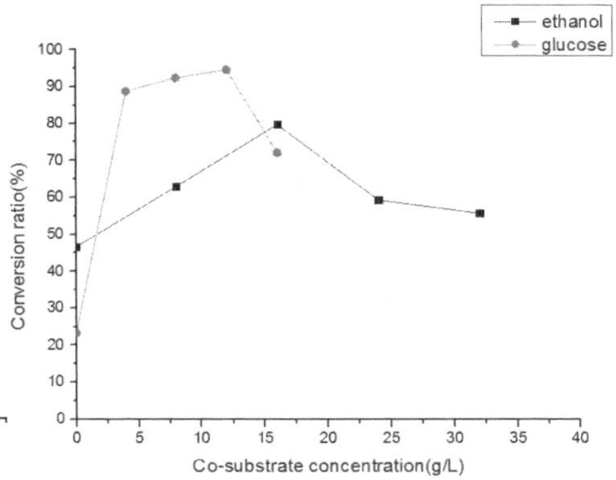

Fig. 5. Effect of co-substrate concentration on reduction.

3.4. Effect of Initial Substrate Concentration on Reduction

The conversion rate decreased with the increase of the initial substrate concentration (Fig.6). Probably, a large amount of substrate is toxic to microorganisms and inhibits the activity of enzyme. Higher conversion rate can be gained when the substrate concentration was lower. Conversion rate can reach 81.53% when initial substrate concentration was 4.8 g/L. Enantiometric excess of (R)-ethyl 3-hydroxy-3-phenylpropionate was 100% at different initial substrate concentration.

3.5. Effect of Biomass on Reduction

The biomass of *C. tropicalis ZJUSS* 4.1.2 used in reduction had significant effect on conversion rate (Fig. 7). More addition of cells can improve the conversion rate. The content of reduction enzyme and co-enzyme NAD(P)H increased with more biomass of cells added in reduction so that the higher conversion rate could be acquired. Enantiometric excess of (R)-ethyl 3-hydroxy-3-phenylpropionate was 100% at different biomass.

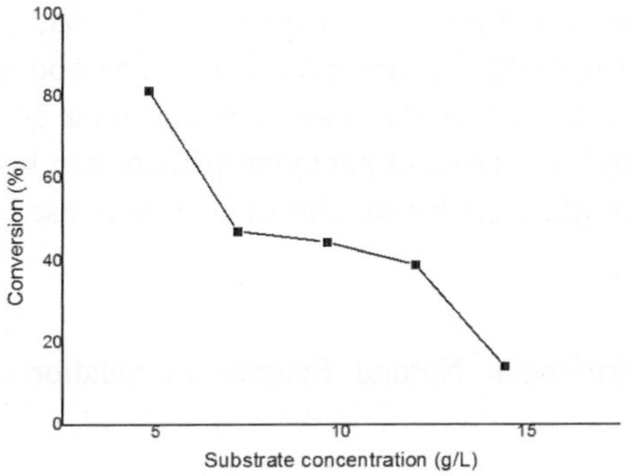

Fig. 6. Effect of substrate concentration on reduction.

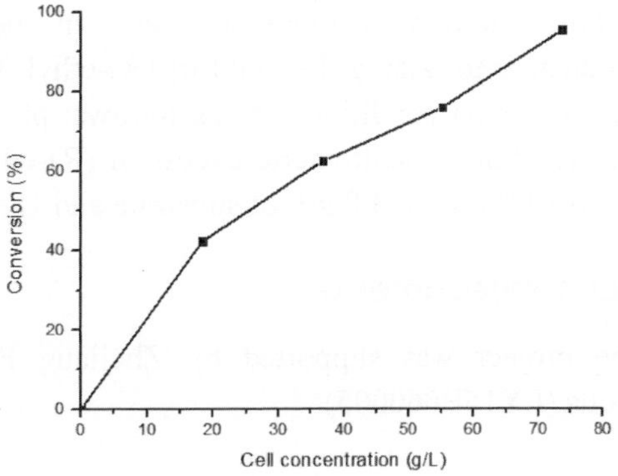

Fig. 7. Effect of cell concentration on reduction.

3.6. Effect of pH on Reduction

The cells of *C. Tropicalis ZJUSS* 4.1.2 should be suspended in optimal buffer to ensure reduction activity effectiveness (Fig. 8). *C. tropicalis ZJUSS* 4.1.2 maintained good reduction ability from pH 5.0 to 6.0. And conversion rate reached 100% when pH was 6.0. Enantiometric excess of (R)-ethyl 3-hydroxy-3-phenylpropionate was 100 % at different pH ofphosphate buffer.

3.7. Effect of Cell Reuse Times on Reduction

The conversion rate was the highest when cells were first used. With the increase in cell reuse times, the conversion rate responsed rapid decline (Fig. 9). Therefore, *C. tropicalis ZJUSS* 4.1.2 is no suitable for reuse many times.

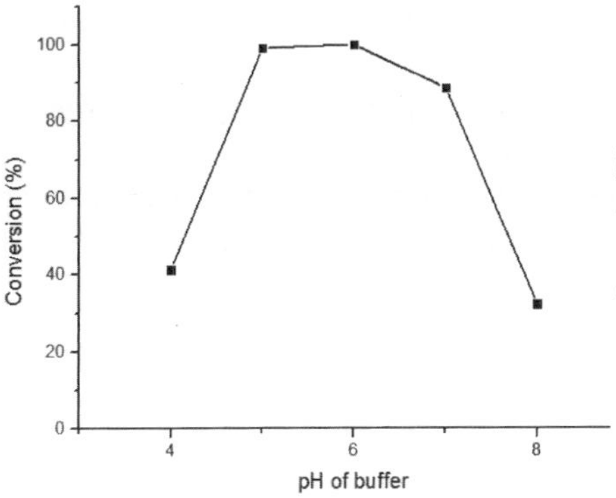

Fig. 8. Effect of pH on reduction.

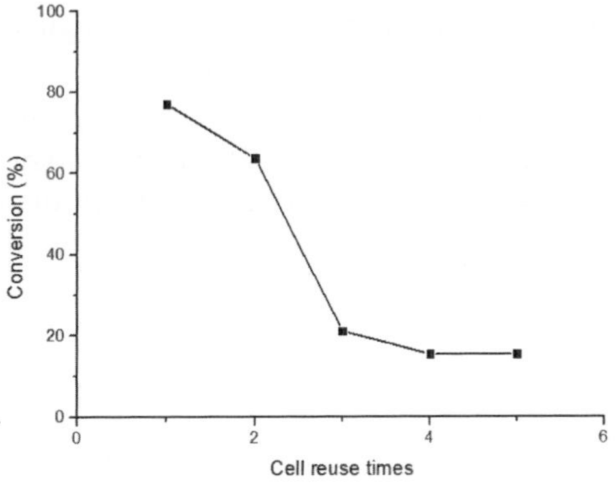

Fig. 9. Effect of cell reuse times on reduction.

4. CONLUSION

(R)-ethyl 3-hydroxy-3-phenylpropionate was synthesized by the asymmetric reduction of ethyl benzoylacetate with *C. tropicalis ZJUTSS* 4.1.2. The conversion rate decreased withthe addition of more substrates in the reduction. The increase of biomass is conductive to raising the yeild of (R)-ethyl 3-hydroxy-3-phenylpropionate. The optimal bioconversion conditions are as follows: pH 6.0, 120 rpm/min and reaction time 24 h. The yield and enantiomeric excess of (R)-ethyl 3-hydroxy-3-phenylpropionate can both get to 100% when 4.8 g/L of substrate and 18.4 g/L (cell dry weight) of cells were used.

Acknowledgments

The project was supported by Zhejiang Provincial Natural Science Foundation of China.(LY15B060005)

References

1. Y. Z. Chen, F. Q. Wen. Drug treatment status and progress of children with Attention Deficit Hyperactivity Disorder. J Chinese *Journal of Pediatrics* 2008; **23**:715–718.
2. Aaron T. Piepmeier, Chia-Hao Shih, Margaret Whedon, et al. The effet of acute exercise on cognitive performance in children with and without ADHD. J *Journal of Sport and Science* 2015; **4**: 97–104.
3. A. Q. Ji, L. Zhou. The separate of intermediates atomoxetine. J *Sichuan Chemical* 2011; **14**: 5–6.
4. L. Wang. The Study on Synthesis duloxetine and atomoxetine. D Shenyang Pharmaceutical University. 2004.
5. H. Shi, Q. M. Ding. Progress in the treatment of ADHD medication atomoxetine. J *Chinese Journal of Hospital Pharmacy* 2010; **30**: 2111–2112.
6. Z. M. Ou, X. Y. Chen. Saccharomyces cerevisiae and it applied in microbial conversion prepared (R) - (-) - β- hydroxy ethyl Application: China, 101709271[P]. 2009-10-23.
7. S. F. Cao, W. C. Li. Research and application of microbial conversion. J *Guangdong Sericultural* 2012; **46**: 45–50.
8. Z. M. Ou, X. Y. Chen. Synthesis of (s)-hydroxyphenyl propionate by microorganism transformation. J *Industrial Microbiology* 2009; **39**: 11–15.
9. X. D. Mao, J. Y. He. Screening of strain producing carbonyl reductase and optimization of its fermentation conditions. J *Industrial Microbiology* 2007; **37**: 21–26.
10. X. X. Shu, L. H. Sun. Strain screening and culture condition optimization for production of carbonyl reductase by Candida sorboxylosa. J *Industrial Microbiology* 2013; **43**: 14–19.

THE IMAGING CHARACTERISTICS OF ACUTE ALCOHOLIC ENCEPHALOPATHY: A CASE REPORT

Mengjiao Li, Haijing Guan, Cuicui Cui, and Ying Xing[a,*]
Department of Neurology, China-Japan Union Hospital of JiLin University, Changchun, 130000, China
E-mail: [a]xingying1970@163.com
[*]Corresponding author

Alcoholic Encephalopathy was originally described as a rare, fatal disease affecting wine drinks, characterized by demyelination and necrosis of the hemispheric white matter. The prevalence of this disease is unclear. Alcoholic Encephalopathy includes a variety of neurologic features such as seizures, ataxia, dysarthria, cognitive dysfunction, and so on. It has presented with acute or chronic alcoholic encephalopathy. Generally speaking, the therapeutic effect was relatively poor. Especially it is difficult to restore the damage presented in image. We report a case of Acute Alcoholic Encephalopathy who has diagnosed early in the disease course with subsequent clinical and radiological recovery by appropriate treatment.

1. THE CASE DETAILS

A 44-Year-old man was admitted to our department because of the abdominal pain associated with nausea and vomiting on March 21, 2015. The patient was acute onset with a background of heavy alcoholism (1L of that day). He had been consuming an average of 400ml of liquor per day for the last 20 years. With examination, BP: 129/82mmHg, HR: 102 times/ min, bilateral papillary diameter and light reflex were normal. And there was no nystagmus. Fundus and extraocular movements were normal. There were no focal or lateralizing neurological signs. The laboratory data revealed his allergic troponin of serum level was high (0.33ng/ml) (0-0.04), and the electrocardiogram presented with sinus tachycardia (102 times/ min) and with the level of S-T (II, III, AVF) elevating about 0.5 mv. On the basis of clinical and laboratory data, acute coronary syndrome was diagnosed on the first day. Intravenous Nitrate yamanashi ester and compound coenzyme were started. But on the second day he suddenly presented with dysphoria, speech blunt, and hemiparesis. On examination, BP: 150/80mmHg, HR: 98 times/ min, He was confused and disoriented without any signs of meningeal irritation. He had a totally type of dysarthria with both scanning and spastic components. And other neurological examinations were not allowed. Findings of brain CT performed after onset were normal (Fig.1) (March 22, 2015). His allergic troponin of serum level was mainly back to normal (0.07ng/ml). His baseline investigation in the form of thermogram, liver function tests, vitamin B12 levels were normal. Cerebral MRI showed multiple, symmetrical, long T1 and long T2 signals, which appeared bilateral periventricular areas (Fig.2) (March 23, 2015). Diffuse and symmetric hyperintensity around bilateral ventricles is main appearance on cerebral DWI (Fig.3A) (March 23, 2015),. Diffuse and symmetric hypointensity around bilateral ventricles is main appearance on cerebral ADC map (Fig.3B) (March 23, 2015). On the basis of clinical and imaging features, Acute Alcoholic Encephalopathy was diagnosed. Intravenous vitamin B complex (including vitamin B1, vitamin B6, and vitamin B12) and dexamethasone therapy (10mg/day for 6 days, next 5mg/day For 3 days) and immunoglobulin therapy (0.4g/kg) were started for 5 days, coordinating with vitamin B1 intramuscular injection (100 mg/day). Based on the mechanism, we also use it in combination with other treatments including clearing free radicals away, decreasing the intracranial pressure and so on. Due to the adequate treatment mentioned above, his neurological disorders alleviated concurrently on the tenth day. Approximately 15 days after the onset, cerebral MRI yielded the signals surrounding bilateral periventricular which showed in the second picture tend to be normal (Fig. 4) (April 1, 2015). In addition, signal - intensity abnormalities on DWI and ADC map had disappeared (Fig.5) (April 1, 2015).

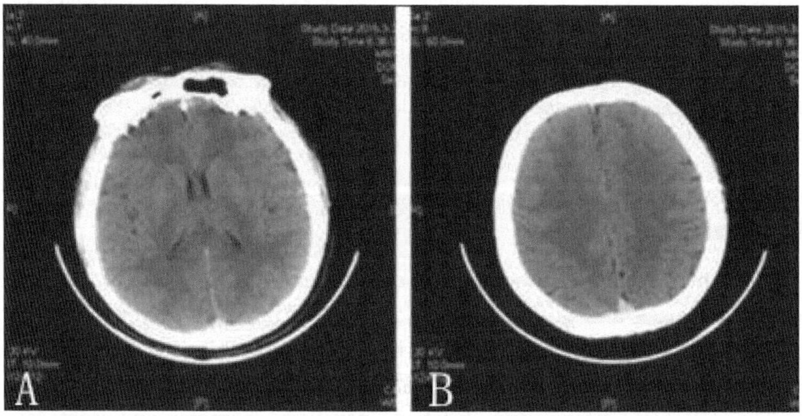

Fig. 1. A: Brain CT showed that the density was normal around bilateral basal ganglia. B: Brain CT showed that the density was normal around bilateral corona radiation. (March 22, 2015).

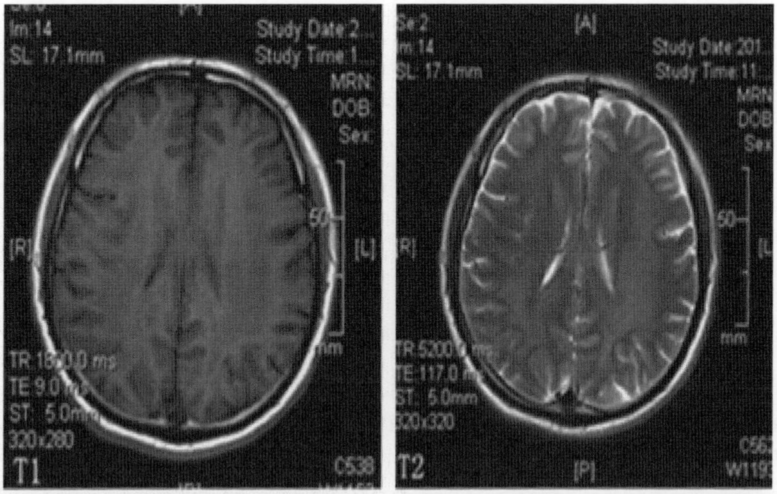

Fig. 2. Brain MRI showed multiple, symmetrical, long T_1 and long T_2 signals, which appeared bilateral periventricular areas (March 23, 2015).

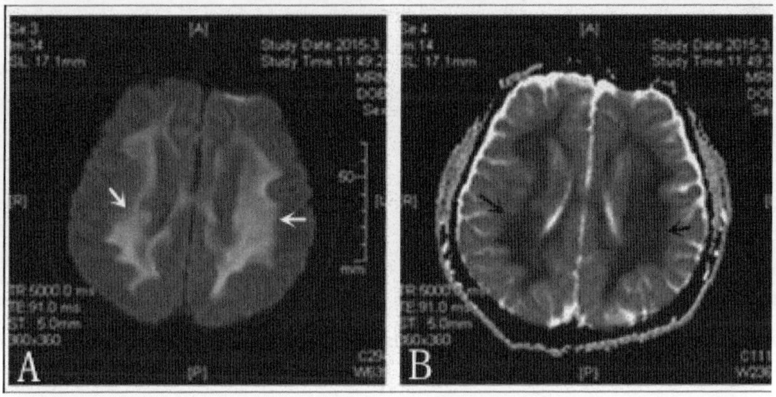

Fig. 3. A: DWI showed diffuse and symmetric hypointensity around bilateral ventricles. B: its ADC map showed diffuse and symmetric hypointensity around bilateral ventricles. (March 23, 2015).

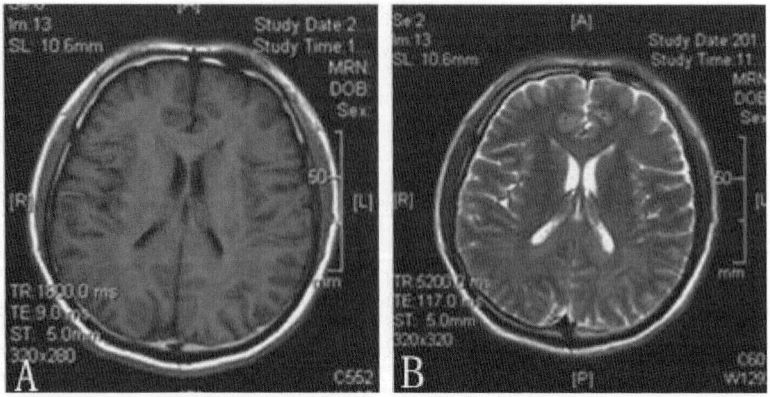

Fig. 4. A: Brain T_1 showed almost normal signal around bilateral ventricle. B: Brain T_2 showed almost normal signal around bilateral ventricle (April 1, 2015).

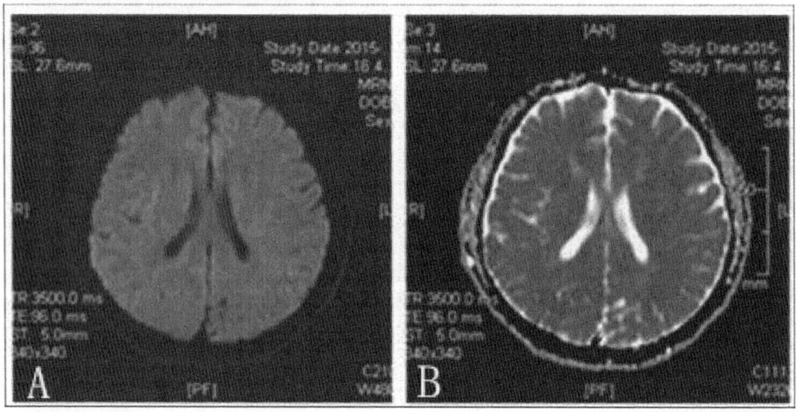

Fig. 5. A: DWI showed almost normal signal around bilateral ventricles. B: its ADC map showed almost normal signal around bilateral ventricles. (April 1, 2015)

2. DISCUSSION

The lesions of alcoholic encephalopathy [1] may be found in the hemispheric white matter, cortex and middle cerebellar peduncles. The characteristic of MRI picture of acute AAE shows that the lesions are hypointense on T_1WI, hyperintense on T_2WI and DWI, and the ADC mapping showed relative hypointensity[2]. The exact pathogenesis mechanism is still unclear, but it is generally attributed to alcohol toxic, vitamin B complex especially vitamin B1 deficiency. On the one hand, toxic substance combined with alcohol and lecithin [3] typically contained inside nerve cell could deposit in the brain tissue, and it can take toll on the nervous system [4]. On the other hand, if the patient suffered gastronintestinal dysfunction which is derived from the long history drink, the absence of the cerebrospinal sheath's components (such as protein, thiamine and so on) resulted in demyelination of the brain [5]. The patient was acute onset with the background of heavy drink, which presented with cognitive dysfunction, dysarthia, hemiparesis. On the basis of the clinical and imaging features, AAE can be diagnosed.

The presentations could be variable in the acute stage: 1) Marchiafava-Bignami disease: the lesions could be found in the splenium of the corpus callosum, and the characteristic of MR exhibited hypointensity on T_1WI and ADC map, hyperintensity on T_2WI and DWI; 2) Wernicke encephalopathy: the lesions could surround the third ventricle, periaqueductal, the thalamus. And the hypointensity presented on T_1WI and ADC map, and the hyperintensity presented on T_2WI and DWI; 3) Leukoencephalopathy: The lesions presented hypointensity on ADC map and hyperintensity presented on DWI. The most likely reason is that the white matter around the ventricles is more sensitive to the alcohol toxic substance. The case characteristic of MR imaging findings which was similar to the third one descriped above was consistented with many literatures reported [6, 7]. There were no specific methods for the treatment of the disease usually. The patients should keep away from alcohol and cooperate with medical treatment at the same time, and intravenous hyperatimentation [8] was initiated with high-dose multivitamins accompanied by support treatment .Some serious patients was ted with high-dose dexamethasone. It was reported [9] that corticosteroid hormone performed a positive effect on the treatment in acute stage. The most likely possibility is that hormones can alleviate edema, inflammation, inhibit demyelination and protect the blood-brain barrier. Most of the patients with AAE exhibited a poor prognosis, and they often died in a short time [10, 11, 12]. There are two cases reported [13] that the neurological disorders alleviated concurrently with the lesions still on MRI. The reason why its MR imaging didn't recover seems to be inadequate dexamethasone and immunoglobulin therapy. We present a patient with AAE showing complete resolution of the MR imaging abnormalities and clinical recovery. This is rare in clinic.

In conclusion, AAE is a rare neurological disease related to heavy alcohol consumption, and is characterized by demyelination and necrosis in the white matter. In addition, the clinical presentations of imaging are variable [14]. But most of them have a poor prognosis, correlated with the increasing availability of brain MRI[15]. It is helpful for the discovery and appropriate treatment of cases of AAE. We regard our sample as representative. It may be significantly reduced the risk of a poor outcome.

References

1. Yan-hui Wang. The clinical and imaging characteristics of alcoholic encephalopathy. *J Guide of China Medicine* 2014, **12** (11): 81-82. (In Chinese).
2. Menegon.P, Sibon.I, Pachai.C,et al. Marchlafava-BIgnami disease: Diffusion-weighted MRI in corpus callosum and cortical lesions. *J Neurology* 2006, **66**: 1286-1285.
3. Anne-Pascale L, Anne-Lise P, et al. Sensitive biomarkers of alcoholism's effect on brain macrostructure: similarities and differences between France and the United States. *J Front Hum Neurosci* 2015, **9**: 354-366.
4. Anonymous. First person account. Life with a mentally ill spouse. *J Schizo Bull* 1994, **20**(1):227-229.

5. Xiao-lan Yan, Kai-bing Liao, Shu-qing Zheng. The Marehiafava—Bignami disease progress of imaging studies. *J Radiologic Practice* 2011, **26**(6):86-88. (In Chinese).
6. Haihel C, Gonnaud PM, Champin S, et aL. Difusion—weighted magnetic resonance imaging in Marchiafava-Bignami disease: follow—up studies. *J Neuroradio1ogy* 2005, **47** (7): 520—524.
7. Ming-yi Kong, Min-min Lv, Qian Ma, et al.Corpus callosum degeneration in 1 case. *J Journal of Medical Imaging* 2014, **24**(5):744-771. (In Chinese).
8. Ashish Aggarwal MD, Ashish Khandelwal MD. A case of marchiafava bignami disease:complete recovery with thiamine. *J Neuropsychiatry Clin Neurosci* 2011, **23** (2):28-28.
9. Yuan Gao, Jing-lun Li, et al. The Marchiafava-Bignami disease research progress. *J The southwest military medical* 2015, **17**(1):86-87. (in Chinese).Yuan Gao, Jing-lun Li, et al. The Marchiafava-Bignami disease research progress. *J The southwest military medical* 2015, **17**(1):86-87. (In Chinese).
10. De Yang,Yu Li, Xiasu Yang, et al. The clinical manifestations and imaging characteristics of acute corpus callosum degeneration. *J Chongqing Medical Journal* 2015, **44**(8):1060-1062. (In Chinese).
11. Gaetana M, Angela De G, Attilio C, et al. MR Imaging Findings in Alcoholic and Nonalcoholic Acute Wernicke's Encephalopathy: A Review. *J BioMed Research International* 2014, 2014:1-12.
12. Alexanden Heinric MD. Clinicoradiologic subtypes of Marchiafava—Bignami diseasel. *J Neurol* 2004, **251** (9): 1050-1059.
13. Sha Xu, Yao Ding, Mei-ping Ding.Corpus callosum degeneration of 6 cases of imagining characteristics and performance. *J Chinese Journal of Neurology* 2013, **46** (9):605-608. (In Chinese).
14. Yon kwon I, Seong su H, and Young Ha p. Acute Marchiafava-Bignami disease: Diffusion- weighted MRI in cortical and callosal involvement. *J Yonsei Medical Journal* 2007, **48**(2): 321-324.
15. Michito N, Yuko N, Imaharu N. Cortical involvement in Marchiafava-Bignami disease can be a predictor of a poor prognosis: A case report and review of the literature. *J Internal Medicine* 2013, **52**: 811-813.

THERMOACOUSTIC IMAGING OF FINGER JOINTS AND BONES: A FEASIBILITY STUDY

Xueliang Xu[1,a], Lin Huang[2,b], Yan Ling[1,c], and Huabei Jiang[3,d,*]
[1]*Teaching Hospital of Chengdu University of T.C.M, Chengdu, China*
[2]*School of Physical Electronics, University of Electronic Science and Technology of China, Chengdu, China*
[3]*Department of Biomedical Engineering, University of Florida, Gainesville, FL, U.S.A*
E-mail: [a]464346844@qq.com, [b]274393582@qq.com, [c]272001703@qq.com, [d]hjiang@bme.ufl.edu
[*]*Corresponding author*

A new application of thermoacoustic tomography (TAT) for human finger joints and bones imaging was investigated in this study. The feasibility of the joints and bones TAT scanner was validated on a volunteer and verified through comparison with MRI. The recovered TAT image revealed apparent microwave absorption difference between the joint cavity and bone, and the TA imaged shape/size of the finger showed consistent with the MRI findings. This initial study suggests that TAT has the potential to be used as a useful auxiliary tool for screening and diagnosing of joints and bones diseases, and treatment monitoring.

1. INTRODUCTION

Thermoacoustic tomography (TAT) is a novel imaging modality, which combines the merits of high ultrasound resolution with high microwave absorption contrast[1-7]. Due to its further non-invasive, low cost, and functional imaging abilities, this method has been applied in breast cancer and foreign body detection, and molecular imaging, etc. [8]. Basically, both photoacoustic/optoacoustic (PA) and TA imaging are the product of thermal acoustic effect found by A.G.Bell in 1880. Thus, TAT has the potential to be used as an auxiliary or preferentially tool in some aspects that PA imaging was widely used, such as breast and cancer detection, brain and finger imaging, and thermal therapy monitoring, etc.[8]. As it reported, diseases related to joints and bones such as osteoporosis and rheumatoid arthritis affecting millions of people worldwide[9]. It is helpful to diagnosing bones and joints related diseases at the early stage with sensitively imaging modalities. Thus, from 2001, our group started to study joints and bones imaging using diffuse optical tomography (DOT),[10] and we are currently evaluating this approach for osteoarthritis (OA) detection and monitoring with a large scale[11]. As PAT booming and fast developing in these days, by knowing the advantages of photoacoustic tomography (PAT), we immediately realized that this technology has the potential for high performance finger disease detection. So after 2 years of studies focusing on validating the finite element (FEM) based PAT approach, in 2005, we reported the first chicken bones and/1 or joints PA images by using our FEM-based PAT method[12]. After that, we further developed our method into quantitative PAT (qPAT) that can quantitatively recover tissue absorption coefficient and physiological/functional parameters (e.g., hemoglobin concentration, blood oxygenation.). Up to now, we have demonstrated that it is possible to detect osteoarthritis in the finger joints and bones with our FEM based qPAT approach by using of various phantom, ex vivo and in vivo experiments[13]. Meanwhile, there are also some other laboratories worldwide interested in the context of arthritis imaging with PAT. For instance, X. Wang and collaborators reported the PA images of rat tail, human cadaver fingers, and recently a healthy human finger[14,15]. On the other hand, after Ermilov et al. presented the large finger blood vessels of a human in three dimensions with its modified three dimensional OA tomography (3D-OAT) system[16]. Recently, van Es et al. showed the small blood vessels of the human index finger with diameter between 1.5 mm and 100 μm[17]. As is all known to us, beyond radiography, computed tomography, magnetic resonance and ultrasound imaging methods being used for bones and joints disease detection, microwave imaging has been used for joint tissues imaging, and has been recognized as a complementary or alternative imaging modality to X-ray, magnetic resonance, and ultrasound imaging for joints and bones diagnosing[18]. While there is some advantages for joints and bones imaging by using of microwave technology, such as non-ionizing and high contrast, the cm level spatial resolution related to microwave wavelength hinder the development of this method for clinical application. Thusas the feasibiliwith PAT and microwave methods has been verified, we proposed to explore finger joints and bones imaging with TAT method in this study.

2. METHODS AND MATERIALS

The schematic of our TAT finger joints scanner is shown in Fig.1. The detail of our TAT system has been introduced in our previous works elsewhere, here, we just get a brief review of the TAT finger joints scanner[6]. Microwave generator produced microwaves propagated to fingers via wave guide and horn antenna assembly. The center frequency and pulse duration of the microwaves interacted with fingers were 3.0 GHz and 0.75 μs, respectively. A 2.25 MHz unfocused transducer fixed on a rotator circularly scanned around the center of the antenna was immersed into transform oil for microwave induced thermal acoustic signal collection. The transducer collected data was converted by a data acquisition card and recorded in a computer for further processing. During data collection, the volunteer stand face to the antenna and with his/her hand being fixed on a homemade jack mildly. With a constant 2.0° interval, our TAT scanner can cover 300°, where the radius for the circular scan is 113.0 mm.

In this study, a female volunteer's index finger (right hand) was thermoacoustically scanned in sagittal section. For cross-validation, MRI of the TA imaged finger was presented for comparison.

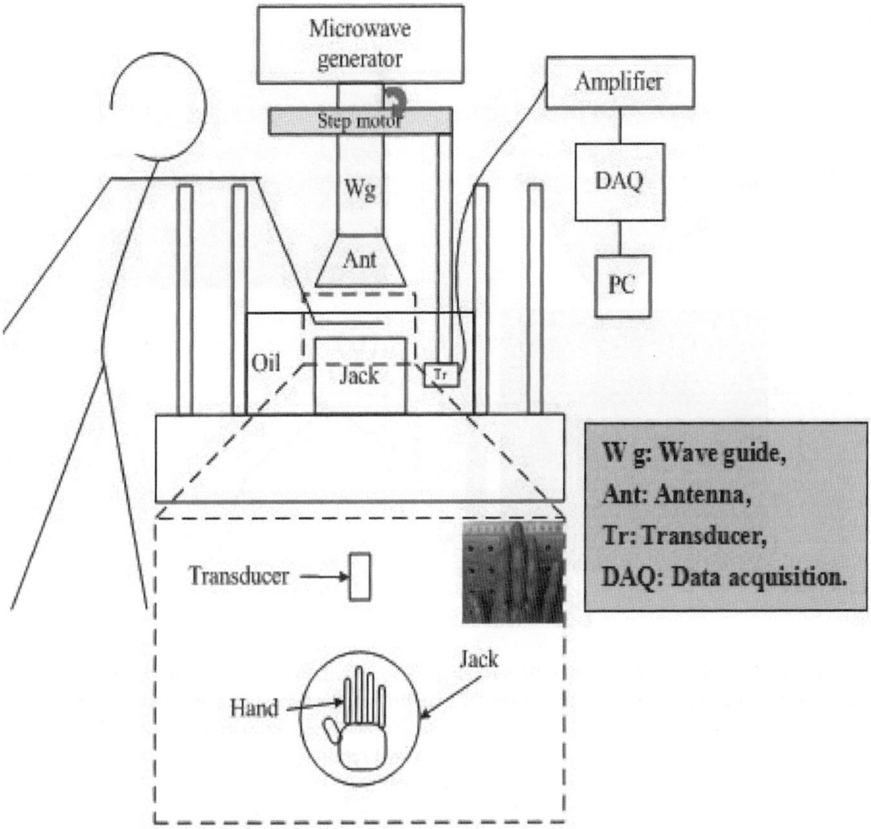

Fig. 1. Schematic of TAT finger joints scanner, the inset picture is a photography of the hand being TA scanned in coronal section.

3. RESULTS

Fig.2(a) shows the picture of the index finger and its imaging plane (sagittal section) being TA scanned during this study. Fig.2(b) displays the the MRI image of two fingers for this study, and indicates the joint space between two bones of the index finger in gray. From Fig.2(b), we estimated that the joint space between the two finger bones of index finger is about 6.8 mm, and the distance between the two joints is about 14.4 mm. Fig.2(c) illustrates the TAT recovered index finger image in sagittal section, which agreed well with the real shape of the index finger as indicated in Figs.2(a) and 2(b). For comparison, Fig.2(d) presents the recovered relative microwave absorption profiles plotted along the red dashed line shown in Fig.2(c). We found that the TA recovered space between the two finger joints is about 6.3 mm, which is in good agreement with the MRI indicated size of 6.8 mm as indicated in Fig.2(b).

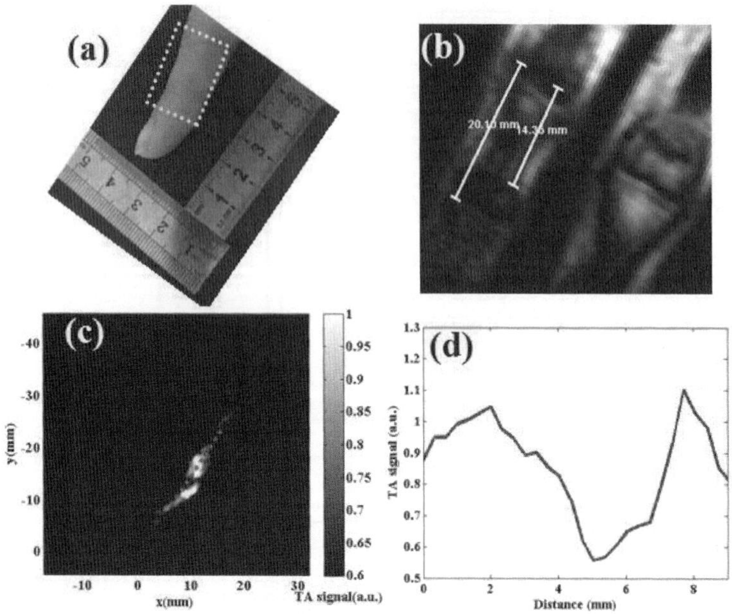

Fig. 2. (a) is the photography of the index finger being TA scanned, the dashed white box in (a) indicates the TA scanned plane. (b) illustrates the MRI image of two fingers for this study, and indicates the joint space between two bones of the index finger in gray. (c) presents TA recovered index finger image of the volunteer. (d) shows the relative microwave absorption profiles along dashed red line indicated in (c). The color bar displays in (c) stands for relative microwave absorption in arbitrary unit (a.u).

4. CONCLUSIONS

In summary, we represent the first attempt to detect finger joints and bones by using of thermoacoutsic tomography, and have demonstrated its feasibility in vivo. However, there is some improvements needed before its application for diseased bones and joints imaging. First, new stabilization method of the fingers during data collection is essential to reduce motion artifacts. A transducer array with multiple elements which can significantly save acquisition time and allow for real-time imaging is also necessary for functional TA joints and bones imaging. In addition, quantitatively recover the microwave absorption related conductivity is helpful for diseased joints and bones diagnosing. With these improvements, a new TA scanner is under investigation in our lab, and some results of diseased joints and bones with this new scanner will be reported in the future.

Acknowledgments

This research was supported in part by the University of Electronic Science and Technology of China and by the J. Crayton Pruitt Family Endowment (To Jiang), the International cooperation project of Sichuan Province Science and Technology Agency (2015HH0043) (to Xu).

References

1. Kruger, R. A., Kiser, W. L. Jr., Miller, K. D., Reynolds, H. E., Reinecke, D. R., Kruger, G. A., Hofacker, P. J. "Thermoacoustic CT: Imaging Principles," Proc. SPIE 2000; **3916**: 150-159.
2. Kruger, R.A., Miller, K.D., Reynolds, H,E., Kiser, Jr W.L., Reinecke, D.R., Kruger, G.A. "Breast cancer in vivo: contrast enhancement with thermoacoustic CT at 434 MHz - feasibility study," Radiology 2000; **216**(l):279-283.
3. Li, C., Pramanik, M., Ku, G., Wang, L. V. "Image distortion in thermoacoustic tomography caused by microwave diffraction," Physical Review E 2008; **77**: 031923.
4. Bauer, D.R., Wang, X., Vollin, J., Xin, H., Witte, R.S. "Spectroscopic thermoacoustic imaging of water and fat composition," APPLIED PHYSICS LETTERS 2012; **101**: 033705.
5. Ji, J., Lou, C.G., Yang, S.H., Xing, D. "Three-dimensional thermoacoustic imaging for early breast cancer detection," *Medical Physics* 2012; **39**(11), 6738-6744.
6. Huang, L., Yao, L., Liu, L.X., Rong, J., Jiang, H.B. "Quantitative thermoacoustic tomography: Recovery of conductivity maps of heterogeneous media," *APPLIED PHYSICS LETTERS* 2012; **101**: 244106.
7. Omar, M., Kellnberger, S., Sergiadis, G., Razansky, D., Ntziachristos, V. "Near-field thermoacoustic imaging with transmission line pulsers," *Medical Physics* 2012; **39**(7): 4460-4466.
8. M. H. Xu and L. H. V. Wang, "Photoacoustic imaging in biomedicine," *Rev. Sci. Instrum.* 2006; **77**(4): 041101.
9. Abramson, S.B., Attur, M., Yazici, Y., "Prospect for disease modification in osteoarthritis," Nat. Clin. Pract. Rheumatol. 2006; **2**: 304-312.
10. Xu, Y., Iftimia, N., Jiang, H.B., Key, L.L, Bolster, M.B. "Imaging of in vitro and in vivo bones and joints with continuous-wave diffuse optical tomography," OPTICS EXPRESS 2001; **8**: 447–451.
11. Yuan, Z., Zhang, Q., Sobel, E., Jiang, H.B., "3D diffuse optical tomography imaging of osteoarthritis: Initial results in finger joints," J. Biomed. Opt. 2007; 12: 034001.
12. Yuan, Z., Zhao, H.Z., Wu, C.F., Zhang, Q.Z., Jiang, H.B. "Finite-element-based photoacoustic tomography: phantom and chicken bone experiments," *APPLIED OPTICS* 2006; **45**(13) 3177-3183.
13. Sun, Y. and Jiang, H., "Three-dimensional photoacoustic tomography of finger joint: from phantom experiment to in-vivo study," *Proceedings SPIE* 2009; 7258: 72584G.
14. Wang, X., Chamberland, D. L., Fowlkes, J. B., Carson, P. L. and Jamadar, D. A., "Photoacoustic tomography of small-animal and human peripheral joints," *Proceedings SPIE* 2008; **6856**: 685604.
15. Xu, G., Rajian, J. R., Girish, G, Kaplan, M. J., Fowlkes, J. B., Carson, P. L., Wang, X. D. "Photoacoustic and ultrasound dual-modality imaging of human peripheral joints," J. *Biomed. Opt.* 2013; **18**(1):10502.
16. Ermilov, S.A., Su, R., Zamora, M., Hernandez, T., Nadvoretsky, V., Oraevsky, A.A. "Optoacoustic angiography of peripheral vasculature," *Proceedings SPIE* 2012; **8223**: 82230D-1.

17. van Es, P., Biswas,S.K., Bernelot, H.J., Steenbergen, M,W,.Manoha,S. "Initial results of finger imaging using photoacoustic computed tomography," J. Biomed. Opt. 2014; **19**(6):060501.
18. Salvador, S.M., Fear, E.C., Okoniewski, M., Matyas, J.R. "Exploring Joint Tissues With Microwave Imaging," Ieee Transactions On Microwave Theory And Techniques 2010; **58**(8): 2307-2313.

ANTI-TUMOR EFFECT OF SHENFUWEIKANG HERBS ON MICE LOADED WITH GASTRIC CARCINOMA

Xiaoping Wang[a,*], Huanping Lin, Qiaoxia Wang, Baoning Qi, Yi Li, Pengfei Zhang, Yan Fang, Xuxu Wei, and Kun Chen

Department of Pathology, Shaanxi University of Chinese Medicine, Xianyang, Shaanxi 712046, China

E-mail:[a]wxpphd@aliyun.com

[*]Corresponding author

To explore the anti-tumor effect of the Shenfuweikang in treatment of gastric cancer. Kunming mice grafted with a mouse gastric adenocarcinoma cell line MFC was used as the experimental model. The mice were divided into four groups, one control and the other three different drug administration experimental groups. Animals in the three experimental groups received Shefuweikang herb over a 60-day period starting at the first day after grafting. Animals received saline as controls. All the mice were sacrificed at 61 days after being grafted. The anti-tumor effect was assessed by two ways: (1) tumor size was periodically measured during the life of the animals and tumor weight was determined by an electron balance immediately after the animals killed. (2) Apoptotic indices (AI) were examined by the terminal deoxynucleotidyl transferase-mediated deoxyuridine triphosphate fluorescence nick end labeling (TUNEL) method and flow cytometric analysis. (3) The expression of VEGF, STAT3 and HIF-1a in tumor tissues was examined by immunostaining and analyzed by Image J analysis system. Compared with controls, tumor growth (size and weight) was significantly inhibited with the Shefouweikang decoction treatment ($P<0.05$). AI in gastric cancer grafted mice was significantly increased in the Shefuweikang decoction group compared with the controls. The expression of VEGF, STAT3 and HIF-1a in tumor tissues were down-regulated after treated with Shefuweikang decoction. The anti-tumor effect of gastric cancer cell growth in vivo by Shefuweikang herbs is related with the induction of the cell apoptosis and down-regulation of VEGF, STAT3 and HIF-1a signal transduction system.

1. INTRODUCTION

In clinic studies, Chinese Shenfuweikang herbs had been found to have effect on pre-malignant lesion, especially on gastric diseases [1,2]. The Shenfuweikang Decoction might inhibit gastric carcinoma cell proliferation and cause tumor cell death. Apoptosis plays a crucial role in the proliferation and turnover of cells in various tumors. It has been clear that its extent is often enhanced in tumor by many anticancer drugs, such as cytotoxic drugs, hormone, or some Chinese herbal medicine [3-5]. Researches indicated that Chinese herbs could enhance apoptosis of human gastric cancer grafted in mice [5,6].

Hypoxia-inducible factor-1a (HIF-1a) mediates hypoxic and non-hypoxic signaling pathways, and plays a crucial role in cancer progression. Over-expression of HIF-1α has been found in gastroenterological cancers and it is implicated in treatment resistance and poor prognosis in the hypoxic region around cancer [7]. Vascular endothelial growth factor (VEGF) has been proved to be a major inducer of angiogenesis, lymphangiogenesis and vasculogenesis in cancer [8]. The JAK/STAT3 signalling cascade play essential roles in promoting cancer cell survival, proliferation, angiogenesis and tumor metastasis [9].

Based on the previous studies, we presumed whether the Shenfuweikang herbs could affect the apoptotic indices of gastric cancer grafted onto mice and the expression of VEGF-STAT3-HIF-1α signal transduction in gastric cancer, further confirming the anti-tumor mechanism of the Chinese Shenfuweikang herbs.

2. MATERIALS AND METHODS

2.1. Mice and Antibody Reagents

Fourty 6-8 weeks old female Kunming mice (weight 18-22 g) and a mice gastric carcinoma cell line MFC were obtained from the Fourth Military Medical University. The mice were subcutaneously grafted with the MFC cells. The tumor transplantation procedure was described previously. Rabbit anti-mouse HIF-1α, STAT3 and VEGF polyclonal antibody were purchased from Santa Cruz Biotechnology, Inc (Santa Cruz, CA, USA). EnVisionTM kits were purchased from Dako Corp (Carpinteria, CA, USA).

2.2. Drugs

The Shenfuweikang decoction consists of Codonopsis pilosula (Franch) Nannf., Atractylodes macrocephala koidz, Poria cocos (Schw.) Wolf, Glycyrrhiza uralensis Fisch. The concentration of the Shenfuweikang decoction was 240 g/L.

2.3. Administration

After grafting the mice were randomly divided into 4 groups, one control and the other three experimental groups assigned to receive the Shenfuweikang Decoction. Each animal in the three experimental groups was given 2.0mL, 1.0 mL and 0.5mL of the Shenfuweikang Decoction by gastric perfusion every day over a 60-day period beginning at 1st day after grafting. The control animals received normal saline according to the same schedule. Animals were killed 61 days after being grafted.

2.4. Assessment of Tumor Growth

The effect of therapy was assessed by two ways: (1) tumor size was measured twice a week by multiplying two perpendicular diameters. (2) tumor weight was determined immediately by electron balance after the animals were killed. The tumor tissues were fixed in 10% formalin, embeded in paraffin and cut into 5μm sections coated on the slides for staining.

2.5. Detection of Apoptosis

For detection of apoptotic cells, apoptotic indices were examined by the terminal deoxynucleotidyl transferase-mediated deoxyuridine triphosphate fluorescence nick end labeling (TUNEL) method and flow cytometry analysis. (1) TUNEL: In situ cell death detection Kit POD (Roche Applied Science, Indianapolis, USA) was used to detect the apoptotic cell. The procedures were referred to the kit protocol. The positive cells were identified, counted and analyzed under the light microscope. Non-necrotic zone was selected in the tissue section and images were sent to computer by AEC camera (ZEISS, Jena, Germany). 10 image at least 1000 cells were selected on the screen, positive ratio analyzed by KS400 Video Image Digital Analysis System (ZEISS, Jena, Germany). (2) Flow cytometry analysis: Propidium iodide (PI) staining was used for flow cytometry detection of apoptosis. 10^6 cells from each of the sample were treated with RNase and stained with PI. The apoptotic cells having DNA strand breaks that had been labeled were measured on a flow cytometer (FACSCalibur, Becton Dickinson, USA). The data from 10^6 cells/sample were collected, stored, and analyzed using CELLQUEST (Becton Dickinson USA) and ModFIT LT for mac V1.01 software (Becton Dickinson, USA).

2.6. Immunostaining Methods

All sections were deparaffinized and rehydrated with graded alcohols. Endogenous peroxidase was then blocked with 3 mL/L H_2O_2 diluted in methanol for 30 min at room temperature. Antigen retrieval was performed by treating the slides in citrate buffer in a microwave for 10 min. The slides were incubated in a moist chamber with HIF-1α, STAT3 or VEGF rabbit polyclonal antibody (1:100) at 4°C overnight. After a complete wash in phosphate buffered saline (PBS), the slides were incubated with horseradish peroxidase labeled goat anti-mouse antibody (1:100) for 45 min at 37°C. After a complete wash in PBS, the slides were developed in 0.5g/L freshly prepared 3,3'-diaminobenzedine solution (DAB, Sigma Co, St.Louis, Mo, USA) for 8 min, and then counterstained with hematoxylin, dehydrated, air dried, and mounted. Normal mouse IgG was used to substitute for the primary antibody as a negative control. Only distinctive intranuclear or intra-cytoplasm immunoreactivity was considered positive. In each case, more than 1000 cells were counted and the percentage of immunoreactivity was independently determined. Image J analysis software was used to acquire the optical density data from the staining sections.

2.7. Statistical Analysis

The results were expressed as x±s and significant difference was assessed by Student's t test.

3. RESULTS

3.1. Inhibition of Tumor Growth by Shenfuweikang Herbs

Compared with the control group, tumor growth (size and weight) was significantly inhibited by treatment with the Shenfuweikang decoction (P<0.05, Table 1). The results showed that the higher the concentration of Shenfuweikang herbs, the less the tumor weight and size. There was a significant difference between the Chinese herbs and control group.

Table 1. Chinese Shenfuweikang herbs-induced effects on gastric cancer (x±s)

Treatment	Tumor weight(g)	Tumor size(mm3)
High-dose Shenfuweikang Decoction	0.52±0.33[a]	260.74±42.56[a]
Middle-dose Shenfuweikang Decoction	0.73±0.27[b]	325.48±37.25[a]
Low-dose Shenfuweikang Decoction	0.98±0.24[c]	576.43±66.54[c]
Control	1.01±0.29	587.32±73.28

[a]P<0.05, [b]P<0.05, [c]P>0.05 vs control group.

3.2. Induction of Tumor Cell Apoptosis by Shenfuweikang Herbs

Apoptotic index (AI) in mice loaded with gastric cancer cells was significantly elevated to 19.23±2.57 % by TUNEL method and 17.76±5.84 % FACScan in the Shenfuweikang decoction treatment group, compared with the controls (TUNEL: 2.93±1.85 %, P<0.05; FACScan: 6.24±2.08 %, P<0.05). But there was no significant difference between low-dose Shenfuweikang decoction treated group and control group by using either TUNEL method or flow cytometry analysis (Table 2).

Table 2. Chinese Shenfuweikang herbs-induced apoptosis on gastric cancer cells (x±s)

Treatment	Apoptotic index (AI)(%)	
	TUNEL	FACScan
High-dose Shenfuweikang Decoction	19.23±2.57[a]	17.76±5.84[a]
Middle-dose Shenfuweikang Decoction	10.76±7.45[b]	9.46±4.25[b]
Low-dose Shenfuweikang Decoction	3.02±1.63[c]	5.39±2.45[c]
Control	2.93±1.85	6.24±2.08

[a]P<0.05, [b]P<0.05, [c]P>0.05 vs control group.

3.3. Expression of VEGF, STAT3 and HIF-1A In Gastric Cancers

HIF-1α, STAT3 and VEGF immunoreactivities were detected in 40 mice gastric tumors. HIF-1α proteins were mainly presenting a nuclei staining, while STAT3 and VEGF mainly stained in the cell cytoplasm. The optical density (OD) value of HIF-1α, STAT3 and VEGF in Shenfuweikang herbs treated groups were significantly lower than that in control (P < 0.05, Table 3).

Table 3. Expression of VEGF, STAT3 and HIF-1α in gastric cancers

Treatment	Optical density (OD)		
	VEGF	STAT3	HIF-1α
High-dose Shenfuweikang Decoction	0.019±0.054[a]	0.025±0.035[a]	0.013±0.020[a]
Middle-dose Shenfuweikang Decoction	0.026±0.015[b]	0.037±0.008[b]	0.021±0.013[b]
Low-dose Shenfuweikang Decoction	0.045±0.009[c]	0.053±0.012[c]	0.036±0.008[c]
Control	0.043±0.035	0.051±0.006	0.035±0.029

[a]$P<0.05$, [b]$P<0.05$, [c]$P>0.05$ vs control group.

4. DISCUSSION

Gastric carcinoma is one of the most common malignant gastrointestinal carcinoma in the world. At present gastric carcinoma is still detected later in most patients throughout the world, and even with curative resection, they remain at a high risk of relapse and mortality. Thus, there is a great need for effective adjuvant therapy for patients with gastric carcinoma. Our previous clinic studies suggested that Chinese herbal recipe Shenfuweikang have therapeutic effects on gastric pre-malignant lesion, with increasing the reversal of the atrophic gastritis, decreasing the recurrence and improving the life quality [1,2]. Because of its lower toxic side-effect compared with chemical therapy, it is worth to make a further research on its anti-cancer mechanism.

Similar to the other malignant tumors, gastric carcinoma is always accompanying with abnormal cell proliferation and differentiation, as well as with abnormal apoptosis [5,6]. The enhanced induction of apoptosis in human gastric carcinoma cells can be observed after treatment with 5-Fluorouracil, Cisplatin, arsenous oxide and so on [5,6,10]. These data suggest that inducing cancer cell apoptosis should be a therapeutic method for gastric carcinoma. The present study indicated that tumor growth was significantly inhibited by treatment with the Shenfuweikang decoction. TUNEL method and cytometry analysis clarified that Shenfuweikang decoction could enhance apoptosis. The results suggest that the mechanism of the inhibition of gastric cancer cells in vivo by Shenfuweikang herbs is related with inducing apoptosis.

The VEGF-STAT3-HIF-1α signal transduction pathway play an important role in the development and progression of gastrointestinal cancers [11]. Our results suggested that Shenfuweikang herbs could down-regulate the expression of VEGF-STAT3-HIF-1α signal transduction proteins in gastric cancer, thus inhibiting the growth of tumor cells.

In conclusion, Shenfuweikang decoction inhibited gastric cancer cell growth. The anti-tumor effect of Shenfuweikang herbs lies in inducing gastric cancer cells apoptosis and down-regulating the expression of VEGF-STAT3-HIF-1α signal transduction protein in tumor cells. The detailed molecular mechanism of Shenfuweikang herbs inhibiting gastric cancer cells still needs further investigation.

Acknowledgments

This work is supported by Scientific Research Program Funded by Shaanxi University of Chinese Medicine and Shaanxi Provincial Education Department (No.2007JK233, 2009xj02, 2010JK484, 14JS025), the Scientific Research Program of Shaanxi Administration of Traditional Chinese Medicine (No.15-SCJH001), the Key Project of Ministry of Education of China (No.205002) and the National Natural Science Foundation of China (No.81172135, 81310108008).

References

1. Lin HP, Li SC. Protective effects of shenfuweikang on acute gastric mucous jnjury on rat. *Shaanxi Journal of Traditional Chinese Medicine* 2007; **28**: 1422–1423. (In Chinese).
2. Li SC, Wang F, Lin HP, Zhou XY. Effects of shenfuweikang on the secretion of serum gastrin and motilin in chronic atrophic gastritis. *Jiangxi Journal of Traditional Chinese Medicine* 2007; **38**: 68–69. (In Chinese).
3. Matsunaga T, Tsuji Y, Kaai K, Kohno S, Hirayama R, Alpers DH, et al. Toxicity against gastric cancer cells by combined treatment with 5-fluorouracil and mitomycin c: implication in oxidative stress. *Cancer Chemother Pharmacol* 2010; **66**: 517–526.
4. Qiao L, Wong BC. Targeting apoptosis as an approach for gastrointestinal cancer therapy. *Drug Resist Updat* 2009; **12**: 55–64 .
5. Zhao AG, Zhao HL, Jin XJ, Yang JK, Tang L. Effects of Chinese Jianpi herbs on cell apoptosis and related gene expression in human gastric cancer grafted onto nude mice. *World J Gastroenterol* 2002; **8**: 792–796.
6. Li J, Sun GZ, Lin HS, Pei YX, Qi X, An C, et al. The herb medicine formula "Yang Wei Kang Liu" improves the survival of late stage gastric cancer patients and induces the apoptosis of human gastric cancer cell line through Fas/Fas ligand and Bax/Bcl-2 pathways. *Int Immunopharmacol* 2008; **8**: 1196–1206.
7. Semenza GL. Hypoxia-inducible factors: mediators of cancer progression and targets for cancer therapy. *Trends Pharmacol Sci* 2012; **33**: 207–214.
8. Pore N, Jiang Z, Gupta A, Cerniglia G, Kao GD, Maity A. EGFR tyrosine kinase inhibitors decrease VEGF expression by both hypoxia-inducible factor (HIF)-1–independent and HIF-1–dependent mechanisms. *Cancer Res* 2006; **66**: 3197–3204.
9. Li N, Grivennikov Sergei I, Karin M. The unholy trinity: inflammation, cytokines, and STAT3 shape the cancer microenvironment. *Cancer Cell* 2011; **19**: 429–431.
10. Korbakis D, Scorilas A. Treatment of gastric cancer cells with 5-fluorouracil/leucovorin and irinotecan induces distinct alterations in the mRNA expression of the apoptosis-related genes, including the novel gene BCL2L12. *Tumour Biol* 2009; **30**: 100–107.
11. Dodd KM, Yang J, Shen MH, Sampson JR, Tee AR. mTORC1 drives HIF-1α and VEGF-A signalling via multiple mechanisms involving 4E-BP1, S6K1 and STAT3. *Oncogene* 2015; **34**: 2239–2250.

APPLICATION OF MULTIPLE-OUTPUT BP NEURAL NETWORK IN AUXILIARY DIAGNOSIS OF CHILDREN'S SHORT STATURE

Jingjun Zhang[1,a], Limin Jiang[1,b], Jian Wang[2], and Ruizhen Gao[1]

[1]*School of Information and Electrical Engineering, Hebei University of Engineering, HanDan, 056038, China*

[2]*HanDan Central Hospital, HanDan, 056001, China*

E-mail:[a]santt88@163.com, [b]18232027370@163.com

As is known to all, traditional approaches to diagnosis of children's short stature have shortcoming, even experienced clinical pediatric specialists often may lead to phenomena of leakage check, missed diagnosis and misdiagnosis because of the professional background, professional direction, clinical experience and other factors. Therefore, with the analysis of the indicators of clinical diagnosis, a computer-aided diagnosis system based on BP neural network is built and trained to make diagnostic factor more comprehensive, diagnostic process more convenient, diagnosis more accurate.

1. INTRODUCTION

The relatively recognized defination of height range of dwarfism in the world is that height of ones is two standard deviation or three percentile lower than groups of same age, gender and ethnicity in the average height[1]. Dwarfism makes a serious threat to children's physical and mental health. In our country, there are about 8 million dwarfism patients with a increasing rate of 161000 people every year .Obviously, prevention and screening and treatment is crucial to dwarfism. However, the probability of errors in diagnosis of Dwarfism is larger, not because the traditional diagnosis process, but also the high requirements for doctors' professional level and quality, in addition, the irrational configuration of medical resources.

The expert system based on neural network not only solves the shortage of traditional expert system and improves the ability of reasoning of information, self-organization and self-learning, which makes information processing faster, considered factors more comprehensive, logical judgment more rigorous[2]. As a result, the computer aided diagnosis system can help doctors shorten diagnosis time, avoid omissions and reduce labor intensity.

2. THE MODEL ESTABLISHMENT OF SHORT STATURE DIAGNOSIS SYSTEM BASED ON BP NEURAL NETWORK

2.1. To Construct Reference Index System for Diagnosis of Short Stature

With analysis of the traditional method of diagnosis of short stature, selection process of diagnostic reference index and diagnosis results is shown in Fig. 1.

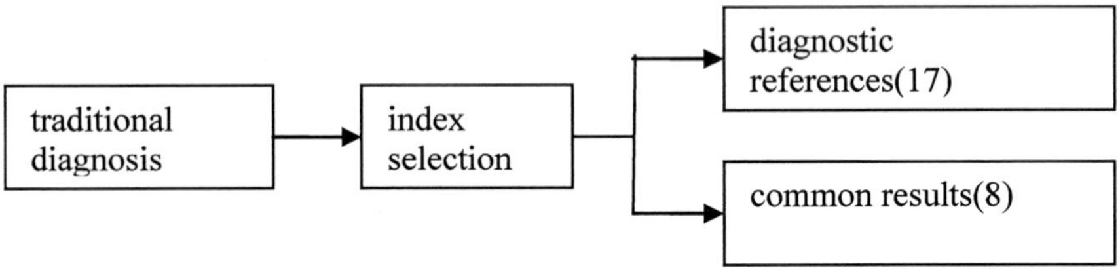

Fig. 1. selection process of diagnostic reference index and diagnosis results.

This paper take long-term monitoring data, from all over the normal hospital in china, from Shijiazhuang Future Technology Co.Ltd to analysis the effects on the height growth of children. After adequately integrated analysis of case data and experience of experts, 17 index eventually get selected as reference factors of computer aided diagnosis as following: gender, actual age, weight, height and bone age, BMI(Body Mass Index)assessment, height of parents, GH (Growth Hormone), TT3 (three serum total triiodothyronine), TT4 (serum total thyroxine), thyroid stimulating hormone (TSH), luteinizing hormone (LH), insulin like growth factor 1 (IGF-1), insulin like growth factor binding protein 3 (IGFBP-3), brain CT, trace elements[3]. Among these factors, take gender, brain CT and trace elements as 0, 1 through data handling. Through analysis of medical pathology of Dwarfism , 8 types of common causes are summarized as expected output patterns of neural network, which can be expressed by binary with 3 nodes in the output layer as following: hereditary dwarf(0 0 0), physical puberty growth retardation(0 0 1), primary GHD(0 1 0), thyroid dysfunction(0 1 1), nutritional dwarf(1 0 0), secondary GHD(1 0 1), precocious puberty(1 1 0), other diseases(1 1 1).

2.2. To Fix the Number of Nodes in the Hidden Layer

Generally, the focus as well the difficulty to fix the neural network structure in BP artificial neural network is selection of layers and the number of nodes in the hidden layer. Considering the system structure simple, too much hidden layers would reduce operation rate. Technically, a hidden layer could meet the demand of operation[4].

Currently, there is no theoretical methods to fix the number of nodes in the hidden layer[5], but generally depends on the empirical formula, as in Eq. (1)~Eq.(3).

$$M = \sqrt{N+L} + \alpha \tag{1}$$

$$M = \log_2 N \tag{2}$$

$$M = \sqrt{NL} \tag{3}$$

Where M represents the neuron number of hidden layer, N is the neuron number of input layer, L is the neuron number of output layer, α is the constant between 1~10.

In this paper, a comprehensive analysis of the corresponding training results of 0.01 precision, with different number of nodes in the hidden layer is made, The results is shown in Table 1, ultimately 10 hidden layers are fixed with relatively less training times and less error of order of magnitude. Moreover, relatively good effect of training can be achieved.

Table 1. The corresponding training results with different number of nodes of hidden layers

Hidden Layers	Training times	Training Errors	Hidden Layers	Training times	Training Errors
1	non-convergence		11	5	0.000474502
2	non-convergence		12	6	0.00157561
3	79	0.00968475	13	4	0.000298972
4	10	0.00990652	14	3	0.00988653
5	8	0.00939664	15	3	0.00207506
6	11	0.00819728	16	3	0.000414254
7	7	0.00559292	17	3	0.00107923
8	5	0.00075854	18	3	0.00319209
9	7	0.00161982	19	3	0.00758851
10	3	0.00599393	20	3	6.91547e-005

So the topology structure and diagnosis system structure of dwarf in BP neural network with a basic structure 17-10-3 of BP neural network is shown in Fig. 2.

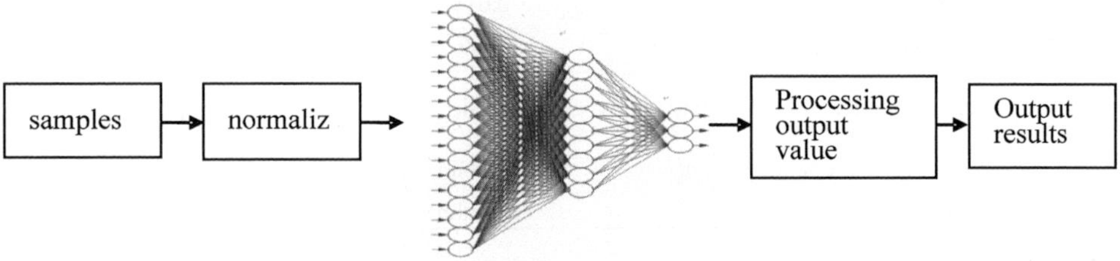

•**Fig. 2.** The diagnosis system structure of dwarf based on BP neural network.

2.3. To Select Training Samples for Model

The collection and sorting of training samples is often limited by objective conditions. And numbers of training samples that properly could reach the need of precision are enough[6]. Therefore, it refers to the rule of experience, as in Eq. (4).

$$P = (5\sim10) \times P_w \tag{4}$$

Where P represents numbers of training samples; P_w is the total of network connection weight, numerically equal to the sum of nodes of input layers and hidden layers. Thus, the training samples in the system should be in range of 135~270. In this paper, 200 samples are selected to train and ten of those are taken to test the system.

In addition, data processing of the selected samples will be conducted so as to improve the convergence speed of BP algorithm and increase stability of the system[7]. Especially, the common method is to make samples normalized, as in Eq. (5) and Eq. (6).

$$Y = \frac{X - X_{\min}}{X_{\max} - X_{\min}} \tag{5}$$

or

$$Y = 2*(X - X_{\min})/(X_{\max} - X_{\min}) - 1 \tag{6}$$

Where, X represents the original sample data, X_{min} and X_{max} indicates the minimum and maximum values of the original sample data, Y represents the normalized data.

Through normalization of the original samples, 17 diagnostic index data of different magnitude to be limited to the interval [0, 1] are chosen to take BP training.

2.4. To Select Training Function and Related Parameters

With the above analysis, the structure of BP neural network can be fixed. Table 2 shows the chosen training function and the relevant parameters.

Table 2. Basic parameters setting of short children intelligent diagnosis model based on BP neural network

Setting items	The value set
The learning rate	0.1
Error bounds	0.01
The number of iterations	1000
Transfer function of hidden layer nodes	tansig
Transfer function of output nodes	purelin
The training function	trainlm

3. STEPS OF ERROR EVALUATION OF INDEX OF MULTI-OUTPUT BASED ON BP

In this paper, the structure of intelligent diagnosis model contains three layers of 17-10-3. The training process of BP neural network is different from one-output BP.

Firstly , set the nodes of input and output and the hidden layers respectively as N, M, L. Assuming the training sample set $\{\xi^P, Y\} \subset R^N \times R^L$, the weight matrix between the input and hidden layers recorded as $V = (v_{mn})_{M\times N}$,where $V_m = (v_{m1}, v_{m2}, \ldots, v_{mn})^T \in R^N$ and m=1,2,…,M;assume the connection weight matrix between the hidden and output layers $W = (w_{lm})_{L\times M}$, where $W_l = (v_{l1}, v_{l2}, \ldots, v_{lm})^T \in R^M$, l=1,2,…,L. In addition, respectively take g and f as the activation function of each node of hidden and output layers. To simplify the process of derivation, take the vector function $G(X)$ for $X=(x_1,x_2,...,x_m)^T \in R^M$, where $G(X)=(g(x_1),g(x_2),...,g(x_m))^T \in R^M$ After the sample $\xi^p \in R^N$ being inputed, the actual output is as in Eq. (7).

$$Y^p = f\left(w_l \bullet G\left(V_m \xi^p\right)\right) \tag{7}$$

Define the error function as in Eq. (8).

$$E(W,V) = \frac{1}{2}\sum_{p=1}^{P}\sum_{l=1}^{L}\left(O^p - f\left(w_l \bullet G\left(V_m \xi^p\right)\right)\right)^2 \tag{8}$$

Objectively, the target of BP training is to compute the W and V so as to minimize the solution of error function $E(W,V)$. Therefore, gradient descent, common and simples taken .To simplify the derivation process[8], set Eq. (9).

$$f(x_{pl}) = \frac{1}{2}\left(O^p - f(x)\right)^2 \tag{9}$$

Then, the error function as in Eq. (10).

$$E(W,V) = \sum_{p=1}^{P}\sum_{l=1}^{L} f_{pl}\left(w_l \bullet G\left(V_m \xi^p\right)\right) \tag{10}$$

And the corresponding gradient function of W and V, as in Eq. (11) and Eq. (12).

$$E_{wl}(W,V) = \sum_{p=1}^{P} f'_{pl}\left(w_l \bullet G\left(V_m \xi^p\right)\right) G\left(V_m \xi^p\right) \tag{11}$$

$$E_{vm}(W,V) = \sum_{p=1}^{P}\sum_{l=1}^{L} f_{pl}\left(w_l \bullet G\left(V_m \xi^p\right)\right) \omega_{lm} g'\left(v_m \bullet \xi^p\right) \xi^p \tag{12}$$

For arbitrary initial values of $W_0 \in R^{L\times M}$ and $V_0 \in R^{M\times N}$, rules to modify the weight of BP learning algorithm as in Eq. (13) and Eq. (14).

$$W_l^{m+1} = W_l^m + \Delta w_l^m, \Delta w_l^m = -\eta_m E_{wl}\left(W^m, \mathrm{V}^m\right) \tag{13}$$

$$V_l^{n+1} = W_l^n + \Delta w_l^n, \Delta v_m^n = -\eta_n E_{wm}\left(W^n, \mathrm{V}^n\right) \tag{14}$$

Where η_i represents the learning efficiency.

For any initial value of $\eta_0 > 0$, η_i changes as the rule, as in Eq. (15).

$$\frac{1}{\eta_i} = \frac{1}{\eta_{i-1}} + \beta, \beta > 0 \tag{15}$$

Where, β is a constant.

4. PERFORMANCE ANALYSIS OF THE DIAGNOSIS MODEL

When BP neural network model has been built, 200 samples with pretreatment are taken into training and learning and the training results as shown in Fig. 3.

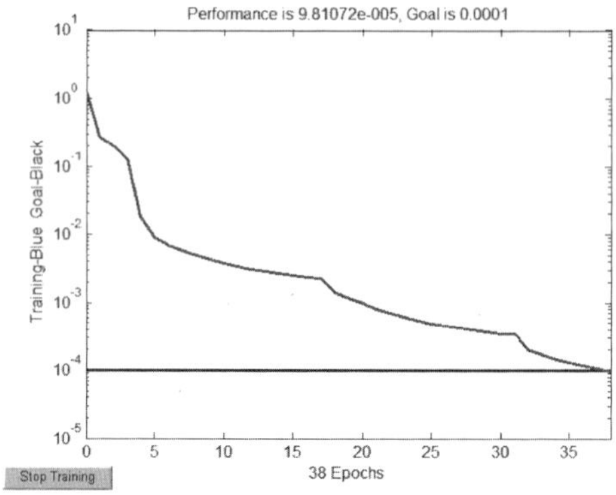

Fig. 3. The training curve of BP neural network of short stature children's diagnostic.

As the figure 3 shown, the training and computing of convergence is achieved and has meets the requirements of accuracy of 0.0001 with 38 training. In addition, through change the accuracy, the system can also effectively fit the intelligent diagnosis in children with short stature.

Table 3. The comparison of the network output results and expected results.

Sample	Expected output value	The actual output value	Consistency
1	(0 0 1)	(-0.0076 0.0057 0.9922)	T
2	(0 0 1)	(0.0058 0.0036 1.0015)	T
3	(0 1 0)	(0.0003 0.9996 0.0017)	T
4	(0 1 1)	(0.0075 1.0042 0.0005)	F
5	(1 1 0)	(0.0043 0.0026 0.0038)	T
6	(0 1 0)	(0.0014 0.9937 -0.0012)	T
7	(0 0 1)	(0.0050 0.0048 1.0019)	F
8	(0 1 0)	(-0.0043 1.0083 0.0073)	T
9	(0 1 1)	(0.0036 0.9938 -1.0004)	T
10	(1 1 0)	(1.0048 1.0066 0.0075)	T

As the network reaches a steady convergence, take the 10 test samples into the trained system and the error between the actual and expected outputs is curved as shown in Table 3 and Fig. 4.

According to Table 3 and Fig.4 about the comparison of the actual and the expected outputs, it shows that the network is most capable of diagnosis in short stature children.

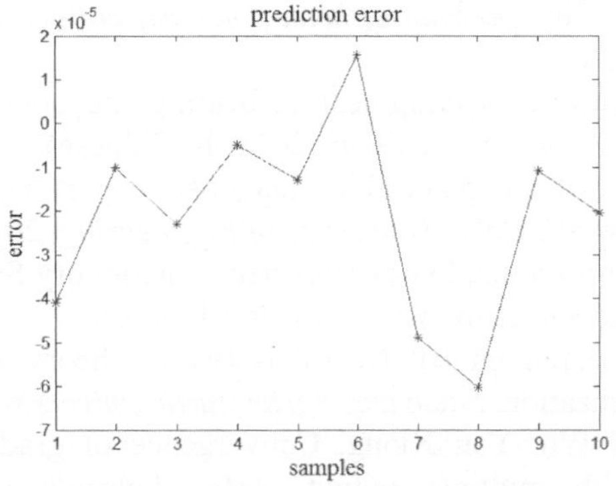

Fig. 4. The error curve of network output and the desired results.

5. CONCLUSION

This paper takes the traditional methods of diagnosis in short stature children as a starting point and an index system of diagnosis based on BP is built and verified with considerate legitimacy and applicability. As for the diagnostic process, the method effectively divorces human factors, and improves the diagnostic reliability, and makes the diagnostic results more effective and objective. Especially, under the circumstances of the seriously uneven distribution of medical resources and the great lack of experienced experts in China, the applicability is more outstanding.

The diagnosis system of short stature children also has some shortcomings and deficiencies in practical application. Firstly, the cause of Dwarf is complex, not just only 8 factors put in this paper. To gain more accurate results in practical medical diagnosis, the rationality of establishing indicators need further proof and perfect. Secondly, BP training is easy to fall into local optimum, which would also affect the accuracy of evaluation results in some extent.

Regardless of the drawbacks, the application of BP model truly, with great development potential, provides a new research idea and tool for the study on diagnosis of Dwarf.

Acknowledgments

This work is supported by the science and technology research and development project of HeBei Province under grant No. 12210148.

References

1. Jun-Feng Liu. Investigate the Etiology and Treatment of 389 Microsomia Children in Rural Areas of One City. *Journal of henan medical college for staff workers* 2014; **26**: 390–392.(In Chinese).
2. Wu-Long Ning. Platform development of bearing life prediction based on neural network. *Southwest Jiaotong University*. 2014.(In Chinese).
3. JianWang. Research of computer-aided diagnosis expert system of short Children based on neural network. *Hebei University of Engineering*. 2013. (In Chinese).
4. Qi Chen. Research on Evaluation of University Laboratory Safety Based on GA-BP Neural Network. *Xiamen University*. 2014.(In Chinese).
5. Ji-Quan Wang. Research on BP Neural Network Theory and Its Application In Agricultural Mechanization. *Shenyang Agricultural University*. 2011.(In Chinese).
6. Feng-qi Zhou. Wei Wu. Yan Xiong. Convergence of gradient algorithm for BP neural network with multiple output units. *Journals of atural sciences of HeiLongJiang university* 2006;**23**(1): 108–113.(In Chinese).
7. Wen-feng Li. BP neural network applications in the Xuchang wheat powdery mildew incidence trends forecast. *The 30th Annual Conference of Chinese Meteorological Society 2013.*(In Chinese).
8. Jian Mao, Hong-Dong Zhao, Jing-Jing Yao. Application and Prospect of Artificial Neural Network . *Electronic design engineer* 2011;**19**(27): 62–65.(In Chinese).

ASYMMETRIC SYNTHESIS OF DULOXETINE INTERMEDIATE (S)-N,N-DIMETHL-3-HYDROXY-(2-THIENYL)-1-PROPANAMINE WITH MICROORGANISMS

Zhimin Ou[a,*] and Ce Yang[b]

Pharmaceuticals College, Zhejiang University of Technology, Hangzhou, Zhejiang, 310014, China

E-mail: [a]oozzmm@163.com, [b]766006181@qq.com

[*]Corresponding author

(S)-N,N-dimethyl-3-hydroxy-(2-thienyl)-1-propanamine is an important intermediate in the synthesis of (S)-duloxetine. (S)-N,N-dimethyl-3-hydroxy-(2-thienyl)-1-propanamine can be gained by reduction of N,N-dimethyl-3-keto-3-(2-thienyl)-1-propanamine with *Candida tropicalis* zjut 22 as catalyst. The optimal bioconversion conditions are as follows: pH 8.0, temperature 35 ^{0}C and reaction time 48h. The yield and enantiomeric excess of (S)-N,N-dimethyl-3-hydroxy-(2-thienyl)- 1-propanamine can get to above 98% and 99% when 10 g/L of substrate and 400 g/L (cell dry weight) of cells were used.

1. INTRODUCTION

Duloxetine with the chemical name (S)-N-methyl-3-(1-naphthoxy)-3-(2-thienyl)-propanamine is a double serotonin and norepinephrine inhibitor. (S)-Duloxetine can inhibit serotonin reuptake in rat synaptosomes two times more potently than (R)-enantiomer. (S)-Duloxetine is not only a potent new antidepressant but also a norepinephrine reuptake inhibitor and a new drug for treating urinary stress incontinence[1-3].

(S)-N,N-dimethyl-3-hydroxy-(2-thienyl)-1-propanamineis ((S)-DHTP) a key chiral building block for the synthesis of (S)-duloxetine and can be obtained by the asymmetric reduction of N,N-dimethyl-3-keto-3-(2-thienyl)-1-propanamine(DKTP) using the biotransformation method[4]. Reaction conditions of biotransformation are mild and environmentally friendly compared with the chemical method [5, 6].

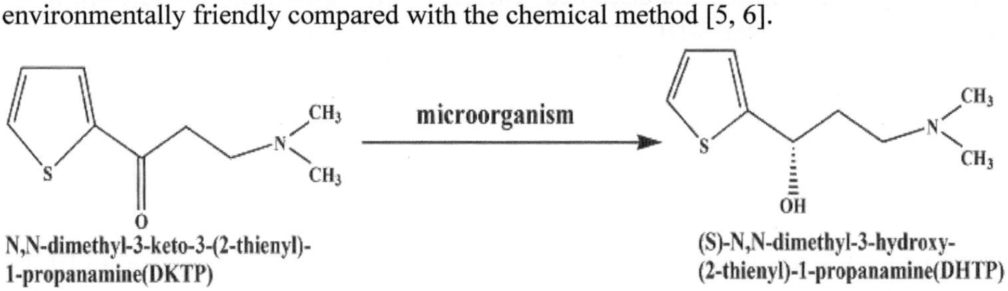

Fig. 1. Transformation of N,N-dimethyl-3-keto-3-(2-thienyl)-1-propanamine(DKTP) to (S)-N,N-dimethyl-3-hydroxy-(2-thienyl)-1-propanamineis ((S)-DHTP).

The asymmetric reduction of N,N-dimethyl-3-keto-3-(2-thienyl)-1-propanamine with a microorganism as the catalyst was recognized as an effective method because conversion can theoretically reach 100% and the microorganism can be cultured on a large scale [7, 8]. Large amounts of oxidoreductase existing in cells can reduce carbonyl compounds stereoselectively.

In this paper, (S)-N,N-dimethyl-3-hydroxy-(2-thienyl)-1-propanamine was synthesized by the asymmetric reduction of N,N-dimethyl-3-keto-3-(2-thienyl)-1-propanamine with *Candida tropicalis* zjut 22. The reaction scheme was shown in Fig. 1. The biotransformation conditions on reduction were investigated in detail.

2. MATERIALS AND METHODS

2.1. Reagents and Instruments

DKTP and Racemic DHTP were supplied by Zhejiang Jiuzhou Pharma Science & Technology Co., Ltd., China. Other chemical reagents were purchased commercially. HPLC (Shimadzu 10AVP) was used in the analysis.

2.2. Microorganism and Cultivation

Microorganisms were preserved in the lab of Zhejiang University of technology. The solid medium for the preservation of the strain was composed of 10 g/L malt juice, 3 g/L yeast extract, 5 g/L peptone, 10 g/L glucose, and 20 g/L agar. The liquid medium for the cultivation of cells on a large scale was composed of 30 g/L glucose, 3 g/L yeast extract, 5 g/L $(NH_4)_2SO_4$, 0.25 g/L $MgSO_4$, 1 g/L $K_2HPO_4 \cdot 3H_2O$, and 1 g/L KH_2PO_4. The strain was inoculated into 100 mL liquid culture medium and cultivated in 30 ^{0}C shaker (120 rpm) for 24 h. The 10 mL cell suspension gained above was inoculated into 100 mL liquid medium. After cultivation for 24 h, the free cells were harvested by centrifugation (4,000 g, 20 min) and used in reduction.

2.3. Biotransformation Process of DKTP

Substrate was added into a 100 mL shaking flask contained 20 mL phosphate buffer (0.2 M, pH 7.0) with cells. The flask was shaken at 120 rpm and 25 °C. Then the cells were collected by centrifugation and discarded. The pH of the supernatant was adjusted to 10 by NaOH solution and the S-DHTP was extracted with ethyl acetate.

2.4. Analytical Methods

DKTP and DHTP concentration were analyzed in a reverse-phase C18 HPLC column (4 mm diam., 25 mm long, 5 μm film; Merck, Germany). The mobile phase was phosphate buffer (50 mM, pH 7.0):acetonitrile:diethyl amine at 85:15:0.02. The flow rate of mobile phase and the UV length wave were 1 mL/min and 241 nm. The retention times of DHTP and DKTP were 7.7 min and 15.0 min, respectively. The enantiomers of DHTP were analyzed using a Chiracel OD-H column (0.46 mm diam, 250 mm long, 5 μm film; Diacel) with hexane:ethanol:trifluoroacetic acid:diethylamineat 95:5:0.2:0.1 at 240 nm. The retention times of S- and R-enantiomers were 24.33 min and 35.06 min, respectively.

3. RESULTS AND DISCUSSION

3.1. Select the Optimum Strain

Seven strains preserved in the lab of Zhejiang University of technology to find our objective strain were used in study. Table 1 shows that *Candida tropicalis* zjut 22 has the highest conversion and enantiomeric excess of (S)-N,N-dimethyl-3-hydroxy- (2-thienyl)-1-propanamine. *Candida tropicalis* zjut 22 was selected as the optimal microorganism and was used in study further.

Table 1. Selection of yeast cells which can transform N,N-dimethyl-3-keto-3-(2-thienyl)-1-propanamine

Strain name	Product configuration	Conversion (%)	Enantiomeric excess (ee%)
Candida pseudotropicalis 104	S	42	30
Candida pseudotropicalis 109	S	38	27
Candida utilis 1257	R	56	34
Candida pseudotropicalis 1145	R	60	85
***Candida tropicalis* zjut 22**	S	99	100
Rhodotorula qlutinis 2.102	R	60	55
Candida famata 2.270	R	30	24

3.2. Effect of pH on Reduction

The cells of *C. tropicalis* zjut 22 should be suspended in optimal buffer to ensure reduction activity effectiveness. Fig. 2 shows that *C. tropicalis* zjut 22 maintain good reduction ability from pH 7.0 to 8.5. The optimal pH of phosphate buffer for reduction is 8.0. Enantiometric excess of (S)-DHTP was 100 % at different pH of phosphate buffer.

3.3. Effect of Temperature on Reduction

Fig. 3 shows that the higher conversion can be obtained at optimal temperature. The catalytic activity of cells decreased when the reduction happened at above 35 ^{0}C. It was because that the deactivated rate of enzyme was accelerated at high temperature. The optimal temperature is 35 ^{0}C. Enantiometric excess of (S)-DHTP was 100 % at different temperature.

3.4. Effect of Transformation Time on Reduction

Fig. 4 showed that the conversion increased as the reaction time prolongs. The optimum reduction time is 48 h. The conversion did not rise too much after 48 h. Probably, the reduction activity of yeast cells was inhibited by product. Enantiometric excess of (S)-DHTP was 100 % at different reaction times.

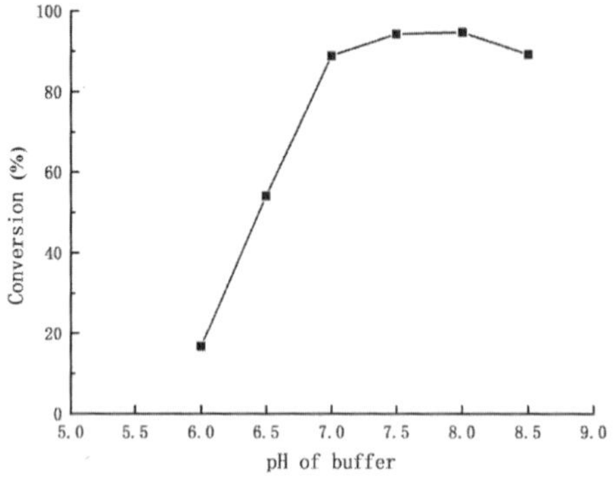

Fig. 2. Effect of pH on the formation of (S)-DHTP ($5\ g{\cdot}L^{-1}$ N,N-dimethyl-3-keto-3-(2-thienyl)-1-propanamine, $167\ g{\cdot}L^{-1}$ cell dry weight, 25 °C).

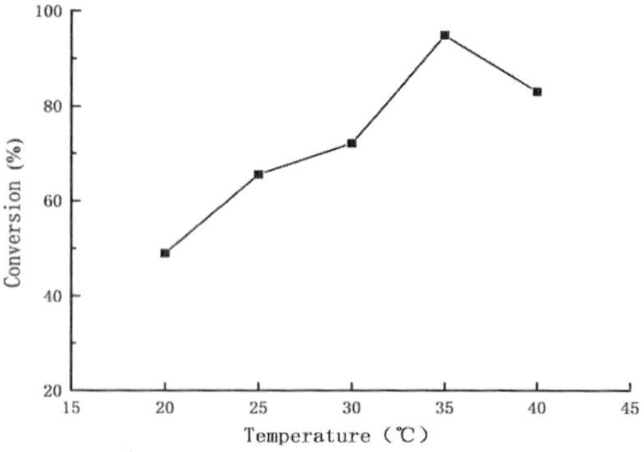

Fig. 3. Effect of temperature on reduction ($8\ g{\cdot}L^{-1}$ DKTP, $267\ g{\cdot}L^{-1}$ cell dry weight).

3.5. Effect of Initial Substrate Concentration on Reduction

Fig. 5 showed that the conversion decreased with the increase of the initial substrate concentration. Probably, a large amount of substrate is toxic to microorganisms and inhibits the activity of enzyme. Higher conversion can be gained when the substrate concentration was lower. Conversion can reach 100 % when initial substrate concentration was 2 g/L. Enantiometric excess of (S)-DHTP was 100 % at different initial substrate concentrations. There was no effect of initial substrate concentrations on enantiometric excess of (S)-DHTP.

3.6. Effect of Biomass on Reduction

Fig. 6 showed the biomass of *Candida tropicalis* zjut 22 used in reduction has significant effect on conversion. More addition of cells can improve the conversion. The content of reduction enzyme and coenzyme NAD(P)H increased with more biomass of cells added in reduction so that the higher conversion could be acquired. Enantiometric excess of (S)-DHTP was 100% at different biomass.

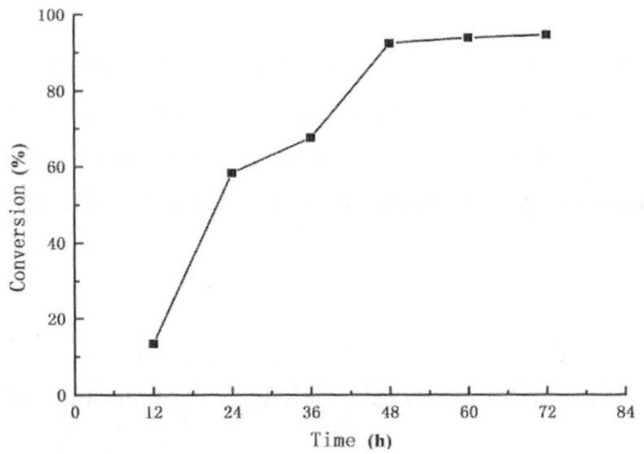

Fig. 4. Time course of reduction (Substrate concentration 8 $g \cdot L^{-1}$, biomass 267 $g \cdot L^{-1}$ (*w*/*v*)).

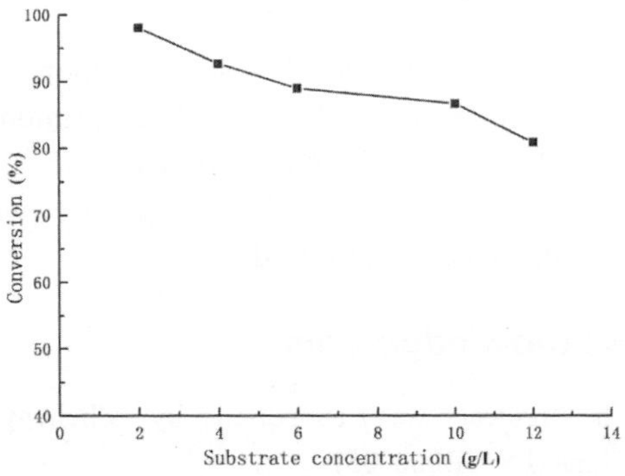

Fig. 5. Effect of the concentration of the substrate on reduction (biomass 87 $g \cdot L^{-1}$ (*w*/*v*)).

3.7. Effect of Co-substrate on Reduction

It was reported that the addition of suitable co-substrate is of advantage to increase yield and can influence the stereoselectivity of reaction. The main function of co-substrate is for coenzyme regeneration. Therefore, different concentrations of methanol, ethanol, isopropanol and glucose were selected as co-substrate to study the effect of co-substrate on reduction. The result showed that addition of methanol, ethanol and glucose can not improve the conversion. Fig. 7 showed that Higher conversion was achieved with 2.5 - 7.5 % isopropanol as co-substrate than without addition of co-substrate. The co-substrate isopropanol can help the regeneration of NADPH in cells and improve the conversion.

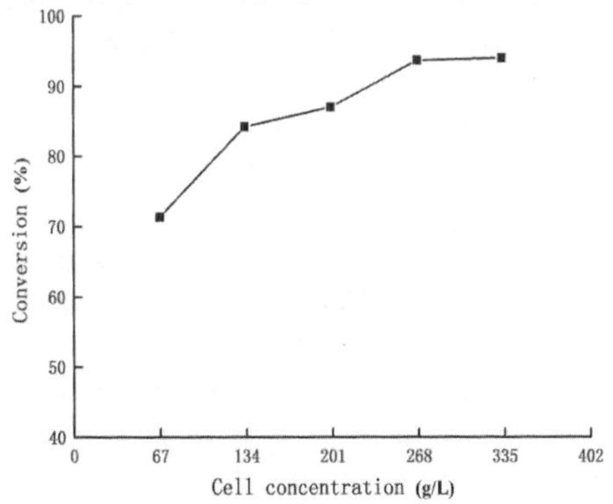

Fig. 6. Effect of biomass on reduction (Substrate concentration 6.67 $g \cdot L^{-1}$).

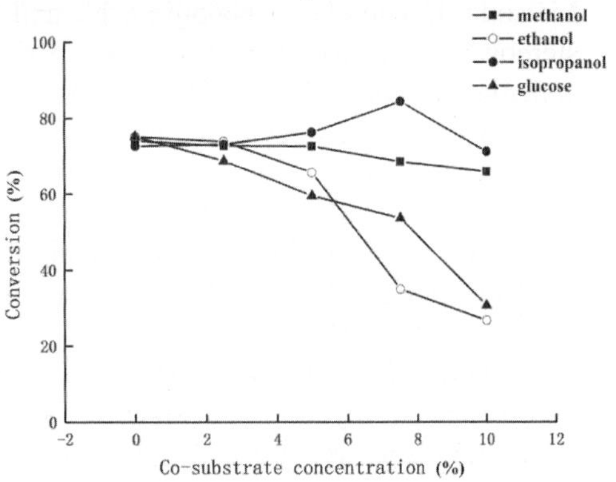

Fig. 7. Effect of the co-substrates on reduction.

4. CONCLUSIONS

(S)-N,N-dimethyl-3-hydroxy-(2-thienyl)-1-propanamine was synthesized by biotransformation with *C. tropicalis* zjut *22*. The conversion decreased with the addition of more substrates in the reduction. The increase of biomass is conductive to raising the yield of (S)-DHTP. The influence of co-substrates on DKTP reduction with *C. tropicalis* zjut *22* as catalyst was investigated. Isopropanol is the best co-substrate and the optimum concentration is 7.5 % (*v/v*). The optimum reaction conditions are as follows: reaction time 48 h, pH 8.0 and temperature 35 ^{0}C. The yield and enantiomeric excess of (S)-DHTP can reach above 98 % and 99 % when 10 g/L of substrate and 400 g/L (cell dry weight) of cells were used.

Acknowledgments

The project was supported by Zhejiang Provincial Natural Science Foundation of China.(LY15B060005).

References

1. Mallinckrodt CH, Prakash A, Andorn AC, Watkin JG, and Wohlreich MM. Duloxetine for the treatment of major depressive disorder: a closer look at efficacy and safety data across the approved dose range. *J Psychiat Res* 2006; **40**: 337–348.
2. Liu H, Hoff BH, and Anthonsen T. Chemo-enzymatic synthesis of the antidepressant duloxetine and its enantiomer. *Chirality* 2000;**12**: 26–29.
3. Perahia DG, Wang F, Mallinckrodt CH, Walker DJ, and Detke MJ. Duloxetine in the treatment of major depressive disorder:a placebo- and paroxetine-controlled trial. *Eur Psychiat* 2006;**21**: 367–378.
4. Soni P, Banerjee UC. Biotransformations for the production of the chiral drug (S)-Duloxetine catalyzed by a novel isolate of Candida tropicalis. *Appl Microbiol Biotechnol* 2005; **67**:771–777.
5. Matsuda T, Harada T, Nakajima N, and Nakamura K. Mechanism for improving stereoselectivity for asymmetric reduction using acetone powder of microorganism. *Tetrahedron Lett* 2000; **41**:4135–4138.
6. Kurbanoglu EB, Zilbeyaz K., and Kurbanoglu NI. Cryptococcus laurentii as a new biocatalyst for the asymmetric reduction of substituted acetophenones. *Tetrahedron Asymmetr* 2011; **22**:345–350.
7. Gelo-Pujic M, Guyader FL, and Schlama T. Microbial and homogenous asymmetric catalysis in the reduction of 1-[3,5-bis(trifluoromethyl)phenyl]ethanone. *Tetrahedron Asymmetr* 2006; **17**:2000–2005.
8. Carballeira JD, Valmaseda M, Alvarez E, and Gago JVS. Gongronella butleri, Schizosaccharomyces octosporus and Diplogelasinospora grovesii: novel microorganisms useful for the stereoselective reduction of ketones. *Enzyme Microb Technol* 2004; **34**:611–623.

CLINICAL EXPERIENCES OF NURSING CARE IN LEUKEMIA PATIENTS

Hongyan Li
Hematological Department, Affiliated Hospital of Beihua University, Jilin, jilin, 132013,China
E-mail:sunhongxiajl@126.com

A general introduction about the nursing contents, measures and key points of nursing for leukemia at all levels is given here through the experiences in the department of hematology after working for more than twenty years.

1. INTRODUCTION

Leukemia is a group of malignant tumor of hematopoietic system characterized by proliferation of widely juvenile white cells appearing mainly in bone marrow and other hematopoietic tissues, which causes patients with anemia, fever, bleeding, liver, spleen, lymph node enlargement at different degrees and a series of symptoms in clinic. As one of the most difficult cancer to be cured, the family members of patients are always with physical and mental fatigue, illness itself, therapeutic side effects and the cost of treatment also pose serious challenges to the patient's body and will, how to help patients defeat disease and tide over the difficulties is the problem which our medical staff should think about constantly.

2. FOR INTENSIVE CARE

2.1. Observation Contents[1]

To observe :a. The bleeding site and scope; b. The bleeding volume and whether the bleeding stopped in visceral hemorrhage; c. The patient's condition and abnormal laboratory tests; d.The changes of skin and mucous membrane; e.Pay attention to the changes of vital signs in patients with large amount of bleeding; f.Note the changes of consciousness, being alert for the occurrence of cerebral hemorrhage; g.Skin mucous ulceration and incision catheter was infected or not; h.The color of urine, sputum and stool.

2.2. Nursing Measures[2]

a. Mark the bedside card and inpatients schedule with red type markers in eye-catching location; b. Establish the nursing care records, planning records and various kinds of records related to illness of critically ill patients, nursing measures are carried out strictly according to the plan. Nursing records are timely, accurate, objective, true and complete. c.Patients will be arranged in resuscitation room or single room and wear gowns inside-out, keeping warm. The patient must stay in bed absolutely in serious bleeding or platelet counting smaller than 20×109/L. Wards are disinfected by UV 2 times daily, 30 minutes every time, visitors are restricted to avoid cross infection; d. Encourage patients to eat high protein, high vitamin diet and digestible or semi liquid diet, fasting too coarse food; e. To observe the disease conditions closely, such as signs of life; consciousness; pupil; skin color,temperature and integrity;bleeding tendency; oral and anal mucous changes; limb functions;color,character and quantity of vomit, urine, feces, sputum, maintain defecation unobstructed, do not be too hard for defecation,glycerine or laxative enema are used for constipation, enema is done to keep the peri-anal and peri-neal cleaning when necessary; f. To maintain the airway open.Coma patients should enable the head to one side, they should be timely aspirated in case respiratory tract secretion and saliva accumulated in larynx to affect respiration. Conscious patients but unable to expectorate sputum maybe considered appropriate suction; long-term bedridden patients should be often transformed posture, button back and be encouraged to keep deep breathing and expectoration; g. Strengthening the clinical nursing of the followings: (1) Oral caring: Coma patients and patients who could not take food by mouth are given routine oral care, symptomatic treatment is given according to oral mucous changes; (2) Skin caring: Patients of long-term bedridden, malnutrition and unable to turn over should be strengthened the skin care, preventing pressure sores. Do not force the static point tapping, do not tie tourniquet tight, blood or static point must be centralized operation to reduce the number of puncture as far as possible; (3) Paying attention to keep the limbs in a functional position: The condition permitting, body massage and passive joint motion might be done. h. To keep the catheter unobstructed. Fix the catheter properly to prevent clogging and falling off. Implement aseptic technique strictly in operation to prevent retrograde infection and record all catheter indwelling time; i. Supplement nutrition and hydration following the doctor's advice to ensure smooth transfusion and ensure the input; j. Caution: patients in delirium, restlessness, loss of consciousness should be provided with the proper use of protecting, constraint zone and bed baffle to prevent falling bed hurt; k. Maintain patients in comfortable and functional position; l. Visit the wards often to observe changes of infusion, disease and whether variety of instruments are running normally, getting to all problems; m. Records of liquid intake and output volume according to the doctor's advice; n. Maintain the normal excretory function. Occlude the catheter every 2--4 hours for Indwelling catheter in patients; o. Psychological nursing. Offer effective psychological support for patients according to the illness, more is accomplished by the non-language communication; p. Prepare various rescue medicines and articles, ready to do a good job with the doctor rescue work; q. Conduct bedside shift.

3. NURSING LEVELS AND MEASURES

3.1. Grade One Nursing

3.1.1. *Observation Points[3]*

a. The location and range of hemorrhage; Patient's condition and abnormal laboratory tests; c .Skin and mucous membrane changes; d. Mucous ulceration of skin and incision catheter was infected or not; e.Observe the color of urine, sputum and stool.

3.1.2. *Nursing Measures*

a.The red triangle is marked in the head of a bed and schedule; b.Establish the nursing care records, planning records and various kinds of records related to illness of critically ill patients, nursing measures are carried out strictly according to the plan. Nursing records are timely, accurate, objective, true and complete; c. Patients will be arranged in resuscitation room or single room and wear gowns inside-out, keeping warm. The patient must stay in bed absolutely in serious bleeding or platelet counting smaller than $20\times109/L$. Wards are disinfected by UV 2 times daily, 30 minutes every time, visitors are restricted to avoid cross infection; d.Encourage patients to eat high protein, high vitamin diet and digestible or semi liquid diet, fasting too coarse food; e.Ward inspection once every hour and observe disease conditions closely; f. Maintain defecation unobstructed, do not be too hard for defecation, glycerine or laxative enema are used for constipation, enema is done to keep the perianal and perineal cleaning when necessary; g.Measure the vital signs timely and observe the reaction and effects after administration; h.Long-term bedridden patients should be often transformed posture, button back and encourage deep breathing and expectoration; i. Strengthening clinical nursing: (1) Oral caring: symptomatic treatment is given according to oral mucosal changes; (2) Patients who cannot take care of themselves must be provided with safety measures. Use of special appliance should be agreed to patient and family members; j. To keep the catheter unobstructed. Fix the catheter properly to prevent clogging and falling off. Strict implement aseptic technique in operation to prevent retrograde infection and record all catheter indwelling time; k.Supplement nutrition and hydration following the doctor's advice to ensure smooth transfusion and ensure the input; l.. Maintain patients in comfortable and functional position; m. Visit the wards often to observe changes of infusion, disease and whether variety of instruments are running normally, getting to all problems; n. Carry out psychological nursing; o. Conduct health guidance according to the different stages of treatment.

3.2. Grade Two Nursing

3.2.1. *Observation Points[4]*

a. Master patient's condition and abnormal laboratory tests, do a good job in the prevention and nursing of hemorrhage; b. Observe the change of skin and mucous membrane; c. Observe the urine, sputum and stool color.

3.2.2. *Nursing Measures*

a. Green triangle is marked to the head of a bed and schedule; b. Keep the bed clean, dry, smooth, no stains, no superfluous items; c.The ward is disinfected with UV two times daily, each time for 30 minutes; d.Patients wear hospital gown and be sure to keep warm; e. Ward inspection is taken once every two hours to observe the disease changes; f. Measure the vital signs timely to observe the reaction and effects after administration; g.Assist patients in life nursing; h.Maintain the smooth transfusion, observe the changes of infusion and disease condition, getting to all problems; i.The patients stay comfortable and functional position; j. Carry out psychological nursing; k. Give health guidance according to the different stages of treatment.

3.3. Grade Three Nursing

3.3.1. *Observation Points*

a.Patient's condition and abnormal laboratory index; b. Change of skin and mucous membrane .

3.3.2. *Nursing Measures[5,6]*

a. Do not mark in the head of a bed or a card; b. Keep the bed clean, dry,smooth, no stains and no superfluous items; c. The ward is UV disinfected 2 times daily, each time for 30 minutes; d. The patients wear hospital gown, being sure to keep warm; e. Ward inspection is taken once every two hours to observe the disease changes; f. Giving timely measurements of vital signs; g. Keep the patients clean; h. Implement the correct treatment and administration measures according to the doctor's treatment; i.Conduct health guidance according to the different stages of treatment.

4. SUMMARY

With the continuous progress of science, technology and treatment, the remission rate and curing rate of leukemia has been significantly improved[7-10]. But various chemotherapeutic agents has no selectivity and its toxic side effects to the body often appear fever, vomiting, nausea, mouth ulcers, diarrhea, constipation and other adverse reactions to the patients which may cause malnutrition, reduced resistance, electrolyte disorders and other adverse reactions or even endangering life[11-14]. So our nursing staff in children should do well in children and their parents' psychological nursing during the period of chemotherapy according to the specific circumstances of patients, eliminating their fear, anxiety,tension and bad mood etc., helping them to keep a good, positive and optimistic attitude, support scientific diet guidance to the children, replenish essential nutrients; Eat food of high protein, high vitamin, high fiber, low sodium in no-chemotherapy period[15]. Food such as oil and fat diet, which is relatively high in fat are forbidden. We must do our best to reduce the adverse reactions that may bring pain to the children to ensure the effectiveness of clinical treatment as far as possible[16].

Reference

1. Hong-Yan Li, Hongxia Sun, Nursing experience of 64 cases of acute leukemia chemotherapy. *Journal of the national min-kang medicine* 2012; **24:** 979-980(In Chinese).
2. Shu-Fang Lu, Hong-Yan Wang, Shu Zhang, Nursing of acute leukemia patients . *China m*edical review 2006; 3:.103(In Chinese).
3. Hai-Xia Yan, Clinical nursing for leukemia patients grass roots. Medical BBS 2014; 18: .3308-3309(In Chinese).
4. Gui-Jing Su, Clinical nursing experience of patients with acute leukemia. Disabled Chinese medicine 2013; 21 : 363-364(In Chinese).
5. Chun-Feng Hou, Ba-Li Liu ,Chun-Yan Zhang, Nursing experience of common symptoms for leukemia chemotherapy. Clinical nursing, August 2014; 12:.326-327(In Chinese).
6. Juan-Juan Li,Yuan-Yin Qu, Jin-Yun Wang. Nursing experience in 35 cases of patients with leukemia. J doctor visit 2013; 11: 245-246(In Chinese).
7. Provan D,Newland A,Bolton-Maggs Pet al.Guidelines for the investi-gation andmanagement of idiopathic thrombocytopenic purpura in adults, children and in pregnancy.*Br J Haematol*,2003; **120:** 574-596.
8. Borst F,Keuning J J,vanHulsteijnHet al.High-dose dexamethasone as a first- and second-line treatment of idiopathic thrombocytopenic purpura in adults .*Ann Hematol* 2004; **83:** 764-768.
9. Cheng Y,Wong RS,Soo YO et al. Initial treatment of immune thrombocytopenic purpura with high-dose dexamethasone. *N Engl J Med* ,2003; **349:** 831-836.
10. Kojouri K,Vesely S K,Terrell D Ret al.Splenectomy for adult patients with idiopathic thrombocytopenic purpura:a systematic review to assess long-term platelet count responses,prediction of response,and surgical complications. *Blood* 2004; **104 :** 2623-2634.
11. Choudhary D R,Naithani R,Mahapatra M,et al.Efficacy of cyclosporine as a single agent therapy in chronic idiopathic thrombocytopenic purpura .*Haematologica* 2008; **93:** 61-62.
12. Howard J,Hoffbrand AV,Prentice HGet al.Mycophenolate mofetil for the treatment of refractory auto-immune haemolytic anaemia and auto-im-mune thrombocytopenia purpura[J].*Br J Haematol* 2002;**117:** 712-715.

13. Bennett C M,Rogers Z R,Kinnamon D Det al.Prospective phase 1/2 study of rituximab in childhood and adolescent chronic immune thrombo-cytopenic purpura. *Blood* 2006; **107:** 2639-2642.
14. McMillan R,Wang L, Tomer Aet al.Suppression of in vitro megakaryocyte production by antiplatelet autoantibodies from adult patients with chronic ITP. *Blood* 2004; **103:** 1364-1369.
15. Cines D B,Bussel J B.How I treat idiopathic thrombocytopenic purpura(ITP) .*Blood* 2005; **106:** 2244-2251.
16. GernsheimerTB.The pathophysiology of ITP revisited:ineffective thrombopoiesis and the emerging role of thrombopoietin receptor agonists in the management of chronic immune thrombocytopenic purpura.*Hematolo-gy* 2008; **1:** 219-226.

DETERMINATION OF RUBIADIN IN *PRISMATOMERIS TETRANDRA* BY HPLC

Licheng Guo, Hongli Teng, and Xiangpei Zhao[a]

Guangxi Institute of Minority Medicine/Traditional Zhuang Hospital of Guangxi Zhuang Autonomous Region, Nanning, 530001, China

E-mail: [a]zhaoxiangpei@126.com

Objective: To develop a method to determine the content of rubiadin in Prismatomeris tetrandra by HPLC. Methods: Agilent XDB C18 (5μm, 4.6mm×150mm) was adopted with mobile phase of Methanol-water-tetrahydrofuran(310:90:3). The assay was carried out at a flow rate of 1.0 mL/min with detection at 265 nm. The column temperature being 25 °C. Results: The formula for the regression line was observed within the range of 0.00503~0.1006 mg/ml (r=0.999 5) and the average recovery was 99.78% with RSD 0.85% (n=6). The rubiadin solution was stable within 12 h at room temperature. The average recovery rate of rubiadin was 98.54%. Conclusion: The HPLC method is convenient, sensitive, and accurate with good linear relationship, can be used for the quality control of Prismatomeris tetrandra.

1. INTRODUCION

Prismatomeris tetrandra(Roxb.) K. Schum is commonly used as Zhuang medicine in Zhuang Autonomous Region. It is mainly distributed in Guangxi, Guangdong, Yunnan, etc[1]. The roots are used medicinally in folk, with many functions such as cleaning heat, stopping bleeding, eliminating blood stasis and promoting tissue regeneration, strengthening bones and muscles. It can be used in the treatment of leukemia, aplastic anemia, gingival bleeding and hepatitis[2]. It is reported that several kinds of chemical components have been isolated from Prismatomeris tetrandra, mainly contains anthraquinone including 2-Methylanthraquinone, rubiadin, rubiadin-1-methylether, damnacanthal, etc[3, 4]. It is also contains Organic aluminium salts [5] and microelement [6], etc. However, no report has been shown the quantitative analysis of rubiadin in Prismatomeris tetrandra. To determine the quality of the Prismatomeris tetrandra, the aim of the present study is to develop for the first time a simple HPLC analytical method for determination of rubiadinin Prismatomeris tetrandra.

2. INSTRUMENTATION AND MATERIALS

An Agilent series1100 HPLC instrument (Agilent, Waldbronn, Germany) equipped with a quaternary pump, a diode-ar-ray detector, an auto sampler and a column compartment was used. The samples were separated on an Agilent XDBcolumn 18 (150 mm×4.6 mm, 5μm, Agilent). Electronic analytical balance (CP2250D, Sartorius). Numerical controlled ultra-sonic cleaning device, model KQ-5200 (manufactured by Jiangsu Kunshan Ultrasonic Apparatus Co., Ltd), CG-16W high speed micro centrifuge (manufactured by medical centrifuge factory of Beijing), acetonitrile (produced by Fisher, USA), Phosphoric acid (≥85%), was purchased from Shanghai Chemical Co. Other chemicals were of analytical grade. All solvents and samples were filtered through 0.45 μm filters before they were injected into HPLC.

The plant used for the present study was collected from Nanning, Guilin,etc. Plant materials were further identified as Prismatomeris tetrandra(Roxb.) K. Schumby Zhu yi-lin, a professor from Guangxi university of Chinese Medicine. Standard sample was as purchased from National Institutes for Food and Drug Control(lot No.: 20120812-1).

3. METHOD AND RESULT

3.1. Chromatographic Conditions

Agilent XDBcolumn 18 (150 mm×4.6 mm, 5μm); The mobile phase was methanol-water-tetrahydrofuran(310:90:3), with the flow-rate of 1.0 ml/min. The detection wavelength was set at 265nm. The column temperature was set at 25°C and the injection volume was 10μl. Theoretical plate number was above 3000 according to rubiadin.

3.2. Preparation of Standard Solution [7]

Some reference rubiadin was weighed accurately and then added into a 10 mL volumetric flask and dissolved with methanol to make a solution containing 0.1993 mg/ml rubiadin. The solution was filtered through 0.45μm micropore filter before HPLC analysis.

3.3. Sample Preparation

A 0.4 g crushed sample of Prismatomeris tetrandrawas accurately weighed and Ultrasonic extracted with 50ml of methanol-water-tetrahydrofuran(310:90:3) for 30 min. After cooling, weighed again, add the extraction solvent to make up the weight loss. Shook well, filtered and took 5ml of the filtrate into 10ml volumetric flask, then brought to volume with the extraction solvent. The sample solutions were filtered through a 0.45 μm micropore filter before HPLC analysis.

3.4. Stability Test

Reference rubiadin solution was analyzed by HPLC for 0, 1, 2, 4, 6, 8, 10h respectively under room temperature. According to above Chromatographic condition, the average of peak area integral values was 2348921. RSD% was 1.30 %. The experimental results indicated that samples were stable in 10 h.

3.5. Determination of The Linear Relationship

Injected 4.0, 6.0, 8.0, 10.0, 12.0μl standard solution into HPLC. Determined the rubiadin peak area integral values according to the above chromatograph condition. rubiadin standard curve: Y = 3.54×10-7 X +0.7458, (x was the rubiadin and y was the peak area integral value of rubiadin). The linear range of rubiadin was between 0.00503~0.1006 mg/mL, with a mean correlation coefficient r=0.99995.

3.6. Precision Test

10μl of rubiadin solution was injected into the HPLC system five times successively. The peak areas were calculated and the average of peak area integral values was 2561213. The RSD was found to be 0.20 %.

3.7. Repeatability Test

Five portions of the same batch of Prismatomeris tetrandrawere sampled and extracted for analysis. The content of rubiadin was determined. The average of rubiadin content was 0.87% and its RSD was 3.3%.

3.8. Recovery Test

We have examined the recovery property of rubiadin by adding certain amount of its standard solution to the plant samples prior to the extraction, and all the samples were processed with the same method as described in Section 3.1. The results were illustrated in Table 1.

Table 1. Result of recovery test of rubiadin

No.	Added amount (mg)	Sample amount(mg)	Found amount (mg)	Recovery(%)	Mean recovery(%)	RSD(%)
1		0.56	1.07	96.22		
2		0.43	0.95	98.11		
3	0.53	0.56	1.09	100.00	98.45	0.85
4		0.48	1.02	101.69		
5		0.50	1.01	96.22		

3.9. Sample Analysis

The content of rubiadin was determined by the above method in Prismatomeris tetrandra from the extraction solution. The results are shown in Table 2, Fig.1. From Table 2, we can conclude that the contents of rubiadin in Prismatomeris tetrandra are 0.5079~0.5684%.

Table 2. Contents of rubiadin in Prismatomeris tetrandra

No.	Sample	Habitat	Time of collection	rubiadin(%)
1	Prismatomeris tetrandra	Yongning	2013.8.11	0.5386
2		Jinxiu	2013.4.9	0.5684
3		Nanning	2013.10.18	0.5183
4		Liuzhou	2013.7.6	0.5285
5		Guilin	2013.4.5	0.5483
6		Hezhou	2013.3.13	0.5079

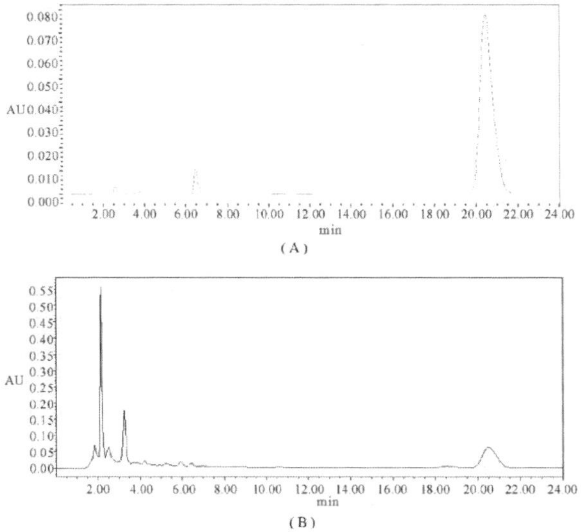

Fig.1. HPLC chromatogram of reference rubiadin.(A) Chromatograms of standard sample; (B) Chromatograms of extraction sample.

4. CONCLUSION

This is the first report on qualification and quantification of rubiadin in Prismatomeris tetrandra. It documents that the verified HPLC method can be successfully applied to quantification of rubiadin in Prismatomeris tetrandra with good sensitivity, repeatability and reproducibility. The results of this study showed that Prismatomeris tetrandra were rich source of rubiadin, as shown in Table 2.Our results suggested that this method can be helpful for the further study on Prismatomeris tetrandra.

Acknowledgments

The National Science-technology Support Plan Projects (20122BAI27B06-1, 2013BAI11B06-7); Guangxi science and technology development projects (14124002-14); Nanning science and technology development projects (20133163).

References

1. Health Bureau of Guangxi. Selections of Chinese material medical used in Guangxi. Nanning: Guangxi People's Publishing House 1974; 17901
2. Editing Group the national assembly of Chinese herbal medicine. National Assembly of Chinese Herbal Medicine. Beijing: People's Medical Publishing House 1996; 32.
3. Likhitwitayawuid K, Dej-adisai S, Jongbunprasert V, et al. Antimalarials from Stephaniavenosa, Prismatomeris sessiliflora, Diospyrosmontana and Murraya siamensis. *Plantamedica*1999; **65**: 754-756.
4. Jing H, Shi XF, Sheng X, et al. Anthraquinone glycosides from the roots of Prismatomeris tetrandra. *Chinese Journal of Natural Medicines* 2011; **9**: 42-45.
5. Tu DJ, Pang ZH, Bi NJ. Chemical constituents from root of Prismatomeris tetrandra. *Acta Pharm Sinica* 1981; **16**: 631-634
6. Zhou JGG, Li SG, Zhou XZ. Comparison of trace elements content of Impatiens balsamina in Prismatomeris connate. *Chinese Traditional and Herbal Drugs* 1986; **17**: 113-115.
7. Guo GMM, Cai L, Liang XY, et al. Content Determination of rubiadin in Morinda office-nails How by HPLC. *Chin Mod Drug Appl* 2013; **7**: 223-224.

QUANTITATIVE METHOD STUDY ON TRANSPORT MECHANISM OF PUERARIN, BERBERINE HYDROCHLORIDE AND PAEONIFLORIN

Xinli Liang[a,*], Wei Dong, Jing Zhang, Guowei Zhao, Yun Luo, and Zhenggen Liao

Key Laboratory of Modern Preparation of TCM of Ministry of Education, Jiangxi University of TCM, Nanchang 330004, China

E-mail: [a]paln7@163.com

[*]Corresponding author

To study the quantitative method evaluated drug transportation across cell membrane. Combined with Fick's law of passive diffusion and Michaelis-Menten equation, with the application of law of mass conservation, to derive equation of drug transportation rate. The overall description for membrane-transportation rate was depicted as the sum of passive diffusion rate and carrier-mediated diffusion rate which were calculated by transportation mass being divided respectively. Puerarin, berberine hydrochloride and paeoniflorin, set as model drugs. The cell membrane transportations for these three drugs were anlysized by fitting into this transportation rate theory on different drug concentration and drug combination to figure out the main biotransportation manner. Transportation rates of model drugs were obtained. Compared with the overall transportation manner, the passive diffusion and carrier-mediated transport were 58.2% and -41.8% respectively of puerarin (167.6μg·mL-1), 73.2% and -26.8% respectively for berberine hydrochloride (98.2μg·mL-1), 57.7% and -42.3% respectively for paeoniflorin (153.9μg·mL-1). At the present of Verapamil or probenecid, the overall transportation rate of puerarin and berberine hydrochloride increased and their ratios converted to positive direction. For paeoniflorin, carrier-mediated transport ratio was changed to positive direction only. The transportation mechanism and the efflux proteins of the three drugs were quantitatively evaluated by the established quantitative methods.

1. INTRODUCTION

Caco-2 cell monolayer model commonly used to study transport mechanism of drugs, almost all the reports that about transport mechanism study of drugs used Caco-2 cell model from the perspective of qualitative angel [1-3]. Currently the research approach focused on the effects of transport time [4], concentration [5-6], temperature [7-8], pH value [9], EDTA [10], cell protein content [11], ATP energy consumption[12], bidirectional transport[13], the presence or absence of protein inhibitors[14-15] and so on factors on the transport rate (Ka) and the apparent permeability coefficient(Papp) to estimate the transport mechanism of drugs. So far, there is no reports that published about the perspective of using quantitative methods to study transport mechanism of drugs by passive diffusion and carrier media accounting for the proportion of the total of transport. If there would be a method that could calculate the proportion of the drugs transport through the carrier medium and passive transport from the chimeric angle of the mathematical mode and use chimeric model parameters quantitatively , we not only could determine the transport mechanism of drugs, but also reflect the proportion of each mechanism more accurately, accordingly selecting strategies to ameliorate the drugs absorption more accurately.

In this paper, taking puerarin, berberine hydrochloride and paeoniflorin as model drugs, to research the theoretical study on mechanism of absorption of small molecule drugs, and try to use mathematical model to chimeric the transport of drugs and then using the model parameters to study the proportions of drug passive diffusion and carrier medium accounting for the total amount of drug transport, and then according to the results to explore transport mechanism of the three drugs, to analysis the transport protein that might involved in transport mechanism.These results can be used to improve strategies for efficient drug transportation.

2. MATERIALS AND METHODS

2.1. Drugs and Regents

Puerarin (batch numbers.101765-200408) and berberine hydrochloride (batch numbers.110713-201212) were purchased from the National Institute for the Control of Pharmaceutical and Biological Products (NICPBP, China). Paeoniflorin (batch numbers.FY15890529) was obtained from Nantong Feiyu Biological Technology Co., Ltd.

Fetal calf serum (FCS) and Dulbecco's modified Eagle's medium (DMEM) were purchased from Hyclone (Thermo Fisher Scientific). Penicillin and streptomycin solutions (10,000 U/mL penicillin and 10,000 μg/mL streptomycin) and Hank's balanced salt solution (HBSS) were purchased from Solarbio (Beijing Solarbio Science & Technology Co., Ltd., China). Non-essential amino acids were obtained from Sigma Chemical Co. (USA). The trypsin-EDTA solution (0.25% (w/w) trypsin/1 mM EDTA) was obtained from Gibco Laboratories (Life Technologies Inc., USA). Verapamil hydrochloride and MK-571 were purchased from Sigma Chemical Co. (USA). Twenty-four-well cell culture plates were purchased from Costar (Corning Incorporated, USA). Millicell® Cell Culture inserts and Millipore Express® PES Membranes (0.22 μm) were obtained from Millipore (USA).

2.2. Cell Cultures and Transport Experiments

Caco-2 was obtained from the American Type Culture Collection (ATCC, USA) ,The cells were cultured according to Zhu et al[16-17]. Briefly, the cells were cultured in medium containing DMEM (high D-glucose 4.5 g/L), 10% fetal bovine serum (inactivation at 56°C for 30 min), 1% non-essential amino acids, 1% L-glutamine, and a penicillin–streptomycin double antibody solution. Cultures were maintained at 37 °C in an atmosphere of 5% CO2, 95% air, and at 95% relative humidity. For transport experiments, the cells were seeded on polycarbonate filter membranes with a pore size of

0.4 mm, filter area 1.12 cm2 in 12-well plates (Transwell, Corning Costar,Cambridge, MA, USA) at a density of 1×105 cells/cm2 and grown for 21 days before the transport experiments.

Using a previously described method (Sun et al.,2006), transepithelial electrical resistance (TEER) and alkaline phosphatase activity were measured. And only monolayers with TEER value >500Ω/cm2 were accepted for the permeation experiments. before the experiment, the monolayer cells were washed twice with 37 °C blank Hank's buffered salt solution (HBSS solution, pH 7.4), the transport experiments were conducted by adding the drug solutions 0.5 mL to the apical (AP) while basolateral side (BL) contained HBSS medium 1.5 mL. After incubated for 120 min at 37 °C, samples were collected from BL side of Caco-2 cell monolayer to study the drug transport.Samples were collected and immediately treated with organic reagents, then preserved below -20 °C for subsequent HPLC analysis.

2.3. Theoretical Research

The intestinal absorption channel of oral drugs included Cell membrane pathways and paracellular transport. The primary transportation mechanisms of small-molecule drugs are simple diffusion and carrier-mediated transport. For small-molecule drugs, passive transport[18-19], is a first-order rate process that obeys Fick's law of diffusion. The equation can be simplified as in Eq. (1):

$$\frac{dc_1}{dt} = KC \tag{1}$$

Where $\frac{dc_1}{dt}$ is the diffusion rate, K is the membrane permeability coefficient, and C is the drug concentration in the gastrointestinal tract.

The carrier-mediated transport process that membrane proteins participated in can be fitted by the Michaelis-Menten kinetics equation as shown in Eq. (2):

$$\frac{dc_2}{dt} = \frac{V_{\max} C}{Km + C} \tag{2}$$

Where $\frac{dc_2}{dt}$ is the carrier-mediated transport rate, V max is the maximum transport rate, Km is the Michaelis constant, and C is the drug concentration.

Diffusion is the external manifestation of molecular movement, while molecular movement is absolute, so all molecules involved passive diffusion. If drug transport occurs in combination with carrier-mediated transport, total absorbed amount of small-molecule drugs should be the sum of the two transport modes, including simple passive diffusion and carrier-mediated transport according to the law of mass conservation states, as shown in Eq. (3):

$$\mathrm{M} = M1 + M2 \tag{3}$$

Therefore, Eq. (4) can be used to model a certain volume of solution (V_{in}) transported through the membrane of a cell over time:

$$\frac{M}{V_{in}} = \frac{M_1}{V_{in}} + \frac{M_2}{V_{in}} \tag{4}$$

Correspondingly, this equation can also be expressed in terms of concentration:

$$C = C_1 + C_2 \tag{5}$$

Where C, C_1, and C_2 are the concentrations of the total amount of drug absorbed, the concentration transported via carrier-mediated transport, and the concentration transported through passive diffusion respectively. The time differential illustrated in Eq. (5) is shown in Eq. (6):

$$\frac{dc}{dt} = \frac{dc_1}{dt} + \frac{dc_2}{dt} \tag{6}$$

The total drug transportation rate is equal to the algebraic sum of the carrier-mediated transport rate and the simple passive diffusion rate, and V is the total transport rate ($\mu g \cdot min^{-1} \cdot cm^{-2}$). It is shown in Eq. (7).

$$V = \frac{V_{\max} C}{K_m + C} + KC \tag{7}$$

To calculate the parameters M_{ax}, K_m, and K using the mathematical fitting software MATLAB 7.5 based on data obtained with different drug concentrations. Because concentration gradient is a driven for simple diffusion, therefore, we can assume that the transportation rate is positive when it is initiated from the brush border side and ended at the basement membrane side. While carrier-mediated transport may simultaneously exist two types of transportation. For one type, the carrier mediated transport net rate is positive when the carrier-mediated transportation rate is greater than the efflux rate. For the other type, the carrier mediated transport net rate is negative when the carrier-mediated transportation rate is less than the efflux rate. The sum of the absolute values of the simple diffusion rate and the carrier-mediated transport rate is the transport ratio of the corresponding transport mechanisms. As a result, the transport ratio of passive transport and carrier-mediated transport can be obtained using the parameters the included in Eq (8) and Eq (9) as follows:

The ratio of simple passive transport can be represented as follows:

$$V_1\% = \frac{KC}{\left|\frac{V_{\max} C}{K_m + C}\right| + |KC|} \times 100\% \tag{8}$$

The ratio of carrier-mediated transport can be represented as follows:

$$V_2\% = \frac{\dfrac{V_{\max} C}{K_m + C}}{\left|\dfrac{V_{\max} C}{K_m + C}\right| + |KC|} \times 100\% \tag{9}$$

According to the ratio of the fitted model and its specific results, two conclusions could be drawn.

(1)If passive diffusion exhibits better fitting than that of the combination of passive and active diffusion, the transport mechanism is simple passive diffusion.

(2)If the combination of passive and active diffusion exhibits better fitting than that of passive diffusion, the transport mechanism is passive diffusion combined with carrier-mediated transport. Carrier-mediated transport can be divided into two types.

(a) When the carrier-mediated transport ratio is positive, transport of drug molecules must involved active transport proteins. It showed that no efflux protein interacts the probe drugs participated in if the rate of the carrier-mediated transport is unchanged after the addition of efflux protein inhibitors. If the ratio of the carrier-mediated transport is increased, the drug efflux phenomenon exists, and the type of efflux protein can be determined according to the drug used.

(b)When the carrier-mediated transport is negative, an efflux protein must be involved in the transport of drug molecules. After the addition of efflux transport inhibitors, if the rate of the carrier-mediated transport is changed, it showed that there had an efflux protein is involved in the process, but not the corresponding efflux protein of the probe drugs. If the carrier-mediated transport rate becomes positive, the type of efflux protein should be determined using a probe drug.

Based on previous quantitative analyses, it not only could explain the transport mechanism and the extent of drug transportation but also could provide an accurate guide to improving the intestinal absorption of a drug. At the same time according to the analysis of case (b), quantitative researches could reveal the existence of potential efflux proteins and may identify factors that influence drug transportation, leading to additional studies of the new efflux protein.

2.4. HPLC analysis

An Agilent 1260 series HPLC system (Agilent Technologies, USA) was used. For puerarin, Reversed-phase chromatography was performed with an analytical Phenomenex Luna5 μC18 column (250 × 4.6 mm, 5 μm). The mobile phase consisted of A (methyl alcohol)/B (0.05% phosphoric acid) (30/70) with a flow rate of 1.0 mL/min, and the detection wavelength was 250 nm. For berberine hydrochloride, Reversed-phase chromatography was performed with an analytical Phenomenex Gemini C18-110A column (250 × 4.6 mm, 5 μm). The mobile phase consisted of A (acetonitrile)/B (0.05

mol/L NaH_2PO_4) (35/65) (pH=2.5), with a flow rate of 1.0 mL/min, and the detection wavelength was 345 nm. For paeoniflorin, Reversed-phase chromatography was performed with an analytical Phenomenex Gemini C18-110A column (250 × 4.6 mm, 5 μm). The mobile phase consisted of A (methyl alcohol)/B (water) (38/62)2, with a flow rate of 1.0 mL/min, and the detection wavelength was 230 nm.

2.5. Data Analysis

The total rate of drug transportation was calculated based on Eq. (10):

$$V = \frac{\Delta Q}{\Delta t A} \tag{10}$$

Where V is the transported amount per unit time and unit area ($mmol\cdot min^{-1}\cdot cm^{-2}$) and A is the surface area of polycarbonate filter membranes (cm^2).
The apparent permeability coefficient (Papp) was caculated based on Eq. (11).

$$P_{app} = \frac{\Delta Q}{\Delta t \times A \times C_0} \tag{11}$$

Where ΔQ is transport rate of drugs ($\mu g\cdot s^{-1}$) in transport times (Δt) , Δt is transport time, A is the surface area of polycarbonate filter membranes, C_o is the initial concentration ($\mu g\cdot mL^{-1}$).

For every experiment there were four samples, and the values are expressed as mean ± S.D. Data were subjected to statistical analysis by Student's t-test, or one-way ANOVA at a significance level of $P < 0.05$ using SPSS 19.0.

2.6. Effect of The Concentration of Drugs on Puerarin, Berberine Hydrochloride and Paeoniflorin Transport and Transport Mechanism Judged

According to the transport experiments methods as "2.2", preparation different concentration of puerarin, berberine hydrochloride and paeoniflorin, respectively. HPLC methods to determinate the drug amount that transported from AP to BL through Caco-2 cell monolayers.The transport rate V and the apparent permeability coefficient Papp value was calculated in accordance with the above corresponding formula. Through two methods to determinate the transport mechanism of three small molecule drugs. One method was traditional qualitative judgment, another methods was the quantitative analysis that established in this paper.

2.7. Effects of P-gp and MRP on the Transport of Puerarin, Berberine Hydrochloride and Paeoniflorin through Caco-2 Cell Monolayers

Puerarin, berberine hydrochloride and paeoniflorin were dissolved and diluted with HBSS, and the diluted drugs were used to dissolve verapamil and probenecid. The P-glycoprotein (P-gp) antagonist verapamil[20]was used at a concentration of 100 μmol/L to determine its effect on the transport rates and Papp of puerarin, berberine hydrochloride and paeoniflorin, respectively. The MRP antagonist probenecid[21] was also used to determine its effect on the transport rates and Papp of puerarin, berberine hydrochloride and paeoniflorin, respectively. Volumes of 0.5 mL of the drug solutions and 1.5 mL of HBSS were added to the AP and BL sides, respectively. The samples were incubated for 2 h, and 1.0 mL samples were then collected from the BL sides. The samples were then processed as described above. The transport rates and Papp of the three drugs were calculated through HPLC analysis, respectively.

3. RESULTS AND DISCUSSION

3.1. Effect of the Concentration of Drugs on Puerarin, Berberine Hydrochloride and Paeoniflorin Transport and Transport Mechanism Judged in Caco-2 Cell Monolayers

The transport results of different concentration of puerarin, berberine hydrochloride and paeoniflorin through Caco-2 cell monolayers was shown in Table 1, Table 2, and Table 3 respectively.

Table 1. Transport of different concentrations of puerarin on Caco-2 cell monolayers($\bar{x}$±sd, n=4)

Group	Transport rate($\times 10^{-2}\mu g \cdot min^{-1} \cdot cm^{-2}$)	Papp($\times 10^{-6} cm \cdot s^{-1}$)
puerarin(36.37$\mu g \cdot mL^{-1}$)	0.35±0.03	1.62±0.13
puerarin(72.74$\mu g \cdot mL^{-1}$)	0.84±0.05	1.92±0.11
puerarin(145.48$\mu g \cdot mL^{-1}$)	1.72±0.10	1.97±0.12
puerarin(290.96$\mu g \cdot mL^{-1}$)	3.20±0.12	1.83±0.07
puerarin(436.44$\mu g \cdot mL^{-1}$)	4.90±0.19	1.87±0.07
puerarin(581.92$\mu g \cdot mL^{-1}$)	6.89±0.41	1.97±0.12

Table 2. Transport of different concentrations of berberine hydrochloride on Caco-2 cell monolayers($\bar{X}$±sd, n=4)

Group	Transport rate($\times 10^{-2}\mu g \cdot min^{-1} \cdot cm^{-2}$)	Papp($\times 10^{-6} cm \cdot s^{-1}$)
berberine hydrochloride (42.75$\mu g \cdot mL^{-1}$)	0.71±0.1	2.76±0.4
berberine hydrochloride (57.81$\mu g \cdot mL^{-1}$)	0.84±0.02	2.43±0.07
berberine hydrochloride (72.98$\mu g \cdot mL^{-1}$)	1.14±0.2	2.6±0.46
berberine hydrochloride (89.98$\mu g \cdot mL^{-1}$)	1.2±0.13	2.22±0.24
berberine hydrochloride (121.18$\mu g \cdot mL^{-1}$)	1.71±0.17	2.35±0.23
berberine hydrochloride (131.65$\mu g \cdot mL^{-1}$)	1.81±0.23	2.29±0.29

Table 3. Transport of different concentrations of paeoniflorin on Caco-2 cell monolayers($\bar{x}$±sd, n=4)

Group	Transport rate($\times10^{-2}\mu g\cdot min^{-1}\cdot cm^{-2}$)	Papp($\times10^{-6}cm\cdot s^{-1}$)
Paeoniflorin (58.63$\mu g\cdot mL^{-1}$)	1.33±0.07	2.64±0.13
Paeoniflorin (88.88$\mu g\cdot mL^{-1}$)	2.09±0.24	2.66±0.31
Paeoniflorin (107.09$\mu g\cdot mL^{-1}$)	2.43±0.1	2.57±0.10
Paeoniflorin (118.55$\mu g\cdot mL^{-1}$)	2.83±0.21	2.70±0.20
Paeoniflorin (148.51$\mu g\cdot mL^{-1}$)	3.47±0.23	2.61±0.17
Paeoniflorin (178.50$\mu g\cdot mL^{-1}$)	4.24±0.28	2.66±0.17

Transport mechanism was judged by two methods as follows:

Traditional qualitative judgment:

From the Table 1, Table 2, and Table 3 we can see that with the concentration of puerarin, berberine hydrochloride and paeoniflorin increased the transport rate increasing, and Papp of the three drugs have no significantly change. We could make a conclusion that the transport mechanism of all the three drugs mainly was passive diffusion.

Quantitative analysis :

Used MATLAB 7.5.0 mathematical fitting soft and combined with the formula (1), formula (2) and formula (7) to make model fitting and statistical analysis on these three tables. The results were shown in Table 4, Table 5 and Table 6. From the tables we can see that with the formula (7) fitting the sum of squares due to error (SSE) and Root mean squared error (RMSE) more smaller, and the R-square relatively larger. It indicates that the formula(7) can get the best results that meet the actual situation. That is to say, the fitting degree of the joint model of simple diffusion and carrier medium transport is the most optimal. It indicated that the transport of puerarin, berberine hydrochloride and paeoniflorin exist in the process of simple diffusion and carrier medium, simultaneously. And these results are consistent with the reported[2-23]results.According to the fitting equation of formula (7), the fitted equations of the two drugs and transport parameters are shown in Table 7.

Table 4. Statistical analysis of puerarin transport chimeric model

Fitting parameters	Transport mode		
	Simple diffusion	Carrier medium	Simple diffusion + Carrier medium
SSE	0.08055	0.1107	0.04683
R-square	0.9975	0.9966	0.9985
RMSE	0.1269	0.1664	0.1249

Table 5. Statistical analysis of berberine hydrochloride transport chimeric model

Fitting parameters	Transport mode		
	Simple diffusion	Carrier medium	Simple diffusion + Carrier medium
SSE	0.03122	0.3318	0.0183
R-square	0.9687	0.9668	0.9817
RMSE	0.07902	0.1052	0.06765

Table 6. Statistical analysis of paeoniflorin transport chimeric model

Fitting parameters	Transport mode		
	Simple diffusion	Carrier medium	Simple diffusion + Carrier medium
SSE	0.01415	0.01709	0.00967
R-square	0.9973	0.9967	0.9982
RMSE	0.0532	0.0665	0.05677

Table 7. The fitting parameters of puerarin, berberine hydrochloride and paeoniflorin transport

Fitting parameters	Drugs		
	puerarin	berberine hydrochloride	paeoniflorin
Fitted equation	$V=-1.624C/(1.139\times10^{4}+C) + 2.523\times10^{-4}C$	$V=-0.0163C/(998.3+C)+1.561\times10^{-4}C$	$V=-3.728\times10^{-3}C/(116.9+C)+2.496\times10^{-4}C$
K ($mL\cdot min^{-1}\cdot cm^{-2}$)	2.523×10^{-4}	1.561×10^{-4}	2.496×10^{-4}
Km ($\mu g\cdot mL^{-1}$)	1.139×10^{4}	998.3	116.9
Vmax ($\mu g\cdot min^{-1}\cdot cm^{-2}$)	-1.624	-0.0163	-3.728×10^{-3}

3.2. The Transport Effects of P-gp and MRP on Puerarin, Berberine Hydrochloride and Paeoniflorin in Caco-2 Cell Monolayers

To further study the type of efflux protein that is involved in the transport process and to determine its role in carrier-mediated transport, the change in the rates of simple diffusion and carrier-mediated transport for the two types of the three drugs and the ratio of each to the sum of the absolute value after the addition of the P-gp inhibitor verapamil and the MRP inhibitor probenecid was investigated. The transport rates of puerarin, berberine hydrochloride and paeoniflorin after the addition of verapamil and probenecid to the AP side were shown in Tables 8, Tables 9 and Table 10.

Table 8. The influence of P-gp and MRP on puerarin in Caco-2 cell monolayers($\bar{x}$±sd, n=4)

Group	puerarin(167.6$\mu g\cdot mL^{-1}$)	puerarin(167.6$\mu g\cdot mL^{-1}$)+ 100$\mu mol\cdot L^{-1}$verapamil	puerarin(167.6$\mu g\cdot mL^{-1}$)+ 100$\mu mol\cdot L^{-1}$ probenecid
Ttotal transport rate ($\times10^{-2}\mu g\cdot min^{-1}\cdot cm^{-2}$)	1.19±0.07	1.82±0.04**	1.81±0.05**
Transport rate of simple diffusion ($\times10^{-2}\mu g\cdot min^{-1}\cdot cm^{-2}$)	4.23	4.23	4.23
Transport rate of carrier media transport ($\times10^{-2}\mu g\cdot min^{-1}\cdot cm^{-2}$)	-3.04	-2.41	-2.42
Proportion of simple diffusion(%)	58.2	63.7	63.6
Proportion of carrier media transport(%)	-41.8	-36.3	-36.4

Note: Compared with puerarin, **P<0.01

Table 8 showed that transport rate of carrier media transport of puerarin was negative, this indicated there was the ratio of the passive diffusion and carrier-mediated transport for puerarin, the results showed that the total direction of the carrier is efflux direction. The rates of carrier-mediated transport and simple diffusion are the absolute values of -41.8% and 58.2%, respectively. This revealed that transport mechanism of puerarin mainly via passive diffusion and involved efflux proteins participated in. The total transport rate, proportion of simple diffusion increased significantly after adding the verapamil and probenecid, while the proportion of carrier media transport decreased from -41.8% to -36.3% and -36.4%,respectively. That is to say transport rate of carrier media all transferred to positive direction after adding the verapamil and probenecid, it suggested that there are P-gp and MRP proteins that participated in in the transport process of puerarin.

But there is a small proportion that reduced of carrier media transport after adding P-gp and MRP inhibitors, the direction was still efflux, we thought that there might be other efflux proteins involved in the transport process of puerarin.

Table 9. The influence of P-gp and MRP on berberine hydrochloride in Caco-2 cell monolayers($\bar{X}$±sd, n=4)

Group	berberine hydrochloride (98.2μg·mL^{-1})	berberine hydrochloride (98.2μg·mL^{-1}) +100μmol·L^{-1}verapamil	berberine hydrochloride (98.2μg·mL^{-1}) +100μmol·L^{-1}probenecid
Ttotal transport rate (×10^{-2}μg·min^{-1}·cm^{-2})	0.97±0.02	1.14±0.05**	1.07±0.02**
Transport rate of simple diffusion (×10^{-2}μg·min^{-1}·cm^{-2})	1.53	1.53	1.53
Transport rate of carrier media transport (×10^{-2}μg·min^{-1}·cm^{-2})	-0.56	-0.39	-0.46
Proportion of simple diffusion(%)	73.2	79.7	76.9
Proportion of carrier media transport(%)	-26.8	-21.3	-23.1

Note: Compared with berberine hydrochloride, **$P<0.01$

Table 9 showed that transport rate of carrier media transport of berberine hydrochloride was negative, the results showed that the total direction of the carrier media was efflux direction. The rates of carrier-mediated transport and simple diffusion are the absolute values of -26.8% and 73.2%, respectively. This suggested that transport mechanism of berberine hydrochloride mainly via passive diffusion and there efflux proteins involved in the transport process of berberine hydrochloride. The total transport rate, proportion of simple diffusion increased significantly after adding the verapamil and probenecid, while the proportion of carrier media transport decreased from -26.8% to -21.3% and -23.1%,respectively. That is to say transport rate of carrier media all transferred to positive direction after adding the verapamil and probenecid, it suggested that there were P-gp and MRP proteins that participated in in the transport process of berberine hydrochloride.But there was a small proportion that reduced of carrier media transport after adding P-gp and MRP inhibitors, the direction was still efflux, we thought that there might be other efflux proteins involved in the transport process of berberine hydrochloride.

Table 10. The influence of P-gp and MRP on paeoniflorin in Caco-2 cell monolayers ($\bar{X}$±sd, n=4)

Group	Paeoniflorin (153.9μg·mL^{-1})	paeoniflorin(153.9μg·mL^{-1}) +100μmol•L^{-1}verapamil	paeoniflorin(153.9μg·mL^{-1}) +100μmol•L^{-1} probenecid
The total transport rate(×10^{-2}μg·min^{-1}·cm^{-2})	1.03±0.02	1.75±0.12**	1.04±0.01
Transport rate of simple diffusion (×10^{-2}μg·min^{-1}·cm^{-2})	3.84	3.84	3.84
Transport rate of carrier media transport (×10^{-2}μg·min^{-1}·cm^{-2})	-2.81	-2.09	-2.8
Proportion of simple diffusion(%)	57.7	64.8	57.8
Proportion of carrier media transport(%)	-42.3	-35.2	-42.2

Note: Compared with paeoniflorin, **P<0.01

Table 10 showed that proportion of simple diffusion of paeoniflorin was 57.7%, while proportion of carrier media transport of paeoniflorin was -42.3%, it suggested that paeoniflorin was absorbed through the intestinal epithelial cells mainly by passive diffusion, at the same time there was carrier media transport involved and the direction of carrier media transport was efflux direction.The total transport rate, proportion of simple diffusion increased significantly after adding the verapamil, while the proportion of carrier media transport decreased from -42.3% to -35.2%, That is to say transport rate of carrier media all transferred to positive direction after adding the verapamil and probenecid, it suggested that P-gp participated in in the transport process of paeoniflorin.There have no significantly changes of total transport rate and proportion of simple diffusion of paeoniflorin when adding probenecid, and the proportion of carrier media transport also has no changed, it suggested that the transport of paeoniflorin did not affected by MRP. Because the rate of carrier media transport was still negative when adding P-gp inhibitor, we thought that there might be other efflux proteins involved in the transport process of paeoniflorin.

4. CONCLUSIONS

Based on the theoretical research on the transport mechanism of drugs, drug absorbed channel was divided to three models, simple diffusion, carrier mediated transport, simple diffusion involved media transport model. A quantitative drug transportation rate model was established, and the role of carrier-mediated transport in overall cell transport was evaluated. From the angle of mathematical fitting equation, set puerarin,berberine hydrochloride and paeoniflorin as model drugs, a quantitative research on transport mechanisms of the three drugs were studied. The results in the present study were consistent with the results reported in the literature using traditional methods, which suggests that the established quantitative method is suitable for the study of drug transport mechanisms.

The present quantitative research study revealed that the transport mechanisms of the three drugs are mainly passive transport together with some carrier-mediated transport. In the presence of efflux protein inhibitors, the drug transport rate and the ratios of passive diffusion and carrier-mediated transport to the total rate of drug transport further revealed that P-gp was substrates of puerarin, berberine hydrochloride and paeoniflorin while MRP was substrates of puerarin and berberine hydrochloride but not paeoniflorin. These results indicated that this quantitative method not only could be used to determine the transport mechanism of a drug, but also could identify the contribution of each transport mechanism and analyze the potential transporters and will therefore guide further follow-up studies.

Acknowledgments

The work was supported by a grant from the Key Laboratory of Modern Preparation of TCM of the Ministry of Education of Jiangxi University of TCM, the Project of the National Natural Science Foundation of China (No. 81303237), The Training Project of Young Scientists in Jiangxi Province (No. 20153BCB23019) and the project of the Department of Health of Jiangxi Province (No. 2012A030).
We confirmed that all of the authors have no conflict of interest.

References

1. Ci XY, Xia YY, Zeng Y, Yin XL, Si RY. Absorption of 3-acetyl-11-keto-β-boswellic acid in Caco-2 cells and MDCK cell models. *Chin Tradit Herbal Drugs* 2014; **44**: 1162-1167.
2. Yu ZC, Tian WW, Ma MH, Liu X, Chen HJ, Chen HZ. Investigation on the mechanism of transmembrane transport of Mycophenolate mofetil and Its active metabolite Mycophenolic acid. *Chin Pharm* 2014; **49**: 322-328.
3. Hu LJ, Guo HL, Hu ZF, Yan BY, Zhao XJ, Xiao XM. Transport Characteristics of α-Cyperone in Caco-2 Cell Model by LC-MS. *Chin J Exp Tradit Med Form* 2014; **22**: 35-38.
4. Saito H, Fukasawa Y, Otsubo Y, Yamada K, Sezaki H, Yamashita S. Carrier-Mediated transport of macrolide antimicrobial agents across Caco-2 cell monolayers. *Pharmaceutical Research* 2000; **6**: 761-765.
5. Chen JY, Xu ZQ, Li YQ. The uptake characteristics of 5-fluorouracil in the Caco-2 model system. *Acta Pharm Sin* 1998; **33**: 222-225.
6. Mizuuchi H, Katsura T, Hashimoto Y, Inui K. Transepithelial transport of diphenhydramine across monolayers of the human intestinal epithelial cell line Caco-2. *Pharmaceutical Research* 2000; **5**: 539-545.
7. Brunet J, Maresca M, Fantini J, Belzunces L. Human intestinal absorption of imidacloprid with Caco-2 cells as enterocyte model. *Toxicology and Applied Pharmacology* 2004; **194**: 1-9.
8. Liang XL, Zhao LJ, Liao ZG, Zhao GW, Zhang J, Chao YC, Yang M, Yin RL. Transport properties of puerarin and effect ofRadix Angelicae Dahuricae extract on the transport of puerarin in Caco-2 cell model. *Journal of Ethnopharmacology* 2012; **144**: 677-682.

9. Petri N, Tannergren C, Rungstad D, Lennernas H. Transport characteristics of fexofenadine in the Caco-2 cell model. *Pharmaceutical Research* 2004; **8**: 1398-1404.
10. Gan LS, Hsyu P, Pritchard JF, Thakker D. Mechanisms of intestinal absorption of ranitidine and ondansetron transport across caco-2 cell monolayers. *Pharmaceutical Research* 1993; **12**: 1722-1725.
11. Me´nez C, Buyse M, Farinotti R, Barratt G. Inward translocation of the phospholipid analogue miltefosine across Caco-2 Cell membranes exhibits characteristics of a carrier-mediated process. *Lipids* 2007; **42**: 229–240.
12. Liang E, Proudfoot J, Yazdanian M. Mechanisms of transport and structure-permeability relationship of sulfasalazine and its analogs in Caco-2 cell monolayers. *Pharmaceutical Research* 2000; **10**: 1168-1174.
13. Zhang SZ, Morris ME. Effect of the flavonoids biochanin a and silymarin on the P-Glycoprotein-Mediated transport of digoxin and vinblastine in human intestinal Caco-2 cells. *Pharmaceutical Research* 2003; **8**: 1184-1190.
14. Yan LC, Liu QC, Zhao JN, Bi YQ, Yi JH. Study on transportation of Xiaochaihu Tang in Caco-2 cell model. *China Journal of Chinese Materia Medica* 2011; **36**: 1087-1090.
15. Yuan ZT, Wang K, Gao PP, Zhao ZH, Hu M. Study on Transport mechanism s of Osthol in Caco-2 cell model. *Chin Pharm J* 2011; **46**: 278-282.
16. Zhu ML, Liang XL., Zhao LJ, Liao ZG, Zhao GW, Cao YC, Zhang J., Luo Y. Elucidation of the transport mechanism of baicalin and the influence of a Radix Angelicae Dahuricae extract on the absorption of baicalin in a Caco-2 cell monolayer model. *Journal of Ethnopharmacology* 2013; **150**: 553-559.
17. Liang XL, Zhao LJ, Liao ZG, Zhao GW, Zhang J, Chao YC, Yang M, Yin RL. Transport properties of puerarin and effect of Radix Angelicae Dahuricae extract on the transport of puerarin in Caco-2 cell model. *Journal of Ethnopharmacology* 2012; **144**: 677-682.
18. Sun, M.J., Sheng, X., Hu, Y.Q. Establishment and validation of Caco-2 cell lines for intestinal epithelial permeability. *Chin. Pharm. J.* 2006; 41: 1431-1434.
19. Sun J. *Absorb and transport of oral drugs*. Beijing: People's Medical Publishing House, 2006:17.
20. Sha XY, Fang XL. Transport characteristics of 9-nitrocamptothecin in the human intestinal cell line Caco-2 and everted gut sacs. *Int. J. Pharm* 2004; **272**: 161-171.
21. Liang XL, Zhu ML, Zhao LJ, Zhao GW, LiaoZG, Cao YC, Yang M. Study on transport mechanism of baicalin in Scutellariae Radix extracts and effect of Angelica Dahurica extracts on transport of baicalin by Caco-2 cell monolayer model.*China Journal of Chinese Materia Medica* 2013; **38**: 2389-2392.
22. Wu AG, Zeng B, Wang CL, Chen JA, Lai XP. Transport of berberine as single compound and berberine in extract in Caco-2 cell model. *Chinese Pharmacological Bulletin* 2011; **27**: 1007-1011.
23. Zhu DF. Transport of Danshensu an Paeoniflorin across Caco-2 Cells Monolays. *Zhejiang University, Hangzhou*. 2006.

COMPARE THE DIFFERENT LAPAROSCOPIC TREATMENT OF VARICOCELE

Yuhan Zhang[1,a], Xinquan Gu[1], and Cao Xia[2,b,*]

[1]*Department of Urology, China-Japan Union Hospital Affiliated to Jilin University, Changchun Jilin, 130033, China*

[2]*School of Pharmaceutical Sciences, Jilin University, Changchun Jilin, 130021, China*

E-mail: [a]837647704@qq.com, [b]caoxia@jlu.edu.cn

[*]*Corresponding author*

Objective: To discuss the treatment choices of varicocele by laparoscopic. Methods: Retrospective analysis clinical datas of 337 patients with primary varicocele above grade II. All of them were treated by operation and followed-up, 38 cases had scrotum pain. 265 cases were infertility after married. History of diseases was between 1-13 years. 249(197 patients with infertility) were treated with laparoscopic high selective ligation of spermatic veins, as Group A. 88(68 patients with infertility) were treated with laparoscopic cluster ligation, as Group B. The clinical data were compared. Results: Three hundreds and thirty-seven cases were followed-up 2 years after operation. The operative times of two groups were(38.0±10.9)min, (19.0±7.5)min, the incidences of hydrocele were 0.4%(1/249),11.3%(10/88);the incidences of testicular atrophy were 0%(0/249),3.4%(3/88); the improving rates of semen quality in 3 months were 77.1%(192/249),62.5%(55/88); the spontaneous pregnant rates of his spouses in 2 years were 56.8%(112/197),39.7%(27/68). There were significant differences among the 2 groups ($P<0.05$). The recurrence rates were 5.6%(14/249),4.5%(4/88),there was no significant difference among the 2 groups($P>0.05$). Conclusion Laparoscopic high selective ligation of spermatic veins in this study had less complications, higher improving rates of semen quality and pregnant rates.

1. INTRODUCTION

Varicocele is a male common disease, the patients that have symptoms or abnormal sperm merge infertility should take the surgical treatment [1]. Laparoscopic ligation of spermatic veins has been widely used in the treatment of varicocele. Between January 2005 and October 2013, our hospital receive 452 patients with varicocele above II degrees , complete follow-up data of 337 cases. Compare the treatment of laparoscopic high selective ligation of spermatic veins and laparoscopic cluster ligation, presently reports as follows:

2. MATERIALS AND METHODS

2.1. General Information

Totally 337 cases. Laparoscopic high selective ligation of spermatic veins as group A, 249 cases, the average age is 26; left side 211cases, both side 38cases; II degrees

199cases, III degrees 50cases; 197 cases was infertility after marriage, course 1 ~ 13 years; 154 cases have scrotal belly pain discomfort. Laparoscopic cluster ligation as group B, 88cases; the average age is 26; left side 74cases, both side 14cases; II degrees 70cases, III degrees 18cases; 69 cases was infertility after marriage, course 1 ~ 11 years, 54 cases have scrotal belly pain discomfort. 2 groups of patients with the above information to compare differences had no statistical significance ($P > 0.05$), comparable.

2.2. Surgical Treatment

Group A: Epidural anesthesia, establish pneumoperitoneum, pressure 14 mmHg (1mmHg=0.133 kPa), edge of umbilical incision in 10 mm casing, insert the laparoscopic, finding the right side of McBurney point to put in 10 mm casing, left side of McBurney point in 5mm casing. Finding the spermatic cord blood vessel bundle, at the avascular area from the spermatic cord blood vessels and vas deferens on 2 ~ 3 cm and cut retroperitoneum, meanwhile reduce the pressure to 5-8mmHg, avoiding blood vessels shrinking. And inject the papaverine 30mg or 2% lidocaine 1ml to prevent vasospasm. Firstly separation obvious enlargement of varicose spermatic vein, then cut off after ligation. Next to separation and ligation the small vein around. Exhaustive check the remaining vein in the adipose tissue and lymphatic tissue. Discharge of CO2 gas, suture the incisions.

Group B: Epidural anesthesia, establish pneumoperitoneum and put in the casing as group A. Don't reduce the pneumoperitoneum pressure and inject the drug, separation varicose spermatic vein and ligation. Then cut off.

2.3. Observation Item

Compare two groups of operation time, postoperative incidence of hydrocele, semen routine change after surgery 3months (Sperm density >20%, class a and class b sperm percentage >5% seems semen quality improvement). 6 months review testicular atrophy (Color doppler ultrasound, volume compared with preoperative < 3 ml is atrophy), recurrence rate in 2 years and spouse natural pregnancy rate.

2.4. Statistical Methods

Using SPSS 10.0 statistical software process the data, operation time comparing application of two independent samples t test, The incidence of postoperative hydrocele, testicular atrophy rate, recurrence rate, semen routine change rate and spouse natural pregnancy rate in 2 years using x^2 test, a=0.05.

3. RESULTS

337 cases are performed successfully, postoperative follow-up of 2 years. 2 groups operation time are (38.0±10.9)min,(19.0±7.5)min, comparative differences between groups was statistically significant ($P<0.05$); 2 groups the postoperative complications,

semen quality changes and spouse natural pregnancy rate was presented in the Table 1. The rate in Group A are higher than Group B and it is statistically significant (P<0.05); the recurrence of two groups was not statistically significant (P>0.05).

Table 1. Results of group A and group B

Group	A	B
cases	249	88
hydrocele	1(0.4)	10(11.3)
testicular atrophy	0(0.0)l	3(3.4)
recurrence	14(5.6)	4(4.5)
Improving semen quanlity	192(77.1)	55(62.5)
Spouse pregant in 2 years	112(56.8)	27(39.7)

4. DISCUSSION

Varicocele is because of the spermatic vein is blocked or valvular function lost cause reflux caused by blood stasis, and the testicles cirsoid venous plexus circuity expansion. After the occurrence of varicose influence of testicular blood circulation, testicular local temperature rise, lead to testicular function impairment and affect sperm [1]. Patients often feel discomfort of lateral scrotal belly pain, and secondary infertility. Literature reported incidence of secondary infertility patients with varicocele as high as 60% ~ 70% [2], 75% ~ 85% patients with varicocele cause secondary infertility [3]. Surgery in the treatment of varicocele can effectively block the venous reflux, eliminate local blood stasis, improve the testicular growth environment, good for sperm functional recovery, and at the same time to eliminate clinical symptoms. There is a lot of varicocele operation method. Laparoscopic ligation of spermatic veins has been widely used in the treatment of varicocele. Compared with open surgery has the advantages of little trauma and quick recovery [4]. Laparoscopic ligation of spermatic veins including Palomo operation of high position ligation of spermatic cord blood vessels of cluster and Ivanissevich operation of retain the spermatic artery [5-6]. The operation of ligation of spermatic cord blood vessels of cluster is simple and the recurrence rate is low. But because at the same time with ligation of spermatic artery and lymphatic vessels,the incidence of postoperative hydrocele was 7% ~ 18%. And have the risk that the testicular atrophy [7]. Student [8] reports on spermatic artery for testicular blood flow within the main sources, save the artery to guarantee adequate testicular blood supply is very important. Grober [9] think studies such as arterial injury although not lead testicular atrophy, but can damage sperm production process, protect arteries to maintain normal sperm production functions is important. 88cases in group B, 10 cases of postoperative hydrocele, accounted for 11.3%; testicular atrophy occurred in 3 patients, accounting for 3.4% were significantly higher than group A. So reserve spermatic artery and lymphatic vessels is very important. Ivanissevieh operation to avoid testicular atrophy that reserve spermatic artery, but the recurrence rate of varicocele was rise to 10%~19%, because of separation of spermatic artery residue and missing the small vein [10]. The 249cases in group A, in addition to operating time extension in the B group, other side were much better than group B. And the recurrence rate is similar. This method is simple, the effect certainly, complication decreased significantly, can be used as an important treatment of varicocele.

References

1. Hongqian Guo, Zeyu Sun, Varicocele // Jieping Wu, *Wu Jieping Department of Urology*. Jinan: Shandong science and technology publishing. 2004:1951-1953.Blute ML, Leibovich BC, Lohse CM, et al.
2. Hongjun Li, Hanzhong Li. Strict control of male infertile patients with varicocele operation indications [J]. *Chin J Urol*. 2010, **31**: 221-222.
3. Shaowen Wu, Varicocele operation // Hua Mei, Lingwu Chen, Xin Gao, *Urology proc.3*, Beijing: People's medical publishing house. 2008:614-618.
4. Zhiqiang Jiang, Xianzheng Jiang, Qing Zeng. The choice of varicocele operation method. *Chinese Journal of Andrology*. 2005, **19**: 44-46.
5. Songlin Tang, Xiaohua Zhang, Yanhong Wang, et al. Laparoscopic ligation of spermatic veins cluster shaped silk varicocele curative effect analysis. Chinese Journal of Endourology, 2008, 2: 40-42.
6. Qizhong Liu, Kai Tian, Yuexi Zhang, et al. Laparoscopic and open surgery spermatic vein high ligation of curative effect analysis. *Chinese Journal of Andrology*, 2009.**15**:625-627.
7. Barqawi A, Caruso A, Meacham R B. Experimental varicocele induces testicular germ cell apoptosis in the rat.[J]. *Journal of Urology*, 2004, **171**(1):501-503.
8. Student V, Zátura F, Scheinar J, et al. Testicle hemodynamics in patients after laparoscopic varicocelectomy evaluated using color Doppler sonography.[J]. *European Urology*, 1998, **33**(1):91-93.
9. Grober E D, O'Brien J, Jarvi K A, et al. Preservation of Testicular Arteries During Subinguinal Microsurgical Varicocelectomy: Clinical Considerations[J]. *Journal of Andrology*, 2004, **25**(5):740-743.
10. Marcus R, Josef O, Mark K, et al. Optimizing the operative treatment of boys with varicocele: sequential comparison of 4 techniques.[J]. *Journal of Urology*, 2003, **169**(2):666–668.

TEXTURE ANALYSIS AND CLASSIFICATION ON XINJIANG KAZAKH ESOPHAGEAL CANCER IMAGES

Suixia Zhang[1,a], Murat Hamit,[1,b,*], Chuanbo Yan[1], Jing Sun[1], and Juan Yao[2]
[1]*Xinjiang medical university, Medical Engineering Technology, Xinjiang, China*
[2]*Department of Radiology, The First Affiliated Hospital, Urumqi, China*
E-mail:[a]1548756091@qq.com,[b]murat.h@163.com

Texture feature and classification are vital clues of image analysis. In this study, we proposed the feature extraction based on gray-gradient co-occurrence matrix (GGCM) and gray level co-occurrence matrix (GLCM) to characterizing the ulcerative and the fungating esophageal cancer. Firstly, converting RGB images to the grayscale intensity images and removing the image noise through median filter, then using adaptive histogram equalization to enhance the contrast. Secondly, GGCM and GLCM were used to extract the texture features of the images, twenty three features were obtained using the two methods and seven most powerful features were selected by principal components matrices (PCA). Finally, the classification ability of the texture features was evaluated by K-Nearest Neighbor (KNN) classifier, conducting the experiment repeatedly by adjusting the optimal K value. Experiment results show that the classification accuracy is the best when optimal K value is 1. This work proved the classification ability of KNN for Xinjiang Kazakh esophageal cancer while verified the significance of parameter selection in classifier, which laid the foundation for the subsequent research.

1. INTRODUCTION

Esophageal squamous cell carcinoma (ESCC), the major histological type of esophageal cancer (EC), is the fourth most frequent cause of cancer deaths in China[1-2]. According to the international agency for research on cancer, there were an estimated 482,000 incident cases of EC with high mortality (84%) around the world in 2008[3]. EC often infiltrates neighboring organs and easily metastasizes to lymph nodes, and as a result, the prognosis of locally advanced patients is extremely poor, with a 5-year survival of 15-34%[4]. Xinjiang, the northwestern part of China, where the Kazakh's ESCC has been reported to be abnormally higher than other minorities[5].

Texture, which as a basic properties of the object surface, is widely spread in the nature world, its extremely important for the description and recognition of the object [6], texture feature is independent of the color and brightness to reflect the homogeneous phenomenon of visual characteristics [7]. Texture feature has been widely used in pattern recognition and computer vision. Stavroula G.M.[8] defined an optimal performing computer-aided diagnosis architecture for the classification of liver tissue from non-enhanced CT images. Doaa Mahmoud-Ghoneim[9] investigated the accuracy of texture analysis results on three color spaces, conventional gray scale, RGB, and Hue-Saturation-Intensity (HSI), at different resolutions[10].

The Classification research of texture images has been widely applied in the field of biomedical engineering, remote sensing and telemetry, industrial product testing. PCA can effectively avoid the dimension disaster when dealing with the high dimensional samples, it is also the small sample size required, which make it has the strong potential to classify the texture feature samples in the high dimension. In this work, feature extraction based on gray gradient co-occurrence matrix (GGCM) and gray level co-occurrence matrix (GLCM) were applied in X-ray images of Xinjiang Kazakh Esophageal cancer due to its particular pathological features. And the classification ability of the texture features was verified by KNN classifier.

2. FEATURE EXTRACTION

2.1. Preprocessing

The preprocessing step aims to enhance the image or to reduce the effect of illumination changes, which can provide the better support for the doctor's diagnosis. Preprocessing can reduce image noise and improve the intelligence and usefulness of computer-aided diagnosis system. For the esophageal cancer X-ray images, converting RGB images to the grayscale intensity images, removing the image noise through median filter, using adaptive histogram equalization to enhance the contrast. Preprocessing results of ulcerative and fungating esophageal X-ray images show in Fig. 1 and Fig. 2, respectively. (a) original image, (b) scale normalized, (c) noise removed, (d) adaptive enhanced. As can be seen from the figures, the contrast of the images has been enhanced distinctly, which lay a good foundation for the subsequent image processing.

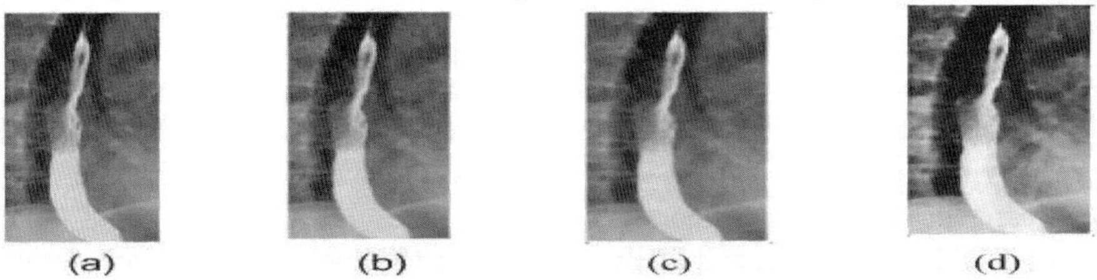

Fig. 1. Preprocessing results of fungating esophageal X-ray image.

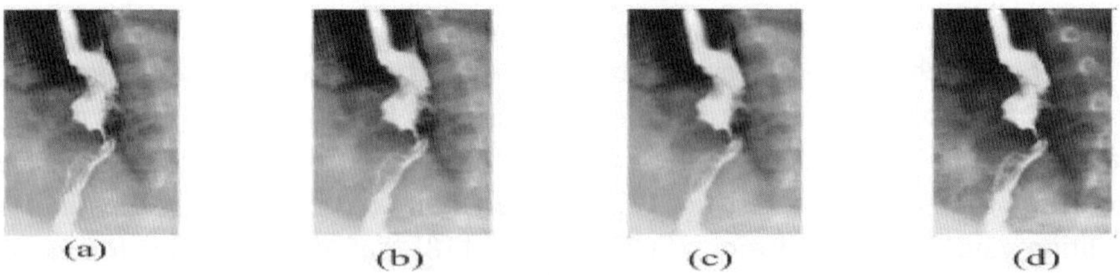

Fig. 2. Preprocessing results of ulcerative esophageal X-ray image.

2.2. Gray Level Gradient Co-Occurrence Matrix

The elements x and y of gray gradient co-occurrence matrix can be defined as the total pixels, which gray value is x and gradient value is y in the normalized grayscale image $\hat{H}(x,y)$ and the gradient image $H(x,y)$ [11-12]. Where L_f is the defined gray level, L_g is the defined gradient level. Gray gradient co-occurrence matrix model reflects the relationship between the gray and the gradient of each pixel in the image. Grayscale is the basis of an image, while the gradient constitutes the edge contour[12]. Therefore, gray gradient co-occurrence matrix not only describes the distribution of the gray and gradient of each pixel in the image, but also reflects the spatial relationship between the pixel and its neighboring pixels. Based on the GGCM, 15 texture features can be defined as shown in Table 1.

Table 1. The definition of texture features from GGCM.

Feature name	Definition	Explanatiom
Small Gradient Advantage	$T_1 = \frac{\sum_{y=0}^{L_g-1}\sum_{x=0}^{L_f-1}\hat{H}(x,y)/(y+1)^2}{\sum_{x=0}^{L_f-1}\sum_{y=0}^{L_g-1}\hat{H}(x,y)}$	Small Gradient Advantage and Big Gradient Advantage measure gray change. A homogeneous image will result in high Small Gradient Advantage value and low Big Gradient Advantage value.
Big Gradient Advantage	$T_2 = \frac{\sum_{y=0}^{L_g-1}\sum_{x=0}^{L_f-1}\hat{H}(x,y)y^2}{\sum_{x=0}^{L_f-1}\sum_{y=0}^{L_g-1}\hat{H}(x,y)}$	
Gray Level Distribution Inhomogeneity	$T_3 = \frac{\sum_{x=0}^{L_f-1}\left[\sum_{y=0}^{L_g-1}\hat{H}(x,y)\right]^2}{\sum_{x=0}^{L_f-1}\sum_{y=0}^{L_g-1}\hat{H}(x,y)}$	Gray Level Distribution Inhomogeneity measures gray level change. If the gray of an image only focuses on some gray levels, Gray Level Distribution Inhomogeneity value will become large.
...	...	....
Gray Level Entropy	$T_{13} = -\sum_{x=0}^{L_f-1}\sum_{y=0}^{L_g-1}\hat{H}(x,y)\log\hat{H}(x,y)$	No matter Gray Level Entropy, Gradient Level Entropy or Mixture Entropy, they all measure the amount of information in the image, also known as randomness of the image texture. A homogeneous image will result in a lower entropy value.
Gradient Level Entropy	$T_{14} = -\sum_{x=0}^{L_f-1}\sum_{y=0}^{L_g-1}\hat{H}(x,y)\log\sum_{x=0}^{L_f-1}\hat{H}(x,y)$	
Mixture Entropy	$T15 = -\sum_{x=0}^{L_f-1}\sum_{y=0}^{L_g-1}\hat{H}(x,y)\log\sum_{y=0}^{L_g-1}\hat{H}(x,y)$	

2.3. Feature Extraction based on GLCM

The gray level co-occurrence matrices (GLCM) [13-14] analyzes the common occurrences of a gray-level value for all of the pairs of pixels in textured images. There are two important parameters in the calculation process: the distance $d = 1$, between the two pixels and the orientation angle of the line connecting these two pixels. Using function in MATLAB, in each ROI, we calculated the GLCM at a specific distance and orientation. Pixels pairs were excluded from the GLCM if the neighbor pixel was outside the ROI. To achieve rotation invariant features, the GLCM of the four directions $\theta = \{0^\circ, 45^\circ, 90^\circ, 135^\circ\}$ were summed to one matrix. After the calculation, all features were standardized to zero mean and unit variance. From those isotropic GLCM, 4 statistical features were calculated resulting in a set of 8 GLCM texture features[15]: Entropy, Inertia Moment, correlation and contrast. Using H1, H2, H3, H4, H5, H6, H7 and H8 represent 8 standardized features.

2.4. Extraction Results of Mixed Texture Features

We selected 200 Kazakh esophageal X-ray images as experimental images, 100 fungating esophageal cancer images, 100 ulcerative esophageal cancer images. Extracted GGCM and GLCM texture features : T1, T2, T3, T4, T5, T6, T7, T8, T9, T10, T11, T12, T13, T14, T15, H1, H2, H3, H4, H5, H6, H7 and H8 consisted of features vector. Table 2 shows the results of feature extraction by GGCM and GLCM.

Table 2. Results of feature extraction by GGCM and GLCM

Type	fungating	fungating	…	fungating	ulcerative	ulcerative	…	ulcerative
T1	0.331	0.348	…	0.349	0.463	0.444	…	0.34
T2	6.624	5.638	…	5.119	3.77	3.975	…	6.509
T3	2151.632	1855.176	…	2173.568	2894.715	2679.627	…	1929.183
…	…	…	…	…	…	…	…	…
T13	3.267	3.265	…	3.162	2.908	2.976	…	3.284
T14	20209.392	21226.621	…	23634.149	23104.987	18533	…	18691.766
T15	0.002	0.003	…	0.005	0.025	0.002	…	0.012
H1	0.023	0.016	…	0.019	0.025	0.024	…	0.029
H2	0.002	0.001	…	0.001	0.003	0.003	…	0.003
…	…	…	…	…	…	…	…	…
H7	0.079	0.059	…	0.058	0.062	0.045	…	0.077
H8	0.004	0.003	…	0.002	0.002	0.002	…	0.003

2.5. Principal Component Analysis (PCA)

Extracted 23 texture features of Kazakh esophageal X-ray images with GGCM and GLCM, using principal component analysis (PCA) [16-17] reduced the dimensionality of texture features and got new PCA principal components matrices. Then, we got seven main components V1, V2, V3, V4, V5, V6 and V7, the cumulative contribution rate of PCA was 93.675%, it indicated that seven main components can be a good indicator of the reaction most of the information of the original variables. Fig. 3 shows accumulative contribution rate of PCA, Table 3 shows component matrix of PCA.

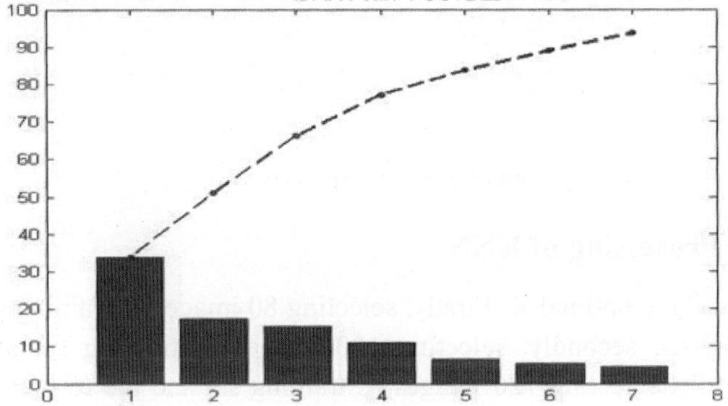

Fig. 3. Accumulative contribution rate of PCA (%).

Table 3. Component matrix of PCA

	V1	V2	V3	V4	V5	V6	V7
T1	-0.25	-0.026	-0.559	0.614	0.189	-0.337	0.164
T2	0.185	-0.315	0.295	0.709	-0.158	0.334	-0.166
T3	0.302	0.025	0.597	-0.614	-0.139	0.355	-0.125
T4	0.454	-0.213	0.74	0.012	-0.074	0.323	-0.117
…	…	…	…	…	…	…	…
H5	-0.099	0.753	-0.325	0.06	0.368	0.198	-0.272
H6	0.671	0.636	-0.004	0.155	0.181	0.098	0.046
H7	0.032	-0.603	-0.431	-0.031	0.371	0.452	0.284
H8	-0.526	0.754	0.163	0.187	0.085	0.18	0.019

3. CLASSIFICATION

3.1. KNN Algorithm

K-Nearest Neighbor Algorithm (KNN) [18-20] is known in the machine learning field as a nonparametric method. It does not use a training set or have a training phase. uses part of the data for classification and the rest of it for testing. Usually, the available data is partitioned in two sets, One part of data is used the classification and the rest of it for testing (Fig. 4(a)). Then, we take a datum at a time from the test set and use the classification set to determine its class. In the KNN method we used as a measure of similarity the distance between data. In our implementation we used the Euclidean distance for it. As shown in the example of Fig. 4(b) the distance to the data we want to classify is measured and sorted, as shown in the table of the same figure. If the number of neighbors (K) is set to 3, as in this example, the data (black dot) will be classified as of type a (white dot).This decision is a kind of majority vote based on distances [21].

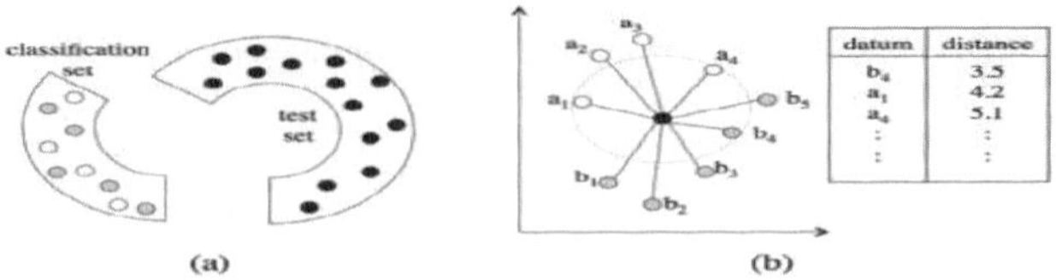

Fig. 4. KNN concept example.

3.2. Trained Processing of KNN

In order to obtained the optimal K. Firstly, selecting 80 images as training set and the rest images as testing set, secondly, selecting 100 images as training set and testing set, respectively. Thirdly, selecting 120 images as training set and the rest images as testing set. We set the number of 1 to 35 as neighbor K. Experimental results of testing set are shown in Fig. 5, the classification accuracy is higher when K is 1. Fig. 5 shows the classification accuracy of three training sets when K vary from 1 to 35. In summary, in order to obtain the optimal K, according to the trend of the three training sets classification accuracy shows: when K=1, ulcerative and fungating have an optimal classification accuracy, The three training set classification accuracy rates of fungating were: 84%, 88% and 90%, ulcerative could bring an accuracy rate of 78%, 74% and 82%.

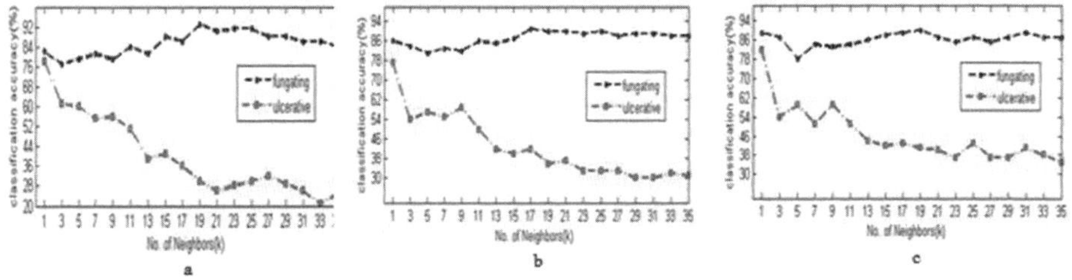

Fig. 5. Three training sets classification accuracy variation.

3.3. Validated Processing of KNN

Comprehensive the results 3 training sets (optimum K value), can be drawn: when K=1, KNN classified on ulcerative and fungating esophageal X-ray images had a high classification accuracy. Then, selected ten different sizes as testing set, which set from 10% to 90% of data (in increments of 10%), validating the optimal K value. Table 4 shows when we got optimal K values, with increasing the amount of validation data classification accuracy increased. In this table, we can assume KNN classifier applied to class fungating and ulcerative esophageal X-ray images, the average classification accuracy rate were high up to 85% and 76%. The accuracy variation of ten classification set sizes when we get optimal K value are showed in Fig. 6.

Table 4. K=1, accuracy variation of ten classification set sizes

Percentage (%)	10	20	30	40	50	60	70	80	90	100	average
fungating	76%	78%	77%	88%	89%	87%	89%	87%	89%	87%	85%
ulcerative	54%	60%	69%	69%	82%	84%	86%	85%	83%	86%	76%

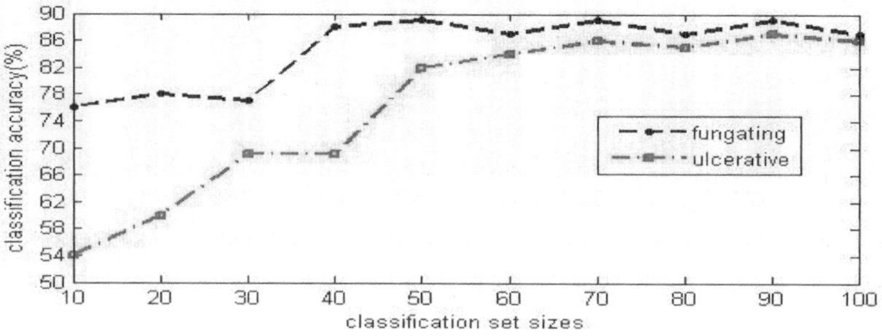

Fig. 6. K=1, accuracy variation of ten classification set sizes.

3.4. Evaluation

The performance measures commonly used are sensitivity, specificity and AUC. The measures are described in as follows:

$$\text{Sensitivity} = \frac{TP}{TP + TN} \quad (1)$$

$$\text{Specificity} = \frac{TN}{TN + TP} \quad (2)$$

$$\text{Accuracy} = \frac{TP + TN}{TP + TN + FP + FN} \quad (3)$$

Where TP=true positive, TN=true negative, FP=false positive, FN=false negative.

An event is said to be TP if a pixel is rightly segmented as a vessel and TN when a pixel is rightly segmented as background. In related development, an event is said to be FN if a vessel pixel is segmented to be a background and a FP when a background pixel is segmented as a pixel in the vessel. Sensitivity measure indicates the ability of a segmentation technique to detect the vessel pixels while specificity measure indicates the ability of a segmentation technique to detect background pixels[22]. The accuracy measure indicates the degree of conformity of the segmented retinal image to the ground truth. A receiver operating characteristic (ROC) curve performance measure is a plot of the rightly classified pixels, referred to as true positive rate (TPR) versus the fraction of the wrongly classified pixels as vessels, referred to as false positive rate (FPR). Area under the curve (AUC) is a performance measure computed from the ROC curve. In this study, when we obtained performances of accuracy on ten classification set sizes. With increasing the amount of validation data classification the values of sensitivity, specificity and AUC are increased[23]. Fig. 7 shows performance of accuracy on ten classification set sizes.

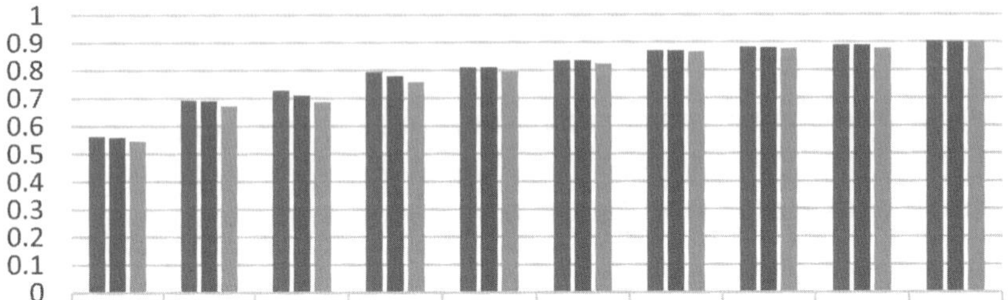

	10%	20%	30%	40%	50%	60%	70%	80%	90%	100%
Sensitivity	0.563	0.692	0.728	0.794	0.81	0.835	0.87	0.882	0.89	0.903
Specificity	0.56	0.69	0.71	0.78	0.81	0.835	0.87	0.88	0.89	0.9
AUC	0.546	0.673	0.686	0.758	0.796	0.824	0.865	0.877	0.878	0.904

Fig.7. Performance of accuracy on ten classification set sizes.

4. CONCLUSIONS

In this paper, we proposed the feature extraction based on GGCM and GLCM to characterize fungating and ulcerative esophageal X-ray image. The taxonomic capacity of the texture features were evaluated by KNN classifier. The experiment results show that classification accuracy varied from different parameters, the classification accuracy is optimum when K is 1. With increasing the amount of validation data classification accuracy increased, which proved the classification ability of KNN for Xinjiang Kazakh Esophageal cancer, while verified the significance of parameter selection in classifier. Preliminary experimental study demonstrated KNN classifier could improve the classification ability and provide a certain basis judgment of Kazakh esophageal cancer classification, which laid the foundation of computer diagnosis system for Kazakh esophageal cancer.

Acknowledgments

The authors would like to acknowledge the support of the first affiliated hospital of Xinjiang medical university of China. This research was partially supported by Nation Science Foundation under grant number (81460281, 81160182, 81560294).

References

1. Wang AH, Liu Y, Wang B, et al. Epidemiological studies of esophageal cancer in the era of genome-wide association studies. *World J Gastrointest Pathophysiol.* 2014; **5**:335-43.
2. Liao YH, Xue Y, Zhang XW, et al. High heat shock factor 1 expression in tumor stroma predicts poor prognosis in esophageal squamous cell carcinoma patients. *Trans Med.* 2015; **13**:338-351.
3. Jemal A, Bray F, Center MM, et al. Global cancer statistics. *CA Cancer J Clin.* 2011; **61**:69-90.
4. Allum WH, Stenning SP, Bancewicz J, et al. Long-term results of a randomized trial of surgery with or without preoperative chemotherapy in esophageal cancer. *J Clin Oncol.* 2009; **27**:5062-67.
5. Liu T, Liu Q, Zheng ST, el al. MicroRNA-21 Promotes Cell Growth and Migration by Targeting Programmed Cell Death 4 Gene in Kazakh'h Esophageal Squamous Cell Carcinoma. *Dis Markers.* Volume 2014, Article ID 232837, 7 pages.
6. GAO CC, HUI XW. GLCM-Based Texture Feature Extraction. *J COMPUT ANAL APPL*, 2010; **19**:195-198.
7. HAO BY, WANG RL, Jun MA, Bing SU, Jian Hua ZHEN. Image retrieval method based on Tamura texture feature improved. *Science of surveying and mapping.* 2010; **35**:136-138.
8. Stavroula G.M, Ioannis K. V, Alexandra N, et al. Differential diagnosis of CT focal liver lesions using texture features, feature selection and ensemble driven classifiers. *Artificial Intelligence in Medicine,* 2007; **41**:25-37.
9. Mahmoud Ghoneim D. Optimizing automated characterization of liver fibrosis histological images by investigating color spaces at different resolutions. *Theoretical biology and medical model,* 2011; **8**: 25-37.
10. Zhang G, Ma ZM. Texture feature extraction and description using gabor wavelet in content-based medical image retrieval. *Beijing: Wavelet Analysis and Pattern Recognition,* 2007; **5**:169-173.
11. Cong R, Gao GP, Fan R X. Based on gray-gradient co-occurrence matrix and fuzzy kernel clustering vibration pattern recognition method. *Vibration and Shock.* 2012;**31**:73-76. (In Chinese).
12. Zhang H, Fan JL. Gray-gradient co-occurrence matrix model Weighted Entropy Threshold. *Computer Engineering and Applications.* 2010; **46**:10-13. (In Chinese).
13. Murat H, Yang F, Kong DW. Classification of Xinjiang Kazakh esophageal cancer image texture feature. *Journal of Xinjiang Medical University.* 2014; **37**:273-276. (In Chinese).
14. Haralick R M. Statistical and structural approaches to texture. *Proceedings of the IEEE.* 1979; **67**: 786-840.
15. Markus B, Huber, Kerstin B. Texture feature ranking with relevance learning to classify interstitial lung disease patterns. *ArtifIntell Med.* 2012; **56**:91-97.
16. Liu ZQ, Lv X, Zhang L. GA-SVM multi-classification model based on highway AID. *Systems Engineering Theory and Practice.* 2013; **33**: 2110-2115. (In Chinese).

17. Wang QQ, Huang ZW, Liu K. Laser-induced breakdown spectroscopy plastic classification based on PCA and Artificial neural networks. *Spectroscopy and Spectral Analysis,* 2013; **32**:3179-3182. (In Chinese).
18. Li R, Ye S W, Shi Z Z. SVM-KNN Classifier: A new method to improve the accuracy of SVM classification. *Journal of Electronics,* 2013; **30**:745-748. (In Chinese).
19. Li GH, Fan P, Yuan L. Continuous K-Nearest Neighbor processing based on speed and direction of moving objects in a road network. *Telecommun Syst.* 2014; 55:403-419.
20. Ahmet A, Gong D H, Kahveci T. Network-based prediction of cancer under genetic storm. *Computational Advances in Cancer Informatics (B).* 2014; 13:14-31.
21. Katsuyoshi O. Detailed Description of the use of the KNN Method for Breast Cancer Diagnosis. *The 2014 7th International Conference on BioMedical Engineering and Informatics.* China: Da Lian, 2014:606-610.
22. Yin Y, Adel M, Bourennane S. Automatic segmentation and measurement of vasculature in retinal fundus images using probabilistic formulation. *Computational and Mathematical Methods in Medicine.* Volume 2013,Article ID260410, 16pages.
23. Mapayi T, Viriri S, Tapamo TR. Adative thresholding technique for retinal vessel segmentation based on GLCM-energy information. *Computational and Mathematical Methods in Medicine.* Volume 2015, Article ID 597475, 11 pages.

RESEARCH OF COMPATIBILITY LAW OF TRADITIONAL CHINESE MEDICINE IN TREATING RHEUMATOID ARTHRITIS

Chun Liu[1,a], Zirong Li[1,a], and Yonglin Liang[2,b,*]

[1]*Information Engineering Institute, Gansu University of Chinese Medicine, Lan zhou, China*
[2]*College of Basic Medicine, Gansu University of Chinese Medicine, Lan zhou, China*
E-mail:[a]7859118@qq.com, [b]875532437@qq.com
Corresponding author

Objective: Through the retrospective research of the prescriptions applied in rheumatoid arthritis (RA) treatment, approaching medicines that are used frequently determine high frequency medicine, so as to discover the law of compatibility and improve theoretical and clinical levels of Chinese medicine. Methods: Articles about RA treatment prescriptions were collected during the past 30 years. Set up a database and analyzing its law according to the efficacy, meridian tropism, taste, property of medicine. Result: According to the analysis, in 376 recipes for RA, 50 medicines were used in the highest frequency accumulated identified as high frequency in 205 herbs. The rheumatism, tonic, activating blood circulation and removing blood stasis, and medicines for relieving exterior syndrome were identified as high-frequency medicines. The forward 3 meridian tropism is liver meridian, spleen meridian, and kidney meridian. The mainly medical taste is pungent, bitter and sweet, and the most commonly used property of medicine is warm property, flat property, and cold property. Conclusions: The high-frequency medicines provide an important basis for the choice of clinical medication.

1. INTRODUCTION

Rheumatoid arthritis[1-2](RA), an autoimmune condition, is a chronic inflammatory polyarthritis, RA causes premature death, disability, and lowers the quality of life in the industrialized and developing world. RA is a systemic inflammatory disease that manifests itself in multiple joints in the body. This inflammation usually affects the lining of the joints (synovial membrane), but can also affect other organs. This inflamed joint lining leads to erosions of the cartilage and bone and sometimes causes joint deformity. Pain, swelling, and redness are common joint symptoms. RA causes are unknown, but it is believed to result from a faulty immune response. RA can begin at any age and causes fatigue and prolonged stiffness after rest. There is no cure for RA, but effective drugs are increasingly available to treat RA and prevent deformed joints. Traditional Chinese Medicine has its own features and interior laws in medicine using in the treatment of RA. We will make a generalization review on the ancient and present series recipe, which is used to treat RA and made a good clinical effect, exploring the kernel drug of RA, analyzing the combination of medicinal, so as to raise the level of theory of Traditional Chinese Medicine , as well as medical practice.

2. LITERATURE AND METHOD

2.1. Medicines Specification

Traditional Chinese medicine has a long history, there are different names of one Chinese medicine, so the first important thing to do is to unify its name. We standardize all of the medicines that has been used in our study according to "Pharmacopoeia of the People's Republic of China" [3]and "Chinese pharmacy"[4].

2.2. Literature Index and Data Collection

We retrieved some large academic database by computer, including Chinese National Knowledge Infrastructure during the past 30 years (CNKI from January 1980 to December 2011), the Chinese biomedical literature database (CBM from January 1978 to December 2011), and science and technology of China journal full-text database (VIP from January 1989 to December 2011). The standard of valid documents are as follows: (1) It should be a publicly published study about the clinical curative of traditional Chinese medicine treatment of RA with formula. (2) The single traditional Chinese medicine treatment of RA, or cooperation with other western medicine, acupuncture and massage treatment of RA, proprietary Chinese medicine injection in the treatment of RA documents are not included. (3) The same authors and the same formula, merge into one record. (4) There are at least 30 cases in his study, and systematically curative effect evaluation should more than 70%.

2.3. Establish Database

We collect all the validation literature, and input the formulas into computer, establish a database about RA. Based on Visual FoxPro, we design a program to complete the statistical information automatically.

3. RESULTS

3.1. High Frequency Medicine

According to analysis of herbs frequency in 376 recipes for RA, there are 205 herbs in total, the total frequency is 4985 times. There are 50 herbs were identified as high frequency in 205 herbs, its cumulative frequency reached 81.2%(4050). The results are shown in Table 1 according to the frequency from high to low.

Table 1. High frequency medicine of RA

No	Medicine	Frequency of usage	Frequency(%)
1	Liquorice Root	200	53.2
2	Chinese Angelica	189	50.3
3	Cassia Twig	174	46.3
4	Milkvetch Root	156	41.5

5	Doubleteeth Pubescent Angelica Root	147	39.1
6	Common Monkshood Mother Root	124	33.0
7	Largeleaf Gentian Root	123	32.7
8	Debark Peony Root	122	32.4
9	Chinese Clematis Root	119	31.6
10	Sichuan Lovage Rhizome	117	31.1
11	Incised Notopterygium Rhizome and Root	116	30.9
12	Coix Seed	110	29.3
13	Divaricate Saposhnikovia Root	109	29.0
14	Suberect Spatholobus Stem	107	28.5
15	Scorpion	96	25.5
16	Fourstamen Stephania Root	93	24.7
17	Black-Tail Snake	90	23.9
18	Medicinal Cyathula Root	82	21.8
19	Centipede	77	20.5
20	Atractylodes Rhizome	75	19.9
21	White Atractylodes Rhizome	73	19.4
22	Prepared Common Monkshood Daughter Root	73	19.4
23	Ephedra	72	19.1
24	Manchurian Wildginger	71	18.9
25	Common Floweringqince Fruit	71	18.9
26	Orientvine Vine	71	18.9
27	Honeysuckle Stem	71	18.9
28	Red Peony Root	65	17.3
29	Earthworm	61	16.2
30	Common Anemarrhena Rhizome	60	16.0
31	Salvia Root	59	15.7
32	Kusnezoff Monkshood Root	58	15.4
33	Chinese Taxillus Herb	57	15.2
34	Eucommia Bark	54	14.4
35	Amur Cork-Tree	53	14.1
36	Kadsura Pepper Stem	53	14.1
37	Unprocessed Rehmannia Root	52	13.8
38	Safflower	51	13.6
39	Himalayan Teasel Root	51	13.6
40	Frankincense	49	13.0
41	Myrrh	48	12.8
42	Indian Bread	45	12.0
43	Mulberry Twig	44	11.7
44	Prepared Rehmannia Root	43	11.4
45	Ground Beetle	39	10.4
46	Herba Siegesbeckiae	38	10.1
47	Epimedium Herb	38	10.1
48	Pangolin Scales	37	9.8
49	Root Leaf or Flower of Common Threewingnut	37	9.8
50	Gypsum	30	8.0

Note: The frequency in Table 1 is frequency of usage in 376 recipes for RA. Frequency (%)=Frequency of usage/376*100%.

The important characters in treating RA among 50 high-frequency medicines are the usage of animal medicine, poisonous medicine, and cany kind of medicine. In detail, the animal medicines are scorpion, black-tail snake, centipede, earthworm, ground beetle, and pangolin scales. The poisonous medicines are common monkshood mother root, scorpion, centipede, prepared common monkshood daughter root, kusnezoff monkshood root, and root leaf or flower of common threewingnut. The cany kind of medicines are suberect spatholobus stem, orientvine vine, honeysuckle stem, kadsura pepper stem,and pangolin scales.

3.2. Efficacy of Medicine

There are 21 kind of efficacy of medicine according to “Chinese pharmacy”. We add up 205 medicines in treating RA by efficacy, the results are shown in Table 2 according to the frequency of usage from high to low.

The high frequency herbal efficacy and its frequencies are: wind-damp-dispelling medicinal (1345), tonic (1059), blood-activating and stasis-resolving medicinal (703), and exterior-releasing medicinal (599).

Table 2. Efficacy of medicine

Efficacy	Count	Frequency of usage	Frequency(%)	Average Frequency
wind-damp-dispelling medicinal	39	1345	26.98	34.5
tonic	38	1059	21.24	27.9
blood-activating and stasis-resolving medicinal	22	703	14.10	32.0
exterior-releasing medicinal	11	599	12.02	54.5
heat-clearing medicinal	30	481	9.65	16.0
liver-pacifying wind-extinguishing medicinal	8	262	5.26	32.8
damp-draining diuretic	13	199	3.99	15.3
interior-warming medicinal	4	107	2.15	26.8
damp-resolving medicinal	4	89	1.79	22.3
resolving phlegm and relieving asthma medicinal	8	45	0.90	5.6
qi-regulating medicinal	5	28	0.56	5.6
digestant medicinal	5	23	0.46	4.6
medicinal for detoxification	3	22	0.44	7.3
hemostatic medicinal	6	13	0.26	2.2
tranquilizing medicinal	3	3	0.06	1.0
resuscitative stimulant	3	3	0.06	1.0
astringent medicinal	1	2	0.04	2.0
purgative medicinal	1	1	0.02	1.0
vermifugal medicinal	1	1	0.02	1.0

Note: Count is the medicine number of the same efficacy. The frequency of usage in Table 2 is the sum of frequency in the same efficacy medicine. Frequency (%)=Frequency of usage/4985*100%. Average frequency= Frequency of usage/Count, The average frequency reflect the concentration of medicinal use.

3.3. Meridian of Medicine

The meridian of medicine belongs to the 12 viscera, the same medicine may belongs to more than one meridian. We add up 205 medicines in treating RA by meridian, the results are shown in Table 3 according to the frequency of usage from high to low.

The high frequency medicinal meridian tropisms and its frequencies are: liver meridian tropism (2958), spleen meridian tropism (1979), and kidney meridian tropism (1615).

Table 3. Meridian of medicine

Meridian	Count	Frequency of usage	Frequency(%)	Average Frequency
liver	127	2958	24.50	23.3
spleen	73	1979	16.39	27.1
kidney	70	1615	13.38	23.1
heart	48	1420	11.76	29.6
lung	58	1336	11.07	23.0
stomach	61	1116	9.24	18.3
bladder	20	980	8.12	49.0
gallbladder	12	276	2.29	23.0
pericardium	6	181	1.50	30.2
large intestine	21	147	1.22	7.0
small intestine	7	44	0.36	6.3
triple energizer	3	22	0.18	7.3

3.4. Flavor of Medicine

Since each kind of medicine may has a variety of flavors. We add up 205 medicines in treating RA by flavor. The results are shown in Table 4 according to the frequency of usage from high to low.

The high frequency medicine flavors and its frequencies are: pungent taste (2733), bitter taste (2442), and sweet taste (2191).

Table 4. Flavor of medicine

Meridian	Count	Frequency of usage	Frequency(%)	Average Frequency
pungent	92	2733	33.27	29.7
bitter	106	2442	29.73	23.0
sweat	89	2191	26.67	24.6
salty	15	334	4.07	22.3
sour	10	303	3.69	30.3
bland	7	188	2.29	26.9
Astringency	5	24	0.29	4.8

3.5. Property of Medicine

Medicinal properties are divided into cold, hot, warm, cool, and normal. We add up 205 medicines in treating RA by property, the results are shown in Table 5 according to the frequency of usage from high to low.

The high frequency herbal properties and its frequencies are: warm property (2405), normal property (1088), and cold property (1046).

Table 5. Property of medicine

Property	Count	Frequency of usage	Frequency(%)	Average Frequency
warm	81	2405	48.24	29.7
normal	46	1088	21.83	23.7
cold	59	1046	20.98	17.7
hot	6	292	5.86	48.7
cool	13	154	3.09	11.8

4. CONCLUSIONS

4.1. High Frequency Medicine

The usage of animal medicine, poisonous medicine, and cany kind of medicine are the important characters in treating RA among 50 high-frequency medicines. The animal medicine has an ability to get rid of the wind from marrow. The poisonous medicine using the principle of "like cures like" toxic medicine to release symptoms. The cany kind of medicine is good for body lesions based on the theory of traditional Chinese medicine "take like than class". So all of them are the representative on treatment of RA.

4.2. Characters of Medicine

Through the statistical analysis of treatment of RA medicine, the wind-damp-dispelling medicinal, tonic, blood-activating and stasis-resolving medicinal, and exterior-releasing medicinal were identified as high frequency medicines. The forward 3 meridian tropism is liver, spleen, and kidney. The mainly herbal flavor is pungent, bitter and sweet, and the most commonly used herbal property is warm herbs. The theory of the etiology and pathogenesis of traditional Chinese medicine science is highly consistent.

The result indicates the coincidence with relevant theories of the traditional Chinese medical science. The high frequency medicines are the representative on treatment of RA, and also give us an important reference for clinical prescription.

Acknowledgments

Supported by Gansu province scientific and technological research of traditional Chinese medicine, (Grant No. GZK-2012-73).

References

1. Xi-Liang KAO. Exploration on recurrent mechanism and clinical syndrome differentiation and treatment of rheumatoid arthritis based on the theory of latent pathogenic factors causing Bi. *China Journal of Traditional Chinese Medicine and Pharmacy*, 2011;**26**(5):1157-1160.
2. Jian-Mei CHEN Jian-mei. Professor JIN Shi's experience in treating active rheumatic arthritis by Guizhi Shaoyao Zhimu Decoction. *China Journal of Traditional Chinese Medicine and Pharmacy*, 2012;**27**(12):3138-3140.
3. The State Pharmacopoeia Committee of PRC. Pharmacopoeia of the People's Republic of China. Beijing: China Medical Science Press, 2010.
4. Xue-Min Gao Xue-min. Traditional Chinese pharmacology. Beijing: Chinese Medicine Press of China, 2007.

GLYCYRRHETINIC ACID INHIBITING HUMAN ESOPHAGEAL CARCINOM ECA9706 CELLS GROWTH THROUGH AFFECTING ON CELL CYCLE AND CELL APOPTOSIS

Sugai Yin[a], Lingli Song, Yulong Chen*, Huihui Wang, Yaosong Wu, and Shanshan Ren

Molecular Biological Experiment Center of Henan University of Traditional Chinese Medicine, Henan, zhengzhou,450046, China

Email: [a]yinsugai@163.com

**Corresponding author*

To Study on the effect and mechanism of glycyrrhetinic acid on growth of human esophageal carcinoma cell. Methods: Inhibitive effect of glycyrrhetinic acid on proliferation of Esophageal Carcinom Eca9706 cells was detected by MTT assay, and cell apoptosis was detected by Annexin-V-FITC/PI assay. Results: In the range of 95-200 μ g/mL, glycyrrhetinic acid with various concentrations effectively inhibited the proliferation of Eca9706 cells in dose dependent manner ($P<0.05$). Compared with the control group, glycyrrhetinic acid in a concentration of 133μg/ml significantly increased the apoptosis rate of Eca9706 cells in 48h ($P <0.01$). Conclusion: glycyrrhetinic acid could significantly inhibit the proliferation of Eca9706 cells through impacting on cell cycle and cell apoptosis.

1. INTRODUCTION

Esophageal cancer is one of the most common digestive tract malignant tumor. Each year about 300000 people die of the disease, the mortality rate of fourth in overall cancer mortality[1]. Chinese medicine has unique advantages in treatment of esophageal cancer and a certain curative effect. Licorice is widely used as a medicinal plant in China since ancient times, its main active components are glycyrrhetic Glycyrrhizic (acid) and acid (licorice acid). In the human body, the hydrolysis of the normal bacteria is decomposed into the acid of the licorice, the latter is a pentacyclic triterpenes,.And then it play a powerful pharmacological effects in the human body[2]. In recent years, research on glycyrrhetinic acid unceasingly found glycyrrhetinic acid, has therapeutic effects,such as human tumor such as digestion tract tumor, gynecological cancer, skin cancer, melanoma tumor had significant inhibition, besides traditional antitussive, antiasthmatic and anti-inflammatory, anti ulcer, anti allergic, antiviral effect [3]. However, the report about the effect of the acid on the growth of esophageal cancer cells is less, and the mechanism is not clear. This study was to observe the inhibitory effect of the acid on the growth of esophageal cancer cell line Eca9706, and to explore the possible mechanism of this study, which provided experimental evidence for clinical treatment of esophageal cancer.

2. MATERIALS AND METHODS

2.1. Materials and Reagents

Human Esophageal Carcinoma Cell Line Eca9706 was kept in Henan University of Traditional Chinese Medicine test center preservation; Glycyrrhetinic acid (Chengdu mansite Biotechnology Co. Ltd., batch number: MUST-13071705), molecular formula:C30H46O4, molecular weight:470.69, 20mg of glycyrrhetinic acid dissolved in 1ml DMSO, with 20mg/ml storage concentration; MTT and pancreatin (cable based company, batch number RPMI1640: T1320-100); medium (cable based company, batch number: 31800-500); fetal bovine serum (Hangzhou, Sijiqing, No. 121005); Annexin V-FITC /PI Apoptosis Detection Kits (KGA512 keygen, batch number: KGA107 apoptosis kit kit, cycle number).

2.2. Methods

2.2.1. *Cell Culture*

RPMI1640 medium containing 10% calf serum, 100 U/mL at 37 °C in a humidified, 5% CO2 atmosphere. The cells were characterized as SMCs by morphology and immunostaining with monoclonal antibody specific for α-SMA. the cells were cultured for 1 times per 2-3D, and Human Esophageal Carcinoma Cell Line Eca9706 in logarithmic phase were cultured in vitro and used in the study

2.2.2. *Detectingrates of Proliferative Inhibition of Glycyrrhetinic to Cells by MTT*

Eca9706 cells were cultured in 96 hole plate, 104 cells / 100 g l per hole.After 24 hours, the culture liquid was discarded. Culture medium was added with different concentrations of glycyrrhetinic acid (5μg/mL, 114μg/mL,166μg/mL, and 200μg/mL). While control group (without glycyrrhetinic acid) and blank group (with DMSO) were taken for comparison. Each group was established 8 holes and cultured in 4 times repeatedly.After 48 h culture, each hole was added with DMSO (150 mu L). After the next 4 hours, each hole was added DMSO 150 mu L again when the liquid discarded. Each hole was added DMSO 150 mu L again. The colorimetry of samples was treated with enzymatic-reader at 570/630n. Cell inhibition rate = (control group OD value of experimental group control group od / OD) * 100%.

2.2.3. *Flow Cytometry Analysis*

Effects of Eca9706 cell apoptosis induced by the acid of licorice was observed. Eca9706 cells were inoculated in the culture dish at the density of 105 /ml per hole. After 24h, RPMI 1640 culture media was added with different dose of glycyrrhetinic acid. The medium without glycyrrhetinic acid and with DMSO were taken as control for common treatment, Each with four replications. The cells were harvested by 0.25% trypsin without EDTA after 48h post-transfection and evaluated by FACS Calibur Flow Cytometrey

2.2.4. *Laser Confocal Microscope*

Effects of Eca9706 cell apoptosis induced by the acid of licorice was observed.Cells were seeded into per laser confocal culture dish with 105 /ml, RPMI 1640 inoculum was incubated with glycyrrhetinic acid of different doses for 48 hours. After being washed two times with PBS buffer, the cell cultures were added with 500 mu 1 of buffer and 5 UL of annexin V-FITC, avoided light for 10 minutes, add PI obstacle light 5 minutes. the cell cultures was dectected by using the laser scanning Confocal microscopy.

2.3. Statistical Analysis

SPSS 17.0 statistical software (SPSS, Inc., Chicago, IL, USA) was used for analysis. The data are presented as the mean ± standard deviation. Repeated measures analysis of variance (ANOVA) was used to evaluate.

3. RESULTS

3.1. Cell Proliferation Inhibition of of Glycyrrhetinic Acid on Eca9706

Glycyrrhetinic acid with different concentration to Eca9706 cells have different degree of inhibition in vitro, in dose dependent, variance analysis, the difference was statistically significant ($P < 0.05$, $n = 4$). Regression analysis to R2 = 0.942, Y = 0.942 x - 9.478, according to the probability method of IC50 = 108.539(Table 1, Fig. 1).

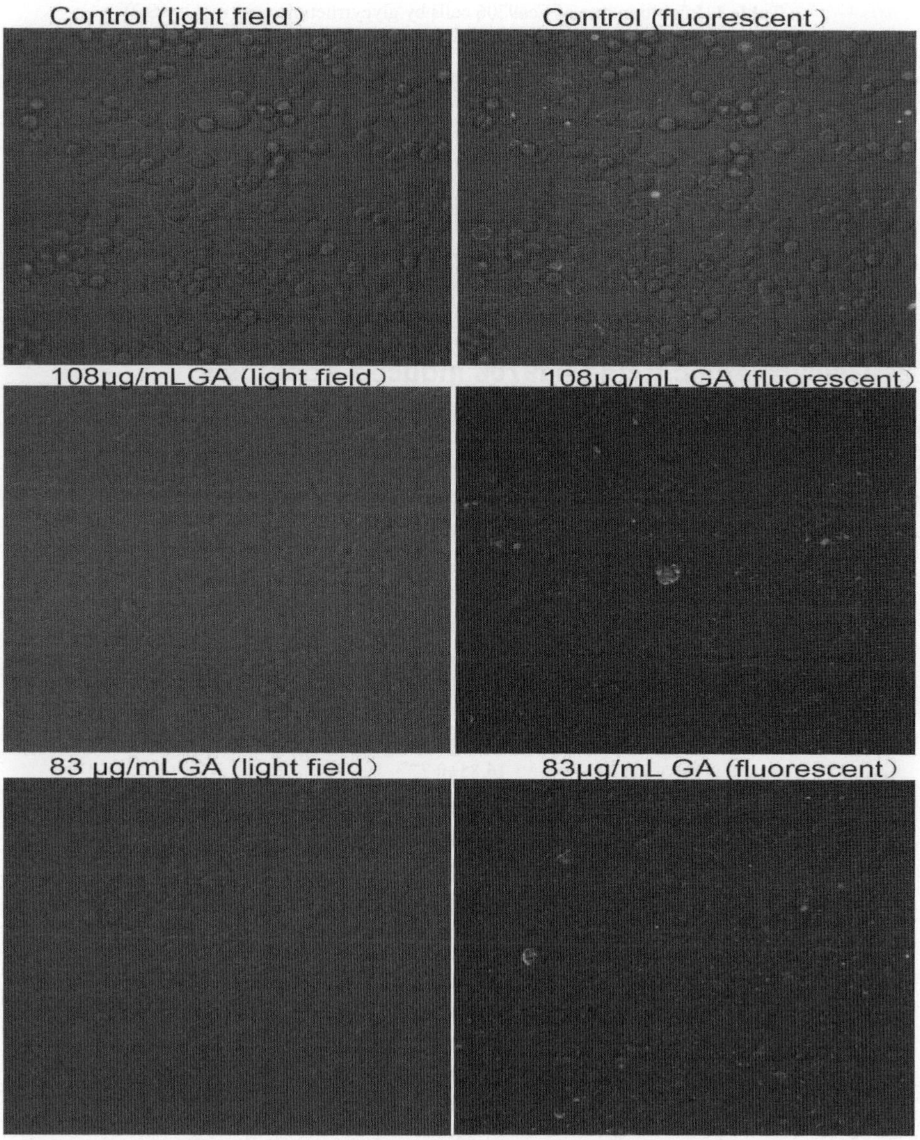

Fig. 1. Cell morphology changes of Eca970 treated with different concentration of GA.

Table 1. Inhibition rate of Eca9706 cells by glycyrrhetinic acid(n=4, $\bar{x}$±SD)

(μg/mL)	OD	suppression ratio(%)
200	0.070±0.009**	95.87
166	0.137±0.054*	58.19
138	0.639±0.128**	20.08
114	0.941±0.021*	18.99
95	1.151±0.037*	8.36
Control	1.984±0.119	----

Note: *P<0.05, **P<0.01, n=4

3.2. Glycyrrhetinic Acid Eca9706 induced Apoptosis

With the cycle, also with glycyrrhetinic acid of high, medium and low three concentration, the inhibition rates were: 30%, 50%, 70%, eca9706 cells after 48 h, the treatment group and control group, apoptosis increased significantly, The difference was statistically significant (P <0.001, n = 4 is used) (Table 2 , Fig. 2).

Table 2. Apoptotic rate of Eca9706 cells after treatment(n=4, $\bar{x}$±SD,%)

(μg/mL)	apoptosis rate
Control	1.77±0.127
High(133μg/mL)	16.85±0.77*
Medium(108μg/mL)	11.99±2.80*
Low(83μg/mL)	5.38±1.32*

*Note:**P<0.001, n=4.

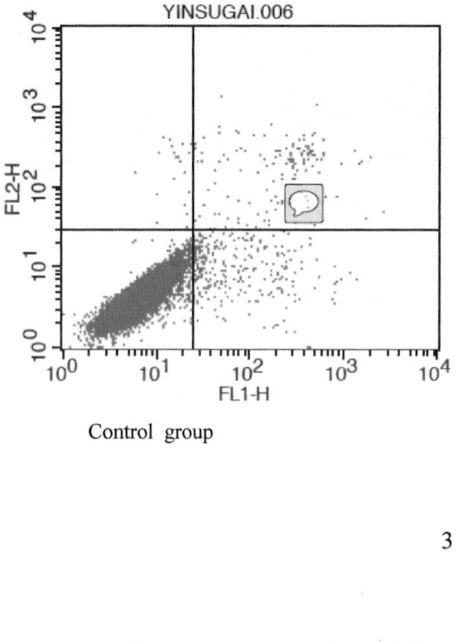

Control group

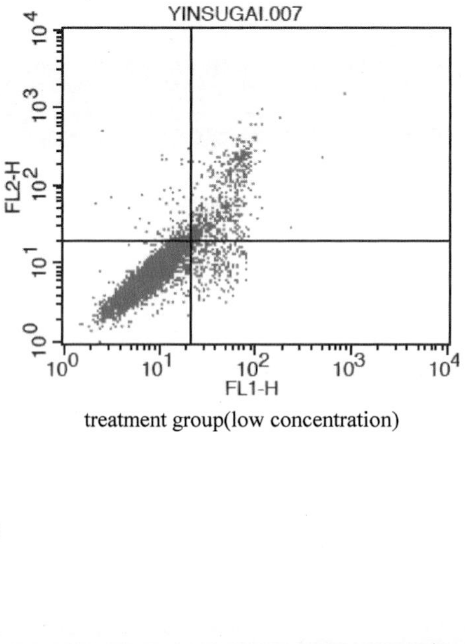

treatment group(low concentration)

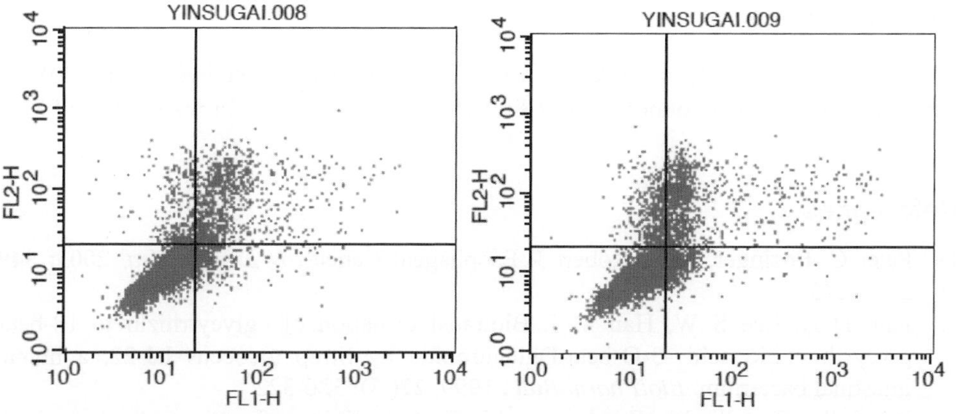

Treatment group(middle concentration) Treatment group(high concentration)

Fig. 2. Apoptosis of Eca9706 cells after treatment.

3.3. Effects of Apoptosis of Eca9706 induced by GA

This study demonstrated that Glycyrrhetinic acid induced Eca9706 cell apoptosis. After the cells were treated with 83, 108μg/mL GA for 24 h, cell apoptosis was detected by confocal laser scanning microscopy. Fig. 1 showed that The volume of cells were smaller than normal, the cell deformability occurred, nucleus was broken in to fraction, and apoptotic body was in the cells with confocal laser scanning microscopy. The apoptosis rate of Eca9706cells was increased with drug concentration.

4. DISCUSSION

In recent years, researchers have sought to find new natural drugs that inhibit the occurrence and metastasis of tumors. [4]. The previous reported revealed that GA has anti-tumor effects to multiple Carcinoma Cells. Malagol et al [5] report that GA has the effect of proliferation inhibition in human HSB-2 acute lymphoblastic leukemia cells.

MTT assay was wildly used for screening of Antitumor Drug and lymphocytotoxicity test. The principle is that the live cells can induce MTT reducible to blue-violet crystallizations which is disposed in the cells. These disposition enable the number of living cells[6].This study revealed that GA can inhibit the esophageal cancer cell proliferation.. This effect was enhanced as concentration and time increased.

Imbalance in the regulation of tumor cells is the main problem,Tumor cells escape normal apoptosis by various mechanism[7,8]. To further investigate that mechanism of GA to cell inhibition. The results of flow cytometry demonstrated that GA had an effect of induced cell apoptosis. This effect was enhanced as concentration and time increased. Its formation mechanism needs further investigation.

Acknowledgments

This article supported by Key research projects of Henan university(15A310020) and national natural science foundation of China (81373535); Thanks for the support of Song Lingli.

References

1. Peter C. Enzinger, M.D. Robert J. Esophageal Cancer. *N Engl J Med,* 2003; **349**: 2241-2252.
2. Kim D H, Lee S W, Han M J. Biotransf ormation of glycyrrhizin to 18-beta-glycyrrh etinic acid -3-O-beta-Dglucuronide by Strep tococcus LJ-22, a human intestinal bacterium. *BiolPharm Bull*, 1999, **22**(3): 320-322.
3. Rossi T, Castelli M, Zandomeneghi G. Selectivity of action of glycyrrhizin derivatives on the growth of MCF-7 and HEP-2 cells. *Anticance Res*, 2003, **23**(5A): 3813-3818.
4. EI-Bayoumy K, Sinha R. Mechanisms of mammary cancer chemoprevention by organoselenium compounds. *Mutant Res*, 2004; **551**(1-2): 181-197.
5. Malagoli M, Castelli M , Baggio A. Effect of glycyrrhizin and its diastereoisomers on the growth of human tumor cells: pre2 liminary finding. *Phytother Res,* 1998;12(2): 952-957.
6. Mosan T. Rapid colorimetric assay for cellular growth and survival: application to proliferation and cytotoxicity assay. *J Immunol Methods*, 1983; **65**(1-2): 55-63.
7. Dibb M, Ang YS. Targeting the cell cycle in esophageal adenocarcinoma: an adjunct to anticancer treatment. *World J Gastroenterol*, 2011; **17**(16): 2063-2069.
8. Hanahan D, Weinberg RA. Hallmarks of cancer: the next generation. *Cell*, 2011; **144**(5): 646-674.

CONTROLLED-RELEASE TABLETS OF DOXAZOSIN MESYLATE AND PHENOXYBENZAMINE IN PREOPERATIVE PREPARATION AND BLOOD PRESSURE CONTROL FOR PHEOCHROMOCYTOMA PATIENTS

Bin Liu[1,a], Xinquan Gu[1,b], and Xia Cao[2,c,*]

[1]*Department of Urology, China-Japan Union Hospital Affiliated to Jilin University, Changchun Jilin, 130033, China*

[2]*School of Pharmaceutical Sciences, Jilin University, Changchun Jilin, 130021, China*

E-mail: [a]120650294@qq.com, [b]guxinq1213@sohu.com, [c]caoxia@jlu.edu.cn

[*]*Corresponding author*

Objective: To compare the effects of Doxazosin Mesylate controlled release (CR) tablets and Phenoxybenzamine in the preoperative management and blood pressure control for pheochromocytoma patients. Methods: Totally 40 patients with left pheochromacytoma were operated on in China-Japan Union Hospital from 1978 to 2006. There were 10 patients in group 1, 15 in group 2 and 15 in group 3. The mean age of patients were (45.1± 24.5) years old in group 1, (43.2 ± 30.1) years old in group 2 and (44.7 ± 23.4) in group 3. The perioperative systolic blood pressure was analyzed retrospectively in these 40 patients. The patients were divided into the group 1 (non-α receptor blocker), group 2 (Phenoxybenzamine) and group 3 (Doxazosin Mesylate CR) according to the preoperative medicine. The systolic blood pressure at five key time points were recorded as SBP1 (before induction), SBP2 (after anesthesia), SBP3 (during tumor exploration), SBP4 (tumor resected) and SBP5 (after operation). Results: Before anesthesia, the diagnostic pressure of the three groups were significantly different (P<0. 01), with the highest level found in group 2, followed by group 1 and group 3. After anesthesia, the diagnostic pressure were decreased in all the three groups, with the lowest level in group 2, followed by group 1 and group 3. During tumor exploration, the pressure was increased in all the three groups, with the lowest level in group 2, followed by group 1 and group 3; there was significant difference between the three groups, but there was no significant difference between group 2 and group 1. After tumor resection, the diagnostic pressure in the three groups was decreased, with no significant difference found in group 2 and group 1, with significant difference between group 1, 2 with group 3 (P<0.01). At the end of operation, the pressures were similar between the three groups. The blood pressure in group 3 was more stable and that in group 2 was not so stable. Conclusion For the patients with hypervascular left adrenal pheochromocytoma, Doxazosin Mesylate CR is better in controlling of systolic blood pressure during the preoperative period, and Phenoxybenzamine is more suitable for intra-operation procedures.

1. INTRODUCTION

Currently, surgery is still the only way to cure pheochromocytoma, but blood pressure fluctuations during the surgery can be the potential fatal risk for patients, so how to maintain the patient's blood pressure stability during surgery is a key issue. Before the surgery, antihypertensive drugs should be ready to control blood pressure, and reduce blood pressure fluctuations during surgery. The First International Pheochromocytoma Expert Forum proposed that, any patients with positive pheochromocytoma during biochemical test should be prepared with appropriate drugs before surgery[1]. α blockers can inhibit the role of catecholamine hormones on the heart and blood vessels in patients, thus lowering blood pressure. Nowadays, Phenoxybenzamine and Doxazosin Mesylate Controlled Release Tablets are most commonly used. In this study, the two drugs will be applied to the preoperative preparation of pheochromocytoma, and then their effects on blood pressure fluctuations during perioperative period will be compared.

2. SUBJECTS AND METHODS

2.1. Study Subjects

216 cases of patients who were diagnosed with pheochromocytoma at China-Japan Union Hospital, Jilin University from 1978 to 2006, including 40 cases of left adrenal pheochromocytoma, the surgical incision is left rib incision retroperitoneal route, and the general anesthetic induced and inhaled by vein is applied. It is divided into 3 group based on the preoperative drug preparation: group 1, 10 patients, with the average age of(45.1 ± 24.5), α blocker was not used before surgery; group 2, 15 patients, with the average age of (3.2 ± 30.1), phenoxybenzamine 20~40 mg / d, took orally at 2 to 3 times, the average taking days was (25.1 ± 24) d; group 3, 15 patients, with the average age of (44.7 ± 23.4), Doxazosin Mesylate Controlled Release Tablets 4~8 mg / d, took at one time, clothing, the average taking days was (10.6 ± 3.9) d. The age difference between the three groups was not statistically significant ($P > 0.05$).

2.2. Methods

Monitor the blood pressure of radial artery, and record the systolic pressure of five time points, namely, before the operative anesthesia induction, after anesthesia, exploring the tumor, after tumor resection and end of surgery. ASBP is adopted to indicate the ratio between the highest and the lowest systolic blood pressure of patients during surgery.

SPSS10.0 statistical software is applied for statistical processing applications. Measurement data is indicated by $X \pm S$, groups comparison was tested $\chi 2$.

3. RESULTS

Before anesthesia, systolic pressure difference between the 3 groups was statistically significant ($P < 0.01$), group 2 was the lowest, followed by group 3, group 1 was the highest. After anesthesia, the systolic pressure of patients in 3 groups was decreased, group 2 was still the lowest, followed by group 1, group 3 was the highest, the difference among the 3 groups was statistically significant ($P < 0.01$). When exploring the tumor, the systolic pressure of patients in 3 groups was elevated, group 2 was still the lowest, followed by group 3, group 1 was the highest, the difference among the 3 groups was statistically significant ($P < 0.01$), but the difference between group 2 and group 3 was not statistically significant ($P > 0.05$). After tumor resection, the systolic pressure of patients in 3 groups was decreased, the difference between group 1 and group 2 was not statistically significant ($P > 0.05$), but the differences between group 3 and group 1, group 2 were statistically significant ($P < 0.01$). At the end of surgery, systolic blood pressure difference in 3 groups was not statistically significant ($P > 0.05$). Please see Table 1 for details. ΔSBP in group 2 and 3 was significantly less than group 1 ($P < 0.01$), while group 3 was significantly less than group 2 ($P < 0.01$).

Table 1. Systolic Pressure Changes in Surgery. ($X \pm S$, p/mmHg)

Group	1 (n = 10)	2 (n = 15)	3 (n = 15)
Pre-anesthesia	156.3 ± 40.5	134.2 ± 23.9	149.1 ± 16.3
Post-anesthesia	91.2 ± 12.4	84.1 ± 16.2	110.2 ± 16.5
Probe the tumor	194.5 ± 12.5	155.2 ± 12.3	163.0 ± 17.9
Tumor resected	71.2 ± 8.4	70.1 ± 4.2	94.2 ± 11.6
End of surgery	112.2 ± 14.5	110.9 ± 13.1	110.2 ± 13.1
ΔSBP	123.3 ± 10.2	85.1 ± 8.3	68.8 ± 13.6

4. DISCUSSION

The scope of this study is limited to patients with left adrenal pheochromocytoma, mainly because the left adrenal has few neighbouring organs, and away from the portal vein, internal celiac artery, superior mesenteric artery and other major blood vessels in enterocoelia, its surgery difficulty is relatively stable, and complications are relatively few. Intraoperative blood pressure fluctuations caused by the surgery and anesthesia is easy to control, facilitate to analysis and observation. On the other hand, the right adrenal is located near the liver, portal vein and inferior vena cava, so the surgery is difficult.

During the process of pheochromocytoma treatment, we have to not only pay attention to the tumor size, site and the local neighbor relations, but also need to pay attention to endocrine hormone changes that caused by the tumor, and its impact on the body's metabolism and vital signs. Pheochromocytoma secretes catecholamine hormones, whose main role is to shrink blood vessels and lead to high blood pressure. Patients with systemic vasoconstriction can make patients' capillary bed in the closed state, and the level of systemic intravascular volume decreased[2-4]. Although patients with pheochromocytoma may have elevated basis blood pressure, their anti-shock ability may be reducted. At present, the method of preoperative rehydration is recommended by most of the clinical centers to improve the patients' anti-shock capability during the operation[1].

Since the blood pressure in patients with pheochromocytoma is fluctuating, some patients' blood pressure is not particularly high, even within the normal range under normal conditions. Therefore, it is not compulsory to control the patient's blood pressure to be in the normal range for preoperative preparation, which makes the processing difficulty of clinicians to increase. Normally, the effect of anesthetic drugs on general patients' blood pressure is not strong, but patients with pheochromocytoma are precisely the exception, as the anesthetic drugs can reduce the response of blood vessels to catecholamine hormones. The results of this study also showed that systolic pressure of 3 group patients after anesthesia is decreased. The main goal for blood pressure monitor during surgery is to stabilize the blood pressure.

Both phenoxybenzamine and doxazosin mesylate are α blockers, the difference is that phenoxybenzamine is a complete non-competitive inhibitor, that can block the binding between catecholamine hormones and α receptor, thus easing the symptoms of pheochromocytoma and lower blood pressure. On the other hand, doxazosin mesylate is a competitive inhibitor, that competes with catecholamine hormones for binding with α receptor[4-6].

The results of this study showed that, the role of phenoxybenzamine is better than doxazosin mesylate in regards of preoperative blood pressure control. The reason is that phenoxybenzamine is noncompetitive, and can block the binding of catecholamine hormones with α receptor, so its effect of blood pressure is better, and consistent with the majority of clinical researches[4-8]. After induction of anesthesia, the body's response to catecholamine hormones declines, the systolic pressure in 3 groups all has a decrease, group 2 has the lowest, followed by group 1. Compared with patients taking phenoxybenzamine, patients who taking doxazosin mesylate controlled release tablets have smaller range of blood pressure decrease; while exploring the tumor, a momentary release of catecholamine hormones may cause severe rise in blood pressure and even lead to severe arrhythmias. In this case, the effect of phenoxybenzamine in controlling blood pressure is better than doxazosin mesylate. After tumor resection, as the level of catecholamine hormones in body was sudden dropped, at the same time, patient's anti-shock capacity was decreased, which will lead to a sharp drop in blood pressure or even requires the use of norepinephrine to increase blood pressure; at this time, the fluctuations in blood pressure in patients used phenoxybenzamine was significantly greater than patients used doxazosin mesylate controlled release tablets. At the end of the surgery, the blood pressure in 3 groups was tended to be uniform.

The results of this study showed that the patients used doxazosin mesylate controlled release tablets for preoperative preparation tended to have more stable blood pressure during the surgical surgery, while the effect of phenoxybenzamine was weaker; patients who didn't have preoperative preparation had the most volatile of blood pressure, and the greatest risk of surgery. In addition, it only takes one week of taking doxazosin mesylate controlled release tablets for preparation, which has greatly reduced the patients exposure time to high concentration of catecholamine hormones, compared with the one month preparation time of Phenoxybenzamine; besides, it can also avoid the occurrence of severe acute vascular complications.

This study can be regarded as a retrospective analysis, some problems can't be avoided, such as the surgeons are not fixed, randomized controlled trial can't be set either. Further studies can design prospective randomized controlled trail with multi-center, so as to provide deeper basis for the perioperative management of pheochromocytoma.

References

1. Pacak K, Eisenhofer G, Ahlman H. Pheochromocytoma: recommendations for clinical practice from the First International Symposium. Nature Clinical Practice Endocrinology & Metabolism 2007; **3**: 92-102.
2. Kinney M A, Narr B J, Warner M A. Perioperative management of pheochromocytoma. J Cardiothorac Vasc Anesth 2002; **16**: 359-369.
3. Prys-Roberts C. Phaeochromocytoma-recent progress in its management. British Journal of Anaesthesia 2000; **85**: 44-57.
4. Prys-Roberts C, Farndon J R. Efficacy and safety of doxazosin for perioperative managementof patients with pheochromocytoma. World J Surg 2002; **26**: 1037-1042.
5. Zhu Yu, Wu Yuxuan, Zhang Chongyu, The Application of Doxazosin Mesylate Controlled Release Tablets in Preoperative Preparation of pheochromocytoma (attached with 25 cases). J Chinese Physician 2006; **8**: 279-280.
6. Young W F Jr. Pheochromocytoma: issues in diagnosis & treatment. Compr Ther, 1997, **23**: 319-326.
7. van der Horst-Schrivers A N, Kerstens M N, Wollfenbuttel B H. Preoperative pharmacological management of phaeochromocytoma. Neth J Med 2006; **64**: 290-295.
8. Brauckhoff M, Gimm O, Dralle H. Preoperative and surgical therapy in sporadic and familial pheochromocytoma. Front Horm Res, 2004, **31**: 121-144.

THE SIGNIFICANCE OF RE-TURBT IN THE TREATMENT OF NON-MUSCLE INVASIVE BLADDER CANCER

Yuhan Zhang[1,a], Xinquan Gu[1], and Xia Cao[2,b,*]

[1]*Department of Urology, China-Japan Union Hospital Affiliated to Jilin University, Changchun Jilin, 130033, China*

[2]*School of Pharmaceutical Sciences, Jilin University, Changchun Jilin, 130021, China*

E-mail: [a]837647704@qq.com, [b]caoxia@jlu.edu.cn

[*]Corresponding author

Objective: To explore the significance of repeated transurethral resection in the treatment of non-muscle invasive bladder cancer(NMIBC). Methods: Selected 90 cases those who was diagnosed with non-muscle invasive bladder cancer with stage Ta(n=30),T1(n=60) after transurethral resection of bladder tumor (TURBT) in China-Japan Union Hospital from 2013 to 2015. All the patients accepted Re-TURBT within 6 weeks after the first time resection and compared the pathologic results of the tumor after the two operation.Results: Pathological findings of Re-TURBT showed that 38(42.2%) patients were found no tumor tissue in the second operation, 36 (40%) patients were found the residual lesions of tumor tissue while 11(12.2%) patients were found missed lesions , the other 5(5.6%) patients were found both residual lesions and missed lesions. Of the 30 patients of the Ta stage at the first TURBT, 18(60%) patients were found no tumor tissue at Re-TURBT, 8(26.7%) patients remained in stage Ta, 3(10%) patients progressed to stage T1 ,1(3.3%) patients progressed to stage T2. Of the 60 T1 patients at the first TURBT, 20(33.3%) patients were found no tumor tissue at Re-TURBT, 9(15%) patients declined to stage Ta, 15(25%) patients remained in stage T1 , 16(26.7%) patients progressed to stage T2. Conclusion: Re-TURBT has the important significance to radical excision and clear pathological stage. And Re-TURBT should be given to the patients those who got tumor in stage T1 at first TURBT, high grade, tumor diameter>3cm, broadly-connected appearance.

1. INTRODUCTION

The incidence of bladder cancer accounted for the first bit ofTumor of urinary system [1],and about 90% are the transitional cell carcinoma[2], 75%-85% [3]were non muscle-invasive bladder cancer (NMIBC). Transurethral resection of bladder tumor (TURBT) is the preferred method for the treatment of non muscle-invasive bladder cancer (NMIBC), but the residual rate is the problem[4].And the new technology which is called Re-TURBT is supposed to solve the problem.Parents acceptd this kind of operation in 4-6 weeks after the first TURBT.This study analyzed the 90 TURBT patients who underwent Re-TURBT between 2013-1 and 2015-1 in China Japan Union Hospital and explored the significance of Re-TURBT for NMIBC, reported as follows.

2. DATA AND METHODS

2.1. Clinical Data

90 bladder tumor patients who underwent Re-TURBT between 2013-1 and 2015-1 at urology of China Japan Union Hospital were included in this research, including 62 males, 28 females; age (58.8±11.6)years. All the patients were diagnosed as NMIBC(Ta,T1) in the first TURBT, 3 of carcinoma in situ(Tis) ,underwent Re-TURBT within 6 weeks after operation. All the patients accepted systemic check-up, laboratory examination, urinary ultrasound, IVP, pelvic CT before operation.

2.2. Methods

The operations had done by the same urologist who was a skilled professor. All the patients took general anesthesia. STORZ resectoscope cutting system was uesd. The 0.9%NaCl as bladder perfusion fluid was about 200ml. All the tumors were cut to the deep muscularis. The tumor tissue, basal part of tumor, muscular layer of bladder wall, edge of tumor tissue were separated sampling pathologic examination. Indwelled catheter and continuous bladder irrigation after TURBT, then did bladder perfusion with pirarubicin 50mg 2hours later. Re-TURBT within 6 weeks with the same methods.

2.3. Observation Index

Record tumor shape, size, site, number, etc during the operation by assistant and histopathological diagnosis report of those 2 TURBT.

2.4. Diagnostic Criteria

The stage of bladder tumor refer to TNM staging criteria of UICC[4], Tumor pathologic stage refer to International classification of tumor histologic grading standards of WHO(the 2004 version)[5].

2.5. Statistical Methods

Using SPSS22.0 statistical software for statistical processing, chi-square test for count data. With $P < 0.05$ for the difference was statistically significant.

3. RESULTS

The pathological of Re-TURBT Pathological findings of Re-TURBT showed that 38(42.2%) patients were found no tumor tissue at the second operation, 36 (40%) patients were found the tumor tissue in primary tumors while 11(12.2%) patients were found tumor tissue in other locations far from the original site, the other 5(5.6%) patients were found both residual lesions and missed lesions.

Secondary postoperative pathologic staging compared with the first Of the 30 patients of the Ta stage at the first TURBT, 18(60%) patients were found no tumor tissue at Re-TURBT, 8(26.7%) patients remained in stage Ta , 3(10%) patients progressed to stage T1 ,1(3.3%) patients progressed to stage T2. Of the 60 T1 patients at the first TURBT, 20(33.3%) patients were found no tumor tissue at Re-TURBT, 9(15%) patients declined to stage Ta, 15(25%) patients remained in stage T1, 16(26.7%) patients progressed to stage T2.

Single factor analysis of pathologic staging progressed to T2 of Re-TURBT stage T1, high grade, tumor diameter>3cm, are related to the pathological staging progressed to T2 of Re-TURBT.(P<0.05, Table 1).

Table 1. Single factor analysis which affect the advance to stage 2 in the Re-TURBT.

Group	Total number	T2 number	P
stage			
Ta	30	1	
T1	60	16	<0.05
Grade			
High grade	32	15	
Low grade	58	2	<0.05
Size			
>3cm	46	13	
<3cm	44	4	<0.05

4. DISCUSSION

One of the important factors that affect prognosis of NMIBC is early recurrence (within 3 months after). However, many of the early recurrence are residual or omission of tumor lesions rather than a true sense of the recurrence, according to reports in the literature, the tumor residual rate is 20%-78% after the first TURBT and 1.4%-64% patients' tumor staging are undervalued[6-9]. 52(57.8%) cases of our group which contains 90 Re-TURBT patients were found residual tumor while 20(22.2%) cases were found that tumor staging were undervalued, similar to the literature results. In order to decrease the residual rate, some scholars proposed application of fluorescence endoscope to better display lesions, but there are some disadvantages, such as clinical practice is difficult to promotion, high rate of false positives, not help solve the tumor staging undervalued etc, and the effect on tumor progression and survival remains to be further clinical research[10-11]. So,doctors are interested in the Re-TURBT, especially for the T1 tumor.

The single factor analysis results in this study suggests that stage T1, high grade, tumor diameter>3cm, broadly-connected appearance are related to the pathological staging progressed to T2 of Re-TURBT, and there was no significant correlation between carcinoma in situ is found for the first time and the second postoperative pathologic staging, may be due to the number of cases(n=3) of carcinoma in situ too little. The precise relationship needs to be studied further large sample to clear.

Re-TURBT has quite important significance for radical resection of bladder tumor and clear tumor pathologic stage, so as to help choose the right follow-up treatment and evaluating prognosis. Although the use of fluorescence endoscope may have to decrease the rate of residual tumor, but dose not help solve the problem of tumor staging undervalued. Suggesting Re-TURBT to the patients those who of stage T1 at first TURBT, high grade, tumor diameter>3cm, broadly-connected appearance. However, whether need Re-TURBT to treat all NMIBC still needs more evidence-based medical evidence to confirm.

References

1. Rodrigue J, Chechil G, A lgaba F, et al. Value of random endoscopic biopsy in the diagnosis of bladder carcinoma in situ [J]. *Eurual*, 1987;**13**(2):150-154.
2. Yinglu Guo, Zhifu He. Urology endoscopic diagnosis and treatment [M].BeiJing; Peking University Medical Press,2004:98.
3. Babjuk M, Oosterlinck W, Sylvester R, et al. EAU guidelines on non muscle-invasive bladder urothelial carcinoma of bladder [J].*Urology*, 2005, **66** (Suppl 6A):4-34.
4. Sobin DH, Witteking CH. TNM classification of malignant tumor [M]. 6th ed. Wiley-liss:New York, 2002:199-202.
5. Sauter G, Algaba F, Amin M, et al. Tumor of urinary system: non-invasive urothelial neoplasia [M] // Eble JN, Sauter G, Epstein JI, Sesterhenn IA (eds). World Health Organization Classification of Tumors: Pathology and Genetics of Tumor of the Urinary System and Male Genital Organs. Lyon (France): IARC Press, 2004:110.
6. Han KS. Results of repeated transurethral resection for a second opinion in patients referred for non muscle-invasive bladder cancer: the referral cancer center experience and review of the literature [J]. *Endourol*, 2008;**22**(12):2699-2704.
7. Sedelaar JPM, Witjes JA. Technique of bladder tumors:value of repeat TUR and random biopsies [J]. *EAU-EBU Update Series*, 2007; **5**(1): 139-144.
8. Divrik RT, Sahin AF, Yildirim, et al. Impact of rountine second transurethral resection on the long-term outcome of patients with newly diagnosed pT1 urotheilal carcinoma with respect to recurrence, progression rate, and disease-specific survival: a prospective randomized clinical trial [J]. *Eur Urol*, 2010;**58**(2):185-186.
9. Cheng L, Neumann RM, Weaver AL. Grading and staging of bladder carcinoma in transurethral resection specimens. Correlation with 105 matched cystectomy specimens [J]. *Clin Pathol*, 2000; **112**(2): 275-279.
10. Dieter J, Herber S, Raphaela. Photodynamic diagnosis in urology: state of the art [J]. *Eur Urol*, 2008;**53**(6):1138-1150.
11. Tritschler S, Scharf S, Karl A, et al. Validation of the diagnosis value of NMP221 bladder chekl test as a marker for bladder cancer by photodynamic diagnosis [J]. *Eur Urol*,2007;**51**(2):403-408.

POLYMORPHISM OF EXON 2 IN BONE MORPHOGENETIC PROTEIN 7 GENE AND CORRELATION ANALYSIS WITH SOW REPRODUCTIVE TRAITS

Wei Zhang[1], Dan Wang[1], Caixia Sun[1], Shuyu Xie[1], Jianshe Zhou[2], Tao Huang[1], Ansi Wang[1], Ningquan Geng[1], Guirong Sun[1,a,*], and Ming Li[1,b,*]

[1]*College of Animal Science and Veterinary Medicine, Henan Agricultural University, Zhengzhou, 450002, China*

[2]*Vegetable Research Laboratory, Tibet Academy of Agricultural and Animal Husbandry Sciences, Lhasa, 850000, China*

E-mail: [a]grsun2000@126.com, [b]leeminghenau@126.com

**Corresponding author*

Bone morphogenetic protein 7 (BMP7) gene is a member of the transforming growth factor β. It has been proposed to be an osteoinductive protein. Recent data have shown that BMP7 also plays a crucial role in the growth and development, and physiological function of reproductive system. To date, studies have shown an association between the BMP gene family and reproduction in many populations, but few studies have completely described this association in sow. Herein, three sow breeds were screened out to investigate the genetic effects of the BMP7 gene. Using polymerase chain reaction-restriction fragment length polymorphism (PCR-RFLP) and DNA sequencing, two polymorphic sites (c.517C>T and c.562C>T) were detected. Correlation analysis with sow reproductive traits revealed that the c.562C>T site had significant effects on litter weight at 21 days, indicating that BMP7 gene could be regarded as a potential candidate gene locus for litter size trait. The data will provide a background for more extensive characterization of the BMP7 gene. And we can choose sows with this gene for swine breeding in order to improve the sow farrowing performance, especially the litter weight at 21 days.

1. INTRODUCTION

BMP7 gene is a member of the transforming growth factor β (*TGFβ*). It encodes one of the bone morphogenetic proteins (*BMPs*) and plays crucial role in osteogenesis[1]. It also has influence on the growth and development and physiological function of the reproductive system. The mutant *FecB* of *BMPRIB* has been reported to have multiplicative effects on ovulation of ewes[2]. Additionally, the families of *TGFβ* proteins are proposed to be the most important growth factors for growth of ovary and differentiation of early ovarian follicles[3]. In females, *BMP7* is associated with the function of the pituitary, ovary and uterus. Systemic treatment of diet-induced obese mice with *BMP7* resulted in increased energy expenditure and decreased food intake, leading to a significant reduction in body weight and improvement of metabolic syndrome. Similar degrees of weight loss with reduced appetite were also observed in

BMP7-treated *ob/ob* mice, suggesting a leptin-independent mechanism utilized by *BMP7*[4]. *BMP7* has stimulative effect on *FSH* secretion and maintains the basic expression of *FSH.* Also, it is associated with the generation of animal follicles, corpus luteum formation and the synthesis and adjustment of steroids. *BMP7* can promote the generation of estradiol and inhibit the synthesis of progesterone[5]. Therefore, it can promote the growth of follicles and transform more primary follicles to be secondary follicles.

Fertility is the main factor that affects the benefits of livestock husbandry[6]. However, fertility is a quantitative character with low heritability which is affected by many factors[7-9]. Ovulation number is one of the important factors that directly influences sow reproductive performance. In recent years, researchers confirmed some genes to be associated with livestock breeding, ovulation and reproductive traits. The genes, such as *FSH, ESR, GDF-9, BMPIB, FecB* gene, etc. playing an important role in porcine reproduction have been reported. Currently great progress has been made towards associations between molecular genetic mechanism and high fecundity of swine. Some research results have shown that *BMPs* are relevant to the embryonic development and the reproduction process, especially to the ovarian function[10,11]. So far, the studies of *BMP7* growth factors mentioned above have been mainly focused on human beings and mice. There is paucity limited of information about *BMP7* growth factor on swine and its relation with porcine reproductive traits. Therefore this study takes *BMP* gene as a candidate gene to investigate its underlying molecular mechanisms in regulation of sow breeding. In recent years, our laboratory has devoted to studying the genetic and molecular biology of *BMP7* gene in livestock. The complete cDNA of *BMP7* gene in domestic rabbit was cloned and the expression of *BMP7* gene was analyzed in tissues. In our former study, three SNPs were identified in intron 2 of the *BMP7* gene and the results showed that *BMP7* is a potential candidate gene for litter size trait[12].

This study is founded on previous findings and it is the further exploration of the related studies. *BMP7* gene was considered as a candidate gene on the purpose to study the association between its polymorphisms and the porcine reproductive traits. Two SNP loci named: c.517C>T and c.562C>T were detected in exon 2 of *BMP7* gene. Association analysis was taken between the two SNP loci and the porcine reproductive traits in order to find valuable molecular genetic marker which may provide scientific theory for swine breeding and reproduction. And the study also offers theoretical basis for exploring the function of the *TGF* superfamily.

2. MATERIALS

Three populations, landrace (n=25), large white (n=148) and duroc (n=85), were investigated in the present study. All these sows were reared under the same environment of Zhengyang breeding farm (Zhengzhou city, Henan province, China). Phenotypic data on sows were collected for the number of piglet born alive (NBA) , litter weight at birth (LBW), total number of piglet born (TNB) and litter weight at 21 days (LW21). Genomic

DNA samples were isolated from the follicles of 258 sows by the phenol-chloroform method[13]. The same amount of DNA sample from each individual was pooled and diluted to a working concentration of 200 ng/ul. Then the samples were stored at -80°C, following the standard procedures.

3. METHODS

3.1. Primer Design and PCR Amplification

The method of the mixed sample sequencing for preliminary screening of *BMP7* gene polymorphisms by building DNA pools was adopted in this research. These sites were genotyping by PCR-RFLP, and then their genetic effects on reproductive traits in Landrace pigs, Duroc pigs and Large White pigs were analyzed. According to the sequencing results and the published pig *BMP7* gene sequence (SEQ: ENSSSCG00000007501) in Ensembl database, Oligo 6.0 was used to design pair primers (F:5'-AACGCTCCCAGCCGTCCGTGAGGGAGG-3', R:5'-CCACTGAGCCACGAC GGGAAATC-3'). The PCR was performed in a 30 μl reaction mixture containing 0.5 μl DNA template, 1.5 μl upstream primer, 1.5 μl downstream primer, 15 μl TaqMix/Taq DNA polymerase and 11.5 μl double-distilled water. The PCR cycling protocol was 5 min at 94°C, 30 cycles of 94°C for 30 s, 64°C for 30 s, and 72°C for 30 s, with a final extension at 72°C for 10 min. The PCR amplification products were sent to the Taihe Biotechnology (Beijing, China) Co.,Ltd. for screening the SNPs. According to the sequencing, two novel SNPs in exon 2 of BMP7 gene were confirmed and typed by Chromas 2.31 and PCR-RFLP. The primers were designed respectively (Table 1). The PCR reactions were carried out in 10 μl of total volume, containing 3.5 μl double-distilled water; 0.5 μl of each primer; 0.5 μl total genomic DNA and 5 μl Taq polymerase. The PCR cycling protocol was 5 min at 94°C, 30 cycles of 94°C for 30 s, 56°C for 30 s, and 72°C for 30 s, with a final extension at 72°C for 10 min.

Table 1. PCR primers for SNP of BMP-7 gene in exon 2.

Primer name	Forward primer sequence(5'-3')	Reverse primer sequence (5'-3')	Size(bp)	Tm(°C)
c.517C>T	5'GGAAGGGGAAGCCGTGACTGGTG3'	5'GTGTCTGCCCC TGCCCCTGTC3'	221	64
c.562C>T	5'GACTACATCCGGGAGCACTTCATT3'	5'GTGTCTGCCCC TGCCCCTGTC3'	178	58

The fragment was successfully amplified in exon 2. After confirming the PCR products with 1% agarose gels stained with ethidium bromide, the PCR products were digested with the corresponding restriction enzyme over-night at 37°C, respectively. All the restriction enzymes were obtained from Taihe Biotechnology (Beijing, China) Co.,Ltd. There were not any natural enzyme sites. So artificial enzyme sites-Ban I and VspI were created by leading in a nucleotide mismatch, which enabled the use of restriction enzymes for discriminating sequence variations (Table 2). The digested products were separated by electrophoresis through 1.5% agarose gel, which was stained with ethidium bromide. Different genotypes of the two SNPs were confirmed by sequencing.

Table 2. The information of created enzyme locus.

Locus	Enzyme	Enzyme site	Recognition site	SNP site
c.517C>T	BanI	G^GYRCC	-4	AGCCGTGACTGgTGCCGAATTCCGCATCTAC
c.562C>T	VspI	AT^TAAT	-3	GGAGCACTTCAtTAATGAGACCTTCCGCATC

Note: Lowercase letters means changed base; underlined letters means mutations.

3.2. Statistical Analysis

Genotype frequencies and alleles were calculated by POPGEN. A $\chi 2$-test was applied to evaluate significance using the SPSS software (Version 16.0). The general linear models procedure of SPSS software was used to analyze the relationship between the genotypes and the reproductive traits in sow. The following linear model was used in the analysis, as in Eq. (1).

$$Y_i = \mu + R + S + P + G_i + E_i. \qquad (1)$$

Yi represents the individual value, u the overall mean, R the effect of sow breed, S the effect of season, P the effect of parity, Gi fixed effect of the single SNP marker genotype, and Ei the random error. Genotype distribution was evaluated for agreement with Hardy-Weinberg equilibrium by the Chi-square test. Genotype data were analyzed with the homozygote of the common allele as the reference group. Both additive and dominance effects were estimated using REG procedure of SAS 8.0 program, where the additive effect was estimated as -1, 0 and 1 for the AA/CC, AB/CD and BB/DD genotypes, respectively; and the dominance effect was represented as 1, -1 and 1 for AA/CC, AB/CD and BB/DD genotypes, respectively[14].

4. RESULTS AND DISCUSSION

4.1. SNP Identification and Genotyping

The 609-bp fragment containing two novel mutations (c.517C>T and c.562C>T) was amplified in exon 2. Two SNPs were found to be polymorphic after products were digested with restriction enzyme (*Ban I* for c.517C>T and *VspI* for c.562C>T). In the present study, compared to the sequencing results of the two homozygous genotypes obtained by PCR-RFLP, substitutions of T→C and T→C were discovered at nucleotide locus 517 and 562 in exon 2 (*BMP7* gene), leading to the variation from allele *A* to allele *B* and allele *C* to allele *D*. The amplified fragment of the c.517C>T is 221 bp in length. The digestion of the PCR fragments of *BMP7* c.517C>T locus by Ban I produced fragments 221 bp for genotype AA, and 221, 201 and 20 bp for genotype AB, and 201 and 20 bp for genotype BB. The amplified fragment of the c.562C>T is 178 bp in length.

The digestion of the PCR fragments of c.562C>T locus using *Vsp I* produced fragments 178bp for genotype CC, and 178, 155 and 23bp for genotype CD, and 155 and 23bp for genotype DD.

4.2. Genetic Polymorphism of Exon 2 in BMP7 Gene

Allele frequencies and value of $\chi 2$, Ne (effective allele numbers), PIC (polymorphism information content), Obs He (observed heterozygosity) and Nei's He (Nei's expected heterozygosity) were shown in Table 3 and Table 4. The $\chi 2$ test showed that the polymorphism of c.517C>T was in Hardy-Weinberg equilibrium (p>0.05), whereas, the c.562C>T deviated from the Hardy-Weinberg equilibrium (p<0.05). The deviation between observed heterozygosity and Nei 's expected heterozygosity was then measured by Fixed index. As shown in Table 4, the efficient allelic numbers were 1.821 and 1.913; the information indexes were 0.643 and 0.670, respectively. All the results were consistent with the computation of Nei 's expected heterozygosity.

Table 3. Genotype frequency and gene frequency of the two SNPs in exon 2

Traits	Genotype frequency			Gene frequency	
c.517C>T	AA	AB	BB	A	B
	0.407 (105)	0.500(129)	0.093 (24)	0.657	0.343
c.562C>T	CC	CD	DD	C	D
	0.213(55)	0.360(93)	0.426(110)	0.393	0.607

Note: AA and CC=homozygous wild-type; AB and CD=heterozygous; BB and DD=homozygous variant genotypes. A and C=wild gene; B and D=mutant gene; The number in the bracket means total number of pigs of the corresponding genotype.

Table 4. Polymorphic indicator of the two SNPs in exon 2

Traits	$\chi 2$	PIC	Obs He	Nei's He	Ne	FIS	I
c.517C>T	2.979 ns	0.349	0.500	0.451	1.821	-0.109	0.643
c.562C>T	15.707**	0.363	0.361	0.477	1.913	0.244	0.670

Note: PIC: Polymorphism information content; Obs He: Observed Heterozygosity Nei's He: Nei's expected Heterozygosity; Ne: Effective number of alleles; FIS Wright's fixation index; I: Shannon's information index I; The values of $\chi 2$ is from the test of the distribution of different genotypes for Hardy-Weinberg equilibrium in different microsatellite locus; Means bearing **differ significantly at p<0.01

Because the genotype frequencies of c.517C>T did not vary significantly among different breeds, we did not perform the Partition of Chi-square test on it. The genetic diversity of SNPs was calculated for each breed. In the genetic diversity of c.562C>T locus, the frequency of allele *C/D* was 0.340/0.660 in the Landrace, 0.490/0.510 in the Large white and 0.241/0.759 in the Duroc. In all three populations the major allele was *D*, with the Large white having the most population, followed by Landrace and Duroc. Overall, the genetic diversity of c.562C>T was highly significant between Large white and Duroc ($P<0.05$). The Large white population also had the highest Ne, with Landrace and Duroc exhibiting successively lower Ne. The PIC and Nei's He of the three populations followed the same pattern as Large white>Landrace>Duroc.

Two mutations were detected in exon 2 of BMP7 gene by DNA pool sequencing in the present study. They both showed polymorphism. The PIC value of c.517C>T and c.562C>T locus were 0.349 and 0.363 respectively, ranged from 0.25 to 0.5, achieving median polymorphism. It revealed that the SNPs had biggish genetic variation and selection potentials. Besides, c.517C>T (0.643) and c.562C>T (0.670) had the higher I (information index), supporting that the heritable variation of experimental population was high. The polymorphism of c.517C>T was in Hardy-Weinberg equilibrium ($p>0.05$), which indicated that this site was less affected by natural selection and artificial selection. Further study of this site was performed, no significant difference was found between the genotypes and reproductive performance. However, the c.562C>T deviated from the Hardy-Weinberg equilibrium ($p<0.05$), indicating that this locus was affected by natural selection and artificial selection. It had certain relevance with porcine reproductive traits judging from correlation analysis.

4.3. Biological Effect of SNPs of exon 2 in BMP7 Gene

Two mutations were detected in exon 2 of BMP7 gene by DNA pool sequencing in the present study. They both showed polymorphism. The PIC value of c.517C>T and c.562C>T locus were 0.349 and 0.363 respectively, ranged from 0.25 to 0.5, achieving median polymorphism. It revealed that the SNPs had biggish genetic variation and selection potentials. Besides, c.517C>T (0.643) and c.562C>T (0.670) had the higher I (information index), supporting that the heritable variation of experimental population was high. The polymorphism of c.517C>T was in Hardy-Weinberg equilibrium ($p>0.05$), which indicated that this site was less affected by natural selection and artificial selection. Further study of this site was performed, no significant difference was found between the genotypes and reproductive performance. However, the c.562C>T deviated from the Hardy-Weinberg equilibrium ($p<0.05$), indicating that this locus was affected by natural selection and artificial selection. It had certain relevance with porcine reproductive traits judging from correlation analysis.

4.4. The Genetic Effect on Reproductive Traits of exon 2 in BMP7 Gene

The association between different genotypes of SNPs and the sow reproduction traits was analyzed by General liner model (GLM) of SPSS16.0. For c.562C>T, the results suggested that it had significant associations with LW21 ($p<0.05$) (Table 5). Additionally the following results were found: sows carrying CD genotype were higher than those with CC and DD genotypes at TNB, NBA and LBW and sows with the DD genotype were the lowest at LW21. Besides, over-dominant effects existed between alleles *C* and

D, the dominant effects on TNB, NBA and LBW were 0.156 head, 0.208 head and 0.210 kg.

Table 5. Associations of different genotypes with sow reproductive traits

SNPs	Traits	SNP genotype (mean±SE)			P value	Additive	Dominant
c.517C>T		AA	AB	BB			
	TNB	8.516±0.600	8.243±0.614	8.319±0.679	0.304	-0.098	0.175
	NBA	8.084±0.607^{a}	7.854±0.621	8.070±0.686	0.418	-0.007	0.223
	LBW	11.401±0.890	11.269±0.911	11.615±1.007	0.732	0.107	0.239
	LW21	46.613±3.179	47.127±3.252	47.070±3.593	0.857	0.229	-0.286
c.562C>T		CC	CD	DD			
	TNB	8.234±0.633	8.517±0.625	8.489±0.601	0.478	0.128	-0.156
	NBA	7.755±0.640	8.123±0.631	8.074±0.607	0.305	0.160	-0.208
	LBW	11.013±0.938	11.431±0.925	11.428±0.890	0.453	0.208	-0.210
	LW21	49.662±3.339^{a}	47.290±3.293^{ab}	46.178±3.166^{b}	0.025	-1.742	0.630

Note: Data in the table are means±standard error. Values in each line with different lower case superscripts are at p<0.05. No marked means no difference. NBA means the number of piglet born alive, LBW means litter weight at birth, TNB means total number of piglet born and LW21 is short for litter weight at 21 days.

For c.517C>T, sows with the homozygous variant genotype AA were the largest at TNB and NBA compared with those of genotypes AB and BB. Sows with the AB genotype had the highest LW21 compared with AA and BB genotypes, and the BB genotype was highest at LBW. However, there were no significant differences among genotypes of SNP c.517C>T and any of the reproductive traits. Correlation analysis was taken among the two SNPs, different genotypes and reproductive traits. There was no significant difference between different genotypes and reproductive traits at c.517C>T site. Whereas, the c.562C>T locus had significant influence ($P = 0.025$) between different genotypes and LW21. Individuals (litter weight) of the CC genotype at 21 days were (7.545%) heavier than that of the CD and DD genotypes ($P < 0.05$). *D* allele was the dominant allele. The DD genotype had the lowest mean value. The mean value of CC genotype significantly differs from DD genotype. The c.562C>T site had an additive effect on LW21. The *D* allele had negative effect on LW21. The additive effect value was -1.742 kg. The dominant effect value was 0.630 kg between the alleles *C* and *D*.

Reproductive traits belong to quantitative character. They are relevant to environmental factors and controlled by multiple genes. Screening out the major genes that affect the porcine reproductive traits would have important implications for the breeding of Chinese local pig breeds. The use of SNPs played a very important role in the livestock genetic breeding. This method could detect a large number of samples in a short time. Shimasaki et al. (2001) [15] designed a specific primer according to the human male hormone receptor gene and a 200 bp fragment of male hormone receptor gene in domestic pig was successfully amplified, besides, a G to A SNP locus was found by using Nae I endonuclease enzyme. Jiang et al. (2000) [16] detected the SNPs of intro 1 of myostatin gene in double muscle rump among three different loci in large white and landrace breeds through PCR-RFLP and PCR-SSCP and found that the allele gene G predominated in the hybrids of large white and landrace. Intracerebroventricular administration of BMP7 to mice led to an acute decrease in food intake, which was mediated, at least in part, by a central rapamycin-sensitive mTOR-p70S6 kinase pathway[4]. Their studies underscored the importance of BMP7 in regulating both food intake and energy expenditure. Combined with researches of recent years, the BMP7 gene can be considered as a candidate gene affecting porcine reproductive traits and further research are needed to explain the special functional mechanism of this gene.

In conclusion, two mutations: a missense mutation locus (c.517C>T) and a synonymous mutation (c.562C>T) locus were detected in exon 2 of porcine BMP7 gene. They both showed polymorphism in landrace, duroc and large white populations. Associated with reproductive traits, the c.562C>T locus had significant influence ($P = 0.025 < 0.05$) in LW21. The site worked mainly by additive effect. Individuals (litter weight) of the CC genotype at 21 days were (3.484kg) heavier than that of the CD and DD genotypes ($P < 0.05$). These results suggest that c.562C>T is a potential candidate gene locus for litter size trait and the BMP7 gene might be associated with the quantitative trait locus (QTL) controlling the litter size. Although, the c. 517C>T locus had no significant effects on reproductive traits, its existence may influence the effects of c.562C>T on reproductive traits. It is hard to confirm the interaction mechanism among the two SNPs simply by statistical analysis. Moreover, our results were based on a limited number of animals. Therefore, this preliminary study should be confirmed with a larger samples size.

Acknowledgments

This work was supported by Modern Agricultural Industry Technology System (Beef Cattle Technology Team) of Henan Province (2013-14).

References

1. Shimasaki, S., Moore, R. K., Otsuka, F., & Erickson, G. F. (2004). The bone morphogenetic protein system in mammalian reproduction. Endocrine Reviews, **25**(1), 72-101.
2. Wang, J. Q., & Cao, W. G. (2011). [Progress in exploring genes for high fertility in ewes]. Yi chuan Hereditas/Zhongguo yi chuan xue hui bian ji, **33**(9), 953-961.
3. Shabir, M., & Ganai, T. A. S. (2012). Nucleotide sequencing and DNA polymorphism studies of BMP 15 gene in Corriedale and local Kashmir valley sheep (Ovis aries). *Gene*, **499**(1), 231-235.
4. Townsend, K. L., Suzuki, R., Huang, T. L., Jing, E., Schulz, T. J., Lee, K., & Tseng, Y. H. (2012). Bone morphogenetic protein 7 (BMP7) reverses obesity and regulates appetite through a central mTOR pathway. The FASEB Journal, **26**(5), 2187-2196.
5. Souza, C. J., MacDougall, C., Campbell, B. K., McNeilly, A. S., & Baird, D. T. (2001). The Booroola (FecB) phenotype is associated with a mutation in the bone

morphogenetic receptor type 1 B (BMPR1B) gene. Journal of Endocrinology, **169**(2), R1-R6.

6. Li, Y.H., Zeng, Y.Q., 2003. Genetic improvement strategy of porcine reproductive traits. *Animal Science and Animal Medicine*. 20, 11-14.
7. Li, F.E., Xiong, Y.Z., Deng, C.Y., 2003. Major gene of reproductive traits of sow and advances of QTL gene. *Pigs and Poultry*. 23, 31-35.
8. Rothschild, M. F., Hu, Z. L., & Jiang, Z. (2007). Advances in QTL mapping in pigs. *International Journal of Biological Sciences*, **3**(3), 192.
9. Onteru, S. K., Ross, J. W., & Rothschild, M. F. (2009). The role of gene discovery, QTL analyses and gene expression in reproductive traits in the pig. *Control Of Pig Reproduction VIII*, 87-102.
10. Shimasaki, S., Moore, R. K., Erickson, G. F., & Otsuka, F. (2002). The role of bone morphogenetic proteins in ovarian function. Reproduction (Cambridge, England) Supplement, 61, 323-337.
11. Yan, G. E., & ZHANG, J. H. (2006). Progress on Bone Morphogenetic Proteins on Animal Reproduction [J]. Progress in Veterinary Medicine, 6, 007.
12. Feng, X., Xie, S. Y., Zhou, J. S., Sun, G. R., Lu, P., & Li, M. (2013). Polymorphisms of the bone morphogenetic protein 7 gene (BMP7) and association analysis with sow productive traits. *Animal reproduction science*, **142**(1), 56-62.
13. ZHOU, J., CHEN, Q., LI, M., XIE, Y., & LIU, X. (2011). An Effective and Fast DNA Extraction Method from Pig Hair-follicle [J]. *Acta Agriculturae Boreali-Occidentalis Sinica*, 8, 002.
14. Liu, B. H. (1997). Statistical genomics: linkage, mapping, and QTL analysis. CRC press.
15. Shimanuki, S., Kobayashi, E., & Awata, T. (2001). A single nucleotide polymorphism in the porcine androgen receptor gene. Animal genetics, **32**(3), 165-166.
16. Jiang, Y. L., Li, N., Wu, C. X., & Du, L. X. (2000). [Analysis on single nucleotide polymorphisms of porcine myostatin gene in different breeds]. Yi chuan xue bao Acta genetica Sinica, **28**(9), 840-845.

APPLICATION OF MULTI CULTURE NURSING IN EMERGENCY WORK

Weiwei Fan[a] and Changping Song[b]

Nursing College, Beihua University, Binjiang East Road No.3999 in Jilin City of Jilin Province, China

E-mail:[a]364760038@qq.com,[b]331848645@qq.com

The serving objects for Emergency Work should be people who come from different countries and different races, and there are some differences in their cultural background, religious belief, language exchange custom etc. This will require that nurses to create a good exchanging environment by using language communication and non-language communication, to recognize culture differences in the expression of sentiment and pain, to respect patients' religious belief and custom, to Esteem different nation's diet habit ,to pay more attention to views of morality and value of patients with different culture practice in Emergency Work. Correctly to implement multi-culture nursing to patients, To meet the patient's physical and mental needs.

1. THE BACKGROUND OF MULTI-CULTURE NURSING

The association and communication among people who are from different regions, even different countries are getting gradually frequent with the deepening of reform and development in economy, and at the same time, a society called multi- culture society in which people with different culture get together has formed. Therefore, the serving objects for nurses should be people who come from different countries and different races, and there are some differences in their cultural background, religious belief, language exchange custom etc. This will require that nurses should refresh their nursing idea and change their roles to meet the need for their nursing works to fit the nursing focused on people's health, namely multi- culture.

In recent years, some western countries have widely put multi-culture nursing into their nursing practices. Essential Health Bureau of United States pointed out that all medical departments should pay more attention to the service for different populations with their different culture backgrounds in1998; Maternal and Childhood Care Bureau emphasized the importance of abilities in culture in serving for different populations with their different culture backgrounds in its concerned documents. Minority Health office, Ministry of Public Health of United States, issued the standard of culture and language adaptability in 2000[1]. Multi-culture nursing was first proposed by lady M·Leininger, an American nurse in 1960s. According to the theory, it is believed that nursing should not be a fixed mode and the particular nursing serving can be provided for people from different race and under different culture backgrounds as long as it is designed to be a rough frame. As soon as it was proposed, the theory was recognized universally. It was in 1995 that the first international symposium on multi-culture nursing perspective was held in our country, and just from that time, the theory, multi-culture nursing, was introduced into China. At present, it is used in clinical practice.

2. MULTI-CULTURE NURSING AND RELATED CONCEPTS

A definite culture is referred to an integrated reflection of the definite history, region, economy, society and politics. Culture is formed in the life of a particular population or society and it is defined as the sum of its members' life styles, including the view of value, language, knowledge, belief, art, law, custom, prevailing practice, living attitude, rule of behavior and related performance manner in substance[1]. Multi-culture is just different cultures which each of multiple races possesses alone[2]. Multi-culture nursing is required to make multi-culture infiltrate into the nursing process based on the diversity of patients' cognition and requirement in health, disease, treatment, nursing, health care and looking after, and the variety of social culture to patients' requirement for their bodies and minds, and social culture[3].

3. THE CHARACTERISTICS OF MULTI-CULTURE NURSING

3.1. Multi-culture Characteristics of Nursing Theoretical System

With the conversion of medicine and nursing modes, the subject which is studied in nursing has changed into a complete or social one from the simply biological human; Nursing, as a subject, is oriented to one which is covert by combining scientific science with social science. Its knowledge does not include only that in scientific science but also that in social science and cultural science. The subjects are interpenetrated to promote the formation of many new cross-subjects and branch-subjects which are comprehensive or peripheral, such as nursing ethics, nursing psychology, nursing aesthetics, nursing communication ect. It follows that what is involved in nursing theory should be extensive and with some cross-culture characteristics.

3.2. The Pluralism of Nursing Client

In present society, nursing clients probably come from different countries or different nations. Their different culture backgrounds lead to their different understandings on health and life, explanation for disease and death, description of symptoms suffered, manifestation of sentiment and sadness. These make them have different demand for nursing, which requires that nursing staff should give people under different culture backgrounds a good holistic nursing service.

3.3. The Pluralism of Nursing Style in Clinic

The modern nursing theory, such as nursing program mode and holistic nursing, is mainly based on the theory of western culture and medicine. China has its own history of civilization for thousands years. Its traditional medicine, Chinese traditional medicine, is still applied and developed to exert its impact on nursing greatly. The unique nursing modes in China are still widely in practice in clinic in China, and they are traditional Chinese medical nursing, and traditional Chinese and western medical nursing. It indicates that cultures in modern time and ancient time or in west and east should be compatible to form a multi-culture nursing mode in clinic.

3.4. The Pluralism of Nursing Role

The enlargement of nursing function and the enrichment of nursing meaning has made a nurse become a pluralistic role, such as caregiver, nursing designer, health consultant, health coordinator, health educator, nursing manager, rehabilitation trainer, nursing researcher.

3.5. The Pluralism of Nursing Matter and Function

A multi-level demand is required for nursing work now just because of the formation of modern medical mode and view of health, and the evolution of multi-culture. The original nursing simply for disease has become an overall and holistic nursing; the past nursing role meeting only the demand of medical department has be extended to be general one serving for hospital and society, and nurses have to move into community, family and society from their hospitals; nursing functions are not only treatment, prevention, care and rehabilitation, but also nursing education, management, communication and research. Due to all described above, the pluralistic nursing matter and function are required.

4. CORRECTLY TO IMPLEMENT MULTI-CULTURE NURSING PRACTICE IN EMERGENCY WORK

Some researchers believe that every one should have his own culture so that a nurse's each contact with her patients will be a multi-culture contact [4]. People with different cultures are different in recognition and demand in health, disease, treatment, nursing, healthcare and looking after. When an Emergency patient comes to a strange environment from his familiar culture environment suddenly, he will develop into an imbalance in psychology and physiology since he is struck by the strange culture, even culture "shock". The culture striking involved in clinical nursing generally includes language, religious belief, eating habit, ethics and morality, and view of value.

4.1. To Recognize Culture Differences in The Expression of Sentiment and Pain

Culture differences can influence the manifestation of sentiment and pain. For example, many British Americans or Indians do not weep while they are painful for fearing that they are regarded as people with puerility or they are self-abandoned, whereas Latinos often express their pains by using their bodies or languages.

4.2. To Create a Good Exchanging Environment by Using Language Communication and Non-language Communication

4.2.1. *Differences of Culture Contents and Communicating Language*

Westerners often take weather and news as their greeting topics since they think that asking others "where are you going", or "what are you going to do" is impolite, and they avoid to be asked about their ages, families and incomes. In China, the word "old" is often used to greet an old man, which shows the respect to him. However, a westerner does not recognize that he has become old and is reluctant to be called "old". Therefore, her words should meet the situation of the patient's culture when a nurse is talking to a patient from a different culture background.

4.2.2. *To Use Body Language as The Primary Manifestation While Communicating in Non-Language Way*

The same body word may have different meaning in a different culture environment. Touching action is the closest way in non-language behavior. In China, it shows the meaning of encouragement or love so as to help to establish a trusting relationship between each other. However, touching a unfamiliar person indicates the meaning of sex. In Thailand, touching other's head is thought that it is an extreme insult. Most nations in the world express their approvals by nodding, but shaking one's head does it in Bulgaria and Nepal. Most races in the western countries think that shrug indicates "having no other way", but shaking one's head with a forced smile means it in China.

4.2.3. *The Distance and Looking While Communicating*

There are some differences in the communicating distance. The distance between the both sides who are talking is very close, but British American likes to keep a distant talking. Populations who speak standard English should gaze at each other, which means they are honest and shows another person a trusted feeling, while in the eastern countries, a proper gaze at the another shows that the listener pays more attention to the topic and he is interested in it to be in favor of keeping the conversation, but very staring at the other side shows a aggressive feeling to him.

4.3. To Pay More Attention to Views of Morality and Value of Patients with Different Culture

Views of morality and value are different for people coming from different nationalities [4]. With the gradual aging in population throughout the world, strengthening the train for elders' self-care abilities to improve their living qualities is an inevitable trend in the society. In the western countries, struggling independently is advocated. For example, supporting elders to go up, go down or give them a help in their lives will make them unpleasant since in these ways, their independent pride and desirability will be hurt. However, in China, the view of filial piety is advocated all the time, elders are looked after carefully and all their things are done by their children since Chinese consider that nothing is more important than the filial piety, which make the elders have dependent mind and neglect the training for their self-care abilities. Since the world is rich and colorful, nurses should enrich their humanistic knowledge, and notice and respect the view of value of patients with different culture.

4.4. To Esteem Different Nation's Diet Habit

With not affecting the treatment and nursing for them as a prerequisite, it would be best to permit patients to choose meal by themselves. The flavoring, such as salt, sugar, pepper powder, may be divided into small packets and then sent to the kitchen together with the meal, and prepared by patients themselves according to their tastes. Manchu and Xibe people in China are forbidden to eat dog meat and Mongolians are forbidden to eat beef. Hui, Uyghur and Tajik nationalities believe in Islam so that pork, dead things and blood are forbidden to be eaten, and the fasting is kept from the dawn to the sunset on their fast days each September. Nurses can meet their demands for nutrition by giving them extra meal and transfusion. Westerners like to eat raw and cold food, since they believe that these can improve their health, but people in the eastern countries regard them as nosogenesis. A nurse should try to understand their food culture and respect their eating habits.

4.5. To Respect Patients' Religious Belief and Custom

Patients from other countries almost have their own religious believes and national customs. The Thai abstain from heading west while sleeping and signing with red ink, as the Thai put their bodies heading west after death, and write the death's name on coffin with red ink. Indonesia and Malaysian are forbidden to pass things by left hand. A Christian fear to listen to "Friday" or "thirteen"; a Buddhist does not eat meat or eat it just at dark; A Mussulman must kneel on a felt pan to pray just before going to bed and having a dinner, he dislikes others to stand or pass nearby while he is praying, and the felt pan is prohibited to be stepped or stridden by feet. Therefore, a nurse should not only try to understand their religious believes and customary habits, but also respect their reading bible silently and worshiping Boddha and praying to create a corresponding environment for them.

Multi-culture nursing is a high-level nursing produced in the new medical mode, a trend with the development of multi-culture, a people' demand for their health, and the necessity for the development of modern nursing. It requires that nursing for diseases and nursing for culture should be implemented simultaneously. It is desired that all nursing fellows should work together, improve our thinking about multi-culture nursing to promote the development in modern nursing.

References

1. Hongmei Zhao,Xiaomin Li. Application of multi culture nursing in the health education in gynecology and obstetrics department. *J Journal of Nurses Training* 2014; **17**(3): 215.
2. Xu Y.N National standards for providing culturally and linguistically appropriate health care: policy implications for nursing. *J Nursing Economics* 2001;**19**(5): 240-241.
3. Wei gu. *Multi-culture nursing* .Beijing: The people's medical publishing house 2011; 2: 7.
4. Marshall. *J International and cross-cultural issues*: six key changes for our professions Folia Phoniatr Logop 2003; **55**(6):329-336.

CHINESE AND WESTERN COMBINED THERAPY OF DEPRESSION

Ruiping Wang

School of Civil Engineering, Qingdao Technology University, Tsingtao, 266033, China

E-mail: ruipingwang@163.com

It is discussed in this paper about the globally usual disease-depression. Combining the Chinese and western study and therapy, the nosogenesis of depression is analyzed with three levels: physical level, psychological level, and spiritual level, particularly from the perspective of oriental culture and traditional Chinese medical science. At the physical level, a new therapy method of depression and its magical curative effect are presented. That is moxibustion. At the psychological level, it is important and necessary that the psychological technique-NLP and the meditation-Vipassana are helpful for depression therapy. At the spiritual level, it is indicated that family constellation is helpful to cure depression caused by the family involvements. The article is valuable and practical.

1. INTRODUCTION

According to WHO statistics, there are about 3.4 billion depressive patients worldwide. The incidence of depression is about 11%. Depression has become the 4th disease damaging human health and its incidence presents the rising tendency with low-age trend. It is predicted that depression will be the second disease next to heart disease and become deadly to human beings. The other name of depression is depressive disorder. It is characterized by clinical features of observable and constant bad mood, being accompanied by hypokinesia, cognitive disorder, inhibition of thought, hypobulia and body symptoms meanwhile, even suicide.

The study of depression deeply about the nosogenesis and therapy is the urgent requirement contemporarily and has notable society values.

2. NOSOGENESIS OF DEPRESSION

There is no clear idea about the pathogenesis of depression. Some clinical research manifested that biological, psychological and social environment factors were involved. Human body is a complicated tissue, and body and mind influence each other so deeply[1]. So the therapy is a complicated process physically to mentally[2]. Every life is a complex of body-mind-spirit. So the nosogenesis of depression is set forth by physical level, psychological level, to spiritual level in this article.

2.1. Physical Level

Nosogenesis of depression is the lesions of Kidney Meridian and Stomach Meridian in traditional Chinese medicine, and with the stagnation of liver-qi[3].

2.1.1. *Lesion of Kidney Meridian*

The symptom of depression caused by the lesion of Kidney Meridian is dull and remote eyes, flustered heart in throat, often hungry but loss of appetite, and terrified as if someone is pursuing him. This is because that the heart-fire can't be restrained which is caused by kidney essence deficiency.

The load of kidney becomes more and more heavy because the fast tempo of modern life, pressure of competition, mental stress, overwork, irregular sleeping patterns, environmental pollution, etc. Another important reason of kidney deficiency and kidney cold is sex acolasia. Dr. Peng Xin, a famous researcher of China Academy of Chinese Medical Science has a profound analysis of the great influence on the human health caused by whoredoms and sex acolasia[4].

2.1.2. *Lesion of Stomach Meridian*

The symptom of depression caused by the lesion of Stomach Meridian is that the patients are very reluctant to contact with outside world. They are afraid of light and even the sound of table pounding with flustered hearts all day long. And they shut the curtains and close the windows at once when they go back home with likes of living in dim environment.

Today the human stomach becomes cold because of the overmuch raw, cold, and cool foods cold in nature, the excessive air conditionings for cool, irregular eating habits, etc. Stomach function is decreasing.

2.1.3. *Stagnation of Liver-Qi*

In traditional Chinese medicine liver, which has the function of catharsis, bearing the dispersive effect plays an important role in the occurrence and regulation of psychological stress. So people will feel depressed with livers-qi are stagnated. This case is more common in female than that in male.

2.2. Psychological Level

Trauma of growth, wrong or negative beliefs and values, and the deficiency of viability cause a series of psychological problems including depression. Today most people are tired and self-blaming. Troubles are produced every day. If the troubles can't be cleaned away timely they would be deposited into subconscious. After days and months multiplying, they will be expressed by depression if great to a certain extent[5].

There are two kinds of fatal poisons in minds in Buddhism: craving and aversion. With the material abundance in modern society the human beings' craving is stimulated maximumly. How much craving, the same as aversion. At the same time, how much aversion, the same as suffering. It is the craving and aversion which push people to a painful world.

In addition, there are significant characteristics of depression patients as perfectionism and holding themselves inordinately. The perfectionist is actually focusing on the negatives and accumulating huge amount of negative energy which can't be cleaned away. That will cause depression. When people hold themselves inordinately, so much unhealthy mood will be buried deep in their hearts, and they will feel depressed to a significant extent some day.

2.3. Spiritual Level

As a human being, every person belongs to many kinds of systems: family system, community system, or any other system. Among them the most important and elementary system is family system. It is discovered by Dr. Bert Hellinger, a German psychotherapist, that there is latent system dynamics-involvements in family which manipulate the relationships between family members and don't operate according to the rules and standards of society and culture. Many body-mind problems including depression are caused by the involvements of family system[6].

3. CHINESE AND WESTERN COMBINED THERAPY

Drug treatment is the main measure for depression patients above medium level in Western medicine. There are quite a few clinical antidepressants so far. These medicines have therapeutic effects at some degree with evident adverse reactions at the same time and can't eradicate the disease with a high recurrence rate.

In traditional Chinese medical science it is stated that the mind and body influence each other greatly. Body diseases should be treated with psychological adjustments and mental diseases should be cured with body treatments. So the therapy of depression may start with body treatments and then the psychological adjustments. Firstly the cold-evils of kidney and stomach should be conquered, and then securing and containing therapy should be done to kidney and stomach in order to recover the functions of them with smoothing the liver and regulating the circulation of qi at the same time.

3.1. Physical Level

There are about three kinds of treatments in traditional Chinese medical science: acupuncture, moxibustion, and Chinese traditional medicine. Which is mainly stated in this article is moxibustion.

Moxibustion is a traditional Chinese medicine treatments using moxa made from mugwort[7]. It is a method of treatment in which moxa is burned to stimulate specific acupoints aiming at health care and healing. Moxibustion, as the oldest and magic Chinese therapy, originates from the ancient times and plays an important role in the traditional medical systems of China. It is called mother's love because it's curative effect by a miracle and warmness[1]. Moxibustion is significant to conquer the cold-evils of kidney and stomach and the stagnation of liver-qi.

- Acupoints in common use for cold-evil of kidney: shenshu, mingmen, shenque, guanyuan, taixi, yongquan.
- Acupoints in common use for cold-evil of stomach: zhongwan, shenque, guanyuan, ganyu, danyu, piyu, weiyu, zusanli, sanyinjiao.
- Acupoints in common use for stagnation of liver-qi: shuifen, shenque, qihai, guanyuan, qimen, zhangmen, taichong, xingjian, yanglingquan, sanyinjiao, ganyu.

It is noteworthy that the quality of moxa is very important. And, different moxibustion technique will lead to different effect. For example, mild moxibustion is supplement and pecking moxibustion is discharge. The main kind in common use is direct moxibustion as well as indirect moxibustion using jars or other partitions such as ginger, salt, etc.

Regular life style, good hobbies and exercises like running are important for depression therapy except treatments.

3.2. Psychological Level

People prefer eliminating physical or mental pains as soon as possible rather than thinking deeply the tremendous hidden values. Ph.D., Wilfred Nelles, an assistant professor of German university of Bonn and an ISCFG at the same time, says that the aims of all accidents are to awaken us[8]. Depression is a reminding of mind and body that the chasing outside should be stopped and inward-looking should be done to see and accept oneself truly with love[9]. Depression can be healed by life itself with good mind and body circumstances. There are many courses for mental growth and healing today.

- NLP- Neuro Linguistic Programming[10]

NLP, as a set of particular and feasible techniques about human behaviors and communication processes, was established in California University in 1976. It can be used to comprehend human beings' experiences and behaviours and transform them although it is not a set of techniques for psychotherapy. It is a set of implicit techniques with rapid, good and powerful effects to enhance human beings' behaviours, abilities, and physical and psychological health widely and permanently.

- Chinese traditional culture-Confucianism, Buddhism, Taoism, Yi-ology, TCM, Military tech, History

Chinese traditional culture is the broad and profound quintessence of national culture of the Millennium. The moral value system is beneficial for human beings' physical and psychological heath.

- Vipassana

Vipassana is one of the oldest meditations in India. Professor S. N. Goenka had taught Vipassana meditation around the world freely from 1969. The distresses and troubles in subconscious can only be covered up and they are difficult to cure through consciousness level. Vipassana means observing reality in a special way, by going beyond apparent truth. It is stated that everything that arises in the mind is accompanied by sensation of the body. And having fully comprehended them, people can become freed from all impurities of distresses and troubles and let their lives full of loving kindness, compassion and wisdom to face up to the ups and downs of lives peacefully[11].

- The power of transforming is not elsewhere or some other time, but here and now[12].

3.3. Spiritual Level

3.3.1. *Family Constellations*

Family constellations are new methods of psychological family therapies. It can explore the root causes, present the hidden influence factors, and show the directions of solutions so as to help people to face up their puzzles[13].

Every person gets birth from parents. So accepting them as they are and loving them to get the power of family system is very important. Love especially with parents is the best remedy[14].

3.3.2. *Religions and Master Teachers*

The cause of sufferings is Karma in Buddhism and original sin in Christian and Catholicism. The method is confession. It is noteworthy that in juvenile years spiritual needs were brought out with physical growth. Parents are not the best persons because so closely or so not. It is very necessary to find a master teacher for guidance.

4. CONCLUSION

Human body is a complicated tissue. Body and mind influence each other deeply. The therapy is complicated and should be done physically to mentally. Firstly, muxibustion, a method of traditional Chinese medicine, is good at the treatment of depression caused by lesions of Kidney and Stomach Meridian. Secondly, Vipassana, an oldest India meditation, is helpful for depression in subconscious. Lastly, family constellations can show the directions of solutions for depression caused by system involvements.

By the way, it is not have to regard the depressed moods as a problem. Some people with a little depressed mood often have great achievements as great litterateurs, artists, and scientists.

Acknowledgments

Thanks for the wise decision of my parents to abandon the antidepressant –Paroxetine Hydrochloride, a drug for depression, which was prescribed mistakenly by doctor when I had a toothache. Thank my brother for his company to help me to overcome from the depression caused by the toothache and the drug. Thanks a lot to my friends and leaders and colleagues of Tsingtao Technology University for their concerns and supports. And thank BMS for the platform of exchange.

References

1. Xu Wenbing, Liang Dong. *Huangdi Neijing·Yifa Fangyi.* Jiang Xi: Jiangxi Science and Technology Press, 2014;106-117.
2. Xu wenbing, Liang Dong. *Huangdi Neijing·Siqi Tiaoshen.* Jiang Xi: Jiangxi Science and Technology Press, 2013;162-166.
3. Qu Limin. *Speaking the health from head to foot.* Wu Han: Changjiang Literature and Art Press, 2008.
4. Peng Xin. *A way from essence injury to enriching.* Peking: Science and Technical Documentation Press, 2014.
5. Xu Wenbing, Liang Dong. *Huangdi Neijing·Shanggu* Tianzhen. Jiang Xi: Jiangxi Science and Technology Press, 2013;5-8.
6. [German]Bert Hellinger, Gunthard Weber, [America] Hunter Beaumon. *Who is in my home: Hellinger's family constellations.* Peking: World Publishing Corporation, 2003;1-132.
7. Fan Changwei. *Fan Changwei's simple Moxibustion.* Peking: Beijing Publishing House, 2010;1-17.
8. Wilfred Nelles. *The truth:A nostrum for spiritual cure.* Hong Kong: Red Publish, 2008.
9. Louise Hay. *You Can Heal Your Life*. California: Hay House Publishing, 1999.
10. Li Zhongying. *Remodel Soul.* Peking: World Publishing Corporation, 2006.
11. William Hart. *VIPASSANA MEDITATION*. New York: Harperone, 1987.
12. Eckhart Tolle. *The Power of Now*. Vancouver: Namaste publishing, 1997.
13. Bertold Ulsamer. *Family constellations for Beginners.* Peking: Chemical Industry Press, 2009.
14. Bert Hellinger. *Orders of Love:The cases of family constellations*. Peking: World Publishing Corporation, 2005;351-513.

INVESTIGATE AND RESEARCH COMPREHENSIVE HOSPITAL NURSING CHARGE ITEMS IN JILIN

Jing Li[a], Weiwei Fan[b], Chenxu Zhao[c], and Changping Song*

Nursing College, Beihua University, Binjiang East Road No.3999 in Jilin City of Jilin Province, China

E-mail:[a]364760038@qq.com,[b]331848645@qq.com

**Corresponding author*

Discuss the current comprehensive hospital nursing charge items in Jilin that is determine to add new nursing charge items. Methods: The questionnaire survey of field investigation on the nursing items and charges.The expert interview method are used to define the nursing charge items which should be added to. The data which be obtained is used for statistical analysis by SPSS13.0 software. Results: A total of 90 projects were carried out in a wide range of nursing services, and the actual nursing services were significantly increased.Charge item 47, no Charge items 43, the proportion of nursing fees low.The expert interview method define the nursing charge items 39 which should be added ($P < 0.05$, the difference is statistically significant)Conclusion:Nursing service fees need to be expanded;By attaching great importance to the nursing work, adjusting the structure of the medical treatment charge and reflecting the labor value of nursing staff in order to promote the development of nursing career.

1. INTRODUCTION

Hospital as an economic entity, it provides health care services should be public welfare, and special products with characteristics of labor services. Hospital charges in our country is mainly based on the contents of doctor's advice, the other is the consumables in nursing activity. Hoever the cost of labor in the process of nursing is often ignored,result in the deviation of nursing charge and nursing cost.That could not reflect the value of nursing services compensation [1]. A study shows that the level of nursing fees and the cost of nursing care has a greater distance [2]. Nursing staff labor value is not reflected,which causes the nurses"" working enthusiasm is not high,and the team is not stable and restrict the professional development of nursing science. This study provide scientific theory basis for the construction of nursing fees system suited to the characteristics of the Jilin Province.

2. REASEARCH OBJECTS

Including 5 therapeutic areas in the Department of internal medicine and 5 surgical treatment areas. Department of internal medicine includes crculatory Department of internal medicine,digestive internal medicine,internal medicine-neurology, respiratory medicine.Surgery consists of 5 therapeutic areas, including department of neurosurgery, department of thoracic surgery, gastrointestinal surgery, department of hepatobiliary surgery, bone surgery, as well as obstetrics and Gynecology, pediatrics, operation room. The above nursing unit covers all nursing items. Study nursing project charging status and put forward more suitable for the local area of nursing cost accounting system.

3. METHOD OF INVESTIGATION

3.1. Historical Review

According to "National Medical Service Price Project Specification (Trial Implementation)" and "Medical Service Price in Jilin Province (Trial Implementation)" content request. List of nursing items, analysis of current nursing fees and charges in Jilin Province, and with the current national standard of care in medical service fees and charges for the project and the standard for comparison. Retrospective investigation on the data of hospitals purchasing price list, medical equipment price list, disposable consumable price list and hospital financial statements ect, in order to understand content and contribution ratio about cost structure and sharing project.

3.2. Field Survey and Questionnaire Survey

Using stratified random sampling method, in a standardized questionnaire form The Affiliated Hospital of the first clinical medical college of Beihua University 13 nursing units field investigation for the project situation, to explore the nursing service items have been extensively developed.

3.3. The Expert Interview Method

A detailed account of the questionnaire about the current nursing service projects have been carried out,Has been carried out but not charging nursing services and has been carried out and should charge nursing service items,To charging but not care services defined theory, principle, discuss and record all kinds of ideas and opinions.Explore the province should add new nursing fees. In a standardized questionnaire form the first clinical medical school affiliated hospital of north China university 13 nursing unit field investigation for the project situation.The definition of 12 experts,which are engaged in clinical nursing management, the average working age of 23.6 years. Professional title, academic qualifications and positions are detailed in Table 1.

Table 1. Participate in defining the new nursing fees experts

Title	senior	10
	intermediate	2
	primary	0
Education	master	2
	undergraduate	10
	junior college	0
Post	Director of nursing department	1
	Finance Director	2
	Head nurses of Department	7

3.4. Data Statistics

With FOXBASE controing collected data entry and input. Statistical analysis using SPSS13.0, propose nursing fee items which Should be added.

4. RESULTS AND DISCUSSION

4.1. Nursing Fee Item Is Low

This research shows that, from 2001 years the ministry of health issued by the national medical service price project specification (try out) "in 3971 the charge in the project ,nursing charge more than 30 items, accounted for only 1% of the total fees; in October 2006, under a rigorous assessment, the National Development and Reform Commission, the Ministry of health and State Administration of traditional Chinese medicine jointly developed a "the national medical service price project specifications"(in 2007", a total of new and revised 345 projects, including 204 new items and related nursing charges 3, accounting for only 1.5 percent of new projects; and nursing related fees 7 of 141 new revised , accounted for about 5%. Therefore, plus original nursing charges, the number but also more than 40 items. At present, the Jilin Province medical institutions charge items and price, still in the Jilin Province medical service price (Trial) "(Jifa change price supervision conjunction words [2006]1507 number, since January 1, 2007 implementation) as the basis, the number of the" medical service price project "is 3964, including hospitals and a total of nursing service related projects is 47, account for 1.2%, nursing fees ratio lower.

4.2. The Actual Nursing Service Projects Increased Significantly

In three general hospitals has been widely carried out nursing service items and charges nursing project survey, the results show that the hospital has been widely carried out nursing service projects a total of 90, and fees strictly to the Jilin Province medical service price (for Trial Implementation) "as the basis, nursing services projects a total of 47, oral administration technology, postural drainage technology, drip eyedrop, terminal disinfection, health education and 43 were not was included in the fees that 43 nursing service project is cost free services, means that 48% of the nursing work despite the need to ask the nurse into the social necessary labor time, but there is no direct economic benefits.

4.3. Chargeable Nursing Service Items

Expert interview to define whether 43 not charging care services should be charge or not charge. With FOXBASE controing collected data entry and input. Statistical analysis using SPSS13.0, $P < 0.05$, the difference is statistically significant, propose 39 nursing fee items which Should be added, accounted for 91%such as chemotherapy for intravenous injection, closed thoracic drainage and so on.

5. CONCLUSIONS

5.1. Nursing Service Fees to Be Expanded

At present, the First Affiliated Hospital of Medical College of Beihua University strictly to the Jilin Province medical service price (Trial) as the basis, the number of nursing service fees is 47.However, a total of this survey display currently the hospital has been carried out nursing service items is 90 There are 43 items of nursing service for cost free service which means that the nursing work despite 48% of the need for nurses into the social necessary labor time, but there is no direct economic benefits. To determine the current international common standards for nursing services specific content is nursing project cost classification (nursing intervention classify, NIC). According to the NIC about the classification and contents of assessing,nursing service items are divided into six categories ,a total of 486: basic physiological needs, physiological needs, out on need for nursing, safety nursing needs, family nursing needs , health care needs.Therefore,the hospital have 47 nursing fees only account for 9.7% of the NIC classification system in the project, illustrate our province nursing service fees need according to the actual situation of national and provincial development and expanding.

5.2. Pay Attention to Nursing Work and Adjust the Structure of Medical charges

Influenced by traditional ideas, the independence of the nursing profession still evident and received by everyone. Because the mistakes that the nurse was fed by doctors,making the nurse in the field of health status is low. As the saying goes "three on the treatment, seven points by raising". With each operation nurses sweat, hospital income all contains nurse labor, but to create economic value is not reflected. The nursing service and nursing project pricing mainly by the administrative department of health, to determine the price. The nature of nursing service with labor productivity, have their specific value. The value of nursing services is the total value[3] of products of the labor value of nursing services and services created by itself.At present, it is generally reflect " To see a doctor is expensive", therefore, it is recommended that the competent department of health supervision, reduce the drug income in the proportion of medical costs, reduce the charge standard of sophisticated instruments and equipment, improving medical services in the appropriate technical standards, not only can control the excessive growth of medical costs, but also can embody the labor value of nursing staff, promote the development of nursing career.

References

1. Chun Xia,ect. The analysis of clinical nursing cost and benefit. *J Nanfang Journal of Nursing* 2005;**12**(8):88-89.
2. Wenqin Ye,ect. Study on actual cost and standard cost of nursing degree insurgical wards of the first grade hoslital in shanghai. *J Chinese Journal of Nursing* 2005; **40**(11):812.
3. Yue-e Chen. Research status of nursing cost accounting in China and thinking of. *J Nursing and Rehabilitation* 2010;**9**(2):109-111.

APPLICATION RESEARCH OF PHYSICAL EDUCATION BASED ON EXERCISE LOAD EVALUATION OF HEART RATE DATA MODEL

Jun Su[1,2], Song Zhang[1], Lin Yang[1,a,*], and Xiaohe Li[3]

[1]College of Life Science and Bioengineering, Beijing University of Technology, Beijing, 100124, China

[2]Youth Sports Department, Beijing Municipal Bureau of Sports, Beijing, 100075, China

[3]ZhongShiHongLi (Beijing) Health Technology Co. Ltd., Beijing, China

E-mail:[a]yanglin@bjut.edu.cn

[*]Corresponding author

Objective: Monitoring and evaluating students' physical education training quality according to students' exercise heart rate data by a teenagers' youth sports load monitoring and evaluation system to realize accurate monitoring and evaluation of each student physical education training quality, realize the process monitoring of students' physical exercise. Methods: The whole motion intensity is divided into three levels: low intensity exercise, medium intensity exercise and high intensity exercise (including extra high intensity exercise) according to the motion of the heart rate collected. Then the movement time of each intensity exercise was calculated. Results: By monitoring the movement of heart rate and duration, in the same class, due to the different levels of effort, the effect of the movement of the students are different. Conclusion: By monitoring the heart rate and movement duration, each students' daily exercise load was visually showed. It provides objective basis and instruction for physical education teachers.

1. INTRODUCTION

With the development of society and the improvement of people's income, the children's physical constitution in China has not been developed. On the contrary, in recent years, the students' physical monitoring data show a gradual downward trend, children's physical health is becoming increasingly prominent. The students' physique growth is facing many difficulties, and the situation is not optimistic. While taking various measures and means of intervention, such as the implementation of the "sunshine sports project", national student physique monitoring once in 5 years, holding the "student physique test match" as one of the methods of evaluation of school physical education work. But these measures and methods can not achieve real-time acquisition and evaluation of students' physical training in the process of exercise intensity and volume. The existing means is difficult to carry out supervision and evaluation of students' physical exercise [1-2]. A survey shows that only 18.3% of middle school students have physical exercise every day for more than 1 hours. In extracurricular activities, the sedentary time and less time activities such as watching TV, computer operation and doing homework accounted for 14.6% of the day time, and outdoor activities, do chores and physical exercise accounted for only 8.4% of the day, far less than the sedentary time [3-9].

Heart rate is an important indicator for students' physical strength and energy consumption. At present, in the research of competitive sports, heart rate data of athletes are real-time acquired and tracked by heart rate watch to compute the motion intensity, effective movement (medium to high intensity) time and proportion, and to calculate energy consumption levels. Monitoring the change of heart rate of students in primary and middle school during sports class is not only one of the effective means of exercise, but also one of the important bases of reflecting the school work of enhancing students' physique [10].

2. METHODS

In this study, the exercise intensity and duration were used for comprehensive evaluation of the exercise load. Combined with domestic and international authoritative children and adolescents physical exercise standards, prescription and recommendations, from the perspective of accuracy and feasibility, the ideal standard and the qualified standard for exercise quality monitoring were established initially [11-13]. The whole motion intensity is divided into three levels: low intensity exercise, medium intensity exercise and high intensity exercise (including extra high intensity exercise) according to the motion of the heart rate collected. Then the movement time of each intensity exercise was calculated. In this study, we evaluated the exercise load from two aspects: exercise intensity and duration.

The maximum heart rate $HR_{max}=220 - age$. If the actual maximum heart rate in the exercise exceeds the calculated heart rate, the actual maximum heart rate was used. A maximum heart rate in a certain period (such as one week) was used. Quiet heart rate (QHR): the heart rate recorded in the quiet state. A quiet heart rate in a certain period (such as one week) was used. Heart rate reserve: HRR= quiet heart rate + (the maximum heart rate - quiet heart rate) * heart rate reserve percentage. All physical activity is divided into low intensity, medium intensity, high intensity, extra high intensity according to the intensity. According to the intensity of the oxygen uptake, the intensity of the exercise was quantitated: the low intensity of the 0-40% reserve aerobic capacity, medium intensity 40-60% and high intensity for 60-85%.

In afternoon physical education of a middle school in December 2014, the heart rate of 12 boys were real-time monitored. The basic information of the 12 boys were showed in Table 1.

Table 1. The basic information of the 12 boys

Numbers	Age (year)	Height (cm)	Weight (kg)	BMI (kg/m2)
12	12.74 ± 0.68	161.15 ± 8.82	59.67 ± 16.83	22.69 ± 4.91

Test instrument is POLAR TEAM2 system, including a notebook computer (the installation of POLAR TEAM2 software), 12 heart rate sensors, 1 base units with 12 heart rate belts and other team system accessories.

Before class, the subjects with POLAR keep quiet and awake for 10 minutes while sitting in the room. The heart rate curve of the subjects was recorded by software, and the heart rate was observed in real time and remained silent for 3 minutes without significant fluctuation. During the student movement, the student's heart rate was recorded and the heart rate curve was formed.

3. RESULTS

Twelve students had monitored the exercise load. The time were from 15:20 ~16:05, which lasted for 40~45 minutes. Because wearing POLAR needs some time, statistical analysis of the research data was intercepted for 30 minutes, which was the main part of physical education. The teaching content was running and skipping, which was mainly in aerobic exercise a lesson. The results were shown in Table 2.

Table 2. The basic information of the 12 boys

Numbers	Average HR	Low intensity duration (percentage)	Medium intensity duration (percentage)	High intensity duration (percentage)	Extra high intensity duration (percentage)	The main component of intensity
Subject 1	92	25:26(85%)	00:35(2%)	00:00(0%)	00:00(0%)	Low
Subject 2	121	18:32(62%)	06:54(23%)	03:46(13%)	00:00(0%)	Low
Subject 3	125	17:29(58%)	04:14(14%)	02:58(10%)	01:22(5%)	Low
Subject 4	130	14:08(47%)	07:58(27%)	03:35(12%)	02:12(7%)	Low
Subject 5	133	10:59(37%)	04:22(15%)	08:11(27%)	03:32(12%)	Low
Subject 6	135	09:29(32%)	09:18(31%)	08:35(29%)	01:52(6%)	Low
Subject 7	124	01:27(5%)	18:56(63%)	05:31(18%)	00:16(1%)	Medium
Subject 8	130	10:06(34%)	14:03(47%)	03:48(13%)	00:26(1%)	Medium
Subject 9	149	04:58(17%)	13:17(44%)	07:15(24%)	04:11(14%)	Medium
Subject 10	134	08:52(30%)	12:51(43%)	04:55(16%)	01:53(6%)	Medium
Subject 11	127	04:55(16%)	12:19(41%)	06:33(22%)	02:32(8%)	Medium
Subject 12	183	01:39(6%)	00:00(0%)	01:13(4%)	23:24(78%)	Extra high

By analysing students' heart rate, average heart rate scope of the 12 boys was 92-183b/min, which was a large span. Among them, Subject 1's average heart rate was 92 b/min in the whole class, which was close to the quiet heart rate. The lesson for him, can not reach the aim of strengthening the body. Subject 12's average heart rate was 183 b/min which was in a high level. If excluding the poor POLAR wearing or other hardware reasons, Subject 12 was considering not suitable for exercise that day.

4. CONCLUSION

By comparing the data of 12 boys in the same physical education class, the result shows that the boys have different performance. By monitoring the heart rate and movement duration, students daily exercise can be visually showed. It provides objective basis and instruction for physical education teachers.

Acknowledgments

This research was financially supported by Projects of Beijing Municipal Commission of Education (PXM2015_014204_500064, PXM2016_014204_500079), Project of Excellent talent of Organization Department of Beijing Municipal Party Committee (2014000020124G044), Project of Excellent talent "RiXin" of Beijing University of Technology.

References

1. Oluyomi AO, Whitehead LW, Burau KD, Symanski E, Kohl HW, Bondy M. Physical activity guideline in Mexican-Americans: does the built environment play a role? *J Immigr Minor Health* 2014; **16**: 244-255.
2. Adamo KB, Langlois KA, Brett KE, Colley RC. Young children and parental physical activity levels: findings from the Canadian health measures survey. *Am J Prev Med* 2012; **43**: 168-175.
3. Zhu YF. Student physical health affecting factors and scholastic physical education. *J Phys Edu* 2006; **13**: 141-144. (In Chinese).
4. Zhu CM, Zhang YM. The advancement and developing trend of physical fitness study in China. *China Sport Sci* 2008; **28**: 25-32. (In Chinese).
5. Yang GH, Yang GQ. Research on the effective evaluation system of the University's Sunshine Sports Movement. *China Adul Edu* 2011; **24**: 131-133. (In Chinese).
6. Xiao SZ, Fang XL, Zhan ZF, Sun XB, Xiao ZF. Research on a new method and index for evaluating cardiac function of the athletes. *China Med Dev Infor* 2002; **4**: 34-36. (In Chinese).
7. Guo XM, Yang Y, Xiao SZ, Xie ML, Wu WZ. Student physical health affecting factors and scholastic physical education. *Chin J Appl Physiol* 2010; **26**: 380-383. (In Chinese).
8. Zhang SL, Zou GZ, Zhang J, Wang J, Yu CL, Wu C, Wang ZP, Zhang J. The design of the evaluation system for the sport technique and the effect to the institute of physical education. *Sports Sci* 2000; **21**: 51-54. (In Chinese).
9. Li R, Jiang CM, Rui R, Wang M. Post-exercise recovery heart rate reflecting heart fitness – limitation of fitness index in evaluation of heart fitness associated with increasing age. *China Sport Sci* 2012; **32**: 81-84. (In Chinese).
10. Ravishankar N, Mujja A, Lewis MG, Nair NS. A tool to measure complexity in public health interventions. *Clin Epidemiol Glob Health* 2014; **2**: 80-86.
11. Bandini LG, Schoeller DA, Fukagawa NK, Wykes LJ, Dietz WH. Body composition and energy expenditure in adolescents with cerebral palsy or myelodysplasia. *Pediatr Res* 1991; **29**: 70-77.

12. Buffart LM, van den Berg-Emons RJ, van Wijlen-Hempel MS, Stam HJ, Roebroeck ME. Health-related physical fitness of adolescents and young adults with myelomeningocele. *Eur J Appl Physiol* 2008; **103**: 181-188.
13. Montirosso R, Morandi F, D'Aloisio C, Berna A, Provenzi L, Borgatti R. International classification of functioning, disability and health in children with congenital central hypoventilation syndrome. *Disabil Rehabil* 2009; 31 Suppl. 1: S144-152.

CAROTID PLAQUES ASSOCIATED WITH CEREBRAL ISCHEMIA: FINDINGS IN HEALTH ASSESSMENT AND PHYSICAL EXAMINATION OF 417 AGED

Le Fang[1], Tianwei Wang[2,a,*], Jingbo Wu[2], Feixiang Zhang[1], and Yang Song[1]

[1]*Neurology Department of China-Japan Union Hospital, Jilin University, Changchun, 130000, China*

[2]*Radiology Department of China-Japan Union Hospital, Jilin University, Changchun , 130000, China*

E-mail: [a]wtw567@sina.com

[*]*Corresponding author*

To investigate the relationship between the nature of carotid plaques and ischemic cerebrovascular disease through the assessing carotid ultrasound in the elderly (>60 years). Methods Individuals(>60 years) living in the same community who had attended in health assessment and physical examination were included in this retrospective research. A total of 417 community-dwelling subjects were recruited. 244 subjects of cerebral infarction and 23 subjects of recurrent TIA were taken as the group of ischemic cerebrovascular disease, and another 150 healthy subjects were taken as the control group. All subjects were grouped based on history, physical examination and imaging of CT and/or MRI. Carotid duplex ultrasonographic assessment was performed in all subjects to determine the ultrasonographic correlates of unstable carotid plaque and cerebrovascular disease. Detailed analyses of carotid plaques included nature, distribution, number, shape and echo characteristics. Results The detection rate of unstable plaque was 83.5% in the cerebrovascular disease group and 52.7% in the control group. The former was significantly higher than the latter ($p<0.01$). Conclusion Unstable carotid plaque, as assessed with carotid duplex ultrasonography, is related to the presence of ischemic cerebrovascular disease. These findings encourage further prospective studies in asymptomatic subjects focused on echographic carotid plaque characteristics as predictors of subsequent TIA or stroke.

1. INTRODUCTION

The importance of carotid artery atherosclerosis in the pathogenesis of cerebral ischemia and infarction has been recognized for many years[1]. Early detection and assessment of carotid plaque and stenosis could help to prevent the occurrence of ischemic cerebrovascular disease. Carotid intima-media thickness and presence of carotid plaque have been found to relate to subsequent ischemic stroke in several large cohort studies mostly performed in whites[1-3] . But according to the report in Asian populations, large artery intracranial occlusive disease is the most common cause of ischemic stroke[4-6]. Thus, we conducted this study to examine whether carotid plaque was correlated with ischemic cerebrovascular disease in community-dwelling health assessment and physical examination project.

2. METHODS

2.1. Object

In total, 417 community-dwelling subjects(>60 years) who had attended in health assessment and physical examination were recruited in this retrospective research from July 2012 to November 2012. 244 subjects of cerebral infarction and 23 subjects of recurrent TIA were collected as ischemic cerebrovascular disease subjects(case subjects). Ischemic cerebrovascular disease was confirmed based on clinical presentation, CT and/or MR imaging. The mean age was 66.1±10.4 years; 214 subjects (80.1%) were male. 150 healthy subjects were recruited as control subjects. The mean age was 64.7±10.1 years; 98 subjects (65.3%) were male, respectively.

2.2. CD Protocol

Carotid duplex ultrasonography(CD) was performed using LOGIQ Ultrasound System (GE, USA) equipped with a 7-MHz high-resolution linear array transducer. An experienced doctor examined all subjects in the supine position. Three scanning angles (anterior-oblique, lateral, and posterior-oblique) were used, focusing on the posterior wall of carotid arteries. Carotid plaques were detected in bilateral common carotid artery (CCAs), bifurcations, and extracranial portion of the internal carotid arteries. It was defined as the presence of focal wall thickening that is ≥50% greater than that of the surrounding vessel wall, or as a focal carotid intima-media thickness (CIMT) >1.5 mm protruding into the lumen[7]. Detailed analyses of carotid plaques included nature, distribution, number, shape and echo characteristics. Histologic classification of carotid plaques is based on their content of fatty, fibrous, and calcific material. This classification allows ultrasonic characterization into soft, dense, or calcified plaques, respectively, based on their echogenic properties. Ulcers or craters which modify the intraluminal surface of the plaque have also been detected. Characteristics of echogenicity divided into four types: Type I, intermediate or echolucent lipid soft plaque; type II, homogeneous intermediate fibrous plaque; type III, heterogeneous echorich plaque with calcification; type IV, heterogeneous plaque with ulceration.

2.3. Statistical Analysis

Ischemic cerebrovascular disease and CD findings were compared between the 2 subgroups in univariate analysis, using χ^2 test, or independent samples t test. All statistical analyses were performed in SPSS Statistics (version 10.0). Two-sided P values of <0.05 were considered statistically significant.

3. RESULTS

Incidence of unstable carotid plaque were different(Table 1) between 2 groups (83.5% in case subjects vs 52.7% in control subjects; p<0.01). The incidence of unstable carotid plaque increased in case subjects significantly. 225 plaques were in the BIF on the left side, and 218 plaques were in the BIF on the right side. The incidences of plaque in the BIF were different (Table 2) between 2 groups. When we analyzed the incidence of plaque for each type(Table 3), we observed high incidences of type Iband type IV in case subjects?while high incidences of type II and type III were observed in control subjects($p<0.01$). The higher total incidence of carotid plaque was observed in case subjects, simultaneously.

Table 1. Incidence of unstable carotid plaque in case and control subjects

	case subjects(n=267)	control subjects(n=150)
unstable carotid plaque	223(83.5%)	79(52.7%)
stable carotid plaque	70(16.8%)	71(47.3%)
$\chi 2$	17.791	
P	<0.01	

Table 2. Position of carotid plaque in case and control subjects

	Left side (plaque)			Right side(plaque)		
	CCA	BIF	ICA	CCA	BIF	ICA
case subjects	93	225	68	57	218	66
control subjects	56	64	37	40	67	36
$\chi 2$	15.296					
P	<0.05					

Table 3. Incidence of each type in case and control subjects

	Type I	type II	Type III	type IV
Case Subjects(n=267)	149	17	33	68
control subjects（n=150）	47	13	54	36
$\chi 2$	20.473			
P	<0.01			

4. DISCUSSION

The importance of carotid plaque in the pathogenesis of cerebral ischemia and infarction has been recognized for many years. The pathogenic role of carotid plaques in the development of stroke is thought to reflect their properties, position, or, more realistically, a combination of both. These plaques can cause stenosis, embolization, and thrombosis with corresponding neurological signs and symptoms.

In this retrospective research, a total of 417 community-dwelling subjects(>60 years) who had attended in health assessment and physical examination were recruited. Carotid duplex ultrasonography (CD) was performed in bilateral common carotid artery (CCAs), bifurcations, and extracranial portion of the internal carotid arteries of all subjects. The results show that the incidence of plaque is higher in case subjects than that of control subjects. When we analyzed the incidence of plaque for each type, we observed high incidences of type I and type IV in case subjects?while high incidences of type II and type III were observed in control subjects($p<0.01$).

With the use of high-resolution ultrasound B-mode imaging, carotid plaques that only poorly reflect the emitted ultrasound signals (echolucent, type I) have been shown to have a higher histological content of lipid and hemorrhage than plaques that strongly reflect ultrasound. Recent studies also indicate that echolucent compared with echo-rich carotid lesions are associated with a higher risk of development of neurological symptoms and brain infarctions[8].

Although the majority of previous pathological work investigating the pathogenesis of carotid artery disease focused on the extent and severity of atherosclerosis (luminal narrowing), recent work seems to indicate that plaque morphology may be important for the occurrence of clinical events. Carotid plaque ulceration(type IV) may be a cerebral embolic source, even when the degree of stenosis is less than 70%[9]. Carotid ulcerations are thought to increase the risk for cerebral embolism by acting as thromboembolic substrates or by allowing plaque components to fracture into the lumen[10]. The results of this research suggested that unstable plaques(type I & IV) place patients at risk of unpredictable ischaemic events($p<0.01$). Results from other researches also reported that high incidences of type I and type IV were observed in symptomatic carotid plaques while high incidences of type II and type III were confirmed in asymptomatic plaques. This finding suggests that unstable plaques may be a significant cause of ischemic cerebrovascular disease.

Carotid atherosclerotic plaque is similar to that found at other arterial sites. The carotid bulb is the major site of involvement in atherosclerotic stenosis[11]. In this research, higher incidences were confirmed both of carotid plaque and unstable carotid plaque in case subjects significantly($p<0.01$). Atherosclerosis in the carotid artery bifurcation has been extensively studied because of its great clinical importance. An inverse relationship exists between the in vivo measured wall shear stress(WSS) and the arterial wall thickness in the carotid arteries. Gnasso et al. measured the peak and mean WSS in the carotid arteries of 23 patients with evidence of plaque in the common carotid or bulb of one side and no plaque in the contralateral carotid tree[12]. The results demonstrate that the atherosclerotic involvement of carotid arteries is usually asymmetrical and that WSS is lower in the carotid arteries where plaques are present than in plaque-free arteries. These findings provide in vivo evidence for a strong association between shear stress and atherosclerotic lesions. Blood flow induced WSS influences many processes affecting vascular remodelling, including tissue regression[13]. Fibrouscapregression induced by high WSS might enhance plaque vulnerability, eventually leading to plaque rupture. A subset of plaques develops into vulnerable plaques in the early phase of atherosclerosis, containing lipids, calcifications, covered by a thin fibrous cap. When mechanical forces in the cap, due to the blood pressure, exceed the maximal strength of the cap, it will rupture leading to thrombus formation, and/or embolisation of thrombus or plaque material into the intracranial circulation. Rupture of the atherosclerotic carotid plaque can therefore result in neurological events like transient ischemic attack (TIA) or ischemic stroke[14].

In this research characteristics of echogenicity divided into four types: Type I, intermediate or echolucent lipid soft plaque; type II, homogeneous intermediate fibrous plaque; type III, heterogeneous echorich plaque with calcification; type IV, heterogeneous plaque with ulceration. we observed high incidences of type I and type IV in case subjects. Carotid atherosclerotic plaque developed into unstable plaques in this group, containing a thin fibrous cap, large lipid-rich necrotic core, increased plaque inflammation, positive vascular remodeling, increased vasa-vasorum neovascularization, and intra-plaque hemorrhage[15]. These characteristics together with the usual hemodynamic pulsating expansion during systole and elastic recoil contraction during diastole contribute to a high mechanical stress zone on the fibrous cap of the atheroma, making it prone to rupture. Same conclusion was reported from similar research in coronary plaques[16]. Complex coronary plaques were associated with intraluminal filling defect consistent with thrombus, ulceration, or irregularity. Complex carotid plaques had one or more of the following features: (a) ulceration, (b) irregular surface, (c) mobile thrombi on plaque surface, (d) predominant echolucency, and (e) heterogeneity with intraplaque echolucent areas.

we observed high incidences of type II and type III were observed in control subjects($p<0.01$). Carotid atherosclerotic plaque developed into unstable plaques in this group, containing fibrous tissue, calcification. The results demonstrate that the incidences of stable and unstable plaques are different in the patients of ischemic cerebrovascular disease and normal elderly.

Unstable plaques place patients at risk of unpredictable ischaemic events and in patients with such lesions, specific preventive treatment beyond long-term antiplatelet therapy can be used to prevent new or recurrent events, and that the choice of management should be individualized.

Unstable carotid plaque, as assessed with carotid duplex ultrasonography, is related to the presence of ischemic cerebrovascular disease. These findings encourage further prospective studies in asymptomatic subjects focused on echographic carotid plaque characteristics as predictors of subsequent TIA or stroke. Future studies are required for understanding the pathophysiology of atherosclerosis and the practical utility of these morphological and biochemical assessments in predicting stroke.

References

1. D'onofrio M, Mansueto G, Faccioli N, et al. Doppler Ultrasound and Contrast enhanced Magnetic resonance angiography in assessing carotid artery Stenosis. *RadiolMed(Torino)* 2006, **111** (1): 93–103.
2. Chambless LE, Folsom AR, Clegg LX, Sharrett AR, Shahar E, Nieto FJ, et al. Carotid wall thickness is predictive of incident clinical stroke: the Atherosclerosis Risk in Communities (ARIC) study. *Am J Epidemiol* 2000;**151**: 478–487.

3. Prabhakaran S, Rundek T, Ramas R, Elkind MS, Paik MC, Boden-Albala B, et al. Carotid plaque surface irregularity predicts ischemic stroke: the northern Manhattan study. *Stroke* 2006; **37**: 2696–2701.
4. Wong LK. Global burden of intracranial atherosclerosis. *Int J Stroke* 2006; **1**: 158–159.
5. Gorelick PB, Wong KS, Bae HJ, Pandey DK. Large artery intracranial occlusive disease: a large worldwide burden but a relatively neglected frontier. *Stroke* 2008; **39**: 2396–2399.
6. Yi L X, Yan C X, Ping C, et al. Correlation of Large Artery Intracranial Occlusive Disease With Carotid Intima-Media Thickness and Presence of Carotid Plaque. *Stroke* 2013, **44**(1): 68–72.
7. Touboul PJ, Hennerici MG, Meairs S, Adams H, Amarenco P, Bornstein N, et al. Mannheim carotid intima-media thickness consensus (2004-2006). An update on behalf of the Advisory Board of the 3rd and 4th Watching the Risk Symposium, 13th and 15th European Stroke Conferences, Mannheim, Germany, 2004, and Brussels, Belgium, 2006. *Cerebrovasc Dis* 2007; **23**: 75–80.
8. Froio A, Deleo G, Piazzoni C, et al. Carotid Plaque Echolucency Measured by Grayscale Median Identifies Patients at Increased Risk of Stroke during Carotid Stenting. The Imaging in Carotid Angioplasty and Risk of Stroke Study. *Noninvasive Cerebrovascular Diagnosis* 2010:149-154.
9. Nakamura T, Tsutsumi Y, Shimizu Y, et al. Ulcerated Carotid Plaques with Ultrasonic Echolucency AreCausatively Associated with Thromboembolic Cerebrovascular Events - Journal of Stroke and Cerebrovascular Diseases. *Journal of Stroke & Cerebrovascular Diseases the Official Journal of National Stroke Association* 2013, **22**(2): 93–99.
10. Grimm J M, Schindler A, Schwarz F, et al. Computed tomography angiography vs 3 T black-blood cardiovascular magnetic resonance for identification of symptomatic carotid plaques. *Journal of Cardiovascular Magnetic Resonance* 2014, **16**(1): 1-10.
11. Weert T T D, Sander C, Groen H C, et al. Atherosclerotic Plaque Surface Morphology in the Carotid Bifurcation Assessed With Multidetector Computed Tomography Angiography. *Stroke* 2009, **40**(4): 1334-40.
12. Gnasso A, Irace C, Carallo C, et al. In vivo association between low wall shear stress and plaque in subjects with asymmetrical carotid atherosclerosis. *Stroke* 1997; **28**: 993-998
13. Groen H C. Atherosclerotic plaque and shear stress in cartotid arteries. *Erkenntnis* 2010, **38**(8):2379-2381.
14. Mughal M M, Khan M K, Demarco J K, et al. Symptomatic and asymptomatic carotid artery plaque. *Expert Review of Cardiovascular Therapy* 2011, **9**(10): 1315-1330.
15. Moreno P R. Vulnerable plaque: definition, diagnosis, and treatment. *Cardiology Clinics* 2010, **28**(1): 1-30.
16. Filippos T, George S, John K, et al. Carotid plaque composition in stable and unstable coronary artery disease. *American Heart Journal* 2005, **150**(4): 782-9.

DESIGN AND PRACTICE OF COMMUNITY-BASED REHABILITATION MODE ON STROKE SEQUELA

Runying Liang[1], Shuai Qiao[3], Meng Lou[3], Yi Meng[2], Xuehui Chang[2], Wen-li Liu [1], Lizhi Shang[1], and Xiaoyan Ma[3]

[1]*Basic Medical School, Henan University of Traditional Chinese Medicine, zhengzhou, 450046, China*

[2]*Henan province hospital of TCM Encephalopathy Division, zhengzhou, 450000, China*

[3]*Henan University of TCM current Master students, Zhengzhou, China*

The rehabilitation mode of stroke sequela was established through the survey into the current situation in community, combined with the advantage of traditional Chinese medicine in recovery. Satisfactory results have been gained after carrying out active exploration and practice in the community. This integrated mode is more effective than single acupuncture modes, leading to the future development of stroke sequela rehabilitation in community.

1. THE CONCEPTION OF COMMUNITY-BASED REHABILITATION ON STROKE SEQUELA

Cerebrovascular accidents are commonly and frequently happened in China among the elders, also one of the leading causes of mortality, particularly in the north [1]. The disability rate of cerebrovascular accident survivors is about 70% to 80%, however, early rehabilitation treatment attaches great importance to the alleviation of sequela, to the reduction of disability rate, and to the activities of daily living (ADL).

Currently specialized health recovery centers are still limited in most parts of China; the vast majority of stroke patients cannot continue to receive formal, long-term, and effective rehabilitation after discharge from hospital, which confines the further recovery of stroke sequela, leaving many sequelas. Since a variety of stroke patients are widely distributed in China, community-based rehabilitation is a very good choice when economic circumstances are limited [2]. Making full use of traditional Chinese medicine's feature combined with modern rehabilitation therapy philosophy, to actively promote the functional recovery of stroke patients, to improve the patients' physical condition, to make psychological adjustment, and to improve their life quality eventually. Studies from abroad showed that patients received early community rehabilitation treatment feel more satisfactory than those from hospital rehabilitation, even though they get the same recovery effect. Besides, it cost less money in community [4]. Economical, effective, convenient, and sustainable, community-based rehabilitation helps the follow-up recovery of the patients.

Using self-designed questionnaires, we went to investigate 23 community health centers in 7 districts of Zhengzhou city. Rehabilitation services were all carried out in these centers, among which acupuncture was the most frequently applied, followed by movement therapy, external therapy with Chinese herbs, herbal administration, and “daoyin” (health preservation method combined respiration with exercise of the limbs). Others like physiotherapy, tuina, massage, moxibustion and cupping were applied sometimes, but not that often. The vast majority of patients accepts community-based rehabilitation, and was eager to know more about the related knowledge [5].

In view of the above situation of stroke sequela in community service centers, combined with the characteristics and advantages of Chinese medicine, we conducted acupuncture and moxibustion, massage with ointment, steaming and washing methods in order to regulate the meridians and collaterals, to activate and harmonize qi and blood, thus the patients can gradually recover in motor function. Chinese food therapy was also used to balance the diet, to adjust to different constitutions of the patients, so as to improve their healthy qi. Only by getting more healthy qi and increasing the resistance against disease, can the patients accelerate their overall function and recovery of the limb function. When motor function was transited from passive movement to active movement, Ba Duan Jin (eight-length brocade exercise) was introduced to the patients as a kind of rehabilitation exercise, which can help their balance, coordination and stability training.

Guided by the principle of low cost, easy to promote, and convenient for nearby rehabilitation, we set up the detailed process based on the previous studies, and designed a mode for the community-based rehabilitation on stroke sequela. Generally the rehabilitation mainly relied on traditional Chinese medicine methods, supplemented by western medicine. After a thorough understanding of the patients’ condition, the most suitable means was selected to promote the patients’ recovery.

Chinese medicine mainly use acupuncture and moxibustion, steaming and washing, massage with ointment, daoyin, and food therapy, complemented by western medication and movement therapy to provide guidance and training for the patients. We followed the patients’ psychological state with interest, and solved their mental problems as soon as possible. For patients with restricted mobility, home service is suggested provided that there are enough personnel. Below is the flowchart of community-based rehabilitation mode on stroke sequela in Fig. 1.

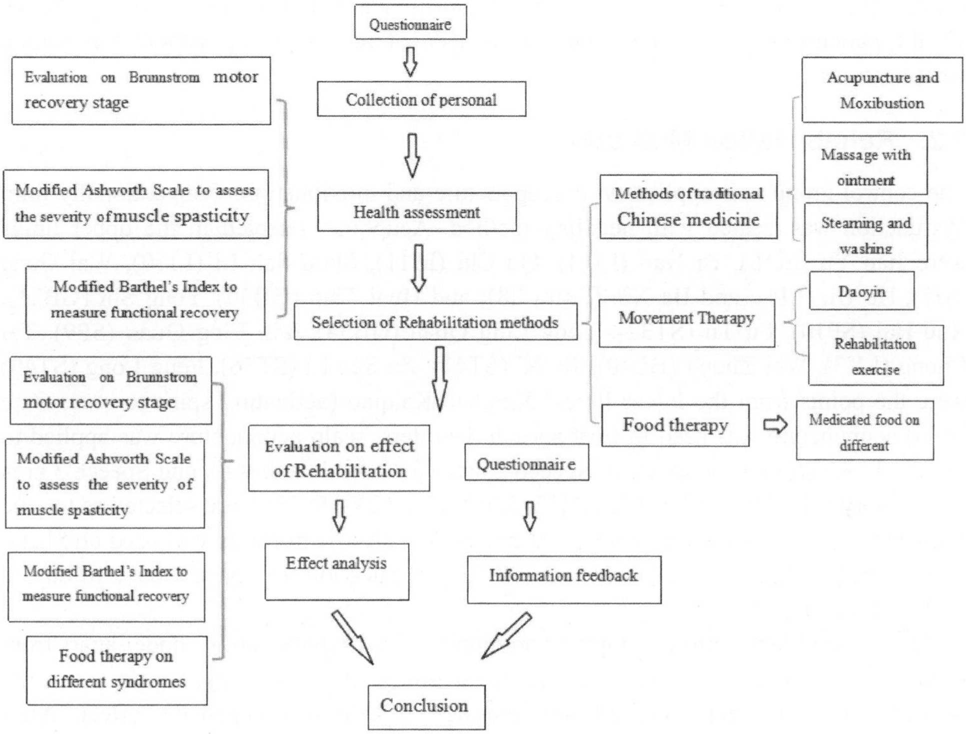

Fig. 1. Flowchart of Community-based Rehabilitation Mode on Stroke Sequela.

2. PRACTICE OF THE COMMUNITY-BASED REHABILITATION ON STROKE SEQUELA

From Oct. 2014 to July 2015, 150 cases with stroke sequela were selected as the objects of research from collaborated community hospitals, all these cases were diagnosed as cerebral hemorrhage or cerebral infarction through brain CT and magnetic resonance imaging (MRI). Cases included were 96 male patients and 54 female patients, from 38 years old to 86 years old, whose mean age was 66±9, and the course of disease varies from 0.5 to 15 years. According to admission time, 150 cases were randomized into rehabilitation group and control group, each with 75 cases. The basic data of the patients between two groups were no statistical significance ($P>0.05$).

2.1. The Entry Criteria

(1) the patients' disease conformed to the diagnostic name in TCM or western medicine; (2) the patients were in sequela period, in need of recovery; (3) without threatening complications.

2.2. Rehabilitation Methods

The control group was applied with acupuncture and moxibustion. Conventionally limb dysfunction was treated with needling method. Acupoints selected in the upper limbs were Jian Yu (LI15), Bi Nao (LI14), Qu Chi (LI11), Shou San Li (LI10), Wai Guan (SJ5), He Gu (LI4), and Ba Xie (Extra 28); and Huai Tiao (GB30), Feng Shi (GB31), Xue Hai (SP10), Fu Tu (ST32), Yang Ling Quan (GB34), Yin Ling Quan (SP9), Tai Chong (LV3), Wei Zhong (BL40), Jie Xi (ST41), Zu San Li (ST36), Feng Long (ST40) were the points from the lower limbs. Xingnao Kaiqiao (activating spirit to resuscitate brain) acupuncture was used to treat speech disorders, scalp acupuncture was applied to Speech 1 area (for motor aphasia), Speech 2 area (for anomic aphasia), and Speech 3 area (for sensory aphasia). Jin Jin (EX-HN12) and Yu Ye (EX-HN13) were selected as tongue acupuncture, and bloodletting with Ya Men (DU15), electric needling was used on Shang Lian Quan (RN23) and Xia Lian Quan (RN23). Application was once a day, 10 days a course of treatment.

The rehabilitation group adopted community-based rehabilitation mode, apart from acupuncture, ointment rubbing, steaming and washing, daoyin therapy, and food therapy on different syndromes were all put into use to gain comprehensive effect. After rehabilitation evaluation, doctors in the community hospital would make a recovery plan to carry out smoothly. Patients were reevaluated in 10 days and 20 days respectively, and the rehabilitation process was altered in accordance with the revised plan.

Massage with ointment: the ointment used in the research was invented by the research team in our university based on the excavation of ancient external therapy of stroke. Patent number is ZL201310055565.5. The ingredients of it were Dang Gui, Chuan Xiong, Qiang Huo, Du Huo, Fang Feng, Xu Changqing, Hua Jiao, Gui Zhi, Xi Xin, Chuan Wu, Fu Zi, Dan Shen, Chi Shao, etc. After apply the ointment to the affected limb, tuina techinique was used on the area. On the one hand, the ointment could penetrate into the body to activate blood circulation, and on the other, tuina technique is a positive stimulus to the patients, which helps the neuromuscular facilitation and the restoration of limbs. During flaccid paralysis stage, tuina technique was required to excite the contractive capacity of the muscle, thus heavy and strong stimuli is better to prevent muscle atrophy and dislocation of the joint; while in spasm stage, muscle tension should be relieved to relax the spasmodic muscle, therefore light and soft technique is desirable to gain relaxation.

Steaming and washing therapy: this method makes use of the heat of decoction to steam and wash on the affected skin. It is suggested to steam first, washing isn't applicable until it cools down. By means of the medicine and heating power, the therapy acts on the body through the skin and mucosa, so that Cou (muscular striae) is dredged, the meridians and collaterals are reconciled, and qi and blood flow smoothly, eventually limb disorder is improved. Four kinds of steaming and washing therapy are available to choose according to the patients' specific condition.

Daoyin therapy: By adjusting the posture and breathing, body-mind relaxation, control of the will, and rhythmic movement of the limbs, daoyin therapy can foster the subjective initiative capacity, arouse the potentials, and regulate the interior function of the body, thereby achieving prevention disease and detoxification purpose. Characterized by slow and soft, coherent, flexible, and elastic, dynamic and static, with physique and spirit, with qi and strength, the modified Ba Duan Jin (eight-lengh brocade exercise) is an appropriate exercise for patients with stroke sequela. According to the degree of stroke sequela, traditional Ba Duan Jin was adapted for standing, sitting and lying posture. In the form of CDs, the patients could learn and practice easily.

Medicated food therapy: As a product of herbal food and cooking method, the effect of healing and nourishment of medicated food is highly stressed by the ancient esteemed physicians. We mainly divided stroke sequela into three traditional syndromes, i.e. wind and phlegm blockage syndrome, collateral stasis due to qi deficiency syndrome, deficiency of the liver and kidney syndrome. Medicated food was applied as adjunctive therapy, and patients could follow our videos to cook their favorite medicated food.

Patients went to the community once a day for acupuncture and massage with ointment, steaming and washing was conducted at home once a day, medicated food was also taken once a day. The exercise of Ba Duan Jin was done at home once a day and patients chose the suitable position according to their own physical conditions. 10 days were a course of treatment, and there were 3 courses in total.

2.3. Criteria of Effect

After 3 courses of comprehensive rehabilitation, subjective symptoms were perceptibly disappeared, enunciation became clearly, paralyzed limbs could move flexibly with force and stability, and able to walk by themselves are the symptoms considered as complete recovery. Subjective symptoms improved, regaining of basic motor function of paralyzed limbs, and disappearing of the majority signs are regarded as marked recovery. Neurological signs and symptoms were slightly improved, affected limbs could move consciously; yet basic self-care ability wasn't possible, all these signs showed the sequelae were improved. Invalid means there were no obvious improvements after the treatment. The overall response rate (ORR) = recovery rate + marked rate + improved rate.

2.4. Statistical Analysis

SPSS Version 20.0 was used for data analysis and $p<0.05$ was taken as significant.

2.5. Result

After rehabilitation therapy, 25 cases were cured, 18 cases got marked effect, 16 cases were improved, and 16 cases stayed unchanged, thus the overall response rate was 79% in the rehabilitation group, while in the control group, the overall response rate was 51%, among which 12 cases were cured, 16 cases got marked effect, 10 cases were improved, and 37 cases got no improvement. Deterioration didn't happen in both groups, $p<0.05$. The result of the research suggests that the comprehensive rehabilitation mode work better than the single acupuncture rehabilitation mode.

We made relevant questionnaires to observe the patients' food, sleep, entertainment, and knowledge on stroke rehabilitation before inclusion and after treatment. The results are shown below in Table 1 and Table 2.

Table 1. Food, Sleep and Entertainment Condition Before and After Inclusion

Groups	Questions	Results of Survey			
...	...	Good	Ordinary	Bad	Very Bad
Before inclusion	Diet	11	31	25	8
After treatment		29	30	16	0
Before inclusion	Sleep	0	8	54	13
After treatment		1	18	56	0
...	...	Many	Some	Few	None
Before inclusion	Entertainment	0	5	44	26
After treatment		2	16	47	10

Table 2. The Recognition and Grasp of the Knowledge on Stroke Sequela

Groups	Questions	Results of Survey			
...	...	Know Well	Know Some	Know a Little	Know Nothing
Before Inclusion	Importance of Rehabilitation	1	29	29	16
After Treatment		15	44	14	2
Before Inclusion	Relevant Knowledge on Stroke Sequela	2	21	31	21
After Treatment		15	41	16	3

The application of comprehensive rehabilitation mode on stroke sequela in the community has better effect than single acupuncture mode. Taking full advantage of the characteristics of traditional Chinese medicine, the comprehensive mode is closely connected with daily life. Not only the functions of the paralyzed limbs are improved, but also promote daily life activities, regulate the patients' overall condition in sleep, diet, and mental state. Community-based rehabilitation plays an important role in improving the patients' recognition and understanding of related rehabilitation knowledge. Thus we predicted that the development of community-based rehabilitation will have a bright future.

Acknowledgments

Funding bodies: The project of science and technology talents in Zhengzhou, Henan province(NO:131PLJRC677).

References

1. Jin-Mei Yang, Jie Li. Effect of Early Rehabilitation Intervention on Extremity Movement Function in the Patients with Cerebral Hemorrhage. *J Journal of Nursing Science*,2001;**16**(2):76.
2. Jing Zhang, Shi-Wen Liu, Zhen-Lan Li, et al. Discussion on the Rehabilitation of Stroke in China. *J Chinese Journal of Rehabilitation Medicine*, 2006; **21**(10):946.
3. Xiao-Feng Jiang, Xue-Yan Hu,Tong Zhang. Study on Community-based Rehabilitation of Stroke. *J Chinese Journal of the Frontiers of Medical Science* (Electronic Version),2013;**5**(10):8-11.
4. Mackenzie AE, Chang AM. Predictors of quality of life following stroke. *J* Disabil Rehabil, 2002;**24**(5):259-265.
5. Run-Ying Ling, Meng Lou, Shuai Qiao. Survey on Current Situation of the Community Rehabilitation of Stroke Sequela. J Chinese Medicine Modern Distance Education of China,2015;**13**(10):126.

SIMULTANEOUS SEPARATION AND DETERMINATION OF SEVEN EFFECTIVE COMPONENTS IN COMPOUND RESERPINE AND HYDROCHLOROTHIAZIDE TABLET BY MICELLAR ELECTROKINETIC CAPILLARY CHROMATOGRAPHY

Suya Gao[a], Yun Zhang, Li Wang, Benquan Hu, and Yan Wang

College of Pharmaceutical Science, Xi'an Medical University, No.1 Xinwang Road, Weiyang District, Xi'an 710021, China

E-mail: [a]18591880376@163.com

A simple and inexpensive high performance capillary electrophoresis was applied to separate and determine seven effective components in Compound Reserpine and Hydrochlorothiazide tablets simultaneously. Two separation modes such as micellar electrokinetic capillary chromatography and capillary zone electrophoresis were investigated in the experiments. The various factors influencing the separation were examined including the pH of the separation system, concentration of surfactants, buffer and organic solvents, etc. The optimal separation was carried out at 20°C and 18kV, using a 10mmol/L monopotassium phosphate (KH_2PO_4) and 20mmol/L borax buffer and 30mmol/L sodium dodecyl sulfate (SDS) buffer (pH9.0).

1. INTRODUCTION

Over the last three decades, high performance capillary electrophoresis(CE) has been proved to be a powerful and highly efficient separation technique[1-4]. In particular, CE offers several advantages for separation including high efficiency (theoretical plate number up to 10^6/m orders of magnitude), short running time, different separation modes and mechanisms, small volumes and low running costs, and a tremendous flexibility for separations because of a wide variety of available additives, etc[5-8]. So CE shows great application prospect in the field of life science.

Compound Reserpine and Hydrochlorothiazide tablet is a kind of commonly used antihypertensive drug. The drug prescription contains vitamin B1, vitamin B6, reserpine, hydrochlorothiazide, dihydralazine sulfate, promethazine hydrochloride, librium, chloroquine phosphate, potassium chloride and magnesium trisilicate. In addition, the concentration of each component in this tablet is very important for our health and safety. Therefore it is very important to separate and determine the components by a simple method. Certainly, in recent years some experiments about this drug have been done by HPLC, CE, etc[9,10]. But it was little that seven components in the medicine were determined simultaneously in literatures. In this paper, by using micellar electrokinetic capillary chromatography (MECC), it was established to separate and determine simultaneously seven effective components such as vitamin B6, reserpine, librium, hydrochlorothiazide, bendazol, dihydralazine sulfate, promethazine hydrochloride, etc.

Satisfactory results were obtained, providing for the quality control and quantitative determination of Compound Reserpine and Hydrochlorothiazide tablet.

2. EXPERIMENTAL

2.1. Apparatus and Reagents

A Beckman MDQ capillary electrophoresis system equipped with diode-array detector (Beckman Instruments Inc., USA). Separations were performed in an uncoated fused-silica capillary (Hebei Yongnian Optical Fiber Factory, China) with a total length of 58cm (50cm to the detector) and 75μm internal diameter. The chromatography software 32Karat (Beckman Instruments Inc., USA) was used to acquire the data and control instrument. SB5200DT ultrasonic instrument (Ningbo Xinzhi Biologic Science and Technology Co. Ltd., Ningbo, China) were used to prepare the sample or mobile phase. Orion pH meter (Orion Research Incorporated, USA) was used to adjust pH of buffer and Sartorius precision electronic balance (Sartorius, German) was used to weigh the sample.

Compound Reserpine and Hydrochlorothiazide tablets were purchased from local drugstores in China, which manufacturer including pharmaceutical factory of Baoji Qinkun, Shandong Xinhua, Changzhou, Shanxi Linfen Jianmin, Shanxi Yabao, respectively. All reference substances were provided from Baoji Qinkun pharmaceutical factory. SDS and beta-cyclodextrin were from Tianjin Hongyan Chemical Reagent Company (Tianjin, China). The buffer and the solution were filtered through 0.45μm microporous membrane, respectively.

2.2. Procedure for CE Analysis

Before the experiment everyday, the capillary was sequentially rinsed for 10min with 0.1mol/L NaOH, water and 0.1mol/L HCl in order to activate the capillary column. Between runs, the capillary was rinsed for 3 min with water and for 5 min with the background electrolyte (BGE) solution. Online diode-array detector was performed from 190nm to 600nm. A voltage of 18kV was applied and the capillary temperature was controlled at 20°C by liquid cooling. Sample solutions were injected by pressure (0.5psi) for 5s. When the components were separated completely, the peaks were designated according to peak height and migration time to respond to increasing the concentration of single-component respectively. The signal was displayed and integrated for 0.1s interval.

2.3. Procedure for Pharmaceutical Formulations

All reference substances were accurately weighted and dissolved respectively in acetone and water up to 10mL. These were stored at 4°C in the refrigerator. Before injection, these stock solutions were diluted and mixed to the desired concentration.

For sample preparation, Compound Reserpine and Hydrochlorothiazide tablet 10 pieces were weighed, then were stripped icing and porphyrized. A suitable amount of powder was accurately weighed, and 5mL acetone was added to dissolve up to 25mL.

The mixture was stirred by a magnetic stirrer for 10 min and extracted by ultrasonication for 15 min. The sample was centrifuged and the liquid supernatant was filtered with 0.45μm membrane before it was injected into the HPCE system.

3. RESULTS AND DISCUSSIONS

3.1. Effect of Applied Voltage and pH of Butter on Separation

In this experiment the voltage was varied from 12kV to 24kV. The migration time of the compounds decreased with increasing voltage. In addition, the intensity of the peaks was also increased. But higher voltages than 18kV could not be used because of the generation of higher currents and noise. Therefore according to the absorption intensity and the migration time, 18kV was selected for further investigation.

The pH value of the 20mmol/L borax buffer was investigated from 4.5 to 11.0. Decreasing pH of the buffer led to a decrease in resolution and even peaks overlapped completely. When pH was less than 9.0, the number of peaks was fewer, the shape and reproducibility were poor. When pH was equal or greater than 10.0, some peaks were dispersed severely although more compounds could be separated exactly. This might be due to a larger difference in the mobility of the buffer and sample solution or a greater adsorption on the capillary wall. When the mobility of buffer was greater than the one of the sample, the peak could be dispersed. Only when pH equaled to 9.0, the number of peaks was maximum, the baseline and reproducibility were good.

3.2. Effect of Composition and Concentration of Buffer

By varying the concentration of the buffer it turned out that the greater the concentration was, the longer was migration time. Higher concentrations could not be used due to the generation of high current values. Using a greater than 50mmol/L sodium tetraborate buffer, peak dispersion occurred clearly. The satisfactory results were obtained from using 20mmol/L borax and 10 mmol/L monopotassium phosphate buffer. However, it was pity that peak-1 and peak-2 was overlapped in the above buffer, which mode was capillary zone electrophoresis (CZE). Therefore further optimization should be investigated for separation conditions. It was found from the experiment the resolution of the peaks could be improved by adding surfactant SDS. Furthermore, when SDS concentration is greater than its critical micelle concentration (CMC, 8.1mmol/L), CZE mode is changed to MECC. Then it turned out that the resolution and intensity of the peaks could be improved after adding 30mmol/L SDS to the buffer solution. In addition, reserpine, librium and hydrochlorothiazide were not easily soluble in water but soluble in acetone. In order to quantitate accurately the components, the sample was dissolved in acetone at first and then metered volume with water. Intensity of peaks was improved. It was worthy of note that two ways were adopted to qualitate the separated peaks, by comparing the migration time of the sample components to the corresponding standard compounds and by adding one standard compound to the sample to observe whether the corresponding peak intensity could be enhanced.

3.3. Determination of Standard Curves

The seven compounds were analyzed simultaneously using gradient analysis. Their standard work curves were determined by the described method with the separation buffer to give serial different concentrations (x, mg/mL), which were injected one by one to obtain peak areas (y). Calibration graphs were plotted by linear regression. The regression equations and linear ranges of seven markers were listed in Table 1. Greater correlation coefficients (*r*) revealed excellent linear relationships in their ranges.

Table 1. Linear regression equations and linear range of components

Components	Regression equations	*r*	Linear range(mg/ml)
Reserpine	$y=1.0\times10^7x+37146$	0.9991	7.2×10^{-4}~1.8×10^{-2}
Hydrochlorothiazide	$y=2.8\times10^6x+43686$	0.9989	1.6×10^{-2}~1.2
Vitamin B_6	$y=1.2\times10^5x+51408$	0.9995	2.2×10^{-2}~2.2
Vitamin B_1	$y=1.4\times10^5x+56302$	0.9993	4.2×10^{-2}~2.1
Chlorodyne	$y=5.0\times10^6x+15364$	0.9997	6.8×10^{-3}~3.4
Dihydralazine	$y=8.0\times10^5x-41972$	0.9961	1.5×10^{-2}~1.2
Promethazine	$y=5.0\times10^6x+15364$	0.9997	6.8×10^{-3}~3.4×10^{-1}

3.4. Method Validation Experiments

In order to assess the possible application of the proposed method, the mixed standard solution containing seven components was repeated injection 6 times in the selected experimental conditions. The relative standard deviation (RSD) of their peak areas for a six-day period was less than 4.0%. Intraday RSDs (n=6) were all less than 3.2%. The sample was injected every hour to check the stability. RSDs (n=6) of the peak areas of seven components were all less than 3.5%, which revealed that the sample solution was stable within six hours. Appropriate concentrations of seven components were injected separately to obtain the determination limits. Within a signal-to-noise ratio greater than 3:1 (S/N=3), the limits of detection (LOD) were less than 0.12 mg/L.

Several standard substances were commonly added into the sample simultaneously to test the recovery. The sample from Qinkun plant was determined accurately at first and the concentration of components was known. Then the mixed standard solutions with an appropriate amount were added into the samples and were processed. Their average recoveries and RSDs (n=3) were in the range of 96.2~104.8% and 0.9~4.1%, respectively.

3.5. Analysis of Real Samples

Following the procedure described previously, the proposed MECC method was applied to determine effective components of the Compound Reserpine and Hydrochlorothiazide tablets from different plants. The determination results of each tablet were listed in Table 2 and the electropherograms were shown in Fig.1.

Table 2. The content of the components in every tablet sample from different plants

Components	Xinhua (mg/g)	Yabao (mg/g)	Qinkun (mg/g)	Jianmin (mg/g)	Changzhou (mg/g)
Reserpine	0.03099	0.03202	0.03195	0.02769	0.02894
Hydrochlorothiazide	2.782	2.649	2.586	2.115	2.994
Vitamin B_6	3.068	3.591	3.927	2.597	3.875
Vitamin B_1	0.9894	1.008	0.9975	0.5036	1.027
Chlorodyne	N.D. [a]	1.582	1.884	N.D.	1.762
Dihydralazine	1.143	1.236	1.287	0.4584	1.215
Promethazine	1.829	1.762	1.916	1.618	1.898

Note: [a]N.D. = not detected.

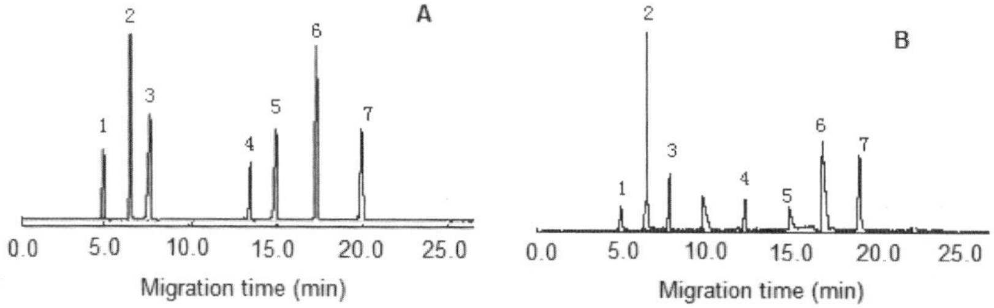

Fig. 1. Electropherograms of standards (A) and Compound Reserpine Tablet (B) by MECC.

1. Reserpine; 2. Hydrochlorothiazide; 3. Vitamin B_6; 4. Vitamin B_1; 5. Chlorodyne; 6. Dihydralazine Sulfate; 7. Promethazine Hydrochloride.

4. CONCLUSIONS

A simple MECC method was established for simultaneous separation and determination seven effective components of Compound Reserpine and Hydrochlorothiazide tablets. Several experimental parameters that influence separation were investigated including pH of the buffer, the concentration of borax, SDS, voltage, and so on. By being optimized process of experiment the best conditions were 20mmol/L borax – 10mmol/L KH_2PO_4 and 30mmol/L SDS buffer (pH 9.0) containing appropriate amount of acetone. A voltage of 18kV was applied and the capillary temperature was controlled at 20°C. The simultaneous separation of seven compounds was obtained in less than 22min with resolution greater than 2.0 when the optimized determination wavelength was performed at 215nm. The method was validated and applied for the quantification of effective components of Compound Reserpine and Hydrochlorothiazide tablets from difference plants. The separation and determination were satisfactory, quick and reliable.

Acknowledgments

This work was supported by the Scientific Research Program Funded by Shaanxi Provincial Education Department (No.14JK1620), the Ph.D. Startup Foundation of Xi'an Medical University of China (No.2012DOC12), Shaanxi Provincial Education Department (No.15JK1630) and the Undergraduates' Innovative Training Program (No.1647).

References

1. Jitka Široká, Daniel N. Polesel, Jose L. Costac, et al. Separation and determination of chlorophenylpiperazine isomers in confiscated pills by capillary electrophoresis. *J. Pharm. Biomed. Anal.*, 2013, **84**: 140-147.
2. Zhang Y J, Huang M X, Zhang Y P, et al. Use of sulfated cyclofructan 6 and sulfated cyclodextrins for the chiral separation of four basic pharmaceuticals by capillary electrophoresis. *Chirality*, 2013, **11**: 735-742.
3. Zhang W H, Lv Z H, Jiang T F, et al. Screening tyrosinase inhibitors from traditional Chinese medicine by capillary electrophoresis with electrophoretically mediated microanalysis. *J. Food Drug Anal.*, 2012, **20**: 159-163.
4. Deng X L, Yuan Y Z, Erwin Adams, et al. Development and validation of a sensitive enantiomeric separation method for new single enantiomer drug levornidazole by CD-capillary electrophoresis. *Talanta*, 2013, **106**: 186-191.
5. He J, Shamsi S A. Application of polymeric surfactants in chiral micellar electrokinetic chromatography (CMEKC) and CMEKC coupled to mass spectrometry. *Methods Mol. Biol.*, 2013, **970**: 319-348.
6. Gu C, Shamsi S A. Evaluation of a methacrylate-bonded cyclodextrins as a monolithic chiral stationary phase for capillary electrochromatography (CEC)-UV and CEC coupled to mass spectrometry. *Electrophoresis*, 2011, **19**: 2727-2737.
7. He J, Shamsi S A. Chiral micellar electrokinetic chromatography-atmospheric pressure photoionization of benzoin derivatives using mixed molecular micelles. *Electrophoresis*, 2011, **10**: 1164-1175.
8. Gao S Y, Li H, Wang L, Yang L N. Simultaneous separation and determination of benzoic acid compounds in the plant medicine by high performance capillary electrophoresis. *J. Chin. Chem. Soc.*, 2010, **6**: 1374-1380.
9. Serena Orlandini, Roberto Gotti, Sandra Furlanetto. Multivariate optimization of capillary electrophoresis methods: A critical review. *J. Pharm. Biomed. Anal.*, 2014, **87**: 290-307.
10. Zheng Cheng, Qing Wan, Yu-yun Guan, et al. Determination of hydrochlorothiazide, promethazine hydrochloride and reserpine in compound reserpine tablets by HPLC. *Anhui Med. Pharm.*, 2010, **14**(1): 38-40. (In Chinese).

APPLICATION OF TASK-BASED LEARNING MODE IN THE TEACHING OF CLINICAL PHARMACOLOGY

He Li[a], Cong Liu[b], Jinghui Sun[c], Chengyi Zhang[d], Hongxia Sun[e], Chunmei Wang[f,*], and Jianguang Chen[g,*]

College of Pharmacy, Beihua University, Jilin City, 132013, China

E-mail: [a]yitonglh@126.com, [b]372131552@qq.com, [c]413437244@qq.com, [d]372300630@qq.com, [e]543328673@qq.com, [f]893716972@qq.com, [g]644257703@qq.com

[*]Corresponding author

Clinical pharmacology is a branch of pharmacology, and a cross discipline to study the action law of drugs in the human body and the interaction between drugs and the human body by taking the human body as the research object. For the teaching design based on task, the first is to make clear the teaching objectives, then determine the contents and ways of tests, namely the teaching assessment, and finally design the teaching activities. The core of task-based learning (TBL) is to focus on the teaching objectives and its characteristic is the teaching assessment prior to the teaching design. In this study, a teaching method based on the learning task was applied in the teaching practice of clinical pharmacology, its effects on the teaching quality of clinical pharmacology were summarized, in order to make the curriculum teaching work closely combined with the clinical practice, and highlight the characteristics connecting basic medicine to clinical medicine, and pharmacy to clinical medicine, ultimately achieving the aim to improve the comprehensive quality of students.

1. INTRODUCTION

Clinical pharmacology is a branch of pharmacology, and a cross discipline to study the action law of drugs in the human body and the interaction between drugs and the human body by taking the human body as the object. On the basis of basic pharmacology and clinical medicine, clinical pharmacology expounds the action law between drugs and the body, including pharmacokinetics and pharmacodynamics (drug action, adverse reactions and mechanism of drug action). Clinical pharmacology is a bridge subject to make the basic theory and methods applied in the clinical practice, and based on basic medicine and clinical medicine, and pharmacy and clinical pharmacy. Clinical pharmacology is a basic course for clinical medical students. The aim to set up clinical pharmacology course is to let students understand the knowledge and research methods of clinical pharmacology, and the ultimate aim is to cultivate the awareness and methods of rational use of drugs, and train students to learn to formulate the reasonable and safe dosing regimens and make a scientific evaluation of the effectiveness and safety of drugs through a continuous learning and accumulation of experiences in the future clinical work[1-4].

For the teaching design based on the learning task (task-based learning), the learning task (teaching objective) must be clear first, then the content and the way of assessment should be determined, namely, the teaching assessment, and finally the learning activities are designed. The core of task-based learning is to focus on the learning task and its characteristic is that the teaching evaluation is prior to the teaching design.

For an effective teaching, a set of precise and concise teaching objectives must be established first. However, in the actual teaching process, many teachers have ignored the establishment of teaching objectives, such as some incomplete teaching objectives, and excessive emphasis on the cognitive or skills but general neglect in the cultivation of emotion, attitude and value. Some teaching objectives are not accurate, such as the lack of earnest research on curriculum standards and teaching materials, and the insufficient understanding of the nature of the subject, and the indifference in the education value implied by the knowledge of subject. Some teaching objectives are not targeted, and lack of in-depth understanding of students' knowledge, life experience, and the practical experience, leading to a randomness and inefficiency in the teaching. In this study, a teaching method based on the learning task was applied in the teaching practice of clinical pharmacology, its effects on the teaching quality of clinical pharmacology were summarized, in order to make the curriculum teaching work closely combined with the clinical practice, and highlight the characteristics connecting basic medicine to clinical medicine, and pharmacy to clinical medicine, ultimately achieving the aim to improve the comprehensive quality of students[5-7].

2. TASK-BASED LEARNING MODE AND CURRENT SITUATIONS

2.1. Task-based Learning

The teaching mode based on the task (task-based learning, TBL) is a teaching design based on the teaching objective, and the teaching based on the curriculum standard is not a teaching method, but an idea, in which on the basis of a profound understanding of the standards, the teachers are required to grasp the general expectations to the students, translate the standards into the grade objectives, and specify them into the learning objectives of each class according to the students' characteristics and the teaching situations.

TBL is a teaching design idea that integrates teaching objectives, teaching evaluation and teaching activities. The first step of this teaching design is to clear the teaching objectives, namely, what the students can achieve finally through the teaching of a class, the importance to highlight the establishment of teaching objectives, and putting the establishment of teaching objectives in the first place; the second step is to determine the contents and methods designed based on the teaching objectives for assessing the teaching, that is, the teachers can understand the students' achievement of three-dimension goal by what kind of evaluation methods; the third step is to determine the teaching activities, the teaching contents and methods to examine the teaching after the teaching objectives and evaluation methods, and the overall planning of the teaching activities.

One of remarkable characteristics of TBL is the teaching evaluation design prior to teaching activities. This design reflects an idea that the evaluation is to promote the teaching, in which the assessment and adjustment of classroom teaching can remedy the classroom teaching to maintain the consistency of teaching objectives and teaching evaluation, and the evaluation of the class can promote the teaching to ensure the effective achievement of three-dimensional goal and be able to effectively avoid "teaching only, no evaluation; content only, no goal; emphasis on the teaching by teachers only, ignoring the learning by students.

The teaching design with TBL is not only a design that clearly targets the teaching objectives, but also one in which the evaluation promotes the achievement of teaching objectives. It is different from the consideration that the teaching evaluation is put after the teaching activity in the traditional teaching, and requires that the teachers should think about the teaching activities with the teaching objectives and problems just at the beginning of the teaching design, to increase the pertinence of teaching activities.

2.2. Thinking on the current situations

Teaching objectives are the starting point and final destination of all teaching activities. Teaching evaluation is a process in which the values how teachers teach and how students learn are studied, and the importance both teachers' teaching and students' learning percent is self-evident, so that teaching evaluation should become the priority among priorities in teaching work. However, the phenomena that emphasize the teaching content, but neglect the teaching objectives and evaluation are not in a minority currently. The following problems should come into notice.

2.2.1. *Weak Consciousness of Teaching Objectives*

Quite a few teachers do not pay enough attention to teaching objectives. When they prepare for classes they often pay more attention to if the teaching is novel, the link are compact, the design of practice can seize the person eyeball, and so on, and just copy other references as their teaching objectives, in which the teachers lack of their own profound interpretation of teaching objectives as a part of teaching plans and the teaching objectives perform practically no function. Through exchanging ideas with the teachers, it is found that while formulating teaching plans, some teachers pay more attention to the teaching contents and time schedule in this semester, what they consider most is the teaching content and teaching methods, but why to teach the content and what are the teaching objectives are not often taken seriously.

2.2.2. *Mere Formal Classroom Assessment*

After the reform of curriculum, although there are more and more teachers have begun to participate in the classroom assessment, there is still a gap between teaching objectives and classroom assessment. The main reason why the classroom assessment becomes a mere formality is that teachers' understanding of classroom assessment is biased. The teachers holding this biased view feel that teacher's task is to complete the teaching of contents contained in textbooks, the assessment is what their superior sectors should do and their superior education sectors will uniformly organize examinations for the assessment, and students' learning is assessed by their scores, so that this kind of one-sided understanding leads directly to the teachers' neglect of the classroom assessment.

3. TBL TEACHING CASES

TBL is not a specific teaching method, but a kind of teaching philosophy, and thus it has no fixed mode of operation and reflects the renewal of teachers' teaching ideas behind their teaching behaviors. The following teaching cases of clinical pharmacokinetics are taken as examples to analyze the thinking way of the TBL teaching design.

3.1. Determination of the Specific Assessable Class Teaching Objectives

Teaching is evaluated and improved based on the behavior and performance of students, and it is important to clearly state the expected outcomes of teaching design before the teaching. The first step of TBL teaching design is to determine the teaching objectives, in which the class teaching objectives can be refined and specified in order to achieve the dynamic integration among the three links, namely the objective, assessment, and teaching.

In the determination of teaching objectives, try not to use some vague expressions, such as "preliminary understanding", "preliminary experience", "further", and so on, and try to avoid use the overall goal of the course as the class objective at the same time, because the application of vague expressions, even using indiscriminately the curriculum standards to replace the class teaching objectives, is the primary reason why the teaching objective and the teaching process can exist in a form of "two pieces of skins in the classroom. The classroom teaching objectives from an "abstract and broad" to a "narrow and specific" content can reasonably subdivide the overall teaching objectives into the teaching of each class, and the teachers can understand the significance of each objective to refine and materialize the teaching objectives, and highlight the materiality and testability of the teaching objectives, which will help to improve the teachers' classroom teaching execution and may have a more important guidance in the design of following class test and teaching links.

3.2. Determination of The Quality and Form of Assessment Contents

As dose the quality of class, the quality of assessment seriously affects the quality of teaching. Through the class assessment, not only the learning outcomes of students can be timely understood and the achievement of teaching objectives can be judged, but also it can be used as the basis to adjust the teaching to clarify the class teaching objectives, and what is closely followed should be the thinking of the class assessment. The content and form of assessment as well as the quantitative standards of partial assessment contents are determined according to the teaching objectives. The assessment is not for the classification and the punishment of students, but for providing a better help in learning for the students, and promoting and guiding the teaching through the assessment.

3.3. Design of Teaching Activities

The basis of teaching activities design is the teaching objectives, but the reorganization of the specific assessment contents, teaching contents and teaching links, the selection of teaching strategies, and the distribution of teaching hours are flexibly designed according to the learning characteristics of the students, the current learning situation and the teachers' personal teaching style.

"Objective", "assessment" and "teaching" are an organic whole, and are connected and integrated with each other, which does not start at the beginning of teaching, and coexists with the teaching activities, runs through the whole teaching from beginning to end, and learning activities, and the arrangement of classroom assessment and teaching activities is always to target the teaching objectives. This kind of teaching design is purposeful and planned, which can provide optimal learning conditions for students' achievement of teaching objectives and promote the development of students' effective learning activities.

4. CONCLUSIONS

Clinical pharmacology is an interdisciplinary subject, involving a variety of knowledge and combining the clinic closely, so that it is necessary for a good job in the teaching of clinical pharmacology to constantly improve the teachers' comprehensive qualities, enrich the teaching contents, embody the characteristics of combination of clinical pharmacology with the clinic, highlight the clinical characteristics, exactly use a variety of teaching methods and means, and develop evidence-based thinking and ideas, to stimulate the students' interest in learning, improve the students' learning enthusiasm and initiative, and achieve the teaching benefits as well as the students for improving the teaching effect and training high-quality compound medical and pharmaceutical talents to meet the clinical need, which may promote the development of the medical and pharmaceutical work, and ultimately improve the medical quality of the Chinese people.

References

1. Quan Ding. Teaching practice of clinical pharmacology and thinking on it [J]. *China Pharmaceuticals*, 2015; **24**(6): 9-11.
2. Qian Zhou. Effective teaching based on the objective. *Bulletin of Biology* [J], 2011; **46**(6): 30-32.

3. Derong Li, Zhandong Zhang. Analysis on the cultivation of students' learning interest in the teaching of clinical pharmacology [J]. *Mdical Information*, 2011, **24**(9): 4935-4936.
4. Juan Li, Shenyin Zhu, Feng Qiu, et al. Exploration on the teaching reform of clinical pharmacology [J]. *China Pharmacy*, 2012; **23**(4): 379-381.
5. Shaoyu Wu, Qiang Wei, Lin Lv, et al. Discussion on the teaching mode reform of clinical pharmacology [J]. *China Foreign Medical Treatment*, 2010; **29**(2): 117.
6. Jingru Meng, Min Jia, Xue Ma, et al. Discussion on the integrated application of various teaching methods in the teaching of pharmacology [J]. *Basic Medical Education*, 2011; **13**(5): 424-426.
7. Junxian Yu, Limin Shi, Rulong Wang. Teaching of clinical pharmacology needs evidence-based thinking [J]. *Journal of Clinical and Experimental Medicine*, 2013; **12**(3): 236-239.

ALZHEIMER'S DISEASE: MECHANISMS OF TAU AND AMYLOID BETA-INDUCED EXCITOTOXICITY

Xiaotong Lin[a] and Tingting Yan[b]

Department of Bioengineering, Harbin Institute of Technology, Weihai, Shandong 264209, China
E-mail:[a]18766312033@163.com,[b]alovelong@aliyun.com

Excitotoxicity is associated with various neurodegenerative diseases, among which Alzheimer's disease (AD) and Parkinson disease (PD) are the typical diseases. AD is a representative degenerative disease with clinical characteristics of senile plaques (SPs), neurofibrillary tangles (NFTs) and neuronal loss. Excitotoxicity has been considered as a mechanism of AD pathogenesis, which may induced by excessive accumulation of glutamate.

1. INTRODUCTION

Glutamate is the major excitatory neurotransmitter in central nervous system (CNS) and excessive glutamate stimulates the glutamate receptors will mediate the excitotoxicity. Increasing evidence show that excitotoxicity is associated with neurodegenerative disease, including stroke, hypoglycemia[1], cerebral ischemia, alcoholism, autoimmune encephalomyelitis, Alzheimer's disease (AD)[2], Parkinson disease (PD), Huntington disease and amyotrophic lateral sclerosis (ALS)[3][4]. Excitotoxicity was firstly proposed by Olney and coworkers in 1969, who found that glutamate treatment induced brain lesions[5][6]. Glutamate, as an excitatory neurotransmitter, participates in various physiological neuronal processes such as nociception, learning and memory. Relatively low concentration of extracellular glutamate prevents excessive activation of glutamate receptors maintaining cellular homeostasis[7]. However, excessive extracellular glutamate in brain could lead to acute neural cell damage[8]. Excessive synthesis or release of glutamate will cause excitotoxicity and then arouse neurotoxicity. Recent clinical and experimental trials have shown that excitotoxicity may occur early in the pathology process of AD and implicate with hyper-phosphorylation of tau and deposition of beta-amyloid (Aβ).

2. ALZHEIMER'S DISEASE (AD)

Alzheimer's disease was first named after the name of Alois Alzheimer, a Bavarian psychiatrist, at a meeting in Munich in 1906[9]. AD is a representative degenerative disease among various neurodegenerative diseases related with CNS, clinical characteristics of which including memory impairment, cognitive dysfunction, disturbing behavior and personality changes. The incidence of AD rises approximately logarithmically with age. According to recent study, every increase in age of five years old, the incidence of AD will be doubled[10]. With the continuous development of aging, Alzheimer's disease is becoming a hot issue of all over the world. However, specific pathogenesis of AD remains unclear at present. Furthermore, revealing the accurate pathogenesis of AD could establish theoretical foundation for its prophylaxis and treatment.

2.1. The Neuropathological Phenotype of Alzheimer's Disease

Preclinical AD, mild cognitive impairment (MCI) and dementia have been considered as major phases of pathological progression of AD[11]. Pathological characteristics of AD mainly include senile plaques (SPs), neurofibrillary tangles (NFTs) and neuronal loss[12].

2.1.1. *Aβ Deposition Formed Senile Plaques*

One of the classic pathological features of AD is characterized by β-amyloid (Aβ) deposition formed senile plaques. Aβ monomer exists by two forms, Aβ40 and Aβ42[13]. Meanwhile, Aβ fiber filaments formed by aggregation of Aβ, with β folded being its main protein second-grade structure. Aβ42, as the main components of senile plaques, aggregates faster than Aβ40, and also the cytotoxicity of Aβ42 is higher than that of Aβ40. Studies reveal that amyloid-producing pathway of AD patients are hyperactive[14].

2.1.2. *Hyperphosphorylation of Tau Protein Cause Neurofibrillary Tangles*

Hyperphosphorylated tau protein is a major component of neurofibrillary tangles, which consist of paired helical filaments (PHFs)[15]. The abnormal phosphorylation of tau protein is considered to be involved in formation of PHFs[16]. Some researches reveal that the number of neurofibrillary tangles in the brain of AD patients was positively correlated with their clinical dementia. The progression of NFTs is classified into six stages based on Break stage[17]. The stages include the transentorhinal (I and II), limbic (III and IV) and neocortical (V-VI) stage groups, corresponding to normal cognition, cognitive impairment and dementia, respectively[18]. Recent evidence indicates that Aβ deposition is correlated with phosphorylated microtubule-associated protein tau[19].

2.1.3. *Neuronal Loss*

Neuronal loss mainly occurs in hippocampus, associating with memory impairment and cognitive dysfunction. Numerous studies have demonstrated that neuronal loss has been considered as one of the crucial evidences in AD. The typical intracellular regulators included in the loss of neuronal involve activated pro-apoptotic protein of bcl-2 family, damaged mitochondria, activated caspase cascade and so on[20][21]. Aβ deposition and phosphorylated microtubule-associated protein tau lead to neuronal loss. Overwhelming evidence has suggested that early diagnosis of AD before massive neuron loss is extremely important for the treatment of AD.

2.2. Pathogenesis of AD

The accurate pathogenesis of AD remains unclear at present. Many researchers hold different theories, such as the amyloid hypothesis, central cholinergic damage hypothesis, mitochondrial function hypothesis and so on. Over the past several decades, excitotoxicity has been considered as a mechanism of AD pathogenesis. Excessive synthesis or release of glutamate or glutamate conversion decreasing would increase the partial concentration of glutamate and then arouse neurotoxicity. This mechanism is considered to be an important cause of AD neurodegeneration.

3. EXCITATORY AMINO ACIDS (EAAS)

Glutamate, a predominant excitatory neurotransmitter in the mammalian central nerve system, has involved in excitatory synaptic transmission and formation of learning and memory. Glutamate receptors (GluRs) were classified into two main sections, ionotropic (iGluR) and metabotropic (mGluR)[22], and they both located on the postsynaptic neurons. Ionotropic (iGluR) include N-methyl-Daspartate (NMDA), α-amino-3-hydroxy-5-methyl-4-isoxazolepropionic acid (AMPA) and kainic acid (KA) according to selective affinity[23]. Under normal circumstances, the concentration of glutamate in the synaptic cleft is low, while excessive glutamate stimulates the glutamate receptors and mediates the excitotoxicity. Excitotoxicity was divided into two forms, acute and chronic. The acute excitotoxicity is associated with cell bodies and dendrites swelling. Glutamate bind with non-NMDA receptors, causing Na^+ influx, and passive influx of Cl^- ions and water and then acute swelling of cell bodies, finally causing neuronal death. This process is called cute excitotoxicity. The chronic excitotoxicity is associated with mitochondrial dysfunction. When the level of energy metabolism decreased, depolarization happens persistently. Excessive glutamate cause NMDA ions channels damaged, and then cause a slow Ca^{2+} and Na^+ influx persistently. Then the level of intracellular Ca^{2+} increased, will stimulate the release of glutamate and activate NMDA receptors, intracellular Ca^{2+} levels continually increasing, eventually leading to intracellular calcium homeostasis and neuronal death. This process is called chronic excitotoxicity [24].

4. EXCITOTOXICITY WITH TAU AND AMYLOID BETA

Glutamate might cause neuronal damage through excitotoxic mechanisms. Excitotoxicity occurs in the late stage of AD, but mounting evidence indicated that it may occur in the early stages of AD and it has been implicated with tau and amyloid beta (Aβ) in the neurodegenerative process. Aβ is produced by proteolytic processing of amyloid precursor protein (APP) through β-secretases (BACE1 and BACE2) and γ-secretase (presenilin-1 [PS-1])[25]. Evidence indicated that Aβ induces glutamatergic transmission and excitotoxicity. Signaling pathway is considered that Aβ disrupt calcium homeostasis, causing Ca^{2+} flux and then leading to excitotoxicity which is mediated through the NMDA receptors[26]. Over-stimulation of NMDA receptors can accelerate Aβ production and neuronal death. Meanwhile, glutamate treatment activates c-Jun N-terminal kinase (JNK) and caspase-3. According to Di's research, glutamate treatment can up-regulate protein levels of inducible nitric oxide synthase (iNOS) and neuronal nitric oxide synthase (nNOS)[27]. Some studies have indicated that blockade of glutamate receptors such as NMDA receptors prevent Aβ-induced neurodegenerative process. By the way, as research continues, new perspectives emerged constantly, such as microglial activation is correlated with the course of AD[28]. Clearing peripheral tissues and organs brain-derived Aβ opens a novel avenue to understand the pathological process of AD[29]. Meanwhile, excitotoxicity lead to temporary dephosphorylation of tau and then sustained hyperphosphorylation of tau that are hyperphosphorylated in AD brain[30]. Recent evidence indicates that decreases glutamate activity rescue tau pathology and memory impairment. Hyperphosphorylated tau was located closely to the postsynaptic site of action. Recent evidence indicated that Aβ increases tau phosphorylation and that preventing this phosphorylation blocked the Aβ induced impairment of long-term potentiation (LTP), suggesting that tau phosphorylation downstream in the Aβ cascade might be involved in regulating NMDA receptor activity[31]. It have showed that glutamate receptor-mediated excitotoxicity is thought to contribute to the development of AD.

5. CONCLUSIONS

Thus far, excitotoxicity has been considered as a mediator of AD pathogenesis and excitotoxic pathway relies on the Aβ-tau cascades. The ultimate purpose is to find a way to prevent cell death and then prevent neuronal function in disease processes.

Acknowledgments

This work was supported by grants from the National Natural Science Foundation of China (31201338 and 31371082), Shandong province key R & D Program project (2015GSF115014), research fund from Harbin Institute of Technology at Weihai [HIT(WH)Y200902] and Weihai Science and Technology Development Program (2009-3-93).

References

1. Lancelot E, Beal MF. Glutamate toxicity in chronic neurodegenerative disease. *Prog Brain Res* 1998; **116**: 331–347.
2. Schaeffer E, Duplantier A. Glutamate and neurodegenerative disease. *Springer Berlin Heidelberg* 2010; **6**: 91–147.

3. Bergmann F, Keller BU. Glutamate, calcium and neurodegenerative disease: impact of cytosolic calcium buffers and their potential role for neuroprotective strategies. *Springer Netherlands* 2004; 365–379.
4. Gao HM, Hong JS. Why neurodegenerative diseases are progressive: uncontrolled inflammation drives disease progression. *Trends Immunol* 2008; **29**: 357–365.
5. Olney JW. Brain lesions, obesity, and other disturbances in mice treated with monosodium glutamate. *Science* 1969; **164**: 719–721.
6. Tian J, Cheng J, Zhang J. Protection of pyruvate against glutamate excitotoxicity is mediated by regulating DAPK1 protein complex. *Plos One* 2014; **9**: e95777.
7. Sheldon AL, Robinson MB. The role of glutamate transporters in neurodegenerative diseases and potential opportunities for intervention. *Neurochem Int* 2007; **51**: 333–355.
8. Ji WT, Tham CL, Israf DA. Neuroprotective effects of biochanin A against glutamate-induced cytotoxicity in PC12 cells via apoptosis inhibition. *Neurochem Res* 2012; **38**: 512–518.
9. Selkoe DJ. Alzheimer's disease: genes, proteins, and therapy. *Physiological Reviews* 2001; **81**: 741–766.
10. Rui SU, Han ZY, Fan JP. Discussion on Alzheimer's disease pathology in Chinese medicine. *China Journal of Traditional Chinese Medicine & Pharmacy* 2010; **25**: 743–744.
11. Awasthi M, Singh S, Pandey VP. Alzheimer's disease: An overview of amyloid beta dependent pathogenesis and its therapeutic implications along with in silico approaches emphasizing the role of natural products. *J Neurol Sci* 2016; **361**: 256–271.
12. Querfurth HM, LaFerla FM. Alzheimer's disease. *N Engl J Med* 2010; **362**: 329–344.
13. Inna K, Kerensa B, Iryna B. Neurotoxicity of Alzheimer's disease Aβ peptides is induced by small changes in the Aβ42 to Aβ40 ratio. *EMBO J* 2010; **29**: 3408–3420.
14. Robakis NK. Mechanisms of AD neurodegeneration may be independent of Aβ and its derivatives. *Neurobiology of Aging* 2011; **32**: 372–379.
15. Kidd M. Paired helical filaments in electron microscopy of Alzheimer's disease. *Nature* 1963; **197**: 192–193.
16. Kondratick CM, Vandré DD. Alzheimer's disease neurofibrillary tangles contain mitosis-specific phosphoepitopes. *J Neurochem* 1996; **67**: 2405–2416.
17. Alafuzoff I, Arzberger T, Al-Sarraj S. Staging of neurofibrillary pathology in Alzheimer's disease: a study of the BrainNet Europe Consortium. *Brain Pathol* 2008; **18**: 484–496.
18. Miyashita A, Kuwano R, Hatsuta H. Genes associated with the progression of neurofibrillary tangles in Alzheimer's disease brains. *Alzheimers & Dementia* 2013; **4**: P475.
19. Ittner LM, Jürgen G. Amyloid-β and tau-a toxic pas de deux in Alzheimer's disease. *Nat Rev Neurosci* 2011; **12**: 65–72.
20. Rui EC, Santos MMM, Glória PMC. Cell death targets and potential modulators in Alzheimer's disease. *Curr Pharm Des* 2010; **16**: 2851–2864.

21. Yili W. Regulation of RCAN1 translation and its role in oxidative stress-induced apoptosis. *FASEB J* 2012; **27**: 208–221.
22. Nakanishi S. Molecular diversity of glutamate receptors and implications for brain function. *Science* 1992; **258**: 597–603.
23. Monaghan DT, Bridges RJ, Cotman CW. The excitatory amino acid receptors: their classes, pharmacology, and distinct properties in the function of the central nervous system. *Annu Rev Pharmacol Toxicol* 1989; **29**: 365–402.
24. Hynd MR, Scott HL, Dodd PR. Glutamate-mediated excitotoxicity and neurodegeneration in Alzheimer's disease. *Neurochem Int* 2004; **45**: 583–595.
25. Matsui T, Ingelsson M, Fukumoto H. Expression of APP pathway mRNAs and proteins in Alzheimer's disease. *Brain Res* 2007; **1161**: 116–123.
26. Johanssen T, Donnelly PS, Nuttall S. Aβ-induced calcium flux is dependent on glutamate and copper. *Alzheimers & Dementia* 2014; **10**: P472.
27. Di X, Yan J, Zhao Y. L-theanine protects the APP (Swedish mutation) transgenic SH-SY5Y cell against glutamate-induced excitotoxicity via inhibition of the NMDA receptor pathway. *Neuroscience* 2010; **168**: 778–786.
28. Su F, Bai F, Zhou H, et al. Microglial toll-like receptors and Alzheimer's disease. *Brain Behavior & Immunity* 2015; **52**: 187–198.
29. Yu JT, Zhang C. Pathogenesis and therapeutic strategies in Alzheimer's disease: From brain to periphery. *Neurotox Res* 2015; 1–4.
30. Liang Z, Liu F, Iqbal K. Excitotoxic damage deregulates tau phosphorylation: Implication to Alzheimer's disease. *Alzheimers & Dementia* 2009; **5**: P396.
31. Boehm J. A 'danse macabre': tau and Fyn in STEP with amyloid beta to facilitate induction of synaptic depression and excitotoxicityMicroglial toll-like receptors and Alzheimer's disease. *Eur J Neurosci* 2013; **37**: 1925–1930.

DNA BINDING PROPERTIES OF SCANDIUM^{III} COMPLEXES DERIVED FROM THREE 8-HYDROXYQUINOLINE-2-CARBOXALDEHYDE-AROYLHYDRAZONES

Yongchun Liu[a,*], Yingying Li, Huili Qi, Huansheng Hu, Ruixia Lei, and Jianning Liu

College of Chemistry and Chemical Engineering, Longdong University & FLUOBON Surfactant Engineering Technology Center, Longdong University, Qingyang, Gansu 745000, China

E-mail:[a]ychliu001@163.com

[*]Corresponding author

Previously, a series of lanthanide-ion complexes were prepared from lanthanide metal ions (III) with 8-hydroxyquinoline-2-carboxaldehyde-aroylhydrazones, and their DNA binding properties were investigated[1−11]. In this study, scandium^{III} complexes derived from three 8-hydroxyquinoline-2-carboxaldehyde-aroylhydrazones were prepared and their DNA binding properties were investigated. Their suggested structures are of $[ScL(NO_3)]$ and they act as non-electrolytes. However, their solubilities in methanol and/or ethanol are better than those of lanthanide-metal-ion (III) complexes derived from the same ligand. It's found that Sc^{III} complexes can bind strongly to calf thymus DNA compared to EB and their ligands, with the binding constants at $10^5 M^{-1}$, and they are probably used as potential antitumor drugs, though the antitumor activities of complex including hydroxyl may be better than the others. Moreover, their DNA binding abilities were slightly weaker than most of lanthanide-ion complexes derived from 8-hydroxyquinoline-2(7)-carboxaldehyde-aroylhydrazones, though all of them have the same binding mode of intercalation.

1. INTRODUCTION

The bioactivities of lanthanides such as antimicrobial, antitumor, antivirus, anticoagulant action, enhancing NK and Macrophage cell activities, prevention from arteriosclerosis, etc., have been explored in recent decades[12–15]. The chemistry of quinoline and its derivatives has also attracted special interest due to their therapeutic properties. Quinolinesulphonamides have been used in the treatment of cancer, tuberculosis, diabetes, malaria, and convulsion[16,17]. It was found out that chemistry of 8-hydroxyquinoline with rare-earth metals is very versatile and a series of different types of complexes (mononuclear, dinuclear, trinuclear) could be observed. Schiff bases are able to inhibit the growth of several animal tumors, and some metal chelates have shown good antitumor activities against animal tumors[18,19]. So, well designed organic ligands enable a fine tuning of special properties of the metal ions. A number of metal chelates, as agents for mediation of strand scission of duplex DNA and as chemotherapeutic agents, have been used as probes of DNA structure in solution[20–22]. In this study, as one of the rare earth metal ions, scandium^{III} is selected to prepare its complexes with three 8-hydroxyquinoline-2-carboxaldehyde-aroylhydrazones and then their DNA binding properties will be investigated.

2. EXPERIMENTAL

2.1. Materials

Calf thymus DNA (CT-DNA), 8-hydroxyquinoline-2-carboxaldehyde and ethidium bromide (EB) were obtained from Sigma-Aldrich Biotech. Co., Ltd. The stock solution (1.0 mM) of the investigated compound was prepared by dissolving the powdered material into appropriate amounts of DMF solution. Deionized double distilled water and analytical grade reagents were used throughout. CT-DNA stock solution was prepared by dissolving the solid material in 5 mM Tris-HCl buffer (pH 7.20) containing 50 mM NaCl. The CT-DNA concentration in terms of base pair L^{-1} was determined spectrophotometrically by employing an extinction coefficient of ε = 13,200 $M^{-1}cm^{-1}$ (base pair)$^{-1}$ at 260 nm[23]. EB was dissolved in 5 mM Tris-HCl buffer (pH 7.20) containing 50 mM NaCl and its concentration was determined assuming a molar extinction coefficient of 5600 $M^{-1}cm^{-1}$ at 480 nm[24].

2.2. Methods

The melting points of the compounds were determined on an XT4-100X microscopic melting point apparatus (Beijing). Elemental analyses of C, N and H were carried out on an Elemental Vario EL analyzer. The metal ion content was determined by complexometric titration with EDTA after destruction of the complex in the conventional manner.The IR spectra were recorded on a Nicolet Nexus 670 FT-IR spectrometer using KBr disc in the 4000−400 cm^{-1} region. ESI-MS (ESI-Trap/Mass) spectra were recorded on a Bruker Esquire6000 Mass spectrophotometer.

Viscosity titration experiments were carried on an Ubbelohde viscometer in a thermostated water-bath maintained at 25.00±0.01 °C. UV-vis spectra were obtained using a Specord 50 (Analytik Jena) UV-vis spectrophotometer with 1 cm quartz cell. Fluorescence spectra were recorded using a RF−7000 spectrofluorophotometer (Hitachi) with 1 cm quartz cell. Both of the excitation and emission band widths were 10 nm. The data of these experiments were treated as literatures [10, 25, 26].

2.3. Preparation of the Sc^{III}Complexes

The synthetic routes for ligands (**1a–c**) are presented in Fig. 1 and their structural characterizations refer to literatures[1−11]. Complex **2a** was prepared by refluxing and stirring equimolar amounts of $Sc(NO_3)_3{\cdot}6H_2O$ and a 40 mL methanol solution of ligand **1a**(0.0582g, 0.2 mmol) on a water-bath. After refluxing for 30 min, triethylamine(0.020 g, 0.2 mmol) was added into the reaction mixture dropwise. Then, the mixture was refluxed and stirred continuously for 8 h. After evaporating part of solvent and cooling to room temperature, the precipitate was centrifugalized, washed adequately with 50% methanol aqueous solution and dried in vacuum over 48 h to give an orange powder. Yield: 0.0486 g, 61.0%. Anal.calcd. for C17H11N4O5Sc: C 51.53, H 2.80, N 15.14, Sc 11.35. found: C 51.60, H 2.79, N 14.12, Sc 11.32. IR (KBr) (υ_{max}/cm^{-1}): 1591 (C=N, azomethine), 1549 (C=N, pyridine), 1491 (υ_1), 1339 (υ_4), 1105 (C–OM), 1072 (υ_2), 839 (υ_3), 744 (υ_5), 581 (MO), 473 (MN), υ_1–υ_4 = 152. UV-vis (λ_{max} nm, $\varepsilon\times10^4$ M^{-1} cm^{-1}, DMF): 295, 3.970; 328, 3.249, 366, 1.204. ESI-MS (DMF) *m/z*: 1586.7535 $[ScL^{1a}(NO_3)]_4\cdot H^+$, 397.0365 $[ScL^{1a}(NO_3)]\cdot H^+$, 334.0298 $[ScL^{1a}]^+$. Λ_M($cm^2\Omega^{-1}mol^{-1}$, DMF): 50.1.

Complex **2b** was prepared from equimolar amounts of $Sc(NO_3)_3\cdot 6H_2O$ with ligand **1b**(0.0614g, 0.2 mmol) as the same way of **2a**. Yield: 0.0606 g, 73.2%. Anal.calcd. for C17H11N4O6Sc: C 49.53, H 2.69, N 13.59, Sc 10.91. found: C 49.58, H 2.68, N 13.62, Sc 10.88. IR (KBr) (υ_{max}/cm^{-1}): 3447 (OH, phenolic), 1607 (C=N, azomethine), 1595 (C=N, pyridine), 1491 (υ_1), 1339 (υ_4), 1256 (C–OH), 1109 (C–OM), 1088 (υ_2), 839 (υ_3), 744 (υ_5), 592 (MO), 457 (MN), υ_1–υ_4 = 152. UV-vis (λ_{max} nm, $\varepsilon\times10^4$ M^{-1} cm^{-1}, DMF): 250, 2.765; 336, 4.369; 379, 3.679. ESI-MS (DMF) *m/z*: 1721.3987 $[ScL^{1b}(NO_3)]_4\cdot DMF\cdot H^+$, 413.0140 $[ScL^{1b}(NO_3)]\cdot H^+$, 350.0218$[ScL^{1b}]^+$. Λ_M ($cm^2\Omega^{-1}$ mol^{-1}, DMF): 45.1.

Complex **2c** was prepared from equimolar amounts of $Sc(NO_3)_3\cdot 6H_2O$ with ligand **1c**(0.0584g, 0.2 mmol) as the same way of **2a**. Yield: 0.0502 g, 62.9%. Anal.calcd. for C16H10N5O5Sc: C 48.38, H 2.54, N 17.63, Sc 11.32. found: C 48.33, H 2.55, N 17.62, Sc 11.35. IR (KBr) (υ_{max}/cm^{-1}): 1636 (C=N, azomethine), 1595 (C=N, pyridine), 1558 (C=N, pyridine), 1491 (υ_1), 1339 (υ_4), 1103 (C–OM), 1074 (υ_2), 849 (υ_3), 752 (υ_5), 577 (MO), 473 (MN), υ_1–υ_4 = 150. UV-vis (λ_{max} nm, $\varepsilon\times10^4$ M^{-1} cm^{-1}, DMF): 250, 2.833; 280, 2.823; 328, 3.616; 368, 2.018. ESI-MS (DMF) *m/z*: 398.0133 $[ScL^{1c}(NO_3)]\cdot H^+$, 335.0228 $[ScL^{1c}]^+$. Λ_M($cm^2\Omega^{-1}mol^{-1}$, DMF): 30.9.

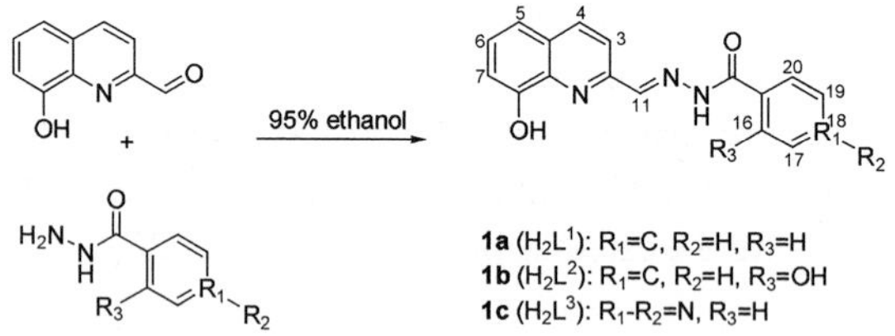

Fig. 1. The synthetic routes for ligands (**1a–c**).

3. RESULTS AND DISCUSSION

3.1. The Powdered Structure of Complexes

Compared with the characteristic IR bands between ligands and powdered Sc^{III} complexes, it is found that phenolate oxygen atom, nitrogen atom of quinolinato unit, C=N of azomethine group, $^-$O–C(C_6H_5)=N– group deprotonated and enolized from

O=C(C_6H_5)–NH– of aroylhydrazones side chain may participate in Sc^{III} complex according to the previous studies[1−11]. Additionally, a nitrate anion bidentately participates in Sc^{III} complex. All these powdered Sc^{III} complexes are orange and stable in air, soluble in DMF and DMSO, moderate soluble in methanol and ethanol, but slightly soluble in acetonitrile, ethyl acetate and acetone, THF and $CHCl_3$, and act as non-electrolyte[27]. However, their solubilities in methanol and ethanol are better than those of lanthanide-metal-ion (III) complexes derived from the same ligand. The melting points of all the powder complexes exceed 300 °C. According to the data of elemental analyses, IR, molar conductivities, ESI-MS and previous studies[1−11], the constitutes of these Sc^{III} complexes may be [ScL(NO_3)] and their suggested structures are shown in Fig. 2. However, there may exist other aggregates.

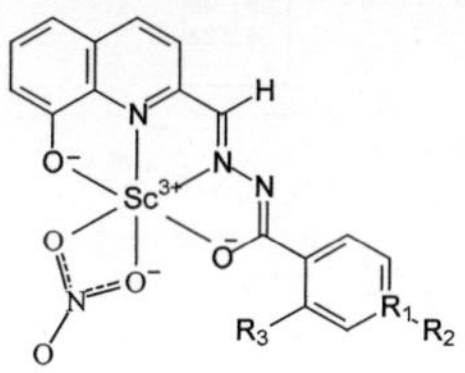

Fig. 2. The suggested structure of Sc^{III} complexes.

3.2. DNA Binding Properties

3.2.1. *Viscosity Titration Measurements*

With the ratio of the investigated compounds to DNA (bps) increasing, the relative viscosity of DNA increases steadily (data not shown), indicating that intercalation takes place between these compounds with DNA helix[24, 28].

3.2.2. *Ultraviolet-visible (UV-vis) spectroscopic study*

The UV-vis spectra of the investigated compounds in the absence and presence of CT-DNA were obtained in DMF:Tris-HCl buffer (5 mM, pH 7.20) of 1:100 containing 50 mM NaCl. The UV-vis spectra of Sc^{III} complexes present two typical bands of λ_{max} at 295 and 328; 250 and 336; 250, 280 and 328, respectively for **2a**, **2b** and **2c**, as well as shoulder peaks at 366~379 nm, which can be assigned to π–π* transition of conjugated aromatic rings and charge transfer from ligand to metal ions (L→Sc^{3+})[29,30]. Upon successive addition of CT-DNA, the UV-vis absorption bands of **2a**, **2b** and **2c** show two types of progressive hypochromisms of 30.0% at 295 nm and 45.7% at 328 nm; 22.4% at 250 nm and 17.8% at 336 nm; 7.06% at 280 nm and 8.03% at 328 nm, respectively, by approximately saturated titration end points at C_{DNA}:$C_{complex}$ = 3.2–4:1. This further indicate the strong stacking interaction between the aromatic chromophore of the compound and base pairs of DNA rather than the non-covalently intercalative binding of compound to DNA helix[31,32]. Fig. 3 shows the plots of [DNA]/(ε_f–ε_a) versus [DNA] for Sc^{III} complexes. Then the binding constants(K_b = 1.439–3.724×10^5 M^{-1}) of DNA and the investigated compounds were determined and the values of K_b are listed in Table 1. Compared with ligands and EB (classical intercalative agent) binding to DNA

($K_{b(EB\text{-}DNA)}$= 0.3068×10^5 M^{-1}, $K_{b(ligands\text{-}DNA)}$ = $0.2767–0.9142\times10^5$ M^{-1}) investigated as the same conditions[10], these Sc^{III} complexes present higher binding abilities to DNA than those of EB and ligands, and complex **2b** present stronger binding abilities to DNA than **2a** and **2c**. However, these Sc^{III} complexes present slightly weaker binding abilities to DNA than most of lanthanide-metal-ion complexes derived from 8-hydroxyquinoline-7-carboxaldehyde-(isonicotinyl)hydrazone(K_b = $9.5–16.7\times10^5$ M^{-1}), 8-hydroxyquinoline-7-carboxaldehyde-(benzoyl)hydrazone(K_b = $2.938–26.13\times10^5$ M^{-1}) and 8-hydroxyquino-line-2-carboxaldehyde-aroylhydrazones(K_b = $1.422–285.3\times10^5$ M^{-1}) [1−11,25,26].

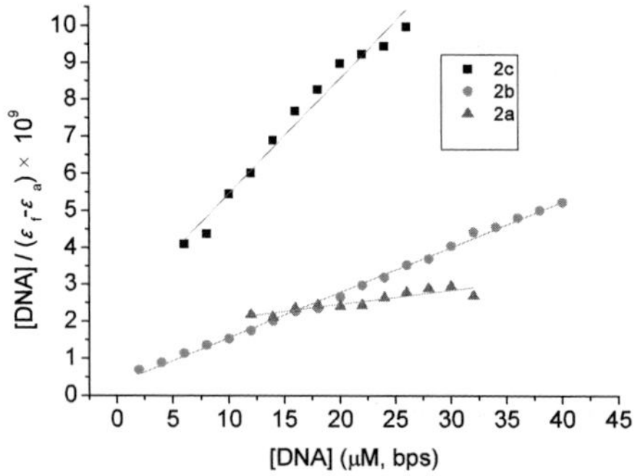

Fig. 3. The plots of [DNA]/(εf−εa) versus [DNA] for Sc^{III} complexes.

Table 1. Parameters of K_b, K_{SV} and FC_{50} for ligands and Sc^{III} complexes

Compound	$K_b \times 10^5$ M^{-1}	$K_{SV} \times 10^4$ M^{-1}	FC_{50} (μM) ($C_{compound}/C_{DNA,\ nucleotides}$)
1a	0.9032 ±0.0444	2.086 ± 0.014	48.18 (12.05:1)
1b	0.9142 ±0.0495	2.352 ± 0.018	38.95 (9.738:1)
1c	0.2767 ±0.0559	1.405 ± 0.013	66.30 (16.58:1)
2a	2.802 ±0.476	1.508 ± 0.027	71.24 (35.62:1)
2b	3.724± 0.417	4.956 ± 0.064	19.08 (9.538:1)
2c	1.439± 0.058	1.964 ± 0.034	50.66 (25.33:1)

3.2.3. *DNA-EB Quenching Assay*

The fluorescence emission intensity of DNA-EB system decreased dramatically upon the increasing amounts of Sc^{III} complexes. The Stern-Volmer plots for Sc^{III} complexes in EB-DNA systems are shown in Fig. 4 and the data of Stern-Volmer quenching constants are collected in Table 1. The loss of fluorescence intensity at the maximum wavelength indicates that most of the EB molecules have been displaced from EB-DNA complex by every quencher at the approximately saturated end point, which indicates further that the intercalative binding takes place between the investigated compound and DNA. More importantly, DNA intercalators have been used extensively as antitumor, antineoplastic, antimalarial, antibiotic, and antifungal agents[24]. FC_{50} value is introduced to denote the molar concentration of a compound that causes a 50% loss in the fluorescence intensity of EB-DNA system. It is interesting that at FC_{50}, all the molar concentration ratios of the investigated compounds to DNA (9.538−35.62:1) are under 100:1, indicating that these investigated compounds are probably used as potential antitumor drugs, and the antitumor activities of **2b** complexes may be better than those of **2a** and **2c**[33]. However, their pharmacodynamical, pharmacological and toxicological properties should be further studied *in vivo*.

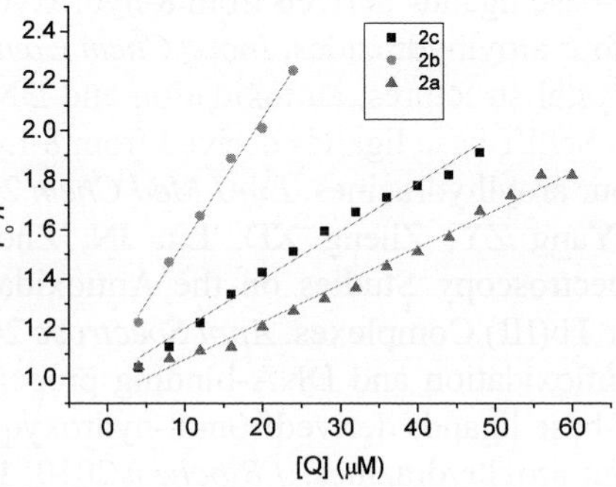

Fig.4. The Stern-Volmer plots for Sc^{III} complexes in EB-DNA systems. λex = 525 nm, λem = 593 nm, 298 K. The concentration of DNA is 4.0 μM (nucleotides) and the concentration of EB is 0.32 μM.

4. CONCLUSION

Scandium^{III} complexes were prepared from three 8-hydroxyquinoline-2-carboxaldehyde-aroylhydrazones. Their suggested structures are of [ScL(NO_3)] and they act as non-electrolytes. Sc^{III} complexes can bind strongly to calf thymus DNA compared to EB and ligands, with the binding constants at 10^5 M^{-1}, and probably be used as potential antitumor drugs. However, their binding abilities to CT-DNA were slightly weaker than most of lanthanide-ion complexesderived from 8-hydroxyquinoline-2(7)-carboxaldehyde-aroylhydrazones, though all of them have the same binding mode of intercalation. Furthermore, the antitumor activities of complex **2b** may be better than those of **2a** and **2c**. Future research efforts will focus on achieving their pharmacodynamical, pharmacological and toxicological properties.

Acknowledgement

The study was supported by Longdong University Doctor Fund (No. XYBY1501 and No. XYBY140210)and Gansu Applied Chemistry Key Subject Fund (No. GSACKS20130113).

References

1. Liu YC, Yang ZY. Crystal structures, antioxidation and DNA binding properties of Yb(III) complexes with Schiff-base ligandsderived from 8-hydroxyquinoline-2-carbaldehyde and four aroylhydrazines. *Biometals* 2009; **22**: 733–751.
2. Liu YC, Yang ZY. Synthesis, crystal structure, antioxidation and DNA binding propertiesof binuclear Ho(III) complexes of Schiff-base ligands derived from

8-hydroxyquinoline-2-carboxyaldehyde and four aroylhydrazines. *J Org Chem* 2009; **694**: 3091–3101.

3. Liu YC, Yang ZY. Crystal structures, antioxidation and DNA binding properties of Eu(III) complexes with Schiff-base ligands derived from 8-hydroxyquinoline-2-carboxyaldehyde and three aroylhydrazines. *J Inorg Biochem* 2009; **103**: 1014–1022.
4. Liu YC, Yang ZY. Antioxidation and DNA binding properties of binuclear Nd(III) complexes with Schiff-base ligands derived from 8-hydroxyquinoline-2-carboxyaldehyde and four aroylhydrazides. *Inorg Chem Commun* 2009; **12**:704–706.
5. Liu YC, Yang ZY. Crystal structures, antioxidation and DNA binding properties of Dy(III)complexes with Schiff-base ligands derived from 8-hydroxyquinoline-2-carboxaldehyde and four aroylhydrazines. *EurJ Med Chem* 2009; **44**: 5080–5089.
6. Liu YC, Jiang XH, Yang ZY, Zheng XD, Liu JN, Zhou TL. Fluorescent and Ultraviolet–Visible Spectroscopy Studies on the Antioxidation and DNA Binding Properties of Binuclear Tb(III) Complexes. *Appl Spectrosc* 2010; **64**: 980–985.
7. Liu YC, Yang ZY. Antioxidation and DNA-binding properties of binuclear Er(III) complexes with Schiff-base ligands derivedfrom 8-hydroxyquinoline-2-carboxaldehyde and four aroylhydrazines. *J Biochem* 2010; **147**: 381–391.
8. Liu YC, Yang ZY, Zhang KJ, Wu Y, Zhu JH, Zhou TL. Crystal Structures, Antioxidation, and DNA Binding Properties of Sm^{III} Complexes. *Aust J Chem* 2011; **64**: 345–354.
9. LiuYC, Xu XY, Lei RX, Zhang XW, Wu Y, Yang ZY. Anti-oxidation and DNA binding properties of La(III) complexes. *Chemical Research and Application (in Chinese)* 2013; **25**: 1475–1483.
10. Liu YC, Li YY, Qi HL, Zhang KJ, Lei RX, Liu JN. Crystal structures and DNA-binding properties of Pr^{III} complexes derived from 8-hydroxyquinoline-2-carboxaldehyde and three aroylhydrazines. *J Coord Chem* 2014; **67**: 3689–3703.
11. Lei RX, Ma DP, Wei XX, Liu YC, Zheng XD, Liu JN, Yang ZY. Binuclear lanthanide complexes derived from aroylhydrazines with 8-hydroxyquinoline-2-carboxyaldehyde and a series of DNA intercalators. In: Wang M, ed. *Medicine Sciences and Bioengineering*, London: Taylor & Francis Group, 2015: 783–791.
12. Parker D, Dickins RS, Puschmann H, Crossland C, Euward JAK. Being excited by lanthanide coordinationcomplexes: aqua species, chirality, excited-state chemistry, and exchange dynamics. *Chem Rev* 2002; **102**: 1977–2010.
13. Albrecht M, OsetskaO, FröhliichR. 2-[(8-Hydroxyquinolinyl)methylene]hydrazine-carboxamide: expandingthe coordination sphere of 8-hydroxyquinoline for coordinationof rare-earth metal (III) ions. *Dalton Trans* 2005; **23**: 3757–3762.
14. Hunter RB, Walker W. Anticoagulant action of Neodymium3-Sulpho-isonicotinate. *Nature* 1956; **178**: 47.
15. Kramsch DM, Aspen AJ, Rozler LJ. Atherosclerosis: prevention by agents not affecting abnormal levels ofblood lipids. *Science* 1981; **213**: 1511–1512.
16. Schmidt LH. Chemotherapy of the drug-resistant malarias. *Ann Rev Microbiol* 1969; **23**: 427–454.
17. El-Asmy AA, El-SonbatiAZ, Ba-IssaAA, Mounir M. Synthesis and properties of 7-formyl-8-hydroxyquinoline and its transition metal complexes. *Transit Metal Chem* 1990; **5**: 222–225.

18. Hodnett EM, Mooney PD. Antitumor activities of some Schiff bases. *J Med Chem* 1970; **13**: 786.
19. Miyadera T, Kosower EM. Improved synthesis of the antibiotic, hexahydrospinamycin. *J Med Chem* 1972; **15**: 339–340.
20. Mahadevan S, Palaniandavar M. Spectroscopic andvoltammetric studies of copper(II) complexes of bis(pyrid-2-yl)-di/trithia ligands bound to calf thymus DNA. *Inorg Chim Acta* 1997; **254**: 291–302.
21. Lippard SJ. Platinum complexes: probes of polynucleotidestructure and antitumor drugs. *AccChem Res*1978; **11**: 211–217.
22. Hecht SM. The chemistry of activated bleomycin. *Acc Chem Res*1986;**19**: 383–391.
23. Zsila F, Bikádi Z, Simonyi M. Circular dichroismspectroscopic studies reveal pH dependent binding of curcumin in the minor groove of natural and syntheticnucleic acids.*Org Biomol Chem* 2004; **2**: 2902–2910.
24. Suh D, Chaires JB. Criteria for the mode of binding of DNA binding agents. *Bioorgan Med Chem* 1995; **3**: 723–728.
25. Liu YC, Zhang KJ, Lei RX, Liu JN, Zhou TL, Yang ZY. DNA-binding and anti-oxidationproperties of binuclear lanthanide(III)complexes of 8-hydroxyquinoline-7-carbaldehyde-(isonicotinyl)hydrazone. *J Coord Chem* 2012; **65**: 2041–2054.
26. Liu YC, Zhang KJ, Wu Y, Zhao JY, Liu JN, Zhou TL. Antioxidation and DNA-Binding Properties of Binuclear Lanthanide(III) Complexes with a *Schiff* Base Ligand Derived from 8-Hydroxyquinoline-7-carboxaldehyde and Benzoylhydrazine. *Chem Biodivers* 2012; **9**: 1533–1544.
27. Geary WJ. The use of conductivity measurements inorganic solvents for the charaterisation of coordination compounds. *Coord Chem Rev* 1971; **7**: 81–122.
28. Palchaudhuri R, Hergenrother PJ. DNA as a target for anticancer compounds: methods to determine the mode of binding and the mechanism of action. *Curr Opin Biotech* 2007; **18**: 497–503.
29. Ismail TMA. Mononuclear and binuclear Co(II), Ni(II),Cu(II), Zn(II) and Cd(II) complexes of Schiff-base ligandsderived from 7-formyl-8-hydroxyquinoline and diaminonaphthalenes. *J Coord Chem* 2005; **58**: 141–151.
30. Moawad MM, Hanna WG. Structural and antimicrobialstudies of some divalent transition metal complexes withsome new symmetrical bis(7-formylanil substituentedsulfoxine) Schiff base ligands. *J Coord Chem* 2002; **55**: 439–457.
31. Barton JK, Danishefsky AT, Goldberg JM. Tris(phenanthroline) ruthenium(II): stereoselectivity in Binding to DNA. *J Am Chem Soc* 1984; **106**: 2172–2176.
32. Lu HL, Liang JJ, Zeng ZZ, Xi PX, Liu XH, Chen FJ, Xu ZH. Three salicylaldehyde derivative Schiff base Zn^{II} complexes: synthesis, DNA binding and hydroxyl radical scavenging capacity.*Transit Metal Chem* 2007; **32**: 564–569.
33. Li ZL, Chen JH, Zhang KC, Li ML, Yu RQ. Fluorescentstudy on the primary screening for nonplatinum type of Schiff-base antitumor complexes. *Science in China Series B: Chemistry* 1991; **21**: 1193–1200.

PET/CT IMAGING AND IMMUNOHISTOCHEMISTRY (IHC) STUDY OF DIABETIC ENCEPHALOPATHIC RATS

Duo Zhang and Heng Meng[a,*]

Department of Radiology, Affiliated Hospital of BeiHua University, Jilin 132011, China

E-mail:[a]beihuamengheng@126.com

[*]Corresponding author

Diabetes mellitus, or simply diabetes, is a group of metabolic diseases in which a person has high blood sugar, either because the pancreas does not produce enough insulin or because cells do not respond to the insulin that is produced[1]. Insulin causes cells in the liver, skeletal muscles, and fat tissue to absorb glucose from the blood. In the liver and skeletalmuscles, glucose is stored as glycogen, and in fat cells (adipocytes) it is stored as triglycerides [2]. Diabetic encephalopathy (DE) is caused by diabetes[3]. The complications of DE include memory loss, dementia, coma, seizures, and finally death.The defects in patients include lethargy, poor judgment and coordination of limbs, dementia, and muscle twitching[4]. Positron emission tomography/computed tomography (PET/CT) uses the positron radioactive nuclide which are labeled trace agent to detect physiological and biochemical changes of normal and abnormal tissues according to the principle of tracing. However, little is known about the PET/CT on metabolic level of diabetic encephalopathic rat. In this study, PET/CT will be applied to detect glycometabolic level in brain of diabetic encephalopathic rat model and Immunology and Histology Chemistry (IHC) method will be utilized to detect the expression condition of main correlation factors in hippocampus of diabetic encephalopathic rats.

1. EXPERIMENTAL ANIMAL

Fifty healthy male Wistar rats of clean degree the 12-week-old （weighting 200-250g) rats are purchased from Experimental Animals Center of Jilin University. Rearing condition after purchasing as following: room temperature 22-28°C, relative humidity 40%-70%, 12h illumination maintenance every day, diurnal cycle, feed with standard feeding blocks with 21% protein content, free water drinking, and raise in separated cages.

2. EXPERIMENTAL METHODS

2.1. Establishment of Diabetic Rat Model

10 rats were randomly selected from 50 Wistar male rats as normal control group, and the rest were model group, after one week adaptive feeding, 12h absolute diet, STZ dissolved in citrate buffer solution (0.1mmol/L, pH=4.6) for preparation of 0.5% solution,

intraperitoneal injection once as per 65mg/kg weight. Collected blood from caudal vein three days later to detect rat's fasting blood glucose, and establishment of diabetic rats model would be successful provided that blood glucose concentration>16.7mmol/L.

2.2. Use Morris Water Maze Experiment to Select Diabetic Encephalopathic Rat Model

Morris water maze is a round water pool with 150cm diameter and 60cm height, the platform with 12cm diameter and 20cm height has 4 quadrants, the surface of platform is 2cm lower than water surface and it is located in the center of one quadrant, water depth is 30cm, and water temperature is controlled at 22-25°C. After 12 full weeks, experiments were started 5 days before PET/CT scanning, training was conducted 4 times per day in the morning and afternoon in the first 4 days, each rat was trained for 2min every time, put rats in water from 4 entering points randomly and record the time taken from entering water to finding platform as test score (escape incubation period).

In case escape incubation period did not have statistical significance comparing with normal control, it would be considered as pure diabetic group (excluding encephalopathy); It would be considered as diabetic encephalopathic patients provided that escape incubation period was obviously extended comparing with normal control group and cognitive function decline was shown. Select 8 rats randomly from normal control group(CON), diabetic group (excluding encephalopathy)(D) and diabetic encephalopathic group(DE) for PET/CT imaging and IHC study.

2.3. PET/CT Imaging of Diabetic Encephalopathic Rats

It was divided into normal control group, diabetic model group (excluding encephalopathy) and diabetic encephalopathic model group. Contrast medium 18F-FDG 0.75mCi/kg was injected according to body weight of rats, and then PET/CT imaging was started after 30min. Use CT image to select the maximum section accurately, apply 18F-FDG PET on PET image to detect glucose metabolism in brains of rats, and use technologies of region of interest to determine quantitative parameter SUV (Standard Uptake Value), and the calculating formula of SUV is as following:

SUV = average radioactivity of the tissue of each gram / radioactivity of nuclide injected / body weight.

2.4. IHC Stain

After PET/CT imaging, cut off the chest quickly to expose the heart, enter aorta ascendens from the left ventricle of heart, cut off the right ventricle of heart in the meanwhile, clean blood off the surface of tissue by ice normal saline, cut off the head and take out brain quickly on ice tray. Take one side of brain tissue, fix with 4% polyoxymethylene and stay overnight for IHC stain.Adopt IHC S-P method (streptomycin avidin - peroxidase) and conduct as per instruction manual of the kit.

3. EXPERIMENTAL RESULTS

3.1. General Condition of Rats of Each Group

In the process of experiments, rats of diabetic group (excluding encephalopathy) had obvious symptoms of polydipsia, polyphagia, diuresis, loss of weight, dull yellow, poor spirit in later period, and slow movement. Besides the conditions of diabetic group mentioned above, rats of diabetic encephalopathic group had worse spirit and slower movement. In the 13th week, compared with normal control group, weight of rats of diabetic group and diabetic encephalopathic group was lost obviously ($p<0.01$) and blood glucose level rose obviously ($p<0.01$)..

3.2. Behavior Test Results

Escape incubation period of rats of each group in the first day did not have statistical difference ($p >0.05$), there was statistical difference in each group from the 2nd day to the 5th day ($p <0.01$ or $p <0.05$). Compared with normal control group, rats were considered as pure diabetic group (excluding encephalopathy) provided that escape incubation period has no statistical significance ($p>0.05$); and rats were considered as diabetic encephalopathic patients provided that escape incubation period was obviously extended compared with normal control group, there was statistical significance ($p <0.01$ or $p <0.05$) and cognitive function decline was shown.

3.3. PET/CT Imaging of Diabetic Encephalopathic Rats

Conduct PET/CT scanning of rats of normal control group, diabetic model group and diabetic encephalopathic model group culculate the SUV's value of each group respectively. Compared with normal control group, SUV value declining trend of rats from diabetic model group is extremely obviously ($p<0.01$); while SUV value of rats from diabetic encephalopathic model group has the same trend and is more obvious ($p <0.01$); there is also obvious difference ($p<0.01$) between SUV values of diabetic model group and diabetic encephalopathic model group; It indicates that glucose metabolism of rats from diabetic model group and diabetic encephalopathic model group declines obviously comparing with normal control group, while glucose metabolism of rats from diabetic encephalopathic model group declines much more obviously by comparing diabetic encephalopathic model group and diabetic model group.

3.4. IHC Stain

(1) IHC detects expression of IGF1R in diabetic encephalopathic model group and diabetic model group. Compared with normal control group, IGF1R content of diabetic encephalopathic model group is obviously more than that of normal control group, while IGF1R content of diabetic model group is less than that of normal control group.

(2) IHC detects expression of IR in diabetic encephalopathic model group and diabetic model group. Compared with normal control group, IR content of diabetic encephalopathic model group and diabetic model group does not change obviously.

(3) IHC detects expression of Glut4 in diabetic encephalopathic model group and diabetic model group. Compared with normal control group, Glut4 content in diabetic encephalopathic model group is less than that of normal control group, while Glut4 content in diabetic model group is more than that of normal control group.

4. DISCUSSIONS AND CONCLUSIONS

PET/CT is one of the widest clinical applications of molecular imaging detection methods at present; it can not only clearly display anatomical structure of body or organs, but also subtly describe physiological pathology on molecular level and biological metabolism process [5]. In this paper, PET/CT scanning is conducted to rats of normal control group, diabetic model group and diabetic encephalopathic model group, it is found that comparing with normal control group, SUV value of rats from diabetic model group declines while SUV of rats from diabetic encephalopathic model group declines more obviously, that it means glucose metabolic rate declines obviously. PET/CT has extremely high sensitivity, functional changes can be found before diabetic encephalopathy has structural change, so that it can confirm partial brain glucose metabolic rate and abnormal condition of brain functional status, and it has greater values to find out and diagnose diabetic encephalopathy.

IR is located on cell membrane, with tetramer structure, includes two α chains, two β chains and disulfide bond in the middle. Combine α subunit of IR on target cell membrane and insulin to change IR conformation, eliminate inhibition of α subunit to β subunit so that tyrosine residue on β subunit has phosphorylation reaction and activates signaling protein of downstream in turn which will urge Glut4 shift, mediate glucose transfer from outside to inside to regulate energy metabolism in cells [6]. Normal running of central nervous system I/IR signal system is the key factor [7] to maintain blood running and normal energy metabolism of brain, cognition impairment and a series of pathophysiological changes will occur in case the system has obstacles.

Relative molecular mass of Glut4 is about 4.5-5.5kD, it contains 12 transmembrane segments (Ml-M12) and 1 extracellular cyclic structural domain at N terminal, it is transmembrane transfer glycoprotein. At the level of organization, Glut4 mainly exists in skeletal muscle, cardiac muscle and adipose cell, the latest research shows that Glut4 has selective expression in brain, for example, expression of Glut4 is found out in hippocampus, sensorimotor area cortex, posterior pituitary and hypothalamus of rats, and the expression in hippocampus tissue is especially rich [8], and its selective expression in brain may be relevant with stimulation of insulin to glucose intake in these brain regions. Glucose is the principal raw material [9] of energy metabolism for brains of mammal animals, it cannot go through lipid bilayer because it is hydrophilic, while glucose going through cell membrane is the precondition for cells producing energy, and this process is mainly fulfilled through Glut4 mediate [10]. For brain tissue organs relying on high level of glucose as source of energy, the effect of Glut4 in energy metabolic process of brain tissue is more important, its level of expression can reflect degree of demand of brain tissue cells to glucose [11].

This experimental result shows that IR expression level among experimental groups is basically same, while rats of diabetic encephalopathic group has increasing IGF1R but declining Glut4, what deserves special attention is that compared with normal control group, only IGF1R expression level in hippocampus of diabetic (excluding encephalopathy) rats declines obviously. This result indicates that compared with IR, IGF1R and Glut4 have more important effect in the process of development and progression of diabetic encephalopathy.

References

1. A. F. Castilho, J. T. Liberal, F. I. Baptista, J. M. "Elevated glucose concentration changes the content and cellular localization ofAMPAreceptors in the retina but not in the hippocampus,"*Neuroscience*, 2012;**219**: 23–32.
2. W. Y. Jeng and N. C. Wang, "High-resolution structures of Neotermeskoshunensis β-glucosidasemutants provide insights into the catalytic mechanism and the synthesis of glucoconjugates," *ActaCrystallographica D*, 2012;**68**(7): 829–838.
3. A. Csibi,D.Communi,N.M¨uller, andS. P. Bottari, "Angiotensin ii inhibits insulin-stimulated GLUT4 translocation and akt activation through tyrosine nitration-dependent mechanisms,"*PLoS ONE*, 2010;**5**(4), Article ID e10070, 2010.
4. M. S. and R. Cripps, "Diabetes mellitus and increased risk of cancer: focus on metformin and the insulin analogs," *Pharmacotherapy*,2010;**30**(11):1159–1178.
5. Dunphy MP, Lewis JS. Radiopharmaceuticals in preclinical and clinical development for monitoring of therapy with PET. J Nucl Med 50, 2009, 106S-121S.
6. Z.L Ma, Jens Sandahl Christiansen, TorbenLaursen, et al . Effects of human insulin and insulin aspart preparations on levels of IGF-I, IGFBPs and IGF bioactivity in patients with type 1 diabetes. *BMC Endocrine Disorder*, 2014; **35**(14):1-11.
7. Dong XC, Park S, Lin XY. Irsl and Irs2 signaling is essential for hepatic glucose homeostasis and systemic growth. *J Clin Invest*, 2006;**1**:32-36.
8. Kobayashi M, Nikami H. Expression and localization of insulinregulatable glucose transporter(Glut4) in rat brain. *NeurosciLett*, 1966;**213**: 103-106.
9. Pardride WM. Brain metabolism: a perspective from the blood-brain barrier. *Physiol Rev*, 1983;**63**: 1481-1535.
10. Steen E, Terry BM, Rivera EJ, et al. Impaired insulin and insulin-like growth factor expression and signaling mechanisms in Alzheimer's disease is this type diabetes. J Alzheimer's Dis, 2005; **7**: 63-80.
11. D. Zhang,S .Jiang, H. Meng, Role of the Insulin-Like Growth Factor Type 1 Receptor in the Pathogenesis of Diabetic Encephalopathy [J] . *International Journal of Endocrinology*, 2014; Article ID 2014-626019.

EXPRESSION OF AQUAPORIN-4 DURING CYTOTOXIC EDEMA AFTER CEREBRAL ISCHEMIC REPERFUSION IN RATS

Songtao Li[1,#], Tianwei Wang[2,#], Bing Han[3], and Le Fang[1,a,*]

[1]Neurology Department of China-Japan Union Hospital, Jilin University, Changchun, 130000, China

[2]Radiology Department of China-Japan Union Hospital, Jilin University, Changchun, 130000, China

[3]Radiology Department of 2nd Hospital of Jilin University, Changchun, 130000, China

E-mail: [a]drfangle@sina.com

[*]Correspondence author

[#]These authors contributed equally to the work

To investigate the rule of the Aquaporin-4 (AQP-4) expression in cytotoxic edema after Cerebral Ischemic Reperfusion (CIR). Methods CIR models were established by occluding unilateral Middle Cerebral Arteries (MCA) of rat with nylon filament.48 SD rats were divided into 2 groups randomly, including false operation group A(n=12)and operation group B (blood flow blocked 1 hours and the reperfusion for 24h,n=24). Both groups were examined with DWI and T2WIMRI.The relative ADC (rADC), relative density(rd), and relative area (rs)of the biggest hyperintensity signal layer on DWI were measured. Then the animals were sacrificed and perfused with the mixture solution consisting of TTC at different time intervals. The AQP-4 expression was measured with the image analyzer. Meanwhile, histologic examination was performed at each group. Results In the operation group,abnormal high intensity was found in DWI of ipsilateral MCA territory at15 min after MCAO, and rADC value decreased quickly within1 h after MCAO ,while the AQP-4 expression, the rd-DWI and the rs-DWI increased rapidly in the stage. With the progress of the time after reperfusion, the rADC value increased slowly till 24h, and the rs-DWI decreased rapidly in the stage. While the AQP-4 expression (ΔS) decreased rapidly in the penumbra but increased still in the center area. Conclusion The high expression of AQP-4 in the acute ischemic cerebral tissues played a significant role in the cytotoxic edema. Certainly, it was also one of the important reasons of the decrease of rADC and hyperintensity on DWI in the early stage of the cerebral ischemia.

1. INTRODUCTION

AQP-4 involved in the reabsorption of cerebrospinal fluid and osmotic adjustment. It is widely accepted that APQ4 plays an important role in the process of vasogenic edema, and it is mostly speculation for the role of AQP4 in cytotoxic edema. At present, MR Diffusion Weighted Imaging (DWI) technology is the most sensitive method to detect cytotoxic edema. In this study, we set up rat model of acute ischemia under the monitoring of DWI, which copied the process of acute ischemia and reperfusion cytotoxic edema as far as possible. The expression of AQP-4 in cytotoxic edema of brain tissue was observed by immunohistochemistry technique. Further analysis of the role of AQP4 in this process was carried out.

2. MATERIAL AND METHODS

48 rats ischemia-reperfusion model after Middle Cerebral Artery Occlusion (MCAO) (A group: 16 rats of sham group; group B: 32 rats of ischemia-reperfusion group, in MCAO 1 h after pulling plug recanalization). 48 SD rats were divided into 2 groups randomly, including false operation group A (n=12)and operation group B (blood flow blocked 1 hours and the reperfusion for 24h,n=24). Measurement included apparent diffusion coefficient (apparent diffusion coefficient, ADC) value, rADC value, rd-DWI value and rs-DWI value, in both brain tissue infarction and contralateral mirror area.

The rats were sacrificed after removing brain tissue perfusion for HE staining and historical examination (AQP-4 monoclonal antibody was purchased from the united states chemical international.inc; immunohistochemistry stain box was purchased from Beijing Zhongshan Biotechnology co., ltd.). Each tissue sample was taken out three slices, and each slice involved ischemic peripheral zone, central zone and contralateral mirror area, each area selected three 400x microscopic field of view, analyzing of immunohistochemistry image, then the color area ratio ΔS (target area / field of statistical area x100%) was calculated under the image analyzer.

All data were processed by SPSS 12.0, one-way analysis of variance (ANOVA) and correlation analysis were used to analysis, $p < 0.05$ was considered statistically significant difference.

3. RESULTS

3.1. MRI Results

Group A of DWI and T2WI showed normal. DWI of Group B showed ischemic areas of higher intensity in 15 minutes after MCAO, but T2WI normal. With the extension of the time, group B DWI signal intensity gradually increased, area is expanded, but rADC decreases rapidly; after recanalization, group B DWI shown high signal is narrower than unplug in varying degrees (Fig.1, Fig.2). Group B and group A have significant differences in each index ($p<0.01$); group B embolization after 1h is compared with reperfusion after 24h, rd-DWI was no obvious difference ($p > 0.05$), rs-DWI and rADC have differences ($p < 0.05$). (Table 1).

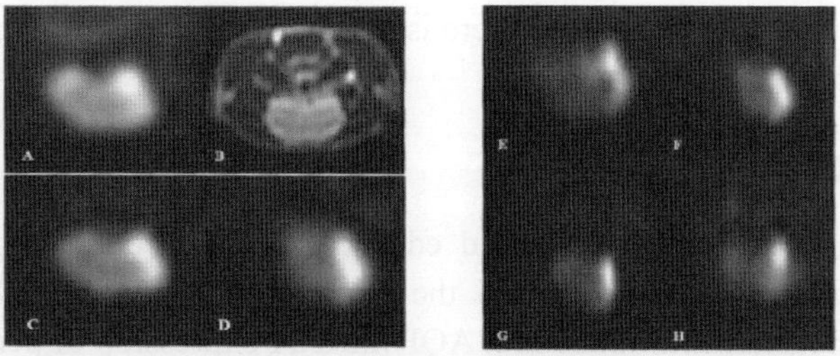

Fig. 1. MRI results at different time point: A.DWI at 15min; B. T2FLAIR at 15min; C. DWI at 0.5h; D. DWI at 1h;E. DWI at 2h; F. DWI at 6h; G. DWI at 12h; H. DWI at 1day.

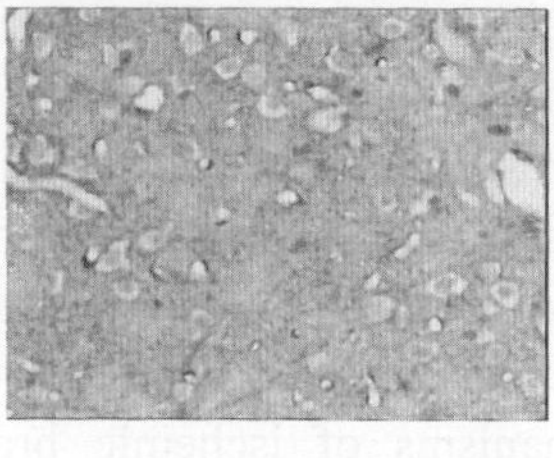

Fig. 2. AQP-4 express (400x)

Table 1. Ischemia reperfusion group at each time point (x±s, n=4)

Time point (h)	Rd-DWI	Rs-DWI(%)	RADC(%)
0	98.6±1.4	0	98.7±0.8
0.25	124.4±16.5	31.45±4.78	63.4±8.9
0.5	138.6±18.6	49.13±7.48	57.6±13.1
1	152.1±20.3	58.67±5.38	47.6±18.5
2	149.9±18.	41.36±8.57	48.7±10.5
6	150.3±16.4	29.40±3.41	54.3±11.8
12	148.5±14.8	23.99±4.39	68.7±11.5
24	149.1±15.1	23.56±4.41	69.3±13.4

3.2. Results of Immunohistochemical Staining

Group A and group B contralateral brain tissue containing a small amount of AQP-4 protein, no significant difference ($p > 0.05$) was found between the two groups. Group B AQP-4 positive nerve cells began to increase from (16.3 ±3.2)% at 15 min after embolism to (51.2 ± 5.4)% at 1h. The amount of positive transitional cells within infarct border zone decreased significantly after reperfusion to 24h, but that of infarct central zones increased slowly. AQP-4 expression was significantly differences at each time point of group B ($F = 17.528$, $p = 0.000\ 1$). Group A has significant difference with Group B ($p < 0.05$).

3.3. Correlation Analysis of AQP-4 Expression, DWI and Time

In group B, both the expression of AQP-4 and DWI high signal area had r linear relation with time (T) before reperfusion (0-1h), but both of them had no linear correlation with time (T) after that(1-24 h) (p> 0.05). There is no linear relation between rADC and DWI high signal density.

4. DISCUSSION

Ischemic brain edema belongs to mixed edema including both vasogenic edema and cytotoxic edema. Taniguch [1] studied the expression of AQP4mRNA after middle cerebral artery occlusion, and found that AQP4mRNA expression of peri-cortex raised at the first day, peaked at the third day, then remained at a high level on the seventh day. This change of expression showed an accordance with the change of brain edema area under MRI, which illustrate AQP4 may be involved in the formation of ischemic brain edema.

Zhou [2] found AQP-4 express in rat cerebral ischemia model increased from 15 min after MCAO, especially increased rapidly within 1h, and showed a "Platform" Period during 1h to 6h, then continued increasing slowly from 6h to 24h. That suggests that AQP-4 expression persists throughout the process in acute cerebral ischemia, and play a key factor in molecular mechanisms of ischemic brain edema, particularly in the formation of intracellular edema.

In our study, model design was refer to Xiao[3]. We set the ischemic time as 1h, while the observing- time was limited within 24h and concentrated in the first of the 6h. So the character of cerebral edema tends to cytotoxic edema status. The expression of AQP-4 protein increased significantly from 15min after MCAO in acute ischemic and the amount of AQP4 positive cells reached 50% at 1h after MCAO. But after the reperfusion, AQP-4 protein expression in transitional zone of the edge of infarction and that in the infarct central region exhibited the opposite pattern: TZ-positive cells was significantly reduced in the edge of infarct zone while increased in the central area. Combined with DWI performance, it suggests that the decrease of AQP-4 protein expression in the edge of infarction transitional zone may be helpful to cytotoxic edema regression and cell function recovery. The expression of AQP-4 protein in infarction central area lead cytotoxic edema increase continually and further induce vasogenic edema, then lead to irreversible cell damage in the end.

Bao[4] observed that the level of AQP4 begins to decrease at 2h after ischemia and reperfusion, down to the lowest point at 24h, then began to recover, and return to normal at the 7th day, with some rats higher than normal. This study on rats with acute cerebral ischemia-reperfusion model found that TZ-positive cells in infarct border zone was still higher than the false operation group and ischemia- reperfusion group contralateral normal brain tissue at 24h after reperfusion, suggesting that AQP4 levels of vary degrees of cerebral ischemia reperfusion brain are not entirely consistent. In early stage of reperfusion injury, AQP4 levels decreasing rapidly in the body's brain injury is a protective response in order to reduce the blood-brain barrier damage and brain edema formation, in particular, can reduce the formation of vasogenic brain edema. With the damage gradually increased, and loss state of protection, AQP4 levels gradually restored and brain edema was also significantly increased during the subsequent period of time.

Considering that early reperfusion injury attributes to cytotoxic edema while lately to vasogenic edema, we suggest that AQP4 level's recovery has some relation with BBB opening. The reduction of the AQP4 level of early stage may be a protective response to injury of the body's self-protection.

In this study, it appears high signal corresponding to the ischemic area on DWI at 15min after obstruction in MCA. On DWI, it showed a high signal resulted by ADC decline, which attribute to restriction of diffusion of water molecules in the cell membrane, and decrease of intercellular spaces of free water. With prolonged obstruction, ADC declines further, DWI high signal area becomes strong gradually, and the boundary of high intensity becomes clearer. The rd-DWI and rADC at 1h after MCAO reached the maximum and minimum individually in our study.

On DWI of early stage of ischemia, ADC in the lesion showed a decreasing trend from periphery to the center, indicating the different extent of cytotoxic edema. It was reported that decrease of the ADC value was linearly related to with a cytotoxic edema degree. The membrane ion pump in reversible damage zone is able to maintain intracellular ionic gradients after reperfusion on time, and return normal. Hasegawa [5] proposed ADC level was possible to distinguish cerebral ischemic injury reversible or not. In this study, rADC of recanalization after ischemia-reperfusion group averages (69.3 ±16.4)% before unplug in reversible ischemic area, while rADC in core area which failed to recover averages (47.6 ±18.5)%. This result consists with the observations of Hasegawa.

Acknowledgments

Medical funding of Health Commission of Jilin Province supported this study.

References

1. Taniguchi M, Ya Mashita T, Kumura E, et al. Induction of aquaporin-4 water channel mRNA after focal cerebral ischemia in rat[J]. *Brain Res Mo* 2000; **28**(1-2):131.
2. Renlan Zhou. The AQP4 expression after cerebral ischemia in rats and proton magnetic resonance spectroscopy [D]. Chongqing Medical University 2005. (In Chinese).
3. Xuehong XIAO, Xiangquan KONG, JIANGLi, et al. An experimental study on acute cerebral ischemia and reperfusion with magnetic resonance diffusion weighted imaging[J]. *Chinese Journal of Radiology* 1999; **10**:662-663. (In Chinese).
4. Shiyao Bao, Xiangqun Shi, Zhilin Zhang, et al. Changes in expression level of aquaporin-4 during cerebral ischemia-reperfusion injury in rats[J]. *Cerebrovascular Diseases Foreign Medical Sciences* 2004; **2**(12):122-125. (In Chinese).
5. Hasegawa Y, Fisher M, Latour L, et al. MRI diffusion mapping of reversible and irreversible ischemic injury in focal brain ischemia1[J].*Neurology* 1994; **44**:1484-90.

HPLC ANALYSIS OF ASTILBIN FROM THE RHIZOME OF THE TRADITIONAL ZHUANG MEDICINE "GAEULANGHAUH"

Xiangpei Zhao[a], Licheng Guo, and Hongli Teng

Guangxi Institute of Minority Medicine/Traditional Zhuang Hospital of Guangxi Zhuang Autonomous Region, Nanning, 530001, China
E-mail: [a]zhaoxiangpei@126.com

AIM: To establish a quantitative analysis method with HPLC for determination of astilbin in the rhizome of the traditional Zhuang medicine "Gaeulanghauh". METHOD: The subsequent HPLC separation and quantification was achieved in 30 min using a Agilent XDB C18 column (5 μm, 4.6 mm×150 mm) and at 290 nm. The mobile phase comprised solvent A (0.2% Phosphoric acid aqueous) and solvent B (acetonitrile) and gradient elution mode was applied. This established method was therefore applied to determine the amounts of astilbin in ten samples collected from different regions in Guangxi province. RESULT: The linear regression analytical data for the calibration plots showed a good linear relationship with r2=0.9999 (astilbin content was 0.16~1.92 μg) within the test ranges. The average recovery was 1.09% indicating good accuracy. It was successfully applied to quantify astilbin in different samples. CONCLUSION: The method was simple, reliable, accurate and can be regard as the quantity control method of Zhuang medicine "Gaeulanghauh".

1. INTRODUCION

"Gaeulanghauh" was a famous Zhuang medicine in Guangxi province, it was the dried rhizome of Smilacaceae plants Smilax gabra Roxb. The theory of Zhuang Medicine was showed that, "Gaeulanghauh" was used for the treatment of "Youhniuj" (Stranguria with Turbid Urine), "Binghbwzdai" (leukorrheal diseases), "Baez" (ulcerative carbuncle), "Binghvaliuz" (syphilis), "Nohnyinzin"(Muscle fasciitis), "Fatvuengh" (Rheumatism bone pain), et.al. "Gaeulanghauh" contained a series of two flavonoid glycosides of hydrogen, and the astilbin was one of the large amount in it. Studies have shown that, astilbin has significant immune inhibitory activity, without the side effects of immune inhibitor, and the pharmacological effects of anti-inflammatory, analgesic and diuretic. In this paper, HPLC was used in the determination of of astilbin content, which is an active ingredient. This method was simple, rapid, accurate.

2. METHODS AND RESULTS

2.1. Chemicals, Reagents and Samples

Among the ten batches of samples, one was collected from Yuenan, another was obtained from Hunan, the others were purchased from nine different regions (Guilin, Xing'an, Liuzhou, Guigang, Pingxiang, Jingxi, Yulin) in Guangxi province.

The reference compounds was purchased from TAUTO BIOTECH (Shanghai, China).

Acetonitile of HPLC was grade from Merck (Darmstadt, Germany); distilled water was further purified by a Milli-Q system (Millipore, Milford, MA, USA); phosphoric acid was purchased from the First Chemical Company of Nanjing (Jiangsu, China); other chemicals were of analytical grade. All solvents and samples were filtered through 0.45 μm filters before they were injected into HPLC.

2.2. Preparation of "Gaeulanghauh" Extracts

Medicinal materials collected from different areas were powdered by an electrical bender and sieved through a 60-mesh sieve. The powder was quantitatively extracted by mixing 0.5 g material with 25 ml methanol in a flask, weighting, and then through ultrasonic extraction at 80 kHz, 45°C for 45 min, adding methanol to weight. The extract was collected and filtered. After filtering through a filter paper and a 0.45 μm membrane filter (Millipore), the solutions of extracts were injected in to HPLC system directly.

2.3. Instrumentation

An Agilent 1100 series HPLC system and Agilent ChemStations software were used for the chromatographic analysis. The separation was carried out on a Agilent XDB C18 column (5 μm, 4.6 mm×150 mm). The mobile phase was composed of solvent A (0.2% Phosphoric acid aqueous) and solvent B (acetonitrile). Linear gradient program at a flow rate of 1.0 ml/min was performed as follows: 0-25 min linear gradient 93%-77% A; 25-26 min, linear gradient 77%-93% A; 26-30min, isocratic 93% A. The injection volume, column temperature, and UV wavelength were set at 10 μl, 35°C and 290 nm, respectively.

2.4. Method Validation

2.4.1. *Calibration Curve*

Standard stock solutions containing astilbin were prepared and diluted to appropriate concentrations for the plotting of calibration curves. The linear range was 0.16 μg-1.92 μg. The linear regression analytical data (Y=2216.1X-0.8018) for the calibration plots showed a good linear relationship with r2=0.9999.

2.4.2. *Precision*

The precisions were determined by analyzing the known concentrations of astilbin in six replicates. The relative standard deviation (RSD) values were 0.51%. Showed that the precision of the instrument is good.

2.4.3. *Stability*

Stability of sample solution extracted from "Gaeulanghauh" collected from Yulin was tested at 0, 4, 6, 8, 12, and 24 h. The relative standard deviation (RSD) values were 0.83%. The results showed that the solution was stable within 24 h.

2.4.4. *Repeatability*

To confirm the repeatability, six replicates of sample solutions describes as above were prepared and analyzed, The relative standard deviation (RSD) values were 0.69%. The results show that the method has good repeatability.

2.4.5. *Average Recovery Rate*

Six samples from known content were contained 2 ml astilbin methanol solution (0.1600 mg/ml) respectively, than were prepared and analyzed as above describes. The average recovery was 1.09% indicating good accuracy, the RSD values were 3.29%.

2.5. Sample analysis

The suitability of the developed analytical method was demonstrated by analyzing "Gaeulanghauh" from different area. The contents of astilbin in "Gaeulanghauh" were showed in Table 1 and Fig. 1.

Table 1. The contents of astilbin in "Gaeulanghauh" from different area

Sample number	producing area	the contents of astilbin(%)
1	Guilin, Guangxi	0.06
2	Xing'an, Guangxi	0.98
3	Liuzhou, Guangxi	1.44
4	Guigang, Guangxi	0.75
5	Pingxiang, Guangxi	2.34
6	Yuenan	0.85
7	Jingxi, Guangxi	1.37
8	Hunan	0.90
9	Yulin, Guangxi	1.31
10	Yulin, Guangxi	0.78

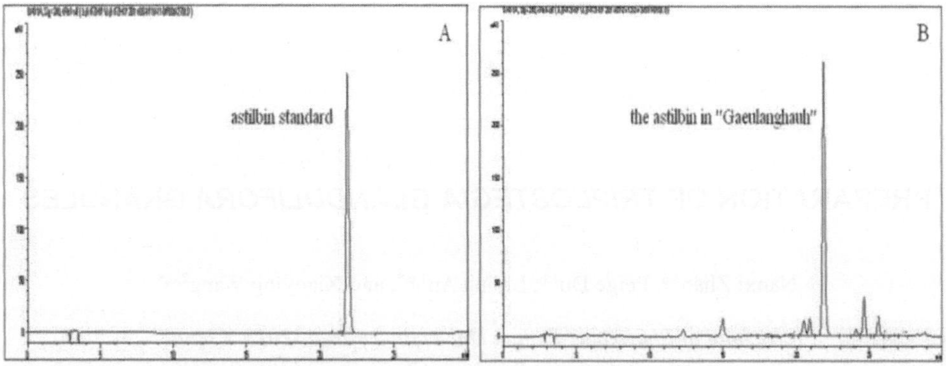

Fig.1. The HPLC figures of astilbin and “Gaeulanghauh” sample.

3. DISCUSSION

The high concentration of astilbin in “Gaeulanghauh” makes it a very important Zhuang medicinal plant, the optimized analytical HPLC method has shown to be useful for the detection of astilbin. This method was fast, sensitive and suitable for quantitative assessment of the astilbin in “Gaeulanghauh”. It could provide experimental basis for the further research of the zhuang medicine.

Acknowledgments

The National Science-technology Support Plan Projects (20122BAI27B06-1, 2013BAI11B06-7); Guangxi science and technology development projects (14124002-14); Nanning science and technology development projects (20133163).

References

1. Guangxi food and Drug Administration. Guangxi Zhuangzu Zizhiqu Zhuangyao Zhiliang Biaozhun Dierjuan (2011 nianban). Nanning: Publishing House Of Guangxi Scientific and Technical.
2. Teng HL, Mei ZN. ZHONG GUO ZHUANG YAO ZI YUAN MING LU. Beijing: Publishing House Of Ancient Chinese Medical Book, 2014.
3. Cai G, Gong CY, Sheng GQ. Simultaneous determination of astilbin and rutin compound tufuling capsules by HPLC. *ZHONG GUO YAO FANG* 2012; **23**: 4486-4488.
4. Bai M, Liu WP, Li SZ. HPLC determination of the different origin of tufuling astilbin and astragaloside content. *YAO WU FENG XI ZA ZHI* 2013; **33**: 1352-1354.
5. Chen ZY, Li Z, Dong QS. Determination of the content of the territory of Guangxi in different areas of soil Fuling astilbin. *ZHONG GUO YAO YE* 2011; **20**: 8-9.
6. Zhang TT, Zhou JS, Wang Q. HPLC Analysis of Flavonoids from the Aerial Parts of Bupleurum Species. *Chinese Journal of Natural Medicines* 2010; **8**:107-113.

PREPARATION OF *TRIPLOSTEGIA GLANDULIFORA* GRANULES

Nanxi Zhao[1,a], Peige Du[1,b], Liping An[1,c,*], and Xiaoying Wang[2,d,*]

[1]School of Pharmacy, Beihua University, Jilin, 132013, China
[2]People's Hospital of Jilin Province, Changchun, 130000, China
E-mail: [a]nancy_z3023@126.com, [b]dupeige2001@126.com, [c]630114130@qq.com, [d]wangxiaoyingzhuren@163.com
[*]Corresponding author

Objective: To study the optimal extraction process of ginsenoside in combination with Acanthopanax senticosus extract as the raw materials for the preparation of Triplostegia Glandulifora Granules. Methods: The auxiliary materials used for the preparation of Triplostegia Glandulifora Granules were screened and the prescription was optimized using orthogonal experiment. Results: The preparation method for Triplostegia Glandulifora Granules included that ginseng dry extract and Acanthopanax senticosus extract (in a mass ratio of 1.2:1) were adopted as the major raw materials, the proportion of them with sucrose and beta-cyclodextrin (β-CD) was 1:1:0.5, 10% CMC-Na was selected as the adhesive, a 16-mesh sieve was used for the granulation the granules were dried at 30°C , and the granules were sized with a 16–mesh sieve. Conclusion: The method used in this experiment can be used for the effective extraction of ginsenoside, and Triplostegia Glandulifora Granules prepared should meet the requirement for the granularity, with a good taste.

1. INTRODUCTION

In ginseng, ginsenoside is the most major active ingredient, followed by ginseng polysugar, volatile oil and some bioactive peptides [1]. Moreover, ginsenoside monomers from different parts of ginsen have different activities due to the difference in their structural properties [2]. Ginsenoside can affect multiple metabolic pathways, characterized by the effect on the central nervous system, significant enhancement of bogy adaptability, strengthening the immunity, improvement on the microcirculation, and anti-tumor, anti-fatigue and anti-aging acitities [3-7]. *Acanthopanax* is commonly used for the treatment of rheumatic arthralgia, aching lumbus and knees, rheumatism, impotent feebleness of feet and knees, and children's retardation of walking. In addition, it can be used in edema, difficult urination, insufficiency of spleen-kidney-yang, body deficiency and lack of power, loss of appetite, and insomnia and dreamful sleep, as well as has efficacies such as recuperating the liver and kidney, strengthening tendons and bones, and resisting fatigue [8, 9], and it is worth noting that the eleutheroside B (syringin) is the main ingredient of acanthopanax. Previous studies has showed that both ginseng and acanthopanax have an excellent anti-fatigue efficacies, and the combined application of these two drugs can show functions such as benefiting vital energy, and tonifying kidney and soothing the nerves, suitable for the insomnia, dreaminess, dizziness, and fatigue caused by the neurasthenia due to the deficiency of the kidney, heart and spleen [5,10]. *Triplostegia Glandulifora* Granules, a compound preparation of triplostegia and glandulifora, is easy to be dissolved, absorbed, carried with and stored, and moreover its preparation technology is simple, its quality of control is not so complicated, and its administration dose is relatively small.

2. INSTRUMENTS AND REAGENTS

AL204 electronic balance (Mettler Toledo Instrument Co., Ltd); air dry oven (Shanghai Boxun Industrial Medical Equipment Plant); *Acanthopanax senticosus* extract (0.8% eleutheroside, Xi'an Acetar Bio-Tech Co. , Ltd); ginseng dry extract (2.6% ginsenoside, self prepared); ethanol (analytically pure, NingQuanRui Reagent Co., Ltd); beta-cyclodextrin (β-CD) (Tuojian Co., Ltd); sucrose (Beijing Dingguo Bio-Technology Co., Ltd); sodium carboxymethylcellulose (Anhui Sunhere Pharmaceutical Excipients Co., Ltd).

3. EXPERIMENTAL METHODS

3.1. Preparation of the Granules

According to the previous studies, the weight ratio of ginsenoside dry extract to *acanthopanax senticosus* extract was set as 1.2:1. The crude drugs and excipients were well mixed for the wet granulation, and the granules were screened with a 16-mesh sieve and dried, and the dried granules were screened through a 16-mesh again and packaged in bags for use.

3.2. Selection of Auxiliary Materials

3.2.1. *Selection of Fillers*

Soluble starch, dextrin, beta-cyclodextrin (β-CD), and PVP were screened respectively as the fillers, and then 80% ethyl alcohol was used as the wetting agent, which was dried at 45°C and used for the breaking of granules.

3.2.2. *Selection of Flavoring Agents*

10% Fructose, 10% sucrose, and 0.1% aspartame were respectively screened as the flavoring agents, and 80% ethyl alcohol was used as the wetting agent, and dried at 45°C for the breaking of granules.

3.2.3. *Selection of Adhesives*

According to the experimental results of the selection of fillers in section 3.2.1 and the selection of flavoring agent in section3.2.2, 10% sucrose and β-CD were selected as the optimal excipients, and 80% ethyl alcohol, water, 10% gelatinized corn starch, and 10% CMC-Na were selected as the optimal wetting agent or adhesive, which were dried at 45°C for the breaking of granules.

3.3. Orthogonal Design of Prescription Formula

Based on the experiment results in section 3.2, the auxiliary materials were mixed with the crude drugs at different ratios, and regarding the granularity (%) of the prepared granules as the investigation index, the orthogonal experiment was conducted according to Table 1.

Table 1. Orthogonal factor level table

Level	Factor		
	A Crude drug: corrective: filler	B Adhesive	C Drying temperature（°C ）
1	1:1:1	water	30
2	1:1:0.5	5%CCNa	45
3	1:2:0.5	10%CCNa	60

4. EXPERIMENT RESULTS

4.1. Selection of Auxiliary Materials

4.1.1. *Selection of Fillers*

The fillers were selected according to the granularity (%) of *Triplostegia Glandulifora* Granules, as shown in Table 2.

Table 2. Results of filler selection

	PVP	Dextrin	Soluble starch	β-CD
Granularity（%）	3.24	3.09	2.78	1.92

According to Table 2, it could be known that the granularity of the granules using β-CD as the filler was best (1.92%), followed by that of the soluble starch and dextrin, and the worst was PVP. Moreover, the prepared granules using soluble starch, dextrin or PVP as the filter led to an excess of fine powders, crispness, and acrid flavor of the granules, so that β-CD was adopted as the filler finally.

4.1.2. *Selection of Flavoring Agent*

The flavoring agent was selected according to the sweetness of *Triplostegia Glandulifora* Granules (Table 3).

Table 3. Results of flavoring agent selection

	10% fructose	0.1% aspartame	10% sucrose
Sweetness	Severe	Slignt	Moderate

It could be known from the results in Table 3 that the sweetness of granules prepared with 0.1% aspartame as the flavoring agent was lower, that with 10% fructose was highest, and that with 10% sucrose was moderate, so that 10% sucrose was adopted as the flavoring agent.

4.1.3. *Selection of Adhesive*

The adhesive was selected according to the granularity (%) of *Triplostegia Glandulifora* Granules (Table 4).

Table 4. Results of adhesive selection

	water	80% ethanol	10% starch paste	10% CCNa
granularity（%）	2.97	1.92	1.67	1.25

As shown in Table 4, the granularity (1.25%) of granules prepared with 10% CMC-Na as the adhesive was best, followed by that with 10% starch paste and that with 80% ethanol, and water was the poorest. Furthermore, CMC-Na as the adhesive could swell automatically, be easy to use, and show a better viscosity in contrast to that of the other adhesives, so that CMC-Na was selected as the adhesive eventually.

4.2. Orthogonal Design of Prescription

Based on the results of auxiliary material selection, β-CD was selected as the filler, sucrose as flavoring agent, and CMC-Na as the adhesive eventually to conduct the orthogonal experiment. The results are shown in Table 5.

Table 5. Results of orthogonal experiment of *Triplostegia Glandulifora* Granules preparation

prescription	Factor A Crude drug:sucrose:β-CD	B adhesive	C drying temperature(°C)	Index Granularity(%)
1	1	1	1	4.19
2	1	2	2	2.28
3	1	3	3	2.05
4	2	1	2	2.74
5	2	2	3	2.62
6	2	3	1	1.78
7	3	1	3	2.23
8	3	2	1	2.11
9	3	3	2	2.24
K1	8.52	9.16	8.08	
K2	7.14	7.01	7.26	
K3	6.58	6.07	6.90	
k1	2.84	3.05	2.69	
k2	2.38	2.34	2.42	
k3	2.19	2.02	2.30	
R	1.94	3.09	1.18	

As shown in Table 5, the value of R was B>A>C, suggesting that the most significant factor should be the adhesive amount, followed by the weight ratio of crude drugs to the auxiliary materials, and the least significant factor should be the drying temperature (°C). Among the granules prepared based on different prescriptions, the granule prepared in line with the No.6 prescription showed a best granularity (1.78%), which meets the requirement for the granularity (less than 15%) prescribed in the appendix IC of the first part of Chinese Pharmacopoeia (2010 edition). According to the results of orthogonal experiment, the prescription of *Triplostegia Glandulifora* Granules was determined as that the ratio of crude drugs to sucrose to β-CD was 1:1:0.5, 10% CMC-Na was adopted as the adhesive, and the drying temperature for the preparation of the granules was 30°C .

5. DISCUSSION

In the selection of auxiliary materials, candidate fillers such as soluble starch, dextrin, PVP, and β-CD were screened, and finally β-CD was selected as the filler according to the granularity (%) of *Triplostegia Glandulifora* Granules; candidate flavoring agents for the preparation of *Triplostegia Glandulifora* Granules, such as 0.1% aspartame, 10%

fructose and 10% sucrose were strictly screened, and finally 10% sucrose as the flavoring agent with a moderate sweetness was adopted according to the sweetness of *Triplostegia Glandulifora* Granules; 80% ethanol, 10% starch paste, and 10%CMC-Na were compared in the selection of them as the adhesive, and finally CMC-Na was proved to be the optimal adhesive based on the requirement for the granularity (%) of *Triplostegia Glandulifora* Granules.

6. CONCLUSION

In this study, the preparation method for *Triplostegia Glandulifora* Granules included that the ratio of crude drugs to sucrose to β-CD was 1:1:0.5, 10% CMC-Na was adopted as the adhesive, a 16–mesh sieve was used for the granulation, the drying temperature was 30°C, and the granules were sized by a 16-mesh sieve. The resulted granules were brown with a granularity of 1.78%. This research may provide reliable data for the industrial production of *Triplostegia Glandulifora* Granules.

Acknowledgements

This project co-funded by Jilin Provincial Pharmaceutical Industry Development Special Fund (20140311043YY), Jilin Provincial NDRC Platfrom (2013G019), Jilin Provincial Science and Technology Agency (20150311029YY), and Science and Technology Bureau of Jilin (201464051).

References

1. Hyeong GK, Sara Y, Hyejung P. Antioxidant effects of Panax ginseng C.A. Meyer in healthy subjects: A randomized, placebo-controlled clinical trial. *Food and Chemical Toxicology* 2011; **49** (9): 2229-2235.
2. Zhi Cao, Yan-di ZHang. Research progress on ginseng active ingredients and the pharmacological action. *Research of Ginseng* 2012; **2**:39-43. (In Chinese).
3. Ying Zhang, Dan-feng Yuan, Li-juan Zhu. Activity study and preparation technology of compound ginseng granules. *Heilongjiang Agricultural Sciences* 2015; **1**: 128-132. (In Chinese).
4. Xiu-li Guo, Shu-lian Gao. Research progess of ginseng chemical component and pharmacology. *Clinical Journal of Chinese Medicine* 2012;**4** (14):26-27.(In Chinese).
5. Hai-ying Huang. Research progress of pharmacological action of ginsenoside-Rg1. *Practical Journal of Traditional Chinese Medicine* 2012; **28** (7): 608-609. (In Chinese).
6. Kang JH, Song KH, Woo JK. Ginsenoside Rp1 from Panax ginseng exhibits anti-cancer activity by down-regulation of the IGF-1R/Akt pathway in breast cancer cells. *Plant Foods for Human Nutrition* 2011; **66** (3): 298-305.
7. King ML, Murphy LL. Role of cyclin inhibitor protein p21 in the inhibition of HCT116 human colon cancer proliferation by American ginseng (Panax quinguefolius) and its constituents. *Phytomedicine* 2010; 17:261.

8. Chun-mei Zhai, Guo-liang Xu, De-qiang Yang. CCD-RSM Method of Optimizing Total Flavonoids of Acanthopanax Senticosus and Macroporous Resin Purification Process. *Information on Traditional Chinese Medicine* 2015; **1**: 14-16. (In Chinese).
9. Ying-hua Zhang, Xue Guan. Research of Oxidation Resistance and Antibacterial Activity of Extracted Flavonoids in Acanthopanax Senticosus Harms. *Journal of Northeast Agricultural university* 2012; **43** (3): 85-90. (In Chinese)
10. Cantarini MV, Bailey CJ, Collins B. The relative bioavailability of gefitinib administered by granular formulation. *Cancer Chemotherapy and Pharmacology* 2007; **62** (2):203-208.

EXPERIENCE ON THE CLINICAL PRACTICE TEACHING OF SURGERY

Shu Jing[a], Cong Liu[b], He Li[c], and Weihai Jiang[d,*]
Beihua University, Jilin City, 132013, China
E-mail: [a]10171812@qq.com, [b]372131552@qq.com, [c]yitonglh@126.com, [d]50598236@qq.com
[*]Corresponding author

Clinical practice is an important stage of clinical medical education and a process in which the theoretical knowledge learned in classroom is combined with the clinical practice, and the results can directly affect the clinical work ability of medical students in the future and even their medical career. In order to further improve the hospital clinical teaching and enable medical students to learn more knowledge of the surgery as much as possible, the characteristics of task-based learning (TBL) and problem-based learning (PBL) methods are discussed and analyzed to design a PBL method for the clinical practice of medical students in general surgery. The proper application of the method in the process of surgical clinical practice teaching may greatly improve the students' surgical theoretical knowledge and clinical practical skills through a good combination of the knowledge learned in classroom with the clinical practical skills learned in hospital, which may lay a solid foundation for the training of students to become qualified clinicians in the future.

1. INTRODUCTION

Clinical practice is an important stage of clinical medical education, a process of the classroom learning theoretical knowledge combined with the clinical practice, and its results have a direct impact on the future clinical work ability of students and may influence the lifetime work of medical students [1-5]. In the traditional teaching mode, generally medical students follow the teacher to make the rounds of wards, change fresh dressing for a wound, participate in a surgery, and be busy with the various clinical operations all day during the surgical practice in clinic, which can result in their superficial understanding of diseases and make the students often formulate treatment plans by blindly imitating what their teachers do, ignoring the cultivation of the students' comprehensive quality, independent thinking and problem-solving abilities. In order to reverse this situation, it is necessary to further improve the hospital clinical teaching and enable students to learn more knowledge of the surgery as much as possible by designing and implementing problem-based learning (PBL) in the clinical practice teaching of general surgery according to the characteristics of task-based learning (TBL) and PBL teaching methods to improve the quality of teaching [6-9]. In this paper, the experiences on the practical teaching of general surgery with PBL method combined with TBL method are discussed.

2. OPERATION MEASURES AND PRACTICES

In accordance with the proposed learning problems, five steps, namely the collection and analysis of clinical data, establishment of hypotheses, the comment on the hypotheses, the comparison, and the summary, are made to complete the clinical practice of representative diseases required by the teaching syllabus. Some important clinical works, such as the medical daily rounds for patients hospitalized newly, the change of diagnosis and treatment options for patients, discussion on intractable cases, and so on, are taken as the vector of PBL clinical teaching. According to the teaching needs and based on the cases, the teachers put forward the preliminary clinical problems, and students of each group collect the partial or complete clinical data, put forward the concrete problems, and then collect more clinical data related to the cases, clear up and analyze the data, establish hypotheses of the diagnosis and treatment options, and discuss the hypotheses. The group leaders are responsible for organizing the discussion on the learning and practice of their own group to further screen, summarize and supplement the data collected by their own group, and the results are collated and summarized in written form for the presentation on behalf of the group during the seminar.

2.1. Problem Design in the PBL Teaching

Problems existing in clinic are the core of PBL teaching. The problems used in the clinical teaching with PBL method must be more complicated. To integrate problems involving multiple disciplines, these problems should have an instructive and representative significance to the future practical work of students although they are only involved in a single case more often than not. For example, before an operation of appendicitis, the students should be informed of the preparation for the operation in advance: what are the signs of appendicitis? How to make a physical examination for the patient? What can cause appendicitis? How to diagnose appendicitis? The students are required to utilize the basic knowledge such as anatomical and pathological knowledge to comprehensively analyze cases, and encouraged to actively preview and collect information through the internet. For example, for a hospitalized patient with melena complicated with dysphagia, the designed problems may be related to the most common anatomical and qualitative diagnosis. For a patient with abdominal pain complicated with vomiting, the designed problems may be what kind of surgical diseases should be differentiated from it. When students propose a series of clinical problems independently in the group, the problems should be studied, respectively, such as the process of pathophysiology, the clinical pathological correlation, the effective examination means, the diagnostic criteria, and the treatment and the curative effect criteria. In order to obtain the data to confirm their hypotheses, the students can freely inquiry the medical history, make the physical examination, and propose the necessary laboratory examinations. During the later stage of peri-operative period, the students are asked to summarize the processes, and the teachers should comment on not only the summary and the discussion of students but also their disagreement, especially on the students' understanding of some basic concepts and thinking methods.

2.2. Guiding Student to Develop a Clear Diagnosis and Treatment Thinking

The transition of hospital clinical practice from the theoretical learning in a medical school is a transformation process from the book knowledge to the clinical skills. Students understand the basic theory knowledge of diseases through the learning of basic knowledge in school and the knowledge should be combined with the clinical practice, which can make the students become qualified doctors in the future. It well known that the knowledge from books is not equal to the actual operation ability and students who can understand what are written in medical books do not mean that they can become qualified doctors, so that teachers should guide their students to have a learning thirst for clinical skills and stimulate the students' interest in learning clinical skills. Taking a patient with jaundice as an example, the first should be based on the icteric sclera and the increased blood bilirubin to determine the diagnosis of jaundice, then the respective characteristics of jaundice in surgery and internal medicine, the differentiation of benign or malignant jaundice if it is a surgical obstructive jaundice, the basis for the diagnosis, the necessary valuable auxiliary examinations, and how to select the examination methods should be introduced and discussed, and finally the procedures of operation, operation types and other matters needing attention in the operation are introduced and discussed.

2.3. Cultivating Students' Holistic View of Disease Diagnosis

Compared with internal medicine diseases, surgical diseases are characterized by clinical manifestations with more typical locality. The students' preliminary understanding on them often is that the diseases are simple and limited, and sometimes they cannot see the wood for the trees. During the practical teaching, the teachers should hierarchically analyze, explain and demonstrate the patient's symptoms and signs to improve the students' overall understanding of the diseases. If students are asked about a patient with bladder cancer, they often reply that the preferred method for the diagnosis of painless hematuria should be the cystoscopy plus biopsy and the therapy should be mainly the surgical resection. At this time, the teachers responsible for the practical teaching should purposefully put forward why the recent recurrence rate of bladder cancer is high after the local excision, and how to explain why some patients still suffer from hematuria. Through these questions, the students are guided to think that bladder cancer is actually a disease of systemic immunodeficiency, the oncogene is expressed after the mutation of tumor suppressor gene, tumors may simultaneously or successively develop in all transitional epithelia from renal pelvis to urethra coexisting in the urine environment, and the local symptoms and signs are the local manifestations of systemic lesion, and in order to improve the treatment effect, the examination on the urinary tract should be carefully made, a comprehensive treatment for it should be taken, and especially the methods for improving the cell immune surveillance function or inhibiting the expression of cancer gene should be found for it [1].

2.4. Guiding the Students to Correctly Understand the Limitation of Operation

When they practice in surgery, students often favor to do operation, and think that surgical diseases can be cured mainly by surgery, and the surgery can removal lesions, relive obstruction, control bleeding, and so on. Due to these reasons, during the practical teaching, it is necessary for the teachers to intend to choose some physiological disorder cases caused by surgeries and take some measures to remind the students to recognize that surgeries have not only therapeutic effects but also adverse effects, such as cough and hoarse voice after the thyroidectomy, asking the patients to describe the feelings of the patients after the operation, and clarifying the adverse consequences that a surgery can bring to the patients. The students should be told that as a doctor, when he or she makes the decision of a surgical plan, he or she must not only take into account of the therapeutic effects of surgery but also should carefully consider the adverse consequences caused by the surgery on the patients, to comprehensively analyze the case and weigh the pros and cons for overcoming the misconception to do a surgery for surgery and strictly choosing a surgery based on the indications of operation. In some cases of obstructive diseases, it is often to think that surgery can relieve the obstruction and not to make a clear distinction between a mechanical obstruction and a dynamic obstruction, which may lead to an exacerbation of the dynamic obstruction disease after the surgery due to the blind surgery and the treatment become more difficult, so we must let students understand these issues from the point of view of physiology and establish a correct concept of surgery.

2.5. Questionnaire Survey of PBL Method

After the end of the clinical practice, the questionnaire survey of PBL was carried out in the students and the results are shown in Table 1 and Fig. 1. The results showed that most of the students believed that the application of PBL in the clinical practice could improve their automous learning ability, expand their knowledge range and depth; cultivate their innovative thinking, problem-posing, problem-analyzing and problem-solving capacity, and language expression and interpersonal communication capacity; not only strengthen their team assistance spirit, but also improve their understanding on the surgery content and their surgery skills significantly, which should be most important.

Table 1. Questionnaire of PBL method

Items	Yes (%)	So so(%)	No(%)
Expanding in knowledge range and depth	90	10	0
Improvement in autonomous learning ability	95	5	0
Improvement in problem-analyzing and solving	90	10	0
Improvement in surgery skills	100	0	0
Improvement in scientific thinking	98	2	0
Improvement in language expression and interpersonal communication	85	10	5

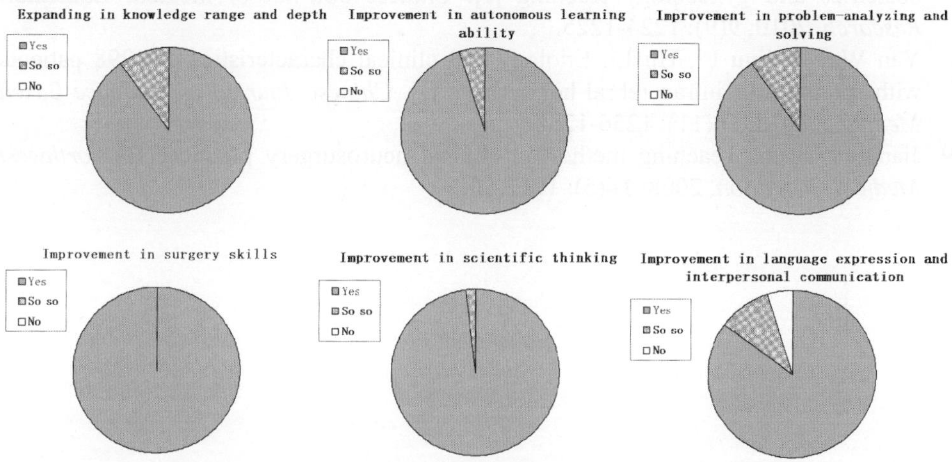

Fig. 1. Questionnaire of PBL method.

3. CONCLUSION

The proper application of PBL method in the process of surgical clinical practice teaching may greatly improve the students' surgical theoretical knowledge and clinical practical skills through a good combination of the knowledge learned in medical school with the clinical practical skills learned in hospital, which may lay a solid foundation for the training of students to become qualified clinicians in the future.

References

1. Baiqing Chen, Zhiqiang Ge, Wang Xin, et al. Application of PBL teaching method in the surgical clinical practice of perioperation [J].*Teaching & Administration*, 2008; **4**: 31-32.
2. Yihua Zhou. Influencing factors and Countermeasures of nursing clinical practical teaching [J]. *Journal of Practical Nursing*, 2003;**19**(1): 69.
3. Fenglin Cao. Mental health status and its coping means and the related research on the social support in nursing undergraduates. [J]. *Journal of Nursing Science*, 2003; **18**(2): 83-85.
4. Jianqiong Xiong, Xianhui Xie, Zhaoxia Deng. To strengthen the training of clinical thinking ability of interns [J]. *Journal of Regional Anatomy and Operative Surgery*, 2006; **15**(1): 40-41.

5. Qing Luo, Runxian Luo, Haiying Huang, et al. Application of clinical teaching approach in the practical teaching of neurosurgery [J]. *Journal of Qilu Nursing*, 2012; **6**(18): 102-103.
6. Quhua Zhu. Application of clinical teaching modes in the teaching of general surgical practice in students. [J]. *Chinese and Foreign Medical Research*, 2013; **18**(11): 145-146.
7. Gang Chen, Weiping Hu. Evaluation on teaching effect of clinical pathway in obstetrics and gynecology teaching [J]. *Chinese Journal of Medical Education Research*, 2010; **9**(9): 1224-1225.
8. Yan Wu, Huaiyu Li, Hu Li. Etiology and clinical characteristics of 1298 patients with spontaneous intracerebral hemorrhage [J]. *Chinese Journal of Evidence-Based Medicine*, 2010; **10**(11): 1256-1258.
9. Jianhong Shen. Teaching method in clinical neurosurgery practice [J]. *Northwest Medical Education*, 2008; **16**(5): 1012-1013.

THE DIFFERENT DOSE EFFICIENCY OF LEPTIN TO NG108-15 CELL LINE AND PRIMARY HYPOTHALAMIC NEURON OF MOUSE

Jiucheng Xia[1,2], Ting Long[2], and Danian Qin[2,*]

[1]*Department of Bio-chemical Engineering, Panzhihua College, Panzhihua, China*

[2]*Department of Physiology, Shantou University Medical College, Shantou, China*

E-mail: dnqin@stu.edu.cn

[*]*Corresponding author*

The NG108-15 cell line is an important neural mode in vitro. When it is exposed for dibutyryl cyclic AMP(dc-AMP), the cell will differentiated as neuron which has some neural morphology character. So many labs use the cell line to research ion channel, membrane receptor and neurotransmitter and so on. In this study, we use NG108-15 cell as a hypothalamic neuron mode to research the response character of the NG108-15 cell to different concentration leptin. The result found that the differentiated NG108-15 cell has no obvious and regular dose effect comparing with the primary hypothalamic neuron of mouse.

1. GUIDELINES

Exploring cell molecular properties such as expression of membrane receptors and neuropeptides is very important for understanding the physiological and pathophysiological functions of neurons. Although acute-isolated primary cell is the optimum choice for pursuing these measurements[1,2,3], cell lines are also served as an appropriate tool for the cell molecular studies, because cell lines provide the advantage of enough homogeneous cells that can make the investigation under easily controlled conditions[4-7]. NG108-15 cell line is a hybrid cell line from the fusion of mouse neuroblastoma cells and rat glioma cells[8]. After differentiation, this cell line presents neurite extension, forms synapses. Therefore, many studies used NG108 cell as the model to investigate cell functions of neurons[4-14].

Leptin is an important cytokine for energy metabolism, when it act on Central nervous system (namely hypothalamic nuclei). Leptin decreases food intake, increases energy expenditure and decreases metabolic efficiency[15]. Beside this, Leptin has wide spectrum of biological functions, such as cytokine secretion, synaptic transmission, phagocytosis, regulating the hypothalamic–pituitary–adrenal axis, hematopoiesis, angiogenesis and so on[16].

So in this study, we want to research the behavior of the NG108-15 cell and primary hypothalamic neuron of mouse, when they were incubated with leptin on different concentration.

2. METHODS

2.1. Differentiation of NG108-15 Cell Lines

The neuroblastoma × glioma NG108-15 cell line was obtained from the American Type Culture Collection (ATCC, Manassas, VA). NG108-15 cells were cultured as previously described. Briefly, cells were cultured at a density of 1 × 104 cells/ml on 12-well plates in Dulbecco's modified Eagle's medium (DMEM, Gibco) with 10% (v/v) fetal bovine serum (Gibco) and 1%(v/v)double antibiotics (Gibco) contained with 104 U/ml penicillin and 104 ug/ml streptomycin. After 3 days, differentiation was induced by culturing the cells in DMEM with 1 mM dc-AMP (sigma). Cells were used for experiments after 7 days of differentiation. Cells were maintained in air incubator at 37°C with humidified atmosphere of 5% CO_2[5,9,11,13].

2.2. Primary Hypothalamic Cell Culture of Mouse

Neonatal brains were extracted (newborn mouse day 1-2), the meninges were removed, and the hypothalamus were dissected immediately on ice. Hypothalamus were digested into single cells by trypsin (Sigma, 25 min, 37°C). Digested Cells were plated at a density of 1×104 cells/ml on 12-well plates. Neurobasal medium with 2% (v/v) B27 supplement (Gibco) and 1%(v/v) L-glutamine (Gibco) with 0.5% (v/v) double antibiotics (Gibco) contained with 104 U/ml penicillin and 104 ug/ml streptomycin was used as culture medium. On the second day in vitro, cells were treated with 10M cytosine arabinoside (Sigma) for 48h to inhibit non-neuronal cell growth. Subsequently, the medium was completely exchanged, and every third day afterward, fresh medium was added. Cells were used for experiments after 10 days of culture[1,2,3].

2.3. Incubate Cell with Leptin

After NG108-15 cells and primary hypothalamic cells of mouse has obvious neural morphology (culture for 10d), use different drug concentration (0pg/ml,100pg/ml,1000pg/ml,10000pg/ml) of leptin (sigma) to incubate the two type cells, each treatment has 3 repeats. Harvested these cells and extracted total RNA using RNAiso plus (Takara) after 12h incubation of leptin.

2.4. Real-time RT-PCR

Cells on each well was aspirated medium out and washed by PBS. Then added 0.5ml RNAiso plus (Takara) to extract RNA following the manufacturer's directions. Purified RNA of each sample was dissolved in 50ul RNase-free water and was conserved at -70°C. cDNA synthesis was performed with a PrimeScript RT reagent kit with gDNA Eraesr (Takara) following the manufacturer's directions. 7ul RNA was added in each RT reaction with total 20ul volume. The cDNA of each sample was conserved at -20°C. The

PCR was performed using specific primers (Table1). Primer sequences were designed using Primer5 software. Amplification was carried out in an ABI7500 system (Applied Biosystems, Foster City, CA) using 40 cycles (95°C, 5s; 60°C, 34s) with SYBR Premix Ex Taq kit (Takara) following the manufacturer's description. Each 25ul reaction volume contained 1ul cDNA as template. The data of mRNA expression level of 5 genes(Lepr, Npy, Agrp, Pomc and β-actin) were analyzed by the 2-ΔΔCt method[9] (β-actin as internal control gene).

Table 1. Sequences of each primer pair and their location in sequences cited in GenBank

Gene name	species	Sense primer	Antisense primer	Amplicom size(bp)	Genebank accession no.
β-actin	mouse	CACTGTGCCCATCTACGA	TGATGTCACGCACGATTT	155	NM_007393.3
Lepr	mouse	GGTCACCCAGCACAATCC	TGGCGAGTCAAGTGAACCTA	230	BC082551.1
Npy	mouse	CGCTCTGCGACACTACAT	CAAGTTTCATTTCCCATCA	161	AF273768.1
Agrp	mouse	GTTCCCAGAGTTCCCAGGTC	ACGGGTCGCAGCAAGGTA	197	U89484.1
Pomc	mouse	CTCCGAGAAGAGCCAGACG	CCTTGGAATGAGAAGACCC	106	NM_001278584.1

3. RESULTS

3.1. Cell Morphology of Two Kinds of Cells

Through different culture procedure, 2 kinds of cell have got. Both of them have neuron character such as axon differentiation, but has some different character of them. The size of NG108-15 cells is bigger than hypothalamic cells of mouse. The hypothalamic cells of mouse has cluster grown, but the NG108 cell has no cluster grown(Fig.1).

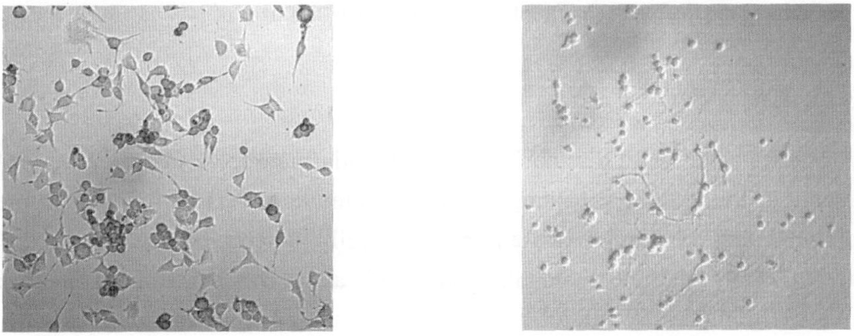

A B

Fig. 1. The Cell morphology of different cells. (A: Differentiated NG108-15 cell lines (100×); B: Differentiated primary hypothalamic cell of mouse (100×)).

3.2. Dose Effect of Leptin with Different Concentration

Through using different concentration of leptin to incubate the primary hypothalamic cells of mouse and NG108-15 cells, respectively. It shows that the mRNA expression level of the receptor of leptin (Lepr), neuropeptide Y (Npy), pro-opiomelanocortin (Pomc) and agouti gene-related protein (Agrp) has obvious and regular dose effect in the primary hypothalamic cells of mouse(Fig.2, A), but the NG108-15 cell line has not this obvious change trend(Fig. 2, B).

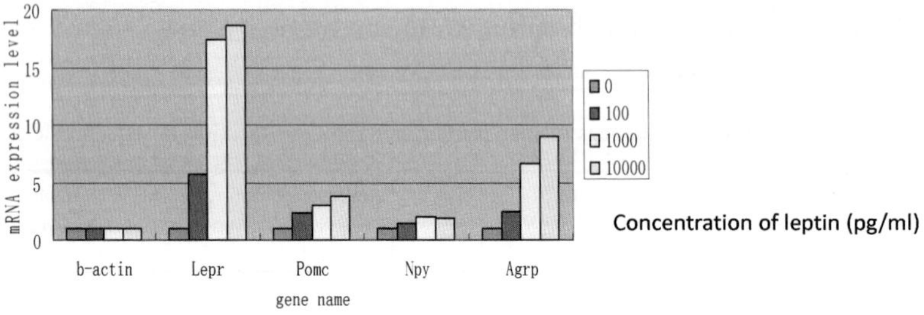

A

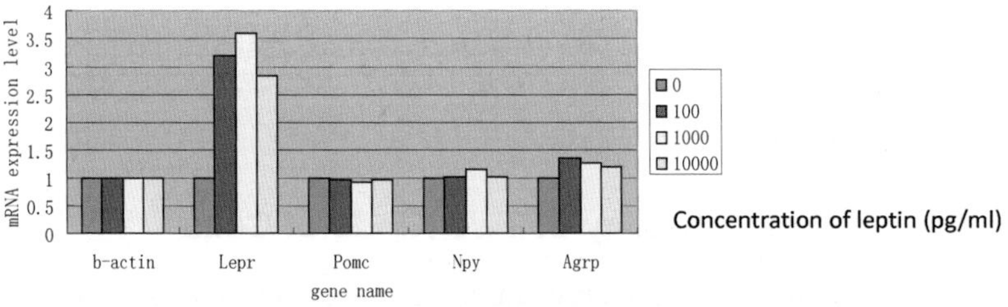

B

Fig. 2. Dose effect of leptin to NG108-15 cell and primary hypothalamic cell of mouse. (A: mRNA expression level of lepr, pomc, npy and agrp of primary hypothalamic cell of mouse incubated with different concentration of leptin for 12h; B: mRNA expression level of lepr, pomc, npy and agrp of NG108-15 cell incubated with different concentration of leptin for 12h).

4. DISCUSSION

Previous studies have found that during the differentiation of NG108-15 cells, dc-AMP is a key factor in the culture medium, which can stimulate NG108-15 cells to present a morphological alterations such as the increase in cell diameter, neurite length[5,9,11,13]. Our present study demonstrated that dc-AMP also can make NG108-15 cell to differentiate (fig1A). Many studies have use differentiated NG108-15 cell as vitro model to research ion channel, neurotransmitter receptor and so on[4-14].

Although acute-isolated primary cell is the optimum choice for understanding the physiological and pathophysiological functions of neurons, but this method has some disadvantage such as waste-time, complicate method, high death rate of neuron and heterogeneous neuron with glial cell[1,2,3]. On the contrary, cell lines provide some advantage: easy method, high efficiency and enough homogeneous cell. So it is served as an appropriate tool for the cellular function studies under easily controlled conditions[8]. But Something need the researcher note that NG108-15 cells cannot cluster grow as primary hypothalamic neurons (Fig.1), so it make cells lack normal communication with synapse liking in vivo. It could affect the experiment result of physiological and pathophysiological functions of NG108-15 cell line in vitro (Fig 2.).

References

1. Freddy D. Jeanneteau, W. Marcus Lambert, Naima Ismaili, et al. BDNF and glucocorticoids regulate corticotrophin-releasing hormone (CRH) homeostasis in the hypothalamus. PNAS 2012, 109;No.4,1305-1310.
2. Lara Marchetti, Matthias Klein, Katalin Schlett, et al. Tumor necrosis factor (TNF)-mediated neuroprotection against glutamate-induced excitotoxicity is enhanced by N-Methyl-D-aspartate receptor activation. *The Journal of biological chemistry* 2004, 279; No.31: 32869–32881.
3. Harald Neumann, Ru¨diger Schweigreiter, Toshihide Yamashita, et al. Tumor necrosis factor inhibits neurite outgrowth and branching of hippocampal neurons by a Rho-dependent mechanism. *The Journal of Neuroscience* 2002; 22, No.3, 854–862.
4. Nidhi Kaushal, Meenal Elliott, Matthew J. Robson, et al. AC927, a receptor ligand, blocks methamphetamine induced release of dopamine and generation of reactive oxygen species in NG108-15 cells. Mol Pharmacol 2012; 81, No.3: 299–308.
5. Dieter Müller, Karen J. Greenland, Robert C. Speth, et al. Neuronal differentiation of NG108-15 cells has impact on nitric oxide- and membrane (natriuretic peptide receptor-A) cyclic GMP-generating proteins. Molecular and Cellular Endocrinology 2010, 320, No.1, 118–127.
6. S. Amina, M. Hashii, W-J. Ma, et al. Intracellular calcium elevation induced by extracellular application of cyclic-ADP-ribose or oxytocin is temperature-sensitive in rodent NG108-15 neuronal cells with or without exogenous expression of human oxytocin receptors. *Journal of Neuroendo crinology* 2010; 22, No.5, 460–466.
7. Katie Leach, Richard E Loiacono, Christian C Felder, et al. Molecular mechanisms of action and in vivo validation of an M4 muscarinic acetylcholine receptor allosteric modulator with potential antipsychotic properties. Neuropsychopharmacology 2010; 35, No. 4, 855–869.
8. Hamprecht B. Structural, electrophysiological, biochemical, and pharmacological properties of neuroblastoma-glioma cell hybrids in cell culture. Int Rev Cytol 1977;**49**, 99–170.
9. Jinxu Liu , Huiyin Tu , Dongze Zhang, et al. Voltage-gated sodium channel expression and action potential generation in differentiated NG108-15 cells. BMC Neuroscience 2012; 13, No.129, 1471-1479.
10. SN Wu, CC. Yeh, HC Huang, et al. Electrophysiological characterization of sodium-activatedn potassium channels in NG108-15 and NSC-34 motor neuron-like cells. Acta Physiol 2012; 206, No.2, 120–134.
11. Akinori Kawaguchi, Hajime Asano, Kayoko Matsushima, Tetsuyuki Wada, Shigeru Yoshida and Seiji Ichida, et al. Enhancement of sodium current in NG108-15 cells during neural differentiation is mainly due to an increase in NaV1.7 Expression. Neurochem Res 2007; 32, No.9, 1469–1475.

12. Geoffrey S Dow. Effect of sample size and P-value filtering techniques on the detection of transcriptional changes induced in rat neuroblastoma (NG108) cells by mefloquine. *Malaria Journal* 2003; 2, No.4, 1-11.
13. Monrudee Sukma, Michihisa Tohda, Hiroshi Watanabe, et al. The mRNA expression differences of RNA editing enzymes in differentiated and undifferentiated NG108-15 Cells. *J Pharmacol Sci* 2005; 98, No.4, 467–470.
14. Jean Chemin, Joel Nargeot and Philippe Lory. Neuronal T-type a1H calcium channels induce neuritogenesis and expression of high-voltage-activated calcium channels in the NG108–15 cell line. *The Journal of Neuroscience* 2002; 22, No.16, 6856–6862.
15. Minokoshi Y, Alquier T, Furukawa N, et al. AMP-kinase regulates food intake by responding to hormonal and nutrient signals in the hypothalamus. Nature 2004; 428, No. 6982, 569-574.
16. Otero M, Lago R, Lago F, Casanueva FF, et al. Leptin, from fat to inflammation: old questions and new insights. FEBS Lett 2005; 579, No.2, 295-301.

CASE STUDY OF CT EXAMINATION SCHEDULING

Chunrong Qin[1,a], Li Luo[2,b], Xian Chen[3,c], and Junwei Yu[1,d]

[1]*Department of Business Administration, College of Mobile Telecommunications Chongqing University of Posts and Telecom, He Chuan, 401520, China*

[2]*Business School, Sichuan University, Chengdu, 610064, China*

[3]*West China Hospital, Sichuan University, Chengdu, 610041, China*

E-mail: [a]jianghu525@126.com, [b]luolicc@163.com, [c]43756107@qq.com,[d]155128196@qq.com

The paper, taking the west China hospital of Sichuan university as an example, tries to do data analysis for CT scheduling status quo by description statistics and parameter test. The analysis result shows that there's a significant mismatch between CT1's examination quantity and reservation quantity; CT1 is under overload operation and some equipment's idle resources has not been fully utilized currently. According to the above problems, our paper gives the following management advice: before service day, taking account of equipment capacity constraints, we should allocate equipment reasonably for each appointment patient on the basis of patient demand; On the service day, considering that both patient demands and equipment capacities are dually uncertain, we should adjust scheduling policy timely to improve the equipment utilization when queue length changes.

1. INTRODUCTION

As an intermediate link to link up outpatient, hospital, surgery and other processes, CT examination is oriented to both patients and other departments in the hospital[1], the service quality has a great influence on the whole hospital system[2]. In addition, CT equipment is very expensive, and the purchase of CTs is strictly regulated by the states[3]. The large public hospitals are facing many dilemmas, for example, CT examination ability is insufficient, patients' waiting time is too long, and so on. At present, according to different illness complexity, hospital staffs arrange patients to the corresponding CTs. In order to serve more respective patients, every CT seldom helps to serve patients for others even though it is idle, this always makes a waste of resource. In addition, due to technical complexity and the trick of models, few scholars focus on the research about resource scheduling of multi-type patients[4]. Due to increasing demand, how to improve equipment utilization and reduce patient waiting time with the assurance of service quality and no additional equipment are important contents of scheduling. To take the west China hospital of Sichuan university as an example, this paper will first do data analysis for CT scheduling status quo, and then give corresponding management proposals.

2. INTRODUCTION OF WEST CHINA HOSPITAL

As an important part of radiology department in west China hospital, CT examination's daily examination quantity is the highest among all examination types, taking about 43.5 percent (CT: 803 patients a day; conventional radiotherapy: 708 patients a day; MRI: 279 patients a day). CT examination involves in a variety of patient types. According emergency degree, patients can be divided into outpatients, inpatients and emergency patients. At the same time, it can have routine scanning patients and contrast scanning patients because of different illness complexity. CT examination scheduling has high complexity. With continuously increasing patients, the pressure of CT examination is more and more great. Without any additional equipment, the yearly output is 161 thousand in 2010 and 199 thousand in 2011, grows about 23.6 percent. The conflicts between the supply and demand become more and more outstanding.

At present, the west China hospital has four CTs, each CT has respective different patients. CT1 can only serve routine scanning patients, while CT2, CT3 and CT4 mainly focus on contrast scanning patients. During the leisure time of each morning, CT2, CT3 and CT4 also serve a small amount of routine scanning patients.

Above all, we can transfer some redundant routine scanning patients to CT2, CT3 and CT4 when the three CTs are idle. This can improve equipment utilization and reduce patient waiting time. Next, we will analyze CT scheduling status quo so as to know the usage of each CT and put forward measures to improve the system.

3. ANALYSIS OF CT EXAMINATION STATUS QUO

3.1. Data Source and Processing

In this paper, we extract about 103248 data in RIS of radiology department from 1 February to 31 July 2012. Through layers of screening, we finally choose 59701 data from 7:30am until 5:30pm.

3.2. Relevant Data Analysis

3.2.1. *Daily Examination Quantity*

The daily examination quantity is shown in Fig. 1. It is observed that the quantity of CT1 is much larger than the other three CTs.

3.2.2. *Utilization Analysis*

By data analysis, we get the unit service time of each CT as shown in Table 1.

Table 1. The unit service time of each CT

Equipment	Mean service time (min)
CT1	2.85
CT2	5.7
CT3	6.14
CT4	8.38
Mean time of CT2, CT3 and CT4	6.54

In each period, let examination quantity times its corresponding unit service time, we can get the estimated total service time as shown in Table 2. Then we can compare it to "30min" by single sample T test, Table 3 and Table 4 are the corresponding results.

Table 2. The estimated total service time in each period (min)

Service period	CT1	CT2	CT3	CT4
7:30—8:00	11.02	7.27	0.80	0.49
8:00—8:30	35.17	23.80	5.93	6.56
8:30—9:00	27.03	21.62	16.33	17.68
9:00—9:30	27.68	24.57	19.97	24.02
9:30—10:00	28.60	25.49	20.63	21.25
10:00—10:30	28.13	26.09	25.74	23.10
10:30—11:00	32.69	32.87	30.39	25.78
11:00—11:30	33.64	33.82	26.28	25.59
11:30—12:00	36.09	35.19	25.64	28.62
12:00—12:30	38.48	34.17	25.28	31.04
12:30—13:00	34.14	35.20	25.63	31.08
13:00—13:30	25.57	21.04	14.81	19.64
13:30—14:00	26.84	24.08	22.50	25.18
14:00—14:30	28.27	24.42	22.02	24.60
14:30—15:00	27.56	24.86	25.31	27.80
15:00—15:30	27.12	24.27	25.15	27.93
15:30—16:00	29.59	25.93	24.85	30.09
16:00—16:30	32.02	28.04	25.90	27.53
16:30—17:00	32.91	27.20	24.94	28.22
17:00—17:30	35.35	31.66	22.93	30.87

Note: Some periods' estimated total service time is greater than "30min". In fact, each period's service time belongs to the individual indicator, there are some fluctuations, it is normal.

Table 3. Single sample statistics

Equipment	N	Mean	Std. Deviation	Std. Error Mean
CT1	20	29.895	5.80935	1.29901
CT2	20	26.5795	6.43734	1.43943
CT3	20	21.5515	7.20876	1.61193
CT4	20	23.8535	7.93623	1.77459

Table 4. Single sample T test

Equipment	Test Value = 30					
	t	df	Sig. (2-tailed)	Mean Difference	95% Confidence Interval of the Difference	
					Lower	Upper
CT1	-0.081	19	0.936	-0.105	-2.8239	2.6139
CT2	-2.376	19	0.028	-3.4205	-6.4333	-0.4077
CT3	-5.241	19	0	-8.4485	-11.8223	-5.0747
CT4	-3.464	19	0.003	-6.1465	-9.8608	-2.4322

From above, we can see that the mean service time is different among the four CTs. Through single sample T test, the CT1's value p is bigger than 0.05, it does not differ significantly from "30min", so CT1 nearly runs at full load status. At the same time, the other three CTs' value p is less than 0.05, it differs significantly from "30min", so the three CTs are obviously idle.

3.2.3. *The Problems and Cause Analysis*

Taking 28400 data of CT1 as an example, we find that the mean working day's reservation quantity (237 patients) is bigger than examination quantity (174 patients). Obviously, the comprehensive center didn't do reasonable appointment (shown in Fig. 2).

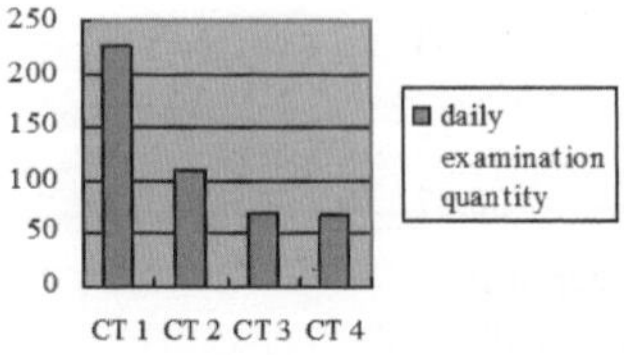

Fig. 1. The daily examination quantity of the four CTs.

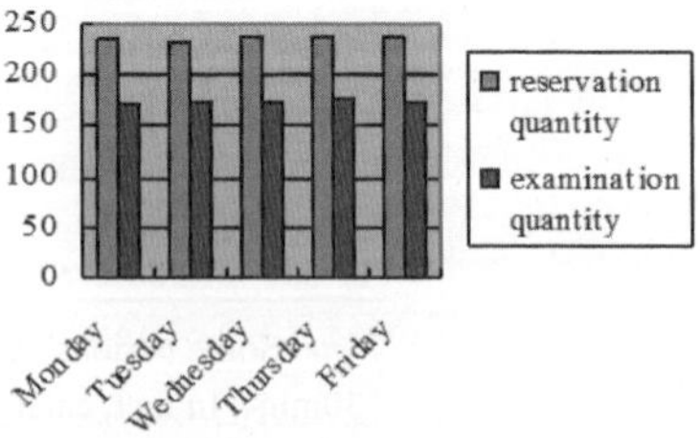

Fig. 2. CT1's reservation quantity and examination quantity in each working day.

Patients cannot be served on schedule, it directly leads to long wait times. Fig. 3 are the patient's mean waiting time on CT1 in each working day.

Then, Hospital staffs can transfer some redundant routine scanning patients to CT2, CT3 and CT4 when the three CTs are idle.

In fact, hospital staffs have done some transfers from CT1 to CT2, CT3 and CT4, it helps to ease the pressure of CT1 to some extent (shown in Fig. 4).

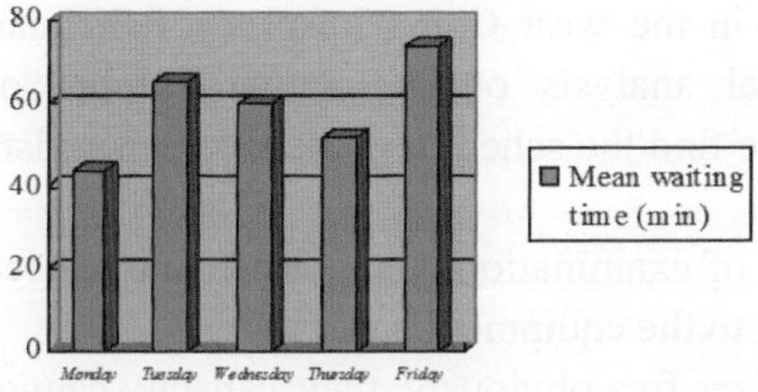

Fig. 3. Patient's mean waiting time on CT1 in each working day.

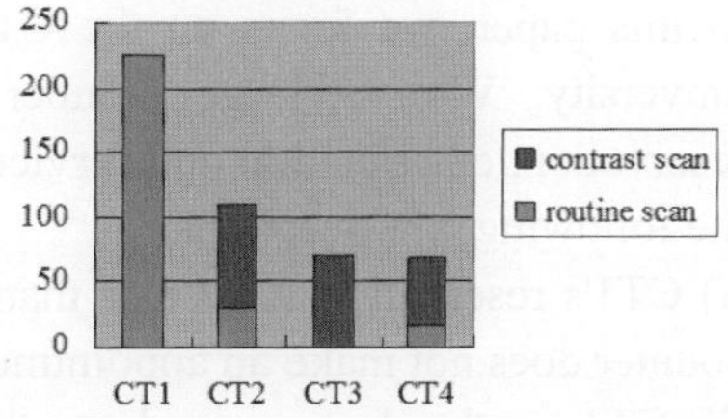

Fig. 4. The quantity of routine scanning and contrast scanning for each CT in each working day.

From Fig. 5, we can find that scheduling from CT1 to CT2 just mainly focuses on 7:30am - 8:30am.

Fig. 5. Analysis for four CTs' actual service quantity.

Through the analysis of four CTs' usage rate (shown in Table 5), CT1 nearly runs at full load status. At the same time, there are certain idle in the other three CTs, they can consider to do part of the routine scan.

Table 5. Utilization of each CT

Equipment	Utilization
CT1	99.65%
CT2	88.60%
CT3	71.84%
CT4	79.51%

Simply put, there is some untapped opportunity for the other three CTs. How to transfer some redundant routine scanning patients to the other three CTs so as to improve equipment utilization and reduce patient waiting time is particularly important.

4. CONCLUSIONS

In this paper, we focus on the CT scheduling in the west China hospital of Sichuan university. With a large number of statistical analysis on the actual number of reservation, examination and service capacity, we find the scheduling management exists the following two questions:
(i) CT1's reservation is greater than the amount of examination, Comprehensive service counter does not make an appointment according to the equipment capacity;
(ii) CT1 overburdens every day, other three devices free obviously, their capacity cannot be made full use of.

In the future, we can go on to optimize the CT scheduling from the following two aspects:
(i) Before the service day, it is uncertain about the requests for appointment, we should make reasonable allocation of equipment to improve the equipment utilization on the basis of patient demand, combining with the equipment capacity constraints;
(ii) On the service day, considering all patients can waiting in line to be serviced, both patient demands and equipment capacities are dually uncertain, we should adjust scheduling policy timely to improve the equipment utilization when queue length changes.

Acknowledgments

This work is supported by the National Natural Science Foundation of China (71131006, 71172197) and the Project Foundation of Chongqing Municipal Education Committee (KJ1502001).

References

1. Yin J. The present situation analysis and the method study on medical examination department cost accounting. Modern Hospital 2006; **6 (12)**: 102-103. (In Chinese).
2. Han YP, Guan HR, Li XC. A overview on hospital department management - the ninth literature: hospital medical examination department management. *Chinese Hospital Management* 1999; **19 (5)**: 59-61. (In Chinese).
3. Green LV, Savin S, Wang B. Managing patient service in a diagnostic medical facility. Operations Research 2006; **54** (1): 11-25.
4. Patrick J, Puterman ML, Queyranne M. Dynamic multipriority patient scheduling for a diagnostic resource. Operations Research 2008; **56** (6): 1507-1525.

FINITE ELEMENT ANALYSIS CENTER OF RESISTANCE OF MAXILLARY DENTITION OF DIFFERENT LOAD MODE IN EXTRACTION SPACE SLIDING CLOSURE

Xiaobo Zhang[a], Hongmei Yao, Yanfei Yin, Li Tian, Peirong Li, Yuanqian Che, Menghan Wang, and Zhenlin Ge*

Department of Orthodontics, School of Stomatology Lanzhou University, Lanzhou 730030, China

E-mail: [a]zxb663@163.com

**Corresponding author*

Objective: Construct a three-dimensional finite element model to investigate the maxillary dentition center of resistance of different load mode on tooth movement in extraction space sliding closure. Methods: DICOM image data including maxilla and upper teeth were obtained through cone-beam CT. The three dimensional finite element model was constructed by Mimics software. Brackets and archwire model were constructed by Creo software. The models were instantiated by Pro/Engineer software. Abaqus software was used to simulate the sliding mechanics by loading 2N force on 6 positions of retraction hook. Predict rotation of the occlusal plane through the initial displacement and stress distribution of tooth. Results The force at the 0mm induced lingual inclination of the maxillary central incisor, gingival movement of the maxillary first molar, and clockwise rotation of the maxillary occlusal plane. In contrast, force application point was located high level of the retraction hook(10mm), the maxillary central incisor was rotated to labial, the maxillary first molar was moved to the occlusion side,and the maxillary occlusal plane was rotated anti-clockwise. When the force at 7.3mm, central incisor, canine and molar vertical displacement distance and the initial rotation angle are the same, very close to 0°. Conclusions: Three dimensional finite element method can be used to predict the center of resistance of maxillary dentition.

1. INTRODUCTION

The movement of teeth restrained by periodontal ligament and alveolar bone when an external force applied to the teeth and all the constraints can be focused on a point on the teeth, this point of constraints is called center of resistance (CRe)[1]. There are many factors influencing the teeth CRe position, including root morphology and number, the height of alveolar bone and different nature of periodontal membrane, etc[2]. However, when micro-implant sliding method applied to close gap of tooth, the CRe of maxillary dentition was also affected by archwire, bracket, micro-implant anchorage position and retraction hook height. In this study, we use three dimensional finite element analysis method analysised the CRe positions of maxillary dentition when implant sliding resistance method applied to close gap of tooth, and make three dimensional finite element analysis method use in clinical treatment.

2. MATERIALS AND METHODS

2.1. The Establishment of 3D Finite Element Model

2.1.1. *Established Model*

Choose a volunteer, using ProMaX3D cone beam CT scanner (Planmeca, Finland) scanned maxilla and dentition. Using Mimics17.0 software (Materialise, Belgium) to read data for three-dimensional reconstruction and the relevant information of alveolar bone and tooth tissue were extracted by threshold operation in Mimics .Finally, the Minics model was established(Fig.1).Establishing entity model of the bracket in Creo3.0 (United States PTC) software according to the axis angle, torque angle and three dimensions of MBT brackets. With $Y=2.619596 \times 10\text{-}5X^4+1.375403\times 10\text{-}2X^2$ as the standard[3], the Creo3.0 software was used to synthesize 0.483 mm×0.635 mm (0.019 inch×0.025 inch) pantograph wire model, and the traction hook was 10 mm high(Fig.2).

2.1.2. *Assembled Model*

CAE (Computer Aided Engineering) model was established in Hypermesh software (American Altair company).Using tetrahedral elements, selecting the basic size of the grid 0.5mm, the total number of nodes of the model is 113564, the total number of units 536998. Due to the force of the left and right sides of the column is basically consistent, in order to reduce the amount of operation, the loss of the model from the middle of the model to separate the half of the whole model to be analyzed [4](Fig.3). The periodontal membrane was expanded from the root of the tooth, and the shell element with a thickness of 0.25 mm was simulated.

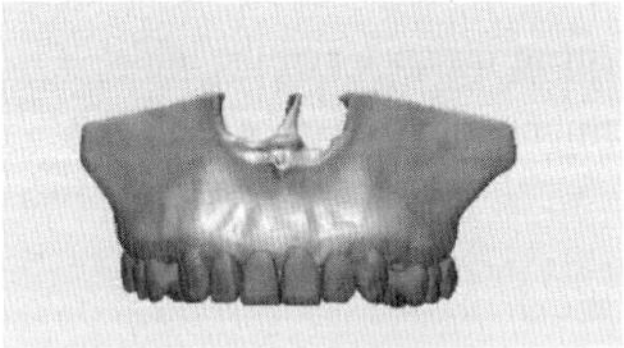

Fig. 1. Mimics model.

Fig. 2. Bracket and archwire model.

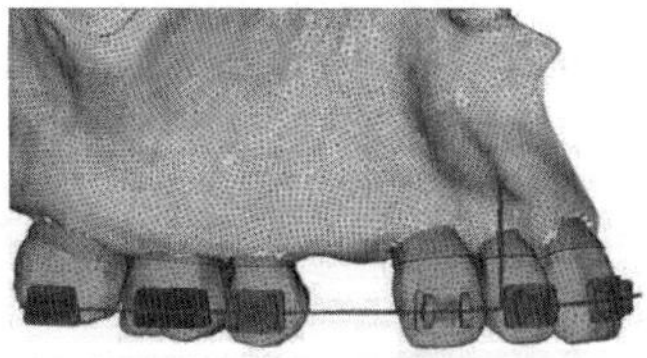

Fig. 3. Loaded positions of hook.

2.1.3. *Material Properties*

The material in the finite element model of teeth including alveolar bone, periodontal ligament, bracket and archwire, retraction hook. The number of nodes in the teeth and alveolar bone are so large that can be set as a uniform structure, which does not distinguish between the enamel and dentin, the cancellous bone and the cortex bone[5]. All kinds of materials are simulated by C3D4 element (a tetrahedral element), and are assumed to be homogeneous, isotropic linear elastic material, material deformation is small deformation, and the material elastic modulus and Poisson's ratio are defined[6,7](Table 1).

Table1. Material parameters of tooth, PDL, alveolar bone, archwire, bracket

Materials	Young's modulus (MPa)	Poisson's ratio
Teeth	20000	0.3
bracket	200000	0.3
Stainless steel	200000	0.3
Periodontal ligament	0.05	0.49
Alveolar bone	2000	0.3

2.2. Mark Point of Displacement Evaluation

With Incisor maxillary margin midpoint CI, apical margin CI, maxillary lateral incisor apical midpoint of the LI, Canine cusp C, apical C, Second anterior molar buccal cusp PM2, root tip PM2, Molar mesial buccal pointed M1, Mesial buccal root M2 as displacement evaluation markers, Central incisor and first molar buccal cusp point connection CIM1 is defined as the maxillary occlusal plane.

2.3. Load Constraint Conditions

In order to study the effect of sliding off the gap when the retraction hook height of the center of resistance of maxillary dentition. According to the clinical practice of orthodontic traction hook is arranged between the maxillary lateral incisors and canines in retraction hook heights were 0mm, 2mm, 4mm, 6mm, 8mm, 10mm,and loaded 2N horizontal orthodontic force.

2.4. Model Coordinate Reference System

To establish a standard coordinate system, the X axis coronary axis, said to represent a buccal and palate, palatal side is positive, negative buccal, Y axis sagittal axis, said to represent a lip and palate, lip side is positive and negative palatal, Z axis for the vertical axis, said direction, positive and negative.

3. RESULTS

When the retraction hook is positioned between the maxillary lateral incisor and canine, the larger force of anterior teeth is mainly distributed in the crown, the Lesser force of posterior teeth is mainly distributed in the root. The force at the 0mm induced 0.069° lingual inclination of the maxillary central incisor, 0.019mm occlusal movement of incisal edge, 0.55° lingual crown rotation and 0.15mm gingival movement of canine, 0.04°mesial rotation and 0.005mm gingival movement of maxillary first molar(Fig.4,

Fig.5) ,and 0.082° clockwise rotation of the maxillary occlusal plane(Fig.6). In contrast, force application point was located high level of the retraction hook(10mm), the maxillary central incisor was rotated to labial 0.033°, incisal edge was moved to gingival side 0.0072mm,the crown of canine was rotated to lingual 0.003° and move 0.0047mm to occlusion, the maxillary first molar was rotated 0.027°to mesial side and moved 0.0018mm to occlusion(Fig.4, Fig.5), maxillary occlusal plane was rotated 0.042°anti-clockwise (Fig.6) . When the force at 7.3mm, central incisor, canine and molar vertical displacement distance and the initial rotation angle are the same, very close to 0°.

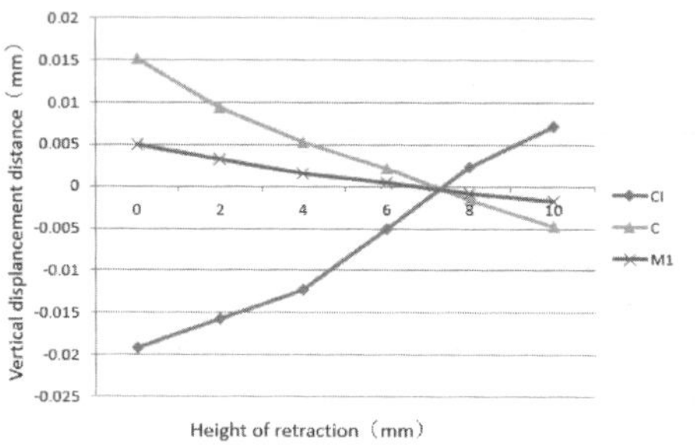

Fig.4. Vertical displacement of the teeth.

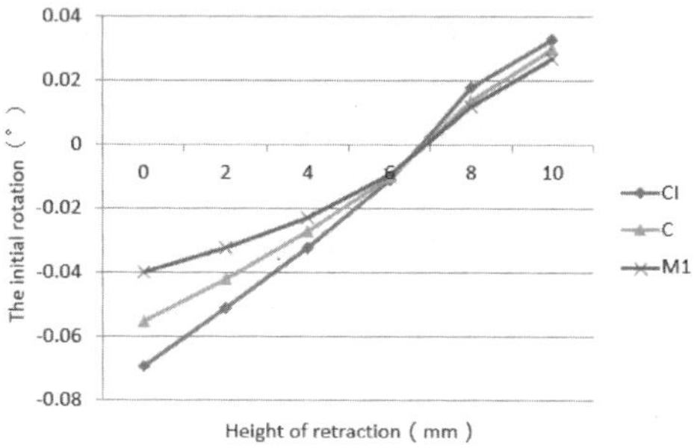

Fig.5. Forward and backward displacement of the teeth.

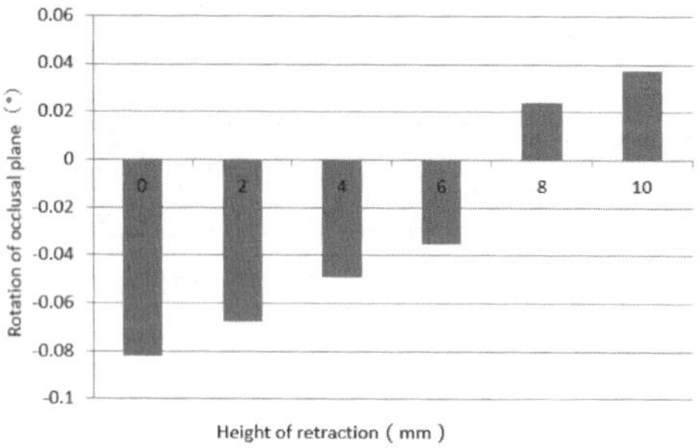

Fig.6. Rotation of occlusal plane.

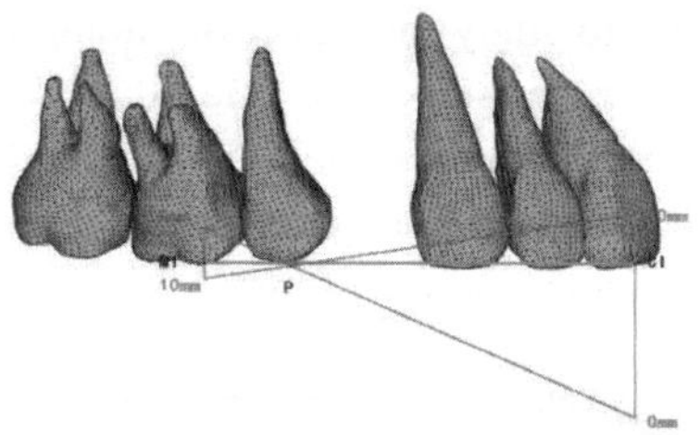

Fig.7. The horizontal position of center resistance.

The distance of maxillary central incisor to mesial buccal cusp of first molar is 32.8mm and to buccal cusp of first premolar is 26.3mm. The vertical displacement of the maxillary central incisor and the maxillary first molar were magnified by 100 times. Measured the data of computer. Analysis of the relationship between the vertical displacement of the occlusal plane when the traction hook is 0mm and 10mm. As the following figure, the distance of point P to CI is 26.7mm(Fig.7).

4. DISCUSSION

The three-dimensional finite element model of this experiment involves the teeth, alveolar bone, periodontal ligament, brackets and archwires and other materials.

According constrained conditions in the experiment to simulate the clinical situation and research trend of the displacement of maxillary dentition. In the process of closing the extraction gap, the control of anchorage is particularly important. The loss of anchorage not only affects the correction effect, but also makes the related research error, which affects the authenticity and the scientific nature of the research results. From the results of this study, when loading in low traction hook, initial displacement mainly lead maxillary anterior crown lingual rotation and margin decline. At the same time, the first premolar and the molar raise to the gum, occlusal plane rotate clockwise. In contrast, when the loading in high retraction hook, initial displacement of anterior teeth mainly is rotation to labial side and the elevation of incisal edge. Occlusal plane counter clockwise rotation. The results of this study are consistent with the results of Kojima Yukio et al [8].

Indicated from Fig.4-Fig.7, When the hook is located at 7.3mm height, vertical displacement and rotation angle of maxillary central incisor, canine and first molar close to 0 ° and the occlusal plane rotation angle is also 0 °. This shows that when the retraction hook height is 7.3 mm overall level mobile maxillary anterior teeth and level to force through the resistance of the maxillary teeth center, so the center of resistance height is 7.3 mm up the archwire. The location of the archwire is up the maxillary incisal edge more than 4 mm, so the center resistance is located in the incisal edge above 11.3 mm. From Fig.7, When the retraction hook height is 0mm and 10mm, the displacement of the P is always the same. The center of resistance is located in the point P impedance occlusal vertical line plane, P distance from maxillary incisor 26.3mm. Therefore, the center of resistance of maxillary dentition of different load mode in extraction space sliding closure is located above the archwire 7.3mm and above the upper central incisors in the upper 11.3mm, the position of 26.3mm after the upper central incisors. Therefore, the implant anchorage lide method to close the gap between the maxillary teeth in the center of the resistance center is located above the arch wire 7.3mm or above the incisal edge of central incisors 11.3mm, and horizontal distance far from the central incisors 26.3mm. This result is consistent with the results that reported by Jeong et al[9], resistance center of the maxillary dentition is located in the central incisors upward 7mm backward 26mm. Yukio et al [10]also obtained the same result that the resistance center of the whole maxillary dentition.

5. CONCLUSIONS

The rotation of the upper occlusal plane is very important in the process of slide extraction space closure of implant anchorage. And the center of resistance influence the movement of occlusal plane. Three-dimensional finite element analysis method can find the center of resistance by analysis Stress distribution of tooth to guide orthodontic.

References

1. Han Y, Mou L, Xu G, et al. [Comparison between one-step and two-step space closing methods of sliding mechanics using three-dimensional finite element].[J]. Zhonghua kou qiang yi xue za zhi - Zhonghua kouqiang yixue zazhi, *Chinese journal of stomatology*, 2015; **50**(3):182-5.

2. Sia S S , Koga Y. Determinating the center of resistance of maxillary anterior teeth subjected to retraction forces in sliding mechanics. Angle Orthodontist,2007; **77**(3) : 999–1003.
3. Zheng H, Zhi C. Three-dimensional Finite Element Modeling of a Maxillary Premolar Tooth Based on the Micro-CT Scanning: A Detailed Description[J]. *Journal of Huazhong University of Science & Technology*, 2013, **33**(5):775-9.
4. Il-Jun Y, Yoon-Ah K, Sang-Jin S, et al. Comparison of tooth displacement between buccal mini-implants and palatal plate anchorage for molar distalization: a finite element study.[J]. European Journal of Orthodontics, 2011, 36(4):cjr130.
5. Reimann S, Keilig L, Jäger A, et al. Biomechanical finite-element investigation of the position of the centre of resistance of the upper incisors[J]. *The European Journal of Orthodontics*, 2007; **29**(3): 219-224.
6. Holberg P C, Winterhalder P, Holberg N, et al. Indirect miniscrew anchorage: biomechanical loading of the dental anchorage during mandibular molar protraction-an FEM analysis.[J]. Journal of orofacial orthopedics, Fortschritte der Kieferorthop die : Organ/official journal Deutsche Gesellschaft für Kieferorthop die, 2014; **75**(1):16-24.
7. Tominaga J Y, Ozaki H, Chiang P C, et al. Effect of bracket slot and archwire dimensions on anterior tooth movement during space closure in sliding mechanics: A 3-dimensional finite element study[J]. *American Journal of Orthodontics & Dentofacial Orthopedics*, 2014; **146**(2):166-74.
8. Kojima Y, Kawamura J, Fukui H. Finite element analysis of the effect of force directions on tooth movement in extraction space closure with miniscrew sliding mechanics[J]. American Journal of Orthodontics & Dentofacial Orthopedics, 2012; **142**(4):501–508.
9. Jeong G M, Sung S J, Lee K J, et al. Finite-element investigation of the center of resistance of the maxillary dentition[J]. Korean Journal of Orthodontics, 2009; **39**(2): 83-94.
10. Yukio K, Hisao F. A finite element simulation of initial movement, orthodontic movement, and the centre of resistance of the maxillary teeth connected with an archwire.[J]. *European Journal of Orthodontics*, 2011; **36**(3):255-261.

THE VARIETY OF ROOT AND CANAL MORPHOLOGY OF MAXILLARY SECOND MOLARS BY CONE-BEAM COMPUTED TOMOGRAPHY IN A NORTHWESTERN CHINESE POPULATION

Miaomiao Liu, Xuexiang Gao, Shuo Wang, Jin Ren, Qi Guo, Yunqing Pang, Yuan Zhao[a,*], and Jing Wang[b,*]

School of Stomatology, Lanzhou University, Lanzhou, 730000, China

E-mail: [a]zhaoy@lzu.edu.cn, [b]lzukqwj@126.com,

[*]Corresponding author

Objective: To investigate variety of root canal system of maxillary second molars in a northwestern Chinese population using cone-beam computed tomography (CBCT). Materials and Methods: 913 maxillary second molars from 496 CBCT images were enrolled. The number of roots and canals, canal configuration, root morphology, location of the second mesiobuccal (MB2) canals and symmetry in homonym teeth were recorded. Canal configuration was classified by Vertucci's classification. Results: Totally 913 maxillary second molars, a higher prevalence of three roots (86.64%) and three canals (77.98%) was present. The incidence of MB2 canals was 10.84%, and the most prevalent configuration of the mesiobuccal (MB) canal when MB2 was present was type II (81.82%). MB2 was located 2.03± 0.42 mm mesially to MB canal, and may be a constant position to locate. The incidence of separate roots, fusion of any two roots, and three fused roots was 69.55%, 21.14%, and 9.31%, respectively. There was a significant difference of the symmetry between different numbers of root on root canal system($p<.001$). Conclusion: The canal system of maxillary second molars was variable in a northwestern Chinese population. The most common form of the variety was MB2 canal, which had a constant position to locate. The variation of four roots may increase the difficulty of endodontic therapy. CBCT could enhance the understanding of root canal anatomy to improve the outcome of endodontic treatment.

1. GUIDELINES

It is commonly acknowledged that the permanent maxillary molars have the most complex and variable root canal morphology, and root canal treatment of the molar teeth has the highest clinical failure rates [1-3]. Numerous literatures have researched the morphological variations of the maxillary permanent molars in different ethnic populations, expounding the presence of root canal number and configuration, especially the second mesiobuccal root (MB2) canal [4-7]. However, previous studies almost focused on the maxillary first molars while ignoring the maxillary second molars. Moreover, because of the special position which is at or near the end of the dentition, detecting all the canals in endodontic treatment is more difficult. Therefore, having a thorough knowledge of the variations of the, maxillary second molars is essential.

Various techniques have been proposed to evaluate the root canal morphology in previous studies, including sectioning [8], canal staining and tooth clearing techniques [9-11], conventional radiography [12, 13], clinical operating microscope [14], and computed tomography scanning [15,16]. Tooth clearing techniques has been generally considered the gold standard for analyzing root canal anatomy in laboratory studies, and in the clinic situation, use of conventional radiography is the most common methods to evaluate tooth anatomy. However, laboratory methods may be inapplicable in clinical operation, and periapical radiography can only provide two-dimensional information yet the teeth were three-dimensional objects. More recently developed, cone-beam computed tomography (CBCT) imaging techniques is taken to evaluation and identify root canal morphology in endodontics which provide a three-dimensional images with a lower radiation dose and a higher resolution and allow a more accurate [17, 18], feasible, and effective studies results and clinical diagnoses [19-21].

CBCT becomes widely used in endodontic applications [22]. Many studies have been conducted to analyze the anatomy of maxillary permanent molars by CBCT. However, there were few studies about detail description of root canal variety and quantitative analysis of the location of second mesiobuccal root canal (MB2) of maxillary second molar teeth in Chinese population.

The aim of this study was to investigate the variety of root and canal morphology and the location of MB2 canal of the maxillary second molars in a northwestern Chinese population by age, tooth position using CBCT images taken for oral examination.

2. MATERIALS AND METHODS

2.1. Patients

Patients, who visited the Hospital of Stomatology Lanzhou University, Lanzhou, China, between January 2013 and May 2015, requiring radiographic examination by CBCT, were enrolled in this study. In total, 913 maxillary second molars from 496 patients were evaluated retrospectively. The mean age of the patients was 39 years, ranging from 16 to77 years. Among all the individuals, 417 patients presented maxillary second molars bilaterally and 79 patients unilaterally which showed a deletion in the other side.

The patients were selected for enrolment in this investigation based on the following criteria: (1) maxillary second molar without periapical lesions; (2) no root canals with endodontic treatment; (3) no root canals with open apices, absorption or calcification; (4) no caries or defect and no filled materials; and (5) CBCT images with good quality.

2.2. CBCT Technique

The patients were scanned by a CBCT scanner (KaVo Dental GmbH, Biberach, Germany) operating at 120KV, 5mA, with an exposure time of 14.7 s. The voxel size was 0.25 mm, the slice thickness was 0.25 mm. Scans were made according to the manufacturer's recommended protocol. All CBCT images were performed by an experienced radiologist.

2.3. Evaluation of the images

Software KaVo 3DeXam Vision 1.9 and Invivo 5 (KaVo Dental GmbH, Biberach, Germany) were used to scan the images. Serial axial, coronal and sagittal CBCT images were evaluated by carefully rolling the toolbar from the floor of the pulp chamber to the apex.

The following observations were recorded: the number of roots and the variety of root morphology, the number of canals and the root canal configuration classified using Vertucci's classification [23], and the location of MB2 canal in relation to the mesiobuccal root, the distobuccal (DB) root and the palatal (P) root [24, 25]. The distance between the MB2 canals and MB canal (MB-MB2) was recorded. The angle □DMP (formed by the DB, MB and P canal orifices) and the angle □DMB2 (formed by the DB, MB and MB2 canal orifices) were evaluated. Straight lines projected between MB-P and DB-MB2 and the point of intersection was T, the angle □MTB2 (formed MB canal orifices, T point, MB2 canal orifices) was analyzed. (Fig.1.)

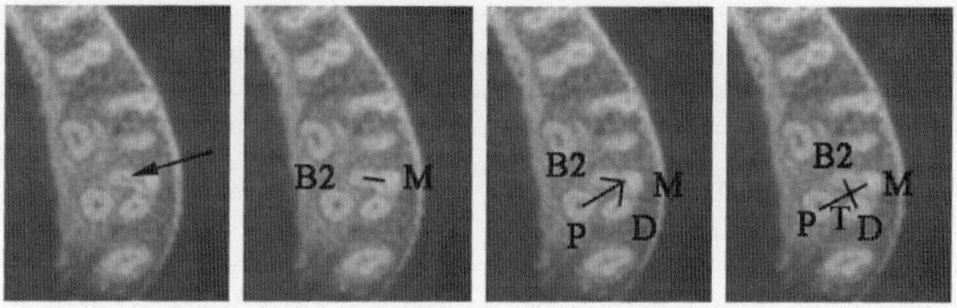

Fig.1. Axial view of maxillary second molar. The location of MB2 canal in relation to the mesiobuccal root (MB), the distobuccal root (DB) and the palatal root (P).

Statistical analysis was performed using SPSS 20.0 software. The relationship of symmetry in 1 root, 2roots, 3roots and 4 roots about root number, root morphology, canal number, and canal configuration were analyzed using chi-square test. In addition, the incidence of additional canals (MB2 canals and 4-rooted) by age was determined using the Kruskal-Wallis test. Difference was considered significant if $P < 0.05$.

3. RESULTS

3.1. The Number of Roots and Canals

The root and canal number of the maxillary second molars shows a variety. The number of roots range from one root to four roots, and the number of canals range from one canal to fives canals. (Table 1) Among 913 maxillary second molars, 791 (86.64%) had three root which was the most frequent pattern, 88 (9.64%) had two roots, 19 (2.08%) had single root, 15 (1.64%) had four roots. The majority 712 (77.98%) had three canals, 110 (12.05%) had four canals, 75 (8.22%) had two canals, 14 (1.53%) had one canal, and only 2 (0.22%) had five canals. When four canals occurred, 97 (10.62%) had an additional MB canal in MB root (MB2 canal), and 13 (1.42%) had an additional canal in an additional root. When five canals occurred, there were an additional canal in MB root and the additional root, respectively.

No difference was found in the occurrence of MB2 canals and 4-rooted in different age periods (p=0.406), and no difference was found between group A and group B (p=0.317). (Table 2)

Table 1. Number and percentage of roots and canals of maxillary second molars

		1-root		2- root		3-root		4-root	
	Total	1-canal	2-canal	2-canal	3-canal	3-canal	4-canal	4-canal	5-canal
Root	913	19(2.08)		88(9.64)		791(86.64)		15(1.64)	
Canal (respectively)	913	14(1.53)	5(0.55)	70(7.67)	18(1.97)	694(76.01)	97(10.62)	13(1.42)	2(0.22)
Canal (total)	913	14(1.53)	75(8.22)		712(77.98)		110(12.05)		2(0.22)

Table 2. Number and percentage of MB2 canals and 4-rooted of the maxillary second molars by age

	A (≤45)			B (>45)		
	15-23	24-35	36-45	46-55	56~	Sig.
MB2 (3-rooted) number/total	22/200	27/213	14/177	21/194	13/129	
Rate (%)	11.00	12.68	7.91	10.82	10.08	>0.05
4-rooted number/total	3/200	2/213	8/177	1/194	1/129	
Rate (%)	1.50	0.94	4.52	0.52	0.78	>0.05

3.2. The Morphology of The Roots

The morphology of the roots showed complicated. (Table 3) 50 (5.48%) two-rooted teeth bifurcated from the pulp chamber or different levels in the roots, and the remaining 38 (4.16%) completely fused. Amongst the three-rooted teeth, 561 (61.45%) formed three independent roots, 83 (9.09%) completely fused, and 147 (16.10%) occurred fusion of any two roots. Amongst the four-rooted teeth, 5 (0.55%) had four separate roots, 8 (0.88%) had two separate roots with two fused roots, and 2 (0.22%) had one separate root with three fused roots.

Table 3. The morphology of the roots of maxillary second molars

	Number	1-root	2- root	3-root	4-root
Separate	635(69.55)	19(2.08)	50(5.48)	561(61.45)	5(0.55)
Two fused	193(21.14)	-	38(4.16)	147(16.10)	8(0.88)
MB+DB		-	-	55	8
MB+P		-	-	74	-
DB+P		-	-	18	-
Three fused	85(9.31)	-	-	83(9.09)	2(0.22)
Total	913	19(2.08)	88(9.64)	791(86.64)	15(1.64)

3.3. The Configuration of Canal System

All patients owned a single canal in distobuccal and palatal root which had a configuration of type I. For 806 three-rooted and four-rooted maxillary second molars, 707 (87.72%) of MB roots had a single canal, and remainder had two canals. When MB2 canal present, the most prevalent canal configuration was type II(81.82%), type III (8.08%), type IV (8.08%), type V (2.02%) were observed in this study. (Table 4).

Table 4. Canal configurations of MB canal of maxillary second molars analyzed by Vertucci's classification

	Total	I(1)	II(2-1)	III(1-2-1)	IV(2-2)	V(1-2)
MB canal (total)	806	707(87.72)	81 (10.05)	8(0.99)	8(0.99)	2(0.25)
MBcanal (MB2 present)	99	-	81 (81.82)	8(8.08)	8(8.08)	2(2.02)

The location of MB2 canal:

We evaluated the location of MB2 canal. (Table 5) The distance between MB and MB2 was 2.03± 0.42 mm. The angle (∠DMP) ranged 49.39± 10.25. The angle (∠DMB2) ranged 76.25± 14.08. The angle (∠MTB2) ranged 88.85± 17.84. The angle (∠DMB2) was larger than the angle (∠DMP), which indicates that the MB2 canal generally locates mesially to an imaginary line between the MB and P canal, and commonly 2 mm to MB canal. The angle (∠MTB2) indicates that imaginary line between MB and P canal had an approximately vertical relation with the imaginary line between DB and MB2 canal in the floor of pulp chamber, which gives us an more specific information about location of the MB2 canal.

Table 5. Measurements were made to characterize the location of MB2 canals

	Percentage	Symmetry	MB-MB2	∠DMP	∠DMB2	∠MTB2
MB2	99(10.84)	34/51(66.67)	2.03± 0.42	49.39± 10.25	76.25± 14.08	88.85± 17.84

3.4. Symmetry in the Bilateral Homonymous Teeth

For 417 patients who had both bilateral homonym teeth in 496 patients, 372 (89.21%) and 363 (87.05%) had the same root number and canal number respectively in bilateral teeth. The morphology of root and canal also had a high symmetry, respectively 312 (74.82%) and 325 (77.94%). (Table 6) This indicates that 14.39% patients had the same root number with different morphology in bilateral teeth, and there were 9.11% patients who owned the same canal number but different configurations in bilateral teeth. In addition, there was a significant difference of the symmetry between different numbers of root ($p<.001$). This indicates that there was a highest symmetry in 3-rooted maxillary second molars, then 2-rooted, 1-rooted, and 4-rooted which has a lowest symmetry.

Table 6. Symmetry in the bilateral maxillary second molar. (n=417)

Symmetry	Total number	Root number	Root morphology	Canal number	Canal configurations
total(%)	417	372(89.21)	312(74.82)	363(87.05)	325(77.94)
1 root	7	4(57.14)	4(57.14)	4(57.14)	4(57.14)
2 root	40	28(70.00)	19(47.5)	27(67.5)	23(57.5)
3 root	358	337(94.13)	287(80.17)	329(91.90)	296(82.67)
4 root	12	3(25.00)	2(16.67)	3(25.00)	2(16.67)
Sig.		<.001	<.001	<.001	<.001

4. DISCUSSION

This study used CBCT to evaluate the root and canal systems of 913 maxillary second molars in 496 northwest Chinese individuals. The results indicate that the maxillary second molars have a more variable root canal system (Fig.2).

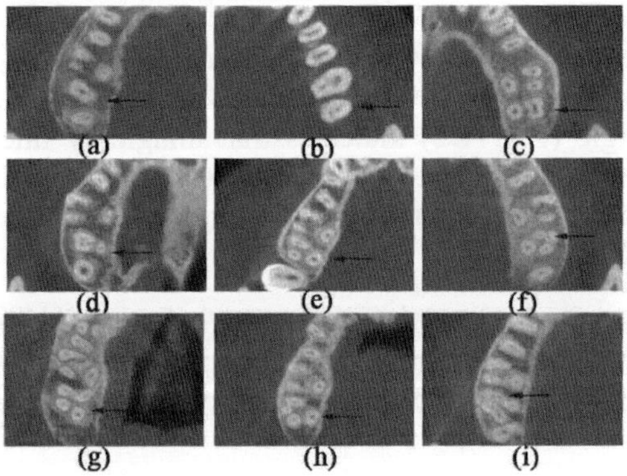

Fig. 2. Maxillary second molars with root and canal variations in the axial sections. (a) One root with a single canal; (b) One root with two canals (a buccal canal and a palatal canal); (c) Two separate roots with two canals; (d) Two separate roots with three canals (a buccal root with two buccal canals); (e) Three separate roots with three canals (conventional model); (f) A additional canal in mesiobuccal root (MB2); (g) Two buccal roots and two palatal roots; (h) Three buccal roots and one palatal roots; (i) Four roots with five canals (fusion of the distopalatal root with the distobuccal root, and an additional canal in mesiobuccal root).

The diversity of the root morphology shows the varieties of root furcation and fusion except for the root number, which was rarely reported in Chinese population. In the present study, we found 278 (30.45%) sample teeth exhibited root fusion, and the result showed a higher prevalence of fusion than in some previous studies by Kim et al. who reported a fusion of 10.71% in Korean population and Rwenyonyi et al. who reported a fusion of 14% in Ugandan people but a lower prevalence than study by Al Shalabi et al. who revealed that the incidence of root fusion was 43% [5, 6, 26]. Moreover, in the present study, for 147 three-rooted maxillary second molars which showed two fused roots, the mesiobuccal root fused to palatal root (50.34%) and mesiobuccal root fused to distobuccal root (37.41%) were the common form, and the fusion in distobuccal and palatal root was lesser(12.24%). In the present study, MB2 canals were present 10.84% in maxillary second molars, which had a lower frequency than previous detection by CBCT [27, 28]. The reason may relate to the isotropic voxel dimensions of CBCT. Bauman et al. who compared the isotropic voxel dimensions at 0.4 mm, 0.3 mm, 0.2 mm, 0.125 mm showed that increasing the resolution of CBCT could improve the detection of the MB2 canal in MB root of maxillary molars and they concluded the highest resolution available as the isotropic voxel dimensions decreased would produce a more accurate indication by root section [29]. In the present study, the patients were examined by using the 0.25 mm voxel size, and this may result in different incidence of detecting MB2 canals in maxillary second molars. However, the radiation dose depends on the voxel resolution because of the exposure time for CBCT. The increase in resolution must be evaluated against this increase in radiation to the patient. Bauman et al. reported that there was not a significant difference ($p<0.01$) for accuracy between the 0.125 mm and 0.2 mm voxel scan, inspite of the otherness of correct identification [29]. We concluded that the highest resolution may be used to provide the most accurate results by CBCT as a standard for laboratory studies and select an appropriate resolution in clinical applications. On the other hand, some researchers showed higher MB2 incidence in their nations generally, and they revealed the incidence was 42.2%, 34.32%, 40.29%, respectively [30, 31, 32]. This may be attributable to the ethnic difference and the sample size.

Regarding the types of canal configurations analyzed by Vertucci's classification, the type II(2 canals with 1 apical foramen) was the most prevalent when MB2 canal was present. This conforms to the finding of some researchers in a Korean population [32, 5], but differs from studies in some other regions, which reported a high incidence of the type IV (2 canals with 2 apical foramen) [7, 33]. Meanwhile, our study and the previous studies proved that the root canal system of the maxillary second molar teeth showed more anatomic variables than the maxillary first molar teeth. Therefore, better understanding the internal morphology of root canal system is essential for successful outcome of endodontic therapy.

We examined incidence of 4-rooted maxillary second molars in this study. 15 maxillary second molars (1.64%) exhibited an extra root, and two of them had an extra buccal root and the rest had an extra palatal root. The result was higher than many previous studies [5, 34, 35]. In the 12 cases of this study, we only found 3 cases with bilateral 4-rooted, all others were unilateral. This indicates that the extra root has a low bilateral symmetry. If an extra root is found on a molar, the possibility of detecting an additional root in the contralateral molar is very low. According to the classification proposed by previous studies, we found that the most form was type I which the two palatal roots were widely divergent which are often longer and more tortuous than the two buccal roots which were less divergent and often ''cow-horn''shaped [36, 37]. In 15 four-rooted teeth, 5 had four separate roots, 8 exhibits fused roots between mesiobuccal and mesiopalatal root, and 2 exhibits fused roots between mesiobuccal, mesiopalatal, and distopalatal roots, whereas the distobuccal root appears to stand alone and may diverge distobuccally. Meanwhile, the morphology of four- rooted maxillary second molars increased the difficulty value of endodontic therapy.

5. CONCLUSIONS

In conclusion, the canal system of maxillary second molar was variable in a northwestern Chinese population. The most common form of the variety was MB2 canal, which had a simplex position to locate. The variation of four roots with four canals or more canals may increase the difficulty value of endodontic therapy. This indicates that increasing sample size and studying in various regions should be the next step program. CBCT images provide a better understanding of the root canal anatomy, which demonstrates potential to improve the outcomes of endodontic treatment.

References

1. Weine FS, Healey HJ, Gerstein H, Evanson L. Canal configuration in the mesiobuccal root of the maxillary first molar and its endodontic significance. Oral Surgery, Oral Medicine, *Oral Pathology* 1969; **28**(3): 419-425.
2. Wolcott J, Ishley D, Kennedy W, Johnson S, Minnich S. Clinical investigation of second mesiobuccal canals in endodontically treated and retreated maxillary molars. *Journal of endodontics* 2002; **28**(6): 477-479.
3. Vertucci FJ, Frank J. Root canal morphology and its relationship to endodontic procedures. *Endodontic topics* 2005; **10** (1): 3-29.
4. Zheng QH, Wang Y, Zhou XD, Wang Q, Zheng GN, Huang DM. A cone-beam computed tomography study of maxillary first permanent molar root and canal morphology in a Chinese population. *Journal of endodontics* 2010; **36**:1480-1484.
5. Kim Y, Lee S, Woo J. Morphology of maxillary first and second molars analyzed by cone-beam computed tomography in a Korean population: variations in the number of roots and canals and the incidence of fusion. *Journal of endodontics* 2012; **38**(8):1063-1068.
6. Al Shalabi RM, Omer OE, Glennon J, Jennings M, Claffey NM. Root canal anatomy of maxillary first and second permanent molars. *International Endodontic Journal* 2000; **33**(5): 405-414.
7. Neelakantan P, Subbarao C, Abuja R, Subbarao C, Gutmann J. Cone-beam computed tomography study of root and root canal morphology of maxillary first and second molars in an Indian population. *Journal of Endodontics* 2010; **36**(10): 1622-1627.
8. Kulild JC, Peters DD. Incidence and configuration ofcanal systems in the mesiobuccal root of maxillary first andsecond molars. *Journal of Endodontics* 1990; **16**(7): 311-317.
9. Barker BCW, Lockett BC, Parsons KC. The demonstrations of root canal anatomy. *Australian dental journal* 1969; **14**(1): 37-41.
10. Vertucci F, Seelig A, Gillis R. Root canal morphology of the human maxillary second premolar. Oral Surgery, Oral Medicine, Oral Pathology 1974; 38(3): 456-464.
11. Yoshioka T, Kikuchi I, Fukumoto Y, Kobayashi C, Suda H. Detection of the second mesiobuccal canal in mesiobuccal roots of maxillary molar teeth ex vivo. *International endodontic journal* 2005; **38**(2): 124-128.

12. Pineda F, Kuttler Y. Mesiodistal and buccolingual roentgenographic investigation of 7,275 root canals. Oral Surgery, Oral Medicine, *Oral Pathology* 1972; **33**(1): 101-110.
13. Pineda F. Roentgenographic investigation of the mesiobuccal root of the maxillary first molar. Oral Surgery, Oral Medicine, *Oral Pathology* 1973; **36**(2): 253-260.
14. Baldassari-Cruz LA, Lilly JP, Rivera EM. The influence of dental operating microscope in locating the mesiolingual canal orifice. Oral Surgery, Oral Medicine, Oral Pathology, Oral Radiology, and Endodontology 2002; **93**(2): 190-194.
15. Park J, Lee J, Ha B, Choi J, Perinpanayagam H. Three-dimensional analysis of maxillary first molar mesiobuccal root canal configuration and curvature using micro-computed tomography. Oral Surgery, Oral Medicine, Oral Pathology, Oral Radiology, and Endodontology 2009; **108**(3):437-442.
16. Somma F, Leoni D, Plotino G, Grande NM, Plasschaert A. Root canal morphology of the mesiobuccal root of maxillary first molars: a micro-computed tomographic analysis. *International Endodontic Journal* 2009; **42**:165-74.
17. Cotton T, Geisler T, Holden D, Schwartz S, Schindler W. Endodontic applications of cone-beam volumetric tomography. Journal of Endodontics 2007; **33**:1121-1132.
18. Patel S. New dimensions in endodontic imaging: Part 2. Cone beam computed tomography. *International endodontic journal* 2009; **42**(6): 463-475.
19. Matherne RP, Angelopoulos C, Kulild JC, Tira D. Use of cone-beam computed tomography to identify root canal systems in vitro. *Journal of endodontics* 2008; **34**(1): 87-89.
20. Domark JD, Hatton JF, Benison RP, Hildebolt CF. An ex vivo comparison of digital radiography and cone-beam and micro computed tomography in the detection of the number of canals in the mesiobuccal roots of maxillary molars. *Journal of endodontics* 2013; **39**(7): 901-905.
21. Cheng L, Zhang R, Yu X, Tian Y, Wang H, Zheng G, Hu T. A comparative analysis of periapical radiography and cone-beam computerized tomography for the evaluation of endodontic obturation length. Oral Surgery, Oral Medicine, Oral Pathology, Oral Radiology, and Endodontology 2011; **112**(3): 383-389.
22. Saberi N. CBCT within endodontics: an introduction. International dentistry African edition; 3(2).
23. Vertucci FJ. Root canal anatomy of the human permanent teeth. Oral Surgery, Oral Medicine, Oral Pathology, Oral Radiology 1984; **58**(5):589-599.
24. Tuncer AK, Haznedaroglu F, Sert S. The location and accessibility of the second mesiobuccal canal in maxillary first molar. *European journal of dentistry* 2010; **4**(1): 12.
25. Betancourt P, Navarro P, Cantín M, Fuentes R. Cone-beam computed tomography study of prevalence and location of MB2 canal in the mesiobuccal root of the maxillary second molar. *International journal of clinical and experimental medicine* 2015; **8**(6): 9128.
26. Rwenyonyi CM, Kutesa AM, Muwazi LM, Buwembo W. Root and canal morphology of maxillary first and second permanent molars teeth in a Ugandan population. *International endodontic journal* 2007; **40**: 679-683.

27. Zhang R, Yang H, Yu X, Wang H, Hu T, Dummer PMH. Use of CBCT to identify the morphology of maxillary permanent molar teeth in a Chinese subpopulation. *International endodontic journal* 2011; **44**(2): 162-169.
28. Han X, Yang H, Li G, Yang L, Tian C, Wang Y. A study of the distobuccal root canal orifice of the maxillary second molars in Chinese individuals evaluated by cone-beam computed tomography. *Journal of Applied Oral Science* 2012; **20**(5): 563-567.
29. Bauman R, Scarfe W, Clark S, Morelli J, Scheetz J, Farman, A. Ex vivo detection of mesiobuccal canals in maxillary molars using CBCT at four different isotropic voxel dimensions. *International endodontic journal* 2011; **44**(8): 752-758.
30. Lee JH, Kim KD, Lee JK, Park W, Jeong JS, Lee Y, et al. Mesiobuccal root canal anatomy of Korean maxillary first and second molars by cone-beam computed tomography. Oral Surgery, Oral Medicine, Oral Pathology, Oral Radiology, and Endodontology 2011; **111**(6): 785-791.
31. Silva EJNL, Nejaim Y, Silva AIV, Haiter-Neto F, Zaia AA, Cohenca N. Evaluation of root canal configuration of maxillary molars in a Brazilian population using cone-beam computed tomographic imaging: An in vivo study. Journal of endodontics 2014; **40**(2): 173-176.
32. Georgia NE, Taxiarchis KG, Nikolaos KP. Suppl 2: M3: Evaluation of the Root and Canal Morphology of Maxillary Permanent Molars and the Incidence of the Second Mesiobuccal Root Canal in Greek Population Using Cone-beam Computed Tomography. *The open dentistry journal* 2015; 9: 267.
33. Alavi AM, Opasanon A, Ng YL, Gulabivala K. Root and canalmorphology of Thai maxillary molars. *International endodontic journal* 2002; **35**: 478-485.
34. Gu Y, Wang W, Ni L. Four-rooted permanent maxillary first and second molars in a northwestern Chinese population. *Archives of oral biology* 2015; **60**(6): 811-817.
35. Peikoff MD, Christie WH, Fogel HM. The maxillary second molar: variations in the number of roots and canals. *International endodontic journal* 1996; **29**(6): 365-369.
36. Christie WH, Peikoff MD, Fogel HM. Maxillary molars with two palatal roots: a retrospective clinical study. *Journal of endodontics* 1991; **17**(2): 80-84.
37. Versiani MA, Pécora JD, de Sousa-Neto MD. Root and Root Canal Morphology of Four-rooted Maxillary Second Molars: A Micro Computed Tomography Study. *Journal of endodontics* 2012; **38**(7): 977-982.

RISK PREDICTION AND EFFECT EVALUATION OF CT ANGIOGRAPHY TO PULMONARY EMBOLISM (PE)

Duo Zhang and Heng Meng[a,*]

Department of Radiology, Affiliated Hospital of BeiHua University, Jilin 132011, China

E-mail:[a]beihuamengheng@126.com

[*]Corresponding author

Pulmonary artery embolism also is known as pulmonary embolism (PE), most of embolus are from deep vein, back-flow with blood, attaches to pulmonary artery lumen to cause a series of clinical syndrome, multiple slice computer tomography pulmonary angiography (MSCTPA) has very high sensitivity and specificity to PE diagnosis, especially, the accuracy to section and above artery is higher [1], can clearly show embolus itself structure and relationship between it and blood vessels.

1. MATERIAL AND METHOD

1.1. Clinical Data

Collect clinical proven 43 PE patients as specific research object, with 26 men, 17 women, 39-77 years old, average age of 63. 43 patients have different degrees of chest tightness, chest pain and breathing difficulties, 17 patients are given priority to paroxysmal chest pain, among them, 3 patients come on at night and progressive exacerbation; 10 patients have the history of cough and expectoration. 6 patients have history of tumor and 13 patients have the history of deep vein thrombosis; 7 patients have history of diabetes and 3 patients have the history of trauma. 5 patients have PE recurrence. All 43 patients are carried out examination for two times and 8 patients are carried out examination for three times.

1.2. Examination Method

Scan by Germany SIEMENS Definition AS 128 layers spiral CT. The scope is from lung top to the bottom, high-pressure syringe: America E-Z-EM Company's EMPOWERCTA high-pressure syringe, contrast medium chooses Bayer Schering nonionic contrast medium (ultravist solution, with concentration of 370mgl/ml), total 60ml, with injection speed of 3.5ml/s. Add 0.9% physiological saline 40ml. Scanning method: first CT plain scan to lung, then enhance scanning, use contrast medium automatically detect trigger technology to determine delay time of scanning.

1.3. Image Diagnosis and Analysis

1.3.1. *Image Processing*

Observe the image by MSCT post-processing technology; main post-processing technologies include: volume representation (VR), maximum intensity projection (MIP), thin layer MIP, curved surface reconstruction (CPR), multi-planar reconstruction (MPR) and CT virtual endoscopy (CTVE), and improve the detection of embolism blood vessels by image magnification technology and CT window setting technique.

1.3.2. *Image Observation*

Adopt double blind method. Two deputy chief physicians respectively study medical film, the observation contents include the characteristics of pulmonary artery embolus and whether own lung disease; record the embolus characteristics, including form, density, position and scope, etc. Carry out imaging evaluation to selected cases, what are divided into three groups: acute PE group (aPE), chronic PE group (cPE) and (acute + chronic) PE group ((a+c) PE); Two physicians will jointly review the films when the diagnosis opinions are not consistent, to reach an agreement finally. Compare the patients who are reexamined after clinical treatment, and record the change before and after treatment.

2. RESULTS

2.1. Pulmonary Artery and Post-processing Technology

Among 43 PE patients, total 658 affected blood vessels are detected, in which, segment and above blood vessels are 575; sub-section blood vessels are 83, accounting for 12.6% of total embolism. Original transverse image displays comprehensive embolus, can detect the situation of distal embolus of sub-section, MPR has better display rate to sub-section embolus (83/83), MIP and thin layer MIP also can greatly detect sub-section embolus (56/83; 59/83); CPR and VR have low rate to sub-section embolus (16/83;18/83); CTVE could not observe distal embolus. In 2D image, CPR has minimum display rate to embolus (270/658), in post-processing technology, the display of MPR and thin layer MIP to embolus is obviously better than VR and CTVE, the difference has statistical significance ($P<0.05$).

2.2. Direct Symptom of PE

2.2.1. *Location of PE*

Divide PE into central type and surrounding type according to the position of embolus of PE: central type means that embolus is located in lobar artery and above lumen, with or without distal arterial embolus; surrounding type means that embolus is only located in section and / or sub-section as well as in distal blood vessels [2]. There is only one surrounding type PE in 43 PE patients of the study.

43 patients are detected 218 lobe PE, in which, total of 73 in right and left pulmonary arteries, including 41 in right pulmonary artery and 32 in left pulmonary artery; total of 145 in lobar artery, in which, 28 in upper lobe of double lung, 94 in lower lobe of double lung and 23 in middle lobe of right lung. Statistics show that, there are more embolisms in lower lobe of double lung, accounting for 43% of lobar artery embolism; the embolism in lower lobe of right lung is more than what in lower lobe of left lung, is about 23.4%. The results show that, embolism in right lung is more than what in left lung, accounting for 61.5% of the group.

2.2.2. *Type and Nature of Embolus*

The basis of CTPA diagnosing PE is the embolus in pulmonary artery lumen [3-5], in CT, it is shown as filling defect of lumen: (1) Central type or off-center type: a. Central type shows 'floating sign', 'target sign' and ' track sign', etc.; b. Wall-attached type: shows 'meniscus sign', it also can be divided into acute angle wall-attached type and obtuse angle wall-attached type; the larger embolus attaches along blood vessel walls, can be changed into a mold; c. Complete filling defect: embolus fills entire arterial lumen, not seen contrast agent all around; (2) Sudden stenosis of lumen; (3) Uneven density of lumen changes, such as honeycombing change, etc. and (4) Embolus calcification. In the group, there are 157 complete filling defect, accounting for 23.9%, 168 central type filling defect, accounting for 25.5%, 137 acute angle wall-attached type filling defect, accounting for 20.8% and 68 obtuse angle wall-attached type filling defect, accounting for 10.3%; sudden stenosis of lumen is 74, accounting for 11.3%, honeycombing is 54, accounting 8.2%, in which, embolus calcification is 8. We can see from above that, in the group, wall-attached type filling defect is more, accounting for 31.1%, in which, acute angle wall-attached type is given priority to, accounting for 20.8% of the group; second is central type and complete filling defect type, accounting for 25.5% and 23.9% respectively.

Based on above direct signs, divide PE into three types [6-7]: (1) aPE; (2) cPE; (3) (c+aPE), observe acute embolus sign on the basis of cPE. The group is detected total 318 acute embolisms, including 112 central type, 114 completely occluded type, 60 acute angle wall-attached type and 32 sudden stenosis of lumen. There are 77 chronic embolisms, in which, 26 obtuse angle wall-attached type, 18 sudden stenosis of lumen and 33 honeycombing change. 263 c + a PE, 56 central type, 77 acute angle wall-attached type, 42 obtuse angle wall-attached type, 43 completely occluded type, 24 sudden stenosis of lumen and 21 honeycombing change.

2.2.3. *Scope of Embolism*

Judgment standard of imaging examines PE [8]: Large area embolism means that PE ≥2 lobar arteries, or ≥7 segment arteries (double lung is based on 20 segment arteries). What inconsistent with large area embolism standard shall be regarded as non-large area PE. In 43 patients of the group, there are 27 patients with large area PE, 15 patients with acute large area embolism, 12 patients with acute and chronic large area embolism, total of 532 blood vessels with embolism; 16 patients with non-large area PE, chronic lung embolism is the most, total of 11, total of 126 blood vessels with embolism.

2.3. Comparison of PE Before and After Treatment

Judgment standard for treatment excellence of PE [4]: CTA cross section shows that the filling defect area in original blood vessels with embolism reduces $\geq$50; completely recanalization means that blood vessels with embolism completely recover and the walls are smooth; invalid performance is that, the area change of blood vessels with embolism is not big before and after treatment, even has worse trend.

Re-examination after thrombolysis. Carry out 128-SCTPA re-examination after 2-4 weeks of thrombolysis treatment (in which, 8 patients shall be carried out CTA re-examination after 24 h of thrombolysis treatment): among 43 patients, 17 with aPE, there are 318 embolisms before treatment and there are 139 after treatment, with effective rate of 56%; 11 with cPE, there are total of 77 embolisms, there are 94 after treatment, the embolisms are not reduced but increased; 15 with (a+c)PE, there are total of 263 embolisms, there are 167 after treatment, with effective rate of 37%. The results show that, the treatment effect of aPE group is obviously better than what of cPE group ($P<0.05$). The comparison of treatment effect between aPE group and (a+c) PE group has no statistical significance ($p>0.05$).

3. DISCUSSIONS

3.1. Factors of Influencing PE Detection

3.1.1. *Scanning Parameters*

The study chooses 128 layers MSCT machine, plain scanning: detector 128x0.6, with pitch 1.2, scanning layer thickness is 5mm; enhancement scanning: detector 128x0.6, with pitch 1.15, scanning layer thickness is 1mm and interlayer spacing is 0.7mm. Scanning time is only 3-4s, the speed is fast and whole lung display is good.

3.1.2. *Post-processing Technology*

Different post-processing technology of MSCT directly determines CTPA image quality and diagnosis rate, 128 layers MSCT adopts 1.0mm thickness to reconstruct volume data, greatly improved the quality of 3D reconstruction. The data of group shows that, original thin layer transverse image can well detect embolus in lumen, especially for surrounding type embolus, e.g. embolus of sub-section pulmonary artery and distal small arteries, has higher sensitivity and specificity (94-100%), the defect is not able to display overall structure of embolus.

During PE diagnosis and on the basis of axial image, flexibly use MPR, MIP and VR, etc. reconstruction method to do comprehensive observation to main pulmonary artery and its branches, to improve the detection rate of embolus.

3.2. Evaluation of 128 Layers MSCT to PE Embolus and Treatment

3.2.1. *Evaluation to PE Embolus*

The direct sign of diagnose PE is filling defect of pulmonary artery lumen, namely, manifestation of embolus. In 43 patients of study group, the embolism in right lung is more than what in left lung, accounting for 61.5% of the group. Embolism in lower lobe of double lung is more than what in upper lobe, and the position is given priority to segment, accounting for 54.3% of the group. The incidence rate of right side is higher than left side, the position of right side pulmonary artery is lower than left side and flow is straight, what is related to thrombus is easy to stay; the not detected embolus in main pulmonary artery may be have a certain relationship with aorta's diameter is wide, blood flow speed is fast, difficult to form embolus and difficult to stay after embolus is formed [9]. PE is easier in segment, especially segment in lower lobe of double lung is more, what is related to pulmonary artery blood flow dynamics and distribution, the blood flow in lower lobe of double lung is rich relatively to upper lobe, and blood flow direction is consistent with lower lobe pulmonary artery. It is consistent with related reports, the causes may be repeated falling off of thrombus or fresh thrombus is broken into multiple small thrombus by mechanical and (or) fibrinolysis function when passing through heart cavity or into pulmonary artery, the blood of right lung and lower lobe of double lung are relatively abundant, so easier to embolism [10-11].

3.2.2. *Evaluation to PE Before and After Treatment*

Acute PE means a series of clinical manifestation caused by embolus obstructs pulmonary artery and its branches in early stage, most embolus is fresh. Chronic PE means that embolus formation time is longer, or long-term repeated accumulation causes obstruction to part or all of pulmonary artery, thus cause clinical change.

In 43 PE patients with re-examination after treatment, there are 17 patients with acute embolism, 12 patients are carried out catheter interventional thrombolysis treatment, 5 patients are treated by drug thrombolysis and anticoagulation, re-examination after 2-4 weeks (among them, 8 patients are re-examined after 24h), among 5 patients with drug thrombolysis, 3 patients have large area embolism and part embolus are disappeared after treatment; all 12 patients with catheter thrombolysis treatment have large area embolism, treated by auxiliary drug, the effect is good. Some literatures said that, drug treatment is the first choice for acute PE [5], however, the patients source of the group is single relatively, so the effect of interventional thrombolysis treatment remains to be further research.

11 patients with chronic thrombus are treated by drug anticoagulation, among them, pulmonary artery embolus of 8 patients has no obvious change after treatment, the treatment effect is not ideal because the embolus time is long or edges are polarized, calcification, however, drug treatment can relatively improve lung ventilation quality of patients.

The judgment of MSCTPA to PE nature can provide basis for clinical choice of treatment options, appropriate treatment option can significantly improve cure rate of patients; the treatment to acute attack of chronic embolism shall be dominated by drug thrombolysis, and for chronic embolism patients, also should give priority to prevent acute attack.

PE will cause a series of pathological and physiological, etc. indirect changes of lung and circulatory system, the observation before and after treatment can assist to judge treatment effect and situations of clinical improvement of patients. In indirect signs of PE, the more important is mosaic sign (13 patients) and pulmonary infarction sign (10 patients), but they are rare, consider that they are related to multiple blood and oxygen supply of lung tissue[5]. Therefore, for flaky fuzzy shadow and basic pathological changes of lung consolidation occurred in lung of PE patient, more is caused by lung parenchyma hemorrhage, exudation and edema with atelectasis.

References

1. Burns PN.n Ultrasound contrast agents in radiological diagnosis. *Radiol Med.* 1994; **87**(5SuPPl l):71-82.
2. Wang Wenhong, Chang Bei and Chang Bin, etc. Application of Spiral CT Vascular Imaging Technology in Diagnosis of PE. *Journal of Practical Radiology*, 2006; **22** (6): 679-682.
3. Remy-Jardin M, Louvegny S, Remy J,et a1.Acute central thromboembolic disease: posttherapeutic follow-up with spiral CT angiography[J].*Radiology*, 1997;**203**:173-180.
4. Remy-Jardin M, Tillie-Leblond I, Szapiro D,et al. CT angiography of pulmonary embolism in patients with underlying respiratory disease:image quality and negative predictive value[J].*Eur Radio*1,2002;**12**(8):1971-1978.
5. Enden T, Klow NE. CT pulmonary angiography and suspected acute pulmonary embolism[J].*Acta Radiol*,2003,**44**(3):310-315.
6. Filipek:MS, Gosselin MV. Multidetector pulmnonary CT angiogra-phy: advances in the evaluation of pulmonary arterial, diseases.[J]. *Semin Ultrasound CT MR*,2004; **25**(2):83-98.
7. Guilabert JP. Manzur DN. Tarrasa MJ, et al. Can multislice CT alone rule out reliably pulmonary embolism, A prospective study. Eur J Radiol. 2007; **62**(2): 220-226.
8. British Thoracic Society Standarss of Care Committee Pulmonary Embolism Guideline Development Group. British Thoracic Society guideline for the management of suspected acute pulmonary embolism[J]. *Thorax*,2003; **58**: 470- 478.

9. Cham M, Yankelevitz D, Shaham D, et al:Deep venous thrombosis: detection using indirect CT venography. The Pulmonary Angiography-Indirect CT venography Cooperative Group[J]. *Radiology*,2000;**216**:744-751.
10. Qin Haiyan, Shen Baozhong and Wang Dan, etc. Diagnosis Value of 64 Layers Spiral CTPA to PE (with 25 cases analysis). Chinese Journal of Medical Imaging Technology, 2008; **24** (1):144-147.
11. Sheng Lei, Ma Deting and Liu Zhiqiang, etc. Diagnosis Value of MSCT Different Thickness Maximum Density Projection in PE [J]. *Journal of Medical Imaging*, 2009; **19** (2): 219-222.

EXPLORATION OF HUMAN ORAL MICROBIOME RELATED WITH HEALTH AND DISEASE BY ILLUMINA HIGH-THROUGHPUT SEQUENCING

Yibin Wang[1], Jianye Zhou[2], Zhiqiang Li[2], Zhenzhen Wang[1], Junping Li[1], Fang Wu[1], Rui Ma[1], Shuang Li[1], Huimin Huang[1], Genyun Yi[1], and Zhanhai Yu[1,*]

[1]*Lanzhou University, Lanzhou, 730000, China*

[2]*Northwest University for Nationalities, Lanzhou, 730000, China*

[*]*Corresponding author*

Taking advantage of high-throughput technologies has produced bacterial profiles and genomic profiles to study the relationships between microbial diversity, genetic variation, and oral diseases. In this study, in order to explore the microbial community structure in the human oral cavity with and without disease, Illumina MiSeq high-throughput sequencing was utilized to analyze 30 salivary samples (10 health, 10 caries and 10 periodontitis). Core microorganisms was verified present in the health group, caries group and periodontitis group, and certain differences of the microbial structures in saliva of all groups at phylum and genus levels are be found. Our findings regarding the characteristics of the microbial communities of the human oral might improve the prevention and cure of oral disease in the future.

1. INTRODUCTION

Oral diseases, such as periodontitis and dental caries, are some of the most prevalent infectious diseases of humans. It is reported that approximately 36% of the world's population have carious lesions in permanent teeth, further more, up to 90% of the population are affected by periodontal disease [1,2]. Oral disease is closely related with the oral microbiome, it have been demonstrated to that bacterial profiles of saliva in subjects with periodontitis and dental caries are different from subjects with health. Oral microbiota is one of most diversity habitat in the human body, lots of hypotheses highlight the impact of ecological variables on microbiota associated with oral disease. Therefore, a comprehensive analysis is needed to explore the relationship between microbial diversity and oral diseases. As the technique developing, high throughput sequencing has become a popular approach applied for oral microbial analysis. This next generation sequencing technique is a cost- effective mothed, it obtains sequences by universal amplification about the 16S rRNA gene, gets thousands of sequence reads generated in a single run. This approach allows an overview of the bacterial communities as a whole, overcoming the limited views that previously employed techniques offered, enabling the analysis about enormous diversity of the oral microbiota to be realized.

2. MATERIALS AND METHODS

2.1. Subjects and Specimen Collection

We collected saliva samples over a one-week period from thirty adult individuals with informed consent, divided them into three groups(ten subjects in each group) and marked as group H (health), group C(caries), group P (periodontitis), all of them were recruited from the Dongxiang baoan autonomous county of Gansu Province in China. Periodontitis was defined as unless two teeth with PD index (Probing Depths) ≥4 mm and AL index(Attachment Loss) ≥3mm. Dental caries was defined as the DMFT index (decayed, missing and filled teeth) ≥4.Saliva samples were collected by unstimulated expectoration into a sterile tube. We mixed 1 mL of each saliva sample with 500 μL of TE buffer (25 mM Tris HCl, 10 mM EDTA, pH 8.0) and kept frozen at −20 °C until processing.

The constitution of oral microorganisms depends on multiple factors, including lifestyle, dietary habit as well as the host's own conditions. In this case, one should try best to reduce errors induced by different external factors when selecting object of study. The objects of study in this research were selected in the same town, a majority of the selected subjects were Ethnic Bonans and Huis who believed in Islam and had basically consistent dietary structure and living habit, reducing the influence of other factors on their oral microorganisms.

Colony structures varied with different sites in the oral cavity. Mager et al[3] analyzed the ratio of 40 kinds of bacteria in different oral cavity sites of 225 healthy subjects through DNA chessboard hybridization, and discovered that there was distinct difference of the bacteria in different sites. It is widely believed currently that as the flow velocity of saliva (the duration from secretion to deglutition) is faster than the growth rate of the bacteria inside, a majority of the contained bacteria derive from dropping of bacteria from other oral cavity sites; as a result, saliva usually provides the information of microflora in all oral cavity sites. In addition, saliva specimens are cheap, easily accessible, and safe to collect, rendering them being commonly used in research on oral microbial community diversity.

2.2. DNA Extraction, Library Construction and Sequencing

Universal Salivary DNA was extracted using the Bacterial DNA Extraction Kit D3350-01 (OMEGA Biotek, USA), according to the method described. To determine the diversity and composition of the bacterial community in each salivary sample, PCR amplifications of the bacterial 16S rRNA V4 region (approximately 240 bp) were performed using the PCR-based protocol described and the forward and reverse primers (5'- AYTGGGYDTAAAGNG-3' and 5'-TACNVGGGTATCTAATCC-3', respectively). In a final reaction volume of 25 μL (9 μL of sterilized water, 5 μL of 5× PCR buffer, 5 μL of 5× PCR GC high enhancer, 2 μL of 2.5 mM dNTP, 2 μL of template 200 ng/μL DNA, 0.25 μL of 5 U/ μL TaKaRa polymerase , 1 μL of 10 μM each primer). The amplification cycle consisted were as follows: an initial denaturation at 98°C for 5 min, followed by 27 cycles at 98°C for 30 s, 50°C for 30 s, and 72°C for 30 s, with a final extension at 72°C for 5 min. All raw sequences were trimmed using the Quantitative Insights into Microbial Ecology (QIIME) toolkit v.1.5.0. Next, the reads that were shorter than 200 bp, contained any ambiguous bases, or exhibited a homopolymer longer than 10 bp were removed. The operational taxonomic units (OTUs) were clustered with a 97% similarity cutoff using UPARSE (version 7.1), and the chimeric sequences were identified and removed using UCHIME. Representative sequences for each OTU were assigned taxonomically using the Ribosomal Database Project (RDP) classifier and the BLAST tool of the Human Oral Microbiome Database (HOMD).

3. RESULTS AND DISCUSSION

Through parallel sequencing and quality control, we obtained 1,046,647 high-quality sequences with a mean read length of 223bp. Clustering of all of the qualified sequences at the 97% similarity level generated a total of 17,796 OTUs, 28 phyla and 506 genera. The average level of Good's coverage per sample was over 99%, suggesting that the 16S rRNA gene sequences identified represent the majority of bacterial taxa inhabiting the saliva in the current study.

3.1. Microbial Community Structure in Each Group

Group H: 25phyla, 389genera, Prevalence>1%: 7phyla, 36genera; Group C: 24phyla, 357 genera, Prevalence>1%: 6phyla, 32genera; Group P: 25phyla, 378genera, Prevalence>1%:9phyla, 34genera. The results are shown in Table 1, Fig. 1 and Fig. 2.

As was shown from the data of the American Human Microbiome Project (HMP), the oral cavity of unrelated individuals possessed larger core sharing biota when compared with other biotopes (such as intestinal tract or skin). Yang et al discovered 0.08% and 0.53% share OTUs from the saliva samples of 19 caries patients and 26 healthy subjects, respectively, through high-throughput sequencing, and consequently verified the presence of core microorganisms.

It could be figured out from table 1 that, the core microbiome was generally the same in the health group and the caries group, and the periodontitis group contained the whole core microbiomes of the former two groups, and had more types. The bacterial genuses that were detectable in all 30 samples were as follows: Actinomyces, Rothia, Streptococcus, Neisseria and Actinobacillus. Thus, we supposed that these five bacterial genuses were the core microorganisms in the oral cavity.

In this research, the bacterial phylum between group C and group H showed no statistical difference which could only be seen among part of the non-dominant bacterial genuses with low detection rate. On one hand, it might be due to the few caries enrolled as the objects of study in the caries group, or certain unknown errors occurred during the research; on the other hand, it might be that one or several bacterial genuses with low abundance acted as the key pathogens and changed the colony structures in the host oral cavities, which might be related to the environment or race and required further investigation. Hajishengallis proposed a key pathogen hypothesis which considered that some pathogenic microorganisms with low abundance might influence the quantities of other bacteria and consequently changed the flora composition, interfered the immune system of the host, and resulted in the occurrence of inflammatory diseases finally.

When compared with group H, the abundance of Actinobacteria, TM7 bacterial phylum was higher in group P, but that of the Bacteroidetes was lower at bacterial phylum level; while that of the Rothia, Megasphaera was higher in group P, but that of the Abiotrophia, Granulicatella and Lactococcus was lower at bacterial genus level.

TM7 bacterial phylum had long been considered as the suspected periodontal pathogens[4-5]. Currently, researchers have isolated and cultured a specific type of TM7 that exists in human oral cavity, which is named TM7x. They discovered that TM7 can attach onto the Actinomyces Odontolyticus (a pathogenic factor that can induce mucosal inflammation) within the oral cavity, inhibit the body's immune reaction induced by Actinomyces Odontolyticus stimulation, and prevent the production of tumor necrosis factor TNFa, leading to periodontal diseases[6].

There have been few reports regarding the role of Megasphaera in periodontitis. Some scholars[7] investigated the microorganisms in the upper gastrointestinal tract of 278 Chinese male that received esophageal cancer screening, and found that the abundance and detection rate of Megasphaera were remarkably higher in smokers than in non-smokers. Given that smoking was a risk factor of periodontitis, we speculated that smoking gave rise to the increased abundance of Megasphaera in the oral cavity, and destructed the otherwise dynamically-balanced oral internal environment, which might be one of the factors that induced the occurrence of periodontal diseases.

The dominant bacterial phyla which had significant difference were Actinomycetes and TM7 bacterial phylum in both group C and group P, with the abundance higher in group P than in group C; while the dominant bacterial genuses were Gemella and Granulicatella, with the abundance in group C higher than that in group P. Gemella and Granulicatella were usually detected in caries-sensitive samples. Some scholars analyzed the composition of the pathogenic bacteria in caries through 16S rDNA-based PGS ion torrent sequencing and discovered that the streptococcus, Rothia, Gemella, and Granulicatella were notably greater in the caries group than in the control group, which was consistent with our research.

Table1. The Prevalence of genera

Prevalence	Group H		Group C		Group P	
	Genera	count	Genera	count	Genera	count;
100%	Actinomyces Rothia Streptococcus Fusobacteria Neisseria Actinobacillus	6	Actinomyces Rothia Streptococcus Veillonella Neisseria Actinobacillus	6	Actinomyces Rothia Streptococcus Veillonella Leptotrichia Fusobacterium Neisseria Actinobacillus	8
51%~99%	Porphyromonas CapnocytophagaGemella Veillonella Leptotrichia	8	Leptotrichia Gemella Fusobacterium	6	Capnocytophaga Gemella	5
21%~50%	Granulicatella Lactobacillus Filifactor Treponema Abiotrophia	10	Granulicatella Abiotrophia Peptostreptococcus Capnocytophaga	5	Porphyromonas Eubacterium Treponema Peptostreptococcus Filifactor Megasphaera Bifidobacterium	8
1%~20%	Corynebacterium Lactococcus Eubacterium Acinetobacter Prevotella Parascardovia Peptostreptococcus	12	Arthrobacter Eubacterium Filifactor Megasphaera Porphyromonas Prevotella CorynebacteriumLactococc us Acinetobacter Prevotella Acinetobacter Solobacterium Parascardovia	14	Prevotella Abiotrophia Acinetobacter Campylobacter Parascardovia Atopobium Parvimonas Solobacterium	13

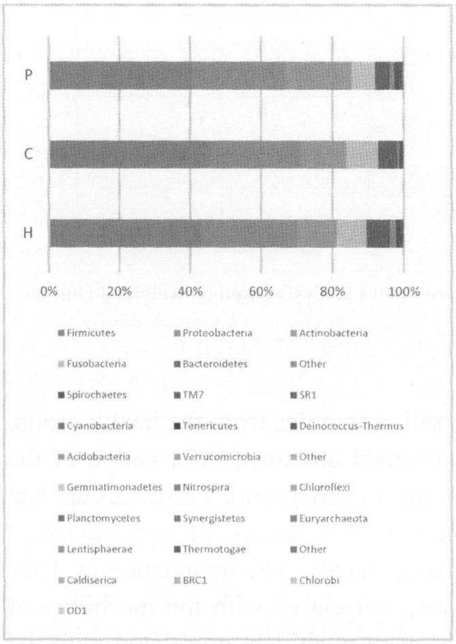

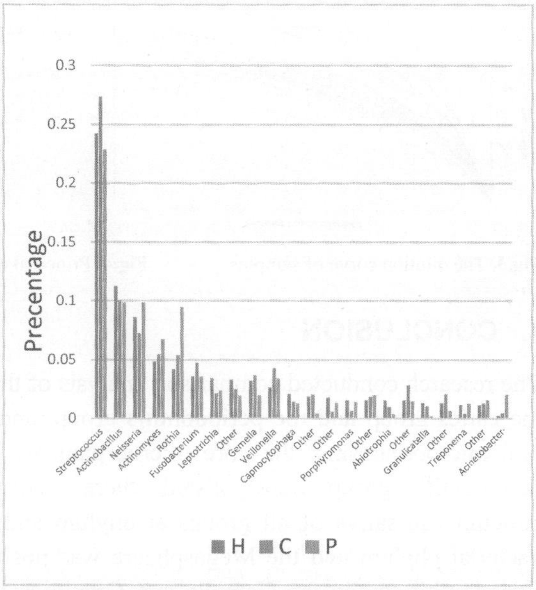

Fig.1. Relative abundance of phyla members **Fig.2.** The genera(at prevalence>1%) presented in Group H, C, P

3.2. Diversity Analysis

α Diversity Analysis: The dilution curve showed that the tail end of the curve of these samples kept on rising(the result is shown in Fig. 3), indicating that there maybe a portion of microorganisms undetected which resulted in certain underestimate of the quantities of the oral microorganisms in the samples. However, the cover degree of all samples was above 99%, it was possible that PCR errors or sequencing errors led to the overestimate of several development types.

β Diversity Analysis: It was shown in the PCoA analysis chart that the 10 samples in the health group were more disperse in distance than another two groups, which indicated that the microbial constitution in the oral cavities of healthy subjects was more diversified, while those of the caries patients and periodontitis patients were relatively single, with several types of bacteria deficiency or decreased abundance. The locations of the C group and P group are separated, but H groups were basically coincided with another two groups, revealing that there are some difference of the microbial community structures between the caries group and the periodontitis group, healthy population distributions are wider than another two group relatively. The result is shown in Fig. 4.

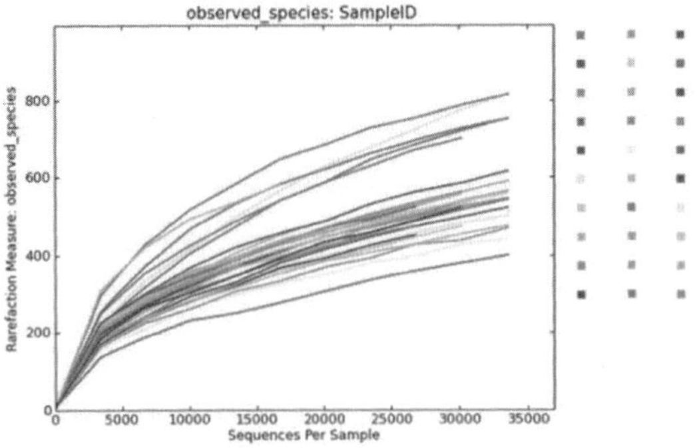

Fig.3. The dilution curve of samples

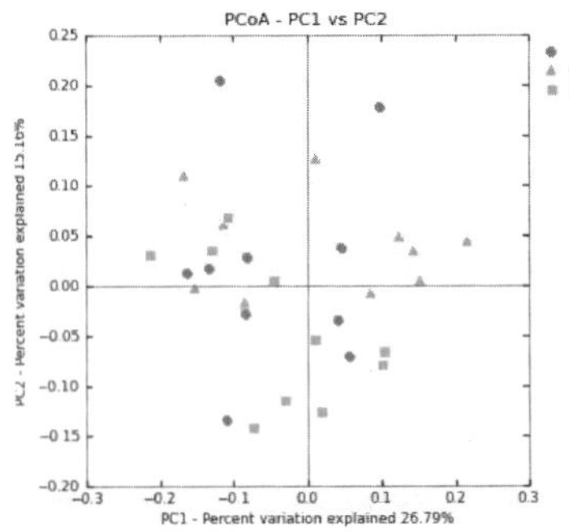

Fig.4. Principal co-ordinates analysis based on weighted UniFrac

4. CONCLUSION

The research conducted comparison analysis of the saliva samples from the health group, the caries group and the periodontitis group, and obtained abundant information of the oral microorganisms: the core microorganisms in the health group, caries group and periodontitis group were existed; there were certain differences of the microbial structures in saliva of all groups at phylum and genus levels. The abundance of TM7 bacterial phylum and the Megasphaera was positively correlated with the morbidity of periodontitis, while the Gemella and Granulicatella were positively correlated with the morbidity of caries. However, the oral microbial environment was extremely complicated; in this case, further in-depth research was required in order to gain more complete microbial information, providing powerful assistance for preventing and treating oral diseases.

References

1. Vos T, Flaxman A D, Naghavi M, et al. Years lived with disability (YLDs) for 1160 sequelae of 289 diseases and injuries 1990 - 2010: a systematic analysis for the Global Burden of Disease Study 2010. Lancet, 2012; **380**(9859):2163 - 2196.
2. Hans-Peter H, Georg C. Diagnosis and anti-infective therapy of periodontitis. Expert review of anti-infective therapy, 2007; **5**(4):703-15.
3. Mager D L, Ximenez-Fyvie L A, Haffajee A D, et al. Distribution of selected bacterial species on intraoral surfaces. *Journal of Clinical Periodontology*, 2003; **30**(7):644 - 654.
4. Bo L, Faller L L, Niels K, et al. Deep sequencing of the oral microbiome reveals signatures of periodontal disease. *Plos One*, 2012; **7**(6): e37919.
5. Rylev M, Bek-Thomsen M, Reinholdt J, et al. Microbiological and immunological characteristics of young Moroccan patients with aggressive periodontitis with and without detectable Aggregatibacter actinomycetemcomitans JP2 infection. *Molecular Oral Microbiology*, 2011; **26**(1):35-51.

6. Xuesong H, Mclean J S, Anna E, et al. Cultivation of a human-associated TM7 phylotype reveals a reduced genome and epibiotic parasitic lifestyle. *Proceedings of the National Academy of Sciences*, 2015; **112**(1):244-249.
7. Vogtmann E, Flores R, Yu G, et al. Association between tobacco use and the upper gastrointestinal microbiome among Chinese men. *Cancer Causes & Control Ccc*, 2015; **26**(4):581-588.

CLINICAL EFFECT OF MODIFIED ROOT CANAL FILLING WITH GUTTAFLOW2

Dandan Wang[1,a], Wen Shi[1,b], and Liya Chen[2,c,*]

[1]Hospital of Stomatology, Lanzhou University, Lanzhou 730000, China
[2]The First Affiliated Hospital, Lanzhou University, Lanzhou 730000, China
E-mail:[a]wangdd2013@lzu.edu.cn,[b]545023246@qq.com,[c]chenchy@lzu.edu.cn
[*]Corresponding author

Objective: To evaluate the clinical effect of modified root canal filling with GuttaFlow2. Methods: seventy-five teeth with single root canal were included in the study. They were randomly divided into three groups of 25 root canals each. The root canals were instrumented and obturated with cold laterally compacted techniques as the control group, GuttaFlow2 technique and modified GuttaFlow2 technique as the treatment groups. The time consuming during root filling, quality of root filling, postoperative pain responses in one week, and the short-term clinical efficacy were chosen as the observation indexes. Results: The root canal filling times of the experimental groups were significantly less than the control group.There were statistically significant differences of postoperative pain response (P=0.003) and root filling quality(P=0.021) in the 3 groups,but no statistical significant differences (P=0.086) were found among the three groups on the short-term clinical efficacy. Conclusion: Root canal filling with Improved GuttaFlow2 showed a superior clinical efficacy compared with cold laterally compacted technique concerning improving work efficiency, alleviating postoperative pain response, improving quality of root filling. Improved GuttaFlow2 system is promising in reducing overfilling of the paste when filling with GuttaFlow2, alleviating postoperative pain, which deserves clinical recommending.

1. INTRODUCTION

GuttaFlow2(Fig. 1) is a further improved new root filling material on the basis of GuttaFlow and combines two products in one[1]. Including gutta-percha in powder form particles and sealants, biocompatible better[2], the edge of the closure and better performance. Silva EJ[3]also believes GuttaFlow2 non-cytotoxic, having the right physical and chemical properties and biological properties of root canal sealer[4], and reliable clinical results. Many [2,5] scholars believe that the guttaflow has the smaller cell toxicity and good biocompatibility, even if beyond the apical foramen would not have obvious periapical reaction and did not show any clinical symptoms.But Endodontic sealers affect the osteoclastic differentiation and activity, which is followed by an adaptive cell response.And the deleterious effect in the bone periapical tissues observed with the root canal sealers might involve, at least partially, a direct effect on the osteoclastic cells[6].Thus, it is necessary to avoid root canal sealer beyond the apical foramen. Studies suggest[7] that using GuttaFlow2 in combination with single gutta percha filling root canal can replace the cold laterally compacted techniques, but its good flowability increased risk of overfilling is still the most concern. From the long-term effects to consider how to reduce the gutta-percha beyond the apical foramen worth exploring in clinical practice.

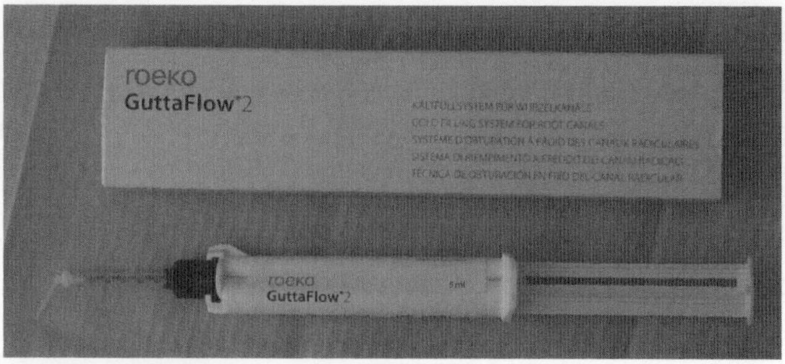

Fig. 1. GuttaFlow 2.

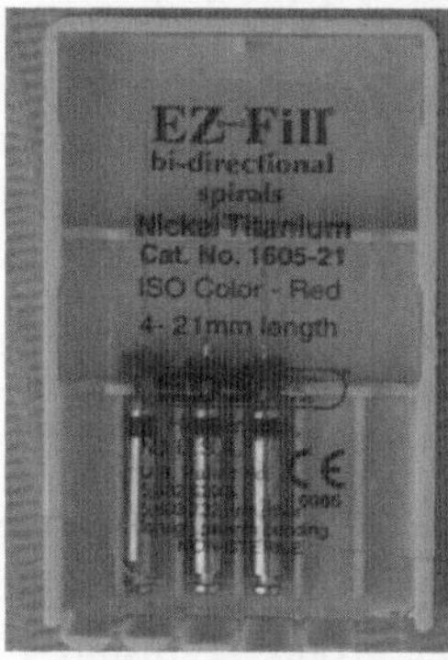

Fig. 2. EZ-FILL Bi-directional spiral plugger.

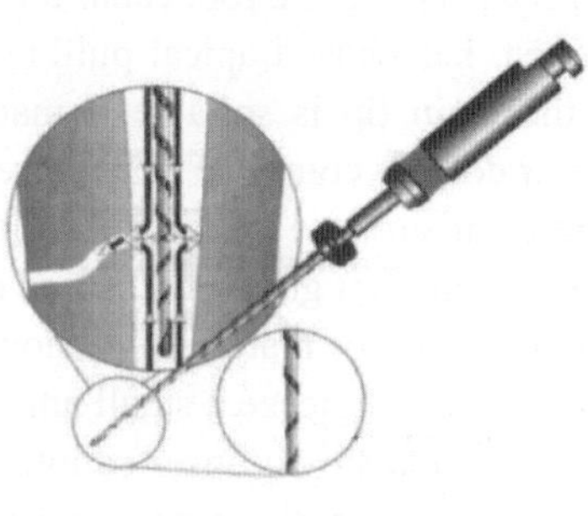

Fig. 3. EZ-FILL plugger.

EZ-FILL Bi-directional spiral plugger (Fig. 2) is a part of the EZ-FILL root canal system, American EDS company's patented design,Its unique reverse spiral tip (Fig. 3) so that the paste flows to the contrary exist thus forced to flow to the side rather than the apical direction, avoid overfilling paste.In the study the cold laterally compacted techniques as the control group, GuttaFlow2 technique and modified GuttaFlow2 technique as the treatment groups. In the modified GuttaFlow2 technique, the paste was introduced using EZ-FILL plugger, then put the main gutta percha into the root canal. To observe the clinical efficacy and provide a reference for clinical application.

2. MATERIALS AND METHODS

Select seventy-five teeth with single root canal and chronic periapical need root canal treatment in our hospital between the ages of 18 and 45.They were randomly divided into three groups of 25 root canals each. Inclusion criteria: No endodontic treatment history; no spontaneous pain and percussion pain, no acute onset of symptoms; apical density of Shadows <2mm; no fistula; no periodontitis; root development is completed and the root canal patency; informed consent to treatment.Exclusion criteria: Calcified root canal, root resorption.

2.1. The Main Experimental Equipments

GuttaFlow2(Switzerland Conte Dental Group), Root canal sealers, ISO 25#EZ-FILL Bi-directional spiral plugger(EDS, USA), Mobile Protaper NiTi instruments(Dentsply, USA), Dentsply X-SMART Root canal treatment using micro-motor, SybronEndo Almighty root canal diagnostic(SybronEndo, USA), 04 tapered gutta percha.

2.2. Clinical Operations

Preoperative shoot apical teeth X-rays, a preliminary understanding of the root morphology, assess periapical PAI index. Open pulp, pulpectomy, small file explore root canal, making length measurement and determine the working length. Hand K file smooth root canal, ProTaper prepared root canal according toSX-F3. Import calcium hydroxide paste, sealing the root canal for two weeks. Select 04 taper gutta be suitable for the protagonist, has caused apical pulling sense of obstruction, making the test piece to determine the main tip is suitable. I made the clinical operations independently, with another senior doctors complete synergy read the piece.

Experimental group 1:Install self-mixing needle, determine the depth of injection, Squeeze a small amount gutta-percha and make slow injection inserted into the root canal, See the small amount of material overflows, back side edge slowly injected gutta-percha to root canal orifice; Squeeze a small amount of gutta-percha to reconcile paper, Suitable main tip dipped gutta slowly inserted into the root canal, Pulling up and down and gently rotate 2-3 times put to the working length. Experimental group 2: Install EZ-FILL bidirectional spiral pluggerand and determine the working length. Injection the gutta-percha to reconcile paper, plugger dipped in an appropriate amount of gutta-percha into the root canal done low and down a circular motion for about 5 seconds. When filling stopper piece does not contact the teeth, from mark a little, so as not to move the catch to the coronal shift leaving the obturator beyond the apical foramen. Repeat the operation until the excess paste overflows from coronal. The method of set in the tip is same to the experimental group 1. The control group: preparation root canal sealers, make main tip as conveyor to lead to an appropriate amount root canal sealers introduced into the root canal and put to the working length, lateral condense gutta-percha until tight root canal filling. After the hot glue broken gutta-percha zinc oxide temporary seal, making X-ray assessment of the quality of root canal, then make the bottom and filling. Let the patient back at 7 days and six months.

2.3. Evaluation Standard

1)Root canal filling quality in accordance with "appropriate filling" "less-filling" "over-filling" evaluated according to the apical X-ray. (i)Appropriate filling: tight root canal filling, and root filling material from the X-ray dental apical 0.5 ~ 2.0mm. (ii)Less filling: transmissive image in root canal or root filling material from the X-ray dental apical> 2mm. (iii)Over-filling: root filling material beyond the X-ray dental apical.

2)Pain assessment standards: Using the previous tests[8] had confirmed the visual analogue scale(VAS) into 0, Level 1, Level 2 and Level 3.

3)Comprehensive clinical evaluation standards: Referring to the European endodontics Annual Meeting 1994[9]The root canal therapy evaluation criteria, after comprehensive assessment of the preoperative and postoperative X-ray examination with clinical results divided into: cured, effective, failed. Effective and failure were classified as not cured. PAI Index in accordance with Orstavik[10] score scoring criteria, divided into five scores, levels 1 and 2 is a normal level, level 3 and above for the inflammatory changes.

2.4. Statistical Methods

Statistical analysis was performed using χ^2 tests and Mean Rank's Post Hoc Multiple Comparisons of Kruskal-Wallis H Test on SPSS22.0, test level P= 0.05.

3. RESULTS

1) The time consuming during root filling: Import root canal sealers after the dried root canal, hot off the gutta percha to date. The average time of control group is 400s, Experimental group 1 is 68s, Experimental group 2 is 90s.

Table 1. The quality of root filling (n. %)

Group	N	Appropriate	Less	Over	Suitable-filling rate(%)
control group	25	17(68.0%)	2(8.0%)	6(24.0%)	68.00
experimental group1	25	22(88.0%)	0	3(12.0%)	88.00
experimental group2	25	24(96.0%)	0	1(4%)	96.00[a]

Note: χ^2 =7.738, P=0.021;

[a] Comparison of the experimental group 2 and control group. P=0.01

Table 2. Postoperative pain responses in one week (n. %)

Group	N	0	1	2	3
control group	25	12(48.0%)	9(36.0%)	4(16.0%)	0
experimental group1	25	20(80.0%)	4(16.0%)	1(4.0%)	0[a]
experimental group2	25	22(88.0%)	3(12.0%)	0	0[b]

Note: H=11.703, P=0.003;

[a]Comparison of the experimental group 1 and control group. P=0.03.

[b]Comparison of the experimental group 2 and control group. P=0.004.

Table 3. The short-term clinical efficacy (n. %)

Group	N	Back(n)	Cured	Not cured	
				effective	failed
control group	25	24	19(79.17%)	5(20.83%)	0
experimental group1	25	22	21(95.45%)	1(4.55%)	0
experimental group2	25	25	24(96.0%)	1(4.0%)	0

Note: χ^2 =4.917, P=0.086

2) The quality of root filling(Table 1):After χ^2 tests, there were statistically significant differences of the root canal suitable-filling rate among the three groups(P=0.021),between Experimental group 2 and the control group(P=0.01), this means compared to cold laterally compacted techniques, improved GuttaFlow2 can improve quality of root filling obviously. Table1 shows the over-filling rate of experimental group2(4%) lower than the experimental group1(12%), may be due to the sake of small sample size, the difference was not statistically significant.

3) The postoperative pain responses in one week(Table 2):After the Kruskal-Wallis H Tests, there were statistically significant differences among the three groups(P=0.003), between Experimental group 1 and the control group(P=0.03), between Experimental group 2 and the control group(P=0.004). But no statistical significant differences (P>0.05) were found between the Experimental groups.

4) The short-term clinical efficacy: the over-filling paste in the control group all had been absorbed, in contrast, in the Experimental group hadn't been absorbed. But patients did not show obvious clinical symptoms. No statistical significant differences (P=0.086) were found among the three groups on the short-term clinical efficacy(Table 3).

4. DISCUSSION

Currently the most commonly used root canal filling techniques includes cold laterally compacted and warm gutta percha techniques. Due to limitations of the material properties and filling technology of gutta filling materials, It's difficult to achieve the desired long-term effect of root canal closed. The introduction of the new filling material as GuttaFlow2, GuttaFlow accordingly encourage the clinical application of single gutta-percha filling techniques[11]. Silicon-based sealers have good closure[12], after curing gutta volume slightly (0.2%) polymerization expansion and the characteristics of not be absorbed, providing effective filling of the apical, ensure long-term apical sealing, prevent microleakage. Uniform structure of Gutta-percha particles and good adaptability to root canal wall make it easy to fill the dentinal tubules. In addition, the materials used in the process does not use heat, shrinkage does not occur. More advantageously ,0.2% volume expansion of the material after solidification. It helps flow gutta more closely with gutta percha and root dentin adapt, thereby enhancing the flow of sealant and adhesion to dentin wall[5]. Good liquidity and compliance, high replication capability to the wall of root, Without the pressure will be able to effectively collateral filling the root canal and dentinal tubules, to achieve the perfect filling three-dimensional effect. Varun K[13] believes GuttaFlow has acceptable performance of apical closure, almost equal to cold lateral condensation, but better than warm gutta percha. Shinde A[1]Scholar used scanning electron microscopy to compare the apical sealing used different sealers with single gutta percha filling techniques, he thought GuttaFlow2 performed better adaptability to root dentin than AH Plus showed , and good sealability.Samiei[14] considered the application under the single gutta percha filling techniques good sealer can replace cold lateral condensation. Rana M[7] believes use GuttaFlow2 combined single master gutta percha root canal filling can be a good substitute of cold lateral condensation filling.In this clinical study, modified GuttaFlow2 technique made clinicians simplify root canal filling process, improved efficacy. Compared with conventional cold lateral condensation filling, root canal filling with Improved GuttaFlow2 showed a superior clinical efficacy concerning alleviating postoperative pain response, improving quality of root filling. Compared with GuttaFlow2 system, Improved GuttaFlow2 system is promising in reducing overfilling of the paste, which deserves clinical recommending.

References

1. Shinde A, Kokate S, Hegde V. Comparative assessment of apical sealing ability of three different endodontic sealers: A scanning electron microscopic study[J]. *Journal of Pierre Fauchard Academy* (India Section), 2014; **28**(3): 78-82.
2. Ebert J, Holzschuh B, Frankenberger R, et al. Sealing ability of different versions of GuttaFlow2 in comparison to GuttaFlow and AH Plus[J]. *RSBO*. 2014 Jul-Sep;**11**(3):224-9.
3. DobynsSilva E J, Neves A A, De-Deus G, et al. Cytotoxicity and gelatinolytic activity of a new silicon-based endodontic sealer[J]. *Journal of Applied Biomaterials & Fundamental Materials*, 2015 Sep 10:0.
4. Silva E J N L, Herrera D R, Rosa T P, et al. Evaluation of cytotoxicity and physicochemical properties of silicone-based endodontic sealer GuttaFlow2 [J]. *Endodontic Practice Today*, 2014, 8(1).
5. Mandal P, Zhao J, Sah SK, et al. In vitro cytotoxicity of guttaflow 2 on human gingival fibroblasts.[J]. *Journal of Endodontics*, 2014, **40**(8):1156-1159.

6. Rodrigues C, Costa-Rodrigues J, Capelas JA, et al. Long-term Dose- and Time-dependent Effects of Endodontic Sealers in Human In Vitro Osteoclastogenesis[J]. *Journal of Endodontics*, 2013, **39**(6):833-838.
7. Rana M, Sandhu G K, Kaur T, et al. New Self Curing Root Canal Filling Material: Gutta flow 2[J]. *Journal of Advanced Medical and Dental Sciences Research*, Vol, 2014, 2(4).
8. Lohbauer U, Gambarini G, Ebert J, et al. Calcium release and pH-characteristics of calcium hydroxide plus points[J]. *Int Endod* J 2005; **38**:683-689.
9. European Society of Endodontology. Consensus report of the European Society of Endodontology on quality guidelines for endodontic treatment[J]. *Int Endod* J,1994;**27**:115-124.
10. Huumonen S, Ørstavik, Dag. Radiological aspects of apical periodontitis[J]. *Endodontic Topics*, 2002;**1**(1):3-25.
11. El Sayed MAAM, Taleb AAA, Balbahaith MSM. Sealing ability of three single-cone obturation systems: An in-vitro glucose leakage study. *Journal of Conservative Dentistry : JCD*. 2013;**16**(6):489-493.
12. De-Deus G, Brandao MC, Fidel RA, et al. The sealing ability of GuttaFlow trade mark in oval - shaped canals:an ex vivo study using a polymicrobial leakage model[J]. *Int Endod* J,2007, **40**(10):794-799.
13. Varun K, Harpreet S, Rajinder B, et al. Qualitative and quantitative comparative evaluation of sealing ability of guttaflow, thermoplasticized gutta percha and lateral compaction for root canal obturation: a cohort, controlled, ex-vivo study[J]. *Oral health and dental management*, 2013, **12**(3): 155-161.
14. Samiei M, Pakdel SMV, Aghazadeh M, et al. Sealing Efficacy of Single-cone Obturation Technique with MTA and CEM Cement: An in Vitro Bacterial Leakage Study[J]. Journal of Dental Research, Dental Clinics, *Dental Prospects*,2014, (2).

CELLS OPTICAL MANIPULATION WITHIN MICROFLUIDIC CHIPS

Wei Zhou[1], Xing Guo[1], Xu Zhang[2], Zhidong Li[1], Xiaochen Wang[1], Yang Li[1], Fengjiao Wang[1], and Sixiang Zhang[1,a,*]

[1]*Research Institute of Modern Measurement and Control Technology, Hebei University of Technology, Tianjin, 300130, China*

[2]*Tianjin University of Technology and Education, Tianjin, 300222, China*

E-mail: [a]zhangsx@hebut.edu.cn

[*]*Corresponding author*

This article describes the method of optical manipulation of cells within microfluidic chips. According to the theory of optical tweezers, we use gradient force and scattering force to manipulate cells. Experimental results show that the moving distance of yeast cells is related to the laser power. Meanwhile, the fiber laser could capture cell in the countercurrent cell solution. It reveals the movement principle of cells under laser beam, and it provides a basis that developing an efficient and non-invasive cell screening method.

1. INTRODUCTION

Early in the 20th century, scientists involved in the field of biology and medicine has already begun the task of cell culture. Cell culture refers to removing tissues from the living body, and then incubating it in vitro physiological environment that simulated in vivo. While traditional cell culture has been trying to create conditions to simulate the vivo situation, but also can not accurately reproduce the actual situation in vivo extracellular microenvironment. The microfluidic chip can precisely control fluid in time and space dimensions, so cell culture on a microfluidic chip has good prospects for development — the extracellular microenvironment may be strictly regulated[1-3], and even single cells were cultured observation[4].

The cell as the basic unit of life activities is the basis of life sciences and biomedical research. Cell selection provides basic information on cell-based drug screening analysis and gene expression studies[5]. The fast, high-throughput, label-free cell laser screening method can not only provide great convenience to biological research, biomedical research but also provide direct assistance to favor of early diagnosis of life processes and major diseases of[6-8]. In this paper, we have analyzed laser force of cell within microfluidic chips, and the result provides a basis to realize the microfluidic chip laser cell screening.

2. OPTICAL RADIATION PRESSURE

Radiation pressure can be analyzed as interactions by either electromagnetic waves or particles (photons). The waves and photons both have the property of momentum, which allows their interchangeability under classical conditions. Electromagnetic radiation is quantized in particles called photons, the particle aspect of its wave–particle duality. Here we using particle model: photons to resolve radiation pressure by reflection. From the perspective of quantum mechanics, although photons are zero-rest mass particles, they have the properties of energy and momentum, thus exhibit the property of mass as they travel at light speed. The momentum of a photon is given by Eq. (1).

$$\mathrm{p} = h/\lambda = mc \tag{1}$$

Where p is momentum, h is Planck's constant, λ is wavelength, m is mass, and c is speed of light in vacuum. Eq. (2) shows the wave–particle duality.

$$E = mc^2 = pc \tag{2}$$

This formula is the mass-energy relationship where E is the energy, as in Eq. (3)

$$\mathrm{p} = E/c \tag{3}$$

The generation of radiation pressure results from the momentum property of photons, specifically, changing the momentum when incident radiation strikes a surface. The surface exerts a force on the photons in changing their momentum by Newton's Second Law. A reactive force is applied to the body by Newton's Third Law.

The orientation of a reflector determines the component of momentum normal to its surface, and also affects the frontal area of the surface facing the energy source. Each factor contributes a cosine function, reducing the pressure on the surface[9]. The pressure experienced by a perfectly reflecting planar surface as in Eq. (4)

$$P_{reflect} = \frac{2E_f}{c} cos^2\alpha (N \cdot m^{-2}\ or\ Pa) \tag{4}$$

Where P is pressure, E_f is the energy flux (intensity) in W/m^2, c is speed of light in vacuum, α is the angle between the surface normal and the incident radiation.

Light pressure produced by ordinary light is faint, it's only 10^{-6}~10^{-7} N/m^2. However, laser with low divergence and high energy density characteristics can have a significant effect mechanics.

3. GRADIENT FORCE AND THE SCATTERING FORCE

Consider the case of parallel light is incident. As shown in the Fig. 1, individual rays of light emitted from the laser will be refracted as it enters and exits the dielectric bead. As a result, the ray will exit in a direction different from which it originated. Since light has a momentum associated with it, this change in direction indicates that its momentum has changed. Due to Newton's third law, there should be an equal and opposite momentum change on the particle. When the bead is displaced from the beam center (b), the larger momentum change of the more intense rays cause a net force to be applied back toward the center of the laser, so we call that gradient force. When the bead is laterally centered on the beam (a), the resulting lateral force is zero. But an unfocused laser still causes a force pointing away from the laser, which we call the scattering force.

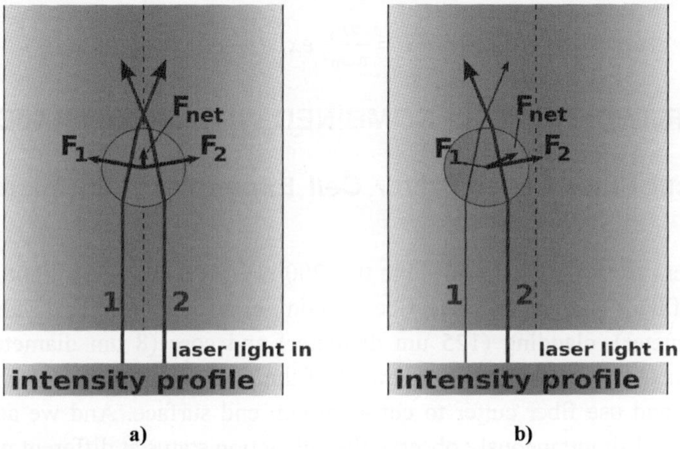

Fig. 1. Beads in symmetric and asymmetric optical field force conditions.

When the laser beam diameter greater than the diameter of the cell, in cells within the beam suffered scattering power is greater than the gradient force.

4. OPTICAL MODEL ANALYSIS

The sum of the forces from all such rays can be split into two components: the scattering force and gradient force. Scattering force is due to the reflection of photons occurs in the surface of the particles, their size is proportional to the incident light intensity, point to the incident direction of the beam. Gradient force is due to the refraction of photons generated inside the fine particles, arising from the gradient of the Gaussian intensity profile and pointing in x-y plane towards the center of the beam. The influence of light on the role of the particles has a lot of factors in laser capture operation, but mainly particle size and its shape[10]. We have known microspheres diameter and the wavelength by Mie scattering theory to calculate the angle of incidence of light.

Any change in the direction of light, by reflection or refraction, will result in a change of the momentum of the light. If an object bends the light, changing its momentum, conservation of momentum requires that the object must undergo an equal and opposite momentum change. This gives rise to a force acting on the object. Microspheres by the force may be decomposed into two components perpendicular to each other, s and g is the unit vector along the scattering force and gradient force direction. These two forces can be expressed as in Eq. (5) and Eq. (6).

$$dF_x = s\frac{n_1}{c}q_x dP \tag{5}$$

$$dF_x = s\frac{n_1}{c}q_g dP \tag{6}$$

Where n_1 is the refractive index of the medium surrounding the microspheres, c is the vacuum speed of light, dP is the power of the incident light. For a Gaussian beam, dP can use the Eq. (7) deduced.

$$dP = \mathrm{I}\cos\theta\, dS = \frac{2P_0}{\pi\omega_0{}^2} exp\left(\frac{-2r^2}{\omega^2}\right)\cos\theta\, dS \tag{7}$$

5. LASER TECHNOLOGY COMBINED WITH MICROFLUIDIC

5.1. Optical Fiber Laser Driver Cell Experiments in Microfluidic Chip

The experiment uses a power adjustable 0 ~ 200 mW 980 nm pump light source with a single-mode fiber. The single-mode fiber is composed of three parts. They are coating (250 um diameter), cladding (125 um diameter) and core (8 um diameter). The fiber pigtail is connected with the output terminal of the pump light source. Then we strip the coating layer and use fiber cutter to cut a smooth end surface. And we adjust the laser output power and simultaneously observe the cell action status at different power laser.

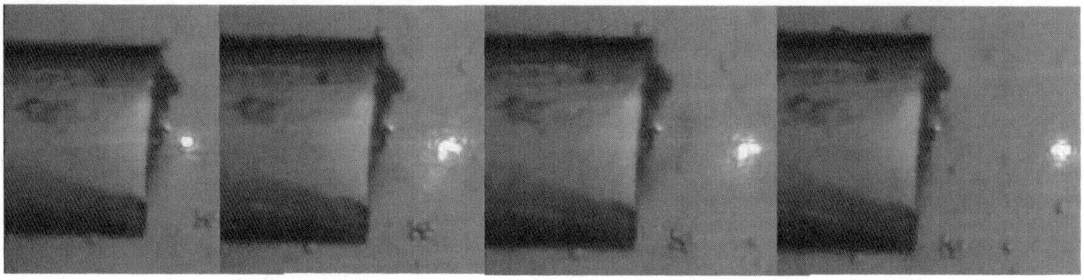

a) **b)** **c)** **d)**

Fig. 2. Yeast cell movement under different laser power role.

We tested the effect of laser on yeast cells in the experiment. The light spot in Figure 2 is a yeast cell driven by laser. Fig. 2 a), b), c), d) Four experiments were 50 mw, 100 mw, 150 mw and 200 mw laser power to the fiber diameter as a reference we can draw yeast cells is driven away from 25 um, 50 um, 71 um, 88 um. Through a lot of experiments, we verify that the yeast cells under static liquid environment can be driven by laser power whose minimum value is 50 mw. The yeast cells within 4 s moved 88 um when laser power adjusted to 200 mw, average speed is 22 um/s.

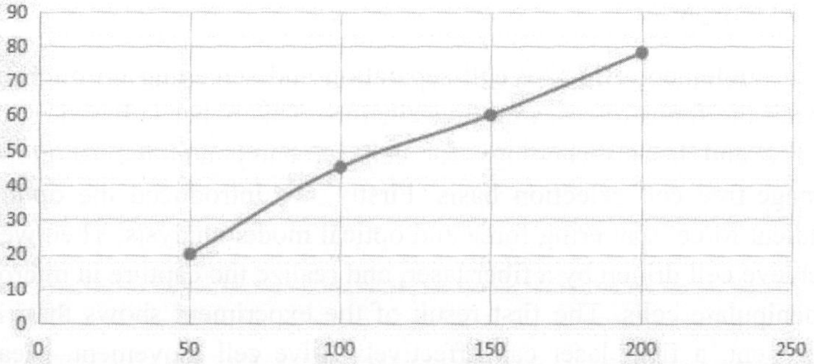

Fig. 3. The laser power and the relationship between cell locomotion distance.

Fig. 3 directly reflects the relationship of the laser power and the driven distance of yeast cells. It verifies that driving action on cells of the laser, also verifies the positive correlation between the laser power and the driven distance.

5.2. Microfluidic Cell Capture

This experiment is designed to channel width 300 um, thickness 200 um, to ensure that the optical fiber can be inserted from the left, using a microinjection pump inject yeast cell at the other end of the tunnel.

Fig. 4 is a fiber laser power 50 mw and 100 mw at capturing yeast cell experiments. Microinjection pump flow set at 5 uL/min, so that we obtain a flow rate within the passage is 1.39 mm/s.

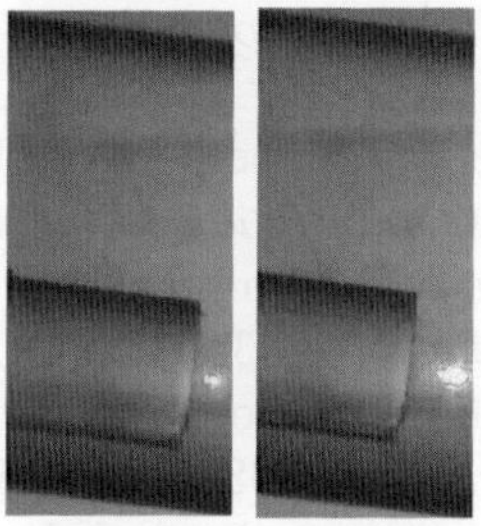

a) b)

Fig. 4. Cells capture.

Through a lot of experiments found that when laser power is about 50 mw, we successfully capture a yeast cell at a position 10 um away from the optical fiber. With the increasing of the laser power, the yeast cells are captured moving away from the fiber end. When the fiber power is increased to 200 mw, it is fixed at a position away from the end face of about 39 um.

6. CONCLUSIONS

The method to implement effective cell separation and screening on microfluidic chips has always been the focus of research investigators. This paper is based on the combine of microfluidics and laser technology for cell screening and the result provides the efficient damage-free cell selection basis. Firstly, we introduced the optical radiation pressure, gradient force, scattering force and optical model analysis. Then we proposed a method to achieve cell driven by a fiber laser, and realize the capture in microfluidic chip method to manipulate cells. The first result of the experiment shows that: in the static liquid environment, a fiber laser can effectively drive cell movement. Meanwhile, the result of the second experiment shows that: in 100 um width of microfluidic chips channel fiber laser can be utilized effectively to capture and manipulate cell in a fluid environment.

Acknowledgments

This work was supported by Hebei Natural Science Fund Project.

References

1. Ashkin A. Optical trapping and manipulation of single cell using infrared laser beams. *Nature* 1987; **33**: 256–267.
2. Ashkin A, Dziedzic J M. Optical trapping and manipulation of viruses and bacteria. *Science* 1987; **235**(4795): 1517–1520.
3. Ashkin A. History of optical trapping and manipulation of small-neutral particle, atoms, and molecules. *IEEE Journal of Selected Topics in Quantum Electronics* 2000; **6**(6): 841–856.
4. Wie Q H, Bechinger C, Leiderer P, Experimental observation of single-file diffusion of brownian particles. *Progress in Colloid and Polymer Science* 1999; **112**: 227–230.
5. Andersson H, Van den Berg A. Microfluidic devices for cellomics: a review [J]. *Sensors and Actuators B: Chemical* 2003; **92**(3): 315–325.
6. KS Bok, YS Youl, S Hyung Jin, KS Soo. Cross-Type optical particle separation in a microchannel. *Analytical Chemistry* 2008; **80**(7): 2628–2630.
7. Nagrath S1, Sequist LV, Maheswaran S, et al. Isolation of rare circulating tumour cells in cancer patients by microchip technology. *Nature* 2007; **450**(7173): 1235–1239.
8. Ophélie I. Berthuy, Loïc J. Blum, Christophe A. Marquette. Cells on chip for multiplex screening. *Biosensors and Bioelectronics* 2016; **76**(15): 29–37.
9. T. Požar, Oblique reflection of a laser pulse from a perfect elastic mirror. *Optics Letter* 2014; **39** (1), 48–51.
10. CB Chang, WX Huang, KH Lee, HJ Sung. Optical separation of ellipsoidal particles in a uniform flow. *Physics of Fluids* 2014; **26**: 062001.

STUDY OF EEG DURING DRIVING MENTAL FATIGUE BASED ON WAVELET PACKET DECOMPOSITION AND APPROXIMATE ENTROPY

Longshun Xu[a,*], Chunlin Zhao, and Min Zhao

Engineering College of Armed Police Force, Xi'an, Shanxi, 710086, China
E-mail: [a]363593659@qq.com
[*]*Corresponding author*

Mental fatigue is an important contributor to some serious transportation crashes. Ten male subjects volunteered in this study by performing 90 min of driving task on a static simulator and the EEG signal was collected. ApEn(Aproximate Entropy) of EEG and the four bands was calculated. ApEn of EEG decreased from 0.70 to 0.41($p<0.01$). ApEn of beta band decreased from 0.63 to 0.45($p<0.01$) and delta band decreased slightly, while theta and alpha bands hardly showed change. The ApEn of original EEG and beta band in Temporal and prefrontal regions reduced the most obviously. ApEn might be an indicator to measure the degree of driving mental fatigue and the change is mainly reflected in the beta band, while Temporal and prefrontal regions are more sensitive. This helps deeply understand the change of EEG when mental fatigue occurs.

1. INTRODUCTION

As the Psychophysiological phenomena during cognitive and behavioral control process of brain, mental fatigue has great negative impact on the safety when driving a car. Research the detection technology of driving fatigue has a very important significance for reducing the frequency of accidents.

Previous studies mainly detect driving fatigue from two aspects. One is analysis vehicle parameters and then fatigue level of the driver is distinguished. The technology is relatively mature and costs lowly, but the detection accuracy is not enough and vulnerable to other influence. Physiological signals and behavior characteristics include facial features, eyes' blinking frequency are also used to determine the fatigue level. The image or video of driver are intercepted through a camera and characteristic of the information can be used to distinguish driving fatigue. But it's useless when the driving environment in low light or driver's head deviates from the normal position [1-3]. It can reflect mental fatigue objectively and effectively through physiological signal features including EEG, ECG and EMG [4-6]. But the physiological signal acquisition devices are relatively expensive and have a greater disturbance to the driver. As portable wireless acquisition devices for physiological signal and the brain-computer interface technology become research hotspots, cost for the devices and resentment of user are greatly reduced.

EEG signal can objectively reflect the real state of the driver's brain activity and show some changes during fatigue state. It usually can be divided into four different bands (delta, theta, alpha and beta). We can determine the status of driver by studying different changes in frequency. The main component of EEG is beta band when driver is in high alert. Otherwise, alpha band will be enhanced. Delta band is intensive during a state of drowsiness [7]. The ratio of energy in different bands changed significantly with time during a simulated driving task [8]. EEG has been thought present a good non-linear characteristic. Complexity parameters such as complexity and ApEn have been used to analyze EEG signal and concerned in the biological medical field [9, 10].

Self-reporting measures from the subjects and the reaction time of Oddball task could be used to verify that 90 min simulated driving task would induce mental fatigue in the subjects. In this paper, ApEn of EEG and the four frequency bands were applied to estimate driving mental fatigue under the simulated condition. The non-linear analysis may helpful to understand the change of EEG during driving mental fatigue.

2. METHOD

2.1. Experimental Design

EEG used in the paper was recorded by Neuroscan system (Synamps2, Scan 4.3, El Paso, TX, USA). Related research proved reliability of driving simulation system [11]. Fatigue can be classified as mental and physical depending on its cause. Physical fatigue is a bodily weakness that can be caused by repetitive muscle activity [12]. But mental fatigue is observed as a reduced efficiency for mental tasks. Boring highway driving tasks was simulated in the experiment in order to make the subjects in a state of driving mental fatigue. To assess the sustaining attention level of drivers, an Oddball task needs to be done at the beginning and end of the driving task. Ten male volunteers from the Xi'an Jiao Tong University carried out the driving task that lasted 90 min. Please consult the tenth reference about the detail of the experiment [13].

2.2. Data Processing

The purpose of the pre-processing was to obtain clear EEG signal and raise computing speed for feature extraction. And then EEG signal was downsampled to 250 Hz. The 30s EEG data segments at the beginning and end of the driving task was chosen for analysis. All the data operations were performed in the matlab2014a. All features were analyzed using SPSS (SPSS19.0.0) for paired t-test and the significance level was defined as 0.05.

2.2.1. *Wavelet Packet Decomposition*

Wavelet packet decomposition (WPD) is extended from the wavelet decomposition (WD). Both the detail and approximation coefficients can be decomposed using WPD [14]. Daubechies ('db4') was used as the mother wavelet and the EEG was decomposed by a 7-level WPD. The minimum frequency resolution can be estimated by the following formula:

$$\Delta f = \frac{1}{2^7} \cdot \frac{f_s}{2} = 0.9766 \qquad (1)$$

Wavelet packet coefficients for each band were added and then the delta band (0.25–4 Hz), theta band (4–8 Hz), alpha band (8–13 Hz) and beta band (13–30 Hz) were reconstructed respectively.

2.2.2. *Approximate Entropy*

Approximate entropy (ApEn) is a statistic technique used to quantify complexity or irregularity of both deterministic and stochastic signals. It was first proposed by Pincus in 1991 [15]. ApEn can characterize incidence of new information of EEG. A non-negative number was used to measure the complexity.

Two parameters that were used when computing ApEn were specified at first. The embedding dimension of the vector to be formed called m usually takes a value of 2. Threshold called r typically takes a value between 0.1 and 0.25 standard deviation (SD) of the original data. In this study, m takes the value 2 and r is o.2 SD. The number of input data points on every eletrode takes a value of 7500. TheApEn of EEG on every eletrode was calculated using computing window lengths of 1250 points and shifted the window (step length) by 25 points. The ApEn was all normalized after computed.

3. RESULTS

3.1. Self-report and Oddball

Subjects began feeling tired after only 30 min of driving and the degree of fatigue gradually deepened. Subjects were expressed feeling tired, bored and drowsy at the end of driving task. According to Li's evaluation score, the mental fatigue scores increased from 1.31 to 5.46 ($p<0.005$). Average reaction time of Oddball tasks increased from 356.35ms 376.65ms and the error rate increased slightly. These symptoms indicated that subjects were mental fatigue after performing a 90 min simulated driving task.

3.2. ApEn of EEG

The values were color-coded and plotted to produce a continuous color map. The mean ApEn of ten subjects is showed in Fig.1. It shows decreasing activity in all regions and extremely obvious in the prefrontal and temporal regions at the end of the driving task. The mean of 30 electrodes decreased from 0.70 to 0.41 ($p<0.01$).

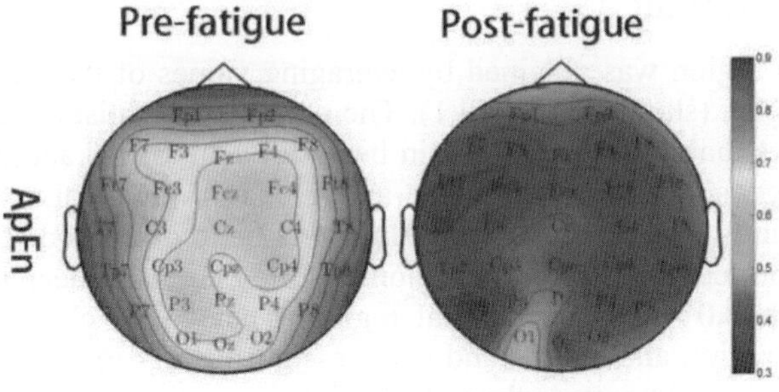

Fig.1. Brain topography of mean ApEn.

The mean ApEn of ten subjects in the four bands is showed in Fig.2. Values decreased obviously in beta band while have no noticeable change in theta and alpha bands. ApEn in delta band decreases slightly after driving task.

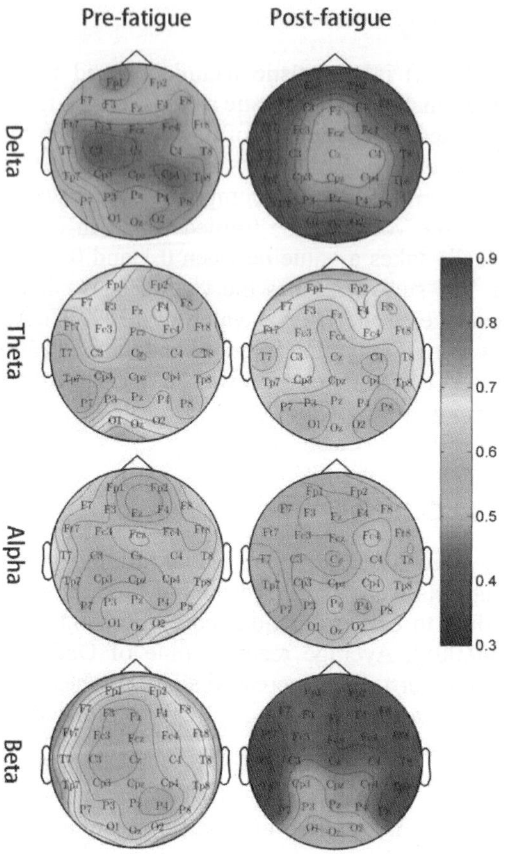

Fig.2. Brain topography of mean ApEn computed by the four frequency bands.

3.3. Statistical Result

The ApEn of each region was obtained by averaging values of the electrodes that were included in this region (showed in Table 1). The p-values of statistical analysis are listed in Table 1. It shows that ApEn decreased in beta and delta bands and the original EEG, but did not change significantly in theta and alpha bands. ApEn of original EEG decreased significantly in all regions ($p<0.01$). Delta band decreased significantly in the frontal ($p<0.01$) and temporal ($p<0.05$) regions, while beta band decreased significantly in other regions ($p<0.01$) except occipital region. Theta band change significantly in occipital ($p<0.05$) region and alpha band change significantly in temporal ($p<0.01$) and central ($p<0.01$) regions. However, the values of ApEn in theta and alpha bands did not increase or decrease obviously.

Table 1. Mean (SD) values of ApEn from EEG and the four bands

Features		Frontal	Temporal	Central	Parietal	Occipital
EEG	1	0.70(0.10)	0.80(0.12)	0.65(0.12)	0.69(0.13)	0.67(0.16)
	2	0.40(0.12)**	0.39(0.13)**	0.42(0.11)**	0.42(0.10)**	0.45(0.14)**
δ	1	0.49(0.11)	0.50(0.09)	0.45(0.06)	0.49(0.07)	0.48(0.08)
	2	0.43(0.10)**	0.41(0.09)*	0.49(0.11)	0.45(0.13)	0.41(0.13)
θ	1	0.64(0.07)	0.61(0.07)	0.62(0.06)	0.62(0.07)	0.67(0.09)
	2	0.65(0.06)	0.64(0.05)	0.64(0.07)	0.61(0.04)	0.61(0.07)*
α	1	0.60(0.07)	0.63(0.06)	0.60(0.11)	0.61(0.07)	0.63(0.10)
	2	0.58(0.07)	0.56(0.08)**	0.59(0.08)**	0.60(0.06)	0.62(0.11)
β	1	0.64(0.11)	0.67(0.12)	0.60(0.16)	0.62(0.15)	0.64(0.14)
	2	0.42(0.10)**	0.40(0.13) **	0.47(0.11) **	0.48(0.10) **	0.56(0.13)

1 means pre-fatigue and 2 means post-fatigue.

*p<0.05, **p<0.01

4. DISCUSSION

According to self-report, reaction time in Oddball task and the decrease of ApEn, it is obvious that subjects were mental fatigue after accomplish the 90 min simulated driving task. The complexity of brain activity decreased with the increasing time of driving simulator task, which would lead to the subjects scatterbrained. The ApEn of original EEG decreased and showed significant differences in the five regions. It meant that subjects were extremely mental fatigue after completing simulated driving task and responded slowly to the change of road and the vehicles around. Enhanced regularity of EEG was embodied in the decrease of ApEn.

It has been studied that EEG showed significant increase of alpha and theta activities and a significant decrease of beta activity in the state of mental fatigue if EEG was evaluated using relative wavelet packet coefficient energy [16]. Previous studies also showed the same result [13]. In this study, ApEn decreased significantly in beta bands and decreased slightly, while showed no noticeable change in theta and alpha bands. The results showed differences compared with the studies from the standpoint of energy mechanism. The reason may be that the nonlinear variation of EEG mainly embodies in the beta band. A decrease of EEG power in beta bands were observed during the transition from increased mental workload to mental fatigue. The complexity of beta band also decreased largely. Although an increase of EEG power in theta and alpha bands were observed in the state of mental fatigue, ApEn did not show significant differences because of the decrease complexity of EEG.

It is widely believed that delta band appeared during deep sleep state [17] and did not show obvious difference between the pre-fatigue state and post-fatigue state [7]. The largest difference between the non-fatigued and fatigued group occurs in the lowest EEG frequency band when analyze the coherence of EEG [18].Mental fatigue is supposed to affect higher-level cognitive processes which are associated with lower EEG frequencies, so delta band may show change when driver are mental fatigue. This may explain why ApEn of delta band decreased in this study. Maybe the signal of blink and body movement were not removed thoroughly and mixed in the delta band. So it showed decrease after subjects were mental fatigue. The change of delta band need to be further researched.

Table 1 show that ApEn of the original EEG and beta band decreased most obviously in temporal region. We can easily know that the features showed more largely difference in prefrontal and temporal regions from Fig.1 and Fig.2. Different regions of the brain are relevant to various functionalities. The frontal region is closely connected with planning, reasoning, movement, emotion and problem solving. The parietal region is involved in movement, recognition, perception of stimuli and temporal region is associated with recognition and perception of auditory stimuli, memory, and speech [6]. The function of temporal and prefrontal regions may be of paramount importance in the driving task, which may explain the larger decrease of ApEn.

5. CONCLUSION

In this study, ApEn of EEG and the four bands were used to measure the mental fatigue caused by simulated driving task. The ApEn of EEG and beta band showed a significant decrease. ApEn can be used to measure the degree of mental fatigue. Beta band show a more decrease than other bands. The ApEn from prefrontal and temporal are more sensitive than other regions. There are several limitations in this study. First, the number of subjects is small to get general results. Second, we did not make forceful constraints that the subjects were not allowed to do some other task totally unrelated to driving.

References

1. Sigari MH, Pourshahabi MR, Soryani M, et al. A Review on Driver Face Monitoring Systems for Fatigue and Distraction Detection. *International Journal of Advanced Science and Technology* 2014; **64**(7): 73-100.
2. Jayanthi D, Bommy M. Vision-based Real-time Driver Fatigue Detection System for Efficient Vehicle Control. *International Journal of Engineering and Advanced Technology* 2012; **2**(1): 238-242.
3. Mbouna RO, Kong SG, Chun MG. Visual analysis of eye state and head pose for driver alertness monitoring. *Transaction on Intelligent Transportation Systems* 2013; **14**(3): 1462-1469.
4. Yang G, Lin Y, Bhattacharya P. A driver fatigue recognition model based on information fusion and dynamic Bayesian network. *Information Sciences* 2010; **180**(10): 1942-1954.
5. Borghini G, Vecchiato G, Toppi J, et al. Assessment of mental fatigue during car driving by using high resolution EEG activity and neurophysiologic indices. *34th Annual International Conference of the IEEE EMBS* 2012: 6442-6445.
6. Kar S, Bhagat M, Routray A. EEG signal analysis for the assessment and quantification of driver's fatigue. *Transportation Research Part F: Traffic Psychology and Behaviour* 2010; **13**(5): 297-306.
7. Borghini G, Astolfi L, Vecchiato G, et al. Measuring neurophysiological signals in aircraft pilots and car drivers for the assessment of mental workload ,fatigue and drowsiness. *Neuroscience and Biobehavioral Reviews* 2014; **44**: 58-75.
8. Jap BT, Lal S, Fischer P, et al. Using EEG spectral components to assess algorithms for detecting fatigue. *Expert Systems with Applications* 2009; **36**(2): 2352-2359.
9. Acharya UR, Sree SV, Chattopadhyay S, et al. Automated diagnosis of normal and alcoholic EEG signals. *International journal of neural systems* 2012; **22**(3): 432-449.

10. Azarnoosh M, Motie NA, Mohammadi MR, et al. Investigation of mental fatigue through EEG signal processing based on nonlinear analysis: Symbolic dynamics. *Chaos, Solitons & Fractals* 2011; **44**(12): 1054-1062.
11. Shechtman O, Classen S, Awadzi K, et al. Comparison of Driving Errors Between On-the-Road and Simulated Driving Assessment: A Validation Study. *Traffic Injury Prevention* 2009; **10**(4):379-385.
12. Tanaka M, Shigihara Y, Ishii A, et al. Effect of mental fatigue on the central nervous system: an electroencephalography study. *Behavioral and Brain Functions* 2012; **8**(2): 262-270.
13. Zhao C L, Zhao M, Liu J P, et al. Electroencephalogram and electrocardiograph assessment of mental fatigue in a driving simulator. *Accident Analysis & Prevention*, 2012; **45**(1): 83-90.
14. Xue J Z, Zhang H, Zheng C X, et al. Wavelet packet transform for feature extraction of EEG during mental tasks, Presented at Proceedings of the Second International Conference on Machine Learning and Cybernetics. *International Conference on Machine Learning & Cybernetics* 2003; **1**(1):360-363.
15. Pincus S. Approximate entropy (ApEn) as a complexity measure. *CHAOS*, 1995; 5(1): 110-117.
16. Jagannath M, Balasubramanian V. Assessment of early onset of driver fatigue using multimodal fatigue measures in a static simulator. *Applied Ergonomics* 2014; **45**(4): 1140-1147.
17. Wang Q, Cheng J, Liu Z W. A novel sleep staging method of EEG signals. Space Medicine & Medical Engineering 2015; **28**(1): 22-27.
18. Ten Caat M, Lorist MM, Bezdan E, et al. High-density EEG coherence analysis using functional units applied to mental fatigue. *Journal of Neuroscience Methods* 2008; **171**(2): 271-278.

Session 3

Food Hygiene, Environment and Human Health

SYNTHESIS AND FUNGICIDAL ACTIVITY OF FURFURYL AMINE DERIVATIVES

Yan Li[a,*] and Geng Li

School of Nuclear Technology and Chemistry & Biology, Hubei University of Science and Technology, Xianning, 437100, China

E-mail: [a]liyan750607@163.com

[*]Corresponding author

Some new furfuryl amine derivatives have been prepared via the reaction of 3-cyano-4-ethoxycarbonyl-5-methyl furfuryl amine and 3-cyano- 4, 5-dimethyl furfuryl amine with aromatic aldehydes, then through reduction with $NaBH_4$. Structures of the products were established on the basis of IR, ^{1}HNMR, EI-MS. The preliminary bioassays indicate that the products possess certain fungicidal activities against five species of fungi at a concentration of 200 mg/L.

1. INTRODUCTION

Rice blast is known as a rice cancer on biological disaster and the main rice diseases in the world, and it has the characteristics of wide distribution, fast epidemic, great harm and difficult to control. It is mainly caused by Magnaporthe grisea. The black pigment of cells attached to the pathogenic bacteria is the essential cell structure in the process of pathogen infection, so it is an effective way to control the disease that inhibit the biosynthesis of the pathogenic bacteria[1-6]. At the present, the chemical methods are mainly on controlling the rice blast, however in our country due to the long-term use of traditional Tricyclazole, Fuji No. 1, Tricyclazole scattered, (ISO) of rice blast net control, the growing resistance of rice blast fungus to become the most prominent problems faced in the current rice production, which requires that pesticide workers are constantly looking for new inhibitors. It has been found that some amide compounds and heterocyclic amino-nitrile groups showed high inhibitory activity[7-10]. On continuing our interest in the biological activities of heterocyclic amino-nitrile group, herein we describe the preparation of new furfuryl amine derivatives. Their structures were characterized by IR, ^{1}HNMR, El-MS, and their fungicidal activities were investigated. The Schiff bases(II)/Schiff bases(V) were formed by the reaction of 3-cyano-4-ethoxycarbonyl-5-methyl furfuryl amine(I)/3-cyano-4, 5-dimethyl furfuryl amine(IV)[8] with aromatic aldehydes, which were reduced by $NaBH_4$ to provide new 3-cyano-4-ethoxycarbonyl-5-methyl furfuryl amine derivatives(III)/3-cyano- 4, 5-dimethyl furfuryl amine derivatives(VI). Synthetic route is shown in Fig. 1 and compounds are shown in Table 1.

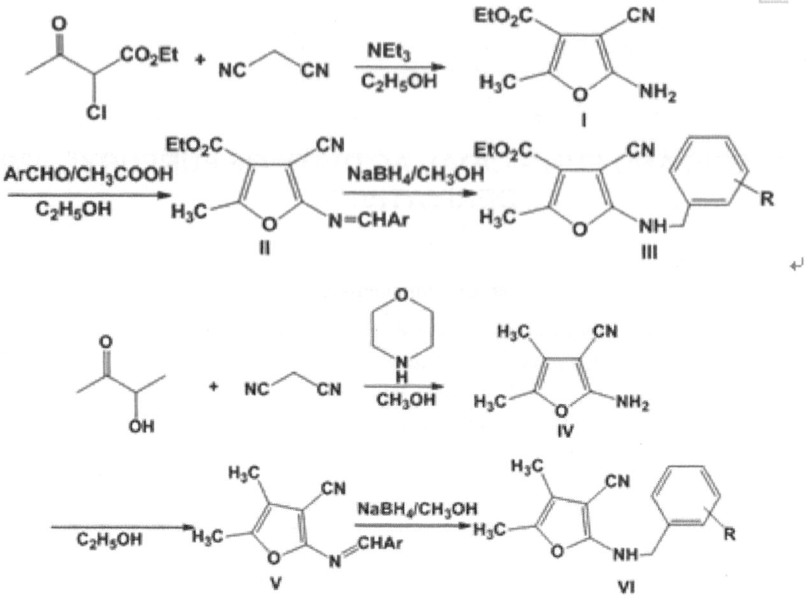

Fig. 1. The synthetic route of furfuryl amine derivatives.

Table 1. Preparation of furfuryl amine derivatives

Compound	R	Compound	R
IIIa	H	IIIf/VIf	*p-t*-Butyl/ *p*-OH
IIIb/VIb	*p*-NO_2	IIIg/VIg	*p*-OCH_3
IIIc/VIc	*o*-NO_2	IIIh/VIh	*o*-OH
IIId/VId	*p*-Cl	IIIi/VIi	*m*-OCH_2O-*p*
IIIe/VIe	*o*-Cl	IIIj	*o*-Cl,*p*-Cl

2. EXPERIENTAL

2.1. Instrumentation

The melting point was determined on a X4 type melting point instrumenta. Infrared spectra were measured on Perkin-Elemer PE-983. Infrared spectrometer (cm^{-1}) was using KBr disks. ^{1}HNMR spectra was measured on Varian Mercury-400 spectrometer (400 MHz), using $CDCl_3$ as a solvent and tetramethylsilane (TMS) (δ) as the internal standard. Mass spectra recorded on Finnigan Tracea instrument at 70 eV. The purity of the synthesized compounds was checked by thin layer chromatography (TLC). The reagents used were domestic (or imported) chemical pure or analytical.

2.2. Synthesis

2.2.1. *3-cyano*-4-ethoxycarbonyl-5-methyl furfuryl amine (I)

According to the reference 8, a mixture of malononitrile (10.00 mmol) and ethyl 2-chloroacetoacetate (10.00 mmol) in absolute ethanol (50 mL) under the catalysis of triethylamine (5 mL) was refluxed under the condition of ice water bath for 3 h. The obtained solids were filtered off and crystallized to give 3-cyano-4-ethoxycarbonyl-5-methyl furfuryl amine(I), which was white solid. The yield was 90% and the melting point was 224~225°C (literature value[8]: 224~225°C)

2.2.2. *3-cyano-4-ethoxycarbonyl-5-methyl furfuryl amine Schiff bases (II)*

A mixture of 3-cyano-4-ethoxycarbonyl-5-methyl furfuryl amine(I) (2 g) and aromatic aldehydes (10.00 mmol) was refluxed for 24 h in absolute ethanol (20 mL) containing of glacial acetic acid (8~10 drop). The obtained solids were filtered off and crystallized by using dichloromethane to give 3-cyano-4-ethoxycarbonyl-5-methyl furfuryl amine Schiff bases (IIa-IIj) which were orange solid in good yield (70~80%), respectively.

2.2.3. *3-cyano-4- ethoxycarbonyl-5-methyl furfuryl amine derivatives(III)*

3-Cyano-4-ethoxycarbonyl-5-methyl furfuryl amine Schiff bases (IIa-IIj) (3 mmol) were reduced by NaBH4 (6 mmol) in anhydrous methanol (10 mL) for 3~4 h. The obtained solids were filtered off and crystallized by using methanol or dichloromethane/petroleum ether to give 3-cyano-4-ethoxycarbonyl-5-methyl furfuryl amine derivatives (IIIa -IIIj), respectively. The details for synthesis of furfuryl amine derivatives (IIIa -IIIj) are shown in Table 1.

2.2.4. *3-cyano-4-5-dimethyl furfuryl amine(IV)*

According to the reference 8, a mixture of malononitrile (10.00 mmol) with 2-butanone-3-hydroxy (10.00 mmol) in anhydrous methanol (50 mL) under the catalysis of Ma Feilin (5 mL) was refluxed under the condition of ice water bath for 3 h. The obtained solids were filtered off and crystallized to give 3-cyano- 4, 5-dimethyl furfuryl amine(IV), which was white solid. The yield was 95% and the melting point was 134~135°C (literature value[8]: 134~135 oC).

2.2.5. *3-cyano-4,5-dimethyl furfuryl amine(V)*

A mixture of 3-cyano- 4, 5-dimethyl furfuryl amine(IV) (1 g) and aromatic aldehydes (8 mmol) in absolute ethanol (20 mL) containing of glacial acetic acid (8~10 drop) was refluxed for 12 h. The obtained solids were filtered off and crystallized by dichloromethane to give 3-cyano- 4, 5-dimethyl furfuryl amine Schiff bases (Vb-Vi), which were orange solid in good yield (80~85%), respectively.

2.2.6. *3-cyano-4,5-dimethyl furfuryl amine(VI)*

3-cyano- 4, 5-dimethyl furfuryl amine Schiff bases (Vb-Vi) (3 mmol) were reduced by NaBH4 (6 mmol) in anhydrous methanol (10 mL) for 3~4 h. The obtained solids were filtered off and crystallized by methanol or dichloromethane/petroleum ether to give 3-cyano- 4, 5-dimethyl furfuryl amine derivatives (VIb-VIi), respectively. The details for synthesis of furfuryl amine derivatives (VIb-VIi) are presented in Table 2.

Table 2. Physical properties of furfuryl amine derivatives

Compounds	Yield (%)	Crystal color	Melting point (°C)
IIIa	75	white	134~135
IIIb	77	——	181~182
IIIc	78	——	101~102
IIId	80	——	104~105
IIIe	75	——	109~110
IIIf	73	——	95~96
IIIg	80	——	93~94
IIIh	79	——	146~147
IIIi	75	——	126~128
IIIj	72	——	112~114
IVb	85	——	125~126
IVc	74	——	160~161
IVd	90	——	90~91
IVe	76	——	103~105
IVf	80	——	136~137
IVg	82	——	89~90
IVh	75	——	106~107
IVi	86	——	82~83

3. RESULTS AND DISCUSSION

3.1. Reaction Conditions

In the literature 8, 3-cyano-4-ethoxycarbonyl-5-methyl furfuryl amine (I) and 3-cyano- 4, 5-dimethyl furfuryl amine (IV) were synthesized using ethanol sodium as catalyst. The reaction is uneasy to control and operate, and the yield was obtained in only 30~40%. By comparison, the reaction is mild and the yield is high (yield up to 95%) using triethylamine as catalyst in the synthesis of (I) and using Ma Feilin as catalyst in the synthesis of (IV).

Intermediate Schiff bases (II)/(V) was reduced by NaBH4 which should be divided into the reaction bottle under stirring conditions at room temperature. In order to prevent hydrogen spillover, plastic bags (or balloons) should be fit in the reaction flask mouth.

After the completion of the reaction solvent is removed and the obtained solids were crystallized by methanol or dichloromethane/petroleum ether to give the target compound (III)/(IV).

3.2. Spectral Characteristics of Target Compounds

Target compounds were confirmed by IR, HNMR and EI-MS. In 1HNMR of target compounds, the chemical shift of hydrogen was on the aromatic ring show multiple peaks between 6.90~8.25, and the chemical shift of hydrogen on the nitrogen is a single peak at about 5.4. In IR spectra of target compounds, The characteristic groups of the products have obvious absorption peaks, the stretching vibration peak of the C=O of the ester is about 1705 cm-1, and the stretching vibration peak of N-H bond is about 3310 cm-1. In MS spectra, target compounds have strong molecular ion peaks, and major fragment ion peaks are in agreement with the expected target molecular structure. These characteristics have confirmed the molecular structure of the target compounds and the details are presented in Table 3 and Table 4.

Table 3. IR and MS dataof furfuryl amine derivatives

Compouds	IR (KBr, ν/cm^{-1})	MS (70eV), *m/z* (%)
IIIa	3303(NH),1711(C=O),1660(benzene C=C), 1351(C-N)	284(M$^+$, 26), 270(7), 245(59), 208(18),180(18), 158(27), 108(100), 80(25)
IIIb	3303(NH),1705(C=O),1660(benzene C=C), 1355(C-N)	329(M$^+$, 26), 277(7), 252(59), 215(18), 187(18), 165(27), 105(100), 77(25)
IIIc		329(M$^+$, 10), 321(13), 273(10), 197(19), 165(36), 136(84), 120(76), 106(100), 78(93), 65(37)
IIId		318(M$^+$, 11), 304(5), 216(24), 165(15), 125(100), 105(98), 89(15), 77(16)
IIIe	3300(NH),1705(C=O),1658(benzeneC= C) , 1 353(C- N)	318(M$^+$, 25), 165(9), 125(100), 105(12), 91(88), 77(7)
IIIf		340(M$^+$,14), 327(4), 150(100), 132(13), 117(14), 105(18), 91(6),77(4)
IIIg		
IIIh		3l4(M$^+$, 19), 269(5), 165(17), 121(100), 106(10), 91(21), 77(40)
IIIi		300(M$^+$, 47), 194(79), 165(75), 149(11), 107(100), 77(61)
IIIj		328(M$^+$, 18), 283(11), 239(13), 165(11), 135(100), 105(24), 77(45)
IVb		352(M$^+$, 6), 309(2), 159(100), 125(4), 89(4)
IVc		269(M$^+$, 11) , 208(9) , 154(24) , 120(15) , 89(100) , 77(98) , 63(16)
IVd	3303(N-H), 1711(C=O), 1660(benzeneC=C),1351(C-N)	269(M$^+$, 11) , 208(9) , 154(24) , 120(15) , 89(100) , 77(98) , 63(16)
IVe	3303(N-H), 1660(benzeneC=C), 1355(C-N)	259(M$^+$, 10) , 197(19) , 165(36) , 136(84) , 120(76) , 106(100) , 78(93)
IVf	3 303(NH) , 1 660(benzeneC= C) ,	

IVg	1 351(C- N)	259(M^+, 10) , 197(19) , 165(36) , 136(84) , 120(76) , 106(100) , 78(93)
IVh	3308(NH), 1658(benzene C= C) , 1 352(C- N)	242(M^+ , 81) , 199(16) , 134(30) , 122(64) , 107(74) , 77(100) , 51(50)
IVi	3300(NH), 1658(benzeneC= C) , 1 350(C- N)	254(M^+ ,47) , 194(79) , 165(75) , 149(11) ,107(100) , 77(61)
	3300(NH), 1658(benzene C= C) , 1 350(C- N)	242(M^+, 100) , 199(71) , 156(17) , 169(35) ,134(38) , 122(16) , 77(50) , 51(26)
	3 430(OH), 3303(NH), 1660(benzeneC=C),1351(C-N)	268(M^+ , 19) , 197(19) , 165(36) , 121(100) , 106(10) , 65(17) , 91(21) , 77(40)
	3 302(NH) , 1 664(benzeneC= C) , 1 360(C- N)	
	3430(OH),3303(NH),1660(benzeneC= C), 1351(C- N)	
	3 304(NH), 1 659(benzene C= C) , 1 353(C- N)	

Table 4. 1HNMR dataof furfuryl amine derivatives

Compouds	^{1}HNMR($CDCl_3$, 400 MHZ, δ)
IIIa	7.26 ~7.39 (m, 5H, Ar-H), 5.02 (s ,1H, -NH-), 4.56~4.57(m, 2H, -CH_2N), 4.30~4.33(m , 2H, -CH_2O), 2.46(s, 3H, -CH_3), 1.36~1.38(m, 3H, CH_3-CH_2-)
IIIb	7.51~8.25(m, 4H, Ar-H), 5.33(S, 1H,-NH-), 4.68~4.69(m, 2H, - CH_2N), 4.30~4.33(m, 2H, -CH_2O), 2.45(S,3H, -CH_3), 1.35~1.37(m, 3H , CH_3-CH_2-)
IIIc	7.51~8.14(m, 4H, Ar-H), 5.70(s, 1H, -NH-), 4.88~4.89(m, 2H, -CH_2N), 4.29~4.33(m, 2H, -CH_2O), 2.45(s, 3H,

	-CH_3), 1.35~ 1.37(m, 3H, CH_3-CH_2-)
IIId	7.25~7.35(m, 4H, Ar-H), 5.21(s, 1H, -NH-), 4.53~4.54(m, 2H, -CH_2N), 4.29~4.33(m, 2H, -CH_2O), 2.45(S, 3H, -CH_3), 1.35~1.37(m, 3H, CH_3-CH_2-)
IIIe	7.26~7.43(m, 4H, Ar-H), 5.40(s, 1H, -NH-), 4.64~ 4.65(m, 2H, -CH_2N), 4.28~4.32(m , 2H, -CH_2O), 2.45(s, 3H, -CH_3), 1.34~1.37(m, 3H, CH_3-CH_2-)
IIIf	7.25~7.40(m, 3H, Ar-H), 4.99(s, 1H, -NH-), 4.52~4.53(m, 2H, -CH_2N), 4.31~4.32(m, 2H , -CH_2O), 2.47(s, 3H, -CH_3), 1.36~1.38(m, 3H, CH_3-CH_2-), 1.32 [s, 9H, -C(CH_3)$_3$]
IIIg	6.89~7.26(m, 4H, Ar-H), 5.00(s, 1H, -NH-), 4.49~4.50(m, 2H, -CH_2N), 4.29~4.33(m, 2H , -CH_2O), 2.46(s, 3H , -CH_3), 1.35~ 1.38(m, 3H, CH_3-CH_2-), 3.81(s, 3H, CH_3O-)
IIIh	6.83~7.26(m, 4H, Ar-H), 5.46(s, 1H, - NH-), 4.54~4.55(m, 2H , -CH_2N), 4.29~4.33(m, 2H, -CH_2O), 2.46(m, 3H, -CH_3), 1.35~1.37(m, 3H, CH_3-CH_2-), 6.09(m, 1H, -OH)
IIIi	6.83~6.84(m, 3H, Ar-H), 5.33(s, lH, -NH-), 4.49~4.50(m, 2H , -CH_2N), 4.34~4.37(m, 2H ,-CH_2O),2.50(s, 3H , -CH_3), 1.40~1.42(m, 3H , CH_3-CH_2-) ,6.0l(s, 2H,-OCH_2 O-)
IIIj	7.25~8.14(m, 3H, Ar-H), 5.38(s, 1H, -NH-), 4.61~4.62(m, 2H, -CH_2N), 4.29~4.31(m, 2H , -CH_2O), 2.45(s, 3H, -CH_3), 1.35~1.37(m, 3H, CH_3-CH_2-)
IVb	7. 26~ 8. 23(m, 4H, Ar-H) , 5. 22(s, 1H, - NH-) , 4. 42- 4. 43(m, 2H, -CH_2N) , 1. 91(s, 3H, -CH_3) , 2. 06(s, 3H, -CH_3)
IVc	7. 27~ 8. 13(m, 4H, Ar-H) , 5. 45(s, 1H, - NH-) , 4. 83~ 4. 84(m, 2H, - CH_2N) , 1.90(s, 3H, - CH_3) , 2. 05(s, 3H, - CH_3)
IVd	7. 24~ 7. 33(m, 4H, Ar- H) ,4. 66(s, 1H, - NH-) , 4. 41- 4. 44(m, 2H, - CH_2N) , 1. 32(s, 3H, - CH_3) , 2. 28(s, 3H, - CH_3)
IVe	7. 26~ 7. 44(m, 4H, Ar-H) , 5. 03(s, 1H, - NH-) , 4. 61~ 4. 62(m, 2H, - CH_2N) , 1.91(s, 3H, - CH_3) , 2. 06(s, 3H, - CH_3)
IVf	7. 15~ 7. 26(m, 4H, Ar-H) , 5. 33(s, 1H, - NH-) , 4. 40~ 4. 58(m, 2H, - CH_2N) ,1.91(s, 3H, - CH_3) , 2. 07(s, 3H, - CH_3) , 6. 12 (s, 1H, - OH)
IVg	
IVh	
IVi	

	6. 89~ 7. 26(m, 4H, Ar-H) , 5. 00(s, 1H, - NH-) , 4. 49~ 4. 50(m, 2H , - CH_2N) ,1.92(s, 3H, - CH_3) , 2. 22(s, 3H, - CH_3) , 3. 81(s, 3H, CH_3O-)
	6. 82~ 7. 26(m, 4H, Ar-H) , 5. 12(s, 1H, - NH-) , 4. 44~ 4. 53(m, 2H, - CH_2N) , 1.90(s, 3H, - CH_3) , 2. 07(s, 3H, - CH_3) , 6. 09 (s, 1H, - OH)
	6.77~6.81(m, 3H, Ar-H), 4.86(s, lH, -NH-), 4.42~4.43(m, 2H , -CH_2N),1.91(s, 3H , -CH_3), 2.21(s, 3H , -CH_3) ,3.98(s, 2H,-OCH_2 O-)

3.3. Fungicydal Activitie

Using the method of containing toxic medium (200 mg/L), the fungicidal activities of the target compounds were initially tested (Table 5) on acetone. The results indicated that the compound (III) show the poor inhibitory activity against 5 species of Rice sheath blight, Rice blast, Cucumber downy mildew, Cucumber powdery mildew and Botrytis cinerea, but the compound(IV) show better inhibitory activity, and the inhibition rate of compound (IVc/IVd) on rice blast could reach to 60%.

Table 5. Inhibition rate of target compounds (III)

Compound	Cucumber downy mildew /%	Cucumber powdery mildew /%	Botrytis cinerea /%	Rice sheath blight /%	Rice blast /%
IIIa	0	23.12	0	0	0
IIIb	10.15	0	0	0	0
IIIc	0	21.87	0	33.17	40.56
IIId	13.73	0	0	0	0
IIIe	0	24.35	20.19	0	24.66
IIIf	0	13.55	0	0	0
IIIg	56.13	34.55	0	15.44	19.87
IIIh	0	0	0	0	30.31
IIIi	0	23.12	0	0	0
IIIj	10.15	0	0	0	0
IVb	42.6	52.6	43.9	48.4	60.6
IVc	40.6	52.1	45.9	53.1	56.0
IVd	35.2	31.9	28.1	30.6	43.1
IVe	26.8	30.2	30.5	33.4	38.2
IVf	28.9	36.3	36.3	40.4	42.6
IVg	28.1	36.1	39.4	40.3	42.6
IVh	30.8	36.2	33.3	40.0	33.3
IVi	41.2	42.2	44.1	43.2	45.1

4. CONSLUSION

In summary, we have developed an efficient and convenient method for preparing furfuryl amine (I) and furfuryl amine (IV). The reaction furfuryl amine(I) or furfuryl amine (IV) with aromatic aldehydes to give Schiff bases (II) and (V), which were reduced by NaBH4 to provide furfuryl amine derivatives (III) and (IV). Their structures were characterized by IR, 1HNMR, EI-MS. The preliminary bioassays indicate that furfuryl amine derivatives (III) possess poor inhibitory activity against five kinds of Rice (sheath blight, Rice blast, Cucumber downy mildew, Cucumber powdery mildew and

Botrytis cinerea) at a concentration of 200 mg/L, but furfuryl amine derivatives(IV) possess better inhibitory activity.

Acknowledgement

This Project was supported by the innovative training program of the provincial college students in Hubei Province (201510927001), the Natural Science Found of Hubei Province (No.2015CFC780), and the Young Teachers Program of Hubei University of Technology and Science (No. KY13040).

References

1. Yamaguchi, Fujimura M. Recent topics on action mechanisms of fungicides[J]. *J Pestic Sci*, 2005, **30**: 67-74.
2. Thompson J E, Basarab G S, Pierce J. 2, 3-Dihydro-2, 5-dihydroxy-4 H-benzopyran-4-one: a nonphysiological substrate for fungal melanin biosynthetic enzymes[J]. *Anal Biochem*, 1998, **256**: 1-6.
3. Jordan D B, Basarab G S, Liao D I. Structure-based design of inhibitors of the rice blast fungal enzyme trihydroxynaphalene reductase[J]. *J Mol Graphics Modell*, 2001, **19**: 434 -447.
4. Basarab G S, Jordan D B, Gehret T C Z. Design of scytalone dehydratase inhibitors as rice blast fungicides: derivatives of norephed-rine[J]. *Bioorg Med Chem Lett*, 1999, **9**: 1613-1618.
5. Gouot J M, Grosjean-Cournoyer M C. Synergistic fungicidal composition comprising a pyridylethyl benzamide derivative and a melanin biosynthesis inhibitor[J]. *EP*, 2005,**15167**:37-42.
6. M.Saeed Butt, Zareen Akhter, M. Zaf ar-uz-Z aman,Synthesis and characterization of some Schiff -base-containing polyimides[J].*Colloid Polym Sci*.2008,**28**(6):1455–1461.
7. Jennings L D, Rayner D R, Jordan D B. Cyclobutane carboxamide inhibitors of fungal melanin: biosynthesis and their evaluation as fungicides[J]. *Bioorg Med Chem*, 2000, **8**: 897-907.
8. Basarab G S, Jordan D B, Gehret T C. Design of inhibitors of scytalone dehydratase: probing interactions with an asparagine carboxamide[J]. *Bioorg Med Chem*, 2002, **10**: 4143-4154.
9. Har Lal Singh, J. B. Singh. Synthesis and characterization of new lead(II) complexes of Schiff bases derived from amino acids[J]. *Business Media*. 2013, **39**(6): 1997–2009.
10. Iffet Sakyan, Synthesis and characterization of four new manganese(III) complexes and amino acid (L -aspartic acid, L -asparagine, L-glutamic acid, L -glutamine) Schiff bases[J].*Transition Metal Chemistry*. 2007, **32**(9): 131-135.

INVESTIGATION OF FOOD SAFETY INFORMATION ASYMMETRY PROBLEM BETWEEN COMPANIES AND CONSUMERS

Minjun Li[a], Zhiyuan Hao, and Jingmin Cheng[b,*]
School of Management, Shanxi Medical University, Taiyuan, 030001, Shanxi, China
E-mail: [a]1604624324@qq.com, [b]72-87@163.com
[*]Corresponding author

Objective: This paper aims at understanding the current situation of information asymmetry between food companies and consumers. Method: This research investigated 511 companies employees and more than 10000 consumers of the country through on-the-spot questionnaire survey and used SPSS20.0 software for dealing with the survey data. Result: There are differences on the understanding of the food safety law, food allergens and genetically modified food between companies and consumers. Conclusion: In the face of the present situation of information asymmetry between food companies and consumers, it is necessary to perfect enterprise information disclosure, improve consumers food safety awareness, and establish a social network information system.

1. PURPOSES AND METHODS OF RESEARCH

In recent years, food safety incidents occur frequently and the information asymmetry in food trade is one of the important reasons which lead to food safety incidents.Among them,the problem of information asymmetry between the food companies and consumers is particularly prominent.This paper aims at understanding the current situation of information asymmetry between the food companies and consumers so that it will provide data support to help solve this problem.

This research investigated 511 companies employees and more than 10000 consumers of the country through on-the-spot questionnaire survey. The author selected three aspects of the data which were understanding of food safety, food allergens, genetically modified food, analyzed by software SPSS20.0.From this analysis, it is concluded that there is asymmetric information problem between the food companies and consumers, so we further analyze the causes and give suggestions of this problems.

2. BASIC SITUATION OF THE INVESTIGATION

We investigated more than 10000 consumers and 511 food businesses which distributed in 23 provinces and cities across the country by the questionnaire survey, field interviews and other forms from March to August in 2014.

The survey for food companies had issued 6200 questionnaires. Among them,5804 questionnaires were retrieved. After the screening, there were 5394 valid questionnaires and the effective recovery rate was 87%.At the same time, 10705 questionnaires related to consumer were issued, and regained 10118 questionnaires. There were 9719 questionnaires available after the screening inspection. The effective recovery rate was 91%.

3. SITUATION ANALYSIS OF INFORMATION ASYMMETRY BETWEEN FOOD COMPANIES AND CONSUMERS

We designed a total of 83 questions about food safety, and picked out the three aspects of the data which illustrates the degree of understanding of the food companies and consumers about food safety law, food allergens, genetically modified food and used the SPSS software to describe and analyze them.

3.1. Knowledge of Food Safety Law

Table 1. The enterprise personnel and consumer's understanding of the food safety law

grouping	enterprise			consumer		
	number	%	cumulative percentage(%)	number	%	cumulative percentage(%)
understand very much	886	16.4	16.4	587	6	6
understand	1771	32.8	49.2	1681	17.3	23.3
generally understand	1903	35.3	84.5	3365	34.6	57.9
do not understand	612	11.3	95.8	3098	31.9	89.8
do not understand completely	222	4.1	100	988	10.2	100
total	5394	100	100	9719	100	100

As you can see from the Table 1, in the case of food safety law,84.5% of the enterprise is "understand very much, understand or general understand", but consumers only accounted for 57.9%.Enterprises which "do not understand or do not understand completely" about food safety law accounted for 15.5%,while consumers accounted for 42.1%.We know chi-square = 1505.068 > 9.49, $P < 0.05$ by chi-square test. And according to the $\alpha = 0.05$ level, it is conducted that the difference of food safety knowledge between enterprises and consumers is statistically significant. The information of food safety law mastered by companies and consumers is different. The understanding of the food safety law of consumers is far below than that of enterprise employees'.

3.2. Knowledge of Food Allergens

Table 2. The enterprise personnel and consumer's understanding of food allergens

grouping	enterprise			consumer		
	number	%	cumulative percentage(%)	number	%	cumulative percentage(%)
understand very much	592	11	11	713	7.3	7.3
understand	1710	31.7	42.7	1917	19.7	27
generally understand	1900	35.2	77.9	3080	31.7	58.7
do not understand	949	17.6	95.5	3067	31.6	90.3
do not understand completely	243	4.5	100	942	9.7	100
total	5394	100	100	9719	100	100

From the Table 2 it can be seen that, taking food allergens for instance, enterprises of "understand very much, understand and general understand" accounted for 77.9%,while consumers accounted for 58.7%. And 22.1% of enterprises "do not understand or do not understand completely" about food allergens but 41.3% of consumers in the same situation. By chi-square test, chi-square = 647.257 > 9.49, $P < 0.05$.According to $\alpha = 0.05$ level, differences of companies and consumers to know about food allergens have statistical significance. The information mastered by companies and consumers about the food allergens is different. Consumers know less about food allergens than enterprises employees do.

3.3. Knowledge of Genetically Modified Food

Looking over Table 3, it's easy to see that 80.10% of enterprises know something about genetically modified food while consumers reach 64.3%. Enterprises which "don't understand or don't understand completely" account for 19.9%,and consumers under similar circumstances attain 35.7%. We draw a conclusion that chi-square = 647.257 > 9.49, $P < 0.05$ by chi-square test. On the basis of the $\alpha = 0.05$ level, differences on the understanding of genetically modified food between enterprises and consumers are statistically significant. The information mastered by companies and consumers about genetically modified food is different. For genetically modified food, understanding of the consumers is much less than enterprises employees do.

Table 3. The enterprise personnel and consumer's understanding of genetically modified food

grouping	enterprise			consumer		
	number	%	cumulative percentage(%)	number	%	cumulative percentage(%)
understand very much	453	8.4	8.4	756	7.8	7.8
understand	1849	34.3	42.7	2319	23.9	31.7
generally understand	2019	37.4	80.1	3168	32.6	64.3
do not understand	851	15.8	95.9	2529	26	90.3
do not understand completely	222	4.1	100	947	9.7	100
total	5394	100	100	9719	100	100

Through the analysis of the above three tables, we conclude that the amount of information which companies and consumers master is different for many problems. The information mastered by the consumers is far below than that of the enterprise employees. They are in a state of obvious information asymmetry.

4. CAUSES AND SUGGESTIONS

4.1. The Poorness of Information Transmission Channel And the Improvement of Enterprise Information Disclosure

Information public platform of companies is unsound and does not have a unified information communication channel. Enterprises lack the appropriate channels to disclose information to consumers, which causes the decrease of information passed to market by companies. Consumers are also lack of convenient way to obtain information of companies. At the same time, interaction between companies and consumers is few. Companies often misunderstand the preferences of consumers.

Enterprises should establish a standard, unified and efficient information communication channel and use standardized information model to distribute information[1]. Enterprises are supposed to establish information platform which is used to communicate with consumers[2]. Strengthen the communication with consumers through this platform so as to understand correctly the preferences of consumers and produce food which meets market demand.

4.2. The Insufficiency of Information Recognition Ability of Consumers and the Improvement of Food Safety Awareness of Consumers

Consumers are at the bottom of the chain of food information so they can gain very little information. Consumers only rely on sensory experience and consumer experience to judge the food safety level. Knowledge of food safety mastered by some consumers is not much so that they are unable to identify fake food. The complicated procedures to safeguard the legitimate rights of themselves make consumers away after the food safety problems.

Improving food safety awareness of consumers becomes particularly important. Consumers should active learning food safety knowledge through television, newspapers, radio and other media so as to improve their food safety attainment. It's purpose is to help consumers to easily identify the truth and reliability of the food certification information [3], and make them simply identify the food quality and know some knowledge and common sense of food[4].

4.3. The Imperfection of Information Supervision And the Establishment of Social Work

The social information supervision is imperfect and lack of adequate supervision. Internal employees of enterprise can easily obtain food information. But they can not really take supervision responsibility for reasons of their own interests. The media is conditioned by various factors, so they do not really speak for consumers. Other nongovernmental organizations, their survival may also depend on financial support of the enterprise or the government, therefore they can not really realize monitoring role.

Strengthen the social work and enrich the supervision power are imminent. Except the enterprise, the consumer to take corresponding measures, the government should also regulate information release, and establish a high quality of the food safety regulatory team[5]. Media should be accurate, timely and fair spread food safety information so as to implement the supervision function of public opinion [6]. The third party power should provide fair and effective information for the market.

Acknowledgments

This work is supported by the fund (grant no. 2201410) hosted by Zhi-yuan Hao for the Youth fund projects at Shanxi medical university. Some students and teachers participated in this project, including Jing-min Cheng and Min-jun Li. Among them, Jing-min Cheng is the corresponding author of this article.

References

1. Fan ZhenHui. Analyze simply purchase behavior of consumers and marketing countermeasures of enterprises under the conditions of information asymmetry-S

vocational and technical college as an example. *Modern business*, 2015; (6): 27-28.(In Chinese).

2. Zou Yanjie. The research of food safety problem result in adverse selection and government regulation. Jiangsu, Nanjing university of Chinese medicine, 2007.(In Chinese).
3. Zhang Man, Tang Xiaochun, Pu Mingzhe, Zhang Jing, Zheng Fengtian. Social work of food safety: enterprise, government and the third party regulatory power. *Journal of food science*, 2014, 35 (13): 286-292.(In Chinese).
4. Zhu Rong. The research of the China food safety information barriers and its elimination in path. Jiangxi, Jiangxi university of finance and economics, 2013.(In Chinese).
5. Cheng Jingmin.*Running situation and countermeasure research of China food safety regulatory system.* Beijing: military medical science press, 2013:193-195.(In Chinese).
6. Deng Ganghong. Build legal logic and path of social work of food safety. *Journal of social sciences in Nanjing*, 2015, (2): 97-102.(In Chinese).

COMPARISON AND EVALUATION OF TWO DIFFERENT COLUMNS FOR DETERMINATION OF THE RESIDUE OF *FUSARIUM* TOXINS (ZEARALENONE, DEOXYNIVALENOL, T-2 AND HT-2 TOXINS) IN FOLOUR

Jianing Zhang[a], Ke Ding[b], Xiangning Chen, and Tao Han
Department of Food Science and Engineering, Beijing University of Agriculture, Beijing, China
Beijing Laboratory of Food Quality and Safety, Beijing, China
Beijing Key Laboratory of Agricultural Product Detection and Control of Spoilage Organisms and Pesticide Residue, Beijing, China
E-mail: [a]s631852@163.com, [b]dingke@tom.com

A simultaneous determination method for four kinds of Fusarium toxins (zearalenone, deoxynivalenol, T-2 and HT-2 toxins) in flour was established using high performance liquid chromatography-tendem mass spectrometry (LC-MS/MS). Samples were cleaned up with MycoSep®226 multifunction purifying columns and immunoaffinity columns DZT MS-PREP separately. The recovery of the Mycosep® 226 multifunction purifying column is 69.42%~111.12%, and the relative standard deviation is 1.46%~13.24%. The recovery of the immunoaffinity columns DZT MS-PREP is 88.01%~107.31%, and the relative standard deviation is 1.09%~14.42%. 7 kinds of flour samples were determined, most of the detection results of the immunoaffinity columns DZT MS-PREP were higher than that of the Mycosep® 226 multifunction purifying column. The immunoaffinity columns provides an alternative effective and fast tool for simultaneous extraction of the residues of four Fusarium toxins from real samples.

1. INTRODUCTION

Fusarium toxins are secondary metabolites produced by Fusarium fungi which is widely distributed in nature[1]. They can colonize a broad range of crops and cereals mainly in the field and/or during storage [2,3]. Zearalenone (ZEA) and trichothecenes, which include deoxynivalenol (DON), T-2 and HT-2 toxins, are major Fusarim toxins[4] and the most common and dangerous mycotoxins in grain products. ZEA and DON were widely detected in feed, and the associative action of ZEA and DON might destroy the intestinal tract function and break the immunization barrier[5-6], T-2 toxin is the most poisoning toxin in trichothecenes and its major metabolite is HT-2[7-8]. Grain products were generally contaminated by the 4 Fusarium toxins above and the detection rate was high. It is essential to have available precise and reliable analytical methods applicable at the regulatory levels for the residue of the four Fusarium toxins.

The main analytical methods for the determination of Fusarium toxins in Grain products are thin layer chromatography (TLC), gas chromatography (GC)[9], gas chromatography-mass spectrometry (GC-MS)[10], high performance liquid chromatography (HPLC)[11-12] and liquid chromatography mass spectrometry (LC-

MS/MS) [13-14]. These methods are mainly aimed at the rapid screening, qualitative or quantitative detection of a certain type of toxin or a class of structurally similar toxins[15-16].

Considering the complexity of matrix and the very low concentration of Fusarium toxins in the real samples, these methods could detect trace levels of target analytes when an extensive pre-treatment step such as solid phase extraction (SPE) was coupled with[17-18]. At present, the multifunction purifying columns and immunoaffinity columns are widely used at home and abroad[18-19]. The study on the effect of two kinds of columns on the actual sample is relatively rare, therefore, the comparision of the two different kinds of columns is very important. The aim of this work was to compare the effect of the two different kinds of columns on the enrichment and purification of the four Fusarium toxins in grain products in order to provide the basis for the later experiments. Finally, it was applied to the extraction of an actual flour sample.

2. EXPERIMENTAL

2.1. Reagents, Solutions and Apparatus

Reagents: DON, T-2, HT-2, and ZEA were obtained from Pribolab (Beijing, China). Methanol and acetonitrile (HPLC grade) were obtained from Dima (Buchs, Switzerland). The other reagents were all of analytical grade.

Solutions: (1) 10 mmol/L ammonium acetate solution; (2) 0.01 mol/L phosphate buffered saline (PBS); (3) 0.1 mol/L NaAc, pH 4.0, containing 0.5 mol/L NaCl; (4) 0.1 mol/L Tris-HCl, pH 8.0, containing 0.5 mol/L NaCl; (5) (6) 0.01 mol/L PBS, pH 7.4; (7) 0.01 mol/L PBS, pH 7.4, containing 0.01% NaN3.

Apparatus: Electronic balance (BSA224S, Sartorius, Beijing, China); Ultraviolet visible spectrophotometer (UV-2300, Techcom Com, Shanghai, China); Mycosep® 226 multifunction purifying column (Pribolab, Beijing, China); immunoaffinity columns DZT MS-PREP (R-biopharm, Darmstadt, Germany); LC-MS/MS (1200-ESI 6410B, Agilent, Beijing, China).

Flour samples: Select 7 different brands of flour samples from different areas, including 5 kinds of wheat flour and 2 kinds of corn powder. Four kinds of them were purchased from large supermarket of China, the other three kinds were purchased from pedlars' market of China.

2.2. LC-MS/MS Parameters

In order to acquire the accurate quantitative results of the four Fusarium toxins extracted from different matrices, the LC-MS/MS method was established. The column was an Agilent ZORBAX Bonus-RP column (50 mm×2.1 mm, 3.5 μm) and was kept at 20°C. A mobile phase gradient programme was used for combining solvent A (Ultra pure water+10 mM ammonium acetate) and solvent B (acetonitrile), the flow rate was 0.3 mL/min, and the total run time was 20 min. In 0~0.1 min, B is 10%; 0.1~2 min, B is from

10% to 50%; 2~10 min, B is from 50% to 80%; 10~15 min, B is 80%; 15~16 min, B is from 80% to 10%; 16~20 min, B is kept 10%. The instrument settings as Table 1.

Table 1. Instrument settings for the four Fusarium toxins

Compound	Capillary voltage (kv)	Ion mode (ESI+/ ESI−)	Ion pairs (m/z)	Fragmentor voltage (V)	CE (V)
DON	4.0	ESI+	319/69※	135	16
			319/283	135	24
HT-2	4.0	ESI+	442.1/263.※	135	17
			442.1/215	135	19
T-2	4.0	ESI+	484.2/245.※	135	20
			484.2/305.1	135	21
ZEA	2.5	ESI−	317.3/131.※	135	40
			317.3/175.2	135	34

2.3. Sample Preparation

2.3.1. *Mycosep® 226 Multifunction Purifying Column*

Weigh 25 g of flour sample into a litre capacity, solvent resistant blender jar, add 100 mL acetonitrile / water (84: 16, v/v) and blend at high speed for 2 minutes. Filter the sample through the filter paper, or centrifuge at 4000 rmp for 10 minutes. Filter the diluted extract through glass microfiber filter paper. Pass 8 mL of filtrate through the Mycosep® 226 column, take 4 mL of post column liquid, dried with N2, set the volume to 1 mL with methanol. Inject 50 μL onto the LC-MS/MS system.

2.3.2. *Immunoaffinity Columns DZT MS-PREP*

Weigh 25 g of flour sample into a litre capacity, solvent resistant blender jar, add 100 mL of 70% methanol (v/v) and blend at high speed for 2 minutes. Filter the sample through the filter paper, or centrifuge at 4000 rmp for 10 minutes. Dilute 2 mL of filtrate with 48 mL of phosphate buffered saline (PBS). Filter the diluted extract through glass microfiber filter paper. Pass 20 mL of filtrate (equivalent to 0.2 g of sample) through the column at a flow rate of 2 mL per minute (or the sample can be allowed to pass through the column by gravity if preferred). A slow, steady flow rate is essential for the capture of the toxin by the antibody. Wash the column by passing 20 mL of water through at a flow rate of approximately 5 mL per minute. Pass air through the column to remove residual liquid. Elute the toxins from the column at a flow rate of 1 drop per second using 1 mL of 100% methanol (v/v) and collect in a 2 mL amber glass vial. Following elution pass 1 mL of water through the column and collect in the same vial to give a 2 mL total volume. Inject 50 μL onto the LC-MS/MS system.

2.4. Quantitative Analysis of The Four Fusarium Toxins

2.4.1. *Calibration Curve*

It is recommended to run at least a 3-6 point calibration curve. In constrcting a suitable curve the levels of the calibration standards should bracket or include the range of expected results. To prepare a five point calibration curve, determine the correlation coefficient of the standard curve and the linear range.

2.4.2. *Quantitative Limit of The Method*

Accurately weigh several copies of 25 g flour samples from each, add standard solution of four Fusarium toxins, and gradually reduce the dosage, reduce flour in 4 kinds of mycotoxins concentration. Extract the sample according to the method 2.3, then inject 50 μL onto the LC-MS/MS system. It continues until when the detected results of signal-noise ratio is 3:1, and there is no peak appeared in the SIM results. At this time the concentration of four Fusarium toxins in flour are the limits of detection (LOD) of each toxin. The same method to calculate the lower limit of quantification (LOQ) when the detected results of signal-noise ratio is 10:1.

2.4.3. *Recovery and Precision of The Method*

Accurately weigh several copies of 25 g flour samples from each, add different concentrations of standard solution of four Fusarium toxins. Extract the sample according to the method 2.3, then inject 50 μL onto the LC-MS/MS system. At the same time, the blank flour sample was detected by the same way. Each sample were made of 3 parallel, repeated 3 times, then calculate the recovery rate and the relative standard deviation (RSD).

2.5. Analysis of the Effect of Different Columns Applied to The Flour Samples

Weigh 25 g of 7 kinds of flour sample and use the two different columns for their pretreatment separately, according to the method 2.3 of extraction, purification and detection. Three parallel should be operated of each sample and repeated three times. Make a comparison of the results of 7 kinds of samples processed by the two different columns.

2.6. Repeatability and Stability Test of The Immunoaffinity Columns

The immunoaffinity columns DZT MS-PREP should be tested to use three or more times to verify the repeatability and stability of them. The column should be regenerated after use. It was washed with 0.1 mol/L NaAc containing 0.5 mol/L NaCl (pH 4.0) and 0.1 mol/L Tris-HCl buffer containing 0.5 mol/L NaCl (pH 8.0) in tunes at least 3 cycles, and the amount of each buffer was fivefold gel volumes. Finally, the column was washed with 0.01 mol/L PBS (pH 7.4) and stored in the same PBS containing 0.01% NaN3 for next use. The stability test required continuous measurement of the recovery for the four Fusarium toxins three or more times.

3. RESULTS AND DISCUSSION

3.1. Detection of The Four Fusarium Toxins by LC-MS/MS

To establish the LC-MS/MS method for simultaneous detection of the four Fusarium toxins, the optimal analytical condition must be firstly confirmed. Other appropriate conditions for LC-MS/MS analysis were mentioned in Section 2.2. Under this optimal condition, the four Fusarium toxins can be completely separated. As shown on Fig. 1, which was acquired from the mixed standard solution, the retention time for HT-2, T-2, DON and ZEA was 5.336 min, 6.105 min, 7.673 min and 7.681 min, respectively.

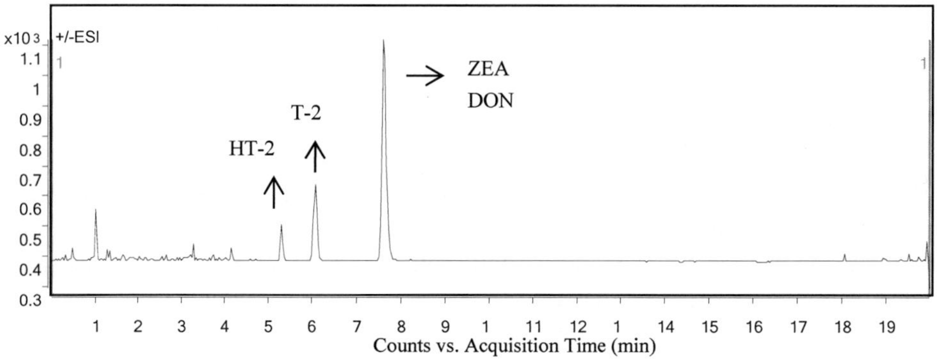

Fig. 1. TIC chromatogram of the four Fusarium toxins standard solutions.

3.2. Quantitative Analysis of the Four Fusarium Toxins

From the standard curves, the liner equation, liner range, correlation coefficient, limit of detection (LOD), limit of quantification (LOQ) were concluded in Table 2. It was seen that the values of LOD of LC-MS/MS for the four Fusarium toxins (DON, HT-2, T-2, ZEA) were 5, 5, 2 and 2 μg/kg, respectively.

Table 2. Linear equation, linear ranges, correlation coefficients, detection limit and quantitative lower limit of four Fusarium toxins.

Compound	Linear equation	Linear range (μg/kg)	Correlation coefficient	LOD (μg/kg)	LOQ (μg/kg)
DON	Y =0.3839 X – 5.9964	5~100	0.9996	5	20
HT-2	Y =1.8281 X – 2.9671	5~100	0.9991	5	20
T-2	Y =4.7100 X – 8.9354	2~100	0.9972	2	5
ZEA	Y =20.930 X – 45.793	2~120	0.9989	2	5

3.3. Recoveries and Relative Standards Deviations of Four Fusarium Toxins

Weigh 25 g of one of the flour sample, add standard solution of the four Fusarium toxins gradiently and use the two different columns for their pretreatment separately, according to the method 2.3 of extraction, purification and detection.The recovery of the Mycosep® 226 multifunction purifying column is 69.42%~111.12%, and the relative standard deviation is 1.46%~13.24%. The recovery of the immunoaffinity columns DZT MS-PREP is 88.01%~107.31%, and the relative standard deviation is 1.09%~14.42%.

The results of the two columns are met the requirements of trace analysis. On the whole, the recovery of the immunoaffinity columns is closer 100% than that of Mycosep® 226 multifunction purifying column as Fig.2. Because the reactions of antigen-antibody are highly specific, the immunoaffinity columns possess great sensitivity and selectivity.

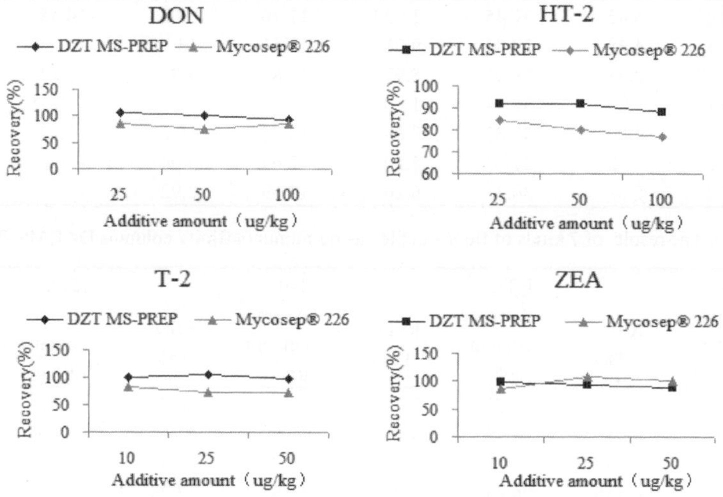

Fig. 2. Comparison of recovery of two different kinds of columns.

3.4. Analysis of The Effect of Different Columns Applied to The Flour Samples

Weigh 25 g of 7 kinds of the flour sample, and use the two different columns for their pretreatment separately, according to the method 2.3 of extraction, purification and detection. The result of the Mycosep® 226 multifunction purifying column showed that the detection rate of DON, HT-2, T-2 were 100% as Table 3, ZEA was detected only in No 1,3,4,5 flour samples, and the average content of the four Fusarium toxins were all the highest of No 4 flour sample. The result of the immunoaffinity columns DZT MS-PREP showed that the detection rate of DON, HT-2, T-2, ZEA were 100% as Table 4, most of the detection results were higher than that of the Mycosep® 226 multifunction purifying column. Considering the complexity of matrix and the very low concentration of the four Fusarium toxins in real samples analysis, the immunoaffinity columns DZT

MS-PREP was the better choice. What's more, the organic solvent used in the purification with the immunoaffinity columns was less than that of with the Mycosep® 226 multifunction purifying column.

In the national standards GB 2761-2005 " Limit of Mycotoxins in food ", there are limits of DON and ZEA in wheat, corn and barley, it shows that DON≤1000 μg/kg, ZEA≤60 μg/kg. However, the results in Table 3 and Table 4 reflect that the content of No 3,4 samples beyond the limits of national standards. So far, the specific standards of HT-2 and T-2 have not been published in China. Based on the application of the actual testing, the limited standards of HT-2, T-2 need to be improved.

Table 3. The results of 7 kinds of flour samples using Mycosep® 226 multifunction purifying column.

Number	DON Average content (μg/kg)	RSD (%)	HT-2 Average content (μg/kg)	RSD (%)	T-2 Average content (μg/kg)	RSD (%)	ZEA Average content (μg/kg)	RSD (%)
1	30.18	5.47	31.45	15.22	11.70	6.02	10.48	7.26
2	27.73	4.68	29.04	9.08	12.51	13.758	-	-
3	115.32	2.51	29.01	8.97	11.81	4.81	68.84	4.30
4	245.14	4.29	34.44	13.09	15.41	6.09	144.162	9.61
5	30.00	4.82	32.85	9.08	13.93	16.20	11.42	7.63
6	17.93	2.18	21.42	5.50	12.69	5.04	-	-
7	29.00	5.58	30.31	6.26	14.69	2.92	-	-

Table 4. The results of 7 kinds of flour samples using immunoaffinity columns DZT MS-PREP.

Number	DON Average content (μg/kg)	RSD (%)	HT-2 Average content (μg/kg)	RSD (%)	T-2 Average content (μg/kg)	RSD (%)	ZEA Average content (μg/kg)	RSD (%)
1	57.45	5.11	16.80	3.41	13.84	2.88	16.52	3.55
2	61.22	0.02	24.22	4.61	18.90	5.06	16.36	8.26
3	129.13	9.00	34.85	2.56	19.72	6.96	61.36	0.81
4	238.23	4.53	26.43	2.92	13.53	13.44	128.06	0.22
5	66.29	1.62	30.16	0.49	19.23	6.69	16.83	0.67
6	67.13	1.46	28.64	5.41	19.44	5.64	15.80	0.09
7	71.25	3.29	21.35	7.12	14.36	13.25	15.92	0.09

3.5. The Repeatability of The Immunoaffinity Columns DZT MS-PREP

The Mycosep® 226 multifunction purifying column can not be reused, however, the immunoaffinity columns are reproducible. The results were shown in Fig.3 after the immunoaffinity columns DZT MS-PREP were used for three times. Compared the recovery after use. It can be concluded that after each use, the recovery of each toxin were reduced, but the reduction was not significant. The immunoaffinity columns can be used repeatedly, which greatly saves the cost and improves its efficiency.

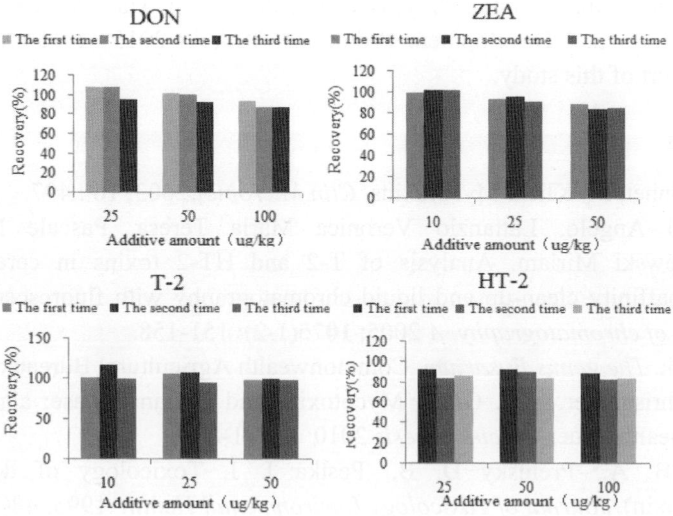

Fig. 3. The recovery after used the immunoaffinity columns DZT MS-PREP for different times.

4. CONCLUSION

In this study, Mycosep® 226 multifunction purifying columns and immunoaffinity columns DZT MS-PREP for simultaneously selective purification of Fusarium toxins (zearalenone, deoxynivalenol, and T-2 and HT-2 toxins) residues from grain products were compared. Applied to actual flour samples, the recovery of the Mycosep® 226 multifunction purifying column is 69.42%~111.12%, and the relative standard deviation is 1.46%~13.24%. The recovery of the immunoaffinity columns DZT MS-PREP is 88.01%~107.31%, and the relative standard deviation is 1.09%~14.42%.

7 kinds of flour samples were determined. The result of the Mycosep® 226 multifunction purifying column showed that the detection rate of DON, HT-2, T-2 were 100% , ZEA was detected only in No 1,3,4,5 flour samples. The result of the immunoaffinity columns DZT MS-PREP showed that the detection rate of DON, HT-2, T-2, ZEA were 100%, most of the detection results were higher than that of the Mycosep® 226 multifunction purifying column.

In a word, the immunoaffinity columns have the advantages of high sensitivity, low matrix interference and low cost, and they are better than the Mycosep® 226 multifunction purifying columns applied to the purification of the four Fusarium toxins (zearalenone, deoxynivalenol, T-2 and HT-2 toxins) residues in real grain products analysis. The immunoaffinity columns provide an alternative effective and fast tool for simultaneous extraction of the residues of four Fusarium toxins from real samples.

Acknowledgments

The authors would like to express their gratitude to the Beijing Natural Science Fund (14L00184), the Importation and Development of High-Caliber Talents Project of

Beijing Municipal Institutions (CIT&TCD20154045) and The Degree and Graduate Education Reform and Development Project of BUA in 2015 (2015YJS034) for the financial support of this study.

References

1. J.W. Bennet, M. Klich. Mycotoxins, *Clin.Microbiol,*2003; 16 : 497.
2. Visconti Angelo, Lattanzio Veronica Maria Teresa, Pascale Michelangelo, Haidukowski Miriam. Analysis of T-2 and HT-2 toxins in cereal grains by immunoaffinity clean-up and liquid chromatography with fluorescence detection. *Journal of chromatography. A* 2005; **1075**(1-2): 151-158.
3. C. Booth. *The genus Fusarium*, Commonwealth Agricultural Bureaux 1971.
4. P.W. Christopher, Y.Y. Gong. Mycotoxins and human disease: alargely ignored global health issue. *Carcinogenesis* 2010; **31**: 71-82.
5. Rotter B. A., Prelusky D. B., Pestka J. J. Toxicology of deoxynivalenol (Vomitoxin). *Journal of Toxicology Environmental Health* 1996; **48**(1): 1-34.
6. Chanemougasoundharam A., Fiona M. D. Trichothecene toxicity in eukaryotes: Cellular and molecular mechanisms in plants and animals. *Toxicology Letters* 2013; **217**(2): 149-158.
7. Visconti Angelo, Lattanzio Veronica Maria Teresa, Pascale Michelangelo, Haidukowski Miriam. Analysis of T-2 and HT-2 toxins in cereal grains by immunoaffinity clean-up and liquid chromatography with fluorescence detection. *Journal of chromatography . A* 2005; **1075**(1-2): 151-158.
8. P.W. Christopher, Y.Y. Gong. Mycotoxins and human disease: alargely ignored global health issue. *Carcinogenesis* 2010; 31 : 71-82.
9. Kotal F, Holadova K, Hajšlová J, et al. Determination of trichothecenes in cereals. *Journal of Chromatography. A* 1999; **830**: 219-225.
10. Eke Z, Kende A, Torkos K. Simultaneous detection of A and B trichothecenes by gas chromatography with flame ionization or mass selective detection. *Microchemical Journal* 2004; **78**: 211-216.
11. Soleimany F, Jinap S, Rahmani A, et al. Simultaneous detection of 12 mycotoxins in cereals using RP-HPLC-PDA-FLD with PHRED and a post-column derivatization system. *Food Additives and Contaminants* 2011; **28**(4): 494-501.
12. Choi C Y, Nakajima-Adachi H, Kaminogawa S, et al. Nivalenol inhibits total and antigen-specific IgE production in mice. *Toxicology and applied pharmacology* 2000; **165**(1): 94-98.
13. Ulrike Brezina, Inga Rempe, Susanne Kersten, et al. Diagnosis of intoxications of piglets fed with Fusarium toxin-contaminated maize by the analysis of mycotoxin residues in serum, liquor and urine with LC-MS/MS. *Archives of Animal Nutrition* 2014; **68**: 425-447.
14. Dall'Asta C, Sforza S, Galaverna G, et al. Simultaneous detection of type A and type B trichothecenes in cereals by liquid chromatography–electrospray ionization

mass spectrometry using NaCl as cationization agent. *Journal of Chromatography A* 2004; **1054**(1): 389-3.

15. Elisabetta De Angelis, Linda Monaci, Michelangelo Pascale, et al. Fate of deoxynivalenol, T-2 and HT-2 toxins and their glucoside conjugates from flour to bread:an investigation by high-performance liquid chromatography high-resolution mass spectrometry. *Food Additives and Contaminants* 2013; **30**: 345-355.
16. Qing Wang, Hua Zhao, Cun-Xian Xi, et al. Determination of chloramphenicol and zeranols in pig muscle by immunoaffinity column clean-up and LC-MS/MS analysis. *Food Additives and Contaminants* 2014; **31**: 1177-1186.
17. Romagnoli B, Ferrari M, Bergamini C. Simultaneous determination of deoxynivalenol, zearalenone, T-2 and HT-2 toxins in breakfast cereals and baby food by high-performance liquid chromatography and tandem mass spectrometry. *Journal of mass spectrometry* 2010; **45**(9): 1075-1080.
18. Vaclavikova M, MacMahon S, Zhang K, et al. Application of single immunoaffinity clean-up for simultaneous determination of regulated mycotoxins in cereals and nuts. *Talanta* 2013; **117**: 345-351.
19. Yun-Mei Li, Bi-Yun Cao, Yan Hong, et al. Development of an immunoaffinity chromatography column forselective extraction of a new agonist phenylethylamine A from feed, meat and liver samples. *Journal of Chromatography B* 2014; 945-946: 178-184.

COMPARE OF DYNAMIC HUMAN MODEL SAR ABOUT DIFFERENT FREQUENCY RADIATION IN BAN

Zhangyong Li[a], Fangfang Liu[b], Dan Su, and Peng Ran

Research Center of Biomedical Engineering, Chongqing University of posts and telecommunications, Chongqing, china

E-mail: [a]lizy@cqupt.edu.cn, [b]719900955@qq.com

Finite-difference time-domain (FDTD) calculations have been performed at 0.9GHZ and 2.4GHZ using two voxel model of an adult male to investigate human exposure to the electromagnetic fields in daily life. The specific absorption rate (SAR) are studied by using the method of finite-difference time-domain and moving human models. Two movements having a different dynamical characteristic each, walking and running, are considered. The half-wave dipole antenna are used to calculate SAR at 0.9GHZ and 2.4GHZ, which would get the SAR under different frequencies and positions. Through the analysis of the result data, it can provide reference for electromagnetic radiation hygiene standards and also have certain significance to medical applications and electromagnetic protection.

1. INTRODUCTION

With the aged tendency of global population, people pay more attention to the healthcare requirements. Due to the advances of wireless communication and medical sensing technologies, many researchers focus on the applications of Wireless Body Area Network(WBAN), which can provide individual healthcare services. WBANs consist of small and intelligent wireless medical sensors, worn or implanted in the human bodies, to sample vital physiological signals and send the records to a base station (usually a portable device such as PDA or mobile-phone) for real-time analysis or remote diagnosis by the Publisher[1-3]. Meanwhile, the implantable equipment make use of the communication channel inside the body tissues. This technology can be applied in medicine and healthcare, sports, fitness, and entertainment.

Under different scenarios and requirements, the sensor nodes can be implanted or deployment on the human body which will inevitably lead to the electromagnetic effect, so the application of the WBAN should ensure that biology is safe. The SAR is a safety assessment index of biological electromagnetic radiation, which works as a general indicator for human tissue exposed to electromagnetic environment, so how to precisely calculate SAR for the human body is a question worth discussing. In addition, the human body is not stationary in our daily life, and the human body posture is also changing along with the daily activities, which could lead to the relative position of sensor nodes changed. The body posture has a certain effect to the sensor nodes of electromagnetic radiation and absorption. So the study of SAR under the human body posture change has an enormous practical significance.

At present, some studies focus on how to carry out the actual SAR measurement of the human body, while another also has focused on the certain part of the human body's SAR calculation, such as head, chest and other parts of human body. A human head model has been established at 0.5~5.5GHZ by the Publisher [4], and the result indicated that the radiation efficiency become lower with the increase of the frequency. The effect of mobile phones on the human body's head has been studied by the Publisher [5], and the result could provide reference to develop the electromagnetic radiation hygiene standards. The SAR values have been calculated in a 10 year old sitting model from exposure to electromagnetic fields at 2.4 and 5 GHz by the Publisher [6,7,8], and this study presents that whole-body and peak localized SAR in the model resulting from both plane wave fields and fields produced by representative WLAN antennas positioned close to the body. Many research show that there are some shortages in the model or research methods. In addition, the most important drawback is that these studies are based on the human body in the static state, which the value of the SAR can't reflect the real situation of the human body absorption of electromagnetic radiation, and even can't provide reference to develop the electromagnetic radiation hygiene standards. Therefore, it is necessary to simulate the real human body environment and study the SAR under motion human body.

2. SAR CALCULATE

In the research on biological effects of organisms, the biological effect of quantitative judgment mainly adopts dosimetry. Experts and scholars generally regard the SAR as its important judgment parameters. As previously mentioned in these columns [9], the maximum power deposition, expressed in SAR, which allowed by the FCC is 1.6 W/kg or 1.6 mW/g in 1g of head tissue from exposure to cellular telephone radiation. The International Commission on Non-Ionizing Radiation Protection (ICNIRP) guidelines stipulate a maximum SAR of 2 W/kg or 2 mW/g in any 10 g of tissue in the head. The SAR is a metric to quantify and register localization of deposited microwave energy. In international standard units, SAR is expressed in units of W/kg. As a dosimetric quantity, it denotes the time rate of microwave energy absorption at a given location inside the tissue. It is derivable from the induced electric field inside tissue, as in Eq. (1).

$$SAR = \frac{1}{2}\frac{\sigma |E|^2}{\rho} W / Kg \tag{1}$$

Where E is the induced electric field (V/m), ρ is the density of tissue (kg/m^3), and σ is the electrical conductivity of tissue (S/m).

The FDTD method was used to calculate SAR values, which is a numerical analysis technique used for modeling computational electrodynamics. Since it is a time-domain method, FDTD solutions can cover a wide frequency range with a single simulation run, and treat nonlinear material properties in a natural way. The time-dependent Maxwell's equations are discretized using central-difference approximations to the space and time partial derivatives[10]. FDTD algorithm directly from Maxwell curl equation adopted Yee type grid structure for each component of the electromagnetic field in different spatial location, which meet the Faraday and Ampere's law in spatial relations. According to the type of Yee grid structure, using central difference scheme, the Maxwell curl equations of each component equations into difference form, Electric field intensity (E), Magnetic permeability(μ), relative permittivity(ε), conductivity(σ), magnetic field intensity(H), which is defined as in Eq. (2-3).

$$\nabla \times E = -\mu \frac{\partial H}{\partial t} \tag{2}$$

$$\nabla \times E = -\varepsilon \frac{\partial H}{\partial t} + \partial E \tag{3}$$

3. HUMAN BODY MODEL

According to the study of SAR with the dynamic human body, the main idea as follows. Firstly, this paper used a 3D animation software for human posing to model the human body motion and produce a set of 9 voxel models (frames) for each movement sequence. Two human movements are considered for dynamic body SAR, including walking and running. Secondly, put the half-wave dipoles antenna on different part of the human body such as head, chest and wrist. Thirdly, exported the three-dimensional human body model into the software of FDTD and set the parameters of the human body model, which would simulate the SAR values at 0.9 GHZ, 2.4 GHZ. Fig. 1 and Fig. 2 show a set of postures for walking and running, respectively. In this paper, the half-wave dipoles antenna has been deployed on head, chest and wrist, the position of those antennas shows as Fig. 3.

Fig. 1. The three-dimensional human body model for a walking man.

Fig. 2. The three-dimensional human body model for a running man.

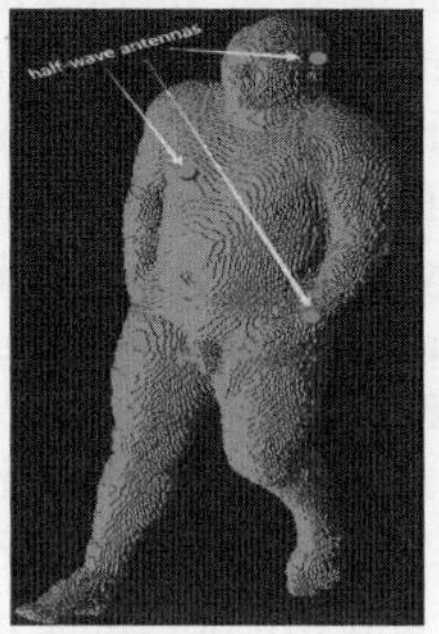

Fig. 3. The position of half-wave dipoles antenna.

Using FDTD method to calculate SAR value in the electromagnetic environment, the size of grid should be taken into consider. The size of grid relate with the wavelength and computer storage space. Therefore, this paper choose 5mm ×5mm×5mm as the size of grid after taking into account all factors. Identical half-wave dipoles are used at 0.9GHZ and 2.4GHZ with a wavelength L=166mm and 62mm respectively. Table 1 and Table 2 show the characteristics of different body tissue at 0.9GHZ and 2.4GHZ, respectively.

Table 1. Characteristic of the simulated tissues at 0.9GHZ

Tissue	Conductivity(S/m)	Relative permittivity	Density(Kg/m3)
Sink	0.86674	41.405	1145
Fat, Bone Marrow	0.051043	5.462	963
Bone Cancellous	0.34	20.788	1074
Blood	1.5379	61.36	1070
Muscle, Duodenum	0.94294	55.032	1046
Cerebellum	1.2628	49.444	1012

Table 2. Characteristic of the simulated tissues at 2.4GHZ

Tissue	Conductivity(S/m)	Relative permittivity	Density(Kg/m3)
Sink	1.319	38.67	1145
Fat, Bone Marrow	0.08082	4.39	963
Bone Cancellous	0.6121	16.43	1074
Blood	2.71528	52.26	1070
Muscle, Duodenum	2.35716	56.75	1046
Cerebellum	2.03481	51.43	1012

4. SIMULATION RESULTS

The safety of electromagnetic radiation is essential to the human body in the special application scenario of body area network. Different body positions have a certain influence on electromagnetic radiation absorption. This paper established two kinds of three-dimensional human body models (walking and running), and each movement process subdivided into 9 frames. Half-wave dipoles are used at 0.9GHZ and 2.4GHZ to simulate the absorption of electromagnetic radiation on human head, chest and wrist. As showing in Fig. 4 and Fig. 5, the two figures show the different SAR value in different part of body at 0.9GHZ and 2.4GHZ. Due to the changing of human body posture, the bones are stretched and the density of each organization are changed, which influenced the organization of electric parameters and also influenced the SAR values of human body. Compared the Fig. 4 and Fig. 5. The results display that with the increase of frequency, the half-wave dipole radiation is also increased. However, no matter in process of walking or running at 0.9GHZ and 2.4 GHZ, the value of SAR is far less than the existing international standards and will not cause harm to human body in this case.

Under the same frequency, the antenna in different locations, different dielectric constant and conductivity of human tissue and different body posture, the SAR value is also different. When the frequency is different at the same time, the antenna is placed on the same position, and the SAR values are also different. The maximum SAR under the process of walking and running at 0.9GHZ and 2.4GHZ are present in Fig. 6 to Fig.9, respectively. Fig. 6 and Fig. 7 shows the half-wave dipole antenna has deployed on the head and chest in the 3 d human body model in the process of walking at 0.9GHZ, respectively. The lift picture of Fig. 6 shows the electromagnetic radiation when the half-wave dipole antenna put on human body head. Color contrast strip can be seen from the figure, the colors are purple, blue, dark green, bright green, yellow, orange and red, with the color from purple to red in the process, half-wave dipole antenna on the human body radiation in turn increases. When the color is red, the electromagnetic radiation is maximum. While the color is violet, the electromagnetic radiation is minimum. When there is 2.4GHZ, whether it is walking or running, human body absorbing the electromagnetic radiation is large than 0.9GHZ. We can also be seen from the figure when the half-wave dipole antenna is placed in a certain part of the body, and the electromagnetic radiation will affect the whole human body. But the place where put the half-wave dipole antenna will absorb more electromagnetic radiation.

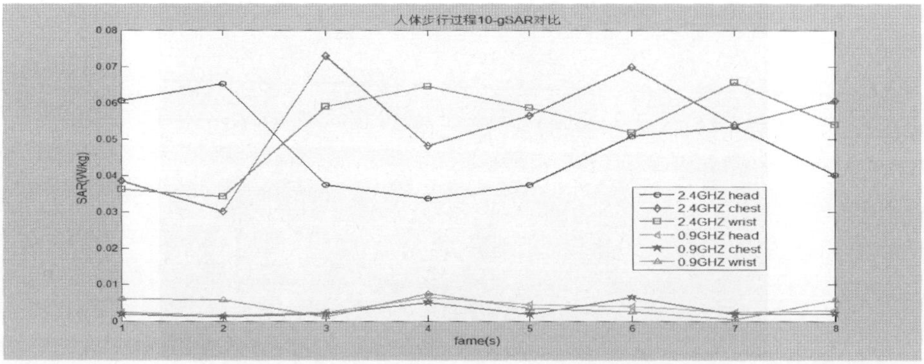

Fig. 4. The 10-g SAR for a walking man at 0.9GHZ and 2.4GHZ.

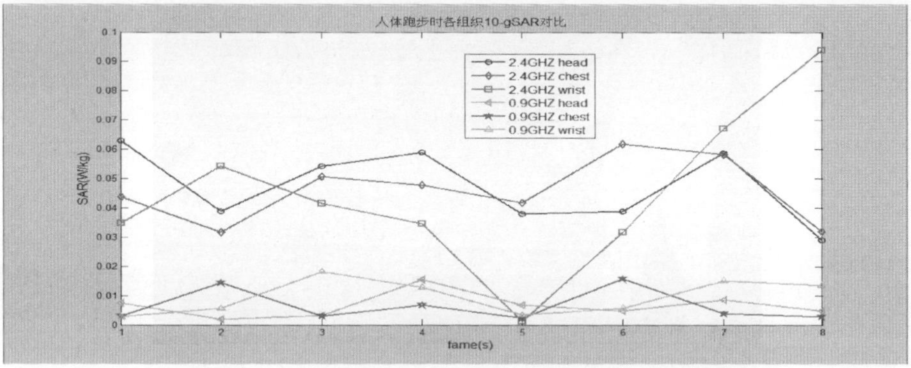

Fig. 5. The 10-g SAR for a running man at 0.9GHZ and 2.4GHZ.

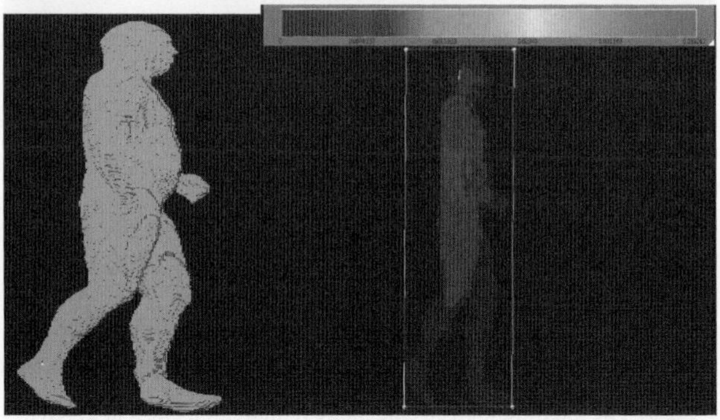

Fig. 6. The maximum electromagnetic radiation on head at 0.9GH.

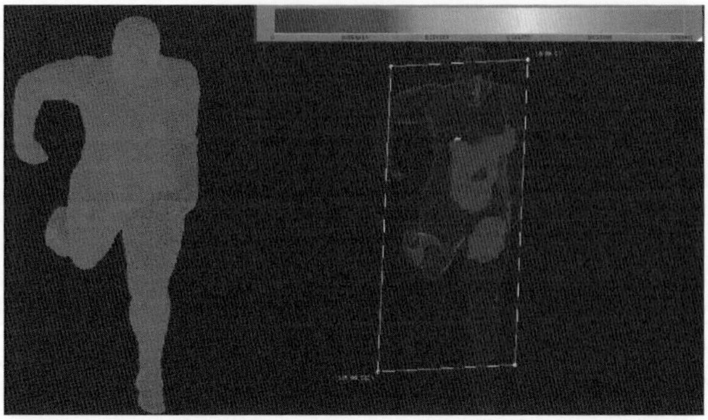

Fig. 7. The maximum electromagnetic radiation on chest at 0.9GH.

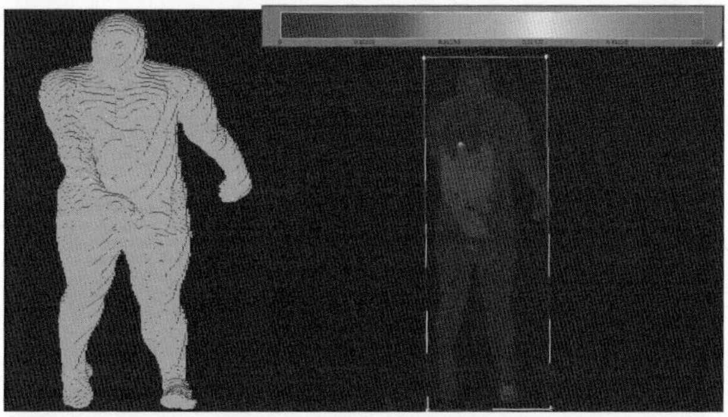

Fig. 8. The maximum electromagnetic radiation on head at 2.4GH.

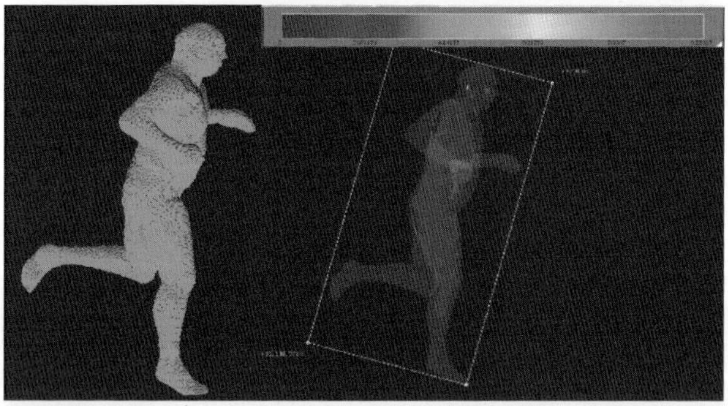

Fig.9. The maximum electromagnetic radiation on chest at 2.4GH.

5. CONCLUSION

At present, the study of SAR is mainly concentrated in the handheld device of electromagnetic radiation to human body in the static human environment rather than the study of SAR in dynamic environment. The specific absorption rate is studied by using the finite-difference time-domain method and dynamic human models. Half-wavelength dipoles are used at 0.9GHZ and 2.4GHZ to simulate the absorption of electromagnetic radiation on human head, chest and wrist. The simulation results show that with the increase of frequency, and the absorption of electromagnetic radiation to human body is increased. While as the change of human body posture, the absorption of electromagnetic radiation to human body is changed, too. But, no matter in the process of walking and running, the SAR values from different parts of the body are in line with the international standards in the United States and Europe, which shows that under a certain frequency, wearable sensors will not cause harm to human body. These data to provide reference data for the establishment of the electromagnetic radiation hygiene standards, it is also significant to medical applications for electromagnetic waves and electromagnetic protection.

Acknowledgments

This project is supported by the National Natural Science Foundation of China (No.61571070), Science Fund of Chongqing China (No.cstc2012jjA10148 and No.KJ1400427).

References

1. Lin, James C. Biological Bases of Current Guidelines for Human Exposure to Radio-Frequency Radiation. *J Antennas and Propagation Magazine*2003; **45**:182-184.
2. Xu L S, Meng, MaxQ. –H, Ren H L. Radiation characteristics of ingestible wireless devices in human intestine following radio frequency exposure at 430, 800, 1200, and 2400 MHz. *J Transactions on Antennas and Propagation* 2009; **57**:2418-2428.
3. Montaser, AM. Design of Multiband PIFA with Low SAR Value for All Commercial Mobile Communication Bands. *J International Journal of RF and Microwave Computer-Aided Engineering* 2015; **25**:194-201.
4. Bakker, JF, Paulides, MM, Christ, A. Assessment of induced SAR in children exposed to electromagnetic plane waves between 10 MHz and 5.6 GHz. *J Physics in Medicine and Biology* 2010; **55**:3115-3130.
5. Fiocchi, S, Markakis. SAR exposure from UHF RFID reader in adult, child, pregnant woman, and fetus anatomical models. *J Bio-electro magnetics* 2013; **34**:443-452.
6. Bonev, IB, Franek, O, Pedersen, GF. Impact of the Hand on the Specific Absorption Rate in the Head. *J Applied Computational Electromagnetics Society Journal* 2014; **29**:470-477.
7. Wang, SM, Shao, Y, Li, SZ. Rapid Local Specific Absorption Rate Estimation for Magnetic Resonance Imaging. *J IEEE Transactions on Electromagnetic Compatibility* 2014; **56**: 771-779.
8. Iqbal-Faruque, MR, Aisyah-Husni, N, Ikbal-Hossain, M.Effects of Mobile Phone Radiation onto Human Head with Variation of Holding Cheek and Tilt Positions. *J Journal of Applied Research and Technology*2014; **12**:871-876.

9. Asayama, R , Wang, JQ, Fujiwara, O. , Statistical Outliers in Voxel SARs and Their Effect of Whole-Body Average SARs in Pregnant Woman and Child for Far-Field Exposure. *J Electronics and Communications in Japan* 2014; **97**:62-69.
10. Bamba, A, Joseph, W, Vermeeren, G. A formula for human average whole-body SAR (wb) under diffuse fields exposure in the GHz region. *J Physics in Medicine and Biology*2014; **59**:7435-7456.
11. Gao, YJ , Yang, HW. A High-Order, Symplectic, Finite-Difference Time-Domain Scheme for Bio electromagnetic Applications within the Mother/Fetus Model. *J PLOS ONE*2014; **9**:e114425.
12. Xu L S, Meng, MaxQ. –H, Ren H L. Effects of dielectric parameters of human body on radiation characteristics of ingestible wireless device at operating frequency of 430 MHz. *J Transactions on Biomedical Engineering* 2009; **56**:2083-2094.
13. Hossain, MI, Faruque, MRI, Islam, MT. A New Design of Cell Phone Body for the SAR Reduction in the Human Head. *J Applied Computational Electromagnetics Society Journal* 2015; **30**:792-798.
14. Faruque, MRI, Islam, MT, Ullah, MH. Biological effect of SAR on the human head due to variation of dielectric properties at 1800 and 2450 MHz with different antenna substrate materials. *J Science and Engineering of Composite Materials*2015; **22**:411-415.
15. Tateno, A, Nagaoka, T, Saito, K. Variability of Specific Absorption Rate of Human Body for Various Configurations of Tablet Computer in Vicinity of Abdomen. *J IEICE Transactions on Communications*2015; **E98B**:1173-1181.
16. Hossain, MI, Faruque, MRI, Islam, MT. Miniature low SAR printed monopole antenna for mobile phone. *J Microwave and Optical Technology Letters*2015; **57**:2471-2475.
17. Zhou YJ, Zhang H, Niu ZQ. Analysis of Electric Stress in Human Head in High-frequency Low-power Electromagnetic Environment. *J Journal of Biomedical Engineering*2015; **32**:295-299.

STUDY ON EXTRACTION PROCESS OF THE FLAVONOIDS FROM SOW THISTLE

Cili Li[a,*], Shiyou Yu, and Qi Zhang

Food and Engineering Department, Harbin University of Commerce, Harbin, No.138 Tongda Street, China

E-mail: [a]foodlcl@163.com

[*]Corresponding author

The flavonoids were extracted from sow thistle by ethanol reflux method. The optimum extraction conditions were determined by single-factor and orthogonal experiment, using the ratio of material to liquid, the volume fraction of ethanol, extraction temperature and extraction time as the independent variables during the extraction process. The experimental results showed that optimal extraction conditions of ethanol reflux method were: material: liquid ratio 1:70g/mL, ethanol-water mixture 40%-volume fraction, extraction temperature 80˚C and extraction time 3h. Under this condition, the extracting rate of flavonoids was 7.23mg·g^{-1}.

1. INTRODUCTION

Perennial sow thistle (Sonchus arvensis) belongs to the families Asteraceae grouped into the subfamily of lettuce[1-2]. Nutrition facts for fresh sow thistle per 100 g contain 5 g carbohydrate, 2.2 g protein, 0.4 g fat, 150 mg calcium, 59 mg phosphorus, iron 52 mg and a large number of trace elements. They are rich in vitamin, including 3.98 mg carotene, 28 mg vitamin C, 0.10 mg vitamin B1, 0.27 mg vitamin B2, 20 mg niacin, etc. Besides that, they also contain tartaric acid, choline, bitter principle and a fistful of phytochemicals including flavonoids[3-7].

As renewable vegetables resources, sow thistle contains abundant flavonoids[8]. Another aspect to consider is the raw materials are cheap, easy to obtain and suitable for industrial mass production. Therefore, extraction of flavonoids from sow thistle offer great potential application in medicine and food[9,10].

In this paper, the influence of process conditions on the flavonoids extraction from sow thistle by ethanol reflux method was investigated, and extraction parameters were optimized. This provide a theoretical basis for utilizing resource advantage of sow thistle and deep processing in our country.

2. MATERIALS AND METHODS

2.1. Materials and Reagents

The fresh sow thistle was sold in Harbin City. Standard rutin was procured from Tianjin science and Technology Co., Ltd. All chemical reagents used were analytical reagent grade, such as sodium nitrite, aluminum nitrate, sodium hydroxide, anhydrous alcohol, ether, etc.

2.2. Method

2.2.1. *Pretreatment*

Initially, the fresh sow thistle was cleaned by DI water to get rid of impurities or muddiness. After rinsing, it was dried at 60°C for 8h in electric thermostat blast oven. And then, dried materials were milled by Chinese herbal medicine and screened to pass through 40-mesh screen. The obtained sow thistle powder was store in a tightly closed bottle.

1g as-prepared dry powder of sow thistle wrapped in filter paper was added into proper amount of ether, and heated under reflux for 12 h in a soxhlet extractor to remove fat-soluble pigment. Then the samples were air-dried in the ventilation cabinet until the solvent is evaporated completely and hermetically closed for the use. Ether, which had been used on the decolorization treatment, was recovered by heating reflux method.

2.2.2. *Determination of Flavonoids Content in Sow Thistle*

(1) Plotting the standard curve

Rutin standard (10.00 mg) was weighed with an analytical balance, and placed into a 100mL volumetric flask, the standard sample was diluted to 0.1 mg·mL-1 with 60 vol.% ethanol as the standard solution.

Firstly, Standard solution (0, 1.0, 2.0, 3.0, 4.0, 5.0 mL) were placed into colorimetric tubes (25mL), then adding 1.0 mL of 5 wt.% NaNO2 solution and standing for 6 min after shaking. Next, 1.0 mL of 10 wt.% Al(NO3)3 were added into colorimetric tubes respectively, then standing for 6 min after shaking again. Finally, 10 mL 4 wt.% NaOH solution were added and a certain concentration was achieved by adding 60 vol.% ethanol to 25 mL. The absorbance value was measured at a wavelength, and the standard curve was plotted (rutin concentration as abscissa, the absorbance value as the vertical axis).

(2) Extraction method of flavonoids

As-prepared dry powder of sow thistle was weighed accurately for the triplicate experiments. After setting the experiment parameters, it was extracted by heating reflux using ethanol solution as the extraction agents. After ethanol extraction, a concentrate was collected by suction filtration, concentration and solution volume was fixed to 25mL by adding 60 vol.% ethanol.

(3) Determination method of flavonoids content in sow thistle

After cooling the extraction solution, prepared as 1.3.3.2 section, was were placed into colorimetric tubes (25mL), then adding 1.0 mL of 5 wt.% NaNO2 solution and standing

for 6 min after shaking. Next, 1.0 mL of 10 wt.% Al(NO3)3 were added into colorimetric tubes respectively, then standing for 6 min after shaking again. Finally, 10 mL 4 wt.% NaOH solution were added and a certain concentration was achieved by adding 60 vol.% ethanol to 25 mL. As the method mentioned above, the absorbance value was measured at a wavelength, the flavonoids concentration was calculated on the basis of the standard curve.
(4) Calculation of flavonoids extraction

$$\text{Flavonids content in the sample } (\text{mg}\cdot\text{g}^{-1}) = \frac{CV_1 \times V_2}{M \times V}$$

C — flavonoids concentration was calculated on the basis of the standard curve. /mg·mL^{-1}, *V* — the volume of extractions added /mL, V_1— the volume of extractions added when measuring /mL, V_2— the volume at fixed volume /mL, *M*— weigh of dry powder of sow thistle /mg

2.2.3. *Study on Extraction Process of Flavonoids from Sow Thistle*

1.00 g dry powder of sow thistle, 30 vol.% ethanol, extracting temperature 70 °C and extracting time 3h was regarded as the basis extraction conditions to determine the best solid–liquid ratio. the obtained solid- liquid ratio was applied for next group of single-factor experiment to determine the best factor levels. And proceed sequentially, the effect of several factors on both flavonoids yield was explored to determine the optimal level settings for each single factor. (1) Effect of material-liquid ratio on the flavonoids yield was investigated: 1:30, 1:40, 1:50, 1:60, 1:70, 1:80, 1:90 g·mL^{-1} material-liquid ratio was selected for relevant single-factor experiment. (2) Effect of ethanol volume fraction on the flavonoids yield was investigated: 20%, 40%, 30%, 50%, 60%, 70%, 80% volume fraction were selected for relevant single-factor experiment. (3) Effect of extraction temperature on the flavonoids yield was investigated: 20°C, 30°C, 40°C, 50°C, 60°C, 70°C, 80°C volume fraction were selected for relevant single-factor experiment. (4) Effect of extraction time on the flavonoids yield was investigated: 1.0h, 1.5h, 2.0h, 2.5h, 3h, 3.5h, 4h extraction time were selected for relevant single-factor experiment.

Four factors during ethanol reflux method influencing the flavonoids yield of sow thistle was studied and analyzed using an L9(34) orthogonal array design. According to the results of single-factor experiment, the factors and levels of orthogonal experiment were determined, as shown in Table 1.

Table 1. Factors and levels of orthogonal experiment of L9 (34) for ethanol reflux method

Level	Factor			
	Material-liquid ratio / g·mL^{-1}	Ethanol volume fraction / %	Extraction temperature / °C	Extraction time / h
1	1:50	30	60	2.5
2	1:60	40	70	3
3	1:70	50	80	3.5

3. RESULTS AND ANALYSIS

3.1. Plotting the Rutin Standard Curve

The horizontal coordinate referred to the rutin concentration (C), and the absorbance value (A) was vertical coordinate. The linear equation was obtained: $A = 24.189C + 0.017$, $R^2 = 0.9991$. The measurement results of the standard curve were shown in Fig. 1.

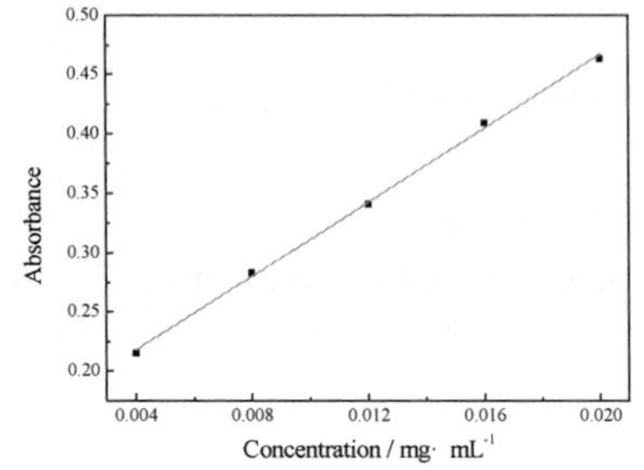

Fig. 1. Linear fitting plot of absorbance and the rutin concentration.

3.2. Extracting Flavonoids from Sow Thistle

According to the experiment, the effect of material-liquid ratio on flavonoids yield by ethanol refluxing was shown in Fig. 2. In the beginning the flavonoids yield increases with the rise of material-liquid ratio and continues to accelerate the rise with range from 1:60 to 1:30 g·mL^{-1}, and the maximum yield (approximately 6.1mg·g^{-1}) is reached on the ratio of 1:60. However, material-liquid ratio exceeds 1:60, flavonoids yield starts to decrease slightly. Therefore, 1:60 g·mL^{-1} is selected as the appropriate material-liquid ratio.

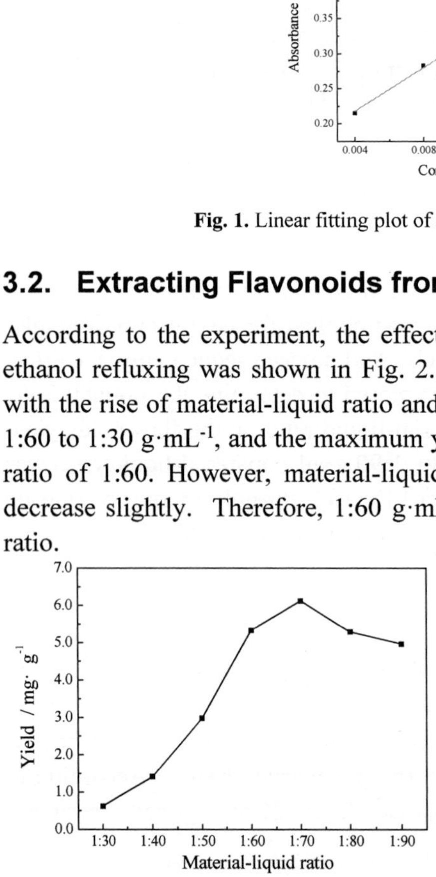

Fig. 2. Relationship between flavonoids yield and material-liquid ratio by ethanol refluxing.

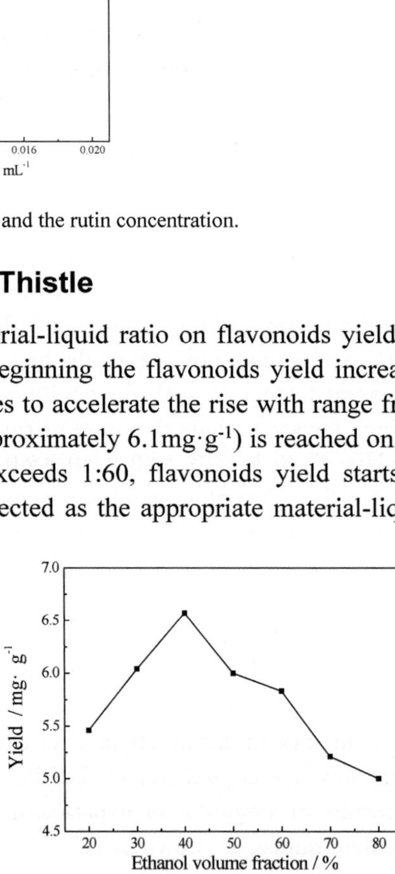

Fig. 3. Relationship between flavonoids yield and ethanol volume fraction by ethanol refluxing.

According to the experiment, the effect of ethanol volume fraction on the flavonoids yield by ethanol refluxing was shown in Fig. 3. The flavonoids yield increases with a shift to higher ethanol volume fraction (20% to 40%), when the ethanol volume fraction attain 40%, the maximum flavonoids are obtained. When using ethanol volume fraction greater than 40%, flavonoids yield began to decrease slightly. Consequently, 40% ethanol volume fraction is more suitable for extracting flavonoids.

According to the experiment, the effect of extraction temperature on the flavonoids yield for ethanol reflux method was shown in Fig. 4. When the temperature rise ranging from 20°C to 70°C, the flavonoids yield goes up steadily and reached to the maximum value at 70°C. When the temperature exceeds 70 0°C, the yield begins to decline. From the point of improving the flavonoids yield, 70°C is chosen in this experiment as the extraction temperature.

According to the experiment, the effect of extraction time on the yield of flavonoids was shown in Fig. 5. At the beginning, the flavonoids yield increases with the increase of time. The flavonoids yield has been rising from 1h to 3h, and reaches the maximum value at the extraction time of 3h. When the extraction time increase more than 3h, a slight decrease of flavonoids yield is revealed. In summary, the optimal extraction time is 3h.

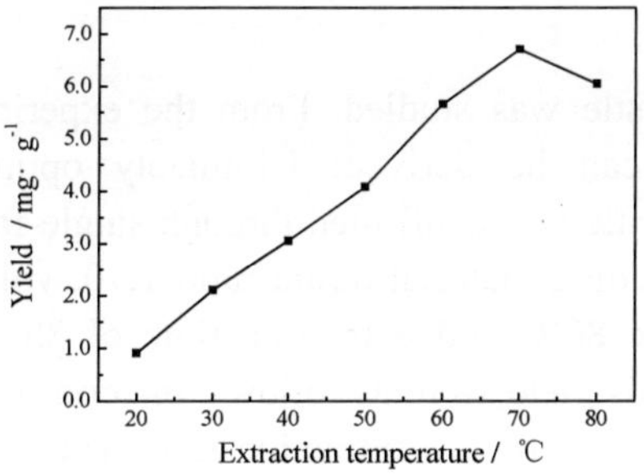

Fig. 4. Relationship between flavonoids yield and extraction temperature by ethanol refluxing.

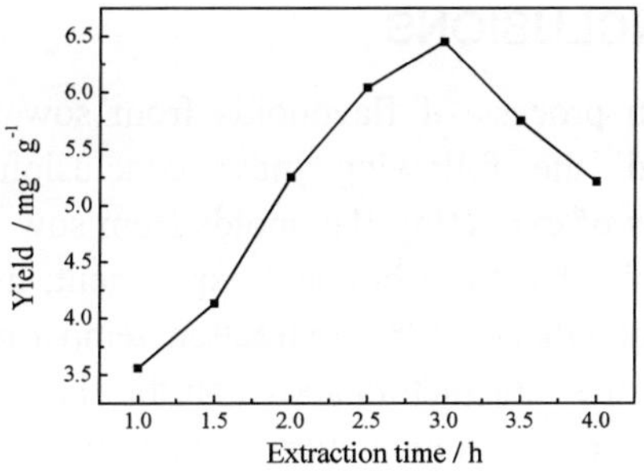

Fig. 5. Relationship between flavonoids yield and extraction time by ethanol refluxing.

With sow thistle as raw material and a mixture of ethanol-water as extraction agent, the results of Orthogonal by ethanol refluxing were shown in Table 2.

From the results in Table 2, it can be concluded that depending on the range, the order of the effect of the different factors on the flavonoids extraction for ethanol reflux method is A>C>B>D (material-liquid ratio > extraction temperature > ethanol volume fraction > extraction time). And determination of the optimal combination of flavonoids extraction is A3B2C3D2. Under this condition, the higher flavonoids yield ($7.23 mg \cdot g^{-1}$) was obtained in triplicate experiments. Therefore, the optimum conditions of flavonoids extraction from sow thistle with ethanol refluxing methods was determined: material-liquid ratio 1:70, volume fraction of ethanol 40%, extraction temperature 80°C and extraction time of 3h.

Table 2. Orthogonal experiment results of the flavonoids extraction process by ethanol refluxing

Number	Factor				Flavonoids yield / mg·g^{-1}
	Material-liquid ratio / g·mL^{-1}	Ethanol volume fraction / %	Extraction temperature / °C	Extraction time / h	
1	1	1	1	1	5.84
2	1	2	2	2	6.38
3	1	3	3	3	5.95
4	2	1	2	3	6.49
5	2	2	3	1	6.95
6	2	3	1	2	6.32
7	3	1	3	2	7.01
8	3	2	1	3	6.46
9	3	3	2	1	6.57
K1	18.17	19.34	18.62	19.36	
K2	19.76	19.79	19.44	19.71	
K3	20.04	18.84	19.91	18.90	
k1	6.06	6.45	6.21	6.45	
k2	6.59	6.60	6.48	6.57	
k3	6.68	6.28	6.64	6.30	
R	0.62	0.32	0.43	0.27	

4. CONCLUSIONS

Extraction process of flavonoids from sow thistle was studied. From the experiment performed, the following main conclusions can be derived: Ultimately optimum conditions of extracting flavonoids from sow thistle was confirmed through single-factor test combined with orthogonal experiment, including material-liquid ratio 1:70, volume fraction of ethanol 40%, extraction temperature 80°C and extraction time of 3h. The effects of the four factors above on the flavonoids yield were as follows: material-liquid ratio > extraction temperature > ethanol volume fraction > extraction time. Under this condition, maximum flavonoids were extracted from sow thistle using ethanol reflux method. a high flavonoids yield of 7.23mg·g-1 was reached.

References

1. Eckersten H, Lundkvist A, Torssell B. Comparison of monocultures of perennial sow-thistle and spring barley in estimated shoot radiation-use and nitrogen-uptake efficiencies. *Acta Agriculturae Scandinavica Section B–Soil and Plant Science* 2010; **60**: 126-135.
2. Chauhan BS, Gill G, Preston C. Factors affecting seed germination of annual sowthistle (Sonchus oleraceus) in southern Australia. *Weed Science* 2009; **54**: 854-860.
3. De-Sheng Jia, Fa-Zeng Li. The revision of the plant of the Shandong lettuce. *Acta Botanica Boreali-Occidentalia Sinica* 2003; **23**: 443-450. (In Chinese).

4. Wei-Guo Cao, Zhi-Qin Liu, Yun Shao, et al. Research progress on pharmacological effects of flavonoids. *Acta Botanica Boreali-Occidentalia Sinica* 2003; **23**: 2241-2247. (In Chinese).
5. Yang Liu, Zhong-Hua Liu. The development prospect of sow thistle tea. *Farm Products Processing* 2014; **6**:44-45. (In Chinese).
6. Zhan-Xin Zhang, Wei-Dong Xie, Ping-Lin Li, et al. Sesquiterpenoids and phenylpropane derivatives from Sonchus uliginosus. *Helvetica Chimica Acta* 2006; 89: 2927.
7. Yang-Jun Xu, Shao-Bo Sun, Li-Mei Sun, et al. Quinic acid esters and sesquiterpenes from Sonchus arvensis. *Food Chemistry* 2008; 111: 92.
8. Jie Yin, Chang-Ling Si, Myeong-Hyeon Wang. Antioxidant Activity of Flavonoids and Their Glucosides from Sonchus oleraceus L. *Journal of Applied Biological Chemistry* 2008; **51**: 57-60.
9. Pei-Zhuo Hu, Chuan-Zong Zou, Ying Zhu. Lipid-soluble Chemical Constituents in sonchus oleraceus. *Acta Botanica Boreali-Occidentalia Sinica* 2005; **25**:1234-1237. (In Chinese).
10. Nancy E, Rawson, Chi TH, et al. Efficacious anti-cancer property of flavonoids from citrus peeis. *Food Science and Human Wellness* 2014; **3**:104-109.

SEMILETHAL TEMPRATURE AND COMPREHENSIVE EVALUATION OF EIGHT KINDS OF GRAPE ROOTSTOCK OF HARDINESS

Pengcheng Li[1,a], Jingjing Wang[1,b,*], Xuede Su[1], Shaojie Guo[1], Ming Li[1], and Xiaofa Fu[2]

[1]Xinjiang Academy of Agricultural and Reclamation Sciences, Shihezi, Xinjiang 832000 China

[2]Xinjiang Academy of Agricultural Sciences, Shihezi, Xinjiang 832000 China

E-mail:[a]lpc830916@163.com,[b]xwangjj530@163.com

[*]Corresponding author

Eight kinds of imported grape rootstock hardiness are evaluated. One year old '5BB', 'Shangheyihao', 'SO4', 'Kangzhensanhao', 'Beida', 'Shuanghong', 'Beibinghong', 'BK-1' grape rootstock branches as test materials, after -40°C~4°C different cold stress treatment 12h, the determination of different varieties of grape rootstock branches of water loss rate, the change of the electrolyte leakage rate, germination rate, and comprehensive evaluation of 8 kinds of grape hardiness in the average of 3 indicators. The results show that electrolyte leakage rate of each type increased significantly with lower processing temperature; Germination rate decreases with the decrease of processing temperature. The cold resistance of 8 kinds of grape was evaluated with the average value of 3 indexes of different strains, and the results are consistent with the comprehensive evaluation results of the membership function method. Cold hardiness of grape variety tested membership function method, comprehensive evaluation, it was concluded that four selected varieties of grape rootstock of hardiness pecking order was: 'BK-1'> 'Shuanghong'> 'Beida'> 'Shangheyihao'> '5BB'> 'Beibinghong'> 'Kangzhensanhao'> 'SO4'.

1. INTRODUCTION

Grape is the high economic value of fruit trees with world's long history of cultivation belonging to families in plant taxonomy and Vitis. Grape industry has been developed rapidly in recent years and become one of the main ways to get rich for many farmers. Chinese grape production was mainly in the northern region, which had abundant sunlight and the temperature difference between day and night so that the grape had high yield and good quality with the production of unique advantages, but the cold winter weather seriously affected the northern grape mass production. North China winter cold weather often makes grapes suffer different degrees of low temperature injury, and grapes were damaged severely, so they need to be buried in cold winter, but it resulted in higher costs and limited the development of northern grape industry[1]. In order to reduce frost damage, cold grafted promotion is one of direct and efficient way[2]. As a result, the study of cold rootstocks and new cold cultivate varieties of grapes is practically significant and valuable.

The identification of cold tolerance and chilling choice on grapes have been reported[3-5]. Researchers worked on the screening study of plant cold tolerance identification[6] in order to provide an effective method for the plant cold resistance identification. However, due to the complexity of the cold hardiness mechanisms and the difference of breeding objectives, the identification of plant hardiness has not unified identification methods and indicators.This study on eight kinds of grapes resistant rootstock membership function[7] by comprehensive comparison, provide a theoretical basis for the Xinjiang region and the development of our Cold rootstock selection and physiological research.

2. MATERIALS AND METHODS

2.1 Test Material

October 2013 20-12 30th, the 1-year-old mature branches with the same cultivation and management are selected substantially as test materials. Eight grape rootstock varieties are selected '5BB', 'BK-1', 'SO4', 'Kangzhensanhao', 'Beida', 'Shuanghong', 'Beibinghong', 'BK-1' (tentative), roughness of 0.6-0.8 cm year old branches are takenas test materials, each species needs 15 branches (2-3 buds per branch).

2.2 Test Processing

After the test material is rinsed several times with tap water and distilled water, it with good plastic bag is placed in 4°C low temperature freezer in the refrigerator to be treated, the test process temperature -15°C, -20°C, -25°C, -30°C, -35°C, -40°C and control (material not cold treatment, remove from the refrigerator directly measured in the refrigerator at 4°C), test the cooling rate and heating rate control 4°C/h, maintained at the set temperature 12h (with JL-1 digital point thermometer monitoring), remove and place in the refrigerator at 4°C refrigerator to thaw, and then measured their physiological and biochemical indicators. Each processing device 3 repeats.

2.3 Index Determination and Cold Resistance

2.3.1 Relative Conductivity The relative conductivity of grape shoots body is determinedby the conductance method[8], each sample was repeated three times.

2.3.2 Germination Rate Germination rate is measured by the growth method[9], Measurement methods are as follows: the fall collection roughness, growth potential, consistent with each rootstock mature branches (each species 50), placed at -10°C low temperature for 24 hours, after removing the water immersion 12h, cut single or double bud cuttings stem segments in the laboratory (18-22°C) for cutting test, 30d after germination rate statistics. Germination rate was calculated as Eq. (1).

$$k(\%) = x / y \times 100\% \tag{1}$$

2.3.3 Water Loss Rate The water loss rate is measured by the method of indoor drying and the [10] of the natural water loss method.

2.3.4 Logistic Equation and Semi-lethal Temperature *(LT50)* Calculation Methods Gai Junyi [11] method, grapes cold study, the relative conductivity of fitting Logistic regression equation in Eq.(2).

$$y=K/(1+ae^{-bx}) \tag{2}$$

Wherein Y represents cell injury rate, x representatives treatment temperature, K is a cell injury rate of saturation capacity, a, b for the equation parameters. In order to determine a, b values, will be linearized equation in Eq.(3).

$$ln\,[(Ky)/y]=lna^{-bx}, \text{ so } y1=ln[(Ky)/y] \tag{3}$$

Then transformed into cell injury rate (y1) and treatment temperature (x) is the linear equation. Obtained by linear regression method a, b values and the correlation coefficient

R, lethal temperature in Eq.(4).

$$LT_{50}=ln[(1/a)]/b \tag{4}$$

2.4 Data processing and Analysis

Using Microsoft Excel and SPSS17.0 for data processing, and the difference is significant indicators Duncan multiple comparisons. While the data collated by fuzzy membership formula[7] quantitative conversion, and then the membership function value of each index averages are compared with each other clones.

Membership function formula in Eq.(5).

$$U(Xi)=\frac{Xi-X\min}{X\max-X\min} \tag{5}$$

If an index with the evaluation result is negative, then the membership function is quantified by reverse conversion.

The formula in Eq.(6).

$$U(Xi)=1-\frac{Xi-X\min}{X\max-X\min} \tag{6}$$

Where: U (Xi) is the membership function value, Xi is a clone indicator measured values, Xmax and Xmin are the maximum and minimum values of certain indicators in all tested clones. The average membership of each grape rootstocks three indicators is calculated according to the above formula.

Referring Liao Jingrong et al.[12], according to the average membership, the method will be divided into five cold resistance: 0.70-1.00 high resistance (High Resistance, HR), defined as grade I; 0.60-0.69 anti (Resistance, R) , as grade II; 0.40-0.59 for the anti-(Middle Resistance, MR), defined as grade III; 0.30-0.39 for the low resistance (Lower Resistance, LR), as grade IV; 0-0.29 is not resistant (Susceptible, S), designated as V level.

3. RESULTS AND ANALYSIS

3.1 Changes of Relative Electrical Conductivity of Branches After Low Temperature Treatment

The Table 1 shows that in low temperature start, all selected grapes material conductivity decreases, but different species begin to rise the temperature of different conductivity. At -15 °C, '5BB', 'BK-1', 'SO4', the conductivity of 'Kangzhensanhao', 'Beida', 'Shuanghong', 'Beibingred' and 'BK-1' is lower than that of control (CK). Under controlled temperature (4°C), the 'SO4' and 'Kangzhensanhao', 'Beida' and 'Shuanghong' has no obvious difference. But the conductivity of 'Kangzhensanhao', 'Shuanghong' and 'Beibingred' is higher at - 20 °Cthan -15 °C, the conductivity of other varieties at -25 °C began to rise, the best frozen temperature of 'Kangzhensanhao', 'Shuanghong' and 'Beibingred' is -20 °C. Overall, -25 °C is good to distinguish the various varieties of cold hardiness, therefore, the processing temperature of -25 °C is as cold stress in this study.

Table 1. The efect of diferent cold temperature on the leakage of electrolytes of grapes' branch

Lines/Cultivars	Lerkage of electriolytes(%)						
	CK	-15°C	-20°C	-25°C	-30°C	-35°C	-40°C
Beida	42.6Dd	41.7Cd	34.0Ef	52.8Bd	66.3BCd	86.9ABb	89.0ABbc
Shanheyihao	45.6Cc	45.4BCc	39.5De	48.9BCe	59.8Cef	76.8Ce	89.0ABbc
SO4	55.5Aa	47.5Bb	52.3Bb	63.0ABb	86.3Aab	87.6ABb	96.5Aa
Kangzhensanhao	57.7Aa	54.0Aa	57.4Aa	68.2Aa	88.9Aa	89.3Aa	94.3Aab
5BB	49.3Bb	43.0Cc	41.3Dc	54.4Bc	69.6Bc	87.1ABb	91.5ABb
Shuanghong	44.2Dd	35.3Ee	37.4DE	49.7BCe	58.6Cf	77.0Ce	87.1Bc
Beibinghong	45.6Cc	37.8De	44.5Cc	50.0Be	61.3Ce	79.9Cd	93.2ABab
BK-1	48.9Bb	46.3Bb	41.1Dc	49.9BCe	62.5Ce	83.8Bc	86.0Bc

Note: Small letters within table means significance at 5% leve1， capital letters means significance at 1% level.

3.2 Establishment of the Logistic Model of Different Varieties Grape Rootstock and Semilethal Low Temperature

Obtaining the Logistic equation, semilethal low temperature and fitting degree (R^2) by the relative conductivity of the materials in different low temperature (Table 2). According the result of fitting statistics of Logistic equation (Table 2), the range of semilethal temperature of 8 grape rootstocks is -23.86 to -33.21°C. The semilethal temperature of 'Beida', '5BB', 'Shuanghong', 'Beibinghong' and 'BK-1' is lower than -30°C, other varieties are higher than -30°C.

Table 2. Logistic equation and semi-lethal temperature(LT50)of eight grape rootstocks

Lines/Cultivars	Treat temperature (□)	Logistic equation	Correlation coefficient (R^2)	Semi-lethal temperature(□)
Beida	-40～4	$y=100/(1+1.5219e^{0.0134x})$	0.935^{**}	-32.18
Shanheyihao	-40～4	$y=100/(1+1.7005e^{0.0147x})$	0.924^{**}	-28.70
SO4	-40～4	$y=100/(1+1.4106e^{0.0143x})$	0.957^{**}	-25.32
Kangzhensanhao	-40～4	$y=100/(1+1.6391e^{0.0131x})$	0.898^{*}	-23.86
5BB	-40～4	$y=100/(1+1.4944e^{0.0137x})$	0.944^{**}	-30.04
Shuanghong	-40～4	$y=100/(1+1.6435e^{0.0130x})$	0.964^{**}	-32.86
Beibinghong	-40～4	$y=100/(1+1.6313e^{0.0149x})$	0.921^{**}	-30.05
BK-1	-40～4	$y=100/(1+1.6012e^{0.0138x})$	0.941^{**}	-33.21

Note: The data in the table are averages.

3.3 The Comprehensive Evaluation of Cold Resistance of Different Grape Rootstocks

From the physiological indices of cold resistance (Table 3), the electrical conductivity of BK-1 is slightly higher than 'Beida' and 'Shuanghong', but its average germination rate is the highest, the dehydrating rate is higher than ‘Beida’ but lower than ‘Shuanghong’, the average subordinative degree is highest, ‘Shuanghong’ and ‘Beida’ second. Contrast the cold resistance level of subordinative degree, ‘BK-1’, ‘Beida’, ‘Shuanghong’ all belong to the type of high cold resistance. Divided into I grade of cold resistance; the average subordinative degree of ‘Shanheyihao’ is 0.68, belong to the type of cold resistance, divided into II grade of cold resistance; the electrical conductivity, germination rate and dehydrating rate of ‘5BB’ and ‘Beibinghong’ in the middle level in all varieties. Belong to the type of middle cold resistance, divided into III grade of cold resistance; the dehydrating rate of ‘Kangzhensanhao’ is minimum, but electrical conductivity was higher, germination rate in the middle level, the average subordinative degree is less, the electrical conductivity of ‘SO4’ is lowest, germination rate is lowest, the average subordinative degree is lowest, ‘Kangzhensanhao’ and ‘SO4’ belong to the type of not cold resistance, divided into IV grade of cold resistance.

Table 3. Different covering methods on grapes under cold Comprehensive evaluation results

Lines/Cultivars	Relative conductivity (%)	Water loss raeio	Percentage of bud bursting(%)	Average of SF	Cold rating
Beida	48.2	1.21	64.3	0.70	I
Shanheyihao	48.8	1.09	54.9	0.68	II
SO4	57.3	1.41	37.8	0.14	V
Kangzhensanhao	54.5	1.43	50.5	0.23	V
5BB	49.3	1.23	53.3	0.52	III
Shuanghong	51.9	1.34	63.7	0.83	I
Beibinghong	49.7	1.64	43.8	0.47	III
BK-1	45.1	1.23	67.4	0.90	I

Note: The data in the table are averages.

3.4 Correlation Analysis of Cold Resistance Indexes

In order to further study the correlation degree between the 3 cold resistance indexes, the correlation analysis of the water loss rate, electrical conductivity, germination rate and average membership degree of the above mentioned materials (Table 4) are made. As can be seen from Table 4, the correlation coefficient between average membership and water loss rate reaches a significant level, and the correlation coefficient between electrical conductivity and germination rate reaches a very significant level, the correlation coefficient is above 0.6, and the correlation between the other 3 indicators is relatively poor, which can be used to represent the other 3 indicators of the size of the grape cold resistance. A significant negative correlation is observed between the electrical conductivity and the germination rate (R=-0.578). There is a negative correlation between water loss rate and germination rate (R=-0.214), but not significant.

Table 4. Analysis of correlation of indexes for cold resistance

Index	Correlation coefficient(*R*)			
	*X*1	*X*2	*X*3	*X*4
*X*1	1.000			
*X*2	0.693*	1.000		
*X*3	-0.578*	-0.214	1.000	
*X*4	0.927**	0.608*	0.891**	1.000

Note: X1.Lerkage of electriolytes;X2.Water loss raeio;X3. Percentage of bud bursting;X4.Average of SF;*Significant eorelation in 5% level;**Significant eorelation in 1% level.

4. DISCUSSION

Identification of cold resistance of Grape Germplasm resources has provided a lot of theoretical basis for the selection of cold resistant grape varieties[13]. Niu Li-xin and other[7], Zhang Wen-e and others[9] have a more systematic study of the cold identification method, the damage degree of low temperature stress on plant tissue in addition to relate to the low temperature, but also with the cooling rate, the duration of the cooling and the speed of the temperature and other factors[7]. In this study, the cooling rate and heating rate are controlled to 4°C/h, and the cooling time is maintained at 12h, and the best freezing temperature of -25°C is used as the germination rate test of each stock.

Electrical conductivity with Logistic equation is used to estimate the semi lethal temperature and compare the cold resistance difference of plant tissue, which is applied in many kinds of plants and got good response[14]. The method has been widely used in the study of the cold resistance of grape populations, grape roots and varieties[15-16]. In this experiment, the cold resistances of introduced and screened eight grape varieties are compared with the Logistic equation of electrical conductivity method, according to the semi lethal temperature, the strong-to-weak sequence of cold resistance of eight varieties are: 'BK-1'> 'Shuanghong'> 'Beida'> 'Shanheyihao'> '5BB'> 'Beibinghong'> 'Kangzhensanhao'> 'SO4'.

The grape cold resistance is a complex physiology and biochemical process. It is difficult to reflect this parameter only depending on a single indicator. There are many defects in the identification of cold resistance of the grape germplasm resources due to the limitation of the previous methods and the resources limitation[17]. The analysis shows that three physiological indexes has a significantly correlation with the average membership degree and the determination coefficients is more than 0.5, so the average membership degree could be used to replace the three physiological indexes. This method is more comprehensive and representative so it could be used in the identification of cold resistant germplasm resources and breeding of new cold resistant varieties. The estimation results of the average membership method is consistent with the Logistic equation of electrical conductivity method for the semi lethal temperature of grape shoots, it is more scientific to use the semi lethal temperature as the evaluation and identification of the cold resistance of grape rootstocks.

The cold resistance is controlled by many factors and there are different mechanisms between different varieties. In order to analyze the cold resistance mechanism of different materials completely, it is necessary to evaluate the parameter in many ways. The cold resistance of plant resources is identified to lay the groundwork for cold resistance breeding.

5. CONCLUSION

Under the condition of artificial low temperature treatment, we selected three key indicators including electrical conductivity, germination rate and the water loss rate to identify the cold resistance of the tested grape materials. Then the cold resistance is evaluated by the membership function method to enhance the identification accuracy of the grape branches, and the results are more reliable. Considering of various measurement indicators we concluded that the strong-to-weak sequence of cold resistance of the varieties/lines are: 'BK-1'>'Shuanghong'> 'Beida'> 'Shanheyihao'> '5BB'> 'Beibinghong'> 'Kangzhensanhao'> 'SO4'.

Reference

1. Yan Xiong, Wan-Min Zhang. Cold Resistance of grapes Overview.*Northern Horticulture* 2007;**6**:69-71. (In Chinese).
2. Jin-Feng Niu, Ji-Cheng Ping, Zhen-Ping Wang, et al. A Comparative Study of cold resistance of several table grape varieties.*Northern Horticulture* 2005;**6**:63-65. (In Chinese).
3. Chen-Huan Liu, Ji-Feng Chen, Qin-Shan Kong. Evaluation of the anti-frost grape germplasm.*Crop Germplasm Resources* 1999;**3**:46-49. (In Chinese).
4. Xiu-Wu Gao, Wang-Heng Fu, Guang-Jie Wang. Studies on cold hardiness of grape roots.*Acta Hprticulturae Sineca* 1989;**1**:17-22.(In Chinese).
5. Jian-Dong Cao, Bai-Hong Chen, Li-Jun Wang, et al. Cold resistance indexes identification and comprehen-sive evaluation of grape varieties. *Acta Bot Boreal-Occident Sin* 2010;**11**:2232-2239. (In Chinese).
6. Jun-Wei Yuan, Ai-hong Ma, Zi-Juan Guo,et al. Discussion grape rootstock resistance identification methods. *Sino-Overseas Grapevine & Wine* 2011;**11**:66-75.(In Chinese).
7. Wen-E Zhang, Fei Wang, Xue-Jun Pan. Comprehensive evaluation on cold hardiness of Vitis species by Subordinate Function (SF). *Journal of Fruit Science* 2007;**6**: 849-853. (In Chinese).
8. Ameguo T, Pigeon D, Archilla O. Adaptation to cold tern-perature and response to freezing in roses.*Acta Hort* 2003;515-520.
9. Li-Xin Niu, Pu-Chao He. Grape cold resistance identified by growth method.*Journal of Fruit Science* 1991;**1**:40-42. (In Chinese).
10. Wen-E Zhang, Xue-Jun Pan, Fei Wang. Moisture content of grape branches and Cold Resistance Identification.*China Fruit* 2007;**1**:14-15. (In Chinese).
11. Jun-Yi Gai. Statistics for experiments. Bering:China Agricultural Press:56-59. (In Chinese)
12. Lin Ai. Table-grape Hardiness research. Urumqi:Xinjiang Agricultural University 2003;44.(In Chinese).
13. Pu-Chao He, Wu-Ji Chao. Cold analysis of vitis wild germplasm resources. *Acta Hprticulturae Sineca* 1982;**3**:17-21. (In Chinese).
14. Yan-Ping Liu, Yan-Lin Zhu, Xian-Yang Kang, et al. Cold resistance determination of different type Magnolia grandiflora with synergistic electrical conductivity

method and Logistic equation.*Journal of Central South University of Forestry& Technology* 2012;**10**: 69-71. (In Chinese).

15. Xiao-He Ma, Xiao-Ping Tang, Zhi-Gan Dong,et al. Comparison of Cold Resistance for 6 Grapewine Cultivars. *Journal of Shanxi Agricultural University(Nat.Sci.Ed.)* 2013;**1**: 1-5. (In Chinese).
16. Qian Zhang, Chong-Huai Liu, Da-Long Guo, et al. Relationship between LT_{50} and cold adaptability of five grape varieties. *Journal of Northwest A&F University (Nat.Sci.Ed.)* 2013;**5**:149-154. (In Chinese).
17. Run-Gang Song, Wen-Peng Lu, Yu-Jie Shen,et al. A New Ice-red Brewing Grape Cultivar 'Beibinghong'. *Acta Hprticulturae Sineca* 2008;**7**:1085. (In Chinese).

THE CHARACTERISTICS, RESEARCH AND DEVELOPMENT OF THE AGED-FOODS

Lidong Wang[a], Weijiang Fan[b,*], Yanping Hu[c], and Deyang Sun[d]
Collage of Food and Drug, Shandong Institute of Commerce and Technology, Jinan, China
E-mail: [a]73653936@qq.com, [b]weijiangfan@126.com, [c]1355929195@qq.com, [d]1324655356@qq.com
[*]Corresponding author

In today, the aging populations were more and more attention from all of peoples. These problems have been focused on how to delay senescence and prevention and treatment of senile diseases etc. The development of elderly food, diet and nutrition in the elderly is an important aspect of anti-aging, prevention-diseases. The food's demands of the elderly were reviewed in this paper. According to the need of nutrition and health of elder people, we put forward the direction of aged food of research and development in the future.

1. CHARACTERISTICS OF AGED-FOODS

According to the physiological characteristics of the elderly, the aged-foods should have the following characteristics: 1, the aged-foods nutrition should show easier to chew, nutritious, digestible, absorbent, and minor irritation. 2, all kinds of nutrients should be adapted to the physiological characteristics and needs of the elderly. 3, all kinds of health functional food should play roles of the prevented, treated and recovered in common diseases of old people. 4, soft and taste pure should be the Characteristics of aged-foods to adapt to the old people's dietary habits. 5, the packages of aged-foods should be simple and convenient for the characteristics of eating more frequent, smaller meals small in the old peoples. The aged-foods should be in line with the "four feet of five low" nutritional basic principles, namely, enough quality protein, enough food fiber, enough vitamin and enough mineral elements, low calorie, low fat, low sugar, low cholesterol and low sodium [1,2].

2. APPLICATION OF SPECIAL FOOD INGREDIENTS IN AGED-FOODS

The material basis is foods for human survival and development. Different groups of people have different requirements for foods. Young people are asked for bodybuilding, and older people are required to live longer. These requirements had been the main direction of research and development in foods.

2.1. Soybean lecithin

Soybean lecithin is the essence of soybean. Lecithin is the basic substance in human cells, which is responsible for the information and energy transfer of cells. The body's metabolism, immunity and life were invigoration if the lecithin is abundant in human body. Therefore, it is important that the lecithin was developed the aged health food.

Firstly, soybean lecithins can anti-aging. The cell membranes are mainly composed of lecithin in body cells. If the soybean lecithin were supplied to the human body, which can repair the damaged cell membrane, increase cell membrane fatty acid saturation, improve membrane function. The human body can improve metabolism, self-healing ability and body tissue regeneration ability after the soybean lecithin was supplied to the human body. The lecithin can primarily to slow the effects of aging for the human body. The lecithin can play an improvement role for the not coordinated action on the elderly people.

On the other hand, soybean lecithin can enhance the memory of the elderly people. The content of lecithin in brain nerve cells is about 20% of that of the brain. Acetylcholine is a kind of transmitter in the brain. Choline is the basic ingredient of soybean lecithin. Choline and acetyl of human body can synthesize acetylcholine. If the lecithin sufficient supply, sufficient transmitter will be ensure in the human body. The transmitter can improve the activity of brain cells, and then improve the memory and intelligence of old people.

2.2. Wolfberry

Wolfberry can be added in the aged-foods as a kind of anti-ageing ingredient. Wolfberry has some physiological functions, as follow:

Firstly, enhancing immune function: modern scientific research shows that wolfberry can enhance the transformation rate of macrophages, macrophages and T lymphocytes, so wolfberry can regulate the function of immune. A study showed that Lycium barbarum polysaccharide from wolfberry can obvious increase the immune of body [3]. The wolfberry has five different kinds of Polysaccharides, which had improve the immune activity of body [4].

Secondly, anti-tumor: Lycium barbarum polysaccharide is the matter of anti-tumor in wolfberry. The result showed that 10mg/kg lycium barbarum polysaccharide had 31% anti-tumor rate for bearing tumor mice [5]. Lycium barbarum polysaccharide can enhance the inhibition of the tumor target cells proliferation in mice macrophage cells of ConA-treated. 20mg/kg lycium barbarum polysaccharide has the most significant inhibition, and the inhibition rates were 85.5% and 63.6% for P388 and P815, respectively [6].

Thirdly, anti-aging: studies showed that wolfberry can increase the products of interleukins by human lymphocytes, while interleukins are one of the main substances to maintain the activity of cells. Therefore, wolfberry in diet can be the anti-aging and longer life. If Lycium extracts were oral administration for 2 months, aging related symptoms can completely disappeared for 95% patients of 56-81 year old [7]. The research showed that the average life of mice and fruit flies can be improved by the wolfberry [8]. The wolfberry can maintain the development of normal cells and promote the aging cells to reverse the direction of the young.

Fourthly, the effect of the hematopoietic system: wolfberry can significantly increase the white blood cells and promote the recovery of hematopoietic function. Wolfberry can promote the proliferation of bone marrow hematopoietic stem cells in normal mice. Wolfberry significantly increase the number of bone marrow cells and promote its differentiation to the myeloid differentiation, meanwhile reduce the inhibition of chemotherapy on hematopoietic system [9].

Fifthly, Hypolipidemic and anti fatty liver: the experimental data show that 10mg/kg.d lycium barbarum polysaccharide were administrated intragastric 10d. Then total cholesterol decreased 3.82mmol/L and triglyceride decreased 1.692mmol/L in serum, respectively, which showed that wolfberry has the effect of hypolipidemic [10].

2.3. Honey

The main ingredients of honey are fructose and glucose, which is about 70% of honey. The other components of honey are a small amount of sugar, maltose, dextrin, and inorganic salts. Honey also contain trace amounts of vitamins, including vitamin A, vitamin C, vitamin D, vitamin B1, vitamin B6, vitamin K and choline, niacin, pantothenic acid, biological essence, folic acid and so on. Nitrogenous compounds of honey included protein, peptone, peptide, amino acids, invertase, catalase, amylase, esterase enzymes and acetylcholine [11]. Organic acid of honey include citric acid, malic acid, malic acid, and acetic acid. In addition, the ash contained Mg, Ca, K, Na, P, Fe, Cu, etc. At present, honey is not only as a family drinks and nutritional supplements, but also applied to the foods [12].

Honey's main physiological functions are:

Firstly, enhancing immune function: a study showed that 5% the linden tree honey were given to mice once every day by gavage treatment. After 7 days, the humoral immune was significantly enhanced in mice [13].

Secondly, the protection of the cardiovascular system: the myocardial metabolic process was improved by honey administration, which can adjust the heart to play the role of the normal physiological function. The active ingredient of honey may be a large amount of sugar and many other nutrients, such as amino acids, enzymes, vitamins, etc. The antihypertensive was acetylcholine in honey [14].

Thirdly, antibacterial: ripe honey would not corrupt at room temperature without any treatment because of the antiseptic. A study has been reported that the antibacterial effect may come from the glucose oxidase of honey. This enzyme is not heat resistant. The most actives of the glucose oxidase were appeared in pH3. Hydrogen peroxide can affect the activity of the glucose oxidase in pollen [15].

Fourthly, the effect on the digestive system: honey has the regulatory effect on the gastrointestinal function. Honey can ensure the stomach acid secretion and adjust the amount of the stomach acid in body. In addition, honey has the good effect on colitis, habitual constipation, the elderly and pregnant women constipation, children with diarrhea, etc [16].

Fifthly, the promotion of tissue regeneration: honey contains the wealth of nutrients, such as vitamins, amino acids and enzymes and so on. Honey is the nutritional supplements of adult, meanwhile the honey can promote children's growth and development and improve the body's resistance to disease. Honey was also accelerated to various delayed healing of the ulcer on the growth of the granulation tissue. In addition, honey has the regulatory function of the nervous system. Honey can improve the sleep of patient, regulate the disorders of nervous system, and improve the activity of mental and physical [16].

In the research and development of aged-foods, some natural special food ingredients or functional factors can not only meet the needs of the elderly, but also can prevent and treat the elderly disease, improve the health of the elderly.

3. THE VARIETIES OF AGED-FOODS

3.1. The Nutrition and Convenience of Aged-Foods

The new nutrition and convenience of aged-foods should be researched and developed in the diet of old people. This kind of aged-food included the convenient staple food and various kinds of non-staple food. The characteristics of this aged-food should be the nutritional balance, cheap, convenient to eat. The nutritional balance of convenience foods were reasonable collocated and produced to the aged-foods of the early, middle and dinner type from the raw materials of ordinary natural animal and plant.

3.2. Functional Aged-Foods

Functional aged -foods should have the function of preventing and curing all kinds of senile diseases, delaying aging and promoting the old people's health. The kind of aged-foods was strengthened the specificity function of elderly, such as the foods of enhancing the immune function of the elderly, the foods of prevention hypertension, the foods of prevention diabetes and metabolic disorders and so on.

3.3. The Aged-Foods for Chinese Diet Structure

The aged-foods should be researched and developed for chinese diet structure. The nutritional status was improved the elderly in China. These extract substances of physiological function would be produced some variety of old formula foods because these extract substances have high quality protein, vitamins and trace elements.

References

1. Wang Jianhui, Liu Shijian. Some advices on aged-foods. China Health Care and Nutrition 2002; **3**: 5-6 (in Chinese).

2. Li Zhemin. Technology Key and Market Prospect of Developing Food for Old Age. Food and Nutrition in China 2004; **4**: 31-32 (in Chinese).
3. Luo Qiong, Yan Jun, Li Jinwei et al. The effects of Pure Lycium Barbarum Polysaccharides on the Immunopharmacological Activity of the mice. Chinese Journal of Gerontology 1999; **19**: 38-41 (in Chinese).
4. Huang Linjun, Lin Ying, Tian Gengyuan et al. Isolation, Purification and Physico-Chemical Properties of Immunoactive Constituents from the Fruit of Lycium Barbarum L. Acta Pharmaceutica Sinica 1998; **33**: 512-516 (in Chinese).
5. Wang Baikun, Xing Shantian, Zhou Jinhuang et al. Effect of Lycium barbarum polysaccharides on the cellular immune and anti tumor in S-180 tumor bearing mice. Chinese Journal of Pharmacology and Toxicology 1988; **2**: 127-131(in Chinese).
6. Zhang Yongxiang, Xing Shantian, Zhou Jinhuang. Effects of Lycium barbarum polysaccharides and combination with Corynebacterium parvum on tumoristatic activity of peritoneal macrophages in mice. Chinese Journal of Pharmacology and Toxicology 1989; **3**: 169-174 (in Chinese).
7. Li Dongyang, Yuan Xiulan, Xia Huifang et al. Preliminary observation for the extract of Ningxia wolfberry in clinically. Chinese Traditional and Herbal Drugs 1989; **20**: 26-28 (in Chinese).
8. Dai Shouzhi, Wang Mudi. Effects of wolfberry and its polysaccharide on the life of Drosophila and mice. Chinese Jounal of Gerontology 1990; **10**: 94-96 (in Chinese).
9. Zhou Zhiwen, Zhou Jinhuang, Xing Shantian. Effect of wolfberry polysaccharide on the proliferation and differentiation of bone marrow hematopoietic stem cells and granulocyte progenitor cells in mice. Chinese Journal of Pharmacology and Toxicology 1991; **5**: 44-46 (in Chinese).
10. He Jin, Zhang Shenghua. Study on the polysaccharide of wolfberry and wolfberry. Food Science 1995; **16**: 14-21 (in Chinese).
11. Xu Qin. Research on the function of milk powder for middle and old age. Master degree thesis of China Agricultural University 2006 (in Chinese).
12. Huang Weikun. Honey and its products . Beijing: China Light Industry Press 1991 (in Chinese).
13. Zheng Jianxian. Functional foods (I). Beijing: China Light Industry Press 1995; 255-258 (in Chinese).
14. Yuan Changqi. Exploitation and utilization of natural drugs. Jiangsu: Jiangsu Science and technology press 2000; 444-448 (in Chinese).
15. Lv Xiaowu. Antibacterial effect of honey. Apiculture of China 1996; **1**: 29-30 (in Chinese).
16. Shi Jinglin. Medical function of honey. Apiculture of China 1995; **1**: 32 (in Chinese).

VARIETIES OF VOLATILE COMPOUNDS IN TREATED FTGSK BY PULSED ELECTRIC FIELDS ESTIMATED USING HS-SPME-GC-MS

Xingfang Li[a], Jie Xing[b], and Songyi Lin[c,*]

College of Food Science and Engineering, Jilin University, Changchun, 130062, China

E-mail:[a]lxf19920530@163.com,[b]xingjie0014@163.com,[c]linsongyi730@163.com

[*]Corresponding author

Pulsed electric field (PEF) as an important non-thermal food processing technology has unique effect on the flavor compounds and health-related phytochemicals of plant-based foods. This paper focuses on estimating the effects of PEF on the volatile compounds of synthetic peptide, Phe-Thr-Gly-Ser-Lys (FTGSK) using the headspace solid phase microextraction (HS-SPME) and gas chromatography-mass spectrometer (GC-MS) combination. HS-SPME was applied to extract the volatile compounds of FTGSK. And the analysis and identification of volatile compounds were performed by GC-MS. The results showed that the major families of identified volatile compounds were hydrocarbons (70.26% for untreated versus 74.68% for PEF-treated), esters (13.25% for untreated versus 9.98 % for PEF-treated) and aldehydes (14.62% for untreated versus 14.85% for PEF-treated). The results illustrated that there were differences of volatile compounds kinds and content between untreated and PEF-treated FTGSK solution. This study provided an effective estimate method for studying the volatile compounds in synthetic peptide and may have applications for promoting peptides quality deterioration studies.

1. INTRODUCTION

As is known, pulsed electric field (PEF) as a sterilization technology was usually used in non-thermal liquid foods processing[1]. PEF can kill most pathogenic and putrefactive microorganisms without significant adverse effects on the flavor, texture, nutrients and heat labile functional food components[2]. A large number of literatures have studied the effects of PEF on the flavor compounds and health-related phytochemicals of plant-based foods[3, 4]. However, there was a lack of knowledge on the effects of PEF on proteins and qualities of protein-based foods. Liu et al.[5] found that PEF could change the physical and chemical properties of protein, including the velocity and foam property, thermal stability, and result in exposure of disulfide bonds and hydrophobic groups. Some researchers have showed that PEF technology could obviously increase the antioxidant activity of peptides including corn peptides, egg-white protein polypeptides and glutathione[6-8]. But there was no research existed on the effects of PEF processing on the changes of volatile compounds of peptide. For that reason, the goal of this paper was to study the influence of PEF on the volatile compounds of synthetic peptide FTGSK.

2. MATERIALS AND METHODS

2.1. Materials and Reagents

Phe-Thr-Gly-Ser-Lys (FTGSK) (Molecular Formula: $C_{24}H_{38}N_6O_8$) with a 98.73% purity was chemically synthesized using Fmoc protected amino acid synthesis methods from Cellmano Biotech Limited (Hefei, China). The MS spectra and molecular graph of synthesis peptide FTGSK were shown in Fig. 1. 2, 2-diphenyl-1-picrylhydrazy (DPPH), 2, 2-azinobis (3-ethylbenzothiazoline-6-sulfonic acid) diammonium salt (ABTS), and potassium bromide (KBr) powder were purchased from Sigma Chemicals Co. (St. Louis, MO, USA). The analytical grade methanol, sodium hydroxide (NaOH), hydrochloric acid (HCl) and all the other chemical reagents required in the experiments were of analytical grade and purchased from Beijing Chemical Plant (Beijing, China).

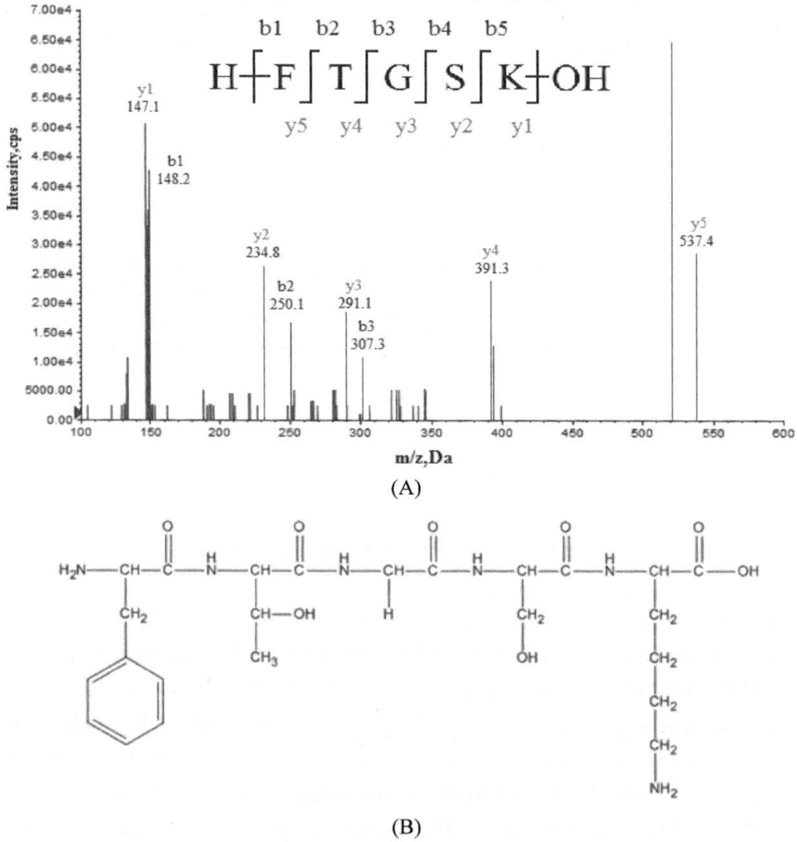

Fig.1. The MS spectra and molecular graph of synthesis peptide FTGSK.

2.2. Treatment by PEF

The PEF system was self-designed by Yongguang Yin and described in the previous study[9]. Synthetic peptide was weighed and dissolved with deionized water to the concentration of 8 mg/mL. Whereafter, the peptide solution was pumped into the PEF system at a flow velocity of 3.2 mL/min and then treated at the electric field strength of 20 kV/cm with the pulsed frequency of 2400 Hz. DPPH and ABTS radical-scavenging activity were tested immediately after the PEF treatment.

2.3. Headspace Solid Phase Microextraction Procedure

SPME fiber assembly and SPME holder for manual sampling were provided from Sigma-Aldrich Co. (St. Louis, MO, USA). An SPME fiber 50/30 μm Divinylbenzene/Carboxen/Polydimethylsiloxane (50/30 μm DVB/CAR/PDMS) was used for analyte extraction. And the fibers were conditioned in the GC injection port for 2 h at 250°C before extraction. The 1.5 mL sample before the PEF treatment and after was transferred to 4 mL brown screw-capped vials, respectively. The fiber was pushed through the sample container cap for exposure to the headspace of the solution, and the analyte was extracted at room temperature (25°C) for 30 min. After extraction, the fiber was immediately inserted into the injection port for 5 min to desorb the analyte. HS-SPME and subsequent analyses were made in triplicate.

2.4. Analysis of Volatile Compounds by GC-MS

GC-MS analysis was performed using a QP-2010 GC-MS (Shimadzu, Japan). The capillary column used was an Rxi®-5ms column (30 m × 0.25 mm, 0.25 μm film thickness, Restek Co., Bellefonte, PA, USA). The injection temperature was 250°C. Sample injection was in the splitless mode. Helium was used as the carrier gas at a flow rate of 1.0 mL/min. The oven temperature was programmed as follows: the initial temperature was kept at 40°C for 2 min, then heated to 90°C at a rate of 6 °C/min, and then increased to 200°C at 4 °C/min. Finally, the temperature was raised to 220°C at 15 °C/min and kept constant at 220°C for 2 min. Electron-impact mass spectra were generated at 70 eV, with ion source temperature of 200 °C and scan range from m/z 35 to 450. The components of the volatile compounds were identified by comparing their mass spectra with those in NIST08 mass spectral library.

3. RESULTS AND DISCUSSES

The volatile compounds were listed in Table 1. 43 volatile compounds were identified in the untreated FTGSK solution by using HS-SPME-GC-MS, while 40 volatile compounds were identified in the PEF-treated FTGSK solution. The major families of identified volatile compounds in the untreated FTGSK solution were hydrocarbons (70.26%), esters (13.25%), aldehydes (14.62%), phenols (1.01%) and heterocyclic compounds (0.86%). On the other hand, the identified volatile compounds in the PEF-treated FTGSK solution were hydrocarbons (74.68%), esters (9.98%), aldehydes (14.85%) and alcohols (0.49%).

As shown in Tab.1, there were 12 volatile compounds identified in the untreated FTGSK solution, butthey are not identified through the PEF treatment. These compounds included 2, 4-Dimethyldecane, naphthalene, ethyl tridecyl ester, 2, 6, 10, 15-Tetramethylheptadecane, 5-methyltetradecane, 6-methylpentadecane, 2, 7, 10-Trimethyldodecane, hexadecane, 1, 4-butanediol diacrylate, 4-methyl-3, 5-dioxatridecane, 2, 6-di-tert-butyl-4-methylphenol, and 4-hydroxybutyl acrylate.

In addition, 9 new compounds were detected in the PEF-treated FTGSK solution. They were 5-methylundecane, 5-isobutylnonane, 4, 4-dimethylundecane, tetradecane, 4-Ethyldecane, 3, 4-dimethylundecane, 2, 6, 11, 15-tetramethylhexadecane, isotridecyl alcohol and nonadecane.

The rest of 31 volatile compounds did not change, but their relative contents changed. The relative contents of some volatile compounds declined, including 2, 4, 6-trimethyloctane, 3-Ethyl-3-methylheptane, 6-methyldodecane, pentadecane, 1-hydroxy-2, 4-di-tert-butylbenzene, etc. Among them, 6-methyldodecane and 1-hydroxy-2, 4-di-tert-butylbenzene changed obviously. Their contents decreased from 2.21% to 1.11% and 11.24% to 9.98%, respectively. Nevertheless, 6 compounds changed a little, such as dodecane and tridecane. What's more, the contents of most compounds increased to some degree, such as 3, 7-dimethyldecane, 4, 6-dimethyldodecane, 8-methylheptadecane, heneicosane, and so on. The content of 8-methylheptadecane decreased from 6.73% to 7.84%, other changes were between 0.14% and 0.74%.

Table 1. Volatile compounds identified in untreated and PEF treated FTGSK solution

	Compound	Molecular Formula	Retention Time (min)	Relative Area (%) Untreated	Relative Area (%) PEF treated
1	2,4,6-Trimethyloctane	$C_{11}H_{24}$	7.20	1.65	1.53
2	2,5-Dimethylnonane	$C_{11}H_{24}$	7.34	0.74	0.74
3	2,5,5-Trimethylheptane	$C_{10}H_{22}$	7.40	2.88	3.02
4	2,4-Dimethyldecane	$C_{12}H_{26}$	8.17	0.61	*nd*
5	3,6-Dimethyldecane	$C_{12}H_{26}$	8.29	6.91	6.4
6	3,3,5-Trimethylheptane	$C_{10}H_{22}$	8.35	1.18	1.35
7	3-Ethyl-3-methylheptane	$C_{10}H_{22}$	8.43	2.39	2.28
8	Undecane	$C_{11}H_{24}$	9.39	0.94	0.98
9	3,7-Dimethyldecane	$C_{12}H_{26}$	9.46	4.24	4.44
10	5-Methylundecane	$C_{12}H_{26}$	9.62	*nd*	1.54
11	2,3,6,7-Tetramethyloctane	$C_{12}H_{26}$	9.78	1.16	0.68
12	5-Isobutylnonane	$C_{12}H_{28}$	10.15	*nd*	0.45
13	2-Methylundecane	$C_{12}H_{26}$	11.14	2.27	2.24
14	4,4-Dimethylundecane	$C_{13}H_{28}$	11.45	*nd*	0.84
15	6-Methyldodecane	$C_{13}H_{28}$	11.48	2.21	1.11
16	Naphthalene	$C_{10}H_{8}$	11.65	0.86	*nd*
17	ethyl tridecyl ester	$C_{13}H_{26}O_{3}$	11.92	0.72	*nd*
18	Dodecane	$C_{12}H_{26}$	12.17	2.01	2.03
19	2,6,10,15-Tetramethylheptadecane	$C_{21}H_{44}$	12.31	1.96	*nd*
20	Tetradecane	$C_{14}H_{30}$	12.33	*nd*	2.01
21	5-Methyltetradecane	$C_{15}H_{32}$	12.48	1.22	*nd*
22	4-Ethyldecane	$C_{12}H_{26}$	12.49	*nd*	1.18
23	2,4-Dimethylbenzenecarboxaldehyde	$C_{9}H_{10}O$	12.59	14.61	14.85
24	3,4-Dimethylundecane	$C_{13}H_{28}$	12.70	*nd*	1.46
25	4-Methyldodecane	$C_{13}H_{28}$	12.81	2.95	3.17

26	Tetradecane	$C_{14}H_{30}$	13.47	1.69	1.74
	Compound	Molecular Formula	Retention Time (min)	Relative Area (%)	
				Untreated	PEF treated
27	Hexadecane	$C_{16}H_{34}$	13.64	1.30	1.55
28	6-Methyldodecane	$C_{13}H_{28}$	13.73	0.86	1.04
29	4,6-Dimethyldodecane	$C_{14}H_{30}$	13.80	1.98	2.3
30	3,8-Dimethylundecane	$C_{13}H_{28}$	13.92	1.23	1.38
31	5-Methyltetradecane	$C_{15}H_{32}$	14.07	4.59	5.33
32	2,6,11,15-Tetramethylhexadecane	$C_{20}H_{42}$	14.38	*nd*	0.57
33	8-Methylheptadecane	$C_{18}H_{38}$	14.56	6.73	7.84
34	Heptadecane	$C_{17}H_{36}$	14.82	2.21	2.18
35	3-Methyl-5-propylnonane	$C_{13}H_{28}$	14.99	2.01	2.42
36	Tridecane	$C_{13}H_{28}$	15.16	0.83	0.84
37	Pentadecane	$C_{15}H_{32}$	15.22	2.30	1.95
38	Isotridecyl alcohol	$C_{13}H_{28}O$	15.36	*nd*	0.49
39	2,3,7-Trimethyldecane	$C_{13}H_{28}$	15.42	1.15	1.39
40	6-Methylpentadecane	$C_{16}H_{34}$	15.85	0.71	*nd*
41	2,6,11-Trimethyldodecane	$C_{15}H_{32}$	15.97	2.70	2.85
42	2,7,10-Trimethyldodecane	$C_{15}H_{32}$	16.24	1.05	*nd*
43	Nonadecane	$C_{19}H_{40}$	16.25	*nd*	0.96
44	Hexadecane	$C_{16}H_{34}$	17.09	0.73	*nd*
45	1,4-Butanediol diacrylate	$C_{10}H_{14}O_4$	17.27	0.70	*nd*
46	4-Methyl-3,5-dioxatridecane	$C_{12}H_{26}O_2$	18.50	0.64	*nd*
47	Heneicosane	$C_{21}H_{44}$	18.90	0.72	0.96
48	Octadecane	$C_{18}H_{38}$	20.12	0.56	0.74
49	Eicosane	$C_{20}H_{42}$	21.16	0.96	1.19
50	2,6-Di-tert-butyl-4-methylphenol	$C_{15}H_{24}O$	21.70	1.01	*nd*
51	1-Hydroxy-2,4-di-tert-butylbenzene	$C_{14}H_{22}O$	21.83	11.24	9.98
52	4-Hydroxybutyl acrylate	$C_7H_{12}O_3$	29.15	0.59	*nd*

Note: Table notes.

[a]The lowercase letters, *nd* is mean: not detected.

4. CONCLUSIONS

The effects on the volatile compounds of FTGSK treated by PEF were studied by HS-SPME-GC-MS. The major families of identified volatile compounds were hydrocarbons, esters and aldehydes. They illustrated that there were differences of volatile compounds kinds and content between untreated and PEF-treated FTGSK solution. The changes of volatile compounds of FTGSK solution are accounted for the PEF treated, but the mechanisms are still unclear and further studied remains to be performed.

Acknowledgments

The authors acknowledge the financial support provided by the Key Projects in the National Science & Technology Pillar Program during the Twelfth Five-Year Plan Period (2012BAD33B03) and the Youth Scientific Innovation Leading Talent and Team Building Project of Jilin Province (20140519014JH).

References

1. Zhanga Q. Engineering aspects of pulsed electric field pasteurization. *J. Food Eng* 1985; **25**: 261–281.
2. Á. Robert Marsellés-Fontanet, Martín-Belloso O. Optimization and validation of PEF processing conditions to inactivate oxidative enzymes of grape juice. *J. Food Eng* 2007; **83**: 452–462.
3. Odriozola-Serrano I, Aguiló-Aguayo I, Soliva-Fortuny R, et al. Pulsed electric fields processing effects on quality and health-related constituents of plant-based foods. *TRENDS FOOD SCI TECH* 2013; **29**: 98–107.
4. Toepfl S, Heinz V, Knorr D. High intensity pulsed electric fields applied for food preservation. *CHEM ENG PROCESS* 2007; **46**: 537–546.
5. Liu YY, Zhang Y, Zeng XA, et al. Effect of pulsed electric field on microstructure of some amino acid group of soy protein isolates. *INT J FOOD ENG* 2014; **10**: 113–120.
6. Ke W, Ying W, Songyi L, et al. Analysis of DPPH inhibition and structure change of corn peptides treated by pulsed electric field technology. *J FOOD SCI TECH* 2015; **52**: 4342–4350.
7. Ke W, Jia W, Bolong L, et al. Effects on DPPH inhibition of egg-white protein polypeptides treated by pulsed electric field technology. *J SCI FOOD AGR* 2013; **93**: 1641–1648.
8. Jia W, Ke W, Ying W, et al. A novel application of pulsed electric field (PEF) processing for improving glutathione (GSH) antioxidant activity. *FOOD CHEM* 2014; **161**: 361–366.
9. Jin Y. Optimization of extraction parameters for trehalose from beer waste brewing yeast treated by high-intensity pulsed electric fields (PEF). *AFR J BIOTECHNOL* 2011; **10**: 19144–19152.

THE STUDY OF CORROSION RESISTANCE OF PURE IRON BY PLASMA NITRIDING

Changzi Chen[a] and Gangyi Gao

College of Mechanical Engineering, JingChu University of Technology, Jingmen, 448000, China

E-mail: [a]ccz198311@163.com

Industrial pure iron used as a new biodegradable vascular stent material was extensively studied. However, the degradation rate was not controlled, and it was the key problem to the stent. In this paper, the surface of industrial pure iron was strength using low-voltage and high-frequency plasma nitriding, the mechanical properties of industrial pure iron was detected at different frequency, and the electrochemical corrosion test was carried out in 0.9%NaCl solution. The result showed that theε-Fe_3N andγ-Fe_4N was formed on the surface of industrial pure iron, and all of the wear resistance was higher than that of pure iron at different frequency. The corrosion resistance was also higher than that of pure iron. When the frequency was about 15.15 KHz, the corrosion resistance was well. The conclusion laid a foundation for the development of advanced stent.

1. INTRODUCTION

Cardiovascular disease is one of the most important reasons for the threat to human health and life in 21 Century, and vascular stent implantation is one of the most common treatment methods [1].Biodegradable metal stent is widely recognized as an effective vascular stent, which has better support and can be used directly in the balloon delivery system for a long time. About 4 grams of iron was in the human body, iron was uptake 10 ~ 15mg each day. Therefore, industrial pure iron is regarded as a new degradable vascular stent material [2]. M.Peuster[3] first reported that the experimental results of the number of sixteen iron (Fe>99.8%)stents implanted in New Zealand rabbit abdominal aorta for 6-18 months, the results show that when placement in 6 months ,the most stent strut is disappear, 6 months, 12 months and 18 months experimental results were not obvious inflammation, hyperplasia and coagulation. Gene chip technology was used to detect the divalent iron ions of iron stent degradation product, which inhibit the ability of proliferation of smooth muscle cells [4]. Pure iron stents, however, there are deficiencies, the degradation rate can not control, the release rate of the metallic elements of the stent should control in local metal ion concentration, no vascular tissue and blood have any negative effects (such as cell toxicity, inflammation, coagulation, etc.).

The current metal materials can not be very good to meet the above conditions, it is necessary to modify the surface of industrial pure iron. Zhu et al[5-6] using plasma immersion ion implantation method to implant nitrogen, carbon and other elements to achieve controlled degradation, and study the corrosion of the sample in simulated body fluid. Other methods were used also, N [7] element implanted by ion implantation technology, Nb[8]implantation improved the surface properties of pure iron. The high frequency and low voltage plasma nitriding technique was used to make the surface of pure iron formation of the different phases of the nitride layer, so as to improve the wear resistance [9-10].

In this paper, the plasma nitriding technique was used at different frequency, the corrosion resistance in 0.9% NaCl solution and mechanical properties of industrial pure iron was studied.

2. EXPERIMENTAL

The pure iron was modified by HL (high frequency and low voltage) plasma immersion and ion implantation at different frequency [11]. Disc samples of about 1mm thick were cut from the incoming bar of 18mm diameter, and wet ground with SiC paper from 240 down to 1200 grit to produce a fine surface finishing. The base pressure of the system was 2.1×103Pa, substrate temperature<30□. The nitrogen was inlet in the vacuum and the plasma was trigger by 800W RF glow discharge with 13.56MHz, the target height is 56cm, then a frequency of 15.15 KHz, 20 KHz and10.87 KHz were applied respectively on the sample for 3h. The nitride samples were identified as1#, 2# and 4# corresponding to the sample treated at 15.15 KHz, 20 KHz and10.87 KHz, and the untreated samples were identified as 0#.the low voltage high frequency plasma nitriding process as shown in Table 1.

Table 1. Low voltage high frequency plasma nitriding process

samole	voltage(V)	frequency(KHz)	N2 flow(sccm)	pressure(Pa)	time (h)
1#	2500	15.15	50	0.5	3
2#	2500	20	50	0.5	3
4#	2500	10.87	50	0.5	3

After plasma nitriding, the composition, microstructure and hardness of the nitrided samples were respectively characterized by metallographic examination, XRD and microhardness tests. Ball-on-disc rotation wear test was performed to evaluate the tribological properties of the nitrided layer, and scanning electron microscopy (SEM) was used to observe the ball-on-disc wear tracks. The friction pair was Si3N4.

The corrosion resistance of the nitrided sample was evaluated using anodic polarization tests in 0.9wt% NaCl water solution. The flat cell was a three-electrode set-up consisting of a saturated calomel reference electrode (SCE), a platinum auxiliary electrode and a working electrode (sample). The sample to be tested was pressed against a Teflon ring at one end of the flat cell, leaving a theoretical circular area of 3 mm^2 on the sample surface in contact with the testing solution through a round hole in the Teflon ring.

3. RESULTS AND DISCUSSION

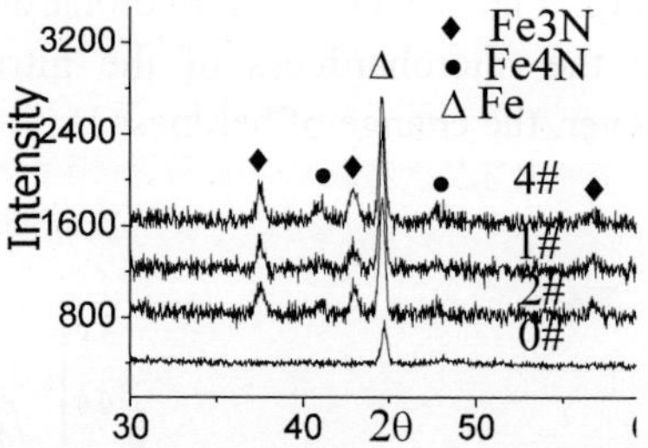

Fig.1. The XRD patterns of the iron treated at different frequency.

The phase composition in the nitrided surface was analysed with XRD using Cu-Karadiation (k = 1.544 A). Fig. 1 showed the microstructure of the sample treated at different frequency. It showed that all of the iron surface modified by plasma nitriding consisted of mainly Fe_3N iron nitride and Fe_4N iron nitride.it should be epsiv phase and gamma phase [12]., the structure does not change with the increase of the pulse frequency.

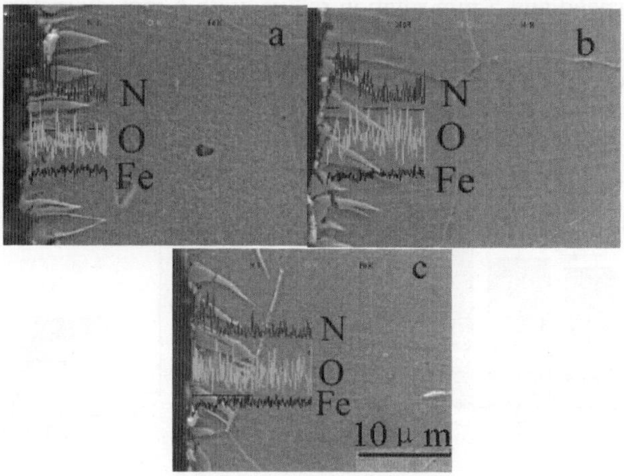

a 4#-10.87KHz b 1#-15.15KHz c 2#-20KHz

Fig.2. The cross section of the iron treated at different frequency.

Fig. 2 showed the cross section and EDX spectrum of the iron treated at different frequency. It showed that acicular structure was formed on the surface of pure iron at different frequency. It scanned the cross from the interface to the center line, these tissues contains N elements, combining with the results of XRD analysis, the acicular structure should be as the epsilon -Fe_3N and gamma-Fe_4N of mixed phase. With the increase of the pulse frequency, the number of acicular structure did not change greatly, and the lamellar structure was not formed. This shows that the pulse frequency has little effect on the change of tissue. Microhardness was measured on the as-nitrided surface with a Vickers indenter at a load of 0.05kgf. Fig. 3 shows the microhardness of the iron at different frequency. It showed that the microhardness of the nitrided layers decreased with frequency increasing. However, the change of hardness is not very obvious.

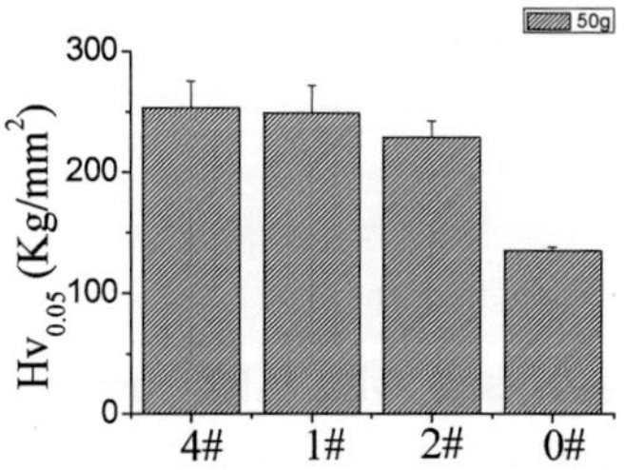

Fig.3. The microhardness of the iron at different frequency.

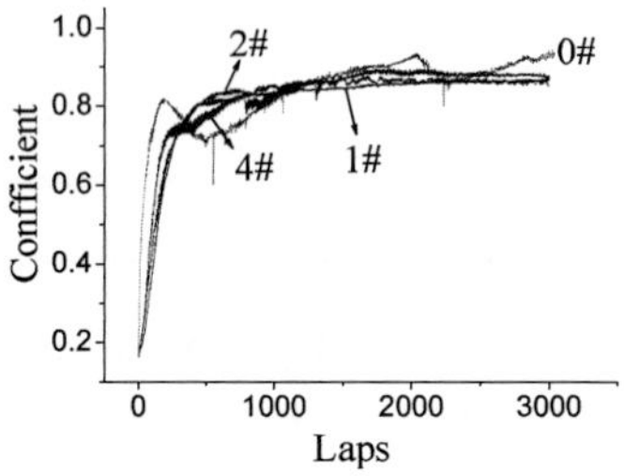

Fig.4. The friction coefficients of the pure iron treated at different frequency.

Dry ball-on-disc tribological tests of samples were performed with a 0.98 N, 100 mm/min rotation speed and 5 mm rotation radius, with 6-mm-diameter Si_3N_4 balls as the counterpart material. Fig. 4 shows the friction coefficients of the pure iron treated at different frequency. It showed that the friction coefficient tends to be stable. Fig. 5 showed that wear width under different pulse frequency.

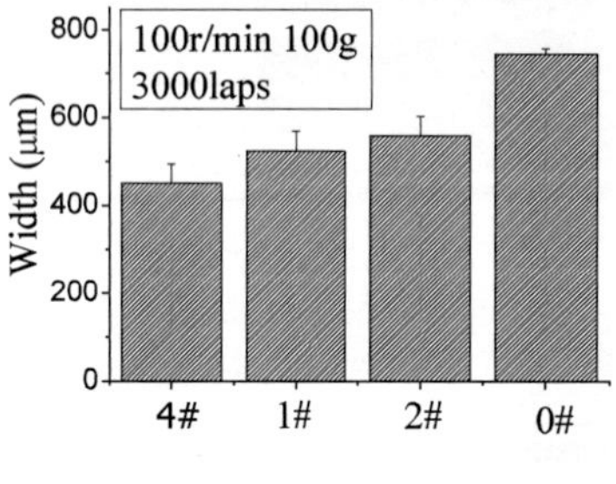

Fig.5. Wear width under different pulse frequency.

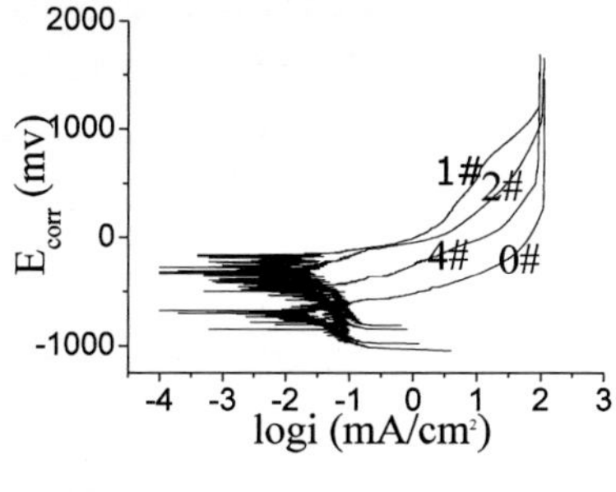

Fig.6. Electrochemical corrosion polarization curves of Industrial pure iron with different pulse frequency.

It showed that the wear resistance of the samples after treatment was better. With the increase of the pulse frequency, the wear scar width has an increasing tendency, and the wear resistance was gradually reduced.

Fig. 6 showed electrochemical corrosion polarization curves of Industrial pure iron with different pulse frequency. it showed that the corrosion potential of the treated sample was higher than that of pure iron, the treated sample surface formed a certain depth of the epsilon -Fe3N and gamma-Fe4N phase can reduce corrosion for Cl ions on metal, also N on the passivity films on metal strengthening and stabilizing role, pitting inhibition of further development, thereby improving the surface corrosion resistance. N element acted a strengthening and stabilizing role to the passivity films on metal, which inhibit pitting. At the same time, it can be seen that with the increase of the pulse frequency, the pitting corrosion resistance of treated sample (15.15 KHz) was the best, and the corrosion resistance was the best.

4. CONCLUSIONS

In this work, the microstructure and properties of pure iron modified by plasma nitriding(HLPIII) at different frequency were studied. The sample was treated by plasma nitriding at different frequency, the properties of industrial pure iron surface is strengthened, the result showed that pure iron surface formed epsilon -Fe_3N and gamma-Fe_4N two mixed phase, so that the wear resistance and hardness is improved.

The result showed that the barrier layer was formed on the treated sample, so the corrosion resistance had been generally improved. The pure iron plasma nitriding at 15.15 kHz has the best corrosion resistance.

Acknowledgements

This research was financially supported by Program for JingChu University of Technology (CXY201503).

References

1. Popma JJ, Califf RM Topol EJ. Clinical trials of in-stent resteno-sis after coronary angioplasty. *Circulation1991*; **84**(3):1426-1436.
2. Kathuria YP. The potential of biocompatible metallic stents and preventing restenosis. *Mater Sci Eng A Struct Mater* 2006; **417** (1-2):40-48.
3. Peuster M, et al, A novel approach to temporary stenting: degradable cardiovascular stents produced from corrodible metal-results 6-18 months after implantation into New Zealand white rabbits. *Heart.*2001; **86**(5):563-9.
4. Mueller PP, et al, Control of smooth muscle cell proliferation by ferrous iron. *Biomaterials.* 2006;**27**(10):2193-200.
5. Y. Oshida, Y. Guven. Biocompatible coatings for metallic biomaterials. *Surface Coating and Modification of Metallic Biomaterials*2015.**10**:287-343.
6. Shengfa Zhu, Li Xu, Hengquan Liu et al. Deposition and Characterization of Biodegradable Fe-O Films. *CHINESE JOURNAL OF VACUUM SCIENCE AND TECHNOLOGY*2008.**29**(3):236-240. (In Chinese).
7. *R. Zenker, A. Buchwalder, K. Rüthrich,et al. First results of a new duplex surface treatment for cast iron: Electron beam remelting and plasma nitriding. Surface and Coatings Technology*2013;**236**: 58-62.
8. *C. Borcz, C.M. Lepienski, S.F. Brunatto. Surface modification of pure niobium by plasma nitriding Surface and Coatings Technology2013;* **224**:114-119.
9. W. P. Tong, C. S. He, J. C. He et al. strongly enhanced nitriding kinetics by means of grain refinement. *Applied Physics Letters*2006; **89**:021918.

10. H. Ferkel, Y. Estrin, C. Blawert, R.Z. Valiev. RF nitriding of severely deformed Armco iron and St2K50. Surface and Coatings Technology2003; **173-174**: 1164-1170.
11. Hengjun Liu, Yongxiang Leng, Jinglong Tang et al., Tribological performance of ultra-high-molecular-weight polyethylene sliding against DLC-coated and nitrogen ion implanted CoCrMo alloy measured in a hip joint simulator. *Surface and Coatings Technology*2012; **206**:4907-4914.
12. Zhenxue Zhang, Xiaoying Li, Hanshan Dong. Plasma-nitriding and characterization of FeAl40 iron aluminide. *Acta Materialia* 2015; **86**:341-351.

PRELIMINARY ASSESSMENT STUDY ON GOUT VULNERABILITY USING *C. ELEGANS* AS A MODEL ORGANISM

Changlu Guo[a], Xipeng Tian[b], Guifeng Liu[c], Hongyuan Wang[d], and Zhizhou Zhang[e,*]
School of Marine Science and Technology, Harbin Institute of Technology (Weihai), Weihai, Shandong Province, China
E-mail: [a]guochanglu@163.com, [b]913826637@qq.com, [c]lgf19861229@163.com, [d]h.wang@ucl.ac.uk, [e]zhangzzbiox@hitwh.edu.cn
[*]Corresponding author

Gout is a group of diseases resulting from increased serum uric acid levels due to the purine metabolism obstacle.Most human gout related genes have corresponding ones in *C. elegans*. By feeding *C. elegans* with mixed extracts of seven kinds of shellfish in three different concentrations (0.1%, 1% and 10% separately), we observed changes of physiological characteristics (shorter lifespan, lower body activity, and more severe body stiffness at later stages) in *C. elegans* fed with either mixed seafood extract or single species of some seafood at 10% concentration. Especially, the *Meretrix meretrix Linnaeus* at 10% concentration raised the general uric acid levels and had the greatest deleterious impact to *C. elegans* in the context of gout-like syndromes.

1. INTRODUCTION

Gout is one of the most common types of inflammatory arthritis in men and affects 1–2% of adults in developed countries. It is resulted from the deposition of monosodium urate crystals in and around the joints. Elevated serum urate levels are a key risk factor for gout onset, though only 10% of people with hyperuricemia develop clinical gout, suggesting that hyperuricemia alone is not enough for the gout arthritis development. By now, People are still looking for gout susceptibility genes [1]. Some of the potential gout genes, such as BCAS3, RFX3 and KCNQ1, are also found involved in the development of other diseases [1-4], indicating the complexity of gout pathology.

Except for genetic susceptibility, food provides another pathological source for gout. In the long run, nutrigenomics will answer most questions on interactions among many different types of food and gout-susceptibility genes. But in the short term, it is needed to set up a model organism to functionally assess gout vulnerability from different food.

C. elegans are model organism that share many biological characteristics of humans. Many of the genes that control cell growth and development in this organism have direct counterparts in humans and therefore it has much significance to understanding basic cell biology in humans. For many years, the worm of *C. elegans* is a good model to study molecullar and cellular mechanisms of lifespan, ageing, development, tumorigenesis, etc. In this study, intake of seven kinds of seafood was undertaken to see if they induce gout-like syndromes (features) on the body of *C. elegans*.

2. MATERIALS AND METHODS

2.1. Maintenance and Cultivation of Nematodes

The *C. elegans* strain for the experiment was wild-type N2. The *C. elegans* worms were cultured onto NGM agar plates at 22°C and fresh living *E. coli* OP50 bacteria were used as the only food source.

2.2. Lifespan Assay

After synchronization L4 larvae were cultured on NGM agar plates (with and without seafood suspensions). The beginning of *C. elegans* hatching was recorded as the day zero, and then nematodes were transferred daily to fresh plates until all died. The survival and death of nematodes were recorded in detail every day [5-8].

2.3. Motility

Motility phenotypes were divided into four levels, where "4" means animals move spontaneously and smoothly, leaving sinusoidal and symmetric tracks; "3" means animals move spontaneously at lower speed; "2" means animals only move slowly when prodded with a platinum wire; "1" means animals only move the mouth or tail. Worms after adulthood were observed daily and their motility was recorded from "4" to "1".

2.4. Uric Acid Assay

The Uric Acid Assay Kit (Shanghai XinFan C012) provides a convenient means for detecting uric acid in biological samples without pretreatment. The principle of the kit lies at the phosphotungstic acid reduced by uric acid under alkaline conditions into tungsten blue, allantoin and carbondioxide. The blue color is proportional to the uric acid concentration measured by spectrophotometry at OD = 690 nm.

3. RESULTS AND DISCUSSION

3.1. High Concentration Seafood Extract Juice Can Decrease Lifespan and Motility of *C.elegans*

There are seven kinds of fresh seafood including *Scapharca broughtonii*, *Cyclina sinensis*, *Scapharca subcrenata*, Cockle, Scallop, *Meretrix meretrix Linnaeus*, and *Ruditapes philippinarum* that were obtained from the supermarket (Weihai). Then we triturated all the seafood and made extract juice. The final concentration of seven seafood mixture extracts was 500.5g/L each (i.e. add 71.5g each kind of seafood to 1L ultrapure water). The final concentration of simplex extract juice of each seafood used in the subsequent experiment was 500g/L (i.e. add 500g each kind of seafood to 1L ultrapure water). 1mL seafood extract juice added to 100mL NGM liquid medium was regarded as "1%" concentration.

To explore the relationship between seafood intake and gout development, we asked whether different concentrations (0.1%, 1%, 10%) of seafood mixture juice have physiological effect on *C. elegans*. Dose-response analysis revealed that worms cultured on NGM plates adding with 10% equally-weighted mixture seafood juice had the most significant lifespan decrease by up to 11.6% (Fig. 1A). The results of another 2 repeated trials were similar to this (data not shown). Subsequently we examined the effect of different kinds of seafood at 10% concentration on *C. elegans*. The results indicated that *Meretrix meretrix L.* reduced mean lifespan by 15.64% (Fig. 1B), and also slowed down the physiological activity (crawling speed, Fig. 1C) of *C. elegans*. Another repeated trial showed the similar result (data not shown).

3.2. Ono Type of Seafood (*Meretrix meretrix Linnaeus*) induced Gout-like Syndrome in *C. elegans*

Followed by *C. elegans* fed with a single extract of 10% concentration in seven kinds of shellfish, we found the *Meretrix meretrix Linnaeus* of 10% concentration had the greatest impact to *C. elegans*, in the terms of shorter lifespan, lower body activity, and even poor reproductive capacity (data not shown). In addition, the body of *C. elegans* had emerged the phenomenon of higher stiffness at later stages. We speculated that the consumption of excessive seafood brings more burden of urate accumulation in the body of *C. elegans*, which might be supported by the results of Table 1.

3.3. Elevated General Uric Acid Levels in Seafood-treated Worms

C. elegans were cultured on NGM plates with and without 10% *Meretrix meretrix L.* extracts at 22°C for 60h. Then nematodes and excreted metabolites on the surface of NGM plates were collected similarly as in the literature [9]. There are four groups of samples as in Table 1. Worms were all washed with saline water for three times under 4000 rpm centrifugation. Finally worms and their excretions were all suspended in 5 ml saline water and subjected to uric acid assay. Uric acid concentrations were detected using Uric Acid Testing Kit (Xinfan, Shanghai), unified with worm numbers (2500~3500 worms each sample) and the results were shown in Table 1.

Table 1. Uric acid levels of worms and their excrement (worm body + excrement = general)

Uric acid concentration (mg/L)	Worms w/o seafood	Excrement of worms w/o seafood	Worms w/ seafood	Excrement of worms w/ seafood
Measurement 1	0.739	0.000	0.080	2.463
Measurement 2	1.531	0.000	2.463	4.337
Measurement 3	0.270	0.270	1.622	1.892
Measurement 4	1.351	0.000	0.273	1.892

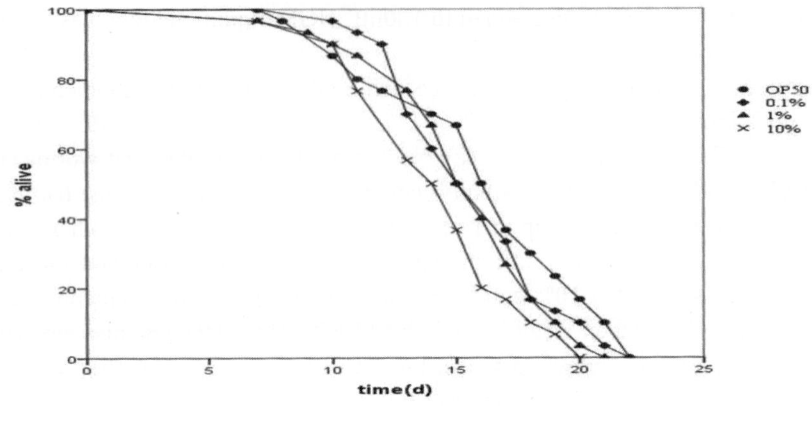

(A)

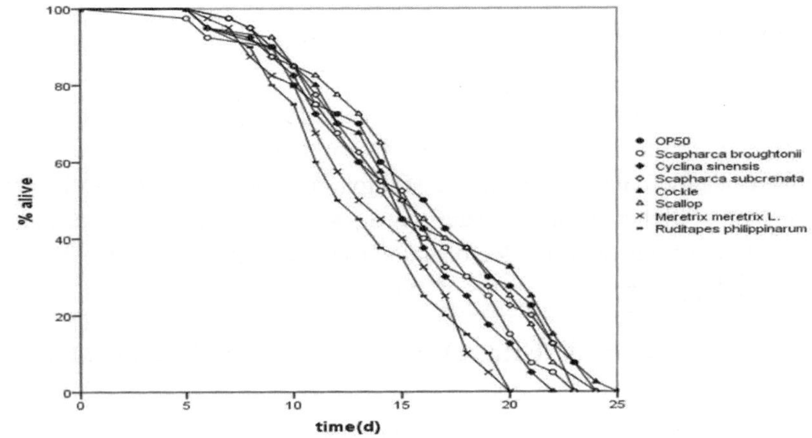

(B)

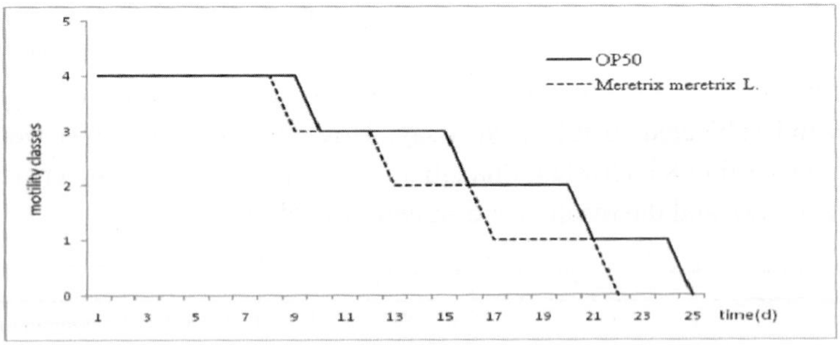

(C)

Fig. 1. Intake of seafood could decrease lifespan and motility of *C. elegans*. A: Kaplan-Meier survival curves for *C. elegans* fed with seafood mixture extract juice at different concentrations. B: Kaplan-Meier survival curves for *C. elegans* fed with 10% separate seafood extract juice. C: *Meretrix meretrix L.* reduced physiological activity of nematodes at the 10% concentration.

4. CONCLUSION

This study employed seven kinds of shellfish seafood to feed *C. elegans* at three different concentrations and found some seafood did induce gout-like features in the worm, such as lower body activity, more body stiffness at late stages, and shorter lifespan. These features are gout-like, not the gout itself, because the key syndrome of gout is monosodium urate crystals in and around joints, which may not be found in *C. elegans*. This preliminary assessment provided us some evidence that *C. elegans* are worthy further investigation whether or not the worm is a good candidate as a "semi-model" organism for gout study, at least in the context of molecular mechanisms of the level elevation of uric acid.

Many questions remain. In the Table 1, it seems that some batch of worms excret uric acid more or faster than others (similar to human), though all exposed to seafood juice and it is known that seafood raises the general uric acid metabolic levels [10].

Acknowledgments

This work was funded by NSFC (No.31071170), GREDBIO fund, HIT (hitwh200904), 985 Fund, HIT-NSRIF (2011101) and HIT fund (WH20140205).

References

1. Li C, Li Z, Liu S, et al. Genome-wide association analysis identifies three new risk loci for gout arthritis in Han Chinese. *Nature Commu* 2015; 6: 7041.
2. Choi HK, Mount DB, Reginato AM, et al. Pathogenesis of gout. *Ann Intern Med* 2005; **143**(7): 499-516.
3. Anzai N, Enomoto A, Endou H. Renal urate handling: clinical relevance of recent advances. *Curr Rheumatol Rep* 2005; **7**(3): 227-234.
4. Hediger MA, Johnson RJ, Miyazaki H, et al. Molecular physiology of urate transport. *Physiology* (Bethesda) 2005; **20**: 125-133.
5. Xiao R, Zhang B, Dong Y, et al. A genetic program promotes C. elegans longevity at cold temperatures via a thermosensitive TRP channel. *Cell* 2013; **152**(4):806-817.
6. Ayyadevara S, Tazearslan C, Alla R, et al. Rec-8 dimorphism affects longevity, stress resistance and X-chromosome non-disjunction in C. elegans, and replicative lifespan in S. cerevisiae. *Front. Genet.* 2014; **5**:1-13.
7. Zhang Y, Shao Z, Zhai Z, et al. The HIF-1 hypoxia-inducible factor modulates lifespan in C. elegans. *PLoS ONE* 2009; **4**(7): e6348.
8. Hekimi S, Burgess J, Bussière F, et al. Genetics of lifespan in C. elegans: molecular diversity, physiological complexity, mechanistic simplicity. *TRENDS Genet* 2001; **17**(12): 712-715.

9. Butler JA, Mishur RJ, Bokov AF, et al. Profiling the anaerobic response of C. elegans using GC-MS. *PLOS ONE* 2012; **7**(9): 1-9.
10. Sha HB, Shengwen HE, Cuiyuan QI, et al. Case control study of risk factors of gout. *Chin J Prev Contr Chron Dis* 2010; **18**(2):147-149.

HOT SPOTS IN CHINESE GENETICALLY MODIFIED RICE RESEARCH: BASED ON CO-WORD CLUSTER ANALYSIS

Kun Ji[a], Dan Guo, and Jingmin Cheng[b,*]

School of Management, Shanxi Medical University, Taiyuan, 030001, China
E-mail:[a]18234118380@163.com,[b]72-87@163.com
[*]*Corresponding author*

The purpose of this study is to gain an overview of hot spots related to research focused on Chinese GM rice from 2001 to 2015. The study collect and min literature data with the keyword "GM rice" published in Chinese periodicals databases (CNKI, Wanfang Data, and VIP) from 2001 to 2015. Using scientific measuring softwares in this study including Bicomb, Excel and SPSS19.0 to try to sort out the hot spots based on co-word cluster analysis of highly-frequent keywords. A total of 1875 publications are indexed in these databases. A total of 30 highly-frequent keywords for more than 20 times' occurrences in the papers about GM rice are extracted for cluster analysis. Thus, a 30×30 co-word matrice is acquired. Through the analysis of the posting and that of high frequency keywords cluster, to find the expansion trend in the research of the GM rice in China, These words are classified into four aspects: biological shapes; safety evaluation; breeding and detection technology; and commercial cultivation.

1. INTRODUCTION

GM is using modern molecular biological technology, transferring gene of certain organisms to other species and transforming the biological genetic material in order to meet people' needs[1]. Genetically modified rice is through GM technology to enrich rice varieties or cultivate insect resistant gene and anti-disease gene. GM rice can provide the means to increase food production and improve food quality. So the study of GM rice is of great significance to the development of food industry.

China is one of the leading states among the developing countries in the area of GM plant research, especially in the researches of GM rice. In order to have a clear understanding of the research hot spots in the field of GM rice in China, this article uses the co-word cluster analysis method to study. Co-word analysis is a type of bibliometric method to identify hot spots and find knowledge in academic literature. Its principle is as follows: if two professional terms expressing a particular research subject appear in the same articles simultaneously, these two terms may have a certain intrinsic relationship. The more frequently these two terms occur in the same articles, the closer their relationship is. According to this "distance", the important keywords of a subject are classified further to sum up the research focus and structure of a discipline by modern statistical techniques[2].

2. MATERIALS AND METHODS

2.1. Materials

Relevant articles with the keyword "GM rice" are identified by searching in Chinese periodicals databases, including CNKI, Wanfang Data, and VIP. Data from January 1st 2001 to December 31st 2015 to ensure the search results are as current as possible.

Based on the search strategy mentioned-above, 2287 publications are found in these databases. The titles and abstracts of the publications are screened according to relevance and selection criteria. The inclusion criteria are: (1) the contents of papers primarily focus on GM rice, and (2) all study designs. The exclusion criteria are: (1)conference report; (2)newspaper; (3)interview; (4)advertisement; and (5)solicit article. Finally, a total of 1875 papers are included in this study. Each publication downloaded from the databases included the following key eligibility items: title, author, institution, country, source, publication year, journal, and keyword. These data are saved as a text file in the format of XML.

From 2001to 2015, the number of published literatures on the study of GM rice in China is shown in Fig. 1.

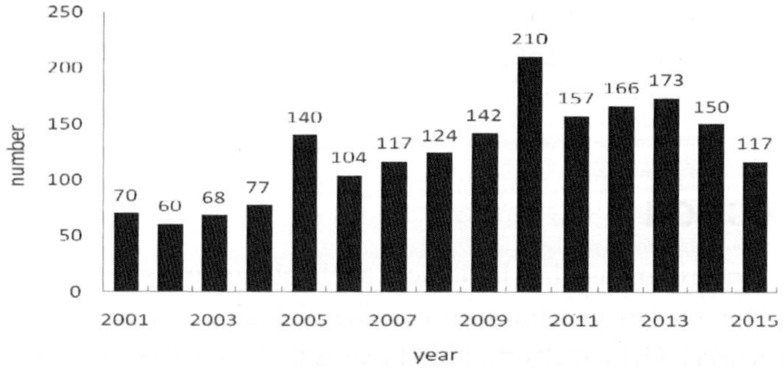

Fig. 1. Number of publications retrieved from 2001 to 2015.

According to the Fig. 1, posting on the GM rice from 2001 to 2015 although experiencing slight fluctuations, basically increased year by year. In 2005 and 2011, there appeared two peaks. Generally, the publication shows a rising trend. Chinese GM rice research shows overall growth trend from 2001 to 2015.

2.2. Methods

Bibliographic Item Co-Occurrence Matrix Builder (BICOMB) provided by professor Lei Cui (China Medical University) and Microsoft Excel are employed to determine the distribution of the publication year, and the frequency ranking of keywords of the included publications.

Considering the papers included in this study having to be related to GM rice, the highly-frequent keyword "GM rice" is deleted for they would be meaningless in the content analysis. The trends of the remaining highly-frequent keywords over time are illustrated visually. The keywords are ranked, and highly-frequent keywords with co-words matrix are gained from the literature database by setting a proper threshold value. Then, the above co-words matrix is used for further hierarchical clustering by using software SPSS19.0 with default options. Dendrogram is drawn for assessing the cohesiveness of the clusters formed and providing information about the appropriate numbers of clusters to keep.

3. RESULT OF CO-WORD CLUSTER ANALYSIS

Highly-frequent keywords are selected according to the occurrence frequency of a descending order in Table 1. The obtained highly-frequent keywords word matrix is imported into SPSS software to cluster analysis; and a co-word clustering dendrogram of highly-frequent keywords is drawn in Fig. 2. According to professional knowledge, the effective group is determined to be four, and finally, it is concluded with the cluster analysis group table in Table 2.

Table 1. The highly-frequent keywords extracted from publications retrieved

No.	Keyword	F.	No.	Keyword	F.
1	rice	303	16	gene drift	34
2	GM food	133	17	gene expression	32
3	GM	104	18	agricultural biotechnology	32
4	commercial cultivation	84	19	transgenic seeds	31
5	GM organisms	50	20	disease resistance	30
6	agricultural GMOs	50	21	agronomic characters	29
7	GM plants	50	22	Brown Planthopper	25
8	genetic transformation	43	23	Greenpeace	25
9	safety evaluation	43	24	security	23
10	Zhang Qifa	40	25	food safety	21
11	bio-safety	37	26	cultivation of new varieties	21
12	GM crops	37	27	insect-resistant	21
13	insect resistant genes	34	28	GM products	21
14	GMO Safety	34	29	rice breeding	21
15	insect resistance	34	30	agrobacterium	21

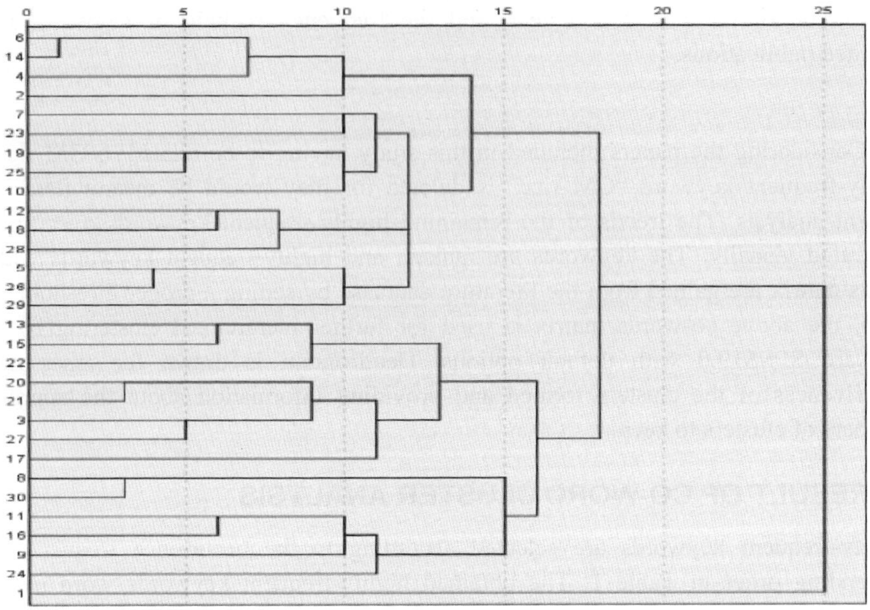

Fig.2. Co-word cluster analysis of the highly-frequent keywords.

Table 2. The cluster analysis group

No.	The Cluster Analysis Group	Serial Number of Keywords
1	biological shapes	13, 15, 22, 20, 21, 3, 27, 17
2	safety evaluation	8, 30, 11, 16, 9, 24
3	breeding and detection technology	7, 23, 19, 25, 10, 12, 18, 28, 5, 26, 29
4	commercial cultivation	6, 14, 4, 2

Through the above analysis, the research focusing on the field of GM rice in China has the following four aspects:

3.1. Biological Shapes of GM Rice

Cluster analysis of highly-frequent keywords number: 13, 15, 22, 20, 21, 3, 27, 1 and 17 showed that research on biological shapes had been a topic of great importance in recent years. The biological characteristics of GM rice mainly studied agronomic traits, plant resistance of the inverse. For example, Hong-Xia Ye Studied the biological properties of pea ferritin GM rice, such as heat resistance, cold endurance and disease resistance[3].

3.2. Safety Evaluation of GM Rice

Cluster analysis for highly-frequent keywords number: 8, 30, 11, 16, 9 and 24 indicated that safety evaluation had been a research focus in recent years. Food safety was often the focus of public attention[4]. The research on safety evaluation of GM rice included two aspects: food safety and ecological safety. The major studies on food safety included nutrition, intracorporal and extracorporeal toxicology, sensitization and exogenous gene transfer, etc. the safety of the ecological environment was also an important topic. GM crops could destroy original biological chain and the ecological structure, thus lead to the diversity of species is affected[5]. The safety evaluation of GM rice should be carried out according to certain principles, for example, principle of science, principle of substantive equality, principle of case analysis, and principle of sequence[6].

3.3. Breeding and Detection Technology of GM Rice

As shown in the cluster analysis of highly-frequent keywords number: 7, 23, 19, 25, 10, 12, 18, 28, 5, 26, and 29, researchers paid more attention to research on breeding and detection technology of GM rice. Studies on GM rice breeding in China was mainly focused on insect resistant GM rice, disease resistant GM rice, herbicide resistant GM rice, male sterility GM rice, in order to improve the quality and yield of rice. Detection of GM rice could be divided into two kinds: one kind was the detection of exogenous gene nucleic acid, the other kind was the detection of exogenous protein.

3.4. Commercial Cultivation of GM Rice

Cluster analysis of highly-frequent keywords number: 6, 14, 4, and 2 showed that commercial cultivation was also a topic of GM rice in China. The research on the commercialization of GM rice in China mainly aimed at the challenges and prospects. Industry researchers had doubts and concerns about the commercialization of GM rice, and they believed that commercial planting of GM rice could increase food production and reduce pests and diseases, but also had existed risks[7-8].

4. CONCLUSIONS

In this paper, based on the co-word cluster analysis method, the research focuses on the study of the GM rice in China during the years from 2001 to 2015. According to the quantity change of literature, the quantity change of the document shows a general upward trend, and the research scale is expanding. Through the high frequency key words co-word cluster analysis, the result shows that the hot spots in Chinese GM rice research as follows: biological shapes of GM rice, safety evaluation of GM rice, breeding and detection technology of GM rice, and commercial cultivation of GM rice.

Acknowledgments

This work is supported by the Fund (grant no. SSKLZDKT2014085)hosted by Dan Guo for the federation of social sciences at Shanxi Province. Some students and teachers participated in this project including Jing-Min Cheng, as the corresponding author, and Kun Ji.

References

1. Jurkiewicz A, Zagórski J, Bujak F, Lachowski S, Florek- Ł uszczki M. Emotionalattitudes of young people completing secondary schools towards genetic modification of organisms (GMO) and genetically modified foods (GMF). Ann Agric Environ Med 2014; **21**(1): 205-11.
2. Jun-ping Qiu, Chun-lei Zhou, Si-luo Yang. Review and prospect of the study of information science in China during the 30 years of reform and opening up (three). Library and Information Science Research 2009; **23** (9): 1. (In Chinese).
3. Hong-xia Ye, Ze-jian Guo, Mei Li, etc. Variation of important biological characteristics of pea ferritin pure line transgenic rice. Journal of Agricultural Biotechnology 2007; **15** (2): 251-256. (In Chinese).
4. Jing-min Cheng. *Administrative Regulation Research of Food Safety*. Beijing: Guanming Daily Press, 2015: 3-5. (In Chinese).
5. Yun-bo Luo, Xiao-yun He, Kun-lun Huang, etc. Food safety evaluation of GM rice at home and abroad situation. *Journal of Food Science* 2010; **23** (6): 689-694. (In Chinese).
6. Cong-min Li. Safety analysis and evaluation of genetically modified food. Modern agricultural science and technology 2014; **12**: 302-304. (In Chinese).
7. Gao-ming Jiang. Commercialization of genetically modified (gm) crops. *Chinese High Technology and New Technology* 2010; **1**: 93-94. (In Chinese).
8. Xiao-mei Lin, Ling-li Zeng, Pei-fen Zhuang. SWOT analysis of the development of commercialization of transgenic rice. *Journal of Taiwan Agriculture Research* 2012; **1**: 50-54. (In Chinese).

PRODUCTION OF ANTI-LISTERIAL BACTERIOCIN PEDIOCIN IN *LACTOCOCCUS LACTIS*

Shanna Liu[1,a], Ye Han[2], and Zhijiang Zhou[2]

[1]*College of Food Science and Bioengineering, Tianjin Agricultural University, 22 Jinjing Road, Tianjin 300384, China*

[2]*School of Chemical Engineering and Technology, Tianjin University, 92 Weijin Road, Tianjin 300072, China*

E-mail:[a]lsn_nxy@aliyun.com

Lactic acid bacteria (LAB) are being used in genetic modification for various novel applications. In this study, pediocin structural gene *pedA* was isolated from *Pediococcus acidilactici* PA003 by polymerase chain reaction (PCR) then cloned into vector pMG36e. Constructed plasmid was electroporated into host strain *Lactococcus lactis* NZ9000. Recombinant pediocin was detected and revealed biological activity by agar diffusion assay taking *Listeria monocytogenes* CVCC1595 as the indicator strain. The results demonstrated that *L. lactis* system could be developed as a useful tool for pediocin production in research or industry.

1. INTRODUCTION

With numerous food poisoning outbreaks and growing consumer refusal of chemical additives in food industry, food safety issue has been placed in ever more important position. Bacteriocins are ribosomally synthesized peptides that could be developed as natural food preservatives due to their antimicrobial activities against food spoilage and pathogenic bacteria. Lactic acid bacteria (LAB) are mostly known for their widespread use and "generally regarded as safe" (GRAS) status in food industry. The potential applications of bacteriocins from lactic acid bacteria (LAB) as antimicrobial alternatives have attracted a strong attention of the academia and the industry with increasing number of LAB bacteriocins have been identified and characterized from wild-type strains[1, 2]. However, there are some limitations when using bacteriocin-producing LAB in situ, such as poor adaptation to the food environment[3], low-level production[4], and negative influence on food property[5]. Thus, interest in the heterologous production of bacteriocin is growing rapidly with advances in genetic technologies and genome sequencing.

Class IIa bacteriocins, also called pediocin-like bacteriocins, are small, heat-stable, and nonmodified peptides that exhibit highly activities against pathogenic *Listeria monocytogenes*[6]. One of the most studied class IIa bacteriocins is pediocin, which is composed of 44 amino acids and predicted to penetrate into cell membrane of sensitive strains and mediate leakage through it[7]. Lactococcal vector pMG36e contains the pWV01 origin of replication. When the gene, deprived of its own promoter sequence, is placed under the transcriptional control of promoter P32 in the pMG36e, it is designed to be functional in *Lactococcus lactis* MG1363. In the present study, we report the cloning of pediocin structural gene *pedA* in pMG36e and expression in *L. lactis* NZ9000, a derivative of MG1363 host strain. Anti-listerial activity of recombinant pediocin was evaluated using *L. monocytogenes* CVCC1595 as indicator strain.

2. MATERIALS AND METHODS

2.1. Bacterial Strains, Plasmids, and Growth Conditions

P. acidilactici PA003, a wild pediocin-producing strain, and host strain *L. lactis* NZ9000, were grown in M17 broth supplemented with 0.5% glucose (M17G) at 30 °C without agitation. *L. lactis* transformants were cultured in M17 broth supplemented with 0.5% glucose and 0.5% sucrose (M17GS). *Escherichia coli* DH5α was grown in Luria-Bertani (LB) medium and incubated at 37 °C with shaking at 150 rpm. When required, erythromycin was used at a concentration of 150 μg ml^{-1} for *E. coli* and 5 μg ml^{-1} for *L. lactis*. The indicator strain used in the bacteriocin assay was *L. monocytogenes* CVCC1595, which was grown in BHI medium. Plasmid pMD18-T and pMG36e were applied to the construction of *pedA* hybrid gene.

2.2. Construction of *L. lactis* Expression Vector

All DNA manipulations were carried out according to procedures described previously[8]. Electroporation of *L. lactis* was performed according to Holo and Nes's method[9]. Total DNA of *P. acidilactici* PA003 was used as the template to amplify the *pedA* gene. The forward primer PedAF (5'-AACCCCTCTAGAATGAAATACTACGGTAATG GGG-3') containing a restriction site for *Xba* I (TCTAGA) and the reverse primer PedAR (5'-CCCGGGCTGCAGTTACTAGCATTTATGATTACCTTG-3') containing *Pst* I restriction enzyme site (CTGCAG) and stop codon marking the end of the *pedA* sequence were used. The PCR amplification was performed as follows: initial denaturation of DNA at 95 °C for 3 min and the amplification programme consisting of 30 cycles of 94 °C for 30 s, 56 °C for 60 s, 72 °C for 60 s, and 1 cycle at 72 °C for another 7 min (final extension). The PCR product was purified and ligated to pMD18-T vector. The T-A cloning product was digested with *Xba* I and *Pst* I restriction endonucleases. The resulted *pedA* gene fragment was purified and ligated into pMG36e vector which was digested with the same restriction enzymes in order to be cloned in frame after the promoter of P32 (Fig. 1). Ligation mixture was transformed into *E. coli* DH5α competent cells. Clones with erythromycin resistance (Emr) were screened in the plate. Recombinant plasmids were analyzed by PCR identification. A positive plasmid confirmed by DNA sequencing was further used to transform *L. lactis* NZ9000.

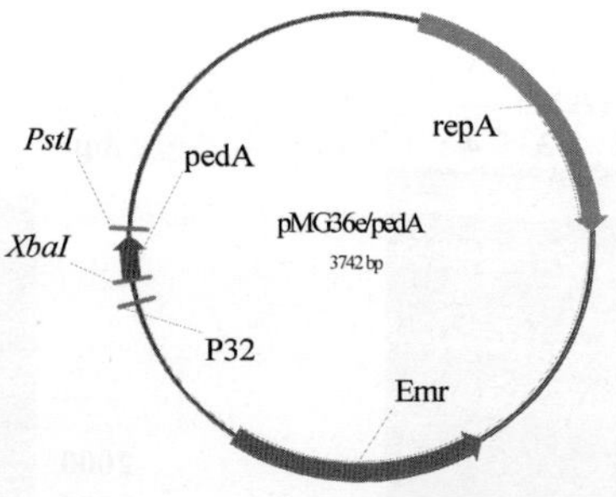

Fig. 1. Construction of pMG36e/pedA.

2.3. Protein Expression and Agar Diffusion Assay

L. lactis NZ9000 (harboring pMG36e/pedA) supplemented with 5 μg ml^{-1} erythromycin was grown at 37 °C in M17G medium without agitation. The overnight culture was centrifuged at 8000×g for 10 min. The supernatant was collected as extracellular fraction. Pellet was kept as whole-cell fraction. Samples with added loading buffer were heated at 100 °C for 10 min before SDS-PAGE.

Cultivated cells were washed twice in 50 mmol L^{-1} Tris-HCl buffer at pH 8.0 and resuspended in the same buffer. The mixture supplemented with 25 μg ml^{-1} lysozyme was incubated at 37 °C for 30 min. Cell lysates were prepared by sonication on ice. The supernatant was kept as the soluble cytoplasmic fraction and kept for further anti-listerial assay.

Agar diffusion assay was adopted to determine the antimicrobial activity of the recombinant product. Agar plate was overlaid with 10 ml of soft agar inoculated with 50 μl of an overnight culture of indicator strain. Samples were added to the wells and the plate was incubated overnight at 37 °C.

3. RESULTS AND DISCUSSION

3.1. Construction of pMG36e/pedA

Wild-type pediocin-producing strain *P. acidilactici* PA003 was isolated from Chinese fermented vegetables in previous work[10] and was used as template to amplify *pedA* gene. Desirable DNA band was detected between 100 bp and 250 bp in gel electrophoresis. The gene was cloned in pMD18-T vector and inserted into pMG36e in accordance with standard techniques. Positive plasmid was selected by PCR identification (Fig. 2) and verified by DNA sequencing. The nucleotide sequence homology of *pedA* gene obtained from *P. acidilactici* PA003 was 100% identity compared with the *pedA* gene that has been published in the National Center for Biotechnology Information (GenBank AY083244).

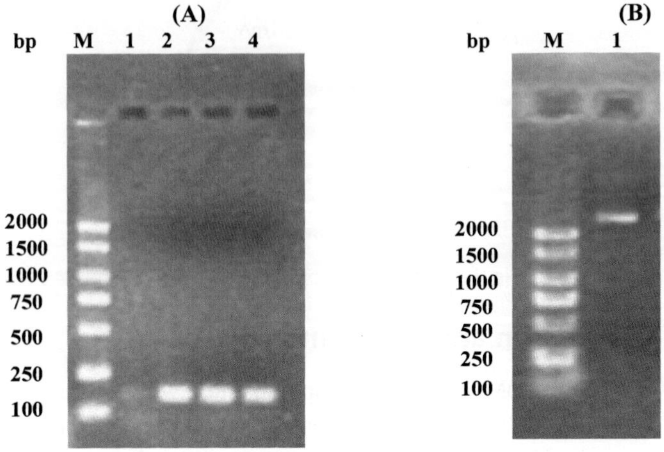

Fig. 2. Identification of pMG36e/pedA. (A) Identification by PCR. M, DNA marker; Lane 1, PCR product using distilled water as the template; Lane 2-4, PCR products using pMG36e/pedA recombinant plasmids as the templates; (B) Identification by molecular weight. M, DNA marker; Lane 1, plasmid isolated from positive recombinant.

3.2. Expression of Recombinant Pediocin

In this study, we selected pMG36e as expression vector since it has the ability of propagation both in *E. coli* and in *L. lactis*. Transformation of competent *L. lactis* NZ9000 with screened positive recombinant plasmid showed evident protein band below 8 kDa after SDS-PAGE of whole-cell lysate which corresponded with the predicted molecular mass of nature pediocin (Fig. 3).

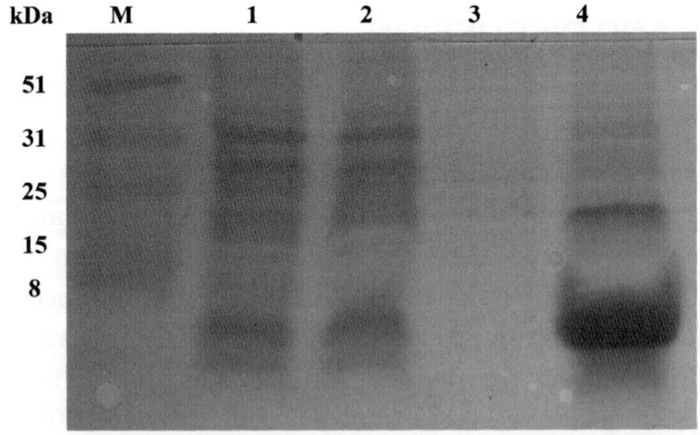

Fig. 3. SDS-PAGE of the expression proteins by *L. lactis* NZ9000. M, polypeptide molecular mass standards; Lane 1, whole-cell lysates of *L. lactis* NZ9000 culture; Lane 2, whole-cell lysates of *L. lactis* NZ9000 (pMG36e) culture; Lane 3, supernatant proteins of *L. lactis* NZ9000 (pMG36e/pedA) culture; Lane 4, whole-cell lysates of *L. lactis* NZ9000 (pMG36e/pedA) culture.

3.3. Antimicrobial Activity Assay

Recombinant pediocin was produced and verified by agar diffusion assay using *L. monocytogenes* CVCC1595 as indicator strain (Fig. 4). Finally, biologically active pediocin was obtained from recombinant *L. lactis* NZ9000. Different yields of bacteriocins by lactococcal strains have been reported before. Usually the yield of pediocin from genetic lactococcal strain was not as high as that from *E. coli* as illustrated by Horn et al.[11]. It may be caused by some unknown metabolic differences in various host strains, promoters used to drive gene expression, recombinant plasmid copy number, higher aggregation and proteolytic activity of host strain[12]. As for this work, strategies used to improve production could be tested by optimizing culture condition, ultrasonic disruption and addition of protecting agent such as glycerol.

The direct utilization of pediocin-producing strain *P. acidilactici* in dairy industry either as starter culture or bacteriocin-producing adjunct culture is an obstacle because of its lack of or slow lactose fermentation[13]. Genetically improved *L. lactis* may have potential in the application of food industry as cell factory by increasing bacteriocin production, leading to a wider antagonistic spectrum (construction of multibacteriocinogenic strains) as well as adaptation of the selected hosts to food environments[14].

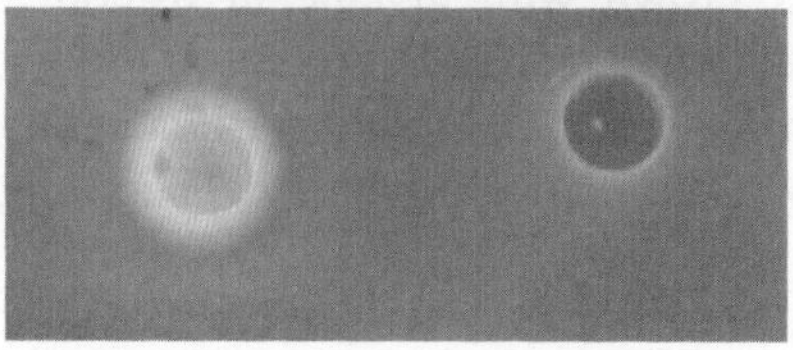

Fig. 4. Agar diffusion assay of *L. lactis* NZ9000 harboring different plasmids.

4. CONCLUSIONS

In conclusion, cloning of structural gene *pedA* under the control of P32 promoter in plasmid pMG36e has permitted the production of pediocin in *L. lactis*. This represents a basic step in the construction of LAB strains applied in food industry. It is anticipated that further study will focus on the optimizing production of pediocin-producing LAB strains with GRAS status and application in different food systems for biopreservation.

Acknowledgments

We thank Professor Mingqiang Qiao (Nankai University) for providing *L. lactis* NZ9000. This work was supported by the National Natural Science Foundation of China (Project 31501583).

References

1. Papagianni M, Anastasiadou S. Pediocins: The bacteriocins of Pediococcus. Sources, production, properties and applications. *Microb Cell Fact* 2009; **8**(3).
2. Roces C, Rodríguez A, Martínez B. Cell wall-active bacteriocins and their applications beyond antibiotic activity. *Probiotic & Antimicro Prot* 2012; **4**: 259-272.
3. Devlieghere F, Vermeiren L, Debevere J. New preservation technologies: Possibilities and limitations. *Int Dairy J* 2004; **14**: 273-285.
4. Yildirim S, Konrad D, Calvez S, Drider D, Prevost H, Lacroix C. Production of recombinant bacteriocin divercin V41 by high cell density *Escherichia coli* batch and fed-batch cultures. *Appl Microbiol Biotechnol* 2007; **77**: 525-531.
5. Galvez A, Lopez RL, Abriouel H, Valdivia E, Omar NB. Application of bacteriocins in the control of foodborne pathogenic and spoilage bacteria. *Crit Rev Biotechnol* 2008; **28**: 125-152.
6. Ji-Eun E, Moon SK, Moon, GS. Heterologous production of pediocin PA-1 in *Lactobacillus reuteri*. *J Microbiol Biotechnol* 2010; **20**: 1215-1218.
7. Lohans CT, Vederas JC. Development of class IIa bacteriocins as therapeutic agents. *Int J Microbiol* 2012; doi:10.1155/2012/386410.
8. Sambrook J, Russell DW. Molecular Cloning: A Laboratory Manual Third Edition. Cold Spring Harbor, NY, USA. 2001.
9. Holo H, Nes IF. High-frequency transformation, by electroporation, of *Lactococcus lactis* subsp. *cremoris* grown with glycine in osmotically stabilized media. *App Environ Microbiol* 1989; **55**: 3119-3123.
10. Zhou ZJ, Han Y, Han X, Zheng F. Isolation of bacteriocin-producing *Pediococcus acidilactici* strain from fermented Chinese cabbage. *Food Science (In Chinese)* 2006; **27**: 89-92.
11. Horn N, Martinez MI, Martinez JM, Hernandez PE, Gasson MJ, Platteeuw C, Alen-Boerrigter I van, Schalkwijk S van, de Vos WM. Food-grade cloning and expression system for *Lactococcus lactis*. *Appl Environ Microbiol* 1996; **62**: 1008-1013.
12. Sanchez J, Borrero J, Gomez-Sala B, Basanta A, Herranz C, Cintas LM, Hernandez PE. Cloning and heterologous production of Hiracin JM79, a sec-dependent bacteriocin produced by *Enterococcus hirae* DCH5, in lactic acid bacteria and *Pichia pastoris*. *Appl Environ Microbiol* 2008; **74**: 2471-2479.
13. Somkuti GA, Steinberg DH. Pediocin production in milk by *Pediococcus acidilatici* in co-culture with *Streptococcus thermophilus* and *Lactobacillus delbrueckii* subsp. *Bulgaricus*. *J Ind Microbiol Biotechnol* 2011; **37**: 65-69.
14. Gutierrez J, Larsen R, Cintas LM, Kok J, Hernandez PE. High-level heterologous production and functional expression of the sec-dependent enterocin P from *Enterococcus faecium* P13 in *Lactococcus lactis*. *Appl Environ Microbiol* 2006; **72**: 41-51.

THE FORMULA AND TECHNOLOGY OPTIMIZATION OF GINSENG WHEY PROTEIN POLYPEPTIDE BEVERAGE

Pingping Hou[1,a], Gang Wang[1,b], Xinhui Zhou[1,c], Ting Zhang[2,d], Dongxue Li[1,e], Xudong Li[1,f], and Tiehua Zhang[1,g,*]

[1]*College of Food Science and Engineering, Jilin University, Jilin Changchun, China*
[2]*Tourism Institution, Jilin Business and Technology College, Jilin Changchun, China*
E-mail: [a]m15754304718@163.com, [b]18844547553@163.com , [c]651065148@qq.com, [d]147834156@qq.com, [e]18843866598@163.com, [f]15688965021@163.com, [g]zhangth@jlu.edu.cn
[*]*Corresponding author*

The ginseng whey protein polypeptide beverage was made mainly from ginseng extract and whey protein hydrolyzate. Single factor experiments and orthogonal experiments were carried out to analyze the formula and technology of ginseng whey protein polypeptide beverage utilizing precipitation rate as study indicator. The results showed the best formula was consisted of 0.2% ginseng extract, 0.7% hydrolyzed WPC-80, 8.0% sugar, 0.1% pectin, and the best secondary sterilization condition was 95°C, 15 min. Under such circumstance, the beverage was rich in nutrients with a pure taste and a good stability.

1. INTRODUCTION

Ginseng is a rare traditional Chinese medicine in China, which belongs to Araliaceae perennial herb[1]. The earliest Chinese medicine book "herbs by Shennong" had listed ginseng as top grade, with the effect of nourishing vitality, calming the nerves, alpiniaoxyphylla, tuning up organs, anti-aging, etc. Ginseng extract is extracted and transformed into ginseng, and the active ingredients are more easily absorbed than the Ginseng, which has the characteristics of quick action, strong activity, etc. [2].

Whey is the production of cheese or casein portion of the remaining liquid after precipitation of casein in milk, which retains 55% of the nutrients in milk. Since whey protein has a high absorbency, complete amino acid composition, low in fat and cholesterol, can anti-aging, promote heart health, cancer, enhance immunity, improve bone mass and control weight ,it is conducive to the promotion of health and is one of the most common protein supplement products today[3-6].

Combined ginseng with whey protein, a unique ginseng whey protein beverage could be developed. In present research paper, we aimed to study the precipitation rate of beverage to optimize the composition.

2. MATERIALS AND METHODS

2.1. Materials and Reagents

Ginseng extract (the total saponin content of 20%) was purchased from Jilin Hongjiu Biotechnology Co., Ltd., China; White sugar in line with GB 317.1-2010 health standards; Citric acid and Danisco pectin were purchased from Henan Fenglu Food Additive Co., Ltd., China; Desalted whey powder was purchased from Galaxy Albert Tianjin Import and Export Co. The company, China; Flavor protease and neutral protease were purchased from Novozymes Biotech Co., Ltd., China.

2.2. Instruments and Equipment

DSY-1-6 shaped hole electric heated water bath was purchased from Beijing Guohua medical instrument factory, China; SW- CJ- 2D type clean bench was purchased from Suzhou Purification Equipment Co., China; TGL- 16G high-speed desktop centrifuge was purchased from Shanghai Anting Scientific Instrument Factory; GYB60- 6S-pressure homogenizer was purchased from Shanghai Donghua high pressure homogenizer Factory, China; TC10KB electronic balance was purchased from Test Instrument Factory Changshu Shuangjie, China.

2.3. Process

Whey powder → rehabilitation → homogenization → pasteurization → enzymolysis →enzyme inactivation→ enzymolysis liquid whey→ mix → homogenization → pasteurization → bottling → sealing → secondary sterilization → cooling → products

2.4. Operating Points

Firstly, the fully hydrated material was homogenized by high-pressure homogenizer, the homogenization pressure was 18-20MPa. And then pasteurization at 85±2°C, sterilization of 30min. 4% neutral protease and 4% flavourzyme were added into the material at 55°C. After 4 hour's enzymolysis[7], it would be heated to 85°C to kill the enzyme for 30min. Then all materials were added in turn and mixed well. The material after mix was homogenized and pasteurized again. Then the liquid material was filled to 2 cm away from the bottle, screw on the lid. Finally, bottling finished products should be sterilized at 90-95°C some time. The products were required to be preserved at 2-28°C.

2.5. Single-factor Experimental Design

2.5.1. *Effect of Whey Type on Product*

The total quality of the product was calculated by 100g, in the case of 0.7% whey, 0.2% ginseng extract, 8.0% sugar, 0.1% citric acid, 0.1% pectin, the secondary sterilization time of 15min, to analyze the effect of WPI, WPC-80 and WPC-34 on product quality utilizing precipitation rate as study indicator.

2.5.2. *Effect of Protein Content on Product*

The total quality of the product was calculated by 100g, in the case of 0.2% ginseng extract, 8.0% sugar, 0.1% citric acid, 0.1% pectin, the secondary sterilization time of 15min, to analyze the effect of WPI content of 0.5%, 0.7% and 1.0% on product quality utilizing precipitation rate as study indicator.

2.5.3. *Effect of Ginseng Extract Content on Product*

The total quality of the product was calculated by 100g, in the case of 0.7% WPI, 8% sugar, 0.1%citric acid, 0.1%pectin, the secondary sterilization time of 15min, to analyze the effect of ginseng extract content of 0.2%, 0.5% and 0.8% on product quality utilizing precipitation rate as study indicator.

2.5.4. *Effect of The Secondary Sterilization Time on Product*

The total quality of the product was calculated by 100g, in the case of 0.7% WPI, 0.2% ginseng extract, 8.0% sugar, 0.1% citric acid, 0.1% pectin, the secondary sterilization time of 10min, to analyze the effect of the secondary sterilization time of 15min, 25min and 35min on product quality utilizing precipitation rate as study indicator.

2.6. Orthogonal Experimental Design

Based on single factor experiments, orthogonal experiment was designed. Whey type (A), whey protein content (B), ginseng extract content (C) and the secondary sterilization time (D) were the main factors affecting the quality of product, performed L_9 (3^4) orthogonal test , carrying out L_9 (3^4) orthogonal experiment, factor level in Table 1.

Table 1. Enzymolysis ginseng whey protein beverage factors level table

Factor	A	B	C	D
	Whey type	Whey protein content	Ginseng extract content	The secondary sterilization time
Level		(%)	(%)	(min)
1	WPI	0.5	0.2	15
2	WPC-80	0.7	0.5	25
3	WPC-34	1.0	0.8	35

2.7. Measurement of Precipitation Rate

Take a sample of product to centrifuge (3000 rpm, 15 min, 4 °C), and discard the supernatant, dry the centrifuge tube and weigh it, according to the following formula to calculate the precipitation rate.

Precipitation rate $X = (W-X) / Z \times 100\%$

Wherein:

X represents the weight of the centrifuge tube, accurate to 0.01g;

Z represents the weight of the material liquid in centrifuge tube, accurate to 0.01g;

W represents the weight of the centrifuge tube after centrifuging, accurate to 0.01g.

2.8. Data Processing

Using SPSS and Origin Software for data processing and mapping.

3. RESULTS AND DISCUSSION

3.1. The Analysis Results of Single Factor Test

3.1.1. *Effect of Whey Type on Product*

Fig. 1 showed that when the whey was WPI, the precipitation rate of beverage was the highest, followed by WPC-34, the lowest was WPC-80 and under such circumstance the drink had the best stability. And the amount of WPC-80 was less in the process of hydrolysis, the gel was not easy to occur, and the flavor of the beverage was good, so the effect of WPC-80 was better.

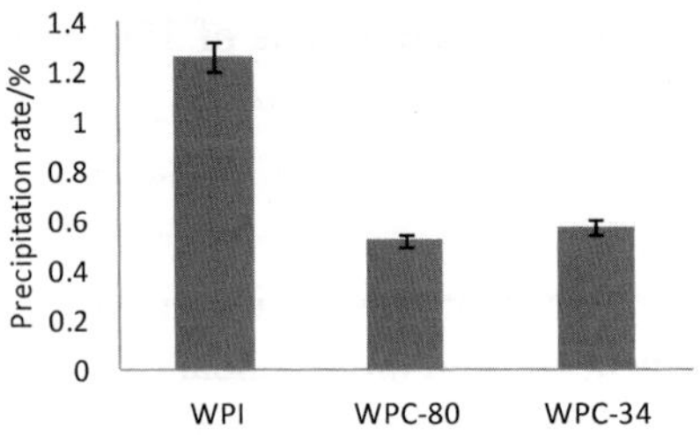

Fig.1. Effect of whey type on product.

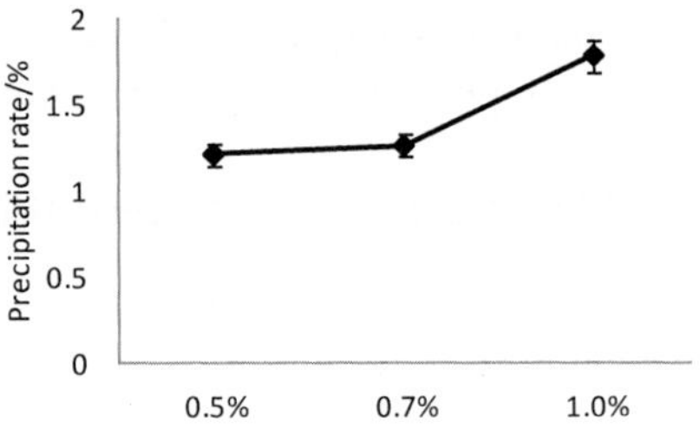

Fig.2. Effect of whey protein content on product.

3.1.2. *Effect of Whey Protein Content on Product*

Fig. 2 showed that with the increase of the whey protein content, the precipitation rate increased gradually. When whey protein content was 0.5% and 0.7%, the precipitation rate had no obvious difference. And when whey protein content was 0.5%, the protein content was very low, couldn't meet the needs of the human body, so the choice of whey protein content for the 0.7% was the best.

3.1.3. *Effect of Ginseng Extract Content on Product*

Fig. 3 showed that with the increase of the ginseng extract content, precipitation rate increased gradually. When the content of ginseng extract was 0.2%, the stability of the beverage was best, under such circumstance, there was a unique aroma of ginseng and the effective components of ginseng extract have been able to work. So the choice of ginseng extract content for the 0.2% was the best.

3.1.4. *Effect of the Secondary Sterilization Time on Product*

Fig. 4 showed that with the increase of the secondary sterilization time, precipitation rate increased gradually. When the secondary sterilization time was 15min, precipitation rate was the lowest. The secondary sterilization is to kill the bacteria on the packaging, but it will affect the stability of the beverage to a certain extent. Tests showed that the best secondary sterilization time was 15min.

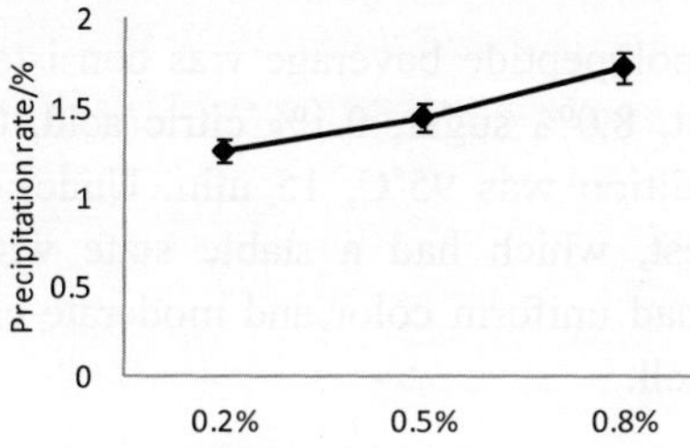

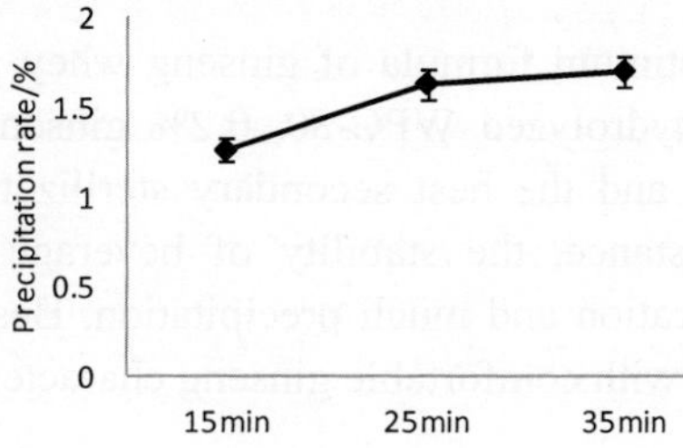

Fig.3 Effect of ginseng extract content on product. **Fig.4.** Effect of secondary sterilization time on .product

3.2. Results and Analysis of Orthogonal Experiment

Based on the conditions of the sugar content of 8%, citric acid content of 0.1%, pectin content of 0.1%, whey types (A), whey protein content (B), ginseng extract content (C) and the secondary sterilization time (D) are the main factors to do the four factors and three levels orthogonal experiment, using L_9 (3^4) orthogonal table to show, the test results were showed in Table 2.

Table 2. The results of orthogonal experimental on Enzymolysis Ginseng Whey Protein beverage

Factor	A	B	C	D	Precipitation rate /%
No.	(1)	(2)	(3)	(4)	
1	1	1	1	1	0.69
2	1	2	2	2	0.73
3	1	3	3	3	1.39
4	2	1	3	3	0.85
5	2	2	1	2	0.78
6	2	3	2	1	1.02
7	3	1	2	2	0.8
8	3	2	3	1	1.11
9	3	3	1	3	0.77
K_1	2.81	2.34	2.24	2.82	
K_2	2.65	2.62	2.55	2.86	
K_3	2.68	3.18	3.35	3.01	
k_1	0.94	0.78	0.75	0.94	
k_2	0.88	0.87	0.85	0.95	
k_3	0.89	1.06	1.12	1.00	
R_j	0.05	0.28	0.37	0.06	
Optimal level	A_2	B_1	C_1	D_1	

According to the range in Table 2, the primary and secondary factors affecting the formula of ginseng whey beverage were C>B>D>A. And the optimal combination is $A_2B_1C_1D_1$. Under such circumstance, the stability of products was the best, but in order to meet the needs of the human body, we chose the content of whey protein was 0.7%, namely the optimal combination was $A_2B_2C_1D_1$.

4. Conclusions

The optimum formula of ginseng whey protein polypeptide beverage was consisted of 0.7% hydrolyzed WPC-80, 0.2% ginseng extract, 8.0% sugar, 0.1% citric acid, 0.1% pectin, and the best secondary sterilization condition was 95°C, 15 min. Under such circumstance, the stability of beverage was best, which had a stable state without stratification and much precipitation. Besides it had uniform color and moderate cream flavor, with comfortable ginseng characteristic smell.

Acknowledgments

This project was funded by Jilin University "Students' Innovation and Entrepreneurship Training Program" (Grant no. 2015831299). The authors thank The National Twelfth Five-Year Grant (2013BAD18B07) for financial support of this study.

References

1. Hongping Wang, Xinbao Yang, et al. Study on the chemical constituents of root and rhizome of Panax ginseng in Jilin in Chinese [J]. *China journal of Chinese meteria medica*, 2013, **9**:2809-2817.
2. Jingpiao XU, Congxin Sun, Guixia Zhuang, et al. RP-HPLC method for simultaneous determination of content of ginseng extract product 20 (S), 20 (R) - ginsenoside Rg2 and 20 (S), 20 (R) -. Ginsenosides Rh1 in Chinese [J] *Journal of Shenyang Pharmaceutical University*, 2011, **9**: 712-715.
3. FULLER M F, MILNE A, HARRIS C I, et al. Amino acid losses in ileostomy fluid on a protein-free diet [J]. *American Journal of Clinical Nutrition*, 1994, 59: 70-73.
4. GAUTHIER S F, POULIOT Y, SAINT-SAUVEUR D. Imunomodulatory peptides obtained by the enzymatichydrolysis of whey proteins[J]. *International Dairy Journal*, 2006, **16**: 1315-1323.
5. DIANE S, SYLVIE F G, YVAN B, et al. Immunomodulating properties of a whey protein isolate, its enzymatic digest and peptide fractions[J].*International Dairy Journal*, 2008, **18**: 260-270.
6. KRUGER M C, POULSEN R C, SCHOLLUM L, et al. A comparison between acidic and basic protein fractions from whey or milk for reduction of bone loss in the ovariectomised rat [J]. *International Dairy Journal*, 2006, **16**: 1149-1156.
7. Luyun Wen, Lijuan Zhang, Bin Zhang, et al. Development of Chromium-rich Germinated Brown Rice Biscuit in Chinese [J]. *Food research and development*, 2015, **4**:91-93.

APPLICATION OF TWO-STAGE AERATION CONTROL STRATEGY TO ENHANCE YELLOW PIGMENTS PRODUCTION BY *MONASCUS ANKA* MUTANT

Bo Zhou[a] and Yuan Tian

School of Food Science and Engineering, Central South University of Forestry and Technology, Changsha, 410004, China

E-mail:[a]1214125436@qq.com

According to the results of the single aeration on *Monascus anka* mutant growth and monascus yellow pigments production in this paper, two stages in the fermentation aeration control strategy has been operated as following: 0.8~1.0 vvm aeration was carried out to improve the *Monascus anka* mutant growth and yellow pigments production before fermentation 96 hours, and then 0.4~0.6 vvm aeration was done to extend the time of yellow pigment production with efficient using substrate. The yield of monascus yellow pigments can arrive with 136.08 *OD* at 102 hours, it was higher 65.75%, 28.07%, 23.93%, 7.43% and 23.85% than that of 0.4 vvm, 0.6 vvm, 0.8 vvm, 1.0 vvm and 1.2 vvm aeration, respectively. The maximum specific production rate of yellow pigments(q_y) could arrive with 0.2852 $OD.g^{-1}.h^{-1}$ at 42 hours and two stage control strategy of aeration could obviously slow down the decreasing speed for yellow pigments production. The maximum specific growth rate (μ_{max}) could arrive with 0.0746 h^{-1} at 6 hours and the decreasing speed for specific growth rate was slowly. The two stage aeration controlling strategy could obviously decrease the aging speed of *Monascus anka* mutant in fermentation anaphase.These results in this paper have very important theoretical and realistic significance for realization the monascus yellow pigments industrial production by submerge culture.

1. INTRODUCTION

Yellow is one of three primary colors along with red and blue. Limited synthetic yellow pigments such as tartrazine and sunset yellow FCF are permitted for utilization in food. However, the toxicity of both synthetic pigments to health is presently still in question. Due to a few safety hazards, natural yellow pigments from animals, plants, or microorganisms have become more attractive in recent years. *Monascus* is an important microbial resource now in use to produce pigments in an industrial level.

Thousands years ago, ancient Chinese had been using Monascus-fermented red rice as a food colorant to make red rice wine, red soybean cheese, meat and fish products and so on [1-3]. Currently, more than 50 patents regarding to utilization of *Monascus* pigments for food have been issued in Japan, the United States, France, and Germany [4]. The pigments produced by original *Monascus* contains three categories of pigments, yellow (monascin and ankaflavin), orange (monascorubrin and rubropunctatin), and red-purple (monascorubramine and rubropunctamine) pigments. These three pigments have two components of polyketide origin in common, which are secondary metabolites possessing an azaphilone skeleton [5, 6]. The pigments produced by *Monascus* are applied widely due to their safety, high protein adhesion, thermal stability, and wide-pH stability [1]. However, *Monascus* yellow pigments need further chemical purification from angkak [6] because both monascin [7] and ankaflavin [8] are minor components in the mixture. Additionally, due to unavailability of microbial species, research on production of *Monascus* yellow pigments is very limited [6, 9]. Scientists in Thailand carried out the breeding of *Monascus* for production of yellow pigment in 2000 and have gained the *Monascus* strain which produced mainly yellow pigment at single peak (λ_{max} 370 nm) [9-11]. Purification of yellow pigments was carried out and three new *Monascus* yellow azaphilones were identified and reported in 2004 [6]. Strong antimutagenic as well as antioxidant properties of this *Monascus* yellow pigment were also investigated [12]. Recently, researchers have focused on *Monascus* yellow pigment area through mutagenesis techniques as well as medium formulation [13] or statistical optimization of nutritional and physical parameters [14].

For the microbial metabolites, the most suitable cultivation method is the liquid-state fermentation.The optimal culture medium and the optimized fermentation conditions for the bench-scale production of monascus red pigments have been pursued. But few studies on the production of monascus yellow pigments in fermenter have been found in the recent literature because of unavailability of microbial genus. Meanwhile, not only microbial genus, but also the environmental conditions play a key influence on monascus yellow pigments production in submerged culture. Environmental conditions include chemical conditions like the type and content of carbon, nitrogen, phosphate, and metals, and physical conditions like mechanical stress, temperature, aeration, agitation and pH [15-17]. Aeration is one of the key control parameters in submerge fermentation for aim metabolites production, and aeration is variation because of different microbial species and aim metabolites. However, the genotypes of strains and environmental conditions play an obviously influence on fungal morphology [18, 19], which not only has a significant impact on mixing and mass transfer, but also determines the overall process productivities and subsequent economics [20, 21]. Thus, fungal morphology is also considered as a key bioprocess parameter for submerged culture.Based on the above, in this work, an investigation of *Monascus* yellow pigments production by *Monascus anka* mutant [13] in a 5 L fermenter by employing different aeration strategies was carried out to enable development of novel applications in the future.

This paper studies the effect of aeration on monascus yellow pigment production and growth of *Monascus anka* mutant strains, culture temperature and agitation speed were carried out with 32 °C and 300 r/min, respectively.

2. MATERIALS AND METHODS

2.1. Organism and Cultivation

Microorganism used in this study is *Monascus anka* mutant, which was screened from physical and chemical combination mutagenesis in our laboratory [13]. Stock cultures of the mutant were maintained on wort agar slants, which contains 15° wort (provided by Guangzhou Zhujiang Beer Co., Ltd, China) and 20 g/L agar (Difco Labatory, Loveton Circle, USA), and subcultivated periodically. Cultures were reactivated by being transferred onto fresh wort agar slants. After cultivation for 2-3 days at 31 °C, spores were collected with 5 mL sterilized water, and the corrected spore suspension was used as inoculum preparation. Spores suspension (0.3 mL) was inoculated in 250 mL Erlenmeyer flasks containing 30 mL of seed culture medium which was composed of 30 g/L of corn flour, 3 g/L of $NaNO_3$, 4 g/L of KH_2PO_4, and 0.01 g/L of $FeSO_4.7H_2O$. The seed culture medium (initial pH 6.0) was cultivated at 31 °C and 200 rpm for 1-2 days and then transferred into a 5 L fermenter (BIOFLO 3000 Batch/Continuous Bioreactor, New Brunswick Scientific Edison, NJ, USA). The submerge fermentation medium ingredients included 10 g/L corn steep liquor, 15 g/L NH_4Cl, 5 g/L KH_2PO_4, 20 g/L glucose and 70 g/L starch. Temperature of 31 °C and agitation of 300 r/min were maintained during fermentation in fermenter. Various aeration from 0.4 vvm to 1.2vvm were applied.

2.2. Determination of Pigments

According to the similar method of Chinese National Standard, GB15961-2005 and some reported articles [13], absorbance was applied to represent the pigments concentration. Five mL of culture broth was mixed with 5 mL of 70% (v/v) ethanol for 1 hour, and then centrifuged at 4,000 rpm for 20 min. The obtained supernatant was filtered through filter paper (45 μm, Xinhua Paper Industry Co., Ltd, Hangzhou, China). The filtrate contained two pigments: yellow pigment and red pigment, whose concentrations were determined by measuring the optical density of the supernatant using a 2802SUV/VIS spectrophotometer (Unicosh Scientific Instrument Co., Ltd, Shanghai, China) at 410 and 510 nm, respectively. Results were expressed as *OD* units per mL of fermented broth. The linearity equation between absorbance and diluting proportions is $y = -0.0054x+1.5462$(R^2= 0.9907), where y is absorbance and x is dilution proportions (in the range from 100 to 300).

2.3. Determination of Dried Cell Weight (DCW)

Fungal biomass was determined by gravimetric analysis after filtration of cell samples through preweighed nylon filters (45 mm diameter, 0.8 μm porosity) and dried to constant weight at 60 °C under partial vacuum (200 mm Hg).

2.4. Determination of Residual Sugar Concentration (RSC)

The residual glucose in the fermentation broth was determined with a spectrophotometer by the standard 3,5-dinitrosalicylic acid (DNS) method [22], and the calibration curve was prepared using glucose.

2.5. Determination of NH_4^+ Concentration

The revised Berthelot reaction [23] was used for determination of ammonium ion.

2.6. Determination of Soluble Starch Concentration

The soluble starch was measured by the modified method of Teng [24]. Briefly, sample (0.5 mL) was hydrolyzed by using 0.5 mL of thermostable α-amylase (EC 3.2.1.1, Novo, Denmark) at 100 ℃ for 60 min and then 0.25 mL of amyloglucosidase (EC 3.2.1.3, Novo, Denmark) at 60 ℃ overnight. The amount of glucose was determined by the DNS method. The starch content was calculated as the amount of glucose×0.9 and expressed as g/L (g starch content/L cultures).

2.7. Data Analysis

The specific growth rate, μ (h^{-1}), was calculated following the equation: $\mu = \left(\frac{1}{X}\right)\left(\frac{dX}{dt}\right)$, where X is the cell concentration (g/L) at time t (h). The specific production rate of yellow pigments, q_y ($OD.g^{-1}.h^{-1}$), was calculated following the equation: $q_Y = \left(\frac{1}{X}\right)\left(\frac{dY}{dt}\right)$, where Y is the yellow pigments value (*OD*) at time t (h). All the data shown in Tables and Figures were expressed as mean of triplicates. The statistical evaluation of all data was performed by OriginPro 8.0(OriginLab Corporation, Northampton, MA, USA).

3. RESULT AND DISCUSSION

3.1. Effect of Aeration on Yellow Pigments Production by Monascus Anka Mutant

As shown as Fig.1A, increasing aeration can made *Monascus anka* mutant produce yellow pigments in advance and the time of the maximum yield of yellow pigments in advance. When the aeration was 1.0 vvm, the maximum yield of yellow pigments can arrive with 126.60 *OD* units at 90 hours, but the decrease speed was quickly in fermentation anaphase. The 1.2 vvm aeration obviously could shorten the time of yellow pigments production and expedited the yellow pigments production, but the maximum yield of yellow pigments was only 109.81 *OD* units with quickly rate of descent in fermentation anaphase. The results above indicated high aeration, such as 1.0 vvm, can expedite the production of yellow pigments in the early culture, but low aeration, such as 0.6 vvm, was good for slow down the rate of yellow pigments descent in fermentation anaphase.

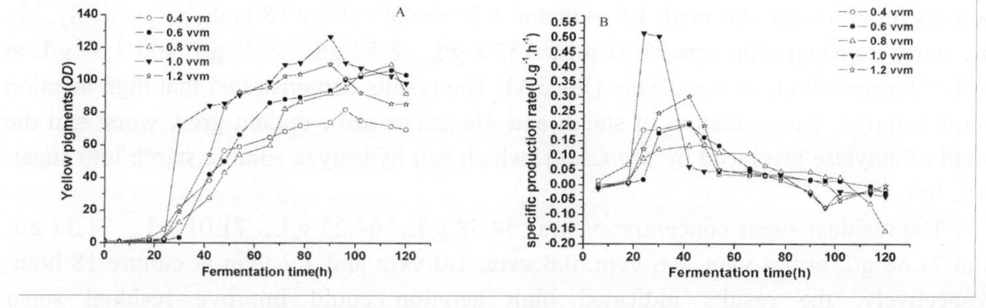

Fig.1. Effect of aeration on *Monascus* yellow pigments production (A) and specific production rate (B).

As shown as Fig.1B, the maximum specific production rate of yellow pigments was 0.5054 $OD.g^{-1}.h^{-1}$ at 30 hours when aeration was 1.0 vvm, under other aeration condition, the maximum specific production rate of yellow pigments was smaller than that of 1.00 vvm aeration and the arrival time for maximum specific production rate of yellow pigments was put off between 42 hours and 48 hours. The small aeration could maintain the high specific production rate of yellow pigments in fermentation anaphase. These results indicated high aeration could improve the maximum specific production rate of yellow pigments and expedited the arrival time, but maximum specific production rate of yellow pigments could maintain for a long time with high value for the low aeration.

3.2. Effect of Aeration on *Monascus anka* Mutant Growth

Aeration has no obviously influence on DCW but *Monascus anka* mutant growth and DCW was obtained from 17 g/L to 20 g/L. Maximum and minimum DCW was 19.98 g/L and 17.40 g/L at 1.0 vvm and 0.4 vvm of aeration, respectively (shown as Fig. 2).

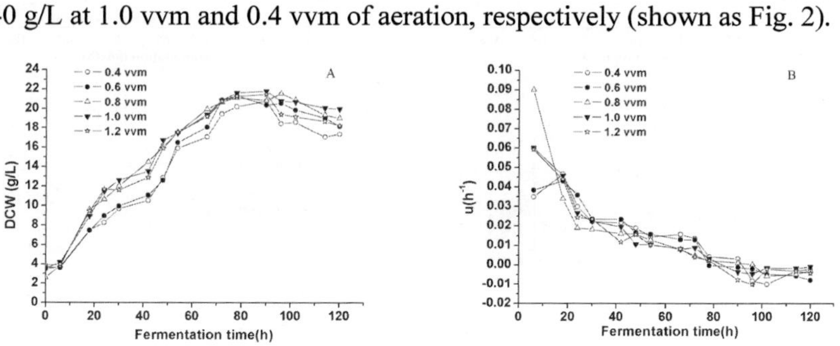

Fig.2. Effect of aeration on DCW (A) and specific growth rate μ(B).

3.3. Effect of Aeration on Substrate Utilization by *Monascus anka* Mutant Growth

The starch concentration was 55.48 g/L, 27.21 g/L, 29.93 g/L, 21.82 g/L and 20.38 g/L at 0.4 vvm, 0.6 vvm, 0.8 vvm, 1.0 vvm and 1.2 vvm at culture 18 hour, respectively, but the starch concentration was 29.93 g/L, 15.28 g/L, 8.52 g/L, 5.21 g/L and 3.59 g/L at end of fermentation, succcessively (Fig.3A). The results demonstrated that high aeration could improve the utilization of starch and *Monascus anka* mutant grew worst and the yield of amylase produced by *Monascus*, which can hydrolyze soluble starch into sugar, was few.

The residual sugar concentration was 54.58 g/L, 64.23 g/L, 71.01 g/L, 71.37 g/L and 71.68 g/L at 0.4 vvm, 0.6 vvm, 0.8 vvm, 1.0 vvm and 1.2 vvm at culture 18 hour, respectively, the results indicated high aeration could improve residual sugar concentration by amylase hydrolyze soluble starch into sugar, which produced by *Monascus*. As shown as figure, the residual sugar concentration was 14.21 g/L at end of culture under 0.6 vvm of aeration, but all residual sugar concentration was below 9 g/L under others aeration in this paper(Fig.3B). Results indicated low aeration (below 0.6 vvm) was not benefit for residual sugar utilization by *Monascus anka* mutant.

As showed as Fig.3c, aeration has no obviously influence on NH_4^+ utilization, the NH_4^+ concentration was maintained from 0.133 mol/L to 0.143 mol/L. but high aeration could improve NH_4^+utilization as far as utilizing speed.

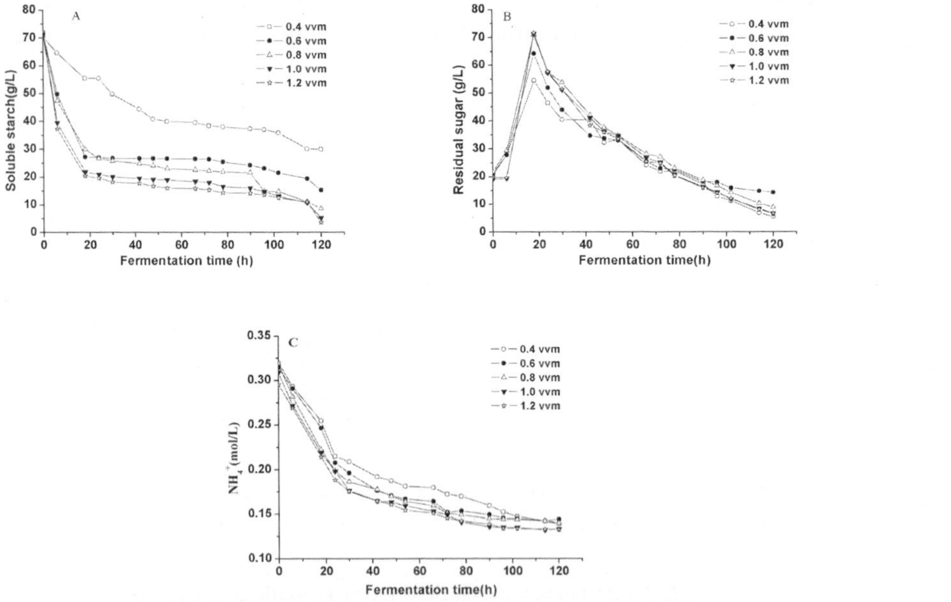

Fig.3. Effects of aeration on soluble starch (A), residual sugar (B) and NH_4^+ (C) utilization by *Monascus anka* mutant.

3.4. Effect of Aeration-shift on Yellow Pigments Production by *Monascus anka* Mutant

From the above experimental results about different aeration on the yellow pigments production by *Monascus anka* mutant, stage-divided strategy maybe can improve the yellow pigments production and benefit for *Monascus anka* mutant growth with efficient using substrate in terms of theory, the stage-divided strategy was as following: 0.8~1.0 vvm aeration was carried out to improve the *Monascus* mutant growth and yellow pigments production before fermentation 96 hours, and then 0.4~0.6 vvm aeration was done to extend the time of yellow pigment production with efficient using substrate.

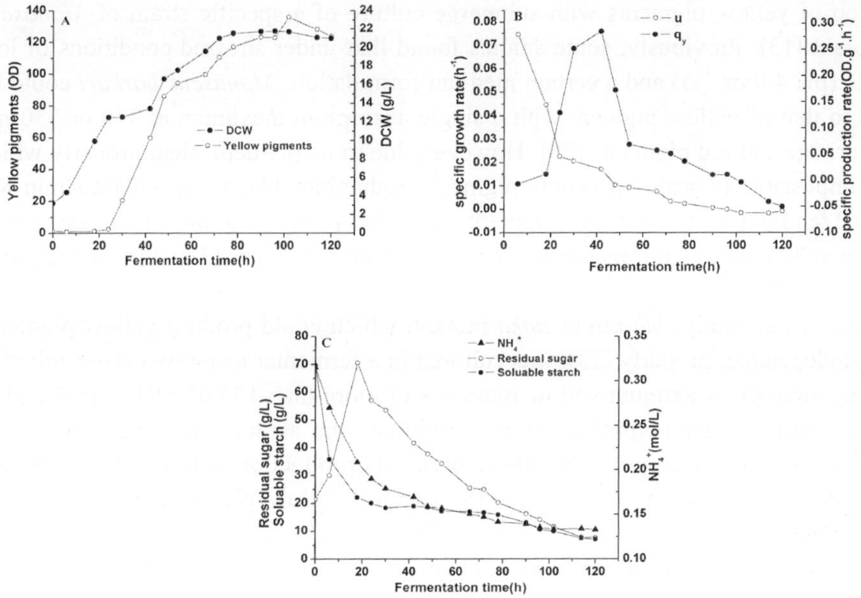

Fig.4. Effects of two stage aeration control on *Monascus* yellow pigments production (A) 、 *Monascus anka* mutant growth (B) and substrate utilization(C).

As shown in Fig.4, using two stage of aeration controlling strategy, the maximum yellow pigments yield could arrive with 136.08 *OD* at 102 hours, it was higher 65.75%, 28.07%, 23.93%, 7.43% and 23.85% than that of 0.4 vvm, 0.6 vvm, 0.8 vvm, 1.0 vvm and 1.2 vvm aeration(as shown in Fig.4A). The maximum specific production rate of yellow pigments could arrive with 0.2852 $OD.g^{-1}.h^{-1}$ at 42 hours, it was smaller than that of 1.0 vvm(0.5165 $OD.g^{-1}.h^{-1}$, 24 hours) and 1.2 vvm (0.30129 $OD.g^{-1}.h^{-1}$, 42 hours)aeration(as shown in Fig.4B), but two stage control strategy of aeration could obviously slow down the decreasing speed for yellow pigments production. The maximum specific growth rate could arrive with 0.0746 h^{-1} at 6 hours and the decreasing speed for specific growth rate was slowly(as shown in Fig.4B). The two stage aeration controlling strategy could obviously decrease the aging speed of *Monascus anka* mutant in fermentation anaphase(as shown in Fig.4C).

Relative to single aeration control, two-stage aeration control could inprove the substrate utilization efficiently to maintain the *Monascus anka* mutant growth and aim metabolites producing, the using the speed and efficiency for substrates was higher than that of single aeration control culture(as shown in Fig.4C).

Normally, yellow, orange, and red pigments are present as a mixture in *Monascus* submerged and solid-state culture. Both monascus red and yellow preparations are commercially available. However, multiple steps of solvent extraction and chemical modification are needed to obtain the yellow fraction [7, 8]. In contrast, our studies led to production of yellow pigments with submerge culture of a specific strain of *Monascus anka* mutant [13]. Previously, some studies found that under stressed conditions of low initial pH (pH 4.0 or 2.5) and a certain medium formulation, *Monascus barkari* could be changed to that of yellow pigment with a single absorption maximum at 370 or 330 nm without orange and red pigments [25]. However, due to its pH-dependent property which made it impossible to scale-up yellow pigment production. *Monascus* mutant strain KB 20 M10.2 could produce yellow pigment with the predominating single absorption maximum at 370 nm in a nitrogen-rich medium in wide pH ranges (2.5–7.0) was reported [9].

In the current study, *Monascus anka* mutant, which could produce yellow pigments with pH-independent property [13], was cultured in a fermenter using two-stage agitation controlling strategy. Maximum yellow pigments of more than 136.08 *OD* were found in submerge culture under optimal aeration conditions: the aeration was 0.8~1.0 vvm to improve the *Monascus* growth and yellow pigments production before culture 96 hour and then 0.4~0.6 vvm was carried out to continue yellow pigment production with efficient substrate utilization. It was higher 65.75%, 28.07%, 23.93%, 7.43% and 23.85% than that of 0.4 vvm, 0.6 vvm, 0.8 vvm, 1.0 vvm and 1.2 vvm aeration, respectively. The μ_{max} and the maximum q_y could reach 0.0746 h^{-1} and 0.2852 $OD.g^{-1}.h^{-1}$, respectively. It showed that the yield of yellow pigments was higher than those of reported by other researchers [7, 8]. High aeration not only has significant influence on monascus yellow pigments yield but could also increase their productivity. The yield of amylase produced by *Monascus*, which can hydrolyze soluble starch into sugar [23], could be increased by the higher aeration speed and therefor the reducing sugar utilization was accelerated. Higher aeration significantly enhanced the carbon source (reducing sugar and soluble starch) consumption by *Monascus anka* mutant (Fig.2A, B). However, NH_4^+ may be only used for *Monascus anka* mutant growth because aeration has no obviously influence on change of NH_4^+ contents in post-fermentation but only in early-fermentation in culture (Fig.2C)). Therefore, *Monascus* growth and yellow pigments production with high speed and longtime could be achieved by two stage aeration control strategy and short fermentation time to improve the yield of yellow pigments. Our *Monascus anka* mutant which produced yellow pigments showed good characteristics for heat resistance and acid resistance [13]. In this study, the citrinin was unable to be detected by HPLC method (The citrinin detection limit was found to be 0.1mg/L in this study)[26], namely, the *Monascus anka* mutant may produce no or just a little citrinin (the detection limit is 0.1mg/L).

In conclusion, to our knowledge, this is the first report on the aeration optimization of *Monascus* yellow pigments production in submerge culture in fermenter as a source of natural yellow pigments. The results from our previously investigation and those reported by Su et al. [27] confirmed that yellow pigments of *Monascus anka* mutant are not toxic. Another UV-absorbing compound, FK17-P2b2, extracted from *Aspergillus* sp. has been patented as an additive in cosmetics [28]. It was found that our yellow pigments showed excellent heat resistance and acid resistance ability, which are very useful for the protection of food products or non-food products, such as cosmetics, from deterioration. Future development on yellow pigments should focus on strain improvement for higher production of yellow pigments as well as process scaling up. Hopefully, the results in this paper have very important theoretical and realistic significance for realizing the industrial production of monascus yellow pigments by submerged culture.

Acknowledgments

This research work was financially supported by the National Natural Science Foundation of China (No.31301550), Scientific research project of hunan province department of education (Nos: 14C1185).

References

1. Blanc PJ, Loret MO, Santerre AL. Pigments of *Monascus*. *J Food Sci* 1994; **9**: 862-865.
2. Ma J, Li Y, Ye Q. Constituents of red yeast rice, a traditional Chinese food and medicine. *J Agr Food Chem* 2000; **48**: 5220–5225.
3. Wild D, Toth G, Humpf HU. New *Monascus* metabolite isolated from red yeast rice. *J Agr Food Chem* 2002; **50**: 3999–4002.
4. Lin TF, Yakushijin K, Buchi GH, Demain AL. Formation of water-soluble *Monascus* red pigments by biological and semi-synthetic processes. *J Ind Microbiol* 1992; **9**: 173–79.
5. Wang TH, Lin TF. Monascus rice products. *Adv Food Nutr Res* 2007; **53**: 123–59.
6. Juzlova P, Martinkova L, Kren V. Secondary metabolites of the fungus *Monascus*. *J Ind Microbiol* 1996; **16**: 163-170.
7. Sato K, Goda Y, Sakamoto SS, Shibata L, Maitami T, Yamada T. Identification of major pigments containing D-amino acid units in commercial *Monascus* pigments. *Chem Pharm Bull* 1997; **45**: 227–229.
8. Suchada, Jongrungruangchok, Prasat, Kittakoop, Busaba, Yongsmith. The fermentation of rice for anka pigment production. *J Ind Microbiol Biotechnol* 2000; **25**: 141-146.
9. Jongrungruangchok S, Kittikoop P, Yongsmith B, Bavavoda R, Tanasupawat S, Lartpornmatulee N, Thebtaranonth Y. Azaphilone pigments from a yellow mutant of the fungus *Monascus kaoliang*. *Phytochem* 2004; **65**: 2569–2575.

10. Kurono M, Nakanish K, Shindo K. Biosynthesis of monascorubrin and monascofla. *J Chem Pha Bull* 1963; **11**: 359-362.
11. Sweeny JG, Estrada-Valdes MC, Lacobucci GA, Sato H, Sakamura S. Photoprotection of the red pigments of Monascus anka in aqueous media by 1,4,6-trihydroxy naphthalene. *J Agric Food Chem* 1981; **29(6)**: 1189–1193.
12. Fielding BC, Holker JSE, Jone DF, Powell ADG, Richman KW, Robertson A, Whalley WB. The chemistry of fungi. Part XXXIX. The structure of monascin. *J Chem Soc* 1961; 4579–4589.
13. Manchand P S, Whalley WB, Chen FH. Isolation and structure of ankaflavin: a new pigment from Monascus anka. *Phytochem* 1973; **12**: 2531–2532.
14. Somboon, Tanasupawat, Nattapat, Lartpornmatulee, Yodhathai, Thebtaranonth. Azaphilone pigments from a yellow mutant of the fungus Monascus kaoliang. *Phytochem* 2004; **65**: 2569–2575.
15. Yongsmith B, Krairak S, Bavavoda R. Production of yellow pigment in submerged culture of a mutant of Monascus sp. *J Ferment Bioeng* 1994; **78**: 223–228.
16. Yongsmith B, Tabloka W, Krairak S, Bavavoda R. Cassava fermentation of yellow pigments and amylolytic enzymes by a mutant of *Monascus* spp. in submerged cultivation, Microb. *Uti Renewable Resour* 1990; **7**: 354–363.
17. Yongsmith B, Kitpreechavanich V, Chitradon L, Chaisrisook C, Budda N. Color mutants of Monascus sp. KB9 and their comparative glucoamylases on rice solid cultures. *J Mol Cat B: Enzym* 2000; **10**: 263–272.
18. Kruawan K, Kangsadalampai K, Yongsmith B, Sriyapai T. Effects of *Monascus* colorants on the mutagenicity of nitrite treated 1-aminopyrene using Ames test. *Thai J Pharmal Sci* 2005; **29(1–2)**: 29–41.
19. Zhou B, Pu YW, Zhu MJ, Liang SZ. Effect of nitrogen sources on *Monascus* yellow pigment production by *Monascus* mutant. *Mod Food Sci and Technol* 2008; **2**: 123–127.
20. Zhou B, Wang JF, Pu YW, Zhu MJ, Liu SM, Liang SZ. Optimization of culture medium for yellow pigments production with *Monascus anka* using response surface methodology. *Eur Food Res Technol* 2009; **228**: 895–901.
21. Yang CL. Fermentation conditions for partial yellow, high color value red rice. *Fujian J Agric Sci* 2009; **4**: 361–364.
22. Jirasatid S, Nopharatana M, Kitsubun P, Vichitsoonthonkul T, Tongta A. Statistical optimization for monacolin K and yellow pigment production and citrinin reduction by *Monascus purpureus* in solid-state fermentation. *J Microbiol Biotechnol* 2013; **23(3)**: 364–374.
23. Dorta B, Bosch A, Ertola RJ. High level of sporulation of Metarhizium anisopliae in a medium containing by products. *Appl Microbiol Biotechnol* 1990; **33**: 712–715.
24. Cui YQ, van der Lans RGJM, Luyben KCAM.. Effects of Dissolved oxygen tension and mechanical forces on fungal Morphology in submerged fermentation. *Biotechnol Bioeng* 1998; **57**: 409-419.
25. Yongsmith B, Tabloka W, Yongmanitchai W, Bavavoda R. Culture conditions for yellow pigment formation by Monascus sp. KB10 grown on cassava medium. *World J Microbiol Biotechnol* 1993; **9**: 85–90.

26. Xu GR, Chen Y, Yu HL. HPLC Fluorescence Method for Determination of Citrinin from *Monascus* Cultures. *Archiv fur Lebensmittelhygiene* 2003; **4**: 82-84.
27. Su NW, Lin YL, Lee MH. Ankaflavin from Monascus fermented red rice exhibits selective cytotoxic effect and induces cell death on Hep G2 cells. *J Agric Food Chem* 2005; **53**: 1949–1954.
28. Takayuki A, Hiroshi S. Novel UV-absorbing compounds TK17-P2a, FK17-P2b1, FK17-P2b2, and FK17-P3 and manufacture of the compounds with *Aspergillus* sp. Patent No. JP6329576 (1994).

ANTIOXIDANT ACTIVITIES AND PHENOLIC COMPOUNDS ANALYSIS OF EXTRACTIONS OF CARICA PAPAYA SEED RESIDUE

Ruoying Ma[a], Youmei Li[b], Xuetian Rao, Jinhong Cai[c,*], Linhe Wu, Zheng Wang, and Bo Zhang

College of Food Science, Hainan University, Haikou 570228, China

E-mail:[a]18289737675@163.com,[b]551907021@qq.com,[c]1275647390@qq.com

[*]Corresponding author

The antioxidant activity of extractions by sub-critical water (SCW) and fractions from papaya seed residue were assessed and the active ingredients were identified by HPLC in this study. The papaya seeds residues(PSR) were extracted by sub-critical water at specified conditions, filtered and dried by vacuum freezing, SCW extract sample was then obtained, identical process similarly, the extract was partitioned with ethyl acetate and n-butyl alcohol, then dried by vacuum freezing, obtained ethyl acetate fraction , n-butyl alcohol fraction and remains samples. Three different methods(DPPH, ABTS and FRAP assay) were used to evaluate the antioxidant activities of SCW extract and fractions from papaya seed meal. The ethyl acetate fraction of papaya seed meal had possessed the best antioxidant activity, the SCW extract took second place. The DPPH and ABTS radical quench activities of the ethyl acetate fraction were better than others, relatively, which implicated the most antioxidant compounds mainly concentrated in the ethyl acetate account for the principle of the dissolution in the similar material structure. Papaya seed meal and the compounds may be acted as natural antioxidant.

1. INTRODUCTION

Numerousness biochemical reactions in human body involve the generation of reactive oxygen species (ROS)[1] mainly covered by hydrogen peroxide(H_2O_2), hydroxyl radicals(OH •) and superoxide anion(O_2-)[2]. Mitochondria[3] are a primary source of ROS with most mammalian cell even in normal circumstance. ROS not eliminated totally by the antioxidant defense system stimulates oxidative damage by attacking macromolecule such as proteins, DNA, and leading to oxidative stress[4] based on imbalance between antioxidant defenses and prooxidant production results in various disorders and chronic diseases such as atherosclerosis, cancer, cerebral thrombosis, macular degeneration, inflammation and emphysema. Free radical had been revealed in the pathology of many diseases[5]. There are a great variety of plant derived components such as vitamin C, vitamin E, flavonoids, β-carotene, triterpenes, tannins, anthocyanins, and other phenolic represent a special group of nutritional supplements with antioxidant activity getting through eliminating free radicals. Food rich in these antioxidants plays an important role in preventing of deleterious oxidations[6].

Papaya, known as pawpaw, a member of the family Caricaceae from Carica genus which originated from tropical and subtropical of America and Africa, is mainly found distributes in the south of China, such as Hainan, Guangdong, Guangxi and Fujian

Province. Papaya is a tropical fruit wealth of dietary antioxidants[7] such as vitamin C, total phenols, tocopherols and bioactive phytochemicals with antioxidant activity. Different parts of papaya are applied for folk medication to cure plenty of disease[8]. Scientific studies testified that papaya play a wide range of therapeutic effects on antibacterial, antifungal and antiviral. These properties mainly counted on the functional components of several secondary metabolites present in the papaya organs. Papaya seed, a conspicuous portion of fruit, wasted from processing and had caused disposal problems[9]. Previous studies indicated[10] that seeds extract contains phenolic compounds, vanillic acid and vitamin C with antioxidant[11] and anticancer activities. Therefore, papaya seed is considered as a good resource of functional ingredients, such as phenolic compounds. Early studies revealed that phenolic compounds[12] can be used as natural antioxidants to improve the stability of food.

The polyphenols used to be extracted by common methods which include solvent extraction, hot water extraction and ultrasound-assisted extraction. However, these techniques require long extraction time and lead to low extraction yield. Lately, more efficient extraction methods such as Subcritical (or superheated) water (SCW) extraction[13] have been used for the extraction of organic compounds from plants. The term subcritical water is used to denote the region of the condensed phase which occurs between 100°Cand the critical point(374°C). Previous study reported that subcritical water[14] is a powerful alternative for extraction of natural products owing to it's enables a rapid extraction[15] and low working temperatures. This avoids the loss of thermolabile compounds because of degradation. Additional positive aspects of the use of subcritical water extraction are its simplicity, low cost and undesirable environmental effects.

Previous studies on the DPPH[16], hydroxyl and superoxide free radical scavenging activities of some tropical fruits and extract fraction indicated strong activities[17]. To the authors' knowledge, rarely researches have explored antioxidant activities of the other extract fraction from papaya seeds. Thus, the aim of this study was to assess the antioxidant activity of papaya seeds. Special attention was paid to assess if the highest antioxidant effect of papaya seed superheated water extract rely on the whole crude extract or on specific sub-fraction. Although many methods are available to evaluate antioxidant activity, it is important to apply a consistent and rapid method. In consideration of each method which used for testing the antioxidant activities in vitro has its own merits and drawbacks, several methods were always used together to evaluate the antioxidant activities of natural products. In this paper, three kinds of methods were used to estimate the antioxidant activities of papaya seeds.

2. MATERIALS AND METHODS

2.1. Chemicals

All chemicals were of analytic grade and purchased from Sigma.

2.2. Sample Description and Preparation

PSR superheated water extract preparation and fractionations. Papaya seeds were collected from the maturity in Hainan, washed and dried by vacuum freezing at -80°C till anhydrous. The waterless seeds were smashed to the powder and achieved uniform particle size and kept in 4°C refrigerator till use. The raw material(PSR), which came from degreased by petroleum ether and deproteinized by phosphate buffer at pH 8.0, were extracted by subcritical water at specified conditions, then filtered and dried by vacuum freezing. Identical process, the extract was partitioned with ethyl acetate and n-butyl alcohol, and then dried by vacuum freezing, obtained ethyl acetate fraction, n-butyl alcohol fraction and remains parts.

2.3. Antioxidant Activities Assay

2.3.1. *DPPH Radical Scavenging Activity*

The molecular 1, 1-diphenyl-2-picryhydrazyl (DPPH) is characterized as a stable free radical by virtue of the delocalization of the spare electron over the molecular as a whole, so that the DPPH doesn't dimerism, as would be the case with most other free radicals[18]. The delocalization of the spare electron give rise to the deep violet color, characterized by an absorption band in ethanol solution centered at about 517nm. When a solution of DPPH is mixed with that a substrate that can donate a hydrogen atom, then this gives rise to the reduced from with the loss of violet color.

In order to assess the antioxidant potential through free radical scavenging by the test sample, the change in optical density of DPPH radicals in monitored. In accordance with Manzocoo et al., the sample extract (1mL) is diluted with water and 3mLof DPPH (25.3mg/L) in ethanol solution. The solutions were vortexes and incubated in the shade for 30min at 37°C. The solvent alone acted as a blank. The decrease in absorbance of the test mixtures (due to quenching of DPPH free radicals) was determined as being 517nm. The percentage of the DPPH radical scavenging is calculated by the equation as in Eq.(1):

$$\%\ \text{inhibitions of DPPH radical} = [1-(As-B)/Ao)] \times 100 \quad (1)$$

Where Ao is the absorbance before reaction, As is the absorbance after reaction has taken place, B is the absorbance of sample background.

The EC50 values were determined as the concentration of the test mixture that gave 50% reduction in absorbance from that of the control blank.

2.3.2. *ABTS radical cation decolorization assay*

In this method, using a diode-spectrophotometer to measure the loss of the color when an antioxidant is added to the blue-green chromophore ABTS • +(2,2-azino-bis(3-ethylbenzthiazoline-6-sulfonic acid))[19]. The antioxidant reduces ABTS • + to ABTS and decolorize it. ABTS radical cations are prepared by adding hydrogen peroxide(2mmol/L) to 0.549g ABTS volume to 100mL,one hour at room temperature till the special blue-green chromophore ABTS • +. Using a 30mmol/L sodium acetate -acetic acid buffer of pH 3.6 diluted to the absorbance 0.7±0.02 at 734nm. Diluted sample (1mL) are mixed with 3mL of ABTS • + radical cation solution, and absorbance is read at 734nm in 6min. The percentage of the ABTS • + radical scavenging is calculated by the equation as in Eq.(2):

$$\% \text{ inhabition of ABTS} \bullet + \text{ radical} = [1-(As-B)/Ao)] \times 100 \qquad (2)$$

Where Ao is the absorbance before reaction, As is the absorbance after reaction has taken place, B is the absorbance of sample background.

The EC50 values were determined as the concentration of the test mixture that gave 50% reduction in absorbance from that of the control blank.

2.3.3. *Ferric Reducing-Antioxidant Power (FRAP) Assay*

The method measured the ability of antioxidants to reduce ferric iron, which based on the reduction of the complex of ferric iron and 2, 3, 5-triphenyl-1, 3, 4-triaza-2-azonizcyclopenta-1, 4-diene chloride (TPTZ) to the ferrous at low pH. This reduction is monitored by measuring the change in ab[illegible]n at 593nm[20]. Briefly, the ferric reducing power(FRAP) reagent containing [illegible] 20mmol/L ferric trichloride, 2.5mL of 10mmol/L TPTZ and 25mL of 300mmol/L sodium acetate-acetic acid buffer at pH3.6. 1mL of sample mixed with 3mLof prepared FRAP regent, the absorbance at 593nm is read after 30min incubated at 37°C. FRAP values can be obtain by comparing the absorption change in the test mixture with those obtained from increasing concentrations of $Fe3+$ and expressed as $Fe2+$ equivalents. A standard calibration curve is constructed for FeSO4 at 0,50,100,200,300,400,800,1000 μ mol concentrations. The antioxidant activity of the sample by the FRAP assay was expressed as FeSO4 equivalent in μ mol/g dry weight of extract.

2.4. Statistical analysis

Experiments conducted in replicates were expressed as means±standard error of the mean. Analysis of variance was performed by the ANOVA produced.

3. RESULTS AND DISCUSSION

3.1. DPPH Radical Scavenging Assay

The DPPH radical scavenging activities of the 4 samples were evaluated by comparing the EC50 of the extract fractions show as Fig. 1 and Table 1. It was found the radical scavenging activities of the positive correlation and diverse solvent extract increasing concentrations, and all the regression equations were significant at $p<0.05$. Thus, all four samples had DPPH radical scavenging activity. On the basis of the EC50 values, the ability to quench DPPH radical of all samples could be ordered by ethyl acetate fraction > SCW extract > n-butyl alcohol fraction > remain fraction. The EC50 values of DPPH radical quench activities of the ethyl acetate fraction, SCW extract, n-butyl alcohol fraction and remain fraction were indicated to be 0.1096 mg/mL, 0.2111 mg/mL, 0.3988mg/mL, and 0.4779 mg/mL, respectively. Based on the previous studied, Kaibing Zhou[21], reported that the EC50 values of DPPH radical quench activities of water extract was 1.6283 mg/mL and other extract fractions such as ethyl acetate fraction, n-butyl alcohol fraction were 6.4 mg/mL,1.093mg/mL from papaya seeds by ethanol extract. The results implicated that the DPPH radical quench activity of ethyl acetate fraction revealed the strongest antioxidant activity which indicated of all extracts from papaya seeds were correlated to the amount of antioxidant constituents extracted.

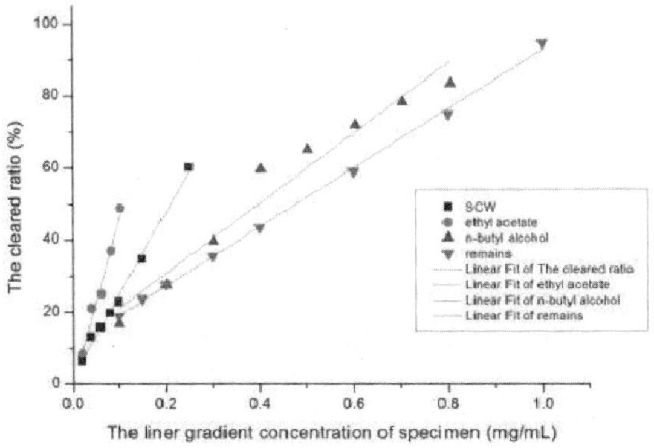

Fig. 1. The regression curves of DPPH.

Table 1. The regression equation of the DPPH radical quench rate on the sample concentration and the EC50 values

Extract and fraction	The regression equations	Adj. R-Square	F Value	Prob>F	EC50 (mg/mL)
SCW	y=2.276x+0.0195	0.99254	799.52194	1.03616E-6	0.2111
ethyl acetate	y=4.6482x-0.0093	0.9732	146.23309	0.00122	0.1096
n-butyl alcohol	y=0.9735x+0.1118	0.94912	131.58588	2.63521E-5	0.3988
remain	y=0.82x+0.1081	0.99837	4291.58535	8.50859E-10	0.4779

3.2. ABTS Radical Cation Decolorization Assay

The ABTS+ quench activities of the four samples were revealed by comparing the EC50 of the extract fractions show as Fig. 2. According to the EC50 values showed as table 2, the ability to quench ABTS+ radical of all samples could be ranked as ethyl acetate fraction > SCW extract > remain fraction>n-butyl alcohol fraction. The EC50 values of ABTS+ radical quench activities of the ethyl acetate fraction, SCW extract, remain fraction and n-butyl alcohol fraction were indicated to be 0.2768 mg/mL, 0.1356 mg/mL, 0.4577 mg/mL, 0.3487 mg/mL, respectively. The ABTS+ quench activity of the ethyl acetate fraction was apparent stronger than that of others, almost twice over SCW extract, quadruple as n-butyl alcohol fraction and 2.5times remains. And the n-butyl alcohol fraction quenched the ABTS+ radical inconspicuous. The results indicated that the extract fraction contained plenty of compounds which were provided with scavenging the ABTS+ radical prefer to ethyl acetate than others. Except ethyl acetate fraction, the SCW extract, n-butyl alcohol fraction and remains had no obviously quenched the ABTS+ radical.

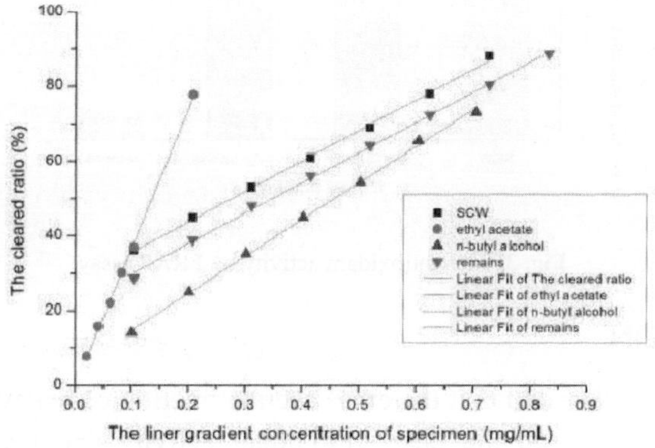

Fig. 2. The regression curves of ABTS.

Table 2. The regression equation of the ABTS radical quench rate on the sample concentration and the EC50 values

Extract and fraction	The regression equations	Adj. R-Square	F Value	Prob>F	EC50 (mg/mL)
SCW	y=0.8175x+0.2737	0.99834	3599.53662	2.43442E-8	0.2768
ethyl acetate	y=3.736x-0.0066	0.99848	3294.09622	5.51824E-7	0.1356
n-butyl alcohol	y=0.9838x+0.0497	0.99823	3389.36429	2.82902E-8	0.4577
remain	y=0.8111x+0.2172	0.99845	4512.74778	7.31924E-10	0.3487

3.3. Ferric Reducing-antioxidant Power (FRAP) Assay

In this assay, reduction of ferric-tripyridyltriazine complex to the ferrous forms a blue-black color which has a maximum absorbance at 593nm. The color depth is correlated with the amount of antioxidant in the extracts. The result of ferric reducing-antioxidant power of different fractions is revealed in the Fig. 3. The antioxidant activity of the SCW extract, ethyl acetate fraction and n-butyl alcohol were 625.71±1.12, 614.29±0.67, 583.09±1.45, 508.38±0.90 as FeSO4 equivalent in μmol/g dry weight of extract, corresponding. Obviously, the antioxidant activity of remains was the poorest. Besides, the antioxidant activity of the SCW extract, ethyl acetate fraction and n-butyl alcohol was remarkable stronger than that of remains, but, inconspicuous difference among themselves. Similar as the results obtained from DPPH and ABTS assay, the ethyl acetate fraction revealed relevantly strong ferric iron-reduction activity.

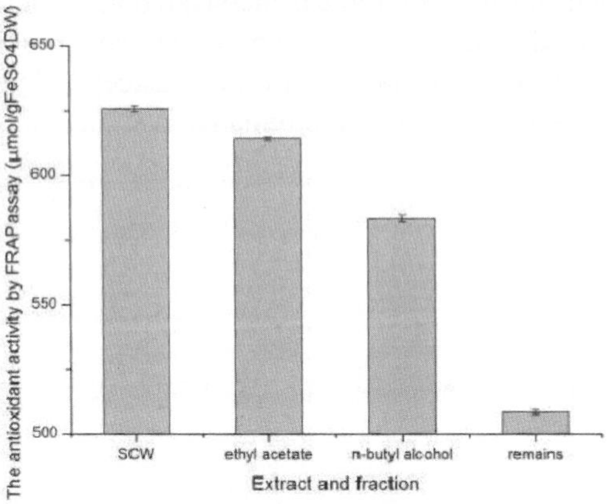

Fig. 3. The antioxidant activity by FRAP assay.

4. CONCLUSION

The date in the paper revealed that the ethyl acetate fraction of papaya seed meal through SCW extract had possessed the best antioxidant activity and the SCW extract took second place. The DPPH and ABTS radical quench activities of the ethyl acetate fraction were better than others, relatively, which implicated the most antioxidant compounds mainly concentrated in the ethyl acetate account for the principle of the dissolution in the similar material structure. Studies revealed that polyphenol , flavone and triterpenes which are widely found in fruits and vegetables have good antioxidant activities and which mainly dissolved in the ethyl acetate. Thus, papaya seed meal and the compounds may be acted as natural antioxidant.

Acknowledgments

This study work was supported by the Natural Science Foundation of the Republic of China (no. 31160325), tropical agricultural products processing in the Midwest, 2012 the national level of College Students' innovation training project (201210589012), 2014 postgraduate student innovation scientific research project of Hainan province (NO.Hys2014-27) and the first funding project of excellent dissertation of postgraduate of Hainan University, 2015 postgraduate student innovation scientific research project of Hainan province (NO.Hys2015-36).

References

1. Balaban RS, Nemoto S, Finkel T. Mitochondria, oxidants, and aging.Cell. 2005;**120**:483-495.

2. Komary Z, Tretter L, Adam-Vizi V. Membrane potential-related effect of calcium on reactive oxygen species generation in isolated brain mitochondria. Biochimica et Biophysica Avta.2010;1797:922-928.
3. Cadenas E, Davies KJ. Mitochondria free radical generation, oxidative stress, and aging. Free Radical Biol Med. 2000; 29:222-230.
4. K.Brieger, S.Schiavone, F.J.Miller, and K.H.Krause. Reactive oxygen species:from health to disease. Swiss Medical Weekly,2012;**142**:1-14.
5. S.A.Stanner, J Hughes, C.N.Kelly. A review of the epidemiological evidence for the "antioxidant hypothesis". Public Health Nutrition.2004;**7**(3):407-422.
6. J.Mursu, K.Robien, L.J.Harnack,et al. Dietary supplements and mortality rate in old women: The Iowa Women's Health Study. Archives of Internal Medicine. 2011;**171**(8):1625-1633.
7. A.Ali, S.Devarajan, M.Waly,et al. Nutritional and medicinal value of papaya. Natural Products and Bioactive Compounds in Disease Prevention.2011;**12**:34-42.
8. P.Jaiswal,P.Kumar,V.K.Singh,et al. Carica papaya Linn: a potential source for various health problems. Journal of Pharmacy Research.2010;**3**(5):998-1003.
9. Y.M. Li, N. Su, H.Q. Yang, X.P. Bai, Q.X. Zhu, H.X. Liu and J.Q. Li. The Extraction and Properties of Carica papaya Seed Oil. *Advance Journal of Food Science and Technology*.2015;**7**(10): 773-779.
10. N.A.Imaga,G.O.Gbenle,V.I.Okochi, et al. Phytochemical and antioxidant nutrient constituents of Carica papaya and parquetina nigrescens extract. Scientific Research and Essay.2010;**5**(6):2201-2205.
11. N.Kothari, S.Seshadi. Antioxidant activity of seed extracts of Annona squamosa and Carica papaya. Nutrition and Food Science.2010; **40**(4):403-408.
12. D'Angelo S, Cimmino A, Raimo M et al. Effect of reddening-ripening on the antioxidant activity of polyphenol extracts from cv. 'Annurca' apple fruits. J *Agric Food Chem*. 2007;**55**:9977-9985.
13. M. Rezaee, Y. Yamini, M. Moradi, A. Saleh, M. Faraji, M.H. Naeeni, Supercriticalfluid extraction combined with dispersive liquid – liquid microextraction as asensitive and efficient sample preparation method for determination of organiccompounds in solid samples, J. *Supercritical Fluids*. 2010; **55**:161-168.
14. Soto Ayala R, Luque de Castro MD.Continuous subcritical water extraction as a useful tool for isolation of edible essential oils. Food Chem. 2001; **75**:109-113.
15. Budrat P, Shotipruk A. Enhanced recovery of phenolic compounds from bitter melon Momordica charantia) by subcritical water extraction. Sep Purif Technol. 2009; **66**:125-129.
16. Li HY, Hao ZB, Wang XL, et al. Antioxidant activities of extracts and fractions from Lysimachia foemun-graecum Hance. Bioresour technol . 2009; **100**:970-4.
17. Kothari.V, Seshadri.S. Antioxidant activity of seed extracts of Annona squamosa and Carica papaya. Nutri. Food Sci.2010;**40**:403-408.
18. Blois.M.S. Antioxidant determinations by the use of a stable free radical.Nature.1958; **181**:1199-1200.

19. Re.R, Pellegrini.R. Antioxidant activity applying an improved ABTS radical cation decolorization assay. Fredd Radical Biol.Med.1999;**1**:1231-1237.
20. Benzie.I.F,Strain.J.J. The ferric reducing ability of Plasma as a measure of 'antioxidant power':The FRAP assay.Anal.Biochem.1996;**239**:487-492.
21. Kaibing Zhou, Hui Wang, Wenli Mei,et al. Antioxidant activity of papaya seed extract. Molecules.2011;**16**:6179-6192.

IMPROVED MYCELIUM BIOMASS BY SUBMERGED CULTURE OF *ANTRODIA CINNAMOMEA* USING PLACKETT-BURMAN DESIGN AND RESPONSE SURFACE METHODOLOGY

Jing Li[1,3], Xiongjie Lin[2], Dongmei Lin[1,3], Xucong Lv[1,3], Jinhua Tong[1,3], and Zhanxi Lin[1,3,a]

[1] *Fujian Agriculture and Forestry University, Fuzhou 350002, China*

[2] *Fruit Research Institute, Fujian Academy of Agriculture Science, Fuzhou 350013, China*

[3] *China National JUNCAO Engineering Research Center, Fuzhou 350002, China*

E-mail:[a]lzxjuncao@163.com

Antrodia cinnamomea (*A.cinnamomea*) fruiting body cultivation has hardly been developed successfully in solid-state fermentation. Thus, to improve its biomass, response surface methodology with a Box-Behnken design was used to predict the optimal conditions in a submerged culture. The most influencing factors of *A. cinnamomea* biomass (Dry weight, DW) were glucose, malt dust, yeast extract, V_{B1} and chitosan ($p < 0.05$). The predicted value of biomass after culturing for 15 days was 1.7843 g/100 mL (DW) under 35.16 g/L glucose, 42.86 g/L malt dust, 20.80 g/L yeast extract, 0.39 g/L V_{B1} and 0.85 g/L chitosan conditions, while the actual biomass yield was 1.6765 ± 0.0103 g/100 mL, which was 2.5 fold more than the control. There was also a significant relation ($p < 0.05$) between the biomass and triterpenoid concentration. Chitosan was thought to have enhanced the growth rate of mycelium and triterpenoid production, and as a result could be used as a natural additive in submerged culture.

1. INTRODUCTION

Medicinal fungus, which is used to enhance longevity in traditional practice and regarded as a functional food [1]. *Antrodia cinnamomea* (*A. cinnamomea*) is well known in Taiwan as "niu zhang ku" in the Polyporaceae, which is brown in heart rot of *Cinnamomum kanehirati* Hay (Lauraceae) and is rare and expensive as medicinal fungi [2]. Some researches have indicated that both mycelium and fruit body are involved in the detoxification of alcohol and drug, and have antianaphylaxis, antidiarrheal, and anti-metastatic effect on carcinoma cell [3-5]. Triterpenoids which exist in both mycelium and fruit body of *A. cinnamomea* are the main components considered for medicinal purposes. It was shown that ethanolic extract of solid state cultured *A. cinnamomea* (EEAC) is an ATPase stimulator and is capable of reducing verapamil-induced ATPase levels [6]. EEAC may be a potent P-gp inhibitor, but higher dosage may be required in subjects carrying variant-types P-gp. Five kinds of triterpenoids substances from *A. cinnamomea* fruit body influence anti-inflammatory activity [7].

However, *A. cinnamomea* seldomly produce basidiomycetes in artificial culture. Therefore, future research and industrial product development shall focus on the mycelium of *A. cinnamomea*, not the fruit body. Submerged cultivation is one of the most efficient fungi culturing method [8,9]. Different physical and chemical factors such as pH, C source, and N source affect the mycelium concentration [10]. In recent years, most studies have investigated the effect of the culture conditions or medium composition on biomass and the formation of triterpenoids and polysaccharides in submerged cultures using natural additives to promote the yield [11,12]. It was found that active ingredients such as peel extracts enhanced significantly the production of crude triterpenoids and biomass for 30 days [13].

Statistical experimental design such as Plackett-Burman (PB) design and response surface methodology (RSM) can collectively optimize all the influencing parameters to eliminate the constraints of a single-factor optimization process. PB design provides an effective way of identifying the important factors among a large number of variables (K), which is a 2-level fractional factorial design used to study K=N-1 factors in N experiments [14]. RSM is a favorable technique for optimizing the complex processes [15] and is successfully used for developing, improving and optimizing processes [16-18]. Box-Behnken design (BBD), one of RSM, only have three levels and fewer experiments than a Central Composite Design (CCD) [19, 20].

In this study, to improve the mycelium biomass of *A. cinnamomea* and illustrate the relation between the biomass and triterpenoid concentration, natural materials including water extract of *Ganoderma lucidum* grass-based spent substrate, alcohol extract of Juncao *Ganoderma lucidum* and chitosan were selected as additives of submerged culture through RSM. The result can promote the development of the mycelium as a functional food and medicinal drug, to investigate the much simpler and easier operation of the culture medium for the improvement of the mycelium biomass, and triterpenoids of the *A. cinnamomea* is essential.

2. MATERIAL AND METHODS

2.1. Microorganism and Culture Conditions

The *A. cinnamomea* (Strain AC001, GenBank No. *KM925002*) was kindly provided by Taiwan Shen-nong Fungal CO., Ltd. Cultures were maintained on potato dextrose agar (PDA) slants and incubated at 28°C for 28 days. The water extract of *Ganoderma lucidum* grass-based spent substrate of (JCSWE) was provided by Chinese National Engineering Research Center of JUNCAO Technology. JCSWE was confirmed as the feed additive that was effective in improving milk yield of a dairy cow [21]. The alcohol extract of Juncao *Ganoderma lucidum* (JCGLAE) was provided by Fujian Jiangle Juncao *Ganoderma* Biotechnology Co., Ltd. JCGLAE was a natural material used to treat various diseases.

2.2. Inoculums and Cultivation

To prepare the seed inoculums, 5 pieces of 0.5 cm sized agar cultures from the PDA slants were inoculated into a 250 mL flask with 100 mL seed medium (25 g/L glucose, 5 g/L peptone, 3 g/L malt dust). The preculture with an inoculum size of 3% was then inoculated into a 250 mL flask containing 100 mL fermentation medium (natural pH) and cultivated at 28°C for 15 days with constant shaking at 120 rpm.

2.3. Biomass of *A. cinnamomea* Mycelium

The mycelium of *A. cinnamomea* was collected after 15 days, dried at 55°C in the oven, and the dry weight was measured.

2.4. Extraction and Measurement of Triterpenoids from Dried *A. cinnamomea* Mycelium

The dried mycelium powder (0.5 g) was extracted with 40 mL 80% ethanol with ultrasonic wave (Kun Shan Ultrasonic Instruments Co., Ltd, Jiangsu, China) at 70°C for 2 h. The slurry was centrifuged at 7,000×g for 20 min. Then, the supernatant was collected and concentrated to 50 mL.

The measurement of total triterpenoid content was determined spectrophotometrically using the modified method of vanillin glacial acetic acid [26]. 0.5 mL sample was dried at 50°C in the scale tubes and mixed with 0.3 mL 5% vanillic acetic acid and 1 mL perchloric acid at 60°C for 20 min. After the tubes had cooled to room temperature, 10 mL acetic acid was added and mixed thoroughly. Total triterpenoid content was determined by measuring the absorbance at 548 nm using UV spectrophotometer (Biochrom Ltd., Cambridge, England) with oleanolic acid as the standard. The concentration of triterpenoids was calculated as follows (Eq. 1):

$$total\,triperpendiods\,(mg/g) = \frac{triperpendiods\ quality\ in\ sample}{\dfrac{weight\ of\ sample}{constant\ volume} \times volume\ of\ sample} \tag{1}$$

2.5. Selection of Significant Variables by PB Design

PB design is an effective way of screening the significant factors from a large number of variables. In this study, a 12-run PB design was applied to evaluate 11 factors (including two dummy variables). Each variable was examined at 2 levels, -1 for the low level and +1 for the high level. The variables and their corresponding levels were designed by experiment experience (Table 1). The PB design assumed that the main factors have no interactions and a first-order multiple regression model is as follow (Eq. 2):

$$Y = \beta + \sum \beta x_i (i = 1, \ldots, k) \tag{2}$$

Where Y is the response function (biomass of *A. cinnamomea* mycelium) and β is the regression coefficient. The response value of biomass of *A. cinnamomea* wasanalyzed and is shown in Table 2. Data were analyzed by analysis of variance (ANOVA). Statistical significance was at $p < 0.05$.

Table 1. Levels of the variables tested in PB design (g/L)

Symbol	Variables	Experimental value	
		Low(-1)	High(+1)
A	Glucose	10	50
B	Malt dust	10	50
C	Yeast extract	5	20
D	Dummy	0	0
E	KH2PO4	0.2	1
F	MgSO4	0.2	1
G	V_{B1}	0.2	1
H	JCSWE	0	1
J	JCGLAE	0	1
K	Dummy	0	0
L	Chitosan	0	1

2.6. Path of Steepest Ascent

Steepest ascent method is a procedure for moving along the maximum increase in the response [22,23]. The direction of steepest ascent is the direction in which the response increases most rapidly by increasing or decreasing the values of the significant factors. The step along the path was determined by the estimated coefficient and practical experience. This steepest ascent point should be near the optimal point and could be used as the center point of RSM [24].

2.7. Experiment Design for RSM

This is a strategical and mathematical approach that was used to find out a suitable relationship between experimental variables and obtained data [25]. For this purpose, the significant variables utilized for the RSM of BBD were as follows: glucose, malt dust, yeast extract, VB_1 and chitosan, which were coded at three levels (-1, 0, +1). A total of 46 trials were conducted, and the biomass was treated as a response. Experiment data were fitted to a second-order polynomial model, and regression coefficients were obtained. The generalized second-order polynomial model used in the response surface analysis of the yield of the mycelium as a function of the independent variables is as follows (Eq. 3):

$$Y = \beta_0 + \sum_{i=1}^{n} \beta_i \chi_i + \sum_{i=1}^{n} \beta_{ii} \chi_i^2 + \sum_{i<j} \beta_{ij} \chi_i \chi_j \quad (3)$$

Where Y is the response, β_0 is constant, β_i β_{ii} and β_{ij} are the regression coefficients for linear, quadratic and interaction coefficients. χ_i and χ_j are the independent variables.

2.8. Data Analysis

Data were determined by analysis of variance (ANOVA) and expressed as means standard errors (SE) of 3 replicates. Design-Expert Version 8.0.6 (Stat-Ease, Inc., Minneapolis, USA) was used to design an experiment; regression model and graphical analysis of the experimental data were obtained.

3. RESULTS AND DISCUSSION

3.1. Screening of Significant Variables using PB Design

The PB design was used to select variables that can be fixed or eliminated in the subsequent optimization process. It showed a wide variation in the biomass of mycelium, which were from 0.4276 to 1.7154 g/100 mL in Table 2. The model F-value of 90.12 implied that the model was significant, and it also suggested that optimization was necessary for improving the biomass of mycelium. It is shown that all the factors had positive effects on mycelium biomass except $MgSO_4$ (F) and V_{B1} (G) in the regression analysis and the t-values of 9 factors (Table 3). The variable with a confidence level above 95% was considered as a significant parameter while those below 95% were considered insignificant and were not included in the steepest ascent and BBD experiments. Glucose (A), malt dust (B), yeast extract (C), V_{B1}, and chitosan (L) were all significant, and therefore, were included in the model. The model equation for biomass (Y) could be written as (Eq. 4):

$$Y = 1.14 + 0.071A + 0.087B + 0.17C + 0.043E - 0.019F - 0.069G + 0.023H + 0.046J + 0.34L \quad (4)$$

Table 2. PB design of variables with biomass as response

Run	A	B	C	D	E	F	G	H	J	K	L	Biomass (g/100 mL, DW)
1	-1	1	1	-1	1	1	1	-1	-1	-1	1	1.5232±0.2956
2	-1	-1	-1	1	-1	1	1	-1	1	1	1	1.0721±0.0558
3	-1	-1	1	-1	1	1	-1	1	1	1	-1	0.9856±0.0353
4	-1	1	1	1	-1	-1	-1	1	-1	1	1	1.7154±0.0729
5	1	1	-1	1	1	1	-1	-1	-1	1	-1	0.8488±0.1622
6	-1	1	-1	1	1	-1	1	1	1	-1	-1	0.7124±0.0655
7	1	-1	-1	-1	1	-1	1	1	-1	1	1	1.2702±0.0402
8	1	-1	1	1	-1	1	1	1	-1	-1	-1	0.8003±0.0915
9	1	1	1	-1	-1	-1	1	-1	1	1	-1	1.0703±0.0685
10	1	-1	1	1	1	-1	-1	-1	1	-1	1	1.7840±0.2085
11	1	1	-1	-1	-1	1	-1	1	1	-1	1	1.5173±0.1321
12	-1	-1	-1	-1	-1	-1	-1	-1	-1	-1	-1	0.4276±0.0349

Table 3. Effects of the variables and statistical analysis of PB design

Variables	Effect	Sum of Squares	df	F-Value	p-Value
A-Glucose	0.071	0.061	1	25.06	0.0377*
B-Malt dust	0.087	0.091	1	37.63	0.0256*
C-Yeast extract	0.17	0.34	1	141.4	0.0070*
E-KH2PO4	0.043	0.023	1	9.31	0.0927
F-MgSO4	-0.019	4.51E-03	1	1.86	0.3062
G-V_{B1}	-0.069	0.057	1	23.64	0.0398*
H-JCSWE	0.023	6.32E-03	1	2.6	0.2482
J-JCGLAE	0.046	0.026	1	10.61	0.0827
L- Chitosan	0.34	1.36	1	558.99	0.0018*
Model		1.97	9	90.12	0.01

* p < 0.05 indicate that model terms are significant

3.2. Path of Steepest Ascent

Experiments were performed along the steepest ascent path until the response showed no further increase (Table 4). The highest response at 28°C was 1.5606 g/100 mL with 40 ˜ glucose, 40 g/L malt dust, 20 g/L yeast extract, of 0.4 g/L V_{B1}, 0.8 g/L chitosan, and ɔ% inoculum size (V/V). The culture medium for steepest ascent result was concluded to be near the optimal point of RSM and identified as the testing formula for further optimization.

Table 4. Effects of the path of steepest ascent (g/L)

Runs	A- Glucose	B-Malt dust	C-Yeast extrac	G- VB1	L- Chitosan	Biomass, g/100 mL, DW
1	10	10	5	1	0.2	0.3869±0.0085
2	20	20	10	0.8	0.4	0.6028±0.0315
3	30	30	15	0.6	0.6	1.0295±0.0571
4	40	40	20	0.4	0.8	1.5606±0.0930
5	50	50	25	0.2	1	1.4208±0.1318

3.3. Analysis of RSM

Based on the result above, there was a total of 46 runs for optimizing the five individual parameters. The point that represented 40 g/L glucose, 40 g/L malt dust, 20 g/L yeast extract, 0.4 g/L V_{B1}, 0.8 g/L chitosan was the central point for 3 levels of variables. The code and actual value are listed in Table 5. The ANOVA result for the BBD is shown in Table 6. The Model F-value of 51.28 implied a good fit of the model, but there was only a 0.01% chance that a "Model F-value" this large could occur due to noise. It showed a good correlation with the experimental data in the model term as values of "Prob > F" less than 0.05 indicates that the model fit efficiency was checked by the coefficient of determination (R^2=0.9762) and adjusted coefficient of determination (Adj R^2=0.9572). It can be seen from the table that the linear coefficients (X_1, X_2, X_3, X_5), the cross product coefficients (X_1X_5, X_2X_4, X_3X_4) and quadratic term coefficients (X_1^2 X_4^2, X_5^2) were significant (p<0.05), while the others were not (P>0.05) (Table 6). In this study, the factors influencing the response value of *A. cinnamomea* mycelium biomass are in the following order: glucose > chitosan > malt dust > V_{B1}> yeast extract. The lack of fit F value of 0.83 implied that the lack of fit was not significantly correlated with the pure error. There was a 66 % chance that a "lack of fit F value" this large could occur due to noise. The data results (Y) and the matrix of RSM experiments used to determine the effects of the five variables including glucose (X_1), malt dust (X_2), yeast extract (X_3), V_{B1} (X_4), chitosan (X_5) were fitted to the following second-order polynomial equation (Eq. 5):

$$\begin{aligned}Y = &-0.94+X_1+0.19X_2+0.018X_3+0.047X_4+0.48X_5-8.425\times10^{-3}X_1X_2+0.059X_1X_3\\&-0.015X_1X_4-0.18X_1X_5-0.012X_2X_3-0.11X_2X_4+0.025X_2X_5-0.012X_3X_4-0.081X_3X_5\\&-0.071X_4X_5-0.32X_1^2-0.33X_2^2-0.26X_3^2-0.3X_4^2-0.4X_5^2\end{aligned} \tag{5}$$

Table 5. The BBD design and experimental responses of the dependent (g/L)

Run	X_1-Glucose	X_2-Malt dust	X_3-Yeast extract	X_4- V_{B1}	X_5-Chitosan	Y-Biomass （g/100 mL, DW）
1	40	40	15	0.4	0.6	0.7264±0.0015
2	40	40	20	0.6	1	0.9488±0.0114
3	30	40	20	0.4	0.6	0.6904±0.0151
4	30	40	25	0.4	0.8	0.9406±0.0004
5	40	40	20	0.4	0.8	1.6079±0.0237
6	50	40	20	0.4	0.6	0.8763±0.0217
7	40	40	25	0.4	0.6	1.1338±0.0163
8	40	40	20	0.4	0.8	1.6007±0.0180
9	30	40	15	0.4	0.8	0.6548±0.0133
10	40	40	25	0.4	1	1.1609±0.0181
11	40	40	15	0.4	1	1.0780±0.0200
12	40	30	15	0.4	0.8	0.7027±0.0161
13	40	40	20	0.4	0.8	1.6992±0.0371
14	30	30	20	0.4	0.8	0.6111±0.0220
15	40	40	25	0.2	0.8	1.3017±0.0216
16	50	40	15	0.4	0.8	1.3732±0.0154
17	50	30	20	0.4	0.8	1.1181±0.0555
18	50	40	25	0.4	0.8	1.4221±0.0235
19	30	40	20	0.6	0.8	0.8153±0.0196
20	40	40	20	0.6	0.6	0.9484±0.0181
21	30	50	20	0.4	0.8	0.9654±0.0342
22	40	50	20	0.6	0.8	1.1766±0.1480
23	40	30	20	0.2	0.8	0.7152±0.0045
24	40	30	20	0.4	0.6	0.7106±0.0013
25	40	40	20	0.2	0.6	0.8219±0.1195
26	40	50	15	0.4	0.8	1.0229±0.205
27	40	50	20	0.2	0.8	1.3471±0.0679
28	40	40	15	0.6	0.8	0.9706±0.0168
29	50	40	20	0.2	0.8	1.1930±0.0652
30	40	50	25	0.4	0.8	1.3447±0.0425
31	40	40	20	0.4	0.8	1.6731±0.0138
32	40	40	25	0.6	0.8	1.2760±0.0771

33	40	50	20	0.4	1	1.2356±0.0618
34	40	30	20	0.6	0.8	1.0031±0.0131
35	50	40	20	0.6	0.8	1.3123±0.0819
36	40	30	20	0.4	1	0.8706±0.0671
37	40	30	25	0.4	0.8	1.0720±0.0686
38	50	40	20	0.4	1	1.5932±0.0700
39	30	40	20	0.4	1	0.6696±0.0179
40	40	40	20	0.4	0.8	1.6435±0.0006
41	50	50	20	0.4	0.8	1.5061±0.0103
42	40	40	20	0.4	0.8	1.7969±0.0418
43	30	40	20	0.2	0.8	0.7574±0.0072
44	40	40	20	0.2	1	1.1053±0.1014
45	40	50	20	0.4	0.6	0.9750±0.0577
46	40	40	15	0.2	0.8	0.9476±0.1925

Table 6. Analysis of variance (ANOVA) and coefficients estimated for biomass yield

Source	Sum of Squares	df	Mean Square	F value	p-value (Prob>F)
Model	4.65	20	0.23	51.28	< 0.0001*
X1	0.52	1	0.52	[illegible]5.76	< 0.0001*
X2	0.034	1	0.034	[illegible]	0.0112*
X3	2.89E-04	1	2.89E-04	0.06	0.8027
X4	2.09E-03	1	2.09E-03	0.46	0.5038
X5	0.22	1	0.22	47.86	< 0.0001*
X1X2	2.84E-04	1	2.84E-04	0.06	0.8044
X1X3	0.014	1	0.014	3.1	0.0907
X1X4	9.43E-04	1	9.43E-04	0.21	0.6523
X1X5	0.14	1	0.14	30.02	< 0.0001*
X2 X3	5.64E-04	1	5.64E-04	0.12	0.7272
X2X4	0.053	1	0.053	11.59	0.0022*
X2X5	2.53E-03	1	2.53E-03	0.56	0.4619
X3X4	5.93E-04	1	5.93E-04	0.13	0.7206
X3X5	0.026	1	0.026	5.81	0.0236*
X4X5	0.02	1	0.02	4.42	0.0458*
X12	0.87	1	0.87	192.72	< 0.0001*
X22	0.93	1	0.93	205.32	< 0.0001*
X32	0.61	1	0.61	134.14	< 0.0001*
X42	0.81	1	0.81	178.63	< 0.0001*
X52	1.37	1	1.37	301.78	< 0.0001*
Lack of fit	0.087	20	4.35E-03	0.83	
Pure Error	0.026	5	5.27E-03		
Cor Total	4.76	45			

R^2=97.62%, Adj R^2=95.72%, C.V.=6.06%

* significant coefficient ($p<0.05$)

The response surfaces (2-D) presented in Fig. 1 was obtained by Design Expert 8.0.6. Two tested variables interactions and the relationships between response and experiment level were provided. The contour plots showed the interactions between the variables. Circular contour plots indicated that it was not a good fit between the independent variables, while elliptical contours predicted a perfect interaction [27,28].

Fig. 1 show the interaction between the independent variables in an elliptical and saddle shape. As can be seen from Fig. 1 a, b and c, the plots for the interaction between glucose and malt dust, glucose and yeast extract, glucose and VB1 showed that glucose had a decrease in biomass yield at the maximum and minimum level, while increasing the malt dust, yeast extract and VB1 led to increasing biomass. However, the three group interactions had no significant effect on the biomass. The glucose and chitosan had a significant effect on the response as shown in Fig. 1d. Increasing the chitosan to 0.8 g/L increased the biomass of *A. cinnamomea* mycelium, which decreased sharply as the chitosan increased further. In contrast, increasing glucose concentration to 40 g/L increased the biomass slightly, possibly up to 1.7969 g/100 mL. The response surface plot of the malt dust and yeast extract had no significant effect (Fig. 1e). The yield was increased while the malt dust and yeast extract increased. The effect of malt dust and VB1 on biomass is shown in Fig. 1f and 2f. The yield was increased when the malt dust and yeast extract were on the 0 level; the biomass increased evidently when VB1 concentration reached to 0.8 g/L. However, the yield slowly increased while VB1 was lower than 0.8 g/L.

Although the combined effects of malt dust and chitosan on the biomass were not significant, there was an obvious but minimal change in the biomass yield when the malt dust and chitosan were at low and high levels (Fig. 1g). The corresponding plot (Fig. 1h) of the yeast extract and VB_1 was not obvious as the surface was flat. It displayed a significant effect on yeast extract and chitosan (Fig. 1i). The biomass yield of *A. cinnamomea* increased significantly with increasing chitosan. A similar result was shown by VB1 and chitosan (Fig. 1j).

Chitosan is a natural material derived from shells, which has an antimicrobial effect on the fruits, vegetables and animals [29]. Chitosan is also used as a growth promoter in plant yield. Liu et al. [30] had indicated that chitosan was a more potent chemical elicitor than CaCl2 in enhancing the biomass on two-stage submerged cultivation of *A. cinnamomea*. Chitosan had not only improved more biomass production in two-stage submerged culture but also produced higher crude triterpenoids at the dosage level of 100 mg/L. In this study, chitosan was the most important factor in the model. The biomass yield was improved sharply when it was influenced by chitosan.

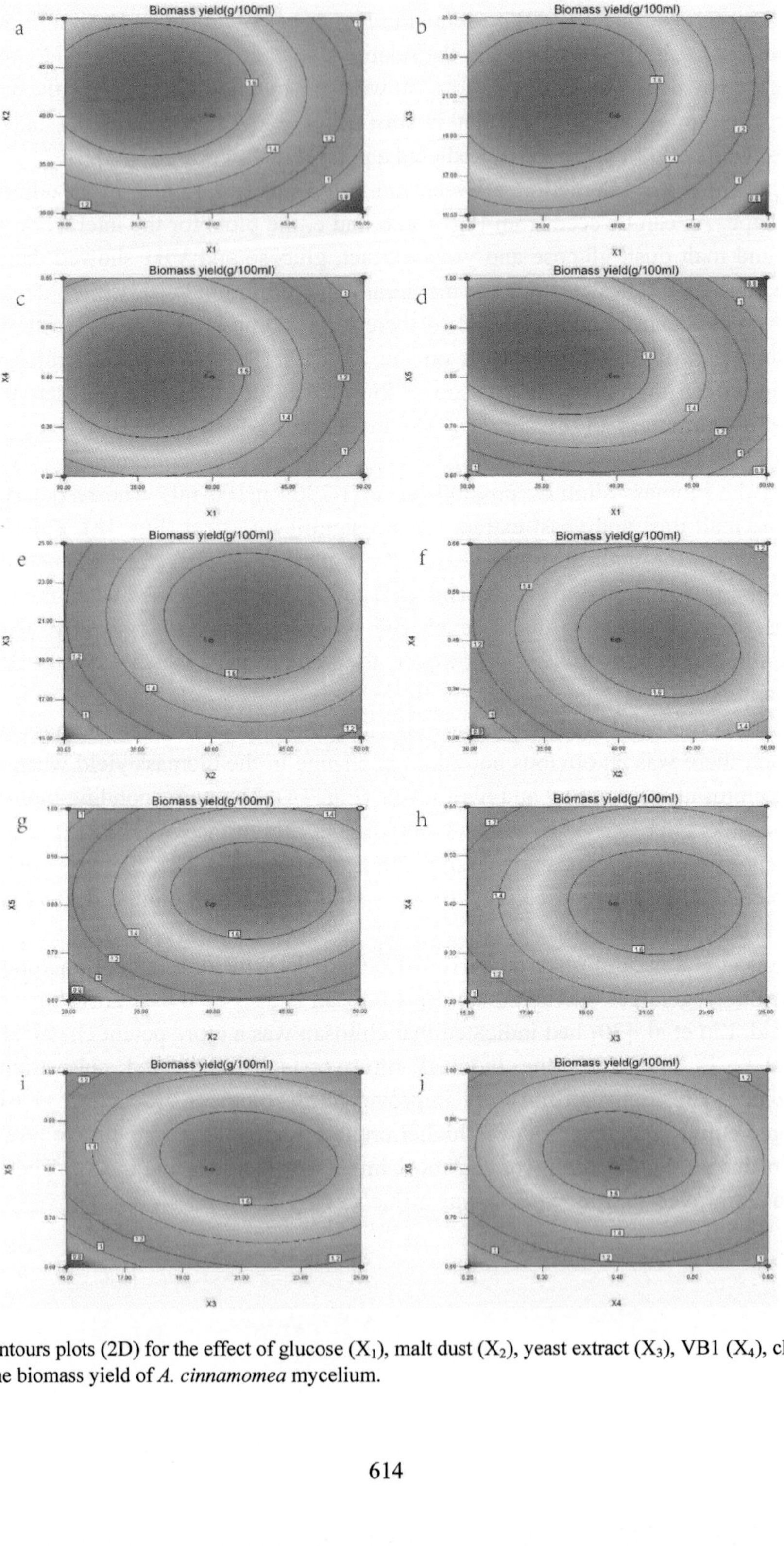

Fig.1. Contours plots (2D) for the effect of glucose (X_1), malt dust (X_2), yeast extract (X_3), VB1 (X_4), chitosan (X_5) on the biomass yield of *A. cinnamomea* mycelium.

3.4. Optimization and Verification

The optimum condition for the biomass yield of *A. cinnamomea* mycelium by submerged culture was as follows: 35.16 g/L glucose, 42.86 g/L malt dust, 20.80 g/L yeast extract, 0.39 g/L V_{B1}, and 0.85 g/L chitosan. The estimated value for biomass was 1.7843 g/100 mL. By recovery experiments (RE), the actual value of biomass yield of *A. cinnamomea* mycelium was 1.6765±0.0103 g/100 mL (n=3). The RE was 2.5 fold more than the control (25 g/L glucose, 5 g/L peptone, 3 g/L malt dust) (Fig. 2).

3.5. Measurement of Triterpenoids of *A. cinnamomea* Mycelium

Fig. 2 shows the relationship between the biomass under different levels and the mycelium triterpenoid concentration. The total triterpenoid concentration reached a stable value at over 1 g/100 mL biomass, signifying a strong association with the biomass. The prediction experiment also had the same result. It was shown that the mycelium content increased to 543 mg/100 mL in an industrial 500-L ALs, and at the same time the ethanol extracts of *A. cinnamomea* mycelium which was 17 μg/mL in 28 days reduced to 10 μg/mL after 42 days [31]. As the mycelium grew, the nutrition of the medium was consumed, and the ethanol extract decreased severely. The culture conditions of *A. cinnamomea* mycelium at 25°C was the chitosan dosage of 100 mg/L, at which the triterpenoid content increased sharply in the preceding 15 days [30].

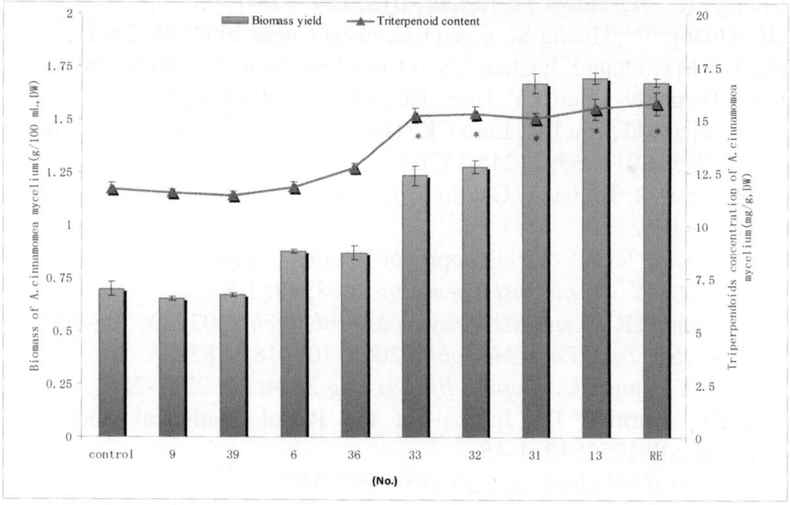

Fig.2. The relation between triterpenoids concentration and biomass of A.cinnamomea mycelium (No. is the runs of the RSM, RE is the recovery experiment, and * significant coefficient, $p < 0.05$).

Since the *A. cinnamomea* fruit bodies are barren, the development of submerged culture can produce more mycelium through the RSM, especially glucose and chitosan. The model showed that the biomass yield of *A. cinnamomea* mycelium improved under the following optimal conditions: 35.16 g/L glucose, 42.86 g/L malt dust, 20.80 g/L yeast extract, V_{B1} 0.39 g/L, and 0.85 g/L chitosan. In addition, it also showed improved triterpenoids content. There was a good fit for mycelium and triterpenoids concentration that may take less processing time and yield more active products. Furthermore, the proposed complex medium will be beneficial to industrial submerged medium for commercial production. However, more research is needed to fully understand the functional activity of triterpenoids under different dosage. Also, more natural materials should be studied for the enhancement of biomass and the active ingredient in the industrial production.

Acknowledgments

Financial support was provided by Collaborative Innovation Center for JUNCAO Ecology Industry (Fujian 2011 Program, K80ND8002). It was also funded by the project of China National Engineering Research Center of JUNCAO Technology (2011FU125X12).

References

1. Liau YJ, Wen L, Shaw JF, Lin CT. *J Biotechnol* 2007; **131**: 84-91.
2. Chang T, Chou W. *Mycol Res*1995; **99**: 756-758 .
3. Lu MC, El-Shazly M, Wu TY, Du YC, Chang TT, Chen CF, Hsu YM, Lai KH, Chiu CP, Chang FR. *Pharmacol Therapeut* 2013; **139**: 124-156.
4. Mau JL, Huang PN, Huang SJ, Chen CC. *Food Chem* 2004; **86**: 25-31.
5. Ho YC, Lin MT, Duan KJ, Chen YS. *J Chin Inst Chem Eng* 2008; **39**: 441-447.
6. Sheu MJ, Teng YN, Chen YY, Hung CC. *PloS one* 9: e89622.
7. Chen LY, Sheu MT, Liu DZ, Liao CK, Ho HO, Kao WY, Ho YS, Lee WS, Su CH. *J Agr Food Chem* 2011; **59**: 11255-11263.
8. Kapich A, Prior B, Botha A, Galkin S, Lundell T, Hatakka A. *Enzyme Microb Tech* 2004 **34**: 187-195.
9. Jordaan J. Doctor Thesis of Philosophy of Rhodes University 2005.
10. Shu CH, Lung MY. *J Chin Inst Chem Eng* 2008; **39**: 1-8.
11. Yang FC, Huang HC, Yang MJ. *Enzyme Microb Tech* 2003; **33**: 395-402.
12. Lin ES, Sung SC. *Int J Food Microbiol* 2006; 108: 182-187.
13. Ma TW, Lai Y, Yang FC. *Bioproc Biosyst Eng* 2014; **37**: 2251-2261.
14. Plackett RL, Burman JP. Journal of the Royal Statistical Society. *Series B (Methodological)* 1951; **13**: 1-45.
15. Kiefer J. *Series B (Methodological)* 1959: 272-319.
16. Xu C, Wang J, Yang T, Chen X, Liu X, Ding X. *Carbohyd Polym* 2014.
17. Lee JW, Kim YH, Lee SM, Lee HW. *Bioresource technol* 2012; **116**: 471-476.
18. Zhong K, Wang Q. *Carbohyd Polym* 2010; **80**: 19-25.
19. Wu S, Gong G, Wang Y, Li F, Jia S, Qin F, Ren H, Liu Y. *Int J Biol Macromol* 2013; **61**: 63-68.
20. Liu YL, Zhao C, Lin DM, Lin H, Lin ZX. *Anim* 2015.
21. Chen XC, Bai JX, Cao JM, Li ZJ, Xiong J, Zhang L, Hong Y, Ying HJ. *Bioresource Technol* 2009; **100**: 919-924.

22. Zhou J, Yu X, Ding C, Wang Z, Zhou Q, Pao H, Cai W. *Journal of Environmental Sciences* 2011; 23: 22-30.
23. Tang XJ, He GQ, Chen QH, Zhang XY, Ali MA. *Bioresource Technol* 2004; **93**: 175-181.
24. Myers RH, Montgomery DC, Anderson-Cook CM. *John Wiley & Sons* 2009.
25. Lu ZM, Tao WY, Xu HY, Lim J, Zhang XM, Wang LP, Chen JH, Xu ZH. *Food Chem* 2011; **127**: 662-668.
26. Guo X, Zou X, Sun M. *Carbohyd Polym* 2010; **80**: 344-349.
27. 28.Liu Y, Gong G, Zhang J, Jia S, Li F, Wang Y, Wu S. *Carbohyd Polym* 2014; **110**: 278-284.
28. Devlieghere F, Vermeulen A, Debevere J. *Food Microbiol* 2004; **21**: 703-714.
29. Liu CJ, Chiang CC, Chiang BH. *Biochem Eng J* 2012; **64**: 48-54.
30. Chiang CC, Chiang BH. *Biochem Eng J* 2013; **73**: 65-71.

STUDY ON MAP KNOWLEDGE DOMAINS OF TRANSGENIC MAIZE: BASED ON CITESPACE

Xintong Li[a], Dan Guo, and Jingmin Cheng[b,*]
School of Management, Shan Xi Medical University, Taiyuan, 030001, China
E-mail: [a]ZnCu_li@163.com, [b]72-87@163.com
[*]Corresponding author

The purpose of this paper is to analyze the hotspot and trend of the studies on global transgenic maize from 2005 to 2015.The paper is mainly based on bibliometric method, taking studies on transgenic maize from 2005 to 2015 which are included in the core journals in the database of Web science as samples, and using Citespace for visual analysis of mapping knowledge domain of studies on transgenic maize. As a result, 2851 pieces of documents have been retrieved and the literatures are in a stable situation and show a tendency of increase. According to Citespace, it has drawn a mapping knowledge domain on scholars, institutions, citation and key words. Therefore, by cluster analysis, it is found that in the study of transgenic maize, Chinese scholars and institutions hold an advantage in quantity while the hotspot of global research is mainly about flora, genes, protein and resistant, etc.

1. MATERIALS AND METHODS

1.1. Materials

Web of Science is developed by the Thomson Scientific from America based on Web, including more than 8000 kinds of one of the most influential journals in the world, after experts' evaluation of quality of journals.

This paper completed data analysis, based on literature of " transgenic maize" which are found on the core journals of Web of Science. Specific search strategy is "subject = ('transgenic maize') ", meanwhile, the range of time is set by "2005-2015",and the retrieval time is set by March 1, 2016.Totally searched 2851 literature, after screening for 2701. Citations saved for other format file, each topic including author, organization, keywords and references and other information.

1.2. Methods

Map Knowledge Domains is a combination of citation analysis with data and information visualization. Citespace is citation visualization software which analyzes the potential knowledge in scientific analysis and is developed under the background of scientometrics, big data and visualization of information [1-2].

Using Citespace to load normalized citations into calculate, the range of time is set as " 2005-2015 ", and time slice is set as 1. Selecting the method of pathfinder in order to reduce the network density and improving the readability of the network. Firstly, screen out repeated citations. Secondly, choose different themes, including the authors, institutions, keywords, cited reference, cited authors, then choose clustering analysis, and finally resize the sizes of individual nodes, the threshold value. Thus we complete the Map Knowledge Domains.

1.3. Posting

A statistical chart about the publication of transgenic maize clearly shows the overall situation of study. As is shown in Fig. 1, from 2005 to 2015, the global study on transgenic maize is keeping steady, published roughly 200 to 300 papers in each year and has a rising trend. Thus, the heat on the study of transgenic maize is relatively stable, and the study has already formed a certain scale and mode while the attention on this study is also increasing. A statistical chart about the publication of transgenic maize clearly shows the overall situation of study. As is shown in figure 1, from 2005 to 2015, the global study on transgenic maize is keeping steady, published roughly 200 to 300 papers in each year and has a rising trend. Thus, the heat on the study of transgenic maize is relatively stable, and the study has already formed a certain scale and mode while the attention on this study is also increasing.

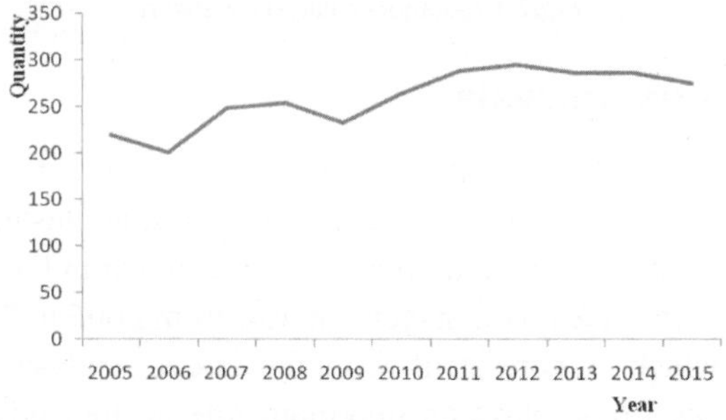

Fig.1. Posting of Transgenic Maize from 2005 to 2015.

2. COLLABORATION NETWORK ANALYSIS

Citespace provides three levels of study, including co-authors, co-institutions, co-countries/ regions; the size of node size represents the number of published papers by individual authors, institutions, countries or regions[3]. This paper analyzes the cooperation network of transgenic maize.

2.1. Co-authors Analysis

Analyzing the network of cooperation between authors can intuitively find the important authors in the field of transgenic maize. Using clustering function analysis citations from 2005 to 2015, the threshold is set to 20, shown in Fig. 2. As shown, Romeis J, writing 40 papers, is the leading figure in the study of transgenic maize, followed by Liu Y, Huang FN, Wang K, Hibbard BE. Besides, the Chinese researchers in this study have advantage of quantity. In addition, various researchers maintained close relate from complex lines, shown that the study of transgenic maize have become interchanged and cooperated normally.

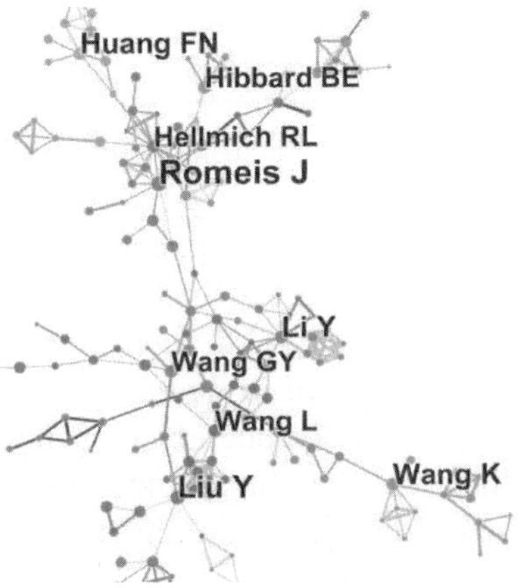

Fig.2. Co-authors Analysis Network.

2.2. Co-institutions Analysis

By analyzing the network of cooperation between institutions, it can intuitively find the important institutions in the field of transgenic maize. Using clustering function analysis citations from 2005 to 2015, the threshold is set as 36, shown in Fig. 3. As shown in table 1, among the top 10 in the study of transgenic maize from 2005 to 2015, China holds half of the institutions, which is consistent with the above analysis results of important authors. It explains that China plays an important role in the study of transgenic maize again.

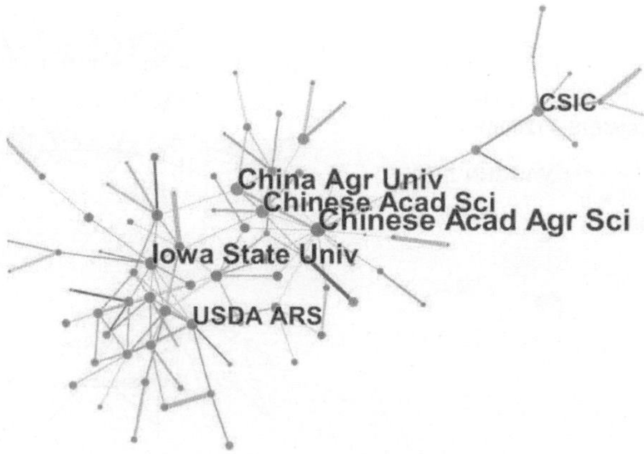

Fig.3. Co-institutions Analysis Network.

Table 1. post on top 10 of institutions

No.	F	Centrality	Institution	No.	F	Centrality	Institution
1	116	0.1	Chinese Acad Agr Sci	6	44	0.03	CSCI
2	86	0.02	China Agr Univ	7	36	0.02	Monsanto Co
3	76	0.07	Iowa State Univ	8	36	0.01	Univ Illinois
4	74	0.06	Chinese Acad Sci	9	35	0.02	Cornell Univ
5	47	0.03	USDA ARS	10	33	0.01	Shandong Agr Univ

3. CO-CITATION NETWORK ANALYSIS

Citation analysis is the method of information measurement that uses mathematical and statistical methods to compare, induct, abstract and generalize, analyses the phenomenon of cited or cite, such as scientific journals, papers, authors, to reveal the quantitative characters and inherent laws [4-5]. This paper mainly does citation analyzes through the author co-citation analysis and literature co-citaion.

By analyzing the situation of citation, it can intuitively find that the of high quality articles in the study of transgenic maize and its development of history and forefront. Using clustering function analysis citations from 2005 to 2015, the threshold is set to 51, shown in Fig. 4. It shows that there is little difference among the size of nodes, explaining situation of cited is similar located in the top. Among them, the articles written by ROMEIS J, published in 2006 and 2008, were cited 86 times and 65 times respectively. It shows that the number of articles written by ROMEIS J is large and the quality is also very high. Using clustering function analysis citations from 2005 to 2015, the threshold is set to 156, shown in figure 5. Among these articles, the largest amount of cited is SAMBROOK J, cited 251 times. By contrast, in term of the number of the important authors and institutional, China has its advantage. In the analysis of cited, this advantage, however, does not display.

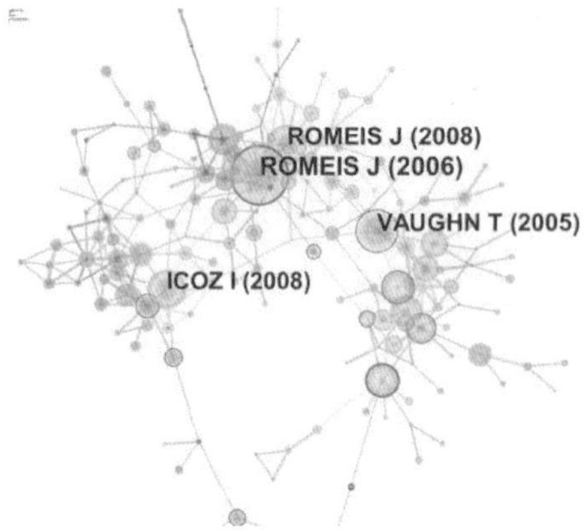

Fig.4. Author co-citation analysis Network.

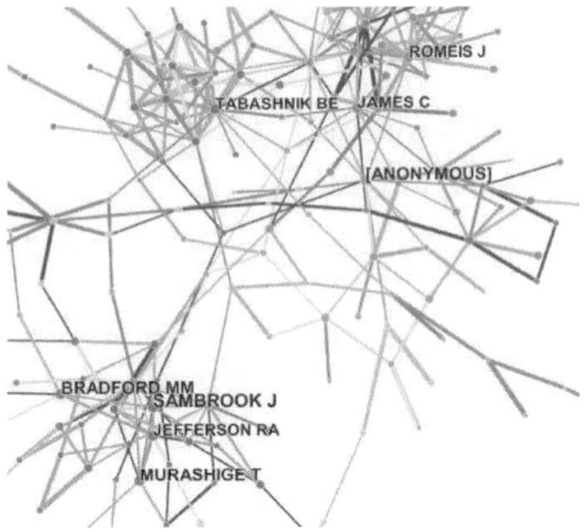

Fig.5. Citation Analysis Network.

4. CO-APPEARANCE NETWORK

Taking keywords as nodes to analyze, it is found that a higher frequency indicates a bigger font. Using clustering function analysis citations from 2005 to 2015, the threshold is set to 287, shown in figure 6. Behind the high heat of keywords on top, the uses of following became smooth and steady. Table 2, the situation of uses of keywords in transgenic maize from 2005 to 2015, shows top 20. " Maize ", " zea may ", and " corn " are synonymous with the term of " corn ", " transgenic maize " with " bt maize " are synonymous with the term of "transgenic maize", "plant" and "crop" are synonymous with "crop". Thus, the study in transgenic maize from 2005 to 2015 mainly revolves in these aspects, such as bacillus thuringiensis, polymerase chain reaction, arabidopsis thaliana, gene, protein, resistant.

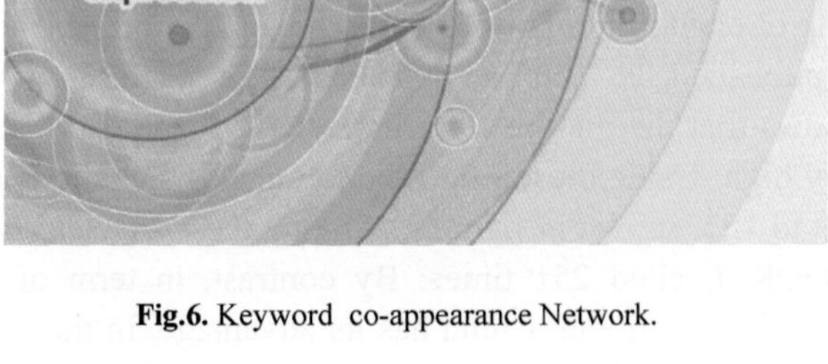

Fig.6. Keyword co-appearance Network.

Table 2. Keywords on top 20

No.	F	Centrality	Keyword	No.	F	Centrality	Keyword
1	896	0.12	maize	11	204	0.05	arabidopsis
2	472	0.18	plant	12	175	0.05	gene expression
3	355	0.19	expression	13	170	0.08	zea may
4	310	0.19	bacillus thuringiensis	14	165	0.17	bt maize
5	287	0.07	arabidopsis thaliana	15	161	0.17	transformation
6	281	0.08	gene	16	155	0.11	corn
7	218	0.24	transgenic maize	17	153	0.16	resistant
8	216	0.16	protein	18	112	0.08	polymerase chain reaction
9	213	0.14	transgenic plant	19	108	0.03	transgenic rice
10	204	0.03	rice	20	104	0.06	crop

5. CONCLUSIONS

This paper mainly uses Citespace, a scientific metrology software, analyzing citation data in the corn of Web of Science about transgenic maize from 2005 to 2015, drawing Map Knowledge Domains. After analysis, we found that Romeis J is one of the main scholars in the study of trandgenic maize. In addition, the researchers and institutions in China have advantage in quantity. Through the analysis in the situation of cited about literature and authors cited, we found that as for the amount of literature being cited, the foreign ones are higher than those of the Chinese. Therefore China should improve the quality of the papers. At the same time, hotspots in recent years mainly converge on bacillus thuringiensis, polymerase chain reaction, arabidopsis thaliana, gene, protein, resistant. As a hot topic in the food safety, transgenic maize has gained a multi-dimensional debate [6]. Capturing the latest development in research on transgenic maize, China should focus on the global hotspot of transgenic maize, expand international cooperation and consistently promote further development of the study.

Acknowledgments

This work is supported by the Fund (grant no. SSKLZDK2014085) hosted by Dan Guo for the the federation of social sciences at Shanxi Province. Some students and teachers participated in this project,including Jing-Min Cheng, as the corresponding author,and Xin-Tong Li.

References

1. Wei-jun Wang, Guo-pei Cai.*Method and Application of Information Analysis*.Beijing Jiaotong University Press,2010;5-10.(In Chinese).
2. Chen C. CiteSpace II: Detecting and Visualizing Emerging Trends and Transient Patterns in Scientific Literature.*Journal of the American Society for Information Science and Technology*2006;**3**:359－377.
3. Jie Li, Chao-mei Chen. Citespace:*Text Mining and Visualization in Scientific Literature*. Capital University of Economics and Business Press,2016;180-182.(In Chinese).

4. Jun-ping Qiu, Xiao-ting Wen, Yan-hui Song. *Knowledge of Metrology*.Science Press, 2014;3-11.(In Chinese).
5. Chun-li Yin. Topological Studies of Science Studies Citation Network. Dalian: Dalian University of Technology,2006.(In Chinese).
6. Jing-min Cheng. *Study on Pre-warning System of Food Safety*.The Economics Daily Press, 2014;13-16.(In Chinese).

FACTORS INFLUENCING THE TISSUE CULTURE OF HYBRID POPLAR *POPULUS DAVIDIANA* DODE × *P. BOLLENA* LAUCHE

Xue Han[a] and Bing Han[b]

[1]College of Accountancy, Harbin University of Commerce, Harbin 150028, China
[2]College of Food Science and Engineering, Harbin University of Commerce, Harbin 150076, China
E-mail: [a]hanxue_home@163.com; [b]hb790118@163.com

Plant regeneration was studied in different media for hybrid poplar *populus davidiana* Dode ×*P. bollena* Lauche. Using leaf as explant, regeneration efficiency on MS containing 0.3 mg/L BA and NAA 0.08mg/L was higher than others. When adventitious microshoots were cultured on MS supplemented with NAA 0.25 mg/L , root formation was obtained after 1 week of culture. The root was length and polyrhizal. This system for micropropagation of *populus davidiana* Dode ×*P. bollena* Lauche is expected to be suitable for studies on transformation of genes.

1. INTRODUCTION

Populus davidiana Dode × P. bollena Lauche is the hybrid cultivar of poplar, which is used *Populus bolleana* as male parent and *Populus davidiana* Dode as female parent. It has several exceptional qualities, such as a high capacity for vegetative propagation and suitability to develop transgenic.

The micro-vegetative propagation and regeneration of poplar has been reported in the recent decades. Different poplar genotypes have been used to perform regeneration experiments [1-10]. Growth and differentiation response is not only controlled by the cultural environment, but is also dependent upon the genotype [11]. In vitro regeneration has been attempted in several poplar species using different explants such as leaves [8,11-14], petioles [15], internodes [8,14], stems [5,16], roots [8,11], and shoot tips [17].

In the present study, we have chosen a hybrid poplar *Populus davidiana* Dode × *P. bollena* Lauche to improve the regeneration efficiency. *Populus* was planted in northeast China and has a well-characterized morphology with cold resistance features. We have optimized the Murashige and Skoog [18] medium with supplementary 6-Benzylaminopurine and Naphthaleneacetic acid to achieve a high regeneration efficiency of *Populus*.

2. MATERIALS AND METHODS

2.1. Plant Material

Populus davidiana Dode × *P. bollena* Lauche leaf explants were obtained randomly from in vitro shoot cultures and then used for regeneration.

2.2 Shoot Regeneration of Leaf Explants

Murashige and Skoog (MS) medium was supplemented with 6-benzylaminopurine (BA) at 0.1 mg/L to 0.5 mg/L and naphthaleneacetic acid (NAA) at 0.01 mg/L to 0.1mg/L shown in Table 1. The leaves were cut into 1cm ×1 cm blade and were cultured in adaxial. The leaves were transferred to fresh regeneration medium at 10-day intervals.

2.2. Rooting of The Shoot

The regenerated shoots (3 cm to 5 cm in length) were excised and rooted on MS medium containing NAA at 0.1, 0.2, 0.25, 0.3, or 0.4 mg/L. All media included 3% (w/v) sucrose and 0.8% (w/v) agar. The pH of the media was adjusted to 5.8 after adding the plant growth regulation, and then autoclaved at 121°C for 20 min.

2.3. Regenerated Plantlet

After 1 to 2 months, the root formation was lignified, and the well-developed plantlets were removed from the media. The plantlets were gently washed under running tap water to remove adhering media. For the hardening process, the plantlets were transferred into plastic cups containing a mixture of sterile compost.

The regeneration experiments were designed with 10 Petri dishes for shoot regeneration and 10 triangular flasks for rooting per treatment, and each dish contained at least six leaf explants per treatment for shoot regeneration analysis, and each flask contained three shoots per treatment for rooting frequency analysis. Experiments were done in triplicate. The cultures were incubated at 25°C ± 2 °C under a 16 h photoperiod (45 μmol m^{-2} s^{-1}).

3. RESULTS AND DISCUSSION

The adventitious shoot induction is influenced by the growth regulators concentration and combination. After 10 days of culture, the leaf explants were curled and expanded. Shoot buds differentiated directly from both the cut blades of leaf explants. It took 15-20 days for bud differentiation. All combinations of different concentrations of BA and NAA induced direct shoot buds from leaf explants, especially in the main vein of leaves blades. A combination of BA (0.1 mg/L) and NAA (0.01, 0.03 mg/L) regenerated minimum leaf explants differention frequency (0%). The combination of BA (0.1 mg/L) , NAA (0.05, 0.08, 0.1 mg/L) and BA (0.5 mg/L) , NAA (0.01, 0.03, 0.05, 0.08, 0.1 mg/L) also regenerated lower shoots than others. BA has vital effect on shoots differentiation, lower or higher concentration lead to the lower percent explant regeneration directly. A combination of BA (0.3 mg/L) and NAA (0.08 mg/L) regenerated maximum shoots (5.87) with 90% explants producing shoots shown in Table 1 and Fig.1 A, B, C.

Table 1. Effect of cytokinins (BA) and auxins (NAA) on shoot regeneration of leaf explants of *populus davidiana Dode* × *P. bollena* Lauche

Medium	BA (mg/L)	NAA (mg/L)	Mean No. Shoots of formed per explant	Explant regeneration (%)
1	0.1	0.01	0	0
2	0.1	0.03	0	0
3	0.1	0.05	0.28±0.41	13.4±1.83
4	0.1	0.08	0.06±0.04	5.51±1.63
5	0.1	0.1	0.03±0.03	2.28±1.72
6	0.3	0.01	3.51±0.37	77.7±6.95
7	0.3	0.03	1.62±0.25	59.8±10.0
8	0.3	0.05	0.32±0.25	16.8±5.69
9	0.3	0.08	5.57±0.79	90.5±8.75
10	0.3	0.1	2.03±0.28	54.6±6.09
11	0.5	0.01	0.58±0.15	14.4±3.85
12	0.5	0.03	1.58±0.64	16.9±5.79
13	0.5	0.05	0.72±0.39	40.8±9.9
14	0.5	0.08	0.03±0.03	3.38±3.31
15	0.5	0.1	0.56±0.09	5.46±3.95

Rooting experiment results showed that formation and growth of the new roots were promoted by NAA at any concentrations that have been setting. The percents of root regeneration were 100%. All the tissue culture derived shoots (2-4 cm in length) developed roots within 10-15 days of culture MS medium supplemented with various concentrations of NAA shown in Table 2. The growth of roots was rapid within 30 days. Of the two concentrations of NAA (0.1 mg/L , 0.2 mg/L) evaluated in the present study, the root of growth and development status were fine root , the number of the root reduces and the root of length shorter. The status of the rooting were the length smaller. The number of the root reduces at the concentrations of NAA (0.3 mg/L 0.4mg/L). NAA at 0.25 mg/L was the best for rooting. The root were length and polyrhizal, e.g. see Figure 1D.

Plantlets were taken out of the culture tubes and hardened after two months when the root induction. 90% of plants survived after the hardening process, e.g. see Fig. 1 E, F.

Table 2. Effect of NAA on root regeneration from shoot explants of *populus davidiana Dode* × *P. bollena* Lauche

Medium	NAA(mg/L)	Root regeneration (%)	Rooting situation
1	0.1	100	The number of the roots reduces.
2	0.2	100	The number of the roots reduces, the root of length shorter.
3	0.25	100	The root were length and Polyrhizal .
4	0.3	100	The root of length smaller and the number of the roots reduces.
5	0.4	100	The root of length smaller and the number of the roots reduces.

Fig.1. High frequency plant regeneration A.B, Regenerated shoots from leaf explants; C, Elongation of regenerated shoots; D, Roots regenerated; E, Growth of regenerated plantlets in pots after 20 d of transfer. F. Growth of regenerated plantlets in pots after six months of transfer.

References

1. Marc DB. Factors Influencing the Tissue Culture and the *Agrobacterium* tumefaciens-Mediated Transformation of Hybrid Aspen and Poplar Clones. *Plant Physiol* 1990; **93**: 1110-1116.
2. Sung HS, Richard BH. Multiple shoot regeneration from root organ cultures of *Populus alba* grandidentata. *Plant Cell, Tissue and Organ Culture* 1990; **20**: 53-57.
3. Thakur DK, Srivastava, DK. High-efficiency plant regeneration from leaf explants of male Himalayan poplar (*Populus ciliata* Wall). *In Vitro Cell Dev. Biol. Plant* 2006; **42**: 144–147.
4. Rakesh Y, Pooja A, Dharmendar K, Dinesh K, Neeraj D, Ashok C. High frequency direct plant regeneration from leaf, internode, and root segments of Eastern Cottonwood (*Populus deltoides*). *Plant Biotechnol Rep* 2009; **3**:175-182.
5. Huang DQ, Dai WH. Direct regeneration from in vitro leaf and petiole tissues of *Populus tremula* 'Erecta'. *Plant Cell Tiss Organ Cult* 2011; **107**:169-174.

6. Fillatti JJ, Sellmer J, McCown B, Haissig B, Comai L. Agrobacterium mediated transformation and regeneration of *Populus*. *Mol. Gen. Genet.* 1987; **206**: 192–199.
7. Dai WH, ChengZM; Sargent W. Plant regeneration and Agrobacterium-mediated transformation of two elite aspen hybrid clones from in vitro leaf tissues. *In Vitro Cell Dev. Biol. Plant* 2003; **39**: 6–11.
8. Silvia F, Dora B, Susana S, Maria SP. Morphogenesis induction and organogenic nodule differentiation in *Populus euphratica* Oliv. leaf explants. *Plant Cell Tiss Organ Cult* 2009; **96**: 35-43.
9. Thakur AK, Sarita S, Srivastava DK. Plant regeneration and genetic transformation studies in petiole tissue of Himalayan poplar (Populus ciliata Wall). *Curr. Sci.* 2005; **89**: 664-668.
10. Tzfira T, Jensen CS, Vainstein A, Altman A. Transformation and regeneration of transgenic aspen plants via shoot formation from stem explants. *Physiol. Plant.* 1997. **99**; 554-561.
11. Nermin G, Kasim B, Zeliha I. (1998) Genotype Differencies in Direct Plant Regeneration from Stem Explants of *Populus tremula* in Turkey. *J. For. Res.* 1998; **3**: 123-126.
12. Kang B, Osburn L, Kopsell D, Tuskan GA, Cheng ZM. Micropropagation of Populus trichocarpa 'Nisqually-1': The genotype deriving the Populus reference genome. *Plant Cell Tiss. Organ. Cult.* 2009; **99**: 251–257.
13. Fillatti JJ, Sellmer J, McCown B, Haissing B, Comai L. *Agrobacterium*-mediated transformation and regeneration of *Populus*. *Mol. Gen. Genet* 1987; **206**: 192-199.
14. Sunilkumar G, Vijayachandra K, Veluthambi K. Preincubation of cut tobacco leaf explants promotes *Agrobacterium*-mediated transformation by increasing vir gene induction. *Plant Science* 1999; **141**: 51-58.
15. Horsh RB. A simple and general method for transferring genes into plant. *Science* 1985; **217**: 1229-1231.
16. Han KH, Bradshaw HD, Gordon MP. Adventitious root and shoot regeneration in vitro in under major gene control in an Fz family of hybrid poplar (*Populus trichocarpa x P. deltoides*). *For Genet* 1995; **1**: 139-146.
17. Han KH, Gordon MP, Strauss SH. High-frequency transformation of cottonwoods (genus *Populus*) by *Agrobacterium* rhizogenes. *Can J For Res* 1997; **27**: 464-470.
18. Murashige T, Skoog F. A revised medium for rapid growth and bioassays with tobacco tissue culture. *Physiol. Plant* 1962; **15**: 473–497.

FUNCTIONAL PROPERTIES OF SOY PROTEIN ISOLATE AND ITS APPLICATION IN FOODS

Deyang Sun[a], Weijiang Fan[b,*], Yuyu Xu[c], and Lidong Wang[d]
College of Food and Drug, Shandong Institute of Commerce and Technology, Jinan, China
E-mail: [a]1324655356@qq.com, [b]weijiangfan@126.com, [c]2313078691@qq.com, [d]736539368@qq.com
[*]Corresponding author

Soybean protein isolate is an important plant protein. As a kind of food additive, soy protein isolate is widely used in all kinds of food, such as baking, dairy, meat and other food. We summarized the processing characteristics and measuring indexes about soy protein isolate, meanwhile the application of the soy protein isolate in food was briefly reviewed in this paper.

1. FUNCTIONAL PROPERTIES OF SOY PROTEIN ISOLATE

1.1. Emulsification

Emulsification is an important functional property for soybean protein isolate (SPI). A protective layer can be from on the surface of oil droplets after emulsification of SPI, which can prevent the accumulation of oil droplets and the failure of the emulsion and maintain the stability of the emulsion [1]. A study showed SPI was modified using the amber and phosphorylation, the emulsification was improved and the stability was increased about the emulsion in SPI [2]. The microwave-treatment can also improve emulsification and emulsion stability of the SPI. The process parameters was the power of microwave 1000W, time 40s [3, 4].

1.2. Hydration

Many groups of the main chain have very good hydration in SPI, so SPI has water absorption, water retention and expansion.

1.2.1. *Water Absorption*

Water absorption is the ability to absorb water. Water absorption is related to many factors, such as protein composition, fat content, pH, temperature, density, diffusion driving force of water molecule (steam pressure difference, convection intensity) and so on [5]. The water absorption of soybean protein is closely related to the water activity, and the water absorption of the soybean protein was the change of fast-slow-fast. The water absorption of SPI had sharply difference, for example, the humidity of the SPI was

84% and 99%, and the water absorption capacity was 20g-35g/100g protein and 40g-60g /100g protein, respectively [5].

1.2.2. *Water Retention*

Soy protein has the ability to maintain moisture in the process. The water retention is a series of physical and chemical reactions, such as protein swelling, viscosity increasing, and formation of gel and so on. The water retention of chicken was significantly increased after addition 2.5% the SPI [6]. An experimental results show that these indexes were minimum values if adding 6% SPI into pork, such as centrifugation loss rate, thawing loss and cooking loss. It indicates that the water retention of pork was maintained the optimum state, meanwhile the yields of primary product were improved in pork products [7].

1.2.3. *Expansion*

Expansion comes from the spreading of the protein in solution. The expansion of the protein had significant changes because of the changes of pH, temperature and salt. The expansion of SPI was increased along with the increases of temperature, and SPI had the best expansion at 80℃. the expansion of SPI was the similar values among 70-100℃.

1.3. Oil Absorption

Oil absorption of soybean protein plays two roles in promoting and controlling of fat absorption.

1.3.1. *Promoting Fat Absorption*

SPI can form the emulsion and gel matrix in meat products, which can prevent the fat moving to the surface and play a role of the promoting absorption and binding for fat in meat. The optimal process conditions were obtained in the oil absorption according to different power of microwave treatment of soybean protein. The optimal process conditions were pH 3, 700W of microwave power, 1:7 of material:liquid, 3 min of the modified time. The oil absorption of soybean protein was 2.36mL/g under the optimal process conditions [8]. The oil absorption increased 27.0% through immobilized enzyme of modified soybean protein [9]. The modified soy protein was higher 64.5% than the normal soy protein in the pig produces [10].

1.3.2. *Controlling of Fat Absorption*

The difference effects of SPI were taking on at fat under the different processing conditions of oil. SPI had the capable to control the absorption of fat under certain conditions. The denatured protein can form the oil lamella in the surface of fried pasta, which prevented too much fat absorption in frying [11].

1.4. Gel properties

SPI had a higher viscosity, plasticity and elasticity because of the gel properties. The gel properties of SPI were mainly influenced by protein concentration, temperature, pH and the presence of thiol compounds in solution [12]. A study showed that the gel strength and gel strength increased from (67.185 ±1.028) g to (150.547±6.374) g because of the addition transglutaminase. The optimum parameters were the addition of 40U/g transglutaminase, the temperature of 40℃, pH 7.5, the action time 2.5 h [13].

1.5. Solubility

The protein solutions can play roles in the function of protein, so the soluble SPI was generally used in food processing [14]. The solubility of soybean protein was impacted on by the heated-treatment materials, water content, pH, the coexistence salt or other conditions. Experimental data showed that the solubility of SPI can be increased under pH 4.5-10.0, temperature$<$70℃, the salt concentration$<$0.8mol/L, guar gum 0-0.06%, xanthan gum 0-1.0% or adding SDS [15]. A research showed high pressure-treatment can change the structure of SPI, which can improve the dissolution and viscoelasticity of SPI [16].

1.6. Foaminess

The foaminess of soybean protein was enlarged in volume during process. As amphoteric substance, soy protein had the better foaminess, moreover the foam of soybean protein is relatively stable and bright whiteness [17]. The protein concentration of SPI was influenced the ability and stability of foam. The foam had the good foam ability and foam height when the protein concentration was higher than 1%. When protein was less than 0.3%, the foam was poor. The foaminess of soybean protein was not significantly changed among 0.6-1.5% of protein concentration [17]. The protein had the best foam ability when the purity of protein was high and fat content was low, pH$>$10, 30-45℃of temperature [18].

1.7. Film Property

The films of porous network structure were formed through the interaction of biological macromolecules (proteins, polysaccharides, and fats). The film is a new type of natural edible packaging material and a barrier of water and aqueous agent. Glycerol as plasticizer, the process parameters of SPI films were: temperature 90 ℃ , pH 7.0, protein:glycerol 215:1 [18]. Adding 0.5% pectin and 0.1% glucose, the films of SPI were effectively improved in the strength of film, and reduced the oxygen and water permeability [19].

2. QUALITY INDEXES OF SPI

2.1. The Sensory Quality

The sensory quality of SPI is as follow: bright color, pale yellow, oyster white powder, no visible impurity [20].

2.2. Physicochemical Indexes

The physicochemical indexes of soybean protein should be in accordance with the follow (Table 1):

Table 1. the physicochemical index of soybean protein [20]

Indexes	Quality
water/% ≤	10.0
Crude protein(stem base)/%	50%≤ soybean protein powder ≤65% 65%≤ soybean protein concentrate ≤90% soybean protein isolate ≥90%
Ash（stem base）/% ≤	8.0
Crude fat (stem base)/%	according to the requirement
Crude fibre (stem base) /%	soybean protein powder ≤ 5.0% soybean protein concentrate ≤ 6.0% soybean protein isolate ≤ 0.5%
Urease activity	—

2.3. Hygienic Indexes

The hygienic indexes of soy protein health as shown in Table 2[20]:

Table 2. the health indicators of soybean protein

Indexes	Requirement
Total plate count (cuf/g)	≤30000
Escherichia coli (MPN/100g)	<30
pathogenic bacterium	not detection
Pb (mg/kg)	≤1.0
As (mg/kg)	≤0.5

3. APPLICATION OF SPI IN FOODS

SPI is a high quality plant protein because of containing a variety of essential amino acids (as shown in Table 3). Added SPI, the food can improve the nutritional quality and increase palatability and chewiness during food process.

Table 3. The protein and essential amino acids of soybean products [21]

constituents	Protein content	Lys	Leu	Val	Ile	Thr	Phe+Tyr	Met+Cys	Trp
FAO/WHO	-	5.5	7.0	5.0	4.0	4.0	6.0	3.5	5.0
FNB	-	5.1	7.0	4.8	4.2	3.5	7.3	2.6	1.1
apolipoprotein	56.0	6.9	7.7	5.4	5.1	4.3	8.9	3.2	1.3
protein concentrate	72.0	6.3	7.8	4.9	4.8	4.2	9.2	3.0	1.5
protein isolate	96.0	6.1	7.7	4.8	4.9	3.7	9.1	2.1	1.4

3.1. Application of SPI in Meat Products

As functional and non functional ingredient, SPI can improve the texture and flavor of meat products, and make full use of the imperfect or incomplete side of meat. Soy protein products as a functional factor can also develop some special functional meat products, such as low fat, low sugar, nutrient fortified of meat products.

Soybean proteins were added in the ham sausage, which can strengthen the nutritional value of the ham sausage, increase the content of protein and increase the amounts of essential amino acids, improve the texture and taste of the ham sausage, and increase the product yields, reduce the cost of products [22].

3.2. Application of SPI in Dairy Products

As a new type of functional additives, SPI was added in the dairy products in order to produce a special taste of dairy drinks. It is a trend in the development of dairy industry. A formula of soybean protein-milk powder was designed in the laboratory. The nutritional compositions were determined for soybean protein-milk powder. These results were the basis of the industrialization of soybean protein-milk powder [23, 24].

3.3. Application of SPI in Flour Products

Added soybean proteins, flour products can be improved the protein content and the quality of products. A study showed bread quality was not significantly changed after soy protein (5%) and dietary fiber (6%) were added in bread. The breads could increase the water retention, prolong the shelf life and improve the quality of bread because of adding protein-fiber powder [25].

3.4. Application of SPI in Other Food Products

After acid hydrolysis, soybean protein can be used as the foam stabilizer of spice and beer [26]. Soybean protein can produce milk of soybean protein, coffee of soybean milk, soy cheese, and fruit juice of soybean protein. These drinks had the good quality and taste.

References

1. Feng Zilong, Yang Zhenjuan, Yuan Baolong et al. Production technology and practice of soybean protein isolate. China Oils and Fats 2004; **29**: 29-30 (in Chinese).

2. Guan Junjun, Qiu Aiyong, Zhou Ruibao. Study on the improvement of emulsifying property and emulsifying stability of soy protein isolate. China Oils and Fats 2003; **11**: 38-42 (in Chinese).
3. Hai Rihan, Chi Yujie. Modifying Soy Protein Isolate by Microwave-Assisted Phosphorylation to Improve Emulsification Properties. Journal of the Chinese Cereals and Oils Association 2009; **6**: 36-40 (in Chinese).
4. Cai Jianrong, Zhang Yinzhi, Sun Xiulan. Effects of Microwave Treatment on Functional Properties of Soy Protein Isolate. Journal of Anhui Agricultural Sciences 2009; **2**: 453-454 (in Chinese).
5. Hua Yufei, Gu Yuxing. Water absorption and water holding capacity of soybean protein. China Oils and Fats 1999; **4**: 64-67 (in Chinese).
6. Fang Hongmei, Chen Conggui, Ma Liliang et al. Effects of Soybean Protein Isolate and Ultra High Pressure on Colour, Water-binding Capacities and Textural Properties of Chicken Meat Gels. Food Science 2008; **29**: 129-132 (in Chinese).
7. Liu Yanxia, Zhao Gaiming, Gao Xiaoping et al. Effects of Isolated Soybean Protein and Modified Starch on Ground Pork Meat Water Holding Capacity. Acta Agriculturae Boreali-Occidentalis Sinica 2009; **18**: 48-51 (in Chinese).
8. Wang Lijuan, Wang He. Improvement for oil absorption of soybean protein concentrate by alcohol extraction through microwave-treatment. Cereals and Oils Processing 2008; **2:** 73-74 (in Chinese).
9. Zhang Chunhong, Zhang Ying, Zhao Fei. Study on promoting the water holding and oil absorption capacity of acid leached soy protein concentrate by immobilized transglutaminase. Science and Technology of Food Industry 2011; **3**: 208-210 (in Chinese).
10. Zhang Chunhong, Zhang Ying, Xu Jiao. Application of acid leached soy protein concentrate modified by immobilized transglutaminase in pork sausage. China Oils and Fats 2011; **36**: 18-20 (in Chinese).
11. Hu Xiaozhong, Wen Guangyuan. Current Situation of production technology of low temperature soy meal at home. Science and Technology of Cereals, Oils and Foods 2004; **12**: 32-34 (in Chinese).
12. Xie Liang, Wang Zhang, Cai Baoyu. Composition and functional properties of soy protein isolate. Journal of the Chinese Cereals and Oils Association 2000; **15**: 6-10 (in Chinese).
13. Yu Guopin, An Jing, Chu Yunbin et al. Effect of transglutaminase on characteristics of soybean protein isolate gel. Journal of Northeast Agricultural University 2010; 41: 100-107 (in Chinese).
14. Xing Xiaopeng, Wu Gaojun. Functional properties of soybean protein isolate. Science and Technology of Food Industry 2000; **4**: 74-76 (in Chinese).
15. Xiong Zheng. Solubility of Soy Protein Isolates. Academic Periodical of Farm Products Processing 2013; **8**: 67-70 (in Chinese).
16. Li Biansheng, Zeng Qingxiao, Peng Zhiying et al. Changes of Solubility and Rheological Property of the Isolated Soybean Protein after High Pressure Treatment and Its Mechanism. Chinese Journal of High Pressure Physics 1999; **13**: 22-29 (in Chinese).

17. Liu Shunhu, Du Maolin. Study on the Effect of Extracted Protein Content on Foaming Properties of Soybean Variety Gaofeng No. 1. Journal of Jining Teachers College 2012; **6**: 48-51 (in Chinese).
18. Deng Yangwu, Tian Shaojun. Characteristics of Soybean Protein Isolate. Journal of Zhengzhou Institute of Technology 2004; **25**: 17-21 (in Chinese).
19. Kang Yujie, Ou Shiyi, Li Mingqing. Effect of polysaccharides and cross-linking agents on properties of soy protein isolate-based edible film. Food Science and Technology 2004; **2**: 66-68 (in Chinese).
20. GB/T 20371-2006, Soy protein for food industry. 2006; 6 (in Chinese).
21. Li Yuzhen, Lin Qinlu, Xiao Huaiqiu. Study on application of isolated soybean proteins in meat produces. Meat Research 2006; **1**: 26-30 (in Chinese).
22. Du Hongyan, Ji Qin. Development of soy protein and its use in food industry. Journal of Zhengzhou College of Animal Husbandry Engineering 2002; **22**: 24-26 (in Chinese).
23. Zhang Zongyan. To explore the whole nutrition of soybean milk. Food Science 1995; **3:** 37-41 (in Chinese).
24. Li Dongling, Yang Guiling. Development of soy protein powder. Soybean Science and Technology 2002; **3**: 21-22 (in Chinese).
25. Zhang Guozhi, Liu Jingshun, He Jian. Application study of soybean protein-fiber powder in bread production . Journal of the Chinese Cereals and Oils Association 1998; **3**: 44-47 (in Chinese).
26. Shufang Lei, Fangfang Zhao, Xuecang Zhang. Soybean Seperating Protein Producing Process and Its Applying in Food Industry. Grain progressing 2004; **29**: 53-56 (in Chinese).

RESEARCH ABOUT THE HEALTH DRINK OF *CORDYCEPS MILITARIS* AND KIWI FRUIT

Ran Ai[1,a], Yanling Quan[1,b], and Shengquan Xie[2,c,*]

[1]*Department of Chemistry and Engineering, Uiversity of Science and Technology Liaoning, Anshan, 114051, China*

[2]*Liaoning Economic vocational Technological institute, shenyang , 110122,China*

E-mail:[a]qyl590@sohu.com,[b]qyl590@163.com,[c]xshq4541@163.com

[*]*Corresponding author*

Cordyceps militaris mycelium and Kiwi fruit were used as main raw material in the test, and the proportion of a mixture of Cordyceps militaris and Kiwi fruit, adding quantity, the amounts of Sucrose and other accessories, and the best formula of the cordyceps militaris - kiwi fruit health drink was determined through experiments. The best formula is that the mix proportion of the Cordyceps militaris and Kiwi fruit juice is 1:5,adding quantity of mixed juice 10%, sugar 8%, citric acid 0.2%, CMC is 0.15% .According to this formula ,the sensorial form of the drink is the best , physical and chemical properties are the most stable. The detection results show that the soluble solids is 13.5%, total acid 0.3%, total sugar content 14%, Vc 0.35% in it.

1. INTRODUCTION

Cordyceps militaris, also known as North-Cordyceps sinensis is edible and medicinal nourishing-fungus. It contains active ingredients such as Cordycepin, Cordycepic acid, Cordyceps Polysaccharide, Mannitol, SOD and a variety of amino acids, trace elements and vitamins, which can invigorate the lung and kidney, resist fatigue, tumor and aging, beautify skin and enhance immunity significantly. Kiwifruit contains not only rich vitamin C, A, E and potassium, magnesium, cellulose, also Carotene-calcium, folic acid, lutein, natural inositol and other amino acids, which also can enhance immunity, regulate glucose metabolism ,prevent diabetes and depression[1-10].

2. MATERIALS AND METHODS

2.1. Materials

2.1.1. *Species*

The experimental strain is saved in experimental center of University of Science and Technology Liaoning.

2.1.2. *Culture medium[5,11]*

Liquid medium
20%potato juice, glucose 1.5%, Peptone 1%
PDA-Culture medium
20%potato juice, glucose 1.5%,peptone 1%, agar 2%

2.1.3. *Kiwifruit*

Buying in market

2.2. Methods

2.2.1. *Preparating Culture Medium*

Preparating Liquid Medium
200g peeled potatoes cutted into small pieces was boiled for about 30min and then the mixture was filtrated with four layers of gauze ,sucrose 15g, peptone 10g were added directly into the filtrate, dissolved completely, then added water to 1000ml. 200ml liquid medium was repackaged into 500ml-Erlenmeyer flask, 6-10 teeth of glass beads joined in each bottle which sterilized under 121°C for 30 minutes.
Preparating PDA-Culture Medium
20g agar was joined into 20% potato juice by the above method, and which was boiled again. When agar was dissolved completely ,filling Water to 1000ml.And then the PDA-Culture medium was sub-packed in test-tubes, sterilized 121°C for 30 minutes , put cant spare.

2.2.2. *Preparating Kiwi Fruit Juice*

Kiwi fruit, which was pure in colour, without plague of insect pest, no mildew, no damage were chosen and peeled after rinsed with water. 400 g peeled after rinsed Kiwi fruit was cut and put into a juicer with 500 mL distilled water and then homogenate treated. We filtered the homogenate and added 100 mL distilled water in the filter residue, and then strained after extracting at the temperature of 80 °C for 30 min, the two mixed filtrate, was kiwi fruit juice.

2.2.3. *Activating Cordyceps Sinensis Spawn*

The strain preserved was inoculated onto the PDA slant culture medium, and cultured at the 25°C. When the mycelium was over the slant culture medium, it was turned over again, we selected thick, white, strong and neat mycelium as feed.

2.2.4. *Cordyceps Sinensis Liquid Fermentation[5,11]*

Take a small piece of activated species in the liquid medium access, static culture at 25°C for 24 hrs, shaking culture (speed of 120 r / min, temperature of 22 °C). After 7d the termination of the culture, get Cordyceps fermentation broth.

2.2.5. *Washing and Homogenizing*

We filtered the Cordyceps fermentation broth with a 100 mesh screen, then rinsed the mycelia with distilled water for three times, put it into a juicer with 100 mL distilled water , labeled as Cordyceps juice.

2.2.6. *Process of Cordyceps-Kiwi Health drink[12]*

Mixing the Cordyceps with kiwi juice→concocting→homogenizating→Filling tank→Disinfecting→ cooling down→ finished product.

Mixing and concocting

Cordyceps juice and kiwi juice as the main raw material were mixed together according to appropriate proportion, then other supplementary materials consisting of white sugar, citric acid, stabilizer were added in it on the basis of the appropriate formulation requirements and while stirring.

Homogenizating

Mixed and concocted juice was homogenized with a homogenizer at the pressure of 20MPa, and under the condition of 55 °C for 1 ~ 2 min.

Filling tank

Packaged the above juice into glass bottles at 75 ~ 80 °C.

Disinfecting

Sterilizing the above glass bottles in the condition of atmospheric pressure for 15 min at 95°C.

Cooling down

Cooling them down rapidly with cold water after the package upside down for 3 ~ 5 min.

The finished product

According to the national standards, the inspection is the finished production.

3. THE DETERMINATION OF THE EXPERIMENTAL FACTORS

3.1. The Mixture Ratio of Cordyceps Militaris Mycelium and Kiwi Fruit Juice

Cordyceps has its own unique effect, kiwi has its unique fragrance, this test will be dubbed Cordyceps juice and kiwi juice in different proportions, selected the best ratio.

3.2. Screening Stabilizer

It was used to test single stabilizers such as CMC, gelatin, xanthan gum on the single factor test.

3.3. Determining the Best Recipe

Orthogonal experiment was carried out according to table 1 to determine the best ratio by the experimental factors such as the amount of mixed juice, sugar, regulating acid agent and stabilizer .The production was made a comprehensive score by some sensory index such as color, smell, taste, organization state, and etc.

4. RESULTS AND ANALYSIS

4.1. Affect to the Health Drink in Different Mixed Proportions of Cordyceps and Kiwi juice (Table 1)

Table 1. The proportion of the mixed juice

The proportion of the mixed juice	Color	Smell	Taste	Organization State
1:7	light green	fruit of kiwifruit thick	sourer	unstable and layered
1:5	soft	fruit of kiwifruit thicker	sweet and sour moderate	stable and no-layered
1;3	colorless	fruit of kiwifruit thin	Sweet lighter	unstable and slightly layered

As can be seen from Table 1, Cordyceps juice: Kiwi juice of 1: 5, the organoleptic state of the product is the best.

4.2. A single Stabilizer's Impact on the Stability of the Product

Table 2. A single stabilizer's stable effect on product

sort	0.10%	0.15%	0.20%
CMC	not stable, layered	not stable,slightly layered	stable, not layered
gelatin	not stable, layered	not stable, slightly layered	not stable, layered
Xanthan gum	not stable, slightly layered	not stable, layered	not stable, layered

As can be seen from Table 2, the stabilizing effect of gelatin and xanthan gum to the product were not good, only CMC stabilizer in an amount of 0.20%, the stabilizing effect was the best.

4.3. Determination of the Best Recipe

Orthogonal experiment was carried out according to table 3,4,5 to do L9(34) orthogonal test to determine the best ratio by the experimental factors such as the amount of mixed juice, sugar, regulating acid agent and stabilizer. The production was made a comprehensive score by some sensory index such as color, smell, taste, organization state, and etc.

Table 3. The level table of orthogonal experimental factor

level	A(amount of mixture)%	B(amount of sugar)%	C(amount of acid agent)%	D(amount of stabilizer)%
1	10	8	0.1	0.10
2	20	10	0.2	0.15
3	30	12	0.3	0.20

Table 4. The standard table of sensory score of the cordyceps militaris & kiwi fruit health drink

grade	color (10)	smell (15)	taste (40)	organization state, (35)
1	soft 10~8	Fruit of kiwifruit thick 15~13	Sweet and sour and lubrication sense 40~35	Stable, not layered 35~28
2	Light green 7~5	Fruit of kiwifruit thicker 12~9	Sweet and sour moderate and throat feeling poor 34~26	More stable, slightly layered 27~16
3	colorless 4~1	Fruit of kiwifruit thin 8~1	Sweet and sour disorders25~1	not stable and stratified precipitation slightly 15~1

Table 5. L9 (34) orthogonal experiment results

No. of test	A(amount of mixture)%	B(amount of sugar)%	C(amount of acid agent)%	D(amount of stabilizer)%	Sense score
1	1	1	1	1	86
2	1	2	2	2	90
3	1	3	3	3	81
4	2	1	2	3	76
5	2	2	1	2	71
6	2	3	3	1	71
7	3	1	3	2	61
8	3	2	1	3	61
9	3	3	2	1	56
K1	85.6	74.3	72.6	71.0	
K2	72.6	74.0	74.0	74.0	
K3	59.3	69.3	71.0	72.6	
R	26.3	5	3	3	
Optimum level	A1	B1	C2	D2	

As can be seen in Table 5 for determination of optimum process conditions of the drink. The main factors affecting the quality of the product were the amount of mixed juice, sugar, citric acid, and CMC. The best combination of process parameters is A1B1C2D2.That is to say when the mixed juice was 10%, sugar 8%, citric acid 0.2%, and CMC 0.15% the quality such as color, smell, taste and organization state were the best.

4.4. The Quality of the Product

The physical and chemical analysis was carried out on the finished product. The results showed that soluble solids content was 13.5% (refractive measurement method); total sugar 14% (film assay reagent); total acid 0.3% (sodium hydroxide titration); Vc 0.35% (2,6 - Dichlorophenolindophenol assay).

References

1. Liu -Jingming, Zhong -Yurong, Yang Zhi, Cui Shulian and Wang Fuhua.Study on the chemical ingredients in cordyceps militaris[J]. *China Journal of Chinese Materia Medica*,1989;**14**(10):32-33.
2. Miyazaki J, Naoko O.H Yamada. Studies on Fungal polysaccharides XX, Galactomannan of Cordyceps sinensis[J].*Chem Pharm Bul*,1997;**25**(12):3324-3328.
3. KihoT, H Tabata, S Ukai. A minor protein containing galactomamian from a sodiumcarbo -mate extract of Cordyceps sinensis[J]. *Carbohydrate Research*,1986;**156**:189-197.
4. Yamada H, Kawaguchi N, Ohmon T, et al. Stractiue and activity of analkali-soluble from Cordyceps ophioglossoides[J].*Carbohydrate Research*,1994;**125**(l):107-115.
5. QUAN Yanling, WANG Lin, XU Yunjian, XIE Shengquan, Production of Cordycepin by liquid fermentation of Cordyces militaris China Brewing, 2013,**32**(2):99-100.
6. Ding-fanglin, Dong-yisheng, Liu-yulan, Development of Ganoderma & fruit health drink Modern Agricultural Science and Technology,2011;(16):87-89.
7. Zhang-sulan, Sun-yunhan and etc. Study on Chinese caterpillar fungus and artificial cordyceps mycelia against Lewis lung cancer of mice[J]. *Bulletin of Chinese Materia Media*,1997,**12**(2):53-54.
8. Liu-jie, chen-yizh, yang-xu. Study on cicadas Cordyceps anti-tumor effect[J]. *Journal of Norman Bethune University of Medical Science*,1992;**18**(5):423-425.
9. Lou-yaqing. Research about cardiovascular pharmacology of artificial Cordyceps mycelium alcohol extract and fermentation broth [J].*Chinese Herbal Medicines*,1996;(5):17-21.
10. Liu-weizhu, Chen-yiping. Preliminary observation on 18 cases of Cordyceps sinensis fungus treatment to chronic renal failure[J]. *Chinese Traditional Patent Medicine*,1995;(8):19-20.
11. Li-liande, Li-zengzhi, Fan-meizhen. Study on Submerged Fermentation of Cordyceps and Cordyceps Polysaccharide[M]. *Journal of AnHui Agricultural University*, 1998;25-36.
12. MIAO Jing-zhi, LI Zhen-yu, DONG Yu-wei. Development of Ganoderma lucidum, mushroom mixed fermented beverage Food Science, 2013;(22):55-57 .

RECOVERY IRON FROM A COPPER FLOTATION TAILING BY NAOH MELTING AND ACID LEACHING PROCESSING

Liangmou Yu[a]

Environmental engineering, Kunming Metallurgy College, Kunming 650033, Yunnan, China
E-email:[a]yullmm613@163.com

An innovate craft of NaOH melting-acid leaching processing was investigated for recovery of iron from a copper flotation tailings. This tailing was of a hazardous industrial waste, which contained 51.34% Fe and high content of S, Cu, Pb, Zn. The results demonstrated that iron can be increased to 62.15% with an iron recovery of Fe 87.64% under the optimal conditions (a melting temperature of 200°C, a melting time of 90 min, a NaOH concentration of 1.5 mol/L and a HCl concentration of 10%), meanwhile, the content of S, Cu, Pb, Zn can be decreased to 0.12%, 0.15%, 0.17% and 0.12% respectively in the final iron concentrate. This craft can be used to utilize copper flotation tailings and produce qualified concentrate as iron-bearing feed for steel industry, which will help to solve the pollution of tailings and extend raw material sourcing for Chinese steel industry.

1. INTRODUCTION

Flotation tailing is a typical industrial wastes when separate valuable mineral form gangue mineral with froth flotation. Generally, tailings was disposed or pile-up. Undoubtedly, the abandoned flotation tailing not only occupy the land, but also cause serious environmental pollution because they usually have high content of heavy metals, such as Cu, Pb, Zn, Au, Ag etc[1,2]. With the continuous and increasing exact demands of environmental protection, comprehensive utilization of tailings to produce of value-added products has been caused everyone's attention. This action will not only prevent pollution of heavy metals but also increase the resource utilization rate.

Recently, many studies have been reported on the use of waste resources such as smelting slags and tailings [3-5]. For the recovery of iron from wastes, gravity separation, froth flotation, high gradient magnetic separation and magnetizing roasting followed magnetic separation are the conventional mineral processing and are performed by many researchers. The results indicated that it is often difficult to recovery the valuable components from this tailings or smelting slags by physical separation, because the valuable components were usually fine grained and disseminated with gangues closely. Sodium salts reduction roasting followed magnetic separation is effective to upgrade iron element and decrease the impurity content, which is potential to processing the smelting slags or tailings, but, its major drawback was the high energy consumption.

In this paper, an innovate craft consist of NaOH melting and acid leaching processing was introduced to recover the iron and remove impurities from a copper flotaiton tailing. Namely, the tailing was firstly treated with NaOH solution under the middle-low temperature to dissolve the impurities, such as silica, alumina etc., resulting in the upgrade of iron element; then, the melted product was leached by sulfuric acid for removal of heavy metals. And a qualified concentrate was obtained. This concentrate obtained from this craft have high grade of iron and low content of impurity, which can be used as iron-bearing feed material for steel industry.

2. MATERIALS AND METHODS

2.1. Materials

A copper flotation tailing was collected from a tailings dam in Yunnan, China. Main chemical composition of the sample was given in Table 1.

Table 1. Main chemical composition of copper flotation tailing (mass-%)

Components	Fe	Cu	Pb	Zn	SiO_2	Al_2O_3	MgO	CaO	P	S
Wt (%)	51.34	0.37	0.59	0.64	12.5	8.2	0.12	0.15	0.016	1.82

From the Table 1, the iron grade is 51.34% and the contents of SiO_2 and Al_2O_3 are 12.5% and 8.2% respectively. These impurities oxides will be detrimental for the iron recovery. Meanwhile, the content of Cu, Pb, Zn and S is high, which make this tailing potential hazardous for stockpiling.

2.2. Experimental Methods

Firstly, a certain amount of sample was mixed with NaOH solution (S/L=1:1) sufficiently in a graphite crucible, melting in the muffle furnace under the middle-low temperature for a period time. After melting, the melted residue was cooled down to room temperature. Powders of melted residue was leached in distilled 60°C water with mechanical stirring (400rpm) for 30minutes for removal of soluble sodium silicate (Na_2SiO_3) and sodium aluminate ($NaAlO_2$). Then, the water leached residue was leached by HCl to remove heavy metals. Finally, the acid leached residue was filtered, washed and dried. The NaOH and HCl used as leaching agent in the experiments were of reagent grade.

3. RESULTS AND DISCUSSION

3.1. Effect of NaOH Concentration on Upgrading of Iron Grade

To investigate the effect of NaOH concentration on the upgrading of iron grade during melting processing, different NaOH concentration experiments were conducted under the following conditions: a melting temperature of 200°C, a melting time of 90 min, and a liquid–solid ratio of 1:1. The results were shown in Fig .1.

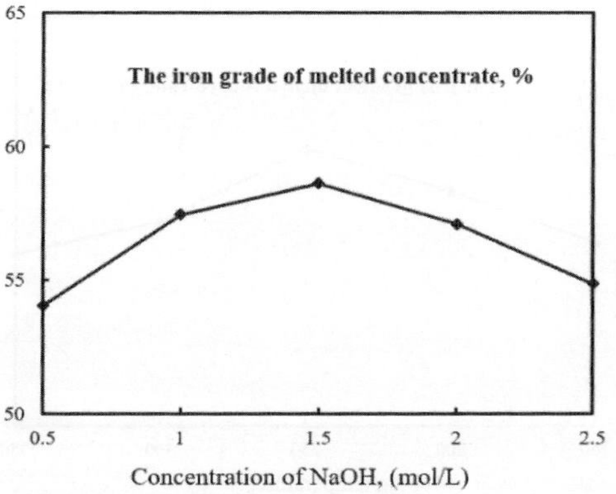

Fig. 1. Effect of NaOH concentration on the upgrading of iron grade in the melted residue.

The results in Fig.1 suggested that the iron grade in the melted residue increased when the NaOH solution concentration increased from 0.5 to 1.5 mol/L. With a further increase in NaOH solution concentration, the iron grade of the melted concentrate began to decline. The reason for this phenomenon was an aluminosilicate hydrate salt of the alkali metal would be precipitated out in the form of gel on the condition of high NaOH solution concentration, resulting in the decrease of iron grade. Typical reactions during NaOH melting processing included the following [6]:

$$SiO_2 + NaOH \rightarrow Na_2SiO_3 + H_2O \quad (1)$$

$$Al_2O_3 + NaOH \rightarrow NaAlO_2 + H_2O \quad (2)$$

$$Na_2SiO_3 + NaAlO_2 + NaOH + H_2O \rightarrow [Na_a(AlO_2)_b(SiO_2)_c \cdot NaOH \cdot H_2O] \quad (3)$$

Hence, the optimal NaOH concentration is recommended at 1.5 mol/L.

3.2. Effect of Melting Temperature on Upgrading of Iron Grade

Melting temperature plays an important role in the NaOH melting processing. The effect of melting temperature on upgrading of iron grade was investigated under the conditions as following: a solid-liquid of 1:1, a melting time of 90 min and a NaOH concentration of 1.5. The experimental results were given in Fig .2.

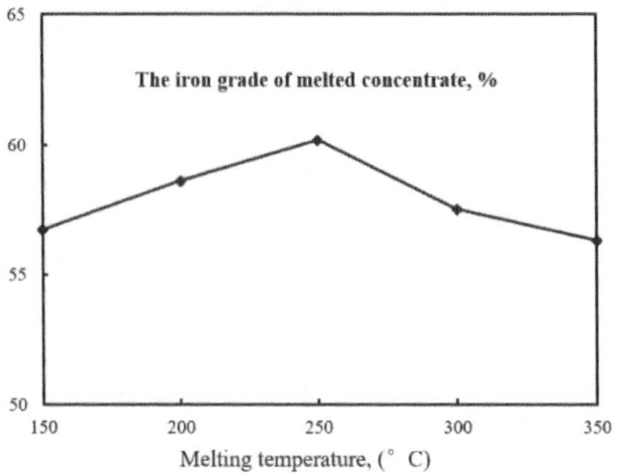

Fig. 2. Effect of melting temperature on upgrading of iron grade.

It can be seen from Fig .2 that increase of the melting temperature was effective in improving the iron grade in the melted concentrate when the temperature was below 250 °C. A possible explanation for this behavior was that increasing temperature would speed up both the reaction and diffusion rate of reactants and reaction products. However, when the temperature was increased further from 250 to 350 °C, part of Fe_2O_3 may be converted into soluble matter. Thus melting temperature of 250 °C was chosen to be optimal. After the process of melting and water leaching, the main chemical composition of melted residue was shown in Table 2 as below. The results indicated the iron grade is 60.17% and the main impurity element contents (S, SiO_2 and Al_2O_3) meet the requirements of iron concentrate, but the content of Cu, Pb and Zn is high.

Table 2. Main chemical composition of the melted concentrate (mass-%)

Components	Fe	Cu	Pb	Zn	S	SiO_2	Al_2O_3	MgO	CaO
Wt (%)	60.17	0.35	0.61	0.68	0.16	3.51	0.35	0.15	0.24

3.3. Effect of Acid Leaching Conditions on Melted Concentrate

In order to removal the heavy metals in the melted concentrate, acid leaching experiments were conducted. Hydrochloric acid (HCl) was used as the leaching agent at room temperature with mechanical agitation at the speed of 400 rpm. The leach residue named iron concentrate was washed and dried. The effect of hydrochloric acid concentration on removal efficiency of heavy metals was investigated under the liquid-to-solid ratio fixed at 2:1. The results were shown in Fig. 3.

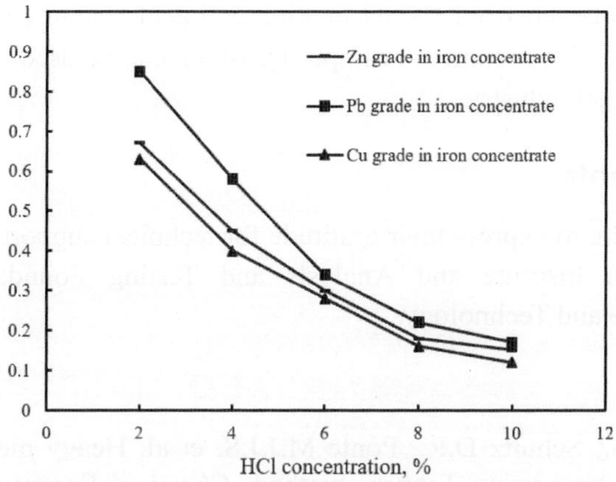

Fig. 3. Effect of HCl concentration on the removal efficiency of heavy metals.

As shown in Fig .3, the content of heavy metals in the iron concentrate decreased as the acid concentration increased. When the concentration of HCl was 10%, the leaching efficiency of Zn, Pb, Cu reached stabilization and maximum nearly. As a result, the final iron concentrate having TFe 62.15% with a total recovery about 87.64% was obtained. The elements analysis results of iron concentrate was shown in Table 3 as below. It can be seen from Table 3 that the content of Cu, Pb and Zn is 0.15%, 0.17% and 0.12% respectively, which can be meet the requirements of iron concentrate. Thus, this concentrate obtained from this craft have high grade of iron and low content of impurity, which can be used as iron-bearing feed material for steel industry.

Table 3. Elements analysis of iron concentrates (mass, %)

Components	Fe	Cu	Pb	Zn	S	SiO_2	Al_2O_3	MgO	CaO
Wt (%)	62.15	0.15	0.17	0.12	0.12	2.37	0.21	0.12	0.20

4. CONCLUSION

On the basis of experimental results, the following can be concluded.

(1) Effect of NaOH concentration on upgrading of iron grade indicates that a medium NaOH concentration during the NaOH melting processing is necessary for the removal of SiO_2 and Al_2O_3, resulting in an increase of iron grade, the recommended NaOH concentration is 1.5 mol/L.

(2) Melting temperature and HCl concentration have a great effect on the iron grade and heavy metals removal efficiency of the present tailing. The optimal melting temperature and HCl concentration are 250 °C and 10% respectively.

(3) Under the optimum operating parameters, an iron concentrate (TFe 62.15%) with a total 87.64% recovery was obtained from a copper flotation tailings. The

concentrate obtained from NaOH melting and acid leaching processing has high grade of iron and low content of impurity, which can be used as iron-bearing feed material for steel industry.

Acknowledgements

The authors would like to express their gratitude for technical support from the Kunming Metallurgy Research Institute and Analysis and Testing Foundation of Kunming University of Science and Technology.

References

1. Kaminari N.M.S., Schultz D.R., Ponte M.J.J.S. et al. Heavy metals recovery from industrial wastewater using Taguchi method. *Chemical Engineering* Journal 2007; **126**:139–146.
2. Abdel Basir S.M., Rabah Mahmoud A. Hydrometallurgical recovery of metal values from brass melting slag Hydrometallurgy (1999); **53**:31–44.
3. Veglio F., Quaresima R., Fornari P., Ubaldini S. Recovery of valuable metals from electronic and galvanic industrial wastes by leaching and electrowinning. Waste Management 2003; **23**: 245–252.
4. Liu J., Wen S.M et al. Process optimization and Reaction Mechanism of Removing Copper From an Fe Rich Pyrite Cinder Using Chlorination Roasting. Journal of iron and steel research, international 2013; **20**(8) :20-26.
5. Jiang K.Q., Guo Z.H., Xiao X.Y., Zhang L. Extraction of metals from a zinc smelting slag using two-step procedure combining acid and ethylene diaminetetraacetic acid disodium. J. *Cent. South Univ* 2012; **19**:1808−1812.
6. He B.B., Tian X. K., Sun Y. et al, Recovery of iron oxide concentrate from high-sulfur and low-grade pyrite cinder using an innovative beneficiating process. Hydrometallurgy 2010; **104**:241–246.

ANALYSIS OF COMMUNITY NURSING SERVICE OF COMMUNITY RESIDENTS

Wenjuan Zhong[a], Linzhi Yu, Men Tian, Xaolong Wei, and Lian Yuan

Department Health Science and Nursing of Wuhan Polytechnic University, China
Email: [a]wjzhong8616@126.com

The objective of the study was to understand the community residents' needs for community nursing service and the characteristics of different groups of community residents' need for community nursing care, and to provide the theory basis for the community nursing care management. We prepared paper questionnaire on 616 community residents in Wuhan, from which we found that their satisfactory needs differ a lot, especially in terms of their degree of education. Results suggested that community nursing services should be carried out according to the characteristics of community residents' needs.

1. INTRODRUCTION

Along with medical system reform and people's health care improvement in China, the community nursing has become into a new orientation for nursing career in new era [1]. Community nursing is a combination of nursing with public health, providing a continuous, dynamic, general nursing service and improving personal health [2]. Due to the further reform of community health services, general practitioners and nurses in community, to a certain extent, take up health education, disease prevention, health care and rehabilitation nursing outside hospitals [3-4]. As an efficient, economic and convenient way of community health services, community nursing is also a new type of nursing way adapted to the biological - psychology - social medical model [5]. The study is made to explore a community nursing mode suitable for China's conditions and provide theoretical basis for management of community health through the investigation on the needs of community nursing by the community residents.

2. OBJECT AND METHODS

2.1. The object of Study

616 community residents were chosen as the research object from Changqin garden, Gushaoshu, Zhanggongdi in Wuhan of Hubei province, including 325 male and 285 female adults. Of which, 244 people are in the Young group(18 to 29 years old), 267 people in the middle age group (30 to 49 years old), 56 people in the menopause group (50 to 59 years old), 38 people in the elderly group (over 60 years old); in terms of

education degree, 24 people are of primary school, 82 people of junior middle school ,153 people of senior high school and 342 people of university or above; and in terms of marriage, 380 people married, 220 people unmarried, 8 people for else.

2.2. Investigation Methods

Self - prepared paper questionnaire on community nursing service, 616 questionnaires were randomly distributed to community residents.

2.3. Statistical Analysis

All analysis was conducted using SPSS, version 12.0.

3. RESULTS

3.1. Community Residents' Satisfactory Requirements for Community Nursing

The results of community residents' satisfactory requirements for community nursing show that the needs of chronic care for elderly patients of community residents was the most wanted.

3.2. Comparison of Community Nursing Contents in Different Groups of Community Residents

There is no significant difference ($P > 0.05$) among community residents in terms of gender, age and marriage for community nursing service content.

Table1. Sample characters of community nursing contents by community residents of different education degree

community nursing contents	Primary school(n=24) Number %	Middle school(n=82) Number %	High school (n=152) Number %	College (n=342) Number %	X^2	P
Psychological nursing	10 41.7	26 31.7	40 26.3	150 43.9	15.332	0.002
Home basic nursing	8 33.3	34 41.5	51 33.3	167 48.8	11.537	0.009
Hospice care	12 50.0	36 43.9	49 32.2	165 48.2	11.381	0.01
Neonatal nursing	15 62.5	34 41.5	79 51.6	206 60.2	11.049	0.01
Monitoring growth and development of children	12 50.0	32 39.0	78 51.0	194 56.7	8.621	0.035
prevention of Common disease for children	19 79.2	55 67.1	83 54.6	225 65.8	8.966	0.03

Table 1 presents sample characters by community nursing contents by community residents of different education degrees. Table 1 showed that there were significant differences in the psychological nursing, home basic nursing, hospice care, neonatal care, monitoring of growth and development of children, and prevention of common disease of children.

4. DISCUSSION

The results showed that the need of chronic nursing service for the elderly patients among community residents is the most wanted. This is mainly due to increasing number of China's elderly population and the elderly patients with hypertension, heart disease, high cholesterol, diabetes and digestive disease, which weakened their ability to care for themselves. At the same time, the rising health care costs in recent years and restrictions on using health insurance also lessened the times the elderly to see a doctor and check-up health. In addition, with the fast pace of life and work, young people have little time to take care of the elderly. So community should be strengthened elderly nursing and establish a perfect community elderly-care system, and more specialists in the field of community elderly nursing [6, 10] are required to meet the needs of the elderly community residents. The results also showed that the community residents of different education degree demand the different contents of community nursing. The higher the education degree of the people, the more needs in psychological nursing, home basic nursing guiding, hospice care, neonatal care, children's growth and development monitoring. Because of economy development, the fast pace of life and complicated interpersonal relationship, the community residents of high education community are inclined to suffer from intensive stress and mental health [7]. Besides, these people have good cognition of the community nursing service[8], and pay great attention to the mental health and pediatric healthcare. In addition, the people who received less education also have higher requirements in psychological care, hospice care and neonatal care, children's common diseases prevention, and they are easily to be involved in confusion due to poor economic condition and unemployment in China. So the community should provide special community nursing services according to the characteristics of different groups. Community nursing will become a new orientation for nursing career in the 21st century. In order to meet the demand of residents for community nursing services and improve the level of the people's health, our country should establish the community nursing services institutions, form a community nursing system [9-10], strengthen the propaganda of community nursing, cultivate high-quality community nursing staff, establish community nursing suitable for China's national conditions and promote the development of community nursing in China [11].

Acknowledgment

The authors wish to acknowledge the support receive from the major project of philosophy and social science in Hubei Province (Research of the interactive development between the higher education and community, 2015z33).

References

1. X.C.Li, P.Ye. The development trend of community nursing in China. *Journal of Nursing Administration*, 2008, **8**(5) :27.
2. L.Yan. Elderly nursing. Beijing: people's medical publishing house, 2003;1- 4.
3. X.S.Liu, S.F.Bai, S.S.Yang etc. Strengthen the construction of the whole family physician team to promote the community health service. *Chinese Journal of Hospital Admininstration*, 2000 (11): 694-695.
4. L.Y.Chen, Z.He, H.Zhao etc. The forecasting research of professional role and qualifications of community health nurses. *Chinese Journal of Nursing*, 2006; **9** (6): 490-493.
5. Y.Y.Zheng, J.B.Qu. Discussion on problems and countermeasures of community nursing in China. *Soft Science of Health*, 2008; **22** (1):68-70.
6. H.Zhang. The situation of elderly population and countermeasures of community geriatric nursing in China. *Journal of community Medicine*. 2008;**6**(3):62-63.
7. W.J.Zhong, J.R.Fang, X.G.Ming. Investigation on mental health status of community residents in Wuhan. *Chinese Journal of Public Health*, 2009; **25** (3):279-280.
8. L.M.Yang, W.J.Zhong, T.Sun etc. Attitude of different groups of reseident toward the community nursing care. *Journal of Public Health and Preventive Medicine*, 2008; **12** (5):66-67.
9. W.Li. The development and it's strategy on community nursing in China. *Journal of Nurse Training*, 2001; **19**(7):551-553.
10. S.P.Lu, Y.J.Xie. The ageing populations and community nursing in China. *Chinese Journal of Gerontology*, 2005; **25** (1):112-113.
11. L.H.Yang, R.L.Yan and H.L.Zhang. Problems and countermeasures of community nursing. *China Foreign Medical Treatment*, 2008, **27** (16):100-111.

MOISTURE DISTRIBUTION AND DYNAMIC CHANGES OF CORN BRAN POLYSACCHARIDES (CBPS) POWDER MONITORED BY LF-NMR

Xuye Liu[a], Shuailing Yang[b], and Songyi Lin[c,*]

College of Food Science and Engineering, Jilin University, Changchun, 130062, China

E-mail: [a]liuxuye007@163.com, [b]yshling826@163.com, [c]linsongyi730@163.com

[*]Corresponding author

The aim of this work is to study the moisture state in the corn bran polysaccharides (CBPS) powder using low field nuclear magnetic resonance (LF-NMR) during the storage process at 25 °C and 4 °C. The CBPS powder was placed under the condition of relative humidity 75 % in a sealed box, and the distribution and dynamic changes were monitored by LF-NMR. The results indicated that mainly three transverse relaxation time fractions existed in CBPS standing for bound water, immobilized water and free water with the initial ratio of 40.35:14.45:45.11. And with the standing time changing, the total moisture content had increased significantly ($p<0.05$) from 6429.586 (Au·ms)/g to 23678.64 (Au·ms)/g at 25 °C and the sample at 4 °C nearly had no obvious change, only inconspicuous transformation occurred in the sample. From the in-depth analysis, we found that the moisture in the sample had an obvious exchange with the external environment at 25 °C and mainly had a dynamic change with itself at 4 °C. This study provided a detail exploration of moisture distribution and dynamic changes of CBPS powder and had a guiding significance to certain for the storage of CBPS powder.

1. INTRODUCTION

Corn bran is an excellent source of polysaccharides with many biological activities, such as immune regulations, intercellular recognitions, intercellular material transport, etc[1]. A large number of facts revealed that polysaccharide is a kind of informational molecule in organism, which take part in the reactions of physiological and pathological procedures. As a matter of fact, polysaccharide is a kind of bioactive macromolecule substance, composed by many monosaccharides, exactly ten or more than ten, linked with glycoside bonds[2]. Nowadays, polysaccharides have been found widespread in animals, plants and fungus as the living matter[3]. More and more functions of polysaccharides have been discovered. As a kind of functional active substances, CBPS is being studied as a hotspot. However, the state of the organizational structure and its biological activity will have a corresponding change during the storage process, moisture content is one of the most vital factors[4]. So having a further understanding of the moisture distribution and dynamic changes in detail is very important[5]. LF-NMR has been widely applied in moisture detection as a nondestructive and non-invasive technique with high precision which requires a small amount of sample[6]. LF-NMR can measure the material through the organization of the length of the hydrogen proton relaxation time to determine the distribution of moisture state[7]. At the same time, LF-NMR can monitor the content of various components of water according to the quality of the peak area has a positive correlation with the moisture content[8].The objective of this study was expected to reveal the distribution and dynamic changes of moisture in CBPS and to provide a positive guiding for the storage of CBPS powder.

2. MATERIAL AND METHODS

2.1. Preparation of CBPS

CBPS powder samples were obtained from the Laboratory of Nutrition and Functional Food, Jilin University and prepared as follows: the corn bran (water content 30 %) was pretreated by microwave drying and ultra micro-pulverization. Corn bran powders were dissolved in distilled water with the liquid solid radio of 40:1 (v/w). The mixture was treated by PEF under the condition of electric field intensity of 25 kV/cm, electric frequency of 2080 Hz. The processed samples were then heated in water bath to the temperature of 59 °C, and 1.5 L Celluclast (enzymatic activity 700 EGU/g) was added to the solution. The ratio of enzyme to material (w/w) of the mixture was adjusted to 1.55 %. Then the mixture was persistently stirred at 59 °C for 2.5 h, using 1 M NaOH to keep the pH within 5.00±0.02. Subsequently the temperature was raised to 100 °C rapidly keeping for 10 min to inactivate the enzymatic activity in order to terminate the enzymatic reaction . The mixture was then cooled to room temperature and centrifuged at 10000 rpm for 15 min. The supernatant was separated and purified for 1.5 h by macroporous adsorption resin to get rid of protein and fat under the condition of pH 6.00, mass concentration 3 g/L, temperature 40 °C. In the end, the collection liquid was taken to free drying for 24 h.

2.2. Measurement by LF-NMR

A MiniMR-40 (magnetic field intensity, 0.4 T; Rf pulse frequency, 1-30 MHz) micro-nuclear magnetic resonance (NMR) analyzer 20-25 (Niumag Co., Ltd.,Shanghai, China) was used to carry out LF-NMR measurements. In order to study the moisture distribution and migration, approximately 0.4 g CBPS powder were separately weighted and then placed into two bottles with the diameter 2 cm and height 4 cm to ensure 1/2 to 2/3 volume of the bottle was covered by CBPS powder to make sure a stable result of the measurement. Then the two bottles were placed into a sealed box with the inside temperature 4 °C and 25 °Crespectively, relative humidity 75 % and saturated sodium chloride solution was used to make a constant relative humidity. The T_2 maps were acquired respectively at 60 min, 120 min, 180 min, 240 min, 300 min, 1440 min and 1980 min. Magnetic resonance imaging of Carr Purcell Meiboo M-gill (CPMG) sequence was used for the acquisition of each sample signal. The CPMG sequence parameter settings was as follows: corresponding resonance frequency (SF) for 18 MHz; offset frequency (O1), 151360.27 Hz; time of echo (TE), 0.3 ms; echo count, 8000; waiting time (TW), 800 ms; sampling number (TD), 239962; pulse widths at 90° (P1), 6 μs; pulse widths at 180° (P2), 9.6 μs, respectively; radio frequency delay time (RFD), 0.05 ms; analog gain (RG1), 30 db; and digital gain (DRG1), set 3[9].

2.3. Data Analysis

The data and distributed exponential curve fitting was analyzed by MultiExp Inv Analysis software (Niumag Co., Ltd.,Shanghai, China). The fitted curve equation was as in Eq. (1) [10].

$$A(t) = \sum_{i=1}^{n} A_{2i} \exp\left(\frac{-t}{T_{2i}}\right) + e(t) \tag{1}$$

where $A(t)$ is the residual magnetic field intensity of the pulse signal intensity values for a settled time t; A_{2i} is the peak area of the i th component; T_{2i} is the time of the i th; $e(t)$ is the residual error of the equation. For a better fitting relationship, the multi-exponential analysis was used for the fitting analysis according to an established model. Each detection was performed for three times, every performance was carried out with 2 min.

The moisture content is normal correlation with the peak area of unit mass, so the peak area of unit mass is selected as the symbol of moisture content, as in Eq. (2) [11].

$$\text{Signal per mass (Au.ms)} = \frac{A_{2i}}{m} \tag{2}$$

where *signal per mass* is peak area of unit mass; A_{2i} is the peak area of the i th component; m is the mass of the samples.

3. RESULTS AND DISCUSSION

3.1. Water Distribution in the CBPS

The changes of moisture have an important influence on the quality of CBPS powder during the storage, even cause spoilage. So the LF-NMR was used to detect the distribution of moisture in CBPS powder at 25 °C and 4 °C respectively. In order to obtain an in-depth understanding of the morphological condition of moisture, the T_2 relaxation time distribution curves at different standing times were analyzed by MultiExp Inv Analysis software. Due to different compositions of moisture show a various binding capacity with the inner material structure, the moisture was identified as three different morphological states: bound water, immobilized water, free water. The simultaneous iterative reconstruction technique (SIRT) method was applied to solve the spectrum. The moisture content was distinguished from each other according to the wave coverage of the area. Fig.1 shows the changes of amplitude curves of CBPS powder with different relaxation time. Three peaks were obviously found in the two-dimensional image for each curve which means there were three kind of moisture in the samples. As shown in Fig.1, T_{21} firstly occurred with the relaxation time almost from 0.1 ms to 1 ms, it is the fastest fraction which occupied a large proportion of 40.35 %. As observed with other materials[12], it was the characteristic spectrum of bound water. T_{22} was the second fraction with the relaxation time nearly from 1 ms to 10 ms which accounted for 14.45 % of the total signal. In other study[13], immobilized water had a T_{22} relaxation time from 1 ms to 20 ms, the T_{22} was in the limited range. And T_{23} was the last fraction of the three proton fractions; the relaxation time was approximately from 10 ms to 100 ms with the proportion of 45.11%. As the results shown in Fig.1, the initial state of the moisture in CBPS powder was bound water, immobilized water and free water with the ratio of 40.35:14.45:45.11. However we still had no exactly understanding of the change of moisture content in the sample as the time change under different temperature.

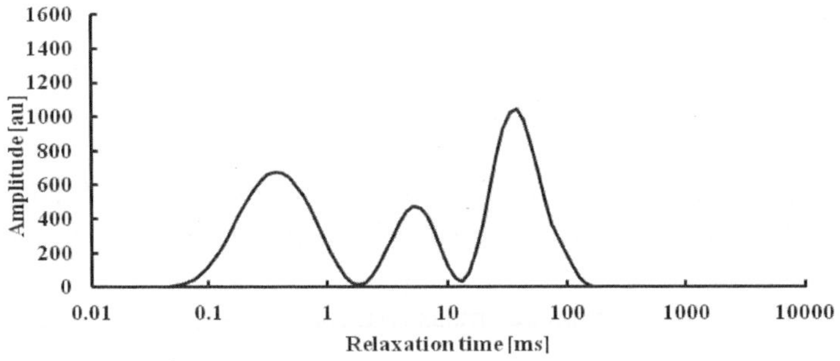

Fig.1. Distribution of the moisture monitored by LF-NMR.

3.2. Moisture State under Different Temperature

For a further understanding of the moisture content in different external storage condition, the moisture state was detected at 25 ˚C and 4 ˚C. The T_2 relaxation time distribution curves at different standing times showed the results of the detection. As the description in previous method 2.2, the original CBPS powder (bound water, immobilized water and free water with the ratio of 40.35:14.45:45.11) was placed in the 25 ˚C and 4 ˚C environment. As the result shown in Fig.2, it was obvious that the change of moisture existed among all the increasing standing times. Fig.2 (a) shows the moisture state of the sample at 25 ˚C. The amplitude reflects the moisture content. The peak existed from 0.01 ms to 1 ms had a significant increase, meaning that the bound water became more and more. Table 1 shows the percentage of each fraction of water in this case. Bound water had the largest proportion among the three fractions from the beginning to the end. The initial content of bound water was 40.35 %, however when the standing time increased to 120 min, it had a significant increase on the percentage by nearly 10.78 %; in the end almost 82.78 % of the water was bound water. At the same time, the percentage of immobilized water and free water reduced from 14.54 % to 9.27 % and 45.11 % to 7.95 % respectively. Fig.2 (b) was the result at 4 ˚C. The moisture had a dynamic change on the whole. Table 2 exactly shows the actual concrete data. From 0 min to 300 min the three fractions had no significant change. However the bound water significantly increase by 30.12 % from 300 min to 1980 min, and free water reduced by 11.28 %, while the immobilized water had no change nearly. From the study, we knew that CBPS powder had a stable state at 4 °C. It had a better guiding for the storage of CBPS powder.

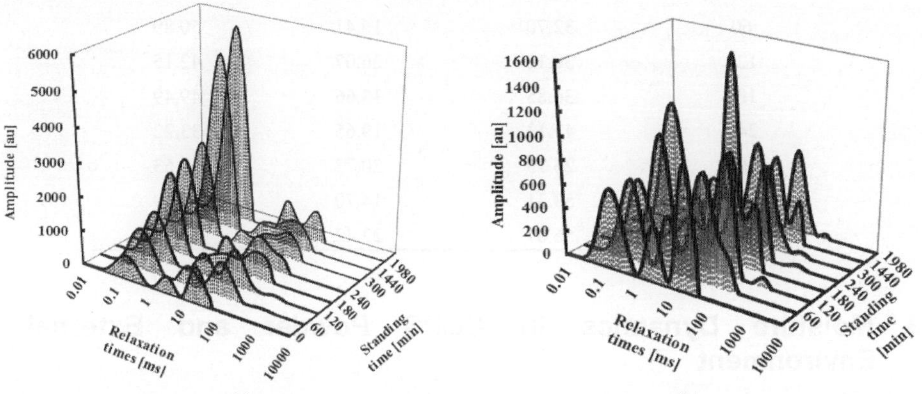

(a) (b)

Fig.2. LF-NMR T_2 relaxation time distribution curves of samples at different standing time. (a).sample at 25°C.(b).sample at 4 °C.

Table 1. The changes of moisture content with different standing time at 25 °C

	Percentage of each water/%		
Standing time/min	Bound water	Immobilized water	Free water
0	40.35	14.54	45.11
60	49.83	15.24	34.93
120	51.13	16.42	32.45
180	61.62	15.86	22.52
240	63.96	13.46	27.56
300	64.99	7.63	27.38
1440	79.99	6.08	13.93
1980	82.78	9.27	7.95

Table 2. The changes of moisture content with different standing time at 4 °C

Standing time/min	Percentage of each water/%		
	Bound water	Immobilized water	Free water
0	43.97	14.91	49.10
60	32.70	16.41	50.89
120	30.12	26.07	42.15
180	36.85	13.66	49.49
240	47.13	19.65	33.22
300	26.75	20.72	52.53
1440	53.31	14.70	31.99
1980	56.87	22.42	20.71

3.3. Moisture Dynamics in CBPS Powder and External Environment

In order to get in-depth understanding of the dynamic changes in CBPS powder and external environment, we created a bar diagram using data from the signal per mass of total water and different moisture fractions and corresponding relaxation times of each group. As Fig.3 (a) shown, the whole moisture in the sample at 25 °C had a significant increase, as time changes the total water had a gradually increase, the 0 min signal per mass was 6429.586 (Au·ms)/g and the maximum signal per mass was 23678.640 (Au·ms)/g at 1980 min which meant the sample absorbed water from the external environment. Compared with 25 °C, during the process of 1980 min, the sample at 4 °C had a stable change, the signal per mass was 6429.586 (Au·ms)/g at 0 min and 5333.158 (Au·ms)/g at 1980 min, the maximum was only 7914.204 (Au·ms)/g had no significant change compared with the original state. The results of signal per mass of each fraction had shown in Fig.3 (b) and Fig.3 (c). Sample at 25 °C had absorbed a large amount of bound water, at the same time, the free water reduced from 2335.260 (Au·ms)/g to 1884.122 (Au·ms)/g and the immobilized water increased from 985.442 (Au·ms)/g to 2196.464 (Au·ms)/g. The result revealed that some free water transferred into bound water and immobilized water. However from Fig.4 (b) we found that the moisture in the sample at 4 °C mainly transferred from one fraction to another fraction itself. It revealed that the moisture in the sample had a obvious exchange with the external environment at 25 °C and mainly had a dynamic change at 4 °C.

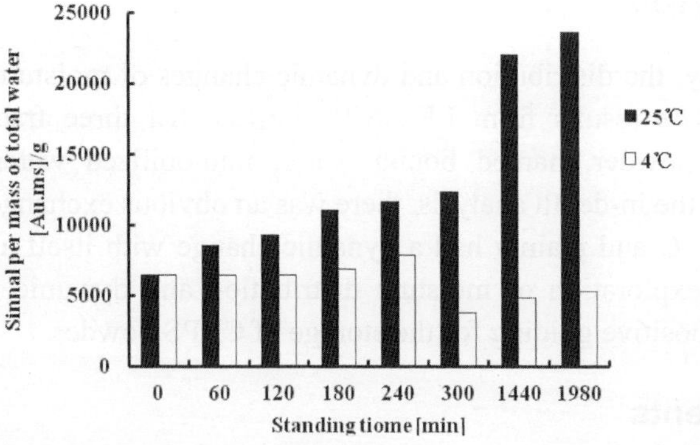

(a)

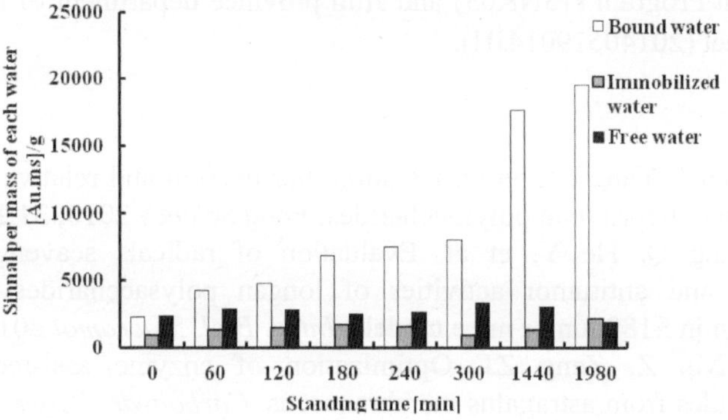

(b)

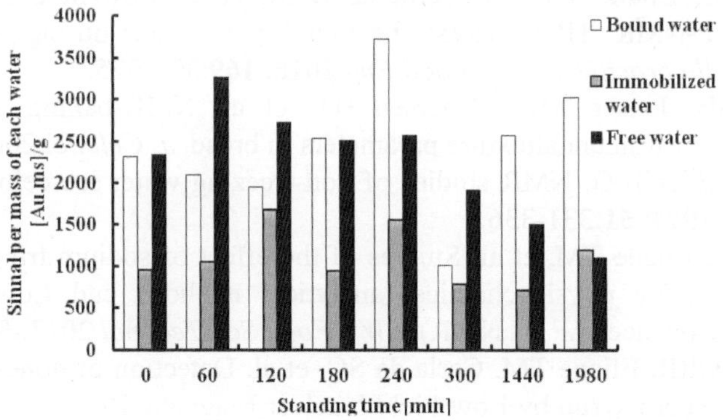

(c)

Fig.3. LF-NMR signal per mass figures of samples at different standing time. (a). signal per mass of total water.(b). signal per mass of each water at 25 ˚C. (c). signal per mass of each water at 4 ˚C.

4. CONCLUSIONS

In the present study, the distribution and dynamic changes of moisture in CBPS powder were monitored. The results from LF-NMR showed that three fractions of moisture existed in CBPS powder, named bound water, immobilized water and free water. Furthermore, from the in-depth analysis, there was an obvious exchange with the external environment at 25 °C and mainly had a dynamic change with itself at 4 °C. This study provided a detail exploration of moisture distribution and dynamic changes of CBPS powder and had a positive guiding for the storage of CBPS powder.

Acknowledgments

The authors acknowledge the financial support provided by Changchun Science and Technology Plan Program (13NK03) and Jilin province department of innovative talent cultivation project (20140519014JH).

References

1. Rong ZY, Yu J, Yang LX, et al. Isolation, purification and relative molecular mass determination of corn bran polysaccharides. *Food Science* 2010; **31**:16–19.
2. Kui Z, Wang Q, He Y, et al. Evaluation of radicals scavenging, immunity-modulatory and antitumor activities of longan polysaccharides with ultrasonic extraction on in S180 tumor mice models. *Int. J. Biol. Macromol* 2010; **47**: 356-360.
3. Guo CH, Xin Z, Zeng ZJ. Optimization of enzyme assisted extraction of polysaccharides from astragalus membranaceus. *Carbohydr. Polym* 2014; **111** :567-575.
4. Carla da SC, Eliane TM , Roberta de ORR, et al. Low-field nuclear magnetic resonance(LF-NMR ^{1}H) to assess the mobility of water during storage of salted fish(*Sardinella brasiliensis*). *J. Food Eng* 2016; **169**:321-325.
5. Engelsen SB, Jensen MK, Pedersen HT, et al. NMR-baking and multivariate prediction of instrument altexture parameters in bread. *J. Cell Sci* 2001; **33**,59-69.
6. Hills BP, LeFloc`h G. NMR studies of non-freezing water ice cellular plant tissue. *Food Chem* 1994; **51**:331-336.
7. Carla da SC, Eliane TM, et al. Studies of the effect of sodium tripolyphosphate on frozen shrimp by physicochemical analytical methods and Low Field Nuclear Magnetic Resonance (LF ^{1}H NMR). *LWT-Food Sci. Technol* 2013; **50**:401-407.
8. Roberta de ORR, Eliane TM, Carla da SC, et al. Detection of honey adulteration of high fructose corn syrup by Low Field Nuclear Magnetic Resonance (LF ^{1}H NMR). *J. Food Eng* 2014; **135**:39-43.
9. Hua SJ, Min DY, Song L. Investigation the effects of protein hydration states on the mobility water and fat in meat batters by LF-NMR technique. *LWT- Food Sci. Technol* 2016; **66**:1-6.
10. Li T, Tu CH, Rui X, et al. Study of water dynamics in the soaking, steaming, and solid-state fermentation of glutinous rice by LF-NMR: a novel monitoring approach. *J Agric Food Chem* 2015; **63**:3261-3270.

11. Yi LS, Ling YS, Fang LX, et al. Dynamics of water mobility and distribution in soybean antioxidant peptide powders monitored by LF-NMR. *Food Chem* 2016; **199**: 280-286.
12. Ling YS, Ye LX, Yan J, et al. Water dynamics in egg white peptide, Asp-His-Thr-Lys-Glu, powder monitored by dynamic vapor sorption and LF-NMR. *J Agric Food Chem* 2016; **64**: 2153-61.
13. Li X, Ma LZ, Tao Y, Kong BH, et al. Low field-NMR in measuring water mobility and distribution in beef granules during drying process. *Adv. Mater. Res*, **500**:3406-3410.

THE SKILLS OF PREOPERATIVE COMMUNICATION FOR SALVAGE OPERATION OF HEAD AND NECK ADVANCED CANCER

Yongjie Kang[1,2,a], Long Xue[1,2], Chenghua Ma[1,2], Bassam Abotaleb[1,3], Lei Shi[1,4], Xueli Zheng[1,2], Ping Chen[1,2], and Lan Yang[2,*]

[1]*Lanzhou University School of Stomatology, Lanzhou, 730030, China*

[2]*Department of Oral and Maxillofacial Surgery Second Hospital, Lanzhou University, Lanzhou, 730030, China*

[3]*Ibb university, faculty of dentistry, department of oral and maxillofacial surgery, Ibb, 70270, Yemen*

[4]*Department of Oral and Maxillofacial Surgery Gansu Provincial Hospital, Lanzhou, 730030, China*

E-mail: [a]kangyj06@foxmail.com

[*]*Corresponding author*

Salvage operation (SO) is usually a surgical operation method to the patients who are treated invalid with all methods except surgical operation method. Head and Neck advanced malignant tumor is a complex tumor with high risk. This article discusses how to make patients have an objective cognition for Salvage operation and have an active cooperation with treatment through the preoperative communication to the patients to reduce medical disputes. It mainly focuses on the following aspects of the preoperative communication for Salvage operation of Head and Neck advanced malignant tumor. psychological factor; preparation of the physician and the patient; the contents and focus of the communication; the skills of Preoperative communication; signature and avoiding medical disputes; preoperative communicative modes and differences; etc.

1. INTRODUCTION

Cancer is a major public health problem in the China and many other parts of the whole world now. More than one in 4 deaths in the China is due to cancer. Head and Neck cancer is a major health care problem, which represents approximately 3% of all malignancies in the body and accounts for more than 274000 newly diagnosed cancers every year worldwide[1].

Salvage operation (SO) is usually a surgical operation method to the patients who are treated invalid with all methods except surgical operation method. Preoperative communication (PC) is a necessary link in the process of diagnosis and treatment of diseases. However, Tieying, Zeng's research[2] find that inappropriate information about cancer easily leads to medical disputes. How to solve the matter like this? It is known to all that the doctor should objectively inform patients (their families or the related person, unit director, etc.) about the patient's condition and the relevant situation in operation before the operation, and answer the question on consulting in order to reach a unified opinion. Still there is not a system and detailed research to solve such an issue[1].

Patients who accept Salvage operation usually have Head and Neck advanced cancer; the tumors often involve important anatomic structures such as the common carotid artery, internal carotid artery, base of skull, cause serious dysfunction, and even be life threatening. The patients were considered not to have surgical indications; Or they were only treated with palliative care even give up previously. It is possible for the patients who with Salvage Operation of Head and Neck Advanced Cancer(SOHNAC)to receive radical resection through SO because of the development of microsurgical technique. But the survival rate of the Patients treated with oral SO is still not ideal: researchers such as Zhang Zhiyuan find that a total of 35 cases of advanced tongue cancer patients accept expand Radical surgery treatment, the 3 years survival rate was 38.7% and the 5 years survival rate only 23.1%[3]. researchers such as Simon N Rogers find that These patients were on average 8 years older with more advanced tambours and overall poorer survival at 5 years, 23% [4] . What can we do to those patients who was diagnosed with Head and Neck Advanced cancer? May be Preoperative communication comes first.

PC is very important for the patients with SOHNAC. However physicians seldom have done some research about PC for SOHNAC in China. The key of PC is to make patients and their family members' ideological work, make SOHNAC related situation, and improve awareness, in order to obtain their consent, understanding and support. But few people did it. At least I did not find anyone who had do some research like this. Combined with personal experience, in this paper we discussed SOHNACPC problems and paid attention to: psychological factor; preparation of the physician and the patient; the contents and focus of the communication; the skills of preoperative communication; signature and avoiding medical disputes; preoperative communicative modes and differences between China and other countries. We hope that such processes would bring about full confidence for surgery and reduce the occurrence of medical disputes. Preoperative Psychological State and Clinical Manifestations of Patients with Head and Neck Advanced Cancer.

2. PSYCHOLOGICAL FACTORS

2.1. Preoperative Psychological State and Clinical Manifestations of Patients with Head and Neck Advanced Cancer

Patient with Cancer psychological status is typically go through five stages: 1)Shock and denial stage; 2)Anger stage; 3)Bargaining stage; 4)Depression stage; 5)Acceptance stage[5]. Most of the patients with cancer were reported feeling with depression, disappointment, nervousness, anxiety, sadness, anger, emptiness or even hopelessness[5,6]. Therefore Clinicians at first need to understand each patient's psychological state and the reason in time, imparting medical knowledge to patients, eliminate the patient's psychological crisis, so that they can carry out various treatments.

2.2. The role of Psychological Intervention in PC

Effective psychological intervention can reduce or eliminate cancer patients with depression, anxiety, tension, fear and despair[7], and may lead to NK cell activity in patients with tumor, increased level of serum IgG, IgM and IL-2 and improve patients' immunity[8] and can get a high survival rate than by despair, helplessness or plain in patient's undergoing coping styles. Therefore, effective psychological intervention can reduce patient's all kinds of bad moods; improve the treatment effect and quality of life of patients. We can give them an Image-guided confidence through psychological counseling, and then told them how to relax such as relax imagination training, massage, and for a walk in the park, listening to music, watching comedy movies and so on.

2.3. The preparation for PC of Patient and Surgeon Physician

Patient and their family should be present at the same time when they talk with doctor. The presence of relatives was necessary to strengthen the confidence of the patients, give them encouragement and courage in the face of surgery, and make them calm in facing operation. Physicians play a key role in the PC. So the physicians must have enough theoretical knowledge and clinical skills, and a positive and sincere attitude is also necessary in our opinion.

3. THE CONTENT AND EMPHASIS OF PC

3.1. The Content of PC

Physicians should make patients and their families understand the preoperative diagnosis of the disease. In other words, patients and their families need to know why this operation is the best choice of treatment for the disease, what consequences will occur, the seriousness of the consequences, the possible risks of operation, its' measures, the expect effect of the operation and the expected costs. The clear opinion of patients and their family members to implement the procedure (Agree with surgery, with the knowledge of the surgery, and understand the operation risk) and sign it indicate the relationship and date through PC.

3.2. The Emphasis of PC

First at all, we need to solve the most urgent problem. Then we need to make the patients have a correct understanding of the risks and major complications in the intraoperative and postoperative SOHNAC for patients usually were seen with one or more complications and even to die. So we must inform SOHNAC patients the possible serious complications, accident, and what we can do to prevent and treat and what about the effect. At last we introduce the current highest level SOHNAC in China and abroad, newest tendency and the level of medical care to the patients or their families and the most important is when the hospital beginning the surgery, the surgery cases, the success rate and complications occurred (how about each proportion), complications, treatment results and effect.

4. THE SKILLS OF PC

4.1. Influence Factors

According to the patient's gender, age, cultural level, psychological quality, the different personality types and national, choosing adaptive way of talking when the physician have a PC with the patients and their relatives.

4.1.1. *Gender*

There is a traditional view that men are superior to women in China since ancient times. The character of the men is more deviate to sane type volitional type or independent type. But women are more deviate to emotional type or obedient type. The men can treat with SO more objective assumed in clinic; while the women are more likely feel wrong. So we should treat them different on the basis of the gender of the patients.

4.1.2. *Age*

Different ages of people have different experience which leads to the difference of physical condition, financial situation and the manner of the life. So we should treat them differently by stages of the patients' ages.

4.1.3. *The Young People (≤ 44)*

Most of them have a Psychological change, they cannot accept the fact of suffering from the cancer, but most of them can seek for therapeutic methods positively after a period of time. So we should give them a positive direction according to the characteristics of the patients with different psychological states in and make the patients to have a correct understanding of these conditions.

4.1.3.1 The middle age (45-59)

These patients' life burden is relatively small, their understanding reached a certain level and the attitude towards life relative to the young man is more optimistic.

4.1.3.2 The aged (≥60)

Those patients' living burden is usually small, they can treat disease objectively because they have more life experiences, but the operating tolerance was significantly decreased, their intraoperative and postoperative risk was bigger, even life-threatening. So the risk must be mentioned in the PC of the old.

The age stages above may not suitable for all patients; the physicians can be adjusted to patients' specific situations.

4.1.4. *The Cultural Level*

The different cultural levels of the patients are a main effect for their PC language comprehension. And to a certain extent, the differences at the knowledge level of the patients affect their psychological status and personality characteristics, etc. Patients with high level of knowledge, the physicians should language fluent, clear and even can use some professional terms. Patient compliance are high that who with low level of knowledge, the physicians should be friendly, expressions should be easy to understand, whenever talk about a problem to ask whether the patients and their families understand, repeat when necessary.

4.1.5. *Psychological Quality*

The psychological factors has a huge impact on the development and prognosis of patients with oral cancer which has a detailed analysis in above. The physicians need to have a basic understanding of the patients' psychological types (Anxiety fear type/ Pessimistic depression type and Optimism active type) through contact with the patients and observe their mood swings. For the patients with Anxiety fear type/ Pessimistic depression type, the physicians can adopt the necessary health protection measures (such as the condition may be given a lighter or advised to hide), but for the families or the unit the need to say the truth to avoid unnecessary medical disputes. For the patients with Optimism active type, the physicians can truthfully inform the condition and operation plan under the premise of triggering the pessimism and despair of the psychological, to arouse the enthusiasm of the active participation of its diagnosis and treatment.

4.1.6. *Ethnic and Religious Customs*

Different ethnic or religious have different habits, customs and believes. Some conservative Muslim patients, for example, it is necessary, however, patients with beard may affect the complete resection of malignant tumor, or postoperative wound infection, etc. so as to imperil their life. Many doctors don't have a better way to persuade such patients to shave their beards. We can start from the creed of the Islam. Such as section 145 chapter 6 of the Koran "forced by potential, the voluntary, edible food or illegally hidden beliefs for the time being"[9]. Seeing in the Koran we can find that faith and survival does not conflict. We can persuade the patient to undergo surgery. Or as far as possible to reduce the patient's resistance, reduce the factors that influence the effect of surgery and make the operation according to the schedule.

4.2. The Language of Artistic Expression

Doctor should combine with intonation, speed, and voice and the conversation, object, ta rget and situation in order to improve the combination of language expression, convincin g and appealing. This not only won the affection and respect of patients, and embodies th e language of art.

5. SIGNATURE AND AVOIDING MEDICAL DISPUTES

The patients sign with the informed consent is a necessary part of the PC in surgical. So PC is a duty that physicians informed patients their consent right, a way to preventing the doctor-patient dispute, but is not the way physicians shirk responsibility. SO it must obtain the consent of the patients, but even if doctors fully fulfilled inform obligation, the signature of the patients agreed to also cannot avoid all medical damage tort liability.

However, the current Chinese laws lag behind this patients' awakening awareness and litigation culture[10]. At present in domestic environment, medical disputes is still a lot, so we should treat every patient serious and be responsible, in the event of medical disputes we should actively seek the protection of the law.

In northwest China, where the medical accident fault under the background of onus probandi inversion, The physicians diagnosis and treatment according to medical procedure are effective to prove their own there are no mistakes, in order to avoid and prevent medical disputes. But to guard against and diminish the responsibility of the physicians, produced many programmatic phenomenon that physicians take a stylized operation for the implementation of the inform obligation and consent of the patients. The physicians should learn more about informed consent, and special methods and guidelines compatible with Chinese culture are urgently required in China[2].

6. CONCLUSION

We should inform the patient's condition and the surgery related to the patients in an orde rly and systematic manner. More important, we should Combining the patients' actual co nditions, including their psychological state, cultural level, Psychological quality and typ es, gender, age, Ethnic and religious customs. When we are talking to the patients, we sh ould use a suitable speed to express the PC content clearly, so that the patients can have a n objective understanding of related conditions and have a positive attitude to deal with t he surgery, thus we can obtain good surgical effect and avoid medical disputes.

References

1. Zheng J W. Clinical epidemiology and risk factors of oral squamous cell carcinoma: An overview. China Journal of Oral & Maxillofacial Surgery, 2007.
2. Zeng T, Huang H, Zhao M, et al. Health professionals' attitude towards information disclosure to cancer patients in China. Nursing Ethics, 2011, 18(3):356-63.
3. Zhang ZY, Zhang CP, Salvage operation of head and neck advanced cancer. Chinese ENT News and Reviews.2007;Vol. 22, No. 6.
4. Rogers S N, Brown J S, Woolgar J A, et al. Survival following primary surgery for o ral cancer. Oral Oncology, 2008; **45**(3):201-11.
5. Kubler-Ross E. On Death and Dying. Southern Medical Journal, 1971; **64**(5):106-10 7.
6. Klikovac T, Djurdjevic A. Psychological aspects of the cancer patients' education: th oughts, feelings, behavior and body reactions of patients faced with diagnosis of

cancer. Journal of B.u.on. Official Journal of the Balkan Union of Oncology, 2010; **15**(1):153-6.
7. Luckett T, Butow P N, King M T, et al. A review and recommendations for optimal outcome measures of anxiety, depression and general distress in studies evaluating psychosocial interventions for English-speaking adults with heterogeneous cancer diagnoses. Supportive Care in Cancer, 2010; **18**(10):1241-1262.
8. Rossman M L. Interactive Guided Imagery as a way to access patient strengths during cancer treatment. Integrative Cancer Therapies, 2002; **1**(2):162-5.
9. Ali M M. Holy Quran Ahmadiyya Anjuman Ishaat Islam Lahore USA, 2011.
10. Dai Q. Informed consent in China: status quo and its future. Medical Law International, 2003; **6**(1):53-71.

ANALYSIS OF RESEARCH FOCUS ON TRANSGENIC SOYBEAN BASED ON CITESPACE

Yanan Bian[a], Yongxu Yuan, Dan Guo, and Jingmin Cheng[b,*]

School of Management, Shanxi Medical University, Taiyuan, 030001, China

E-mail: [a]673643727@qq.com, [b]72-87@qq.com

[*]*Corresponding author*

The purpose of this study is to analyze research focus in the field of transgenic soybean and to summarize the research trend of this field. The study retrieve relevant literature from 2001 to 2015 from CNKI database and Web of science core collection,and analyze literature with the help of information visualization tool-CiteSpaceIII. The global researches on transgenic soybean research keep a steady development, while literature research in China is changing from the rapid growth into slow development. They both have similarities but also have their own focuses. The global researches focus on gene expression of transgenic soybean, however, the researches in China concentrate on transgenic soybean growing. We should strengthen international cooperation in order to promote the deeper research of transgenic soybean.

1. INTRODUCTION

Transgenic soybean refer to those specific soybeans which are developed using transgenic technology, through leading by gene engineering method[1]. While researching transgenic technology, transgenic soybean gradually entered each link of the human food chain. In this paper, the research achievements of transgenic soybean are teased and analyzed comparatively by using CiteSpaceIII. If the research focus in the field is understood, and the differences and similarities are found out, it'll provide theoretical basis for the further research of transgenic soybean.

2. DATA SOURCES AND RESEARCH METHODS

2.1. Data Sources

In this study, the global data sources are from Web of Science, setting “transgenic soybean” as subject, “from 2001 to 2015” as time span, “SCI-EXPANDED” as index, and retrieving 1732 articles finally. Chinese data sources come from CNKI, the retrieval method is that setting “genetically modified” and “soybean” as subject, “from 2001 to 2015” as time span, “core journals” as journal source, “journal” as literature type and “Refworks” as way of downloading data. Having finished these steps through necessary cleaning and processing, 1658 references are found out. They have different retrieval themes because if the theme of WOS database is “transgenic and soybean”, only one article could be get without solving any problem. That is to say, they can't make a comparison with Chinese databases. In this paper, we develop the study on the above two kinds of data.

2.2. Research Methods

This study chooses information visualization software-CiteSpaceIII developed by professor Chen Chaomei in Drexel University as a research tool, and sets up the time of analyzing software from 2001 to 2005. The time division is for one year, according to the following three time zones respectively, threshold is set for (2, 2, 20), (4, 3, 20), (4,3,20) in the study , at which the field of transgenic soybean could be analyzed.

3. ANALYSIS OF THE NUMBER OF TRANSGENIC SOYBEAN RESEARCH

Through calculating the number of published papers in the field of transgenic soybean every year, the research status and development trend of this field can be found intuitively from the overall. Fig. 1 is the number of papers about transgenic soybean in CNKI and Web of science database over years.

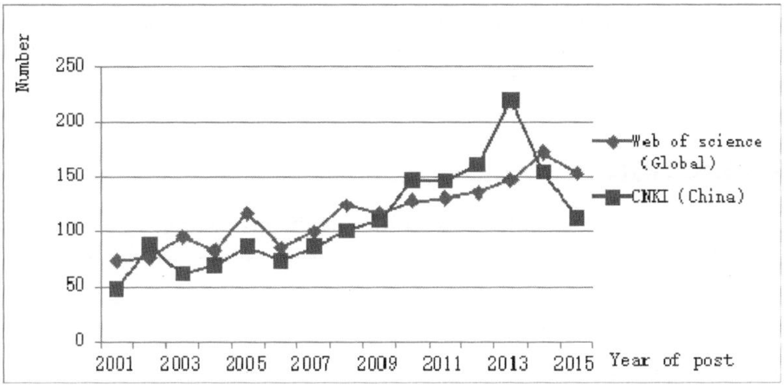

Fig. 1. Statistical analysis of research papers on transgenic soybean.

It can be seen from Fig. 1 that the number of articles about transgenic soybean tends to increase generally, but is different.Global data of papers about transgenic soybean continue to increase, which reflects the mature in the research of this field.However, Chinese data shows a trend of rising wave[2], and began to slow down after reaching its peak in 2013(219 papers), which reflects the Chinese transgenic soybean research is changing from the rapid growth to slow development. It not only results from the lag of recording articles in database,but also may be associated with the agreement of importing three transgenic soybean from Chinese Ministry of agriculture.

Before 2010, researching transgenic soybean in WOS database was hotter than in CNKI, because the pouring imports of American transgenic soybean made China lose the market in soybean market according to the datum. After 2010, more and more Chinese researchers paid attention to transgenic soybean research, because it turned into a heated debate about whether transgenic crops have been industrialized and scholars have produced a lot of research focuses on transgenic soybean.

4. ANALYSIS OF THE RESEARCH FOCUS BASED ON KEYWORDS

The researches on transgenic soybean can be showed in the atlas below. Each node in this atlas respects a keyword, and the more frequency of occurrence, the greater the nodes are. The coarser the connection between the nodes are, the higher co-occurrence among the keywords will be. The shorter distance from the center, the more important the key nodes will be such as: key words in the CNKI database "transgenic soybean".

4.1. Analysis of Research Focuses in CNKI Database

Selecting "keywords" as the network nodes, running CiteSpaceIII, 376 nodes and 1262 connections can be obtained. Fig. 2 is a biggest clusters of keywords from Chinese transgenic research focuses from 2001 to 2015 appearing frequently in the knowledge map.

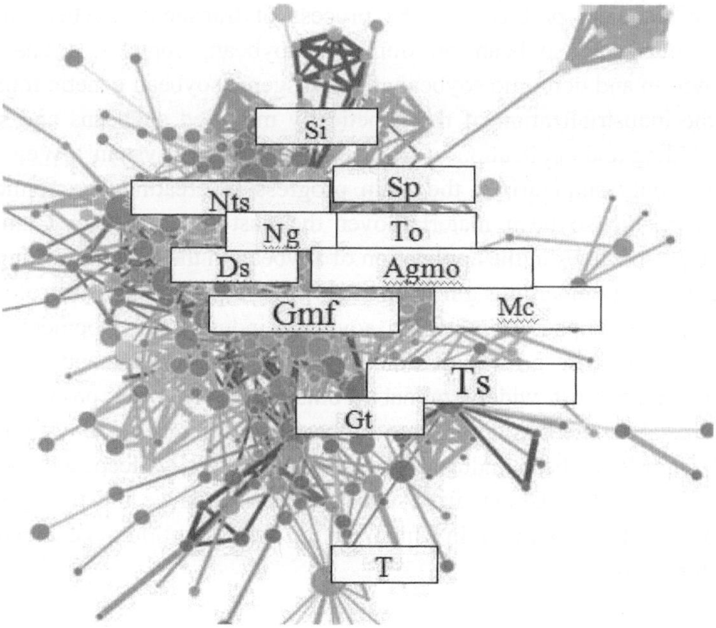

Fig. 2. The co-occurrence graph of transgenic soybean research focuses in China from 2001 to 2015.

(Because labels of the figure are in Chinese, I use English instead of here.)

(Si: Soybean Industry; Nts: Non-transgenic soybean; Sp: Soybean planting; Ng: Non-gmo; To: Transgenic organism; Ds: Domestic soybeans; Agmo: Agricultural genetically modified organisms; Gmf: Genetically modified food; Mc: Monsanto Company; Ts:Transgenic soybean; Gt: Genetic transformation; T: Transgene)

Table 1. Ranks of keywords in Chinese transgenic soybean literature with high frequency

Freq	Capacity	Level
639	0.28	Transgenic soybean
221	0.23	Genetically modified food
208	0.12	Soybean planting
181	0.09	Transgene
180	0.09	Non-transgenic soybean
140	0.09	Domestic soybeans
114	0.06	Genetic transformation
107	0.08	Monsanto Company
102	0.02	Soybean industry
96	0.05	Non-gmo

It can be indicated from Fig. 2 and Table 1 that there exists a mainstream of knowledge composed by research focuses in the field of transgenic soybean. We can see that the frequencies of the first 10 high-frequency keywords are quite different, which range from 639 times of the transgenic soybean to 96 times of non-gmo. High -frequency keywords will be the theme after the combination,and it can be summarized as the following: the technical problems in the process of transgenic soybean planting, the influence of transgenic soybean on domestic soybean, contrastive analysis of the transgenic soybean and non-gmo soybean, the transgenic soybean genetic transformation, upgrade of the industrialization of the genetically modified soybeans and so on. “The large-scale building and application of transgenic technology system” (Wen-Sheng Hou, etc. 2014), not only summarized the main progress in creating agronomic traits and improving transgenic soybean material over the past few years in China, but also conducted the perspectives in the application of soybean in the genome editing and vertex integration technology emerging gm technology[3], “The countermeasure of transgenic soybean under the impact of China's soybean industry development” found the contradiction between the growth of domestic rigid soybean demand and the limited domestic supply.We must widely monarch outside the soybean production base,and accelerate the establishment of a non-gmo soybean independent price system in China to revitalize of China's soybean industry[4]. The topics greatly deepened the transgenic soybean research level in China, from which we can analyze the transgenic soybean researches focus on the cultivation in China, paying little attention to gene expression and genetic transformation.

4.2. Analysis of Research Focuses in WOS Database

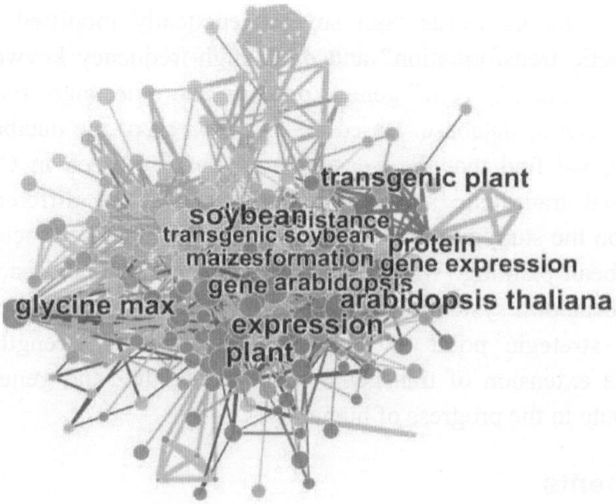

Fig. 3. The co-occurrence graph of global transgenic soybean research focuses from 2001 to 2015.

Table 2. Rank of high-frequency keywords in global gm soy literature

Freq	Centrality	Keyword
245	0.21	Soybean
216	0.11	Plant
203	0.11	Expression
189	0.11	Glycine max
166	0.13	Arabidopsis thaliana
150	0.18	Transgenic plant
127	0.05	Protein
124	0.2	Gene
109	0.06	Arabidopsis
101	0.13	Gene expression

Through the analysis of the keywords in Fig. 3, Table 2 and the summary of the literature of global gm soy, it can be summed up as the following topics: the transgenic soybean planting, the transgenic soybean gene expression, transgenic soybean nutrition ingredient and so on.Since we entered into commercial cultivation of genetically modified crops, the United States has been a developed country using transgenic technology and biological technology all the time.And application of biotechnology to improve soybean fatty acid composition is A successful example of soybean genetic engineering breeding, delta 2 fatty acid dehydrogenase gene (FAD2-1) soybean [5-6] and other related quality genetic improvement research[7] is being further conducted. Some species have been promoted and began to cultivate in large area. The research on transgenic soybean planting and industrialization has gone into mature, while genetically modified focuses on the studies of gene expression and genetic transformation.

5. CONCLUSIONS

This paper selects the keywords "gm soy", "genetically modified food", "soybean cultivation", "genetic transformation" and other high-frequency keywords from CNKI database and "soy", "plant", "gene" gene expression and other high-frequency keywords from the Web of science database. Based on the research of the database of transgenic soybean literature, we find that the transgenic soybean research in China follows the trend of the global transgenic soybean, but also shows the differences. The world researches focus on the study of gene expression, but in China they focus on the research of transgenic soybean planting. We should follow the international pace, and optimize the genetic transformation system of transgenic soybean to increase the food safety to a national security strategic point of view[8]. We ought to strengthen international cooperation in the extension of transgenic soybean planting and gene expression ,and moreover, contribute to the progress of humanity.

Acknowledgments

This work is supported by the Fund (grant no. SSKLZDKT2014086) hosted by Dan Guo for the federation of social sciences at Shanxi Province, Jing-Min Cheng, as the corresponding author, Yong-Xu Yuan and Ya-Nan Bian involved.

References

1. Jin-zhuan Zhong, Wen-liang Wu, You-fu Xia. Development of transgenic soybean and countermeasures of Chinese soybean industry. *Journal of China Agricultural University* 2005;**4**:43-50. (In Chinese).
2. Jun-xiang Zhang, Xiao-feng Li. The analysis and reflection on Chinese development of genetically modified food. *Forum on Science and Technology in China* 2011;**11**:13-135. (In Chinese).
3. Wen-sheng Hou, Kang-xue Lin, Pu Chen, Zhi-wei Jia , Yang Zhou, Yang Yu, Yan-hua Liu. The large-scale building and application of transgenic technology system. *Scientia Agricultura Sinica* 2014;**21**:4198-4210. (In Chinese).
4. Qin Xiao, Jian-ping Li, Dong-mei Liu. The countermeasure of transgenic soybean under the impact of China's soybean industry development. *Forum on Science and Technology in China* 2015;**6**:137-153. (In Chinese).
5. Kinney A J. Development of genetically engineered soybean oils for food applications. *Food Lipids* 1996;**3**:273—292.
6. Bnhr T, Sato S, Ebrabim F. Ribozyme termination of RNA transcripts down-regulate seed fatty acid genes in transgenic soybean. *The Plant Journal* 2002; **30**(2):155-163.
7. Cahoon E B, Ripp. Formation of conjugated delta 8, delta 10 double bonds by dehal2-oleic acid desaturase related enzymes: Biosynthetic origin of calendic acid. *Journal of Biological Chemistry* 2001;**276**: 2637-2643.
8. Jing-min Cheng. Study on pre-warning system of food safety. Beijing: Economic Daily Press 2014;**9**: 180-200. (In Chinese).

PUBLIC HEALTH IN CHINA: THERE IS STILL A LONG WAY TO GO

Zhongzhong Wu[a], Dan Guo, and Jingmin Cheng[b,*]
School of Management, Shanxi Medical University, Taiyuan, 030001, China
E-mail: [a]wuzhongzhong828@163.com, [b]72-87@163.com
[*]Corresponding author

The problem of public health has become a topic which could not be avoided by the Chinese government. This paper backtracks and summarizes the historical stages of development and experience of the public healthin China, especially the influence of SARS. For the present China's public health problems and thecrisis whichwill meetin the future, the Chinese government stillshould do a lot of work in terms of public health.

1. INTRODUCTION

Nowadays, more and more attention has been paid to the problem of public health all over the world. Since enter the 21st century, the severe acute respiratory syndrome (SARS), west Nile, H1N1, Ebola, H7N9 virus had caused one after another worldwide's epidemic. The rising of tuberculosis and HIV/AIDS prevalence, the possible outbreaking of plague and cholera, the changingof international political and economic order after the end of the cold war led to extreme nationalism and religious power growth that increased risk of bioterrorism, which have caused concern of the governmentsin the world. The problem of public health has not only confined to a country's health system, but also penetrated into every field of the international political, economic and cultural life, public healthis becoming a serious challenge to the survival and development of human society [1].

2. THE DEVELOPMENT CHARACTERISTICS OF PUBLIC HEALTH IN CHINA

For China, public health is also an issue that cannot be ignored. For the past few years, the Chinese government was seeking a development paththat suit for China's national conditions of medical and health career. "Taking history as a mirror, can make clear gain or loss", this is an old adage in China, for China's public health development, it is alsonecessary to review and summarize the history, making progress better in the experience for reference.

Since the founding of new China, based on the basic national conditions of China at that time, the Chinese government established the health work policy of "geared to the needs of workers, peasants and soldiers, prevention first, unitethe Chinese and western medicine", the whole society participate public health campaigns at that time, medical workers prevent activelywhile treating diseases, the prevention of infectious disease been put in the first place, andmade remarkable achievements in the improvement of the public health level. By the end of 1956, 29 provinces ,cities and its subordinatecounty of China, has established a health epidemic prevention station, railway and large corporations also have their own independent health epidemic prevention station, at the same time a large number of professional prevention agencies also established in succession, such as kala-azar, plague, malaria prevention and control institutions, TB prevention, infectious hospital, etc., and epidemic prevention station,theseinstitutions and stationformed the original disease defense system in China. To the early 1960s, China has wiped out the plague, cholera and smallpox [2]. During this period, the countryside also gradually formed a three-level network of medical treatment, prevention,with county level institutions, township hospitals and village clinics, and established the rural cooperative medical care system. These measures have effectively promoted the work of disease prevention and control in rural areas, and improve the public health in rural China [3].As the government of the People's Republic of China, with the domestic impoverished circumstances at that time, public health result is positive; the experience is also should be summarized and learned by the government today.

After the ten years “cultural revolution” from 1966 to 1976, the development of China's public health improved rapidly. By 1982, China prevention center (later renamed the Chinese Institute of Preventive Medicine) was established, which marked China established the disease prevention and control system that from country to county and department of health epidemic prevention station, to the township hospitals and village clinics, and at various levels and specialized prevention station (clinic). In addition,the legal system construction gradually improved in the field of public health, the government has introduced a 9 special laws and 23 special regulations to ensure the smooth operation of health system, which involves the drinking water, food hygiene, labor health, radiation protection and food cosmetics health and quarantine method, and also formulated a series of hygienic regulation and safety evaluation procedures and methods, basically meet the need of the health prevention and supervision. In order to adapt to the development of the society better, in 2001 China's ministry of health issued “the several opinions about implement health supervision system reform” and “the guidance on reform of the system of disease prevention and control”,these document put forward reform goals of disease prevention and control and institution setting clearly, guide and promote the reform of health supervision system and disease prevention and control system, formed the system of disease prevention and control, health supervision system, maternal and child health care system and the public health organizations such as patriotic public health organizations, food and drug administration.

Public health recovered from the "cultural revolution" and organization structure reform was completed in 2002, in general, Chinese public health development has obtained certain achievements during this period, such as in world development reportthe health investment (1996), Chinese action, that provided one - hundredth world's health input for basic medical and health service of twenty-two percent population all over the world, was commended. The action played a significant role in improving people's health with less investment and high efficiency. However, after the reform and opening-up, we took economic construction as the center of the development direction, and problems of public health were emerged under such background. For example, some department heads of local governments thought that public health work just need "no accident" or "no big problem", and less investment put into public health. As a result, public health team was pushed into market economy, on the one hand, they should finished their task, on the other hand, they need earn money to fed themselves, then fewer people majored in prevention and control of disease. Areas under-developed economy usually had more public health problems, but got less investment. Heavy pressure of institution of disease prevention and control to fed themselves contributed to less work of public health .In addition, the public health team emerged a phenomenon: On the one hand, professional technologist with abundant knowledge of public health left seriously because of low wages, living benefits; on the other hand, many workers who have no professional knowledge and medical training in poor quality poured into this group for "iron rice bowl". For a long term, such phenomenon brought destructive change to the Chinese public health team and its functions [4].Of course, it also had created conditions for ineffective prevention and control after the outbreak of SARS.

3. SARS: THE TURNING POINT OF THE DEVELOPMENT OF PUBLIC HEALTH IN CHINA

The outbreak of SARS could be thought as a milestone in the development of Chinese public health.The SARS suddenly broke out in Shunde Guangdong in 2002, and quickly spreaded to the southeast Asia and the world, epidemic situation was gradually controlled until the middle of 2003. According to the relevant data of WHO, The global tide of SARS affected 32 countries and regions, there were cumulative total of 8422 cases of SARS and 919 deaths, nearly 11% fatality rate.

In response to the sudden disaster, Chinese government and department of health bared weaknesses of emergency management, SARS as a mirror, reflected Chinese public health ignored problems overnight. Before the outbreak of SARS, China had no preparedness to deal with public health emergency, no emergency identification and classification standard, also had no corresponding command system, no grading system and no post responsibility system. Meanwhile, the government did not make clear regulations for health emergency information publication and news spread, multi-sectoral coordination, social mobilization, major control measures and so on. Even early in the SARS epidemic, policy makers emphasized clinical treatment and etiology research, epidemiology investigation and prevention and control countermeasures' research was not enough, and missed the best opportunity to control in the early. Because of the lack of timely and accurate information communication and improved emergency management system, the SARS epidemic caused huge losses to Chinese development. According to statistics data, the SARS epidemic accumulated 5327 cases in mainland, and 349 deaths. In addition to people's life and health was threated, Chinese aviation transportation, food and beverage retail, tourism, foreign business, construction, manufacturing processing and business services has been impacted unprecedentedly, the impact on the employment level was greater than the impact on the economy, lead to position decreases about 1.27 million in 2004[5].

Once there was an expert said, "No SARS, no recent development of public health in China."Indeed, SARS is a disaster, but it is also an opportunity to accelerate the development of China's public health. After SARS, there was an unprecedented large-scale construction of Chinese public health system.The government spent 11.7 billion Yuan on hardwares to equip centers for disease control [6]. In a few years after the outbreak of SARS, led by the ministry of health, China has formed a nationwide public health monitoring and early warning system and a set of rigorous information reporting institution.Among them, the most striking is the disease network reporting system. This covers the central, provincial, municipal, county, township 5 administration information system, which can directly report to the central, including infectious disease outbreak, public health emergencies of China's 31 provinces. In addition, a number of laws and regulations related to public health got to refresh, including the law of the People's Republic of China on the prevention and control of infectious diseases, the emergent public health event emergency ordinance, the national emergency preplans for public health emergency and other laws and regulations. Those make China rules-based when they respond to emergencies. In recent years, China's public health has also seen the H1N1, hand, foot and mouth disease (HFMD), H7N9 etc.. In terms of epidemic preparedness, emergency disposal, department coordination, information release and so on various aspects, the construction of public health has made great progress, comparing with the SARS outbreak.

4. CHINESE PUBLIC HEALTH STILL HAS A LONG WAY TO GO.

Don't forget the pain once the wound is healed.Currently there are still many problems of public healththat need to be vigilanted and resoled by China government. As the world's largest developing country, Chinese government has got a lot of great achievements on the economic in recent years. In first time, China's GDP is more than Japan's which is only behind United States' in the world in 2010. However, the health input seems to have no corresponding achievementsin the same year, which accounts for only 4.98% of GDP, and the total health expenses had not reached the GDP minimum limit of 5% of the WHO standard. By 2011, China's total health expenditure is 5.1%, but compared with developed countries at the same period, such as 17.7% in United States, 11.6% in France, 11.3% in Germany and 10.0% in Japan, and there is a big gap in terms of health expenditure input levels for China. Especially public health has been in a relatively weak position in China, and the idea that "medical treatment is more important than prevention" limits the rapid development of China's public health. In addition, the shortcoming of public health workers' configuration and management is more and more serious in China's different regions and groups.

It is not too late to mend the fold after the sheep have been stolen. Therefore, for China, it not only needs to mend the hole of Chinese public health, but also inspect the loopholes systematically with the public health perspective, to do it for a rainy day. With the change of ecological environment, the social development of economy with high speed, the development of technology, the collision of legal and ethical, and the biological terrorist risk increases, the public health risk, with various new problems will be more complex, all kinds of vulnerable groups will be hurt easily. To improve the ability of public health of China, it is important to ensure adequate and sustained attentions. Capital input is also necessary, so that a comprehensive response to infectious diseases, chronic non-communicable disease and accident harm to people's health threat to the public; In the longer run been contemplating, you should spread the notion of health promotion, make the "prevention first" is not just a slogan. Public health in China, there is still a long way to go.

Acknowledgments

This work is supported by the Fund (grant no.KXKT1405) of Shanxi Association for Science and Technology, hosted by Jingmin Cheng. Some students and teachers participated in this project, including Jingmin Cheng and Zhongzhong Wu. Among them, Jingmin Cheng is the corresponding author of this article.

References

1. Yun Zhang. Global governance of the global health problem: a case study of emergent public health events [D].*Shandong University*, 2006: 2-4. (In Chinese).
2. Peng-fei Zhao. Public Health Expenditure, Health and Economic Development—the Empirical Analysis Based on China's Statistics [D], *Beijing Jiaotong University*, 2012: 44-47. (In Chinese).
3. Yu-kuanGuo. Rethinking the public health system in China [J].*China Reform*, 2005;(5):50-53. (In Chinese).
4. Le Wang.ZengGuang: SARS10 years, public health's rapid development [N].*Wenhui Daily*, 2013-3-15:1-5. (In Chinese).
5. Na-zhengChen, Zheng Zhang. From SARS to H1N1 Flu: Promotion on Emergency Management Ability for Outburst Public Health Events in China[C].*EBM*,2010:5820-5821. (In Chinese).
6. Ru-pengCai.SARS is a mirror that reflects the problem of Chinese public health[N].*China News Weekly*,2013-3-4:45-48. (In Chinese).

